Synthesizing and Characterizing Plant-Mediated Biocompatible Metal Nanoparticles

Susanta Das
Ajeenkya D.Y. Patil University, India

Shankar Mukundrao Khade
Ajeenkya D.Y. Patil University, India

Debanjali Barman Roy
Ajeenkya D.Y. Patil University, India

Khushbu Trivedi
Ajeenkya D.Y. Patil University, India

IGI Global
Publishing Tomorrow's Research Today

Vice President of Editorial	Melissa Wagner
Managing Editor of Acquisitions	Mikaela Felty
Managing Editor of Book Development	Jocelynn Hessler
Production Manager	Mike Brehm
Cover Design	Phillip Shickler

Published in the United States of America by
IGI Global Scientific Publishing
701 East Chocolate Avenue
Hershey, PA, 17033, USA
Tel: 717-533-8845
Fax: 717-533-8661
Website: https://www.igi-global.com E-mail: cust@igi-global.com

Library of Congress Cataloging-in-Publication Data

Names: Das, Susanta, 1975- editor. | Khade, Shankar, 1988- editor. | Barman Roy, Debanjali, 1992- editor. | Trivedi, Khushbu, 1989- editor.
Title: Synthesizing and characterizing plant-mediated biocompatible metal nanoparticles / edited by Susanta Das, Shankar Mukundrao Khade, Debanjali Barman Roy, Khushbu Trivedi.
Description: Hershey, PA : IGI Global Scientific Publishing, [2025] | Includes bibliographical references and index. | Summary: "This book will delve into diverse synthesis processes for environmentally benign bionanoparticles utilizing various medicinal plants. It will explore associated parameters, such as extracts from different plant parts (e.g., seeds, peels, roots, and flowers), as well as physical and thermal conditions (e.g., temperature, pH levels, and concentrations) to regulate reaction kinetics. Characterizations (including crystal structure, shape, size, surface roughness, etc.) of these biosynthesized particles using various spectroscopic analytical techniques will be discussed in detail"-- Provided by publisher.
Identifiers: LCCN 2024053169 (print) | LCCN 2024053170 (ebook) | ISBN 9798369362402 (hardcover) | ISBN 9798369362419 (paperback) | ISBN 9798369362426 (ebook)
Subjects: LCSH: Plant biotechnology. | Nanobiotechnology. | Nanostructured materials. | Medicinal plants.
Classification: LCC TP248.27.P55 S96 2025 (print) | LCC TP248.27.P55 (ebook) | DDC 660.6--dc23/eng/20250116
LC record available at https://lccn.loc.gov/2024053169
LC ebook record available at https://lccn.loc.gov/2024053170

British Cataloguing in Publication Data
A Cataloguing in Publication record for this book is available from the British Library.

Table of Contents

Detailed Table of Contents

are typically used to manufacture these nanoparticles. Manufacturing methods like nanoprecipitation, ionic gelation, microemulsion, and coacervation are discussed in this context. They are proven effective as food additives, vehicles for nutrient retention, packaging matrices, and functional food formulations. For wider application of this tool, testing in different systems to develop safe, effective, and degradable nano-delivery vehicles must be undertaken.

Hammad Ali Khan, Government College University, Faisalabad, Pakistan
Abdul Malik, University of Agriculture, Faisalabad, Pakistan
Qasim Waheed, Kohat University of Science and Technology, Pakistan
Muhammad Shahjahan, Government College University, Faisalabad, Pakistan
Behram Khan Ajat Khel, Abdul Wali Khan University, Mardan, Pakistan
Samrin Gul, University of Sargodha, Pakistan
Muhammad Huzaifa Mahmood, University of Turin, Italy
Ayesha Hameed, University of Gujrat, Pakistan
Aamir Raza, University of Agriculture, Faisalabad, Pakistan
Samreen Anjum, University of Engineering and Technology, Lahore, Pakistan

This chapter explores the use of plant-mediated production of biocompatible metal nanoparticles (MNPs) for sustainable agriculture. It highlights the benefits of this green synthesis method, including its environmental friendliness, affordability, and renewable resource use. The chapter discusses the processes by which phytochemicals in plant extracts act as reducing and capping agents, enabling the creation and stabilization of MNPs with specific features. It also discusses the characterization methodologies used to evaluate the dimensions, morphology, electrical charge, and compatibility with living organisms of the synthesized MNPs. The chapter also discusses the potential uses of MNPs in sustainable agriculture, such as targeted nutrient delivery, plant development promotion, and environmentally friendly pest control. The chapter explores the use of IoT technology for immediate monitoring and accurate implementation of plant-mediated MNPs in agriculture, highlighting the financial implications, cost reductions, and market opportunities for environmentally friendly solutions.

Santosh Kumar Sethi, School of Biotechnology, Gangadhar Meher University, Sambalpur, India
Smita Rani Dalei, School of Biotechnology, Gangadhar Meher University, Sambalpur, India
Augustine Okpani Oko, Department of Biotechnology, Ebonyi State University, Abakaliki, Nigeria
Asish Kumar, Department of Biotechnology, Guru Ghasidas University, India
Arati Kindo, School of Biotechnology, Gangadhar Meher University, Sambalpur, India

Nanofertilizers represent a groundbreaking advancement in agricultural techniques, poised to transform how nutrients are delivered and managed in the interaction between plants and soil. Unique characteristics of nanoparticles facilitate improved nutrient absorption, precise delivery, and enhanced availability, potentially revolutionizing conventional fertilization methods. Numerous studies demonstrate significant enhancements in growth metrics, seed yield, and overall plant vigor following nanofertilizer application.

The escalating threat of antibiotic resistance poses a formidable challenge to the global healthcare system, necessitating novel approaches to combat multidrug-resistant bacterial infections. The synthesis of nanoparticle antibiotic conjugates emerges as a promising avenue. Nanoparticle antibiotic conjugates represent a groundbreaking strategy that harnesses the unique properties of nanoparticles and antibiotics to enhance antimicrobial efficacy. This chapter delves into the innovative fusion of nanoparticles and antibiotics through green synthesis techniques, aiming to address the pressing challenges of antibiotic resistance. By exploring these hybrid nanostructures' synthesis, characterization, and therapeutic potential, this chapter sheds light on their significance in revolutionizing infectious disease management. This article highlights potential mechanisms of action of Nanoparticle antibiotic conjugates as next-generation antimicrobial agents and future implications and challenges.

Over the past decade, there has been a growing interest in particles with sizes between 1 and 100 nm because of their many uses in fields including drug delivery, cancer diagnosis, and therapy. There are many different chemical compositions of nanobioparticles, ranging from micelles to metal oxide, from synthetic polymers to large biomolecules. Characterizing nanoparticles is crucial after they are synthesized. Size, shape, dispersity, and surface area are the main characteristics of nanoparticles. Numerous methods have been employed to assess the dimensions, crystallographic arrangement, elemental composition, and several other physical attributes of nanoparticles. Multiple techniques can be used to evaluate physical attributes in various cases. Each technique has different strengths and limitations, which complicate the choice of the most suitable method. The common techniques for the characterization of nanoparticles are as follows: XRD, UV-Vis, FTIR, DLS, SEM, TEM, HRTEM, & TGA. This chapter aims to review the various types of techniques for the characterization of nanoparticles.

Nanotechnology will emerge as the most significant technological breakthrough of the twenty-first century. The rapid development of nanotechnology has led to the creation and application of nanoparticles across various fields, including medicine, materials science, agriculture and environmental engineering. It is imperative to characterize these nanoparticles to comprehend their characteristics, actions and possible consequences. Depending on the intended usage, an array of techniques such as SEM, TEM, ZP, DLS,

demands and promote environmental sustainability and human health by utilizing AI-driven technologies and biogenic nanoparticles through ethical innovation and interdisciplinary collaboration.

Preface

In an era where sustainability and environmental consciousness are increasingly prioritized, the exploration of green technologies has become more critical than ever. Among these innovations, plant-mediated biocompatible metal nanoparticles offer a promising and eco-friendly alternative to traditional nanoparticle synthesis. Nanoparticles synthesized through plant-based methods not only reduce the environmental impact of conventional chemical processes but also open doors to safer applications in fields ranging from medicine to agriculture. As the world faces growing challenges in sectors like healthcare, environmental remediation, and food production, this green approach to nanotechnology provides sustainable solutions with minimal environmental disruption.

This book, *Synthesizing and Characterizing Plant-Mediated Biocompatible Metal Nanoparticles*, serves as a comprehensive exploration of the synthesis, characterization, and application of plant-based nanoparticles. The focus is on green methods that leverage plant extracts to create nanoparticles with diverse applications, ranging from healthcare to agriculture and environmental management. As such, this work positions itself at the intersection of nanotechnology, sustainability, and applied research, addressing both scientific curiosity and practical industrial needs.

The target audience for this book includes researchers, scholars, and professionals working in the fields of nanotechnology, green chemistry, materials science, and environmental engineering. Additionally, students and educators engaged in academic research, especially those interested in sustainable innovations and the interdisciplinary applications of nanoparticles, will find this volume invaluable. By providing insights into the synthesis processes, characterization techniques, and potential uses of biocompatible nanoparticles, this book aims to serve as both an educational resource and a reference guide for those looking to advance their understanding of the field.

The target audience for this book includes researchers, scholars, and professionals working in the fields of nanotechnology, green chemistry, materials science, and environmental engineering. Additionally, students and educators engaged in academic research, especially those interested in sustainable innovations and the interdisciplinary applications of nanoparticles, will find this volume invaluable. By providing insights into the synthesis processes, characterization techniques, and potential uses of biocompatible nanoparticles, this book aims to serve as both an educational resource and a reference guide for those looking to advance their understanding of the field.

Each chapter of this book offers a deep dive into a specific aspect of plant-mediated nanoparticle synthesis. The first chapter, *Green Innovations: Evaluating Plant–Derived Metal Nanoparticles and Their Superior Properties Over Traditional Methods*, provides an introduction to the field, comparing traditional nanoparticle synthesis methods with plant-derived approaches. This chapter establishes the framework for understanding why green methods are increasingly favored.

In *Eco-friendly Synthesis And Characterization Of Gold Nanoparticles Using Leaf Extract Of Eucalyptus Globulus* (Chapter 2), readers will delve into specific methodologies of nanoparticle synthesis, with a focus on the eco-friendly production of gold nanoparticles using the leaf extract of *Eucalyptus globulus*, highlighting its advantages in terms of sustainability and efficiency.

Green Synthesis Methods for Metallic Zinc Oxide Nanoparticles - A Sustainable Angle (Chapter 3) continues the discourse by exploring zinc oxide nanoparticles, emphasizing their green synthesis and the potential these methods hold for sustainable development.

Wastewater treatment is a critical issue in today's world, and Chapter 4, *Green Nanoparticles - Innovative Solutions for Sustainable Wastewater Treatment*, showcases how plant-based nanoparticles are being used to address water contamination in an environmentally friendly manner.

The application of nanoparticles in the food industry is covered in Chapter 5, *Biopolymeric Nanoparticles: A Novel Ingredient for Application in Food Industry*, where the potential for nanoparticles to improve food preservation, safety, and quality is discussed.

In *Plant-Mediated Synthesis of Biocompatible Metal Nanoparticles for Sustainable Agriculture* (Chapter 6), we examine how nanoparticles can be applied in agriculture, specifically through the development of nano-fertilizers to enhance crop yield and reduce environmental damage.

Chapter 7, *Effect of Nano-fertilizer on Crop Plant Growth and its Effect on Rhizospheric Microorganism and its Application: A Comprehensive Review*, offers an in-depth review of the interactions between nano-fertilizers, plants, and rhizospheric microorganisms, providing insights into how these technologies can support sustainable agricultural practices.

The therapeutic applications of nanoparticles are explored in Chapter 8, *Harnessing the Therapeutic Potential of Metallic Bionanoparticles: Integrating Medicinal Plant as Green Factories*, where the role of medicinal plants in synthesizing therapeutic bionanoparticles is discussed.

Moving forward, *Plant Biogenic Nanoparticles as Antifungal Agents: Synthesis, Characterization, and Mechanistic Insights* (Chapter 9) investigates the potential of plant-based nanoparticles in combating fungal infections, an area of growing concern in health care.

Chapter 10, *Green Synthesis of Nanoparticles and Their Conjugations with Antibiotics for Treating Bacterial Infections*, bridges the gap between nanotechnology and modern medicine, presenting the use of plant-synthesized nanoparticles in conjunction with antibiotics to treat bacterial infections more effectively.

The development of sophisticated instruments for nanoparticle analysis is an essential part of the research process. Chapters 11 and 12, *Development of Advanced Instruments for Spectroscopic Analyses of Nanoparticles/Nanobioparticles* and *Advanced Instrumentation for Characterization of Nanoparticles: A Discourse*, focus on the spectroscopic and instrumental advancements that enable the precise characterization of nanoparticles, essential for understanding their properties and behavior.

Chapter 13, *Plant-Based Inorganic Quantum Dots: Synthesis, Characterization, and Applications*, takes the exploration into quantum dots, offering an insightful look into their synthesis from plant sources and their wide array of applications, including in electronic devices and medical diagnostics.

Finally, in Chapter 14, *Mapping the Landscape of Biogenic Nanoparticles in Bioinformatics and Nanobiotechnology: AI-Driven Insights*, we explore the intersection of artificial intelligence and biogenic nanoparticles, shedding light on how AI is driving innovation and offering new insights into nanoparticle applications.

This book is poised to make a significant impact on the field by compiling cutting-edge research on plant-mediated nanoparticle synthesis and presenting it through a sustainability lens. The integration of green chemistry with advanced nanotechnology offers a roadmap to innovations that are not only scientifically promising but also crucial for creating a more sustainable future. We hope that this book serves as both a foundational text and a source of inspiration for future research and practical applications in nanotechnology and beyond.

We express our sincere gratitude to all the authors and contributors who have made this work possible and to the institutions that supported us in the development of this book.

Susanta Das
Ajeenkya D.Y. Patil University, India

Shankar Mukundrao Khade
Ajeenkya D.Y. Patil University, India

Debanjali Barman Roy
Ajeenkya D.Y. Patil University, India

Khushbu Trivedi
Ajeenkya D.Y. Patil University, India

Acknowlegment

We are thankful to all the authors and reviewers for their valuable contributions to the development of this book.

Our sincere gratitude goes to our friends, colleagues, and the entire fraternity of Ajeenkya DY Patil University for their continuous support and encouragement.

We are also thankful to IGI-Global, USA for giving us the platform to contribute to the society through this book.

Last but not the least, our heartfelt thanks go to our family members for their relentless support and motivation throughout the journey, without which it would have been impossible for us to go from writing the proposal to completing the project on time.

Susanta Das
Ajeenkya D.Y. Patil University, India

Shankar Mukundrao Khade
Ajeenkya D.Y. Patil University, India

Debanjali Barman Roy
Ajeenkya D.Y. Patil University, India

Khushbu Trivedi
Ajeenkya D.Y. Patil University, India

Introduction

The book *Synthesizing and Characterizing Plant-Mediated Biocompatible Metal Nanoparticles* serves as a comprehensive resource aimed at guiding various stakeholders through the dynamic field of biocompatible metal nanoparticle synthesis, using eco-friendly, plant-mediated methods. This work explores the synthesis, characterization, and multifaceted applications of plant-mediated nanoparticles in areas such as environmental sustainability, nanotechnology, drug discovery, and cancer treatment. In addition to addressing technical and scientific aspects, the book examines societal and technological challenges, assisting scientists, policymakers, academic institutions, and industries in making informed decisions in this rapidly advancing field.

Nanoparticles, particularly noble metal nanoparticles, have garnered significant attention due to their small size—ranging from 1 to 100 nanometers—and their unique physical, chemical, and biological properties. These attributes have led to an explosion of interest in various applications. Traditional methods of nanoparticle synthesis, however, rely on physical and chemical techniques, which often involve hazardous chemicals and complex, expensive processes. These methods also present significant risks, including environmental contamination and health hazards, as nanoparticles can be released into the environment from household items or industrial processes.

In response to these challenges, the green synthesis approach—also known as biological synthesis—has emerged as an eco-friendly alternative. This method uses plants and microorganisms to produce nanoparticles that are more biocompatible and stable than those generated by traditional means. Plant extracts, rich in bioactive compounds such as polyphenols, play a crucial role in reducing metal ions to nanoparticles and stabilizing their size and shape. As a result, this green method is not only more sustainable but also offers cost and time efficiencies.

The book delves into the use of various medicinal plants in synthesizing biogenic nanoparticles . The field of *Phytonanotechnology*, which highlights the integration of plant-derived materials in nanotechnology, stands at the forefront of modern research, offering vast potential for applications ranging from environmental cleanup to targeted drug delivery for life-threatening diseases.

In addition to synthesis methods, this book provides detailed insights into the characterization of these biosynthesized nanoparticles using advanced spectroscopic techniques. Techniques such as Ultra-Violet-Visible (UV-Vis) spectrometry, X-Ray Diffraction (XRD), Fourier Transform Infra-Red (FTIR), Scanning Electron Microscopy (SEM), and more are discussed in depth to provide readers with a thorough understanding of the physical and chemical properties of these nanoparticles.

Moreover, the book explores the integration of artificial intelligence (AI) in advancing nanoparticle applications, particularly in targeted drug delivery and medical imaging. It examines the potential of biogenic nanoparticles in fields such as wastewater treatment, agriculture, antimicrobial treatments, cancer therapy, and wound healing, among others.

By focusing on both current advances and future opportunities in nanoparticle research, the book *Synthesizing and Characterizing Plant-Mediated Biocompatible Metal Nanoparticles* provides a vital framework for understanding the ethical considerations, challenges, and opportunities that define this rapidly evolving discipline, ultimately offering readers a roadmap to navigate this burgeoning area of scientific inquiry.

Chapter 1
Green Innovations:
Evaluating Plant-Derived Metal Nanoparticles and Their Superior Properties Over Traditional Methods

Nehir Arik
https://orcid.org/0000-0002-6724-9463
Turkish Patent and Trademark Office, Turkey

ABSTRACT

Metal nanoparticles (MNPs) have the potential to be employed in both therapeutic and diagnostic applications. Despite the plethora of techniques employed for the formulation of MNPs, it is evident that there is an urgent need to develop a non-toxic, environmentally friendly MNP formulation technology, namely plant-derived MNPs. This section reviews the formulation techniques of plant-derived MNPs, their application areas and a summary of the studies in the literature. First, comprehensive details on plant-derived MNP formulation techniques are provided. Then, literature review is made on the formulation and application areas of different plant-derived MNPs, and overview of the studies is provided. The following sections of this article examine the environmental impact, sustainability and industrial applicability potential of plant-derived MNPs in more detail, while also addressing the limitations of these techniques. Plant-derived MNPs have been a subject of significant interest among researchers and continues to be an endless potential area of research with ongoing studies.

1. INTRODUCTION

Thanks to nanotechnology, nanoparticles (NPs) smaller than 100 nm have been produced and are open to exponential development. These NPs, which were born because of these developments, continue to play a vital role in human life (Malik et al., 2023; Sahu, 2023).

The formulation of these nanoparticles is based on many techniques, as shown in Figure 1. However, the processing of toxic chemicals into NPs formulated with physicochemical techniques from traditional production methods results in the formation of highly toxic waste products. In addition, laborious and costly production conditions such as exposure to radiation, high temperature, and pressure requirements also make these techniques negative (Bloch et al., 2021). In addition, defects occur in the morphology

DOI: 10.4018/979-8-3693-6240-2.ch001

of metal NPs (MNPs) formed with the top–down approach. These defects in morphology and surface structures limit the application areas of MNPs (Bao et al., 2021). In chemical techniques, which are another production technique of NPs, NPs have superior features in terms of controlling morphology and structural parameters, easy, cost-effective, and high-efficiency production. Despite this, when the necessity of disposal of wastes resulting from production and the damage it will cause to the environment as a result of biological accumulation are taken into consideration, it is seen that the effectiveness of the technique decreases (Karunakaran et al., 2023; Szczyglewska et al., 2023). Due to these negativities, less harmful, alternative methods are needed for nanoparticle formulation. The technique formed by the combination of nanotechnology and biotechnology is revealed as environmentally friendly technology, also known as green technology (Kulkarni, 2024).

Figure 1. Different techniques for formation of NPs.

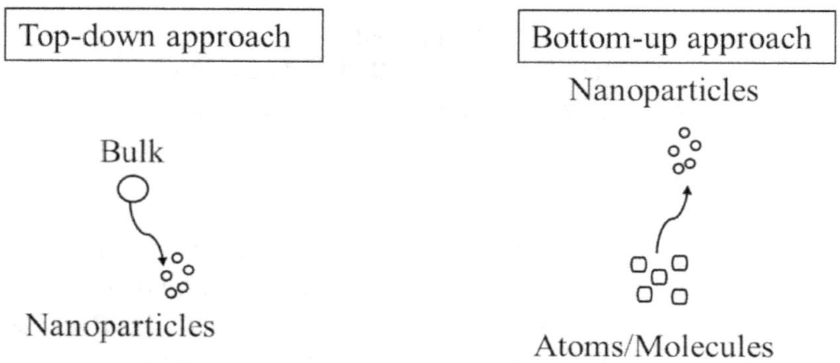

Top-down approach

- Physical Technique
  - Chemical etching
  - Laser ablation
  - Milling process
  - Sputtering
  - Lithography

Bottom-up approach

- Chemical Technique
  - Chemical vapor deposition
  - Sol-gel
  - Laser prolysis
  - Spray method
  - Molecular condensation
  - Aerosol process
  - Thermal decomposition
- Biological Technique
  - Virus
  - Bacteria
  - Fungus
  - Algae
  - Plant

The advent of green technology has provided a means to meet the recent need for less harmful, environmentally friendly, and inexpensive techniques for the formulation of MNPs for potential pharmaceutical and biomedical applications. MNPs have come to the forefront of research thanks to their extraordinary characteristics, including a high surface–to–volume ratio and stability under a range of conditions, such as optical, mechanical and chemical environments (Chandrakala et al., 2022). Plant–derived MNPs with unique features are utilized in wastewater treatment, electronics, optoelectronics, food industry and cosmetics. In addition, MNPs are used in bioengineering and biomedical applications for antibacterial and anticancer purposes and have gained a place in various sectors, both medical and non-medical.

Compared to traditional formulation techniques, plant–derived MNPs offer low cost, scalability, stability and environmentally friendly hazardous waste. It has many advantages in terms of being able to be disposed of in a friendly way (Abuzeid et al., 2023). Studies have shown that there is a 30% reduction in energy consumption and approximately 40% reduction in production costs by using plants in MNP synthesis. Moreover, in addition to these, there is a decrease in product yield A 50% increase was recorded (Osman et al., 2024).

2. ENVIRONMENTAL IMPACT AND SUSTAINABILITY OF PLANT–BASED NANOPARTICLE SYNTHESIS COMPARED TO TRADITIONAL METHODS

The synthesis of plant–derived MNPs is basically based on standard techniques including filtration and/or chromatography as purification methods after the extraction stage. Plant–derived MNPs are formulated by adding plant extract to the prepared metal salt solution under appropriate conditions (temperature and pH). To prove the impurity of these MNPs, centrifugation is performed immediately before washing and the resulting solution is removed. There is absolutely no use of microbial, chemical or toxic substances in the synthesis of plant–derived MNPs. For this reason, in addition to being biocompatible, it is known as a cheaper, more accessible and reliable process, which makes it different from other traditional production techniques (Singh et al., 2023). For example, in the first stage of the co-precipitation technique used in the formulation of MNPs, acidic materials (nitric acid (HNO_3), hydrochloric acid (HCl) and sulfuric acid (H_2SO_4)) are used. After the formulation is completed, a precipitation process is applied with basic materials (such as MgO or NaOH) to eliminate agro–industrial wastes. During this formulation, plant–derived MNPs can be synthesized in the desired size and morphology by changing the basic parameters (pH, temperature and mixing speed).

The damage caused by millions of tons of agro–industrial biowaste to the environment we live in every year is a great threat to the health of living beings. Biological wastes not only pollute the air we breathe and all our water resources, including drinking water, but also cause global warming and the formation of harmful gases (Yaashikaa et al., 2022). Post–harvest incineration is traditionally used for their disposal. However, this method is associated with the release of many harmful gases such as CO_2, CH_4 and NO, which are known to have adverse effects on health. It is known that developing alternative strategies is essential to achieve sustainable cycles in the industry. In this context, using agro–industrial biological waste in MNP formulation is an attractive approach. Moreover, thanks to this approach, MNPs can be formulated with lower cost and energy compared to those synthesized by traditional techniques (physical and chemical). Natural biomolecules found in biological wastes act as a coating and stabilizing reducing agent. These non-toxic biomolecules reduce the need for chemical additives (Osman et al., 2024). When biological wastes are used in the MNP formulation, the wastes are both easily disposed of and recycled. This contributes to the sustainability of the ecosystem (Razak et al., 2022). Finally, renewable and unlimited resources such as plants are also evaluated.

3. FORMULATION MECHANISMS OF PLANT–DERIVED MNPs

The use of plants in medical applications has been widespread since ancient times. Accordingly, plants are among the most suitable raw materials for the synthesis of MNPs to be used in the medical field (Mikhailova, 2021). In recent years, due to their excellent physico–chemical properties, interest in plant-derived MNPs used for different applications has increased. When the current literature is examined, it is seen that there are many studies investigating MNPs formulated from various natural sources such as gold, silver, palladium, platinum, cobalt, zinc, copper, titanium oxide, magnetite and nickel (Ali and Mohammed, 2021; Fahmy et al., 2020; Zuhrotun et al., 2023). Recently, it has been discovered that plant–derived MNPs, whose potential for use in nanomedicine has been on the rise, possess superior properties to synthetic NPs. Alcoholic or aqueous extracts are used in the formulation stages of plant–derived MNPs and finalized by techniques such as percolation, maceration or infusion, as shown in Figure 2. The variety of phyto–biomolecules in the plant to be used in the study to obtain plant–derived MNPs varies from plant to plant. The process of extraction is not a simple one, but rather a complex one that plants must undergo in four distinct stages. The stages are, in order, the selection of the requisite materials, the washing and sterilization of these materials, the drying of the materials, and finally the grinding of the materials. The initial step involves subjecting the selected material to boiling, a traditional method for the extraction process. It is crucial to consider the parameters that influence the extraction efficiency, including temperature, time, and pH. In the final stage, the resulting plant extracts will undergo stabilization and closure processes to form MNPs, as well as undergo functional group modifications to align with the intended purpose of the MNPs (such as antioxidant, antimicrobial, or anticancer applications) (Antonio-Pérez et al., 2023).

Figure 2. Schematic representation of the preparation of plant extracts for the formulation of plant-derived MNPs (Antonio-Pérez et al., 2023).

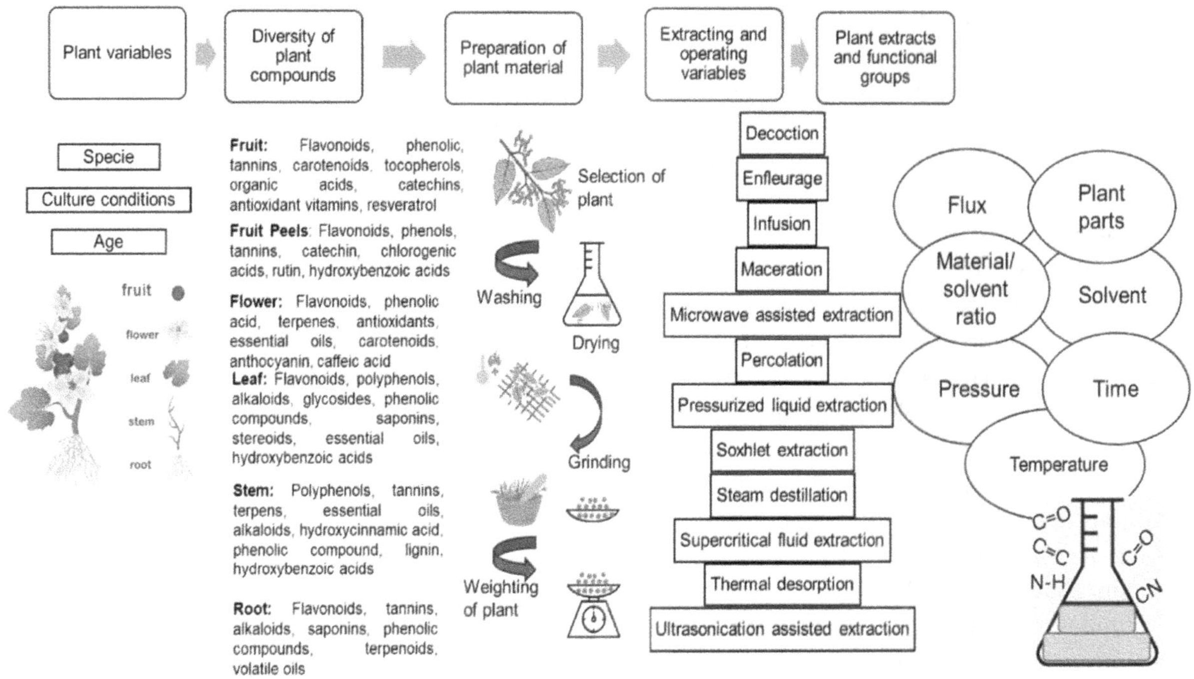

The reducing, capping, and stabilizing properties of bioactive compounds/phytochemicals in plants facilitate the conversion of metal sources into NPs (Khan & Khan, 2023). Due to their inherent reducing properties, plants can participate in the synthesis of MNPs through the involvement of various endogenous reductases, dehydrogenases, flavonoids, and extracellular electron carriers, as well as external reductants, such as lemon salt (citric acid), vitamin C (ascorbic acid), and other compounds that facilitate the green synthesis of MNPs. Flavones, organic acids and quinones are naturally occurring reducing agents that are used in the formulation of MNPs (Roy & Das, 2015).

The formulation of plant–derived MNPs is achieved through the utilization of biological techniques derived from various plant parts. MNP derivatives obtained from plant sources, including stems, roots, seeds, bark, fruits, calluses, flowers, and leaves, can be synthesized in a multitude of forms and sizes (Mekuye & Abera, 2023). It can be argued that such differences are related to the biomolecules found in plants acting as stabilizing and degrading agents (Susanti et al., 2022). Notable plant biomasses employed in the formation of gold (Au) NPs include *Medicago sativa* (alfalfa) (Kokina et al., 2024) and *Pelargonium graveolens* (geranium) (Nancy & Elumalai, 2019). Conversely, Shankar et al. (2004) employed the leaves of *Azadirachta indica* (neem) in the formulation of bimetallic gold, silver and bimetallic gold core–silver shell NPs (Shankar et al., 2004). In a study conducted by Chandran et al. (2006), an aqueous extract of aloe vera was used to form spherical silver NPs and gold nanotubes It is stated that this formulation is achieved because of the gradual reduction of gold ions and the subsequent formation of carbonyls in the extract (Chandran et al., 2006). Also, Harris and Bali (2008) used *Brassica juncea* and *Medicago sativa* plants to synthesize silver NPs (Harris and Bali, 2008). Conversely, plants such as

Brassica juncea, *Helianthus annuus* and sunflower are also employed in the synthesis of other MNPs, including nickel (Ni), cobalt (Co), zinc (Zn) and copper (Cu) (Saeed et al., 2021).

There are three principal techniques for formulating plant–derived MNPs: (i) intracellular (within the plant), (ii) extracellular (using plant extracts) and (iii) using individual phytochemicals. Indeed, numerous plant species have been observed to accumulate metals intracellularly, whereby they synthesise homoaggregate and heteroaggregate MNPs from these accumulated metals in a sequential manner (Wang & Liu, 2022). A wide variety of biomolecules, including amino acids, alkaloids, aldehydes, flavones, ketones, proteins, phenolics, polysaccharides, saponins, tannins, terpenoids, and vitamins, can serve as the building blocks for the biosynthesis of MNPs. The specific biomolecules utilized in this process, as well as the nature of the plant extract employed in the MNP formulation, can influence the resulting nanoparticle properties. It can be concluded, therefore, that the phenolic acids present in the extract are responsible for the reduction of metal ions and the subsequent formation of the NP.

Proteins and enzymes which are part of the extracts that play a key role in biogenic synthesis can help in formulation of MNPs by using metal ions. Since they have high reducing power they can be attracted to specific sites of the molecule of interest and help in formulation of MNPs without reducing the metal ions. Also, their chelating power is not excessive. Removal of protons from amino acids or other molecules present in the system results in formation of resonance structures with oxidation capacity. Then metal ions are reduced actively, and plant–derived MNPs are formed.

Glucose another molecule that is part of the formulation of plant–derived MNPs can act as a reducing agent since it can be oxidized to a carboxyl group by addition of hydroxyl groups.

Polyphenols that act as ligands in the formulation of MNPs have notable activity. The aromatic hydroxyl groups in polyphenols can bind to metal ions and then get oxidized to carbonyl groups. These groups help in stabilizing MNPs and formation of stable complexes. This complex is directly decomposed at high temperatures, resulting in the release of MNPs. Another extract of particular importance is that of flavonoids, which are a type of polyphenol. In addition to their ability to actively chelate metal ions, they also have the capacity to reduce them. These flavonoids, which produce protons that can reduce metal ions, are involved in the nucleation, formation and further aggregation stages of plant-derived 3+

Flavonoids, which belong to the polyphenol group, are another plant extract that has a very important function in biogenic synthesis. For example, Au^{3+} ions can form a complex with the flavonoid radical and thus can be oxidized to the keto form and reduced to Au–NPs. It is thought that this process occurs due to internal mechanism that converts the ketones found in flavonoids into carboxylic acids. Some flavonoids can chelate metal ions by the effect of their carbonyl functional groups or π-electrons. For example, quercetin can form chelates at three positions, both carbonyl and hydroxyl groups at the C3 and C5 positions and catechol groups at the C3' and C4' positions. Additionally, these groups can chelate several metal ions including $Fe^{2+,}$ $Fe^{3+,}$ $Cu^{2+,}$ Zn^{2+}, $Al^{3+,}$ $Cr^{3+,}$ Pb^{2+} and $Co^{2+.}$ Thanks to their chelating ability, flavonoids take an active role in the nucleation process and aggregation step and initiate the MNP formulation (El-Seedi et al., 2019; Makarov et al., 2014). Because flavonoids are adsorbed on the surface of MNPs during bioreduction (Kordy et al., 2022). In one study, anisotropic gold NP and silver NP were formulated from apigenin glycoside (apiin) obtained from *Lawsonia inermis* extracts. FTIR results showed that apigenin glycoside was bound to MNPs by the presence of carbonyl groups. The findings show that flavonoids and flavonoid glycosides can be used as inducing agents in MNP formulation (Kasthuri et al., 2009). Apart from this, phenolic compounds are known to have a significant effect on MNP formulation. For example, salicin facilitates the formation of gold NPs via hydroxyl group and glycoside bonds. It has also been shown that salicin increases the reduction of Au^{3+} to Au^0 and the stabilization of gold NPs. Terpenoids, one of the phenolic compounds, reduce metal ions by oxidation of aldehyde groups in molecules to form carboxylic acids (Mikhailova, 2021). Polyphenols are another class of phenolic compounds that facilitate the reduction of metal ions by the presence of hydroxyl and carboxylic acid groups. Thanks to these biomolecules that act as capping agents, MNPs can maintain their stability in solutions with different pH or at high temperatures (Al-Zahrani et al., 2021).

It has been reported by Kuppusamy et al. (2016) that the most vital parameters in obtaining MNPs with different sizes and shapes are the metal concentration used and the volume of the plant extract (Kuppusamy et al., 2016). Besides, an increase in reaction temperature results in a corresponding acceleration in reaction rate. This permits the formation of nucleation centres to be enhanced and the generation of MNPs with considerably greater efficiency. Furthermore, a linear proportionality exists between temperature and MNP size. Conversely, it has been documented that when continuous mixing is conducted throughout the reaction, the discernible change in colour occurs and MNPs with a uniform distribution of sizes are synthesised (Darwish et al., 2022; Sharma et al., 2024). Also, the synthesis of MNPs with homogeneous distribution sizes has been reported to be feasible through ultrasonic treatment (Glaubitz et al., 2022). Similarly, an increase in the number of nucleation centers was observed in formulations performed at high pH values. Unlike these, it has been proven that MNPs are formulated at acidic or neutral pH values. In long reaction times, there is a longer interaction between the reactants. This allows for higher efficiency and larger sizes of MNPs to be obtained (Álvarez-Chimal and Arenas-Alatorre, 2023). The diversity of plant–derived MNPs increases their potential for use in many therapeutic applications. Plant–derived MNPs can convert insoluble and poorly absorbed bioactive compounds into more easily dispersible substances. Therefore, they are preferred in both pharmaceutical and biomedical applications (Khan and Khan, 2023).

This section provides an overview of the natural plant sources used in the formulation of various plant–derived MNPs and the studies conducted.

3.1 Formulation of Gold Nanoparticles

Gold NPs (Au–NPs) have been the focus of researchers in the field of nanomaterials due to their superior properties. These superior properties are stability, adjustable core size, biocompatibility, and large surface-to-volume ratio. In addition, Au–NPs have high adsorption capacity, inertia, and high catalytic activity (Mikhailova, 2021; Panwar et al., 2024). Since non–cytotoxic Au–NPs can be easily functionalized with various bioactive molecules, they are a good alternative for targeted drug delivery (Milan et al., 2022). Au–NPs have many applications in the treatment of drug–resistant diseases due to their antibacterial properties, in cancer treatment (Chen et al., 2021; Panwar et al., 2024) and gene therapy (Niu et al., 2023) due to their anti-inflammatory properties, in autoimmune disorders and chronic inflammation (Minnelli et al., 2022) due to their anti-inflammatory properties. In addition, their potential in biological imaging (Shamloo et al., 2022) and optical research (Shanmugam, 2022) are some of the areas where much progress has been made.

Various chemical components found in biogenic complexes can facilitate the reduction of gold metal ions, leading to the creation of Au–NPs. Biomolecules such as flavonoids, phenols, and proteins are crucial in the synthesis of Au–NPs through the reduction of Au ions. In a study conducted by Ahmad et al. (2019), the potential of polyphenolic compounds from the aqueous extract of *Elaeis guineensis* leaves as agents for Au–NPs synthesis was examined. The findings indicated a reduction in the levels of total phenolic compounds and flavonoids during the reaction, which suggests that the hydroxyl groups of the biomolecules were converted to carbonyl groups as Au–NPs formed. Ultimately, the concentration of these biomolecules attached to the Au–NPs surface diminished, leading to the generation of Au–NPs via the reduction of Au ions (Ahmad et al., 2019).

As shown in Figure 3, the reduction of chloroauric acid ($HAuCl_4$) to Au–NPs can be expressed as: $H+Au^{3+}4Cl–4H_2O$ + plant molecule (OH, COOH, etc.) → Au^0 NPs. As a result, the aqueous solution acquired a ruby red color, which is enough to prove the presence of Au–NPs in the solution.

Figure 3. Generation of plant–derived Au–NPs.

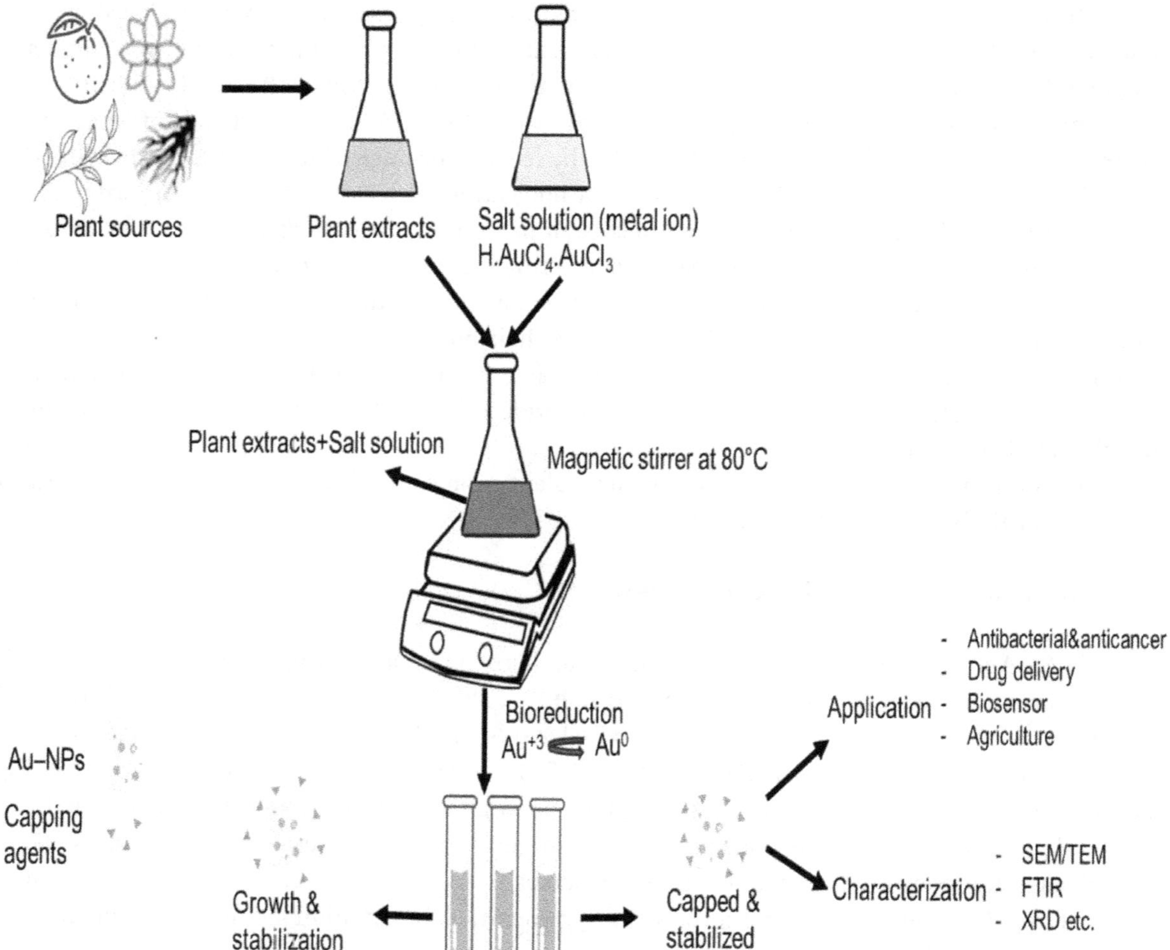

Rodriguez Leon et al. (2019) demonstrated that hydroxyl groups in polyphenolic molecules undergo deprotonation during the synthesis of Au–NPs derived from *Mimosa tenuiflora* extract. They found that the initial step of the deprotonation process involves electrons transferring from the protonated hydroxyl group to Au^{3+} ions. This leads to the reduction of Au^{3+} ions to metallic gold atoms (Au^0), and simultaneously, the oxidation of the polyphenolic ring occurs (Rodríguez-León et al., 2019). Furthermore, other studies have highlighted that tannins, alkaloids, and polyols play significant roles in the biosynthesis of gold nanoparticles from plant sources (Anbu et al., 2020; Huang et al., 2010). During the conversion of Au^{3+} to Au^0, the oxidation of hydroxyl groups in polyols to carboxylate groups was noted. Additionally, the reduction sites on the polysaccharides found in plant extracts reveal amino groups that enhance the stability of metal nanoparticles (MNPs), allowing for strong bonding of both amino groups and carbohydrates to the hydrophilic surface of Au–NPs. The oxidation of polysaccharide hydroxyl groups to carbonyl groups also aids in the reduction of Au^{3+} to Au^0 (Liu & Corma, 2018). Additionally, it has been demonstrated that high molecular weight proteins are involved in the synthesis of Au–NPs derived from plant sources (Kamble, 2009; Lee et al., 2020). In the studies conducted by Patra et al. (2015, 2016), plant–derived Au–NPs were formulated using citrulline, a molecule obtained from watermelon peels. It was discovered that Au^{3+} ions were reduced and stabilized by citrulline in the formulation. This discovery clarified that in addition to conventional molecules in Au–NP synthesis, there was also the potential of molecules that had never been used before. It is similarly important to consider the parameters at the synthesis stage. For instance, Armendariz et al. (2004) examined the formation of Au–NPs derived from Avena sativa at varying pH levels. The researchers observed that at relatively high pH values (pH 3–4), Au–NPs of smaller sizes were formed, whereas at low pH values (pH 2), aggregation occurred in the experimental results. These findings showed that the pH of the medium has a significant effect on the morphology and dimensions of the synthesized MNPs (Armendariz et al., 2004). Therefore, in an intensely acidic environment, the reduction of atoms and the primary nucleation process of MNP aggregation are more. Because, at lower pH values (pH 3–4), many of functional groups that bind, and nucleate tetrachloroauric acid ions are active. Thus, it is assumed that metal ions are involved in fewer nucleation events, resulting in aggregation of the metal (Selvakannan et al., 2004).

Some of the remarkable results obtained from studies conducted with plant–derived Au–NPs are presented in Table 1. Jing et al. (2019) synthesized Au–NPs using Aloe vera leaf extract. The findings of the study show that Au–NPs are protected from agglomeration and oxidation by the plant extracts they are coated with. Furthermore, it was emphasized that the temperature during the reaction and the effect of both chloroauric acid and Aloe vera leaf extracts played a significant role in the formation process (Jing et al., 2017). In 2017, the biological reduction of chloroauric acid ($HAuCl_4.3H_2O$) with a 5% leaf extract of *Azadirachta indica* (neem) was employed for the formulation of Au–NPs. The researchers concluded that the phenols and proteins present in the plant extract play an important role in the green synthesis process (Mishra et al., 2017). In 2015, Au–NPs were formulated by green synthesis using an aqueous *Eucommia ulmoides* bark extract. It was observed that the plant–derived Au–NPs exhibited favorable stability and a narrow size distribution. Furthermore, the natural plant source employed in the formulation was deemed to be straightforward, cost–effective and non–toxic (Guo et al., 2015). In a 2012 study, Sumit et al. employed a green synthesis technique to prepare plant–derived Au–NPs utilizing an array of diverse plant and spice extracts, as outlined in their research paper. The extracts were found to be self-stabilized in a specific crystalline phase by the reduction of aqueous $HAuCl_4.3H_2O$ to Au^0. Yellow auric chloride was used to confirm the presence of synthesized Au–NPs. The results of the studies showed that the used plant and spice extracts served as effective and low-cost reducing agents in the formation of Au–NPs (Lal & Nayak, 2012).

Table 1. Some plant–derived Au–NPs found in the literature and natural reducing agent used.

Reducing agent (Plant source)	Parts of plant	Properties	Reference
Zingiber zerumbet	Aqueous extract	Potential for cosmetic and biomedicine application	(Liao et al., 2024)
Sphagneticola trilobata (L.) J. F. Pruski,	Flower and Leaf	Anticancer	(Mardina et al., 2024)
Allium cepa L. and Allium sativum L.	Aqueous extract	Antioxidant, anticancer	(Tan et al., 2023)
Platycodon grandiflorum	Aqueous extract	Antipathogenic	(Anbu et al., 2020)
(Petroselinum crispum	Leaf extract	Antioxidant, anticancer, antibacterial and photocatalytic	(Borady et al., 2020)
Jasminum auriculatum	Leaf extract	Anticancer, antibacterial	(Balasubramanian et al., 2020)
P. integerrima galls	Gall extract	Antifungal, antinociceptive	(Islam et al., 2019)
Ziziphus zizyphus	Leaf extract	Antibacterial, antifungal	(Aljabali et al., 2018)
Dodonaea viscosa	Leaves	Anticancer	(Anandan & Prabu, 2018)
Eclipta prostrata	Leaf extract	Antibacterial, antioxidant, and cytotoxic activity	(Rajakumar et al., 2016)

This green synthesis method for the formation of plant–derived Au–NPs is known for its simplicity, straightforwardness, and environmentally friendly nature due to the absence of toxic reagents and neutral pH conditions. As a result, it has recently attracted the attention of an increasing number of researchers.

3.2 Zinc Oxide Nanoparticles

In contemporary nanotechnology, considering the constant contact of MNPs with the biological environment, nontoxic and safe formulation techniques have become a research priority.

In recent years, zinc oxide NPs (ZnO–NPs) have become a focus among researchers because they have application potential in various fields. ZnO–NPs are among the most prevalent metal oxides in the natural world, characterised by their low production costs, simplicity of synthesis and extensive applicability. Additionally, its physical and chemical properties may be readily modified to suit the specific requirements of any given application area. Different formulations or materials are used to change the morphology of ZnO–NPs. Thanks to their optical, electrical and photocatalytic properties, they can be used in solar cells, photocatalysis and the chemical sector. Their harmlessness, high ultraviolet absorption and antimicrobial features make them attractive in applications such as nano(bio)medical, drug and gene delivery and biosensors. Furthermore, the solubility of ZnO–NPs in acidic media makes them a promising candidate for use in drug delivery systems (Lakshmipriya & Gopinath, 2021).

Table 2 presents a selection of significant findings pertaining to the green synthesis of ZnO–NPs utilising plant biomasses. Recent studies have demonstrated that biocompatible and biodegradable ZnO–NPs have begun to be formulated through a green synthesis approach, thereby becoming increasingly specific for biomedical applications. The inherent biomolecules in the plant extracts used in this approach not only assist in stabilizing ZnO–NPs, but also fulfill the requirement for toxic and costly reagents used in chemical techniques (Dikshit et al., 2021). In his study, Acar (2021) formulated ZnO–NPs using an aqueous extract of *Achiella millefolium* L. by the green synthesis approach. The experimental results demonstrated that the ZnO–NPs exhibited cytotoxic activity on A549 lung cancer cells and HT29 colon adenocarcinoma cells. Additionally, it was observed that the toxicity of the ZnO–NPs was less pronounced than that of the NPs produced by chemical techniques (Acar, 2021).

Edible plant fruits are a popular subject of study due to their innocuous nature and capacity to reduce the metals required for MNP synthesis. In a study conducted in 2019, the researchers formulated ZnO–NP and TiO_2–NP by mixing lemon fruit extract with ZnO and TiO_2 solutions. The findings of the study demonstrate that the synthesised NPs (NPs) exhibited inhibitory effects against the growth of *Dickeya dadantii* bacteria, whereas the suspended metal solutions demonstrated no antimicrobial activity (Hossain et al., 2019).

Plant–derived ZnO–NPs are an attractive solution to combat bacterial infections, considering their antimicrobial properties. Supporting this, Fahimmunisha et al. (2020) formulated ZnO–NPs using *Aloe socotrina* leaf extract in their study. As a result of the study, ZnO–NPs showed antimicrobial activity against four bacterial pathogen biofilms studied. This supports the claim that the formulated ZnO–NPs may play a key role in combating drug–resistant bacterial infections (Fahimmunisha et al., 2020).

Table 2. Some plant–derived ZnO–NPs found in the literature and natural reducing agent used

Reducing agent (Plant source)	Parts of plant	Properties	Reference
Punica granatum L.	Waste Peel Extract	Antioxidant, catalytic activity	(Meshkatalsadat et al., 2023)
Limonium pruinosum L. Chaz	Aqueous extract	Anticancer, antibacterial, antifungal, antioxidant	(Naiel et al., 2022)
Calotropis gigantea (Giant Milkweed)	Leaf, flower, and stem extracts	Antibacterial, antifungal, enzyme-inhibition activities	(Farooq et al., 2022)
Calotropis gigantea	Leaf powder	Synthesis of 3-amino thymoqui-none connected 1,4-dihyropyridine derivatives	(Gobinath et al., 2022)
Saponariaofficinalis L	Root	Antibacterial	(Kamçı et al., 2022)
Achiella millefolium L.	Aqueous extract	Anticancer	(Aydın Acar, 2021)
Aloe socotrina	Leaf extract	Antimicrobial	(Fahimmunisha et al., 2020)
Eucalyptus globules	Whole plant	Antifungal	(Ahmad et al., 2020)
Cassia fistula and Melia azadarach	Leaf extract	Antibacterial	(Naseer et al., 2020)
Carica papaya fruit	Seed extract	Electrochemical sensing	(Sharma et al., 2018)

3.3 Copper Nanoparticles

Copper NPs (CuO–NPs), another one of the plant–derived MNPs known as bioengineered products, are employed in several biomedical applications, as well as in the medical field and elsewhere. They are of interest due to the large surface area, antimicrobial activity, as well as the fact that they are harmless to nature, sustainable, highly efficient and biocompatible. CuO–NPs can be formulated via cheap and biocompatible natural reducing agents. The synthesis reaction of CuO–NPs is summarized in the chemical equation given below (Priya et al., 2023).

$Cu^{2+}[CuSO_4 \cdot 5H_2O] + \text{Extract} \rightarrow [Cu - \text{Extract}]^{2+}$

$[Cu - Extract]^{2+} + NaOH \rightarrow [Cu(OH)_2 - Extract]$

$[Cu(OH)_2 - Extract] + Stirring \rightarrow [CuO]\text{–}Precipitate \rightarrow Heating \rightarrow CuO\text{–}NPs$

In their study, Cheirmadurai et al. (2014) employed the use of extracts derived from *Lawsonia inermis* (henna) leaves as reducing and capping agents to facilitate the formation of CuO–NPs. The researchers demonstrated that the extracts obtained during the synthesis reaction could reduce Cu^{2+} salt ($CuSO_4$–$5H_2O$), thereby confirming the successful completion of the synthesis. Vitamin C, also known as ascorbic acid, has been observed to react directly or indirectly with serum copper protein, ceruloplasmin, via direct or indirect pathways. Moreover, ascorbic acid, when employed as an organic-based reductant, facilitates the formation of CuO–NPs (Khan et al., 2016). Priya et al. (2023) employed the use of *Morinda citrifolia* leaf extract in the synthesis of CuO–NPs, capitalising on the reducing property of ascorbic acid (Priya et al., 2023).

Table 3. Some plant–derived CuO–NPs found in the literature and natural reducing agent used

Reducing agent (Plant source)	**Parts of plant**	**Properties**	**Reference**
Morinda citrifolia	Leaf extract	Antifungal, antibacterial	(Priya et al., 2023)
Moringa Oleifera	Leaf extract	Supercapacitors, Energy storage application	(Ikhioya et al., 2023)
Piper nigrum fruit	Aqueous extract	Antioxidant, antibacterial, cytotoxic	(Kiranmayee et al., 2023)
Rhatany *(Krameria sp.)*	Root extract	Antifungal, antioxidant, antibacterial	(Alshammari et al., 2023)
Carica papaya	Leaf extract	Antioxidant, anticancer	(Dulta et al., 2021)
Punica granatum L. (Pomegranate peel)	Aqueous extract	Electrical conductivity	(Johnson et al., 2023)
Sesbania aculeata	Leaf extract	Antimicrobial, enhance plant growth	(Elakkiya et al., 2022)
Beta vulgaris L.	Aqueous extract	Antibacterial, anticancer	(Chandrasekaran et al., 2020)
Galeopsidis herba	Aqueous extract	Antioxidant, catalitic activity	(Dobrucka, 2018)
Azadirachta indica	Leaf extract	Characterization experiments	(Nagar & Devra, 2018)
Lawsonia inermis (Henna)	Leaf extract	Lousicidal activity, electrical activity	(Cheirmadurai et al., 2014)

In a recent research study, Kiranmayee et al. (2023) successfully developed metallic CuO–NPs derived from the fruit of *Piper nigrum*, commonly known as pepper. The alkamides and piperine alkaloids, which are biomolecules contained in the plant, facilitated the reduction of CuO and thereby provided stability. Evidence of the formation of CuO–NPs was observed in a color change from brown to blue. Chandrasekaran et al. (2020) achieved the synthesis of plant–derived CuO–NPs using the aqueous extract of *Beta vulgaris* L. Their experiments demonstrated that synthesized NPs could be developed as nanomedicines for infectious diseases and tumor research, due to the antibacterial effect of CuO–NPs. Dobrucka et al. (2018) successfully biosynthesized CuO–NPs using *Galeopsidis herba* extract. Experimental results proved that plant extracts took an active role in the reduction and stabilization of CuO–NPs. Moreover, the formulated CuO–NPs also showed enhanced catalytic activity in the reduction of malachite green.

3.4 Palladium Nanoparticles

Palladium (Pd), a platinum group metal, has high conductivity and corrosion resistance (Vijilvani et al., 2020). Pd–NPs can be used in many areas thanks to their physical, chemical, optical and thermodynamic properties. There are studies in the literature in basic fields such as biotechnology, medicine and electronics, where they are used as sensors, catalysts or drug delivery systems (Alamdari et al., 2022; Morriss et al., 2022; Yanik et al., 2023). There are difficulties in the formulation of Pd–NPs with traditional synthesis techniques due to their size control and their tendency to agglomerate during both synthesis and storage (Tao et al., 2021). Due to these difficulties, Bankar et al. (2010) suggested that biological materials such as plant extracts can also be used in the formulation of Pd–NPs. MNPs formulated with green synthesis technique using plant extracts can be synthesized in a highly controlled and simple manner. Moreover, their morphological properties (size and shape) can be controlled. Petla et al. (2012) succeeded in synthesizing Pd–NPs using a protein–rich Glycine max (soybean) leaf extract. It has been shown that proteins and amino acids present in soybean leaf extracts play an active role in the reduction of palladium ions. Moreover, it has been observed that these amino acids overcome the agglomeration problem present in conventional synthesis techniques of Pd–NPs. The authors also suggested that these plant-derived Pd–NPs they formulated have the potential to be used as catalysts for the degradation of azo dyes (Petla et al., 2012). Singh et al. (2023) emphasized that plant–derived Pd–NPs can have an important place in biomedical applications since they are formulated with natural stabilizers and reducing agents. Notwithstanding these properties, the rate of utilization of Pd–NPs in studies remains relatively low, particularly in the context of MNPs employed in medical applications. A summary of recent studies can be found in Table 4. Sampatkumar et al. (2024) employed a green synthesis technique to formulate Pd–NPs embedded on graphitic carbon nitride using banana leaf extracts. This resulted in a 3–step reaction. The active phytochemicals present in banana leaf extract not only facilitated the reduction of Pd^{2+} to Pd^{0} but also stabilized Pd-NPs. According to the study results, Pd–NPs have antibacterial activity against both gram (-) and gram (+) bacterial strains. Vijilvani et al. (2020) successfully synthesized Au–NPs, Pd–NPs and silver NPs from the leaves of *Solanum nigrum* plant. According to the catalytic activity test results, the reduction time of the silver catalyst was reported to be longer than the Au and Pd catalysts. Also, the antibacterial effect of silver NPs was reported to be greater than the other two (Vijilyani et al., 2020)

Table 4. Some plant–derived Pd–NPs found in the literature and natural reducing agent used

Reducing agent (Plant source)	Parts of plant	Properties	Reference
Banana	Leaf extract	Catalytic efficiency, antibacterial activity	(Sampatkumar et al., 2024)
Allium fistulosum, Basella alba & Tabernaemontana divaricate	Leaf extract	Antimicrobial, antifungal, and antioxidant	(Vinodhini et al., 2022)
Peganum harmala seed	Alkaloid fraction.	Anticancer, antioxidant	(Fahmy et al., 2021)
Solanum nigrum	Leaf extract	Antimicrobial	(Vijilvani et al., 2020)
Anogeissus latifolia (Gum ghatti)	Extract	Catalytic activity	(Kora & Rastogi, 2018)
Pimpinella tirupatiensis	Extract	Photocatalytic activity dye degradation	(Narasaiah et al., 2017)
Prunus × yedoensis tree.	Leaf extract	Antimicrobial	(Manikandan et al., 2016)

3.5 Silver Nanoparticles

Silver NPs (Ag–NPs) formulated from one of the basic elements, silver (Ag), also have the potential to be used as spectrally selective coatings for solar energy absorption, interlayer material for electric batteries, and catalysts in chemical reactions. It can also be used in various materials including textiles, packaging, and plastics. It is also a promising option for use as a food additive. Apart from these, the antimicrobial activity of Ag ions, their capacity to facilitate wound healing and minimize scar formation make them valuable in both medical and cosmetic industries (Singh et al., 2022).

The reasons for preferring plant –derived Ag–NPs are their non-toxicity, compatibility with ecological balance, simplicity of the one-step production process, and sustainability. Various biomolecules found in plants provide the reduction and stabilization of Ag ions. A summary of the synthesis of Ag–NPs can be summarized as follows (Roy & Das, 2015):

(i) Selected plant parts are washed with distilled water, then dried in the dark and ground into powder.
(ii) Subsequently, the resulting powder is boiled with deionized distilled water to obtain the desired extracts. Subsequently, the solution is filtered and combined with an $AgNO_3$ solution.
(iii) As a consequence of the excitation of surface plasmon vibrations in Ag–NPs, the color of the prepared mixture changes to brown, which serves as proof that Ag^+ ions are reduced to Ag^0 (Adrianto et al., 2022).

The morphological properties of MNPs synthesized using a green approach have been shown to vary depending on a few factors, including temperature, pH, reaction time, metal salt concentration in the solution, and the quantity of plant extracts used. As an illustration, an increase in temperature will result in a reduction in the required reaction time for the synthesis of Ag–NPs. The reaction at room temperature is considerably slower. However, the critical temperature is between 30 and 100°C, as below this temperature the concentration of Ag^+ ions decrease, which may result in homogeneous nucleation of Ag nuclei. Moreover, the size of Ag–NPs formulated at high temperatures decreased and their stability increased (Ansari et al., 2023). In the study conducted by Makarov et al. (2014), it was shown that the synthesis of Ag-NPs obtained using *Curcuma longa* (turmeric) tuber powder and extract increased at alkaline pH levels. This can be attributed to the fact that the extracts at alkaline pH levels have a significant amount of negatively charged functional groups that have the potential to reduce and bind Ag ions (Makarov et al., 2014). As mentioned before, these multiple functional groups in the content of flavonoids facilitate their reduction to MNPs by forming stable complexes with metal ions. It has been postulated that the tautomeric transformations of flavonoids from enol–form to keto–form may facilitate the release of a reactive hydrogen atom, enabling the synthesis of MNPs through metal ion reduction. Ahmad et al. (2010) suggested that in extracts of *Ocimum basilicum* (sweet basil), the enol form of luteolin facilitates the release of hydrogen, which is essential for the reduction of Ag^+ to Ag^0. Additionally, the keto form of rosemarinic acid also promotes the release of reactive hydrogen. However, the same researchers also reported that this complex formed was inherently unstable due to the presence of two hydroxyl groups on a carbon atom. They observed that the complex eventually returned to the keto form. It was thus highlighted that the release of reactive hydrogen was a key process in the synthesis of Ag–NPs in both cases (Ahmad et al., 2010). A summary of the plants and parts thereof employed in the synthesis of Ag–NPs is provided in Table 5.

Table 5. Some plant–derived Ag–NPs found in the literature and natural reducing agent used.

Reducing agent (Plant source)	Parts of plant	Properties	Reference
Rubus discolor	Leaf extract	Multi–drug resistant bacteria and cancer cells	(Ghasemi et al., 2024)
Elaeocarpus serratus (Ceylon olive)	Fruit extract	Antimicrobial	(Fernando et al., 2024)
Uvaria narum	Leaf extract	Antibacterial, Antiangiogenic, Anticancer, Catalytic Properties	(Ajaykumar et al., 2023)
Moringa oleifera	Leaf extract	Antibacterial	(Asif et al., 2022)
Capsicum chinense	Root, body, leaf extract	Characterization experiment	(Lomelí-Rosales et al., 2022)
Moringa oleifera	Aqueous extract	SPR based sensor	(Adrianto et al., 2022)
Eugenia roxburghii DC	Leaves	Inhibit biofilm formation	(Giri et al., 2022)
Euphorbia rigida	Leaf extract	Antimicrobial	(Kocazorbaz, 2021)
Mandragora autumnalis	Extract	Antioxidant, antimicrobial	(Dağlioğlu & Öztürk, 2021)
Clerodendrum inerme	Leaf extract	Antimicrobial, antioxidant	(Khan et al., 2020)
Tectona grandis	Seed extact	Antibacterial	(Rautela et al., 2019)
Shikakai, Reetha	Leaf extract	SERS active substrate	(Kumar Sur et al., 2018)
Cinnamon zeylanicum	Bark Extract	Characterization experiment	(Saliem et al., 2018)

Most plants contain saponins, which are surface active molecules with wetting, foaming and grease distribution capacity. The hydrophilic component of these molecules consists of sugars and the hydrophobic part consists of sapogenic acid and oleanolic acid. MNP synthesis can also be carried out through plant–derived surfactants (biosurfactants). These provide alternatives for MNPs to be biocompatible and environmentally friendly. Biosurfactant molecules present in plant extracts play an important dual role as both stabilizers and reducing agents. Kumar Sur et al. (2018) biosynthesized Ag-NPs from plant extracts of *Shikakai* and *Reetha*. The findings indicate that saponins in plant extracts fulfill the roles of stabilizers and reducing agents. Moreover, it has been shown that these Ag–NPs can be used for dye molecule synthesis and bacterial detection purposes (Kumar Sur et al., 2018). Ajaykumar et al. (2023) synthesized Ag–NPs using phytochemicals found in the leaves of *Uvaria narum* plant as reducing and stabilizing agents. Antibacterial, antiangiogenic, anticancer and cytotoxic properties of the synthesized Ag–NPs were tested in three fish lines (Ajaykumar et al., 2023). Adrianto et al. (2022) synthesized Ag–NPs with extracts obtained from Moringa oleifera plant. They also formulated Ag-PVP-NPs using polyvinylpyrrolidone (PVP) as a stabilizer. According to the localized surface plasmon resonance (LSPR) behavior test of these Ag–PVP–NPs, the addition of PVP caused a shift in the LSPR angle. Accordingly, the microstructural and LSPR behavior of Ag–NPs can be regulated using PVP, thus paving the way for optimizing the performance of advanced SPR–based sensors (Adrianto et al., 2022). Khan et al. (2020) formulated Au–NPs and Ag–NPs using compounds from leaf extracts of plant as both reducing and capping agents. During formulation, these leaf extracts were shown to utilize reduced Ag^{+} and Au^{3+} and cap them with many functional groups. It was noted that these synthesized MNPs have antimicrobial, antifungal and cytotoxic effects. They also reported that they possessed antioxidant properties according to the DPPH free radical scavenging test (Khan et al., 2020).

The primary drawback of synthesizing plant–derived MNPs using phytochemicals is that the reaction time is longer compared to conventional methods that utilize chemicals. To overcome this problem, researchers have combined the green synthesis technique with the microwave (MW)–mediated synthesis technique. Thanks to this solution, they have noted that toxic substances will not be formed in the chemical reduction and high efficiency has been achieved (Ajaykumar et al., 2023). Saidu et al. (2019) formulated Ag–NP using the aqueous extract of Clammy cherry (*Cordia obliqua* Willd.). However, besides the green synthesis reaction steps of plant–derived NPs, they also used MW and irradiated the reaction mixture under a household MW oven to synthesize Ag–NP with high stability, both quickly and inexpensively. In addition to testing the optical properties of the formulated Ag–NPs, they also evaluated their catalytic activity. Also, it has been reported that it has highly effective antibacterial properties (Saidu et al., 2019). In another study, Francis et al. (2017) bioformulated Ag–NP supported by MW using *Elephantopus scaber* leaf extracts to act as phytoreductive agents. It was reported that the Ag–NPs they synthesized cleaned up the damage caused to nature using dyes such as eosin Y in a very short time. In addition, these Ag-NPs, which have antimicrobial activity, also have anticancer properties.

3.6 Nickel Nanoparticles

The use of nickel oxide NPs (NiO–NPs) has become widespread in biomedical and medical fields due to their antifungal, antibacterial, anticancer, and cytotoxic effects, as well as their usability as enzyme inhibitors (Bhoye et al., 2023; Haritha et al., 2022; Lv et al., 2023; Punitha & Mary Saral, 2024; Taş et al., 2021). In addition, NiO–NPs have also become a subject of scientific interest due to their exceptional properties, including electrochemical stability, high coercive forces, magneto–crystalline anisotropy, high specific capacitance, durability, and antiferromagnetic properties. Notwithstanding the numerous properties exhibited by NiO–NP-based films formulated through conventional techniques, most of these materials demonstrated an insufficient affinity. Moreover, the synthesis of ultraporous NiO–NP materials with a diameter < 20 nm has proven to be particularly challenging. To address the issue of limited cycle stability, the green synthesis of plant–derived NiO–NPs has been proposed as a potential solution. The formulation processes of NiO–NPs synthesized with the green synthesis technique were evaluated in terms of energy and were found to be efficient and environmentally benign (Hussain et al., 2023). NiO–NPs have the potential to be employed as catalysts in a number of processes, including hydrogen autotransfer and transfer hydrogenation reactions, low-temperature oxidation of CO and propane, diffusion bonding of stainless–steel surfaces, and hydrogenation of nitrobenzene. Additionally, they have applications as supercapacitor electrode materials. Consequently, the specific surface characteristics of these NiO–NPs result in an increased contact surface area of conductive carbon materials, thereby enhancing the conductivity of the system. To enhance conductivity, this process prevents the adsorption or dispersion of active materials. It is acknowledged that the aggregation of NPs during the process is minimized by the replacement of carbon molecules present in plant extracts, which constitute a primary component of the green synthesis technique, with the aforementioned materials.

Table 6. Some plant–derived NiO–NPs found in the literature and natural reducing agent used.

Reducing agent (Plant source)	Parts of plant	Properties	Reference
Fragaria ananassa (strawberries), *Vitis vinifera* (grapes), *Citrus paradisi* (grapefruits)	Aqueous extract	Removal of pollutant dyes (Photocatalytic activity)	(Nejo et al., 2024)
Orange	Leaf extract	Antibacterial	(Khodair et al., 2022)
Acacia nilotica	Leaf extract	Electrochemical, antibacterial haemolytic activity	(Hussain et al., 2023)
Piper betle	Leaf extract	Antibacterial, photocatalytic	(Singh et al., 2022)
Aloe vera	Gel extract	Antimicrobial	(Ahmad et al., 2022)
Peumus boldus	Aqueous extract	Antibacterial	(Taş et al., 2021)
Agathosma betulina	Aqueous extract	photovoltaic/photodiode response in IR.	(Thema et al., 2016)
Stevia	Leaf aqueous extract	Antibacterial	(Srihasam et al., 2020)
Ananas comosus.	Leaf extract	Photocatalytic activity	(Olajire & Mohammed, 2020)
Calotropis gigantea	Leaf extract	Catalytic activity, antibacterial	(Din et al., 2018)
Aegle marmelos	Leaf extract	Cytotoxic, antibacterial, photocatalytic	(Ezhilarasi et al., 2018)
Neem	Leaf extract	Magnetize, coercivitic antibacterial	(Helan et al., 2016)

In 2016, Thema et al. pioneered the synthesis of NiO–NPs using extracts of the Agathosma betulina plant via a green synthesis technique. The extracts primarily serve to form complexes with various Ni-derived salts, including Ni (II) hexahydrate ($NiCl_2.6H_2O$), N i(II) nitrate hexahydrate ($Ni(NO_3)_2.6H_2O$, soluble at 56°C), Ni(II) sulfate hexahydrate ($NiSO_4.6H_2O$, soluble at 100 °C), and Ni(II) sulfate hepta-hydrate ($NiSO_4.7H_2O$) (Thema et al., 2016). A summary of the studies conducted on the synthesis of plant–derived NiO–NPs in the literature is provided in Table 6. Nejo et al. (2024) achieved the synthesis of NiO–NPs from the extracts of *Fragaria ananassa* (strawberries), *Vitis vinifera* (grapes) and *Citrus paradisi* (grapefruits), which exhibited notable antioxidant activity. They tested the catalytic activity of the synthesized NiO–NPs on methyl red dye. They concluded that NiO–NPs are effective in environ-mental remediation (Nejo et al., 2024). Hussain et al. (2023) also synthesized NiO–NPs from *Acacia nilotica* leaf extracts (water and ethanol) using nickel nitrate (Ni $(NO_3)_2.6H_2O$) and sodium hydroxide (NaOH) (Hussain et al., 2023). In their study, Srihasam et al. (2020) successfully synthesized NiO–NPs using stevia leaf juice. They showed that these NiO-NPs exhibited antioxidant and *in vitro* antimicrobial activity against both Gram (+) and Gram (-) bacteria. (Srihasam et al., 2020).

3.7 Iron Oxide Nanoparticles

The focus of attention of researchers is shifting towards iron oxide NPs (IO–NPs) due to their ex-traordinary thermal, optical, electronic and superparamagnetic capabilities. Fe_3O_4–NPs, which occur naturally and can be formulated using a variety of techniques, are particularly intriguing due to their distinctive magnetic properties, including saturation magnetization, coercivity and permanent magneti-zation. Additionally, their biocompatibility has rendered them an attractive subject of investigation for disparate research teams (Laurent et al., 2010; Nguyen et al., 2021). As with other MNPs, the green synthesis technique is preferred to enhance the production conditions and augment the efficiency of IO–NPs formulated with traditional techniques (Patiño-Ruiz et al., 2021). Patino-Ruiz et al. (2021) made a comparison between the green chemistry and the conventional coprecipitation technique employed in the synthesis of IO–NPs as part of their study. To this end, the researchers employed a green chemistry technique utilizing *Cymbopogon citratus* extract and sodium carbonate (Na_2CO_3) as stabilizers. In contrast with the preceding technique, the coprecipitation method involved the utilization of Fe (II) and NaOH, respectively. The study revealed that plant–derived IO–NPs resulted in less toxic substance release compared to that synthesized by other technique. It is argued that these released toxic substances are related to the formulated process which involves land occupation and excessive water usage. As a result, the production process of plant-derived nanoparticles has been proven to be environmentally friendly (Patiño-Ruiz et al., 2021). Rehman et al. (2023) synthesized IO–NPs from the extract of the *Parthenium hysterophorus* plant and as a result, they developed many chemically modified carbon adsorbents (Rehman et al., 2023). Irum et al. (2020) were the first research group to investigate the effects of IO–NPs on the induction and shoot regeneration of the resistant chickpea crop (Cicer arietinum L.), which is of significant economic importance. They employed *Cymbopogon jwarancusa* leaf extracts that they had synthesised (Irum et al., 2020) as a means of conducting their experiments. The MW technique, when employed in conjunction with the green synthesis technique, is not the preferred option due to its ability to be heated rapidly, lack of detrimental impact on the natural environment, high yields, relatively short reaction times and cost–effectiveness. Furthermore, the capacity to regulate the morphological characteristics of the nanoparticles (NPs) undergoing synthesis renders this integrated system advantageous. Buarki et al. (2022) synthesised α–Fe_2O_3–NPs from *Hibiscus rosa sinensis* (hibiscus) flower extract, which they used as reducing and stabilising agent, by green synthesis technique. In addition to the green synthesis technique, they also received support from the MW technique (Buarki et al., 2022).

4. CHARACTERIZATION TECHNIQUES OF PLANT–DERIVED METAL NANOPARTICLES

Once plant–derived MNPs are formed, their characterization is invaluable to determine their potential for therapeutic applications and to assess their biocompatibility. A wide range of basic physicochemical techniques, such as UV-Visible absorption spectroscopy, Fourier Transform Infrared Spectroscopy (FTIR), Raman spectroscopy, Scanning Electron Microscopy (SEM), Transmission Electron Microscopy (TEM) and Powder X-ray Diffraction (XRD), have been used to characterize plant-derived MNPs. These techniques are used to ascertain the size, morphology, chemical composition, crystallography and surface area of the MNPs (Meyns et al., 2022; Tugarova et al., 2018). In addition to the traditional characterisation tests, recent studies have employed a range of novel techniques, including X-ray photoelectron spectroscopy (XPS), dynamic light scattering (DLS), energy dispersive X-ray spectroscopy (EDX), zeta potential and field emission scanning electron microscopy (FE–SEM) (Matouk et al., 2022; Titus et al., 2019).

4.1 Ultraviolet–Visible Absorption Spectroscopy

Ultraviolet–visible (UV–vis) absorption spectroscopy is a simple technique that measures the amount of light absorbed and scattered by the sample, indicating the absorption of the peak. The strength of the bands at the peak provides an idea about the stability of plant-derived MNPs. The optical properties of plant –derived MNPs are highly sensitive to many factors, including reaction time, temperature, pH, and suspension concentration. These properties are affected by numerous additional factors, including the size, shape, concentration, aggregation state, and refractive index of the MNPs themselves. The sensitivity of UV–vis absorption spectroscopy allows the characterization of MNPs and contributes to scientific studies through the results it provides (Nisar Ahmad et al., 2022; Xulu et al., 2022). In essence, the identification and characterization of MNPs with sizes between 2 and 100 nm requires the use of wavelengths between 200 and 800 nm (Bagheri et al., 2023). It has been reported that MNPs with surface plasmon resonance (SPR) peaks between 350 and 500 nm, such as Ag–NPs observed in the wavelength range of 400-450 nm and Au–NPs observed in the wavelength range of 500–550 nm in UV-visible absorption spectroscopy (Soliman et al., 2023), have sizes smaller than 100 nm. Goh et al. (2014) investigated the relationship between the size of ZnO–NPs and their absorption spectra in the UV–vis range. The findings showed that the UV absorbance increased with the size of MNPs between 15 and 40 nm. However, it was also emphasized that the particle concentration number is an important factor. Because it was stated that the absorbance decreased with the decrease in the concentration of larger sized MNPs (>70 nm) (Goh et al., 2014).

4.2 X–Ray Crystallography and X–Ray Diffraction

The X–ray crystallography technique allows for the characterisation of the crystal structure and phase identification of plant–derived MNPs. In this technique, X–rays are directed at the surface of the crystal and interact with the atoms within it, thereby measuring the degree of crystallinity. It has been demonstrated that plant-derived MNPs form crystal structures with particle sizes within the range of 25 to 30 nm (Karade et al., 2018; Nguyen et al., 2021).

X–ray diffraction (XRD) is employed to characterise the crystal structures and phases of formulated plant-derived MNPs. The fundamental objective of this technique is to examine the crystal structures of synthesised nanomaterials and to identify the phases present. It has been demonstrated that the formation of plant-derived MNPs results in the generation of a robust diffraction pattern (Ghoohestani et al., 2024; Ramesh et al., 2018). In their study, Ibrahim et al. (2023) formulated Cu–NP using avocado seed extracts and demonstrated that Cu–NPs have a crystalline nature, as evidenced by XRD results (Ibrahim et al., 2023).

4.3 Fourier Transform Infrared Spectroscopy

Fourier transform infrared (FTIR) spectroscopy is a technique primarily employed for the characterisation of the chemical structures of samples, including the identification of functional groups (such as amide, hydroxyl and carboxyl groups) and bond parameters (Arik et al., 2023). The technique can provide extensive information about the surface chemistry of plant–derived MNPs by detecting the bioactive agents and functional groups attached to the surface of plants or plant extracts used in the green synthesis of MNPs. One of the reasons why the FTIR technique is preferred is to identify the changes that occur in the secondary structures of protein molecules (Al-Zahrani et al., 2021; Albeladi et al., 2020).

4.4 Microscopic Techniques

The most fundamental techniques employed in the characterisation of plant-derived MNPs are high-resolution scanning electron microscopy (SEM), transmission electron microscopy (TEM) and atomic force microscopy (AFM). These are utilised to ascertain the morphology of the MNPs in question. In essence, two–dimensional images can be generated by sectioning samples of specified dimensions and employing the SEM technique to ascertain their dimensions and morphology (Arik et al., 2022). Furthermore, another technique, namely field-emission scanning electron microscopy (FE-SEM), can be employed to obtain more detailed images of the bioactive agents present in plant-derived MNPs. Another imaging method, TEM, can provide superior resolution for the dimensions and morphology of plant–derived MNPs (Thilakan et al., 2022). In their study, Amjad et al. (2021) used *Fortunella margarita* leaves in the formulation of Cu-NPs and reported that they formed Cu–NPs with sizes in the range of 51–57 nm according to the characterisation test results performed with the SEM technique (Amjad et al., 2021). Furthermore, the surface roughness of the samples can be determined qualitatively in the AFM technique, which offers superior resolution compared to both SEM and TEM, which are used to obtain three-dimensional images (Bellotti et al., 2022; Menon et al., 2017). Wu et al. (2019) achieved the successful formulation of Cu–NPs using *Cissus vitiginea* leaves and obtained three–dimensional images of spherical Cu–NPs clustering at 5–25 mm magnifications in the AFM technique used to analyse the surface structure and topography roughness (Wu et al., 2020).

4.5. Light Scattering Methods and Zeta Potential (ζ)

The most basic objective of dynamic light scattering (DLS), a light scattering technique employed to ascertain the physicochemical attributes of plant–derived magnetic nanoparticles (MNPs), is to furnish qualitative data. Furthermore, the diameter and stability of plant-derived MNPs dispersed in colloidal suspensions under different conditions can be determined. Concurrently, this technique is employed to ascertain the surface charge and dimensions of MNPs. According to the spreading motion of MNPs, small MNPs are observed to move faster than large ones (M. F. Khan & Khan, 2023).

In another technique, zeta potential (ζ) information can be obtained regarding the surface charge and its relations with other molecules on the surface. Furthermore, the stability of plant-derived MNPs can be ascertained, and it can be inferred that an elevated ζ value correlates with enhanced MNPs stability. For instance, despite MNPs typically exhibiting low stability within the 0–5 mV range, they demonstrate the greatest stability when the ζ potential is >50 mV (Takma et al., 2023; Yasmin et al., 2023).

5. SCALABILITY AND INDUSTRIAL APPLICABILITY OF PLANT-DERIVED NANOPARTICLE SYNTHESIS

Plant–derived MNPs, particularly the most renowned Ag–NPs and Au–NPs, are highly sought after in the commercial sector due to their extensive range of applications in industrial, electronic, environmental, bioenergy, biosensor and, most prominently, medical fields. The plant–derived MNPs do not have the potential to cause significant adverse effects on human health, in contrast to those formulated using chemical techniques. Consequently, they are employed in a multitude of applications. Some of these applications are pharmacological and medical, due to the antimicrobial, antioxidant and anti-inflammatory activities of the MNPs. Furthermore, they can be utilized in the agricultural sector for applications such as the detection and treatment of plant diseases, crop protection, bioremediation, and the controlled release of agricultural chemicals. Plant-derived MNPs, which are widely used in applications such as food science and technology and bioengineering, also have significant potential for use in the field of cosmetics, for example as a raw material for sunscreen, anti-ageing and hair care creams (Aboyewa et al., 2021; Hano & Abbasi, 2022). The industrial–scale synthesis of plant-derived MNPs is more advantageous than the synthesis of other micro or macro raw materials due to the simplicity of the process (Verma et al., 2021). Nevertheless, significant challenges must be addressed to facilitate the transition to this larger scale. The reproducibility of plant–derived MNPs produced in different volume reservoirs is relatively low. It is hypothesised that this issue can be addressed through the utilisation of automated batch or flow–type reactors. Nevertheless, the development of disparate equipment is imperative, as not all devices are compatible (Bharali et al., 2023). It is evident that further studies are required to enhance their efficiency and scalability, which will facilitate the development of additional bioactive agents derived from MNPs for incorporation into pharmaceutical and nutraceutical applications. It is also anticipated that they will be applicable in water disinfection for phytopathogen treatment in agriculture or environmental cleaning.

6. LIMITATIONS

A considerable body of research has recently emerged on the potential applications of MNPs, as previously outlined. Despite the numerous advantages of plant–derived MNPs, which have recently garnered attention and become a focal point in research, their large–scale synthesis and some applications remain constrained. For instance, the high phytochemical compound content of plant extracts precludes the control of MNPs in terms of size and morphology. This is because detailed optimization of both the composition ratios of plant extracts and synthesis parameters, such as temperature and pH, is required. This is attributed to the fact that some factors, including product quality control, synthesis parameters, material selection and product application, are believed to act as constraints (Dikshit et al., 2021; Zuhrotun et al., 2023). It has been demonstrated that the quality of MNPs is affected by the type of extract used, resulting in alterations to their morphology and the acquisition of inadequate properties. The observed changes in morphology are a consequence of the diverse methodologies employed in the synthesis of MNPs via green techniques. These methodologies result in the formation of MNPs with varying morphologies, including spherical and/or non-crystalline, hexagonal, or irregular clusters. This constrains their potential applications and complicates the establishment of a mass production network. Additionally, the yields and conversion rates of plant–derived MNPs are relatively low, indicating that the efficiency of converting metallic ions into NPs is less than 50%. Despite the use of elevated metal ion concentrations in the synthesis process, the formulation of a substantial number of magnetic NPs remains unfeasible. Consequently, the commercial potential of this approach is limited. Furthermore, the vast number of plant species involved in the selection of materials represents a significant challenge. It has been proposed that it is nearly impossible to produce MNPs derived from endemic plants on a global scale. Additionally, the time constraints associated with MNPs synthesized using plant flowers, coupled with the fact that the flowering cycle of the plant in question varies across countries and even regions, introduce further limitations. The second to last limitation is the specific application area and form of plant–derived magnetic nanoparticles (MNPs). For instance, it has been proposed that MNPs employed in the remediation of heavy metals in wastewater do not consistently demonstrate the same level of efficiency and may even prove ineffective in field studies (Borji et al., 2020). The final limitation is the lengthy reaction times inherent to the formulation process of plant-derived MNPs, coupled with the substantial energy expenditure involved. Furthermore, the green chemistry technique, which may necessitate elevated temperatures, may also result in prolonged reaction times. In such instances, both a considerable amount of energy is consumed, and ecologically detrimental outcomes may ensue. Furthermore, the storage conditions of plant extracts used in synthesis require low temperatures, which presents a similar challenge to that previously discussed, namely excessive energy consumption (Alsaiari et al., 2023; Dikshit et al., 2021).

Deciphering the steps in the reaction processes of plant–derived MNPs using the green chemistry technique remains a topic of ongoing research. To fully comprehend the underlying mechanisms and address the limitations, further investigation is necessary.

7. FUTURE PERSPECTIVES

In recent years, the database of plants used in the formulation of various plant–derived MNPs has been the subject of scrutiny in the context of studies in literature. The synthesis of plant-derived MNPs using environmentally friendly and cost–effective methods has been the subject of extensive research. These green methods aim to reduce the environmental impact and cost of current physical and chemical approaches to MNPs synthesis. Nevertheless, further research is required to fully exploit its potential. It is evident that further studies are required to enhance their efficiency and scalability, thereby facilitating the utilisation of plant–derived MNPs in the synthesis of other bioactive agents, including pharmaceuticals and nutraceuticals, and potentially in diverse applications. Furthermore, it is anticipated that they will be employed in the disinfection of water for the treatment of phytopathogens or the cleansing of the environment in agricultural contexts. It is acknowledged that the synthesis of plant-derived MNPs has become a popular area of research, and it is predicted that this will continue to grow exponentially in the future. However, there is a lack of research in the literature showing the possibility of adapting them to an industrial and larger scale. If plant–derived MNPs are adapted to the industrial scale, it is thought that plant resources, which are thought to be insufficient to meet long-term demand, will also be increased. Nevertheless, the lack of sufficient clarification regarding the aforementioned factors precludes any definitive conclusion regarding the feasibility of scaling plant–derived MNPs to industrial levels, which continues to represent an unknown scientific curiosity (Peralta-Videa et al., 2016).

It is established that application–specific MNPs must be synthesized for utilization in disparate applications. This is a subject that requires further investigation to facilitate resolution. It is thus imperative that future studies elucidate the structure of the phytochemicals that play a role as reducing and stabilizing agents in MNP synthesis and actively participate in biosynthesis in future studies. The issue of metabolic profile inhomogeneity resulting from abiotic stress, herbicide utilisation or genetic accumulation has led to a lack of reproducibility in plant–derived MNPs. To address this issue, future research may consider the use of cultivated plants or plant tissue culture. Further research should focus on the use of seasonal plants, the value of the raw material and the difficulties experienced during the extraction of plant extracts, with the aim of reducing production costs and making the synthesis process more applicable. Additionally, future studies should consider the use of evergreen plants globally, the use of agricultural wastes, which also increase sustainability, as mentioned in the section, or the development of easier techniques for extraction. It was highlighted that plant–derived MNPs offer a more energy–efficient solution. Nevertheless, to further reduce the energy expenditure resulting from prolonged reaction times, elevated temperatures, and high pressures, in addition to designing experiments where plant–derived MNPs can be synthesized at room temperature, it is imperative to conduct studies that further elucidate the issue of energy conservation in the future. It is evident that to enhance the efficacy and reliability of MNP, while also reducing the reaction time, a key focus should be placed on optimizing the reaction parameters. Furthermore, there is a paucity of studies on the production of MNP as a commercial product using green chemistry techniques. Consequently, it is envisaged that a futuristic approach can be achieved by prioritizing studies aimed at increasing the production scale. Not only that, although it is established that the toxicity of plant–derived MNPs is less than that of MNPs produced by chemical techniques, the extent of this difference is also unclear. This may influence the interest of pharmaceutical manufacturers planning to produce on an industrial scale, potentially leading to a reduction in their enthusiasm. For plant–derived MNPs to realise their potential in a range of application areas, it is essential that future studies prioritise in vivo experimentation and undertake a comprehensive evaluation of the issue of toxicity. For MNPs to realize their potential for use in a variety of application areas, it is essential to conduct thorough human and animal experiments. Furthermore, the removal and purification of MNPs from the reaction mixture after synthesis represents a crucial aspect that requires attention.

8. CONCLUSION

In the last few years, nanobiotechnology has become an important scientific research area that offers promising opportunities to develop new materials using various natural resources. This section presents the parameters affecting the synthesis of MNPs, including the formulation of plant–derived MNPs and the plant species used, as well as an evaluation of their advantages over other conventional techniques. The plant species mentioned in the preceding section, along with numerous others, are elucidated in detail as potential raw materials for the green synthesis of MNPs. Phenolic, amino, carboxyl and hydroxyl groups as well as proteins and amino acids found in extracts of plant parts such as leaves, flowers and fruits are used as metal reducing agents in MNP biosynthesis. It is possible to obtain MNPs in a sustainable manner using bioresources, in a reliable and reproducible way that does not involve high costs. Furthermore, the use of water and ethanol solvents during synthesis allows for the formulation of environmentally benign products. The diversity of plant species and from variations in process parameters allows for the synthesis of plant–derived MNPs with varying morphologies, sizes, and properties, thereby expanding their potential applications. Furthermore, the coating of plant extracts on the surface of MNPs renders them more biocompatible without the necessity for additional processing, in comparison to those synthesized through traditional techniques.

REFERENCES

Abdul Razak, N. A., Othman, N. H., Mat Shayuti, M. S., Jumahat, A., Sapiai, N., & Lau, W. J. (2022). Agricultural and industrial waste-derived mesoporous silica nanoparticles: A review on chemical synthesis route. *Journal of Environmental Chemical Engineering*, 10(2), 107322. DOI: 10.1016/j.jece.2022.107322

Aboyewa, J., Sibuyi, N., Meyer, M., & Oguntibeju, O. (2021). Green Synthesis of Metallic Nanoparticles Using Some Selected Medicinal Plants from Southern Africa and Their Biological Applications. *Plants*, 10(9), 1929. DOI: 10.3390/plants10091929 PMID: 34579460

Abuzeid, H. M., Julien, C. M., Zhu, L., & Hashem, A. M. (2023). Green Synthesis of Nanoparticles and Their Energy Storage, Environmental, and Biomedical Applications. *Crystals*, 13(11), 1576. DOI: 10.3390/cryst13111576

Adrianto, N., Panre, A., Istiqomah, N., Riswan, M., Apriliani, F., & Suharyadi, E. (2022). Localized surface plasmon resonance properties of green synthesized silver nanoparticles. *Nano-Structures & Nano-Objects*, 31, 1–12. DOI: 10.1016/j.nanoso.2022.100895

Ahmad, B., Khan, M. I., Naeem, M. A., Alhodaib, A., Fatima, M., Amami, M., Al-Abbad, E. A., Kausar, A., Alwadai, N., Nazir, A., & Iqbal, M. (2022). Green synthesis of NiO nanoparticles using Aloe vera gel extract and evaluation of antimicrobial activity. *Materials Chemistry and Physics*, 288, 126363. DOI: 10.1016/j.matchemphys.2022.126363

Ahmad, H., Venugopal, K., Rajagopal, K., De Britto, S., Nandini, B., Pushpalatha, H. G., Konappa, N., Udayashankar, A. C., Geetha, N., & Jogaiah, S. (2020). Green Synthesis and Characterization of Zinc Oxide Nanoparticles Using Eucalyptus globules and Their Fungicidal Ability Against Pathogenic Fungi of Apple Orchards. *Biomolecules*, 10(3), 425. DOI: 10.3390/biom10030425 PMID: 32182874

Ahmad, N. (2022). Fozia, D., Jabeen, M., Ul-Haq, Z., Ahmad, I., Wahab, A., … Yahya Khan, M. (2022). Green Fabrication of Silver Nanoparticles using Euphorbia serpens Kunth Aqueous Extract, Their Characterization, and Investigation of Its In Vitro Antioxidative, Antimicrobial, Insecticidal, and Cytotoxic Activities. *BioMed Research International*, •••, 1–11.

Ahmad, N., Sharma, S., Alam, M. K., Singh, V. N., Shamsi, S. F., Mehta, B. R., & Fatma, A. (2010). Rapid synthesis of silver nanoparticles using dried medicinal plant of basil. *Colloids and Surfaces. B, Biointerfaces*, 81(1), 81–86. DOI: 10.1016/j.colsurfb.2010.06.029 PMID: 20656463

Ahmad, T., Bustam, M., Irfan, M., Moniruzzaman, M., Asghar, H., & Bhattacharjee, S. (2019). Mechanistic investigation of phytochemicals involved in green synthesis of gold nanoparticles using aqueous Elaeis guineensis leaves extract: Role of phenolic compounds and flavonoids. *Biotechnology and Applied Biochemistry*, 66(4), 698–708. Advance online publication. DOI: 10.1002/bab.1787 PMID: 31172593

Ajaykumar, A. P., Mathew, A., Chandni, A. P., Varma, S. R., Jayaraj, K. N., Sabira, O., Rasheed, V. A., Binitha, V. S., Swaminathan, T. R., Basheer, V. S., Giri, S., & Chatterjee, S. (2023). Green Synthesis of Silver Nanoparticles Using the Leaf Extract of the Medicinal Plant, Uvaria narum and Its Antibacterial, Antiangiogenic, Anticancer and Catalytic Properties. *Antibiotics (Basel, Switzerland)*, 12(3), 564. DOI: 10.3390/antibiotics12030564 PMID: 36978431

Al-Zahrani, S., Astudillo-Calderón, S., Pintos, B., Pérez-Urria, E., Manzanera, J. A., Martín, L., & Gomez-Garay, A. (2021). Role of Synthetic Plant Extracts on the Production of Silver-Derived Nanoparticles. *Plants*, 10(8), 1671. DOI: 10.3390/plants10081671 PMID: 34451715

Alamdari, S. G., Amini, M., Jalilzadeh, N., Baradaran, B., Mohammadzadeh, R., Mokhtarzadeh, A., & Oroojalian, F. (2022). Recent advances in nanoparticle-based photothermal therapy for breast cancer. *Journal of Controlled Release*, 349, 269–303. DOI: 10.1016/j.jconrel.2022.06.050 PMID: 35787915

Albeladi, S., Malik, M., & Al-thabaiti, S. (2020). Facile biofabrication of silver nanoparticles using Salvia officinalis leaf extract and its catalytic activity towards Congo red dye degradation. *Journal of Materials Research and Technology*, 9(5), 10031–10044. DOI: 10.1016/j.jmrt.2020.06.074

Ali, N. H., & Mohammed, A. M. (2021). Biosynthesis and characterization of platinum nanoparticles using Iraqi Zahidi dates and evaluation of their biological applications. *Biotechnology Reports (Amsterdam, Netherlands)*, 30, e00635. DOI: 10.1016/j.btre.2021.e00635 PMID: 34094893

Aljabali, A. A. A., Akkam, Y., Al Zoubi, M. S., Al-Batayneh, K. M., Al-Trad, B., Abo Alrob, O., Alkilany, A., Benamara, M., & Evans, D. J. (2018). Synthesis of Gold Nanoparticles Using Leaf Extract of Ziziphus zizyphus and their Antimicrobial Activity. *Nanomaterials (Basel, Switzerland)*, 8(3), 174. DOI: 10.3390/nano8030174 PMID: 29562669

Alsaiari, N. S., Alzahrani, F. M., Amari, A., Osman, H., Harharah, H. N., Elboughdiri, N., & Tahoon, M. A. (2023). Plant and Microbial Approaches as Green Methods for the Synthesis of Nanomaterials: Synthesis, Applications, and Future Perspectives. *Molecules (Basel, Switzerland)*, 28(1), 463. DOI: 10.3390/molecules28010463 PMID: 36615655

Alshammari, S. O., Mahmoud, S. Y., & Farrag, E. S. (2023). Synthesis of Green Copper Nanoparticles Using Medicinal Plant Krameria sp. Root Extract and Its Applications. *Molecules (Basel, Switzerland)*, 28(12), 4629. DOI: 10.3390/molecules28124629 PMID: 37375184

Álvarez-Chimal, R., & Arenas-Alatorre, J. Á. (2023). *Green Synthesis of Nanoparticles: A Biological Approach* (Shah, K. J., Ed.). IntechOpen.

Amjad, R., Mubeen, B., & Syed, S. (2021). Green Synthesis and Characterization of Copper Nanoparticles Using Fortunella margarita Leaves. *Polymers*, 13(24), 4364. DOI: 10.3390/polym13244364 PMID: 34960915

Anandan, M., & Gurumallesh Prabu, H. (2018). Dodonaea viscosa Leaf Extract Assisted Synthesis of Gold Nanoparticles: Characterization and Cytotoxicity Against A549 NSCLC Cancer Cells. *Journal of Inorganic and Organometallic Polymers and Materials*, 28(3), 932–941. DOI: 10.1007/s10904-018-0799-6

Anbu, P., Gopinath, S., & Jayanthi, S. (2020). Synthesis of gold nanoparticles using Platycodon grandiflorum extract and its antipathogenic activity under optimal conditions. *Nanomaterials and Nanotechnology*, 10, 1–9. DOI: 10.1177/1847980420961697

Angel Ezhilarasi, A., Judith Vijaya, J., Kaviyarasu, K., John Kennedy, L., Ramalingam, R. J., & Al-Lohedan, H. A. (2018). Green synthesis of NiO nanoparticles using Aegle marmelos leaf extract for the evaluation of in-vitro cytotoxicity, antibacterial and photocatalytic properties. *Journal of Photochemistry and Photobiology. B, Biology*, 180, 39–50. DOI: 10.1016/j.jphotobiol.2018.01.023 PMID: 29413700

Antonio-Pérez, A., Durán-Armenta, L. F., Pérez-Loredo, M. G., & Torres-Huerta, A. L. (2023). Biosynthesis of Copper Nanoparticles with Medicinal Plants Extracts: From Extraction Methods to Applications. *Micromachines*, 14(10), 1882. DOI: 10.3390/mi14101882 PMID: 37893319

Arik, N., Elcin, E., Tezcaner, A., & Oktem, H. A. (2023). Optimization of whole-cell bacterial bioreporter immobilization on electrospun cellulose acetate (CA) and polycaprolactone (PCL) fibers for arsenic detection. *Environmental Monitoring and Assessment*, 195(6), 666. DOI: 10.1007/s10661-023-11227-4 PMID: 37178337

Arik, N., Horzum, N., & Truong, Y. B. (2022). Development and Characterizations of Engineered Electrospun Bio-Based Polyurethane Containing Essential Oils. *Membranes*, 12(2), 1–16. DOI: 10.3390/membranes12020209 PMID: 35207129

Armendariz, V., Herrera, I., peralta-videa, J. R., Jose-yacaman, M., Troiani, H., Santiago, P., & Gardea-Torresdey, J. L. (2004). Size controlled gold nanoparticle formation by Avena sativa biomass: Use of plants in nanobiotechnology. *Journal of Nanoparticle Research*, 6(4), 377–382. DOI: 10.1007/s11051-004-0741-4

Asif, M., Yasmin, R., Asif, R., Ambreen, A., Mustafa, M., & Umbreen, S. (2022). Green Synthesis of Silver Nanoparticles (AgNPs), Structural Characterization, and their Antibacterial Potential. *Dose-Response*, 20(2), 15593258221088708. DOI: 10.1177/15593258221088709 PMID: 35592270

Aydın Acar, Ç. (2021). Green Synthesis Of Zinc Oxide Nanoparticles Using Aqueous Extract Of Achiella Millefolium L.: In Vitro Anti-Cancer Potential On Lung And Colon Cancer Cells. *Turkish Journal of Health Science and Life*, 4, 40–45.

Bagheri, A. R., Aramesh, N., Hasnain, M. S., Nayak, A. K., & Varma, R. S. (2023). Greener fabrication of metal nanoparticles using plant materials: A review. *Chemical Physics Impact*, 7, 100255. DOI: 10.1016/j.chphi.2023.100255

Balasubramanian, S., Kala, S. M. J., & Pushparaj, T. L. (2020). Biogenic synthesis of gold nanoparticles using Jasminum auriculatum leaf extract and their catalytic, antimicrobial and anticancer activities. *Journal of Drug Delivery Science and Technology*, 57, 101620. DOI: 10.1016/j.jddst.2020.101620

Bankar, A., Joshi, B., Kumar, A. R., & Zinjarde, S. (2010). Banana peel extract mediated novel route for the synthesis of palladium nanoparticles. *Materials Letters*, 64(18), 1951–1953. DOI: 10.1016/j.matlet.2010.06.021

Bao, Y., He, J., Song, K., Guo, J., Zhou, X., & Liu, S. (2021). Plant-Extract-Mediated Synthesis of Metal Nanoparticles. *Journal of Chemistry*, 2021, 14. DOI: 10.1155/2021/6562687

Bellotti, R., Picotto, G. B., & Ribotta, L. (2022). AFM Measurements and Tip Characterization of Nanoparticles with Different Shapes. *Nanomanufacturing and Metrology*, 5(2), 127–138. DOI: 10.1007/s41871-022-00125-x

Bharali, A., Deka, B., Sahu, B. P., & Laloo, D. (2023). Major challenges and probable scientific solutions toward the large-scale production of plant-based metallic nanoparticles: A systematic review. *Nanotechnology for Environmental Engineering*, 8(4), 933–941. DOI: 10.1007/s41204-023-00347-4

Bhoye, M., Pansambal, S., Basnet, P., Lin, K.-Y. A., Gutierrez-Mercado, K. Y., Pérez-Larios, A., Chauhan, A., Oza, R., & Ghotekar, S. (2023). Eco-Friendly Synthesis of Ni/NiO Nanoparticles Using Gymnema sylvestre Leaves Extract for Antifungal Activity. *Journal of Composites Science*, 7(3), 105. DOI: 10.3390/jcs7030105

Bloch, K., Pardesi, K., Satriano, C., & Ghosh, S. (2021). Bacteriogenic Platinum Nanoparticles for Application in Nanomedicine. *Frontiers in Chemistry*, 9, 624344. DOI: 10.3389/fchem.2021.624344 PMID: 33763405

Borji, H., Ayoub, G., Bilbeisi, R., Nassar, N., & Malaeb, L. (2020). How Effective Are Nanomaterials for the Removal of Heavy Metals from Water and Wastewater? *Water, Air, and Soil Pollution*, 231(7), 330. DOI: 10.1007/s11270-020-04681-0

Buarki, F., AbuHassan, H., Al Hannan, F., & Henari, F. Z. (2022). Green Synthesis of Iron Oxide Nanoparticles Using Hibiscus rosa sinensis Flowers and Their Antibacterial Activity. *Journal of Nanotechnology*, 2022, 1–6. DOI: 10.1155/2022/5474645

Chandrakala, V., Aruna, V., & Angajala, G. (2022). Review on metal nanoparticles as nanocarriers: Current challenges and perspectives in drug delivery systems. *Emergent Materials*, 5(6), 1593–1615. DOI: 10.1007/s42247-021-00335-x PMID: 35005431

Chandran, S. P., Chaudhary, M., Pasricha, R., Ahmad, A., & Sastry, M. (2006). Synthesis of Gold Nanotriangles and Silver Nanoparticles Using Aloevera Plant Extract. *Biotechnology Progress*, 22(2), 577–583. DOI: 10.1021/bp0501423 PMID: 16599579

Chandrasekaran, R., Alagar Yadav, S., & Sivaperumal, S. (2020). Phytosynthesis and Characterization of Copper Oxide Nanoparticles using the Aqueous Extract of Beta vulgaris L and Evaluation of their Antibacterial and Anticancer Activities. *Journal of Cluster Science*, 31(1), 221–230. DOI: 10.1007/s10876-019-01640-6

Cheirmadurai, K., Biswas, S., Murali, R., & Thanikaivelan, P. (2014). Green synthesis of copper nanoparticles and conducting nanobiocomposites using plant and animal sources. *RSC Advances*, 4(37), 19507–19511. DOI: 10.1039/c4ra01414f

Chen, J., Li, Y., Fang, G., Cao, Z., Shang, Y., Alfarraj, S., Ali Alharbi, S., Li, J., Yang, S., & Duan, X. (2021). Green synthesis, characterization, cytotoxicity, antioxidant, and anti-human ovarian cancer activities of Curcumae kwangsiensis leaf aqueous extract green-synthesized gold nanoparticles. *Arabian Journal of Chemistry*, 14(3), 103000. DOI: 10.1016/j.arabjc.2021.103000

Darwish, M., Mafla-Gonzalez, C., Kolenovic, B., Deremer, A., Centeno, D., Liu, T., Kim, D.-Y., Cattabiani, T., Drwiega, T. J., Kumar, I., Li, C., & Traba, C. (2022). Rapid Synthesis of Metal Nanoparticles Using Low-Temperature, Low-Pressure Argon Plasma Chemistry and Self-Assembly. *Green Chemistry*, 24(20), 8142–8154. DOI: 10.1039/D2GC02592B PMID: 37153712

Dikshit, P. K., Kumar, J., Das, A. K., Sadhu, S., Sharma, S., Singh, S., Gupta, P., & Kim, B. S. (2021). Green Synthesis of Metallic Nanoparticles: Applications and Limitations. *Catalysts*, 11(8), 902. DOI: 10.3390/catal11080902

Din, M. I., Nabi, A. G., Rani, A., Aihetasham, A., & Mukhtar, M. (2018). Single step green synthesis of stable nickel and nickel oxide nanoparticles from Calotropis gigantea: Catalytic and antimicrobial potentials. *Environmental Nanotechnology, Monitoring & Management*, 9, 29–36. DOI: 10.1016/j.enmm.2017.11.005

Dobrucka, R. (2018). Antioxidant and Catalytic Activity of Biosynthesized CuO Nanoparticles Using Extract of Galeopsidis herba. *Journal of Inorganic and Organometallic Polymers and Materials*, 28(3), 812–819. DOI: 10.1007/s10904-017-0750-2

Dulta, K., Ağçeli, G. K., Chauhan, P., & Chauhan, P. K. (2021). Biogenic Production and Characterization of CuO Nanoparticles by Carica papaya Leaves and Its Biocompatibility Applications. *Journal of Inorganic and Organometallic Polymers and Materials*, 31(4), 1846–1857. DOI: 10.1007/s10904-020-01837-7

Egan Morriss, C., Kimber, R., Powell, N., & Lloyd, J. (2022). Biotechnological synthesis of Pd-based nanoparticle catalysts. *Nanoscale Advances*, 4(3), 654–679. DOI: 10.1039/D1NA00686J PMID: 35224444

el Borady, O., Ayat, M., Shabrawy, M., & Millet, P. (2020). Green synthesis of gold nanoparticles using Parsley leaves extract and their applications as an alternative catalytic, antioxidant, anticancer, and antibacterial agents. *Advanced Powder Technology*, 31(10), 4390–4400. DOI: 10.1016/j.apt.2020.09.017

El-Seedi, H. R., El-Shabasy, R. M., Khalifa, S. A. M., Saeed, A., Shah, A., Shah, R., Iftikhar, F. J., Abdel-Daim, M. M., Omri, A., Hajrahand, N. H., Sabir, J. S. M., Zou, X., Halabi, M. F., Sarhan, W., & Guo, W. (2019). Metal nanoparticles fabricated by green chemistry using natural extracts: Biosynthesis, mechanisms, and applications. *RSC Advances*, 9(42), 24539–24559. DOI: 10.1039/C9RA02225B PMID: 35527869

Fahimmunisha, B. A., Ishwarya, R., AlSalhi, M. S., Devanesan, S., Govindarajan, M., & Vaseeharan, B. (2020). Green fabrication, characterization and antibacterial potential of zinc oxide nanoparticles using Aloe socotrina leaf extract: A novel drug delivery approach. *Journal of Drug Delivery Science and Technology*, 55, 101465. DOI: 10.1016/j.jddst.2019.101465

Fahmy, S., Mamdouh, I., Saleh, B., Issa, M., Bakowsky, U., & Azzazy, H. (2021). Green Synthesis of Platinum and Palladium Nanoparticles Using Peganum harmala L. Seed Alkaloids: Biological and Computational Studies. *Nanomaterials (Basel, Switzerland)*, 11(4), 965. DOI: 10.3390/nano11040965 PMID: 33918743

Fahmy, S. A., Preis, E., Bakowsky, U., & Azzazy, H. M. E.-S. (2020). Platinum Nanoparticles: Green Synthesis and Biomedical Applications. *Molecules (Basel, Switzerland)*, 25(21), 4981. Advance online publication. DOI: 10.3390/molecules25214981 PMID: 33126464

Farooq, A., Khan, U. A., Ali, H., Sathish, M., Naqvi, S. A. H., Iqbal, S., Ali, H., Mubeen, I., Amir, M. B., Mosa, W. F. A., Baazeem, A., Moustafa, M., Alrumman, S., Shati, A., & Negm, S. (2022). Green Chemistry Based Synthesis of Zinc Oxide Nanoparticles Using Plant Derivatives of Calotropis gigantea (Giant Milkweed) and Its Biological Applications against Various Bacterial and Fungal Pathogens. *Microorganisms*, 10(11), 2195. DOI: 10.3390/microorganisms10112195 PMID: 36363787

Fernando, K. M., Gunathilake, C. A., Yalegama, C., Samarakoon, U. K., Fernando, C. A. N., Weerasinghe, G., Pamunuwa, G. K., Soliman, I., Ghulamullah, N., Rajapaksha, S. M., & Fatani, O. (2024). Synthesis of Silver Nanoparticles Using Green Reducing Agent: Ceylon Olive (Elaeocarpus serratus): Characterization and Investigating Their Antimicrobial Properties. *Journal of Composites Science*, 8(2), 43. DOI: 10.3390/jcs8020043

Francis, S., Joseph, S., Koshy, E. P., & Mathew, B. (2018). Microwave assisted green synthesis of silver nanoparticles using leaf extract of elephantopus scaber and its environmental and biological applications. *Artificial Cells, Nanomedicine, and Biotechnology*, 46(4), 795–804. DOI: 10.1080/21691401.2017.1345921 PMID: 28681662

Ghasemi, S., Dabirian, S., Kariminejad, F., Koohi, D. E., Nemattalab, M., Majidimoghadam, S., Zamani, E., & Yousefbeyk, F. (2024). Process optimization for green synthesis of silver nanoparticles using Rubus discolor leaves extract and its biological activities against multi-drug resistant bacteria and cancer cells. *Scientific Reports*, 14(1), 4130. DOI: 10.1038/s41598-024-54702-9 PMID: 38374139

Ghoohestani, E., Samari, F., Homaei, A., & Yosuefinejad, S. (2024). A facile strategy for preparation of Fe3O4 magnetic nanoparticles using Cordia myxa leaf extract and investigating its adsorption activity in dye removal. *Scientific Reports*, 14(1), 84. DOI: 10.1038/s41598-023-50550-1 PMID: 38168136

Giri, A. K., Jena, B., Biswal, B., Pradhan, A. K., Arakha, M., Acharya, S., & Acharya, L. (2022). Green synthesis and characterization of silver nanoparticles using Eugenia roxburghii DC. extract and activity against biofilm-producing bacteria. *Scientific Reports*, 12(1), 8383. DOI: 10.1038/s41598-022-12484-y PMID: 35589849

Glaubitz, C., Rothen-Rutishauser, B., Lattuada, M., Balog, S., & Petri-Fink, A. (2022). Designing the ultrasonic treatment of nanoparticle-dispersions via machine learning. *Nanoscale*, 14(35), 12940–12950. DOI: 10.1039/D2NR03240F PMID: 36043853

Gobinath, P., Packialakshmi, P., Hatamleh, A. A., Al-Dosary, M. A., Al-Wasel, Y. A., Balasubramani, R., Surendrakumar, R., & Idhayadhulla, A. (2022). Calotropis gigantea Assisted Synthesis of Zinc Oxide Nanoparticle Catalysis: Synthesis of Novel 3-Amino Thymoquinone Connected 1,4-Dihyropyridine Derivatives and Their Cytotoxic Activity. *Journal of Nanomaterials*, 2022(1), 1–10. DOI: 10.1155/2022/9697057

Goh, E. G., Xu, X., & Mccormick, P. (2014). Effect of particle size on the UV absorbance of zinc oxide nanoparticles. *Scripta Materialia, s 78–79*, 49–52.

Guo, M., Li, W., Yang, F., & Liu, H. (2015). Controllable biosynthesis of gold nanoparticles from a Eucommia ulmoides bark aqueous extract. *Spectrochimica Acta. Part A: Molecular and Biomolecular Spectroscopy*, 142, 73–79. DOI: 10.1016/j.saa.2015.01.109 PMID: 25699695

Hano, C., & Abbasi, B. H. (2022). Plant-Based Green Synthesis of Nanoparticles: Production, Characterization and Applications. *Biomolecules*, 12(1), 31. DOI: 10.3390/biom12010031 PMID: 35053179

Haritha, V., Gowri, S., Janarthanan, B., Faiyazuddin, M., Karthikeyan, C., & Sharmila, S. (2022). Biogenic synthesis of nickel oxide nanoparticles using Averrhoa bilimbi and investigation of its antibacterial, antidiabetic and cytotoxic properties. *Inorganic Chemistry Communications*, 144, 109930. DOI: 10.1016/j.inoche.2022.109930

Harris, A. T., & Bali, R. (2008). On the formation and extent of uptake of silver nanoparticles by live plants. *Journal of Nanoparticle Research*, 10(4), 691–695. DOI: 10.1007/s11051-007-9288-5

Helan, V., Prince, J. J., Al-Dhabi, N. A., Arasu, M. V., Ayeshamariam, A., Madhumitha, G., Roopan, S. M., & Jayachandran, M. (2016). Neem leaves mediated preparation of NiO nanoparticles and its magnetization, coercivity and antibacterial analysis. *Results in Physics*, 6, 712–718. DOI: 10.1016/j.rinp.2016.10.005

Hossain, A., Abdallah, Y., Ali, M. A., Masum, M. M., Li, B., Sun, G., Meng, Y., Wang, Y., & An, Q. (2019). Lemon-Fruit-Based Green Synthesis of Zinc Oxide Nanoparticles and Titanium Dioxide Nanoparticles against Soft Rot Bacterial Pathogen Dickeya dadantii. *Biomolecules*, 9(12), 863. DOI: 10.3390/biom9120863 PMID: 31835898

Huang, X., Wu, H., Liao, X., & Shi, B. (2010). One-step, size-controlled synthesis of gold nanoparticles at room temperature using plant tannin. *Green Chemistry -. Green Chemistry*, 12(3), 395–399. DOI: 10.1039/B918176H

Hussain, S., Ali Muazzam, M., Ahmed, M., Ahmad, M., Mustafa, Z., Murtaza, S., Ali, J., Ibrar, M., Shahid, M., & Imran, M. (2023). Green synthesis of nickel oxide nanoparticles using Acacia nilotica leaf extracts and investigation of their electrochemical and biological properties. *Journal of Taibah University for Science : JTUSCI*, 17(1), 1–12. DOI: 10.1080/16583655.2023.2170162

Ibrahim, F. M., Najeeb, D. A., & ThamerSadeq, H. (2023). Green preparation of Cu nanoparticles of the avocado seed extract as an adsorbent surface. *Materials Science for Energy Technologies*, 6, 130–136. DOI: 10.1016/j.mset.2022.12.006

Ikhioya, I., Onoh, E., Nkele, C., Abor, B., Obitte, B., Maaza, M., & Ezema, F. (2023). The The Green Synthesis of Copper Oxide Nanoparticles Using the Moringa Oleifera Plant and its Subsequent Characterization for Use in Energy Storage Applications. *East European Journal of Physics*, 1(1), 162–172. DOI: 10.26565/2312-4334-2023-1-20

Irum, S., Jabeen, N., Ahmad, K. S., Shafique, S., Khan, T. F., Gul, H., Anwaar, S., Shah, N. I., Mehmood, A., & Hussain, S. Z. (2020). Biogenic iron oxide nanoparticles enhance callogenesis and regeneration pattern of recalcitrant Cicer arietinum L. *PLoS One*, 15(12), 1–19. DOI: 10.1371/journal.pone.0242829 PMID: 33259506

Islam, N. U., Jalil, K., Shahid, M., Muhammad, N., & Rauf, A. (2019). Pistacia integerrima gall extract mediated green synthesis of gold nanoparticles and their biological activities. *Arabian Journal of Chemistry*, 12(8), 2310–2319. DOI: 10.1016/j.arabjc.2015.02.014

Kamble, P. (2009). Protein-mediated synthesis of gold nanoparticles. *Materials Science and Engineering B*, 163(2), 93–98. DOI: 10.1016/j.mseb.2009.05.013

Kamçı, H., Celebioglu, H. U., Taş, R., & Köroğlu, E. (2022). Biogenic Synthesis of Zinc Oxide Nanoparticles Using Saponaria officinalis L., Characterisation and Antibacterial Activities. *European Journal of Science and Technology*, 35, 227–234.

Karade, V. C., Dongale, T. D., Sahoo, S. C., Kollu, P., Chougale, A. D., Patil, P. S., & Patil, P. B. (2018). Effect of reaction time on structural and magnetic properties of green-synthesized magnetic nanoparticles. *Journal of Physics and Chemistry of Solids*, 120, 161–166. DOI: 10.1016/j.jpcs.2018.04.040

Karunakaran, G., Sudha, K. G., Ali, S., & Cho, E.-B. (2023). Biosynthesis of Nanoparticles from Various Biological Sources and Its Biomedical Applications. *Molecules (Basel, Switzerland)*, 28(11), 4527. DOI: 10.3390/molecules28114527 PMID: 37299004

Kasthuri, J., Veerapandian, S., & Rajendiran, N. (2009). Biological synthesis of silver and gold nanoparticles using apiin as reducing agent. *Colloids and Surfaces. B, Biointerfaces*, 68(1), 55–60. DOI: 10.1016/j.colsurfb.2008.09.021 PMID: 18977643

Khan, A., Rashid, A., Younas, R., & Chong, R. (2016). A chemical reduction approach to the synthesis of copper nanoparticles. *International Nano Letters*, 6(1), 21–26. DOI: 10.1007/s40089-015-0163-6

Khan, M. F., & Khan, M. A. (2023). Plant-Derived Metal Nanoparticles (PDMNPs): Synthesis, Characterization, and Oxidative Stress-Mediated Therapeutic Actions. *Future Pharmacology*, 3(1), 252–295. DOI: 10.3390/futurepharmacol3010018

Khan, S. A., Shahid, S., & Lee, C.-S. (2020). Green Synthesis of Gold and Silver Nanoparticles Using Leaf Extract of Clerodendrum inerme; Characterization, Antimicrobial, and Antioxidant Activities. *Biomolecules*, 10(6), 835. DOI: 10.3390/biom10060835 PMID: 32486004

Khodair, Z. T., Ibrahim, N. M., Kadhim, T. J., & Mohammad, A. M. (2022). Synthesis and characterization of nickel oxide (NiO) nanoparticles using an environmentally friendly method, and their biomedical applications. *Chemical Physics Letters*, 797, 139564. DOI: 10.1016/j.cplett.2022.139564

Kiranmayee, M., Rajesh, N., Vidya Vani, M., Khadri, H., Mohammed, A., Chinni, S. V., Ramachawolran, G., Riazunnisa, K., & Moussa, A. Y. (2023). Green synthesis of Piper nigrum copper-based nanoparticles: In silico study and ADMET analysis to assess their antioxidant, antibacterial, and cytotoxic effects. *Frontiers in Chemistry*, 11, 1–22. DOI: 10.3389/fchem.2023.1218588 PMID: 37736256

Kocazorbaz, E. (2021). Green Synthesis, Optimization, and Characterization of Silver Nanoparticles from Euphorbia rigida Leaf Extract and Investigation of Their Antimicrobal Activities. *Bilecik Şeyh Edebali Üniversitesi Fen Bilimleri Dergisi*, 8(2), 512–522. DOI: 10.35193/bseufbd.843005

Kokina, I., Plaksenkova, I., Jankovskis, L., Jermaļonoka, M., & Galek, R. (2024). New Insights on Biosynthesis of Nanoparticles Using Plants Emphasizing the Use of Alfalfa (Medicago sativa L.). *Journal of Nanotechnology*, 9721166, 14. DOI: 10.1155/2024/9721166

Kora, A. J., & Rastogi, L. (2018). Green synthesis of palladium nanoparticles using gum ghatti (Anogeissus latifolia) and its application as an antioxidant and catalyst. *Arabian Journal of Chemistry*, 11(7), 1097–1106. DOI: 10.1016/j.arabjc.2015.06.024

Kordy, M. G. M., Abdel-Gabbar, M., Soliman, H. A., Aljohani, G., BinSabt, M., Ahmed, I. A., & Shaban, M. (2022). Phyto-Capped Ag Nanoparticles: Green Synthesis, Characterization, and Catalytic and Antioxidant Activities. *Nanomaterials (Basel, Switzerland)*, 12(3), 373. DOI: 10.3390/nano12030373 PMID: 35159718

Kulkarni, D.-S. (2024). Combinations of Biotechnology and Nanotechnology in Industrial Wastewater Treatment. In Mishra, B. K. (Ed.), *Sustainable Science and Intelligent Technologies for Societal Development* (pp. 96–106). DOI: 10.4018/979-8-3693-1186-8.ch006

Kumar Sur, U., Ankamwar, B., Karmakar, S., Halder, A., & Das, P. (2018). Green synthesis of Silver nanoparticles using the plant extract of Shikakai and Reetha. *Materials Today: Proceedings*, 5(1), 2321–2329. DOI: 10.1016/j.matpr.2017.09.236

Kuppusamy, P., Yusoff, M. M., Maniam, G. P., & Govindan, N. (2016). Biosynthesis of metallic nanoparticles using plant derivatives and their new avenues in pharmacological applications – An updated report. *Saudi Pharmaceutical Journal*, 24(4), 473–484. DOI: 10.1016/j.jsps.2014.11.013 PMID: 27330378

Lakshmipriya, T., & Gopinath, S. C. B. (2021). 1 - Introduction to nanoparticles and analytical devices. In S. C. B. Gopinath & F. B. T.-N. in A. and M. D. Gang (Eds.), *Nanoparticles in Analytical and Medical Devices* (pp. 1–29). Elsevier.

Lal, S., & Nayak, P. L. (2012). Green synthesis of gold nanoparticles using various extract of plants and spices. *International Journal of Science Innovations and Discoveries*, 2, 325.

Laurent, S., Bridot, J.-L., Vander Elst, L., & Muller, R. (2010). Magnetic Iron Oxide Nanoparticles for Biomedical Applications. *Future Medicinal Chemistry*, 2(3), 427–449. DOI: 10.4155/fmc.09.164 PMID: 21426176

Lee, K. X., Shameli, K., Yew, Y. P., Teow, S.-Y., Jahangirian, H., Rafiee-Moghaddam, R., & Webster, T. J. (2020). Recent Developments in the Facile Bio-Synthesis of Gold Nanoparticles (AuNPs) and Their Biomedical Applications. *International Journal of Nanomedicine*, 15, 275–300. DOI: 10.2147/IJN.S233789 PMID: 32021180

Liao, W., Liao, C., Lin, Y.-C., & Lien, C.-Y. (2024). Characterization of Gold Nanoparticles Synthesized with Zingiber zerumbet Extracts. *Journal of Chemistry*, 2024, 1–8. DOI: 10.1155/2024/1358495

Liu, L., & Corma, A. (2018). Metal Catalysts for Heterogeneous Catalysis: From Single Atoms to Nanoclusters and Nanoparticles. *Chemical Reviews*, 118(10), 4981–5079. DOI: 10.1021/acs.chemrev.7b00776 PMID: 29658707

Lomelí-Rosales, D. A., Zamudio-Ojeda, A., Reyes-Maldonado, O. K., López-Reyes, M. E., Basulto-Padilla, G. C., Lopez-Naranjo, E. J., Zuñiga-Mayo, V. M., & Velázquez-Juárez, G. (2022). Green Synthesis of Gold and Silver Nanoparticles Using Leaf Extract of Capsicum chinense Plant. *Molecules (Basel, Switzerland)*, 27(5), 1692. DOI: 10.3390/molecules27051692 PMID: 35268794

Lv, Y.-C., Esmaeili Shahri, E., Mahmoudi, A., Keifi Naughabi, R., Abbaspour, S., & Tayebee, R. (2023). Bioinspired nickel oxide nanoparticle as an efficient nanocarrier in the delivery of doxorubicin as an anti-bladder cancer drug. *Inorganic Chemistry Communications*, 152, 110650. DOI: 10.1016/j.inoche.2023.110650

Makarov, V. V., Love, A. J., Sinitsyna, O. V., Makarova, S. S., Yaminsky, I. V., Taliansky, M. E., & Kalinina, N. O. (2014). "Green" nanotechnologies: Synthesis of metal nanoparticles using plants. *Acta Naturae (English Ed.)*, 6(1), 35–44. DOI: 10.32607/20758251-2014-6-1-35-44 PMID: 24772325

Malik, S., Muhammad, K., & Waheed, Y. (2023). Nanotechnology: A Revolution in Modern Industry. *Molecules (Basel, Switzerland)*, 28(2), 661. DOI: 10.3390/molecules28020661 PMID: 36677717

Manikandan, V., Velmurugan, P., Park, J.-H., Lovanh, N., Seo, S.-K., Jayanthi, P., Park, Y.-J., Cho, M., & Oh, B.-T. (2016). Synthesis and antimicrobial activity of palladium nanoparticles from Prunus × yedoensis leaf extract. *Materials Letters*, 185, 335–338. DOI: 10.1016/j.matlet.2016.08.120

Mardina, V., Fadlly, T. A., Harmawan, T., Sufriadi, E., Iqramullah, M., Umar, H., & Ilyas, S. (2024). Green synthesis of gold nanoparticles from the aqueous extracts of Sphagneticola trilobata (L.) J.F Pruski as anti-breast cancer agents. *Journal of Advanced Pharmaceutical Technology & Research*, 15(2), 75–80. DOI: 10.4103/japtr.japtr_410_23 PMID: 38903547

Matouk, Z., Islam, M., Gutiérrez, M., Pireaux, J.-J., & Achour, A. (2022). X-ray Photoelectron Spectroscopy (XPS) Analysis of Ultrafine Au Nanoparticles Supported over Reactively Sputtered TiO2 Films. *Nanomaterials (Basel, Switzerland)*, 12(20), 3692. DOI: 10.3390/nano12203692 PMID: 36296882

Mekuye, B., & Abera, B. (2023). Nanomaterials: An overview of synthesis, classification, characterization, and applications. *Nano Select*, 4(8), 486–501. DOI: 10.1002/nano.202300038

Menon, S., Agarwal, H., Kumar, S., & Kumar, V. (2017). Green synthesis of silver nanoparticles using medicinal plant Acalypha indica leaf extracts and its application as an antioxidant and antimicrobial agent against foodborne pathogens. *International Journal of Applied Pharmaceutics*, 9(5), 42. DOI: 10.22159/ijap.2017v9i5.19464

Meshkatalsadat, M. H., Momeni, A., & Abdollahzadeh, M. R. (2023). Biosynthesis of Zinc Oxide Nanoparticles Using Punica granatum L. Waste Peel Extract, and Assessment of Antioxidant and Catalytic Activity. *Nano Biomedicine and Engineering*, 15(4), 378–388. DOI: 10.26599/NBE.2023.9290030

Meyns, M., Dietz, F., Weinhold, C.-S., Züge, H., Finckh, S., & Gerdts, G. (2022). Multi-feature round silicon membrane filters enable fractionation and analysis of micro- and nanoplastics with Raman spectroscopy and nano-FTIR. *Analytical Methods*, 15(5), 606–617. DOI: 10.1039/D2AY01036D PMID: 36644945

Mikhailova, E. O. (2021). Gold Nanoparticles: Biosynthesis and Potential of Biomedical Application. *Journal of Functional Biomaterials*, 12(4), 70. DOI: 10.3390/jfb12040070 PMID: 34940549

Milan, J., Niemczyk, K., & Kus-Liśkiewicz, M. (2022). Treasure on the Earth—Gold Nanoparticles and Their Biomedical Applications. *Materials (Basel)*, 15(9), 3355. DOI: 10.3390/ma15093355 PMID: 35591689

Minnelli, C., Moretti, P., Laudadio, E., Gerelli, Y., Pigozzo, A., Armeni, T., Galeazzi, R., Mariani, P., & Mobbili, G. (2022). Tuning curvature and phase behavior of monoolein bilayers by epigallocatechin-3-gallate: Structural insight and cytotoxicity. *Colloids and Surfaces. B, Biointerfaces*, 209, 112171. DOI: 10.1016/j.colsurfb.2021.112171 PMID: 34736221

Mishra, D., Rajurkar, S., Mishra, N., Jadhav, N., & Ballurkar, B. (2017). Green Synthesis of Gold Nanoparticles by Azadirachta indica Leaf Extract and Coating with Morinda citrifolia Fruit Extract: Their Characterization. *International Journal of Livestock Research*, 7, 205–219. DOI: 10.5455/ijlr.20170306091736

Nagar, N., & Devra, V. (2018). Green synthesis and characterization of copper nanoparticles using Azadirachta indica leaves. *Materials Chemistry and Physics*, 213, 44–51. DOI: 10.1016/j.matchemphys.2018.04.007

Naiel, B., Fawzy, M., Halmy, M. W. A., & Mahmoud, A. E. D. (2022). Green synthesis of zinc oxide nanoparticles using Sea Lavender (Limonium pruinosum L. Chaz.) extract: Characterization, evaluation of anti-skin cancer, antimicrobial and antioxidant potentials. *Scientific Reports*, 12(1), 20370. DOI: 10.1038/s41598-022-24805-2 PMID: 36437355

Nancy, B. A., & Elumalai, K. (2019). Synthesis of silver nanoparticles using Pelargonium graveolens essential Oil and anti-fungal activity. *International Journal of Pharmacy and Biological Sciences*, 9, 176–185.

Narasaiah, P., Kumar Mandal, B., & Sarada, N. C. (2017). Green synthesis of Pd NPs from Pimpinella tirupatiensis plant extract and their application in photocatalytic activity dye degradation. *IOP Conference Series. Materials Science and Engineering*, 263, 22013. DOI: 10.1088/1757-899X/263/2/022013

Naseer, M., Aslam, U., Khalid, B., & Chen, B. (2020). Green route to synthesize Zinc Oxide Nanoparticles using leaf extracts of Cassia fistula and Melia azadarach and their antibacterial potential. *Scientific Reports*, 10(1), 9055. DOI: 10.1038/s41598-020-65949-3 PMID: 32493935

Nejo, A., Adetona, A., & Lawal, A. (2024). Green Synthesis of Nickel Oxide Nanoparticles and its Application in the Degradation of Methyl Red. *Lafia Journal of Scientific and Industrial Research*, 2, 54–60. DOI: 10.62050/ljsir2024.v2n2.328

Nguyen, M. D., Tran, H.-V., Xu, S., & Lee, T. R. (2021). Fe(3)O(4) Nanoparticles: Structures, Synthesis, Magnetic Properties, Surface Functionalization, and Emerging Applications. *Applied Sciences (Basel, Switzerland)*, 11(23), 11301. DOI: 10.3390/app112311301 PMID: 35844268

Niu, G., Zhao, L., Wang, Y., & Jiang, Y. (2023). PDA/gold nanorod-based nanoparticles for synergistic genetic and photothermal combination therapy for cancer treatment. *ChemPhysMater*, 2(1), 83–89. DOI: 10.1016/j.chphma.2022.07.001

Olajire, A. A., & Mohammed, A. A. (2020). Green synthesis of nickel oxide nanoparticles and studies of their photocatalytic activity in degradation of polyethylene films. *Advanced Powder Technology*, 31(1), 211–218. DOI: 10.1016/j.apt.2019.10.012

Osman, A. I., Zhang, Y., Farghali, M., Rashwan, A. K., Eltaweil, A. S., Abd El-Monaem, E. M., Mohamed, I. M. A., Badr, M. M., Ihara, I., Rooney, D. W., & Yap, P.-S. (2024). Synthesis of green nanoparticles for energy, biomedical, environmental, agricultural, and food applications: A review. *Environmental Chemistry Letters*, 22(2), 841–887. DOI: 10.1007/s10311-023-01682-3

Özkan Dağlioğlu, Y. (2021). Green Synthesis of Silver Nanoparticles Using Mandragora autumnalis; Its Characterization, Antioxidant and Antimicrobial Activities. *Erzincan Üniversitesi Fen Bilimleri Enstitüsü Dergisi*, 14(3), 1039–1054. DOI: 10.18185/erzifbed.915107

Panwar, C., Phillips, A., Sangal, A., & Gupta, D. (2024). Synthesis of Gold Nanoparticles and Their Applications in Cancer Therapy. *Biointerface Research in Applied Chemistry*, 14, 12.

Patiño-Ruiz, D. A., Meramo-Hurtado, S. I., González-Delgado, Á. D., & Herrera, A. (2021). Environmental Sustainability Evaluation of Iron Oxide Nanoparticles Synthesized via Green Synthesis and the Coprecipitation Method: A Comparative Life Cycle Assessment Study. *ACS Omega*, 6(19), 12410–12423. DOI: 10.1021/acsomega.0c05246 PMID: 34056392

Patra, J. K., & Baek, K.-H. (2015). Novel green synthesis of gold nanoparticles using Citrullus lanatus rind and investigation of proteasome inhibitory activity, antibacterial, and antioxidant potential. *International Journal of Nanomedicine*, 10, 7253–7264. PMID: 26664116

Patra, J. K., Das, G., & Baek, K.-H. (2016). Phyto-mediated biosynthesis of silver nanoparticles using the rind extract of watermelon (Citrullus lanatus) under photo-catalyzed condition and investigation of its antibacterial, anticandidal and antioxidant efficacy. *Journal of Photochemistry and Photobiology. B, Biology*, 161, 200–210. DOI: 10.1016/j.jphotobiol.2016.05.021 PMID: 27261701

Peralta-Videa, J. R., Huang, Y., Parsons, J. G., Zhao, L., Lopez-Moreno, L., Hernandez-Viezcas, J. A., & Gardea-Torresdey, J. L. (2016). Plant-based green synthesis of metallic nanoparticles: Scientific curiosity or a realistic alternative to chemical synthesis? *Nanotechnology for Environmental Engineering*, 1(1), 4. DOI: 10.1007/s41204-016-0004-5

Petla, R. K., Vivekanandhan, S., Misra, M., Mohanty, A., & Satyanarayana, N. (2012). Soybean (Glycine Max) Leaf Extract Based Green Synthesis of Palladium Nanoparticles. *Journal of Biomaterials and Nanobiotechnology*, 03(1), 14–19. DOI: 10.4236/jbnb.2012.31003

Priya, M., Venkatesan, R., Deepa, S., Sana, S. S., Arumugam, S., Karami, A. M., Vetcher, A. A., & Kim, S.-C. (2023). Green synthesis, characterization, antibacterial, and antifungal activity of copper oxide nanoparticles derived from Morinda citrifolia leaf extract. *Scientific Reports*, 13(1), 18838. DOI: 10.1038/s41598-023-46002-5 PMID: 37914791

Punitha, U., & Mary Saral, A. (2024). Nickel oxide nanoparticles from Sargassumwightii: Synthesis,Characterization, and biomedical applications. *Results in Chemistry*, 7, 101289. DOI: 10.1016/j.rechem.2023.101289

Rajakumar, G., Gomathi, T., Abdul Rahuman, A., Thiruvengadam, M., Mydhili, G., Kim, S.-H., Lee, T.-J., & Chung, I.-M. (2016). Biosynthesis and Biomedical Applications of Gold Nanoparticles Using Eclipta prostrata Leaf Extract. *Applied Sciences (Basel, Switzerland)*, 6(8), 222. DOI: 10.3390/app6080222

Ramesh, A. V., Devi, D., Battu, G., & Basavaiah, K. (2018). A Facile plant mediated synthesis of silver nanoparticles using an aqueous leaf extract of Ficus hispida Linn. f. for catalytic, antioxidant and antibacterial applications. *South African Journal of Chemical Engineering*, 26, 25–34. DOI: 10.1016/j.sajce.2018.07.001

Rautela, A., & Rani, J., & Debnath (Das), M. (. (2019). Green synthesis of silver nanoparticles from Tectona grandis seeds extract: Characterization and mechanism of antimicrobial action on different microorganisms. *Journal of Analytical Science and Technology*, 10, 5. DOI: 10.1186/s40543-018-0163-z

Rehman, A., Naeem, A., Ahmad, I., Fozia, F., Almutairi, M. H., Aslam, M., Israr, M., Almutairi, B. O., & Ullah, Z. (2024). Synthesis of Plant-Mediated Iron Oxide Nanoparticles and Optimization of Chemically Modified Activated Carbon Adsorbents for Removal of As, Pb, and Cd Ions from Wastewater. *ACS Omega*, 9(1), 317–329. DOI: 10.1021/acsomega.3c05299 PMID: 38222602

Rodríguez-León, E., Rodríguez-Vázquez, B. E., Martínez-Higuera, A., Rodríguez-Beas, C., Larios-Rodríguez, E., Navarro, R. E., López-Esparza, R., & Iñiguez-Palomares, R. A. (2019). Synthesis of Gold Nanoparticles Using Mimosa tenuiflora Extract, Assessments of Cytotoxicity, Cellular Uptake, and Catalysis. *Nanoscale Research Letters*, 14(1), 334. DOI: 10.1186/s11671-019-3158-9 PMID: 31654146

Roy, A., Pandit, C., Gacem, A., Alqahtani, M. S., Bilal, M., Islam, S., Hossain, M. J., & Jameel, M. (2022). Biologically Derived Gold Nanoparticles and Their Applications. *Bioinorganic Chemistry and Applications*, 2022(1), 8184217. DOI: 10.1155/2022/8184217 PMID: 35959230

Roy, S., & Das, T. (2015). Plant Mediated Green Synthesis of Silver Nanoparticles-A Review. *International Journal of Plant Biology & Research*, 3, 1044.

Saeed, M., Haq, A. U., Muneer, M., Usman, M., Naqvi, S. A. R., Adeel, M., & Nisar, A. (2021). assisted green synthesis of CoO and Ag-CoO and evaluation of their catalytic activities toward photodegradation of crystal violet dye. *Environmental Progress & Sustainable Energy*, 40, e13591. DOI: 10.1002/ep.13591

Sahu, S. C. (2023). Impact of Nanotechnology in the Modern World. *Annals of Clinical Case Studies*, 5, 43–44.

Saidu, F. K., Mathew, A., Parveen, A., Valiyathra, V., & Thomas, G. V. (2019). Novel green synthesis of silver nanoparticles using clammy cherry (Cordia obliqua Willd) fruit extract and investigation on its catalytic and antimicrobial properties. *SN Applied Sciences*, 1(11), 1368. DOI: 10.1007/s42452-019-1302-x

Saliem, A., Ibrahim, O., & Sarwa, I. (2018). Biosynthesis of Silver Nanoparticles using Cinnamon zeylanicum Plants Bark Extract . *Kufa Journal For Veterinary Medical Sciences*, 7(1), 51–63. DOI: 10.36326/kjvs/2016/v7i14294

Sampatkumar, H. G., Antony, A. M., Trivedi, M., Sharma, M., Ghate, M., Baidya, M., Dateer, R. B., & Patil, S. A. (2024). In situ biosynthesis of palladium nanoparticles on banana leaves extract-coated graphitic carbon nitride: An efficient and reusable heterogeneous catalyst for organic transformations and antimicrobial agent. *Biomass Conversion and Biorefinery*, 14(9), 10045–10066. DOI: 10.1007/s13399-022-03222-5

Selvakannan, P. R., Mandal, S., Phadtare, S., Gole, A., Pasricha, R., Adyanthaya, S. D., & Sastry, M. (2004). Water-dispersible tryptophan-protected gold nanoparticles prepared by the spontaneous reduction of aqueous chloroaurate ions by the amino acid. *Journal of Colloid and Interface Science*, 269(1), 97–102. DOI: 10.1016/S0021-9797(03)00616-7 PMID: 14651900

Shamloo, H. B., Jafarzadeh, S., Jouyban, A., & Soleymani, J. (2022). Synthesis and application of concanavalin A-conjugated green luminescent gold nanoparticle/fluorescent polydopamine nanoparticles for specific differentiation of cancer cells from normal cells using glycan bioreceptors. *Materials Chemistry and Physics*, 288, 126344. DOI: 10.1016/j.matchemphys.2022.126344

Shankar, S. S., Rai, A., Ahmad, A., & Sastry, M. (2004). Rapid synthesis of Au, Ag, and bimetallic Au core–Ag shell nanoparticles using Neem (Azadirachta indica) leaf broth. *Journal of Colloid and Interface Science*, 275(2), 496–502. DOI: 10.1016/j.jcis.2004.03.003 PMID: 15178278

Shanmugam, P. (2022). Green route synthesis of alpinia calcarata functionalized gold nanoparticles for nonlinear optical applications. *Heliyon*, 8(8), e10409. DOI: 10.1016/j.heliyon.2022.e10409 PMID: 36082338

Sharma, D., Sabela, M. I., Kanchi, S., Bisetty, K., Skelton, A. A., & Honarparvar, B. (2018). Green synthesis, characterization and electrochemical sensing of silymarin by ZnO nanoparticles: Experimental and DFT studies. *Journal of Electroanalytical Chemistry (Lausanne, Switzerland)*, 808, 160–172. DOI: 10.1016/j.jelechem.2017.11.039

Sharma, G., Das, S., Kumar, P., & Stangar, U. L. (2024). Parametric effect of continuous synthesis of metal nanoparticles within microchannels and application: A review study. *Chemical Engineering Research & Design*, 209, 142–156. DOI: 10.1016/j.cherd.2024.07.058

Singh, A., Goyal, V., Singh, J., Kaur, H., Kumar, S., Batoo, K. M., Gaur, J., Pal, M., Rawat, M., & Hussain, S. (2022). Structurally and morphologically engineered single-pot biogenic synthesis of NiO nanoparticles with enhanced photocatalytic and antimicrobial activities. *Journal of Cleaner Production*, 343, 131026. DOI: 10.1016/j.jclepro.2022.131026

Singh, H., Desimone, M. F., Pandya, S., Jasani, S., George, N., Adnan, M., Aldarhami, A., Bazaid, A. S., & Alderhami, S. A. (2023). Revisiting the Green Synthesis of Nanoparticles: Uncovering Influences of Plant Extracts as Reducing Agents for Enhanced Synthesis Efficiency and Its Biomedical Applications. *International Journal of Nanomedicine*, 18, 4727–4750. DOI: 10.2147/IJN.S419369 PMID: 37621852

Singh, M., Thakur, V., Kumar, V., Raj, M., Gupta, S., Devi, N., Upadhyay, S. K., Macho, M., Banerjee, A., Ewe, D., & Saurav, K. (2022). Silver Nanoparticles and Its Mechanistic Insight for Chronic Wound Healing: Review on Recent Progress. *Molecules (Basel, Switzerland)*, 27(17), 5587. DOI: 10.3390/molecules27175587 PMID: 36080353

Soliman, M. K. Y., Salem, S. S., Abu-Elghait, M., & Azab, M. S. (2023). Biosynthesis of Silver and Gold Nanoparticles and Their Efficacy Towards Antibacterial, Antibiofilm, Cytotoxicity, and Antioxidant Activities. *Applied Biochemistry and Biotechnology*, 195(2), 1158–1183. DOI: 10.1007/s12010-022-04199-7 PMID: 36342621

Srihasam, S., Thyagarajan, K., Korivi, M., Lebaka, V. R., & Mallem, S. P. (2020). Phytogenic Generation of NiO Nanoparticles Using Stevia Leaf Extract and Evaluation of Their In-Vitro Antioxidant and Antimicrobial Properties. *Biomolecules*, 10(1), 89. DOI: 10.3390/biom10010089 PMID: 31935798

Susan Johnson, M., Nair, R. S., & Nambiar, A. (2023). Electrical conductivity of green synthesized copper nanoparticles by P.granatum. *Materials Today: Proceedings*, •••, 1–5. DOI: 10.1016/j.matpr.2023.02.175

Susanti, D., Haris, M. S., Taher, M., & Khotib, J. (2022). Natural Products-Based Metallic Nanoparticles as Antimicrobial Agents. *Frontiers in Pharmacology*, 13, 895616. DOI: 10.3389/fphar.2022.895616 PMID: 35721199

Szczyglewska, P., Feliczak-Guzik, A., & Nowak, I. (2023). Nanotechnology–General Aspects: A Chemical Reduction Approach to the Synthesis of Nanoparticles. *Molecules (Basel, Switzerland)*, 28(13), 4932. DOI: 10.3390/molecules28134932 PMID: 37446593

Takma, D. K., Bozkurt, S., Koç, M., Korel, F., & Nadeem, H. Ş. (2023). Characterization and encapsulation efficiency of zein nanoparticles loaded with chestnut fruit shell, cedar and sweetgum bark extracts. *Food Hydrocolloids for Health*, 4, 100151. DOI: 10.1016/j.fhfh.2023.100151

Tamil Elakkiya, V., Meenakshi, R. V., Senthil Kumar, P., Karthik, V., Ravi Shankar, K., Sureshkumar, P., & Hanan, A. (2022). Green synthesis of copper nanoparticles using Sesbania aculeata to enhance the plant growth and antimicrobial activities. *International Journal of Environmental Science and Technology*, 19(3), 1313–1322. DOI: 10.1007/s13762-021-03182-9

Tan, G., Tevlek, A., & Aydin, H. M. (2023). Comparison of garlic and onion extract-derived gold nanoparticles: Characterization and anticancer activity. *Journal of Drug Delivery Science and Technology*, 84, 104542. DOI: 10.1016/j.jddst.2023.104542

Tao, R., Kang, K., Li, X., Li, R., Huang, R., Jin, Y., Qiu, L., & Zhang, W. (2021). Controlled Synthesis of Palladium Nanoparticles with Size-Dependent Catalytic Activities Enabled by Organic Molecular Cages. *Inorganic Chemistry*, 60(16), 12517–12525. DOI: 10.1021/acs.inorgchem.1c01723 PMID: 34320317

Taş, R., Köroğlu, E., & Celebioglu, H. U.Extract and Antibacterial Activity. (2021). Green Synthesis of Nickel Nanoparticles Using Peumus Boldus Koch. Extract and Antibacterial Activity. *International Journal of Innovative Engineering Applications*, 5(2), 152–155. DOI: 10.46460/ijiea.929625

Thema, F. T., Manikandan, E., Gurib-Fakim, A., & Maaza, M. (2016). Single phase Bunsenite NiO nanoparticles green synthesis by Agathosma betulina natural extract. *Journal of Alloys and Compounds*, 657, 655–661. DOI: 10.1016/j.jallcom.2015.09.227

Thilakan, D., Patankar, J., Khadtare, S., Wagh, N. S., Lakkakula, J., El-Hady, K. M., Islam, S., Islam, M. R., Khan, M. S., Alafaleq, N. O., & Tarique, M. (2022). Plant-Derived Iron Nanoparticles for Removal of Heavy Metals. *International Journal of Chemical Engineering*, 2022, 1517849. DOI: 10.1155/2022/1517849

Titus, D., James Jebaseelan Samuel, E., & Roopan, S. M. (2019). Chapter 12 - Nanoparticle characterization techniques. In A. K. Shukla & S. B. T.-G. S. Iravani Characterization and Applications of Nanoparticles (Eds.), *Micro and Nano Technologies* (pp. 303–319). Elsevier.

Tugarova, A. V., Mamchenkova, P. V., Dyatlova, Y. A., & Kamnev, A. A. (2018). FTIR and Raman spectroscopic studies of selenium nanoparticles synthesised by the bacterium Azospirillum thiophilum. *Spectrochimica Acta. Part A: Molecular and Biomolecular Spectroscopy*, 192, 458–463. DOI: 10.1016/j.saa.2017.11.050 PMID: 29220816

Verma, M. L., Dhanya, B. S., Thakur, M., Jeslin, J., & Jana, A. K. (2021). Chapter Eight - Plant derived nanoparticles and their biotechnological applications. In Verma, S. K., & Das, A. K. B. T.-C. A. C. (Eds.), *Biosynthesized Nanomaterials* (Vol. 94, pp. 331–362). Elsevier. DOI: 10.1016/bs.coac.2021.01.011

Vijilvani, C., Bindhu, M. R., Frincy, F. C., AlSalhi, M. S., Sabitha, S., Saravanakumar, K., Devanesan, S., Umadevi, M., Aljaafreh, M. J., & Atif, M. (2020). Antimicrobial and catalytic activities of biosynthesized gold, silver and palladium nanoparticles from Solanum nigurum leaves. *Journal of Photochemistry and Photobiology. B, Biology*, 202, 111713. DOI: 10.1016/j.jphotobiol.2019.111713 PMID: 31760373

Vinodhini, S., Vithiya, B. S. M., & Prasad, T. A. A. (2022). Green synthesis of palladium nanoparticles using aqueous plant extracts and its biomedical applications. *Journal of King Saud University. Science*, 34(4), 102017.

Wang, T., & Liu, W. (2022). Emerging investigator series: metal nanoparticles in freshwater: transformation, bioavailability and effects on invertebrates. *Environmental Science. Nano*, 9(7), 2237–2263. DOI: 10.1039/D2EN00052K PMID: 35923327

Wu, S., Shanmugam, R., Madasamy, M., & Vanaja, M. (2020). Green synthesis of copper nanoparticles using Cissus vitiginea and its antioxidant and antibacterial activity against urinary tract infection pathogens. *Artificial Cells, Nanomedicine, and Biotechnology*, 48(1), 1153–1158. DOI: 10.1080/21691401.2020.1817053 PMID: 32924614

Xulu, J. H., Ndongwe, T., Ezealisiji, K. M., Tembu, V. J., Mncwangi, N. P., Witika, B. A., & Siwe-Noundou, X. (2022). The Use of Medicinal Plant-Derived Metallic Nanoparticles in Theranostics. *Pharmaceutics*, 14(11), 2437. DOI: 10.3390/pharmaceutics14112437 PMID: 36365255

Yaashikaa, P. R., Senthil Kumar, P., & Varjani, S. (2022). Valorization of agro-industrial wastes for biorefinery process and circular bioeconomy: A critical review. *Bioresource Technology*, 343, 126126. DOI: 10.1016/j.biortech.2021.126126 PMID: 34673193

Yanik, S., Emre, D., Alp, M., Algi, F., Yilmaz, S., Bilici, A., & Ozkan-Ariksoysal, D. (2023). A novel electrochemical biosensor based on palladium nanoparticles decorated on reduced graphene oxide-polyaminophenol matrix for the detection and discrimination of mitomycin C-DNA and acyclovir-DNA interaction. *Journal of Pharmaceutical and Biomedical Analysis*, 234, 115524. DOI: 10.1016/j.jpba.2023.115524 PMID: 37320972

Yasmin, H., Giwa, S. O., Noor, S., & Aybar, H. Ş. (2023). Influence of Preparation Characteristics on Stability, Properties, and Performance of Mono- and Hybrid Nanofluids: Current and Future Perspective. *Machines (Basel)*, 11(1), 112. DOI: 10.3390/machines11010112

Zuhrotun, A., Oktaviani, D. J., & Hasanah, A. N. (2023). Biosynthesis of Gold and Silver Nanoparticles Using Phytochemical Compounds. *Molecules (Basel, Switzerland)*, 28(7), 3240. DOI: 10.3390/molecules28073240 PMID: 37050004

KEY TERMS AND DEFINITIONS

Green (Chemistry) Synthesis Technique: The formulation of plant-derived metal nanoparticles is underpinned by an environmentally friendly approach, with a particular emphasis on sustainability and environmental friendliness.

Plant Extract Mediation: The utilization of natural resources (plant extracts) in the synthesis process is underscored by the emphasis placed on the ecologically compatible and renewable resources of plant-derived metal nanoparticles in nanomaterial production. The plant extracts employed in the formulation are clearly identified.

Metal Nanoparticles: The synthesis of nanoparticles is examined in detail, with particular attention paid to their structure, properties and potential applications.

Biomedical Applications: It suggests a wide range of potential areas of use in the medical field, such as drug delivery systems, biosensors, wound healing and various biomedical technologies, and demonstrates the types of applications these nanoparticles can be employed in.

Biogenic Metal Nanoparticles: Metal nanoparticles formulated with biological steps are of great interest due to their nano size and high surface area–to–volume ratio.

Chapter 2
Eco-Friendly Synthesis and Characterization of Gold Nanoparticles Using Leaf Extract of *Eucalyptus globulus*

Samadhan Waghmode
https://orcid.org/0000-0001-7747-0703
Elphinstone College, Dr. Homi Bhabha State University, Mumbai, India

Sushilkumar Ghadage
Changu Kana Thakur Arts, Commerce, and Science College, New Panvel (Autonomous), India

Anup Sonawane
Institute of Chemical Technology, Matunga, India

Rahul Zanan
https://orcid.org/0009-0005-6646-7046
Elphinstone College, Dr. Homi Bhabha State University, Mumbai, India

ABSTRACT

An essential component of nano-biotechnology is the production of reliable and eco-friendly routes for synthesizing nanoparticles. Plant extracts are an affordable, cost-effective and eco-friendly material for the synthesis of nanoparticles. In the current study, gold nanoparticles (AuNPs) were bio-synthesized utilizing E. globulus leaf extract. Rapid reduction of the gold ions into AuNPs is seen once the gold ions are exposed to an aqueous solution of leaf extract. TEM, EDX and UV-Vis spectroscopy methods were used to investigate these bio-synthesised AuNPs. The aqueous media containing AuNPs had a peak in the UV-visible spectrum at 540 nm. According to TEM investigation, AuNPs were evenly dispersed and had spheres with spherical and triangular forms that were between 15 and 30 nm in size. By using the agar well diffusion method, the antibacterial activity of the biosynthesized AuNPs was confirmed against prevalent human pathogens such S. aureus (NCIM 5021) and P. aeruginosa (NCIM 5029). This process uses cheap, easily accessible biomaterial to synthesize metallic nanoparticles.

DOI: 10.4018/979-8-3693-6240-2.ch002

1. INTRODUCTION

Nanotechnology is a broad field of science and technology that encompasses many disciplines. In fact, many different industrial and scientific fields exploit the chemical, physical and biological properties of materials smaller than 100 nanometers. Since nanotechnology has so many uses across a wide range of industries, it is a keystone of contemporary scientific advancement.

Geometry, structure and morphology of the nanomaterials have a significant impact on how well they function in various applications. Nanoparticles (NPs) have attracted a lot of attention recently due to their valuable features and possible uses in a variety of industries, including sensor, electrical, optical, catalytic and pharmaceutical (Nikam et al. 2019). The use of gold nanoparticles (Au-NPs) in biomedical research techniques like magnetic resonance imaging and X-ray computed tomography is one of the most important uses of nanotechnology (Siddique et al. 2020). The optical characteristics of NPs support cancer nanotechnology for photothermal therapy, medication delivery and cancer diagnostics (Zhao et al. 2022; Ghosh et al. 2008). Green-mediated NPs synthesis and characterization have grown to be a key area of nanotechnology, particularly for precious metals including gold, silver, platinum and palladium (Waghmode et al. 2021). Moreover, the chemical or physical pathways are usually employed to synthesize metal NPs because of their inherent advantage in producing well-defined NPs with quite governable shapes and sizes (Iravani et al. 2014). A more sophisticated substitute for hazardous chemicals and expensive physical processes could be the synthesis of NPs using biological materials (Rahimi & Doostmohammadi, 2019). The production of metal NPs employing bacteria, fungi, actinomycetes, yeast and viruses has received a lot of attention (Bahrulolum et al. 2021). In addition to chemical and physical methods, phytosynthesis that exploit parts of whole plants as biological mill to synthesize metallic NPs are under exploitation and is not only advantageous but also profitable approach (Thakkar et al. 2010). Unlike microorganisms, the phytosynthesis process is relatively quick and inexpensive and it can be scaled up simply for bulk production. It also does not require complex and multi-step operations such as microbial isolation, cultivation and maintenance (Iravani, 2011). Moreover, it has been observed that the use of plants instead of microorganisms accelerates the synthesis of nanoparticles and results in more stable NPs (Akhtar et al. 2013). Plants are known to have a diverse range of metabolites. However, their potential for producing metallic NPs has not yet been completely harvested (Jha et al. 2009). Metal NPs can be produced on a large scale using plant tissue culture techniques and improved downstream processing (Ramrakhiani & Ghosh, 2018). Due to its environmental friendliness, the usage of plant components has been regarded as a safe and reliable method for the biosynthesis of NPs (Das et al. 2017). It is clear from the literature that plants have been effectively used to produce noble metal NPs quickly and extracellularly (Dudhane et al. 2017).

The integration of nanotechnology in various fields has led to remarkable advances. Nanotechnology covers the manipulation and use of materials with dimensions below 100 nanometers, including their chemical, physical and biological characteristics (Bayda et al. 2020). In nanotechnology research, considerable attention has been paid to the use of metal NPs, especially silver nanoparticles (Ag-NPs). Ag-NPs have attracted much attention due to their unique properties, such as high surface area-to-volume ratio, excellent adsorption capacity, reactivity and electrical conductivity (Zhang et al. 2016). These properties make Ag-NPs adapted to a wide range of applications. In recent times, they have found applications in textiles, cosmetics, paintings, conductive inks, photonics, electronics, catalyzing, biocapper, water purification, air filters, containers food storage and biomedical fields (Verma & Kumar, 2022). However, traditional synthesis methods, including chemical or physical approaches, have raised environmental and health concerns associated with the use of toxic substances. Correspondingly, there is an increasing need to develop safe, environmentally friendly and cost-effective synthesis methods (Moond et al. 2022). The conventional methods applied to the synthesis of nanomaterial involve the use of hazardous chemical compounds and/or physical procedures with high energy requirements (Kumar et al. 2021). Green chemistry approaches have been applied to the creation of nanomaterials to prevent these negative effects. Unlike the current methods of synthesis, the 'green synthesis of nanoparticles' involves processes, which are clean, safe and friendly to the environment (Álvarez-Chimal & Arenas-Alatorre, 2023). One of the green routes of synthesis of NPs involves the use of vegetal extracts. The publication by Gardea-Torresdey et al. (1999), which is one of the pioneering research results on the production of Au-NPs from alfalfa plant biomass. Green synthesis methods using natural reducing agents such as plant extracts are a promising direction. *Eucalyptus* is widely used as a medicinal plant due to its rich content of phytochemical constituents and has attracted attention in the synthesis of NPs. Extracts from various *Eucalyptus* species are abundant and have demonstrated potential for green synthesis (Sadiq et al. 2023).

Aqueous extract of *Eucalyptus* leaves has high antioxidant activity therefore usually used as reducing agent for synthesis Ag-NPs (Alghoraibia et al. 2019). Because of its broad range of biological activities, mainly linked to the diversity of phytochemical elements in plant parts, *Eucalyptus* is one of the most commonly used medicinal plants (Shala & Gururani, 2021). Decoctions made from E*ucalyptus* leaves are applied topically, rinsed wounds, applied as compresses in trauma cases and used for rheumatism and upper respiratory tract pain. It can also be inhaled for bronchitis and other upper respiratory tract ailments. (Balčiūnaitienė et al. 2022). The biologically active compound complex found in *Eucalyptus* leaves determines a particularly strong antioxidant effect of their extracts (González-Burgos et al. 2018). The most important phenolic compounds that largely determine the antiradical and reductive activity of *Eucalyptus* leaves are phenolic acids, quercetin glycosides and other groups of phenolic compounds of various structures. Like sage leaf extract, *Eucalyptus* leaf extract has extremely strong antibacterial and antifungal activity (Gullón et al. 2020). In general, the chemical, physical, mechanical and antimicrobial properties of Ag-NPs depend on their chosen precursor and synthesis method (Zhang et al. 2016). *Eucalyptus* plant has a wide range of medicinal uses to treat tuberculosis, aromatherapy, reduce the risk of malaria, cure joint pain, respiratory problems, common bacterial infection, cough, cold etc. (Vecchio et al. 2016). The environmental benefit of the *Eucalyptus* tree is high because of its desirable traits such as being fast-growing sources of wood (Sunder, 1993). The leaves, bark and extracts of the *Eucalyptus* tree are rich sources of bioactive compounds, including phenolic compounds, terpenoids and essential oils.

The technique for the plant extract-mediated synthesis of Au-NPs is disclosed that is economical, cost-effective and eco-friendly (Abdelkhalek et al. 2020). This study involves the reduction of $HAuCl_4$ solution utilizing aqueous *E. globulus* leaf extracts. UV-Visible spectroscopy was used to monitor the bio-reduction process, while EDX was used to look into the crystal structure. TEM was used to determine the nanostructure and size of the produced Au-NPs. In addition, antibacterial efficacy against common bacterial pathogens has been discussed. Numerous methods can be employed for the synthesis of NPs (Pinto et al. 2017). These methods encompass reduction in solutions (including distilled water, ethanol, hexane, toluene, ethylene glycol and others), electrochemical and sonochemical approaches (Nam & Luong, 2019), microwave-assisted technique (optimized for power, duration and medium), radiation-assisted method and biological routes coupled with chemical reactions (Dhand et al. 2015). Generally, NPs are crafted using a diverse range of chemical and physical procedures. However, these conventional approaches prove costly and environmentally detrimental due to their utilization of hazardous and toxic chemicals, posing biological and health risks (Abuzeid et al. 2023). Nevertheless, Ag-NPs find extensive application in healthcare and medicine owing to their robust antibacterial properties (inhibiting DNA replication, enzyme functions etc.), antiviral attributes (blocking virus attachment to cell walls), antifungal characteristics (cell membrane destruction) and other activities (Naganthran et al. 2022). Therefore, there is an urgent need to develop a safe, cost-effective and feasible method to synthesize Ag-NPs to maximize the potential benefits to society (Tsilo et al. 2024). NPs derived from their bulk counterparts often exhibit new and improved physical, chemical, electrical, mechanical and optical properties depending on factors such as size, shape, distribution and morphology (Joudeh & Linke, 2022). Gold as a material exhibits changes in these properties depending on its size, shape and concentration. Au-NPs have emerged as an interesting material for a variety of biological and biomedical applications due to their unique and diverse properties. Different types of NPs that have been biosynthesized using leaf extracts from *Eucalyptus* sp. are presented in Table 1.

Table 1. The various types of nanoparticles biosynthesized with the use of leaf extracts of Eucalyptus sp.

Plant species	Precursor	NPs	Reducing agent	Characterization technique	UV-Visible spectra	TEM/SEM Size	Application	References
Eucalyptus sp.	$AgNO_3$	Ag	Leaf extracts	UV-Vis, FTIR, SEM, XRD	370 nm	~68 to 106 nm	Antimicrobial activity	Gohil et al. 2024
Eucalyptus sp.	$FeCl_3$, $PdCl_2$	Fe, Pd	Leaf extracts	FTIR, Zeta analysis	--	--	Removal of oily wastewater	Hassan et al. 2024
Eucalyptus sp.	$FeSO_4.7H_2O$	Fe	Leaf extracts	UV-Vis, FTIR, XRD, SEM, EDX	360 nm	--	Photocatalytic activity	Katta et al. 2024
Eucalyptus globulus	$ZnSO_4$	ZnO	Leaf extracts	UV-Vis, FTIR, Particle size analyzer	430 nm	--	Antibacterial & Mosquito repellency test	Majeed et al. 2024
Eucalyptus globulus	$AgNO_3$	Ag	Leaf extracts	UV-Vis, FTIR, SEM, XPS, EDS, XRD, TGA, DLS	470 nm	42.9 to 142.9 nm	Antibacterial, Photocatalytic activity	Rocha et al. 2024
Eucalyptus teriticornis	$HAuCl_4$	Au	Leaf extracts	--	--	--	Antioxidant effect	Faiz et al. 2023
Eucalyptus sp.	$AgNO_3$	Ag	Leaf extracts	UV-Vis, FTIR, SEM, TEM, EDX, XRD	419 nm	5–20 nm	To enhance the shelf life of banana	Nayab & Akhtar, 2023
Eucalyptus globulus Labill.	$ZnSO_4$	ZnO	Leaf extracts	UV-Vis, FTIR, SEM, EDX	370 nm	25 nm – 151 nm	--	Shahid et al. 2023
Eucalyptus globulus	$AgNO_3$	Ag	Leaf extracts	UV-Vis, FTIR, TEM, EDS	438 nm	17.5 ± 5.89 nm	Antioxidant, Antimicrobial activity	Balčiūnaitienė et al. 2022
Eucalyptus globulus Labill	Chitosan	CNPs	Leaf extracts	UV-Vis, FTIR, EDX, TEM, TGA, XRD, DSC, Zeta potential	295 nm	6.92 - 10.10 nm	Antibacterial activity	El-Naggar et al. 2022
Eucalyptus sp.	$AgNO_3$	Ag	Leaf extracts	UV-Vis, AFM, SPM, Zeta Sizer, FTIR	430 nm -440 nm	57.94 nm	Catalytic degradation	Rabeea et al. 2021
Eucalyptus camaldulensis	$AgNO_3$	Ag	Leaf extracts	UV-Vis, SEM, EDX, XRD, FTIR, DLS	450 nm	28 nm	Antioxidant activity	Alghoraibi et al. 2020

continued on following page

Table 1. Continued

Plant species	Precursor	NPs	Reducing agent	Characterization technique	UV-Visible spectra	TEM/SEM Size	Application	References
Eucalyptus globulus Labill	$Zn(NO_3)2{\cdot}6H_2O$	ZnO	Leaf extracts	UV–Vis, XRD, FE-SEM, EDX, BET, Zeta potential, DLS, DSC, FTIR	257 nm	27-35 nm.	--	Barzinjy & Azeez, 2020
Eucalyptus sp.	NH_4Cl	Carbon	Twigs extract	FE-SEM, XPS, FTIR	360 nm	40–50 nm	Sensing, Bioimaging, Cell viability cytotoxicity	Damera et al. 2020
Eucalyptus globulus	$HAuCl_4{\cdot}4H_2O$	Au	Leaf extracts	UV-Vis, SEM, XRD, SPR	525 ~ 540 ± 2 nm	21 - 30 nm	--	Ibraheem et al. 2020
Eucalyptus tereticornis	$AgNO_3$	Ag	Leaf extracts	UV-Vis, SEM, TEM, XRD, EDS	569 nm	35 nm	Antibacterial, Antioxidant, Anticancer studies	Kiran et al. 2020
Eucalyptus Concoction	$HAuCl_4{\cdot}4H_2O$	Au	Bud Extract	UV-Vis, FTIR, XRD, FESEM, AFM	550 nm	<100 nm	Antimicrobial, Cytotoxic Studies	Muthiah et al. 2020
Eucalyptus obliqua	$AgNO_3$	Ag	Leaf extracts	UV, SEM, TEM	420 nm	10 nm	--	Rahmatyian & Dadmehr, 2020
Eucalyptus globulus	$AgNO_3$	Ag	Leaf extracts	Particle size distribution, Zeta potential	259.5 nm	--	Anthelmintic activity	Sankpal et al. 2020
Eucalyptus robusta	$(NH_4)_2$ Fe (SO_4) $6H_2O)$	Fe	Leaf extracts	UV-Vis, AFM, DLS	280 nm	--	Antibacterial, Antioxidant activity	Vitta et al. 2020
Eucalyptus Corymbia	$AgNO_3$	Ag	Leaf extracts	UV, FTIR, TEM, EDX	425 nm	18-20 nm	--	Sila et al. 2019
Eucalyptus globulus	$HAuCl_4{\cdot}4H_2O$	Au	Leaf extracts	UV,TEM, EDS, ATR-FTIR	520–570 nm	12.8 ± 6.3 nm	--	Dzimitrowicz et al. 2019
Eucalyptus globulus	$Mg(NO_3)_2{\cdot}6H_2O$	MgO	Leaf extracts	UV–Vis, FTIR, TEM	340 nm	6–8 nm	--	Jeevanandam et al. 2018
Eucalyptus Mellidora	$ZnSO_4$	ZnO	Leaf extracts	XRD, SEM	350 nm	30-50 nm	Antibacterial activity	Rafiee et al. 2018
Eucalyptus urograndis	$HAuCl_4$	Au	Leaf Extract - oil	UV-Vis	566 - 580 nm	103 - 171 nm	Antioxidant activity	Thesing et al. 2018
Eucalyptus globulus	$HAuCl_4.3H_2O$	Au	Bark extracts	UV–Vis, TEM, DLS	525 nm	25.8 ± 5.1, 20.1 ± 3.4, and 20.1 ± 3.3 nm	--	Pinto et al. 2017
Eucalyptus oleosa	$HAuCl_4$	Au	Leaf extracts	UV, TEM	552 nm	28 nm	--	Pourmortazavi et al. 2017
Eucalyptus globulus	Zn $(NO_3)_2{\cdot}6H_2O$	ZnO	Leaf extracts	UV-Vis, FTIR, SEM, TEM, EDX, XRD, PSA, Zeta potential, Raman spectroscopy	361 nm	11.6 nm	Photocatalytic, antioxidant activity	Siripireddy & Mandal, 2017
Eucalyptus globules	$AgNO_3$	Ag	Leaf Extracts	UV, XRD, AFM	375 nm	15-71 nm	Antibacterial activity	Thakur et al. 2017

continued on following page

Table 1. Continued

Plant species	Precursor	NPs	Reducing agent	Characterization technique	UV-Visible spectra	TEM/SEM Size	Application	References
Eucalyptus globulus	$AgNO_3$	Ag	Leaf Extracts	UV,TEM, XRD, FTIS	425 nm	30-36 nm	--	Balamurugan & Saravanan, 2017
Eucalyptus sp.	$FeSO_4{\cdot}7H_2O$, $Ni(NO_3)2{\cdot}6H_2O$	Fe/Ni	Leaf extracts	GC-MS, FTIR, TG, SEM, EDS, XRD, XPS	--	20 to 50 nm	Degradation of methyl orange	Weng et al. 2017
Eucalyptus globulus	$AgNO_3$	Ag	Leaf extracts	UV-Vis, XRD, TEM, SEM-EDX, FTIR, TGA	428 nm	1.9 - 4.3 nm, 5 - 25 nm	Antibacterial, Antibiofilm activity	Ali et al, 2015
Eucalyptus Camaldulensis	$AgNO_3$	Ag	Leaf extracts	UV-Vis, XRD, AFM, EDX, SEM, SPR	425 nm	110 - 250 nm	--	Bashir & Qureshi, 2015
Eucalyptus camaldulensis	$AgNO_3$	Ag	Leaf extracts	UV-Vis	420 nm	--	Antibacterial property	Mohammed, 2015
Eucalyptus Macrocarpa	$AgNO_3$, $HAuCl_4$	Au, Ag	Leaf extracts	UV-Vis. XRD, EDS, TEM, FESEM	430 nm (Ag), 570 nm (Au)	10 to 100 nm (Ag), 20 to 100 nm (Au)	--	Shah et al. 2015
Eucalyptus urophylla (EU), *Eucalyptus citriodora* (EC) *and Eucalyptus robusta* (ER)	$AgNO_3$	Ag	Leaf extracts	UV–Vis, FE-TEM, XRD, FTIR	447, 445, 482 nm	4 – 60 nm	--	Ya-yuan Mo et al. 2015
Eucalyptus oleosa	$AgNO_3$	Ag	Leaf extracts	UV, FTIR, SEM, EDX	460 nm	21 nm	--	Pourmortazavi et al. 2015
Eucalyptus Globulus	$FeCl_3$	Fe	Leaf extracts	UV–Vis, XRD, SEM, EDS, TEM, FTIR	350 nm	30-36 nm	--	Balamurugan et al. 2014
Eucalyptus sp.	$FeSO_4$	Fe	Leaf extracts	FTIR, SEM, EDS, XRD	--	20–80 nm	Wastewater treatment	Wang et al. 2014
Eucalyptus globulus	$AgNO_3$	Ag	Leaf extracts	UV-Vis, TEM	530 nm	9.18-32.39 nm	Antibacterial activity	El-Rahman et al. 2013
Eucalyptus macrocarpa	$AgNO_3$	Ag	Leaf extracts	UV-Vis, XRD, TEM, SEM	430 nm	10 - 100 nm	--	Poinern et al. 2013
Eucalyptus sp.	$FeCl_3$	Fe	Leaf extracts	XRD, TEM, XAS, Cyclic Voltammograms	--	40–60 nm	Adsorption flocculation capacity	Wang, 2013
Eucalyptus hybrida	$AgNO_3$	Ag	Leaf extracts	UV–Vis, XRD, SEM, EDX	412 nm	50 - 150 nm	--	Dubey et al. 2009
Eucalyptus camaldulensis	$HAuCl_4$	Au	Leaf extracts	UV/Vis, TEM, EDS	534.4 nm	1.25 – 17.5 nm	--	Ramezani et al. 2008

2. MATERIAL AND METHODS

2.1 Preparation of Plant Extracts

The plant extract from *E. globulus* was used to make Au-NPs. Fresh leaves of plants were collected from Botanical garden of Lokmangal Biotechnology College, Wadala, North Solapur. The plant leaves were surface-washed with distilled water and air-dried. The dried leaves were ground into a fine powder and 20 g of that powder was mixed with 180 mL of distilled water before being heated for 15 minutes in a water bath. The content underwent a 10-minute centrifugation to remove cell debris. The aqueous leaf extracts in the supernatant were filtered using filter paper (Whatman No. 1). The resulting filtrate was kept at 4 °C and further used for the preparation of Au-NPs (Dudhane et al. 2017).

2.2 Bio-Synthesis of Au-NPs

Synthesis of biogenic Au-NPs was performed with $HAuCl_4$ solution and aqueous leaf extract. 5 mL of *E. globulus* leaf extract was added to 95 mL of aqueous auric chloride solution (1 mM) in an Erlenmeyer flask. The flasks were kept at 40 °C on a rotary shaker at 150 rpm. The change in color of the solution after the reaction completed acted as a visual sign for the success of the synthesis process (Ghosh et al. 2012). The biosynthesis procedure and characterization of Au-NPs using leaf extract is illustrated in Fig. 1.

Figure 1. Graphical abstract of Au-NPs synthesis using Eucalyptus globulus leaf extract.

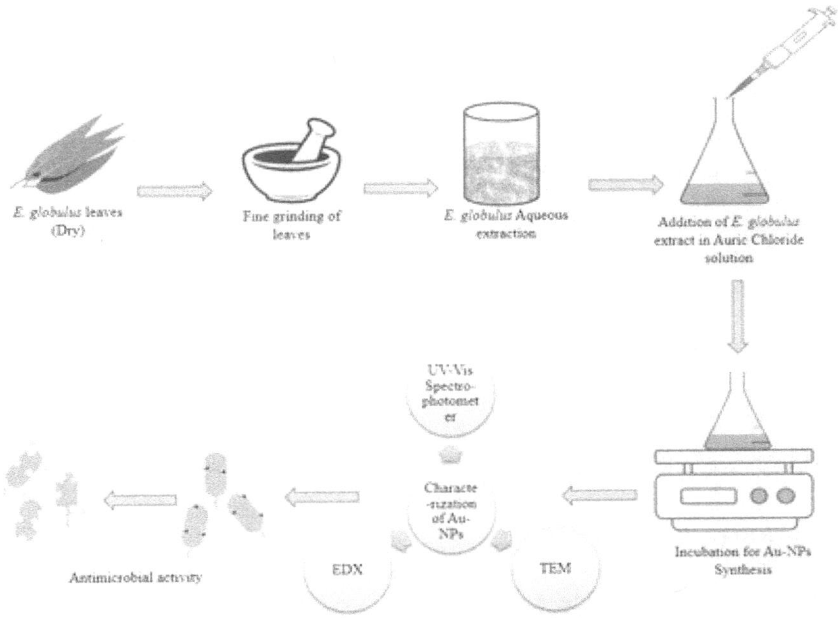

2.3 Characterization

2.3.1 UV–Vis Spectroscopy

The formation of Au-NPs in aqueous solution was confirmed by using UV-Vis spectrophotometer. UV-Vis spectroscopy examination was performed on the mixture of leaf extract and $HAuCl_4$ solution during the reduction process. The bio-reduction of pure Au^+ ions with the leaf extract of *E. globulus* was monitored. The optical absorbance of Au-NPs suspended in distilled water was recorded on UV-Vis Spectrophotometer (UV-Vis 1601 Schimodzu spectrophotometer (Kyoto, Japan) at 200-800 nm wavelength range (Waghmode et al. 2021).

2.3.2 Energy Dispersive Spectroscopy and Transmission Electron Microscopy

The size and morphology of biosynthesized Au-NPs prepared using *E. globulus* were studied by TEM analysis. TEM analysis of the sample was done using an instrument functioned at an accelerating voltage of 200 kV with a resolution of 0.23 nm. Additionally, EDX analysis was used to assess the elemental makeup of the synthesized Au-NPs. EDX analysis was done on a thin-film sample made with Au-NPs on copper grid. The single drop of sample was applied using a micropipette (100 μL) and dried for 30 minutes at room temperature (Dzimitrowicz et al. 2019; Ghosh et al. 2012).

2.4 Antibacterial Activity

The antibacterial activity of the plant-mediated Au-NPs was evaluated against human pathogens such as *Staphylococcus aureus* (NCIM 5021) and *Pseudomonas aeruginosa* (NCIM 5029). On sterile nutrient agar plates, the pure bacterial cultures were uniformly distributed. Using a sterile cork borer, 10 mm diameter wells were created on the nutrient agar plates. The 100 μL solution of green synthesized Au-NPs was added in the well whereas *E. globulus* leaf extract was used as a negative control. After 24 hrs, inhibition zones around the wells were measured in millimeter (mm) (Kiran et al. 2020).

3. RESULTS

3.1 Synthesis of Au-NPs

The change in color from the yellow color of Ruby Red indicates the formation of Au-NPs. The appearance of red ruby red in two minutes of incubation confirms the decrease in gold chloride in Au-NPs.

3.2 Characterization of Au-NPs

3.2.1 UV-Vis Spectra of Au-NPs

UV-Vis spectroscopy is one of the most important methods to measure the formation of Au-NPs in aqueous solutions. The production and stabilization of the reduced Au-NPs in the colloidal solution was monitored by a UV-Vis spectrophotometer. The Au-NPs are known to exhibit at maximum in the range of 400-700 nm. The Au-NPs synthesized by *E. globulus* are positioned at 540 nm in good agreement with previous reports (Fig. 2) (Kiran et al. 2020; Dzimitrowicz et al. 2019).

3.2.2 Transmission Electron Microscopy

The shape, size and morphology of the synthesized Au-NPs were elucidated with the help of TEM. According to the earlier findings, the TEM image further confirms that the Au-NPs are primarily spherical and triangular in shape and range in size from 15 to 30 nm (Fig. 2) (Sadiq et al. 2019). Some of the NPs have been discovered in small aggregates with larger dimensions. The nano triangles and sphericals clearly showed the presence of bioreduced NPs, which are comparable to the Au-NPs that have been previously reported to be produced from *Gnidia glauca* flower extract (Ghosh et al. 2012).

3.2.3 Energy Dispersive Spectroscopy Studies

The EDX spectrum reveals the elemental composition profile of the green synthesized Au-NPs. For Au-NPs, the peak was attained at an energy of 2.12 keV. Some of the weaker peaks, including Fe, Cl, Si and O are also present. The sample contained a significant peak of copper, which comes from the copper substrate. The reduction of gold ions is shown by the emission energy at 2.12 keV (Fig. 2). The successful bioreduction of the Au^{+} ions was confirmed by Ghosh et al. (2012) a comparable signal of energy emission of Au in EDX spectra was observed.

Figure 2. UV-Visible, TEM and EDX spectra of Au-NPs synthesized using Eucalyptus globulus leaf extract

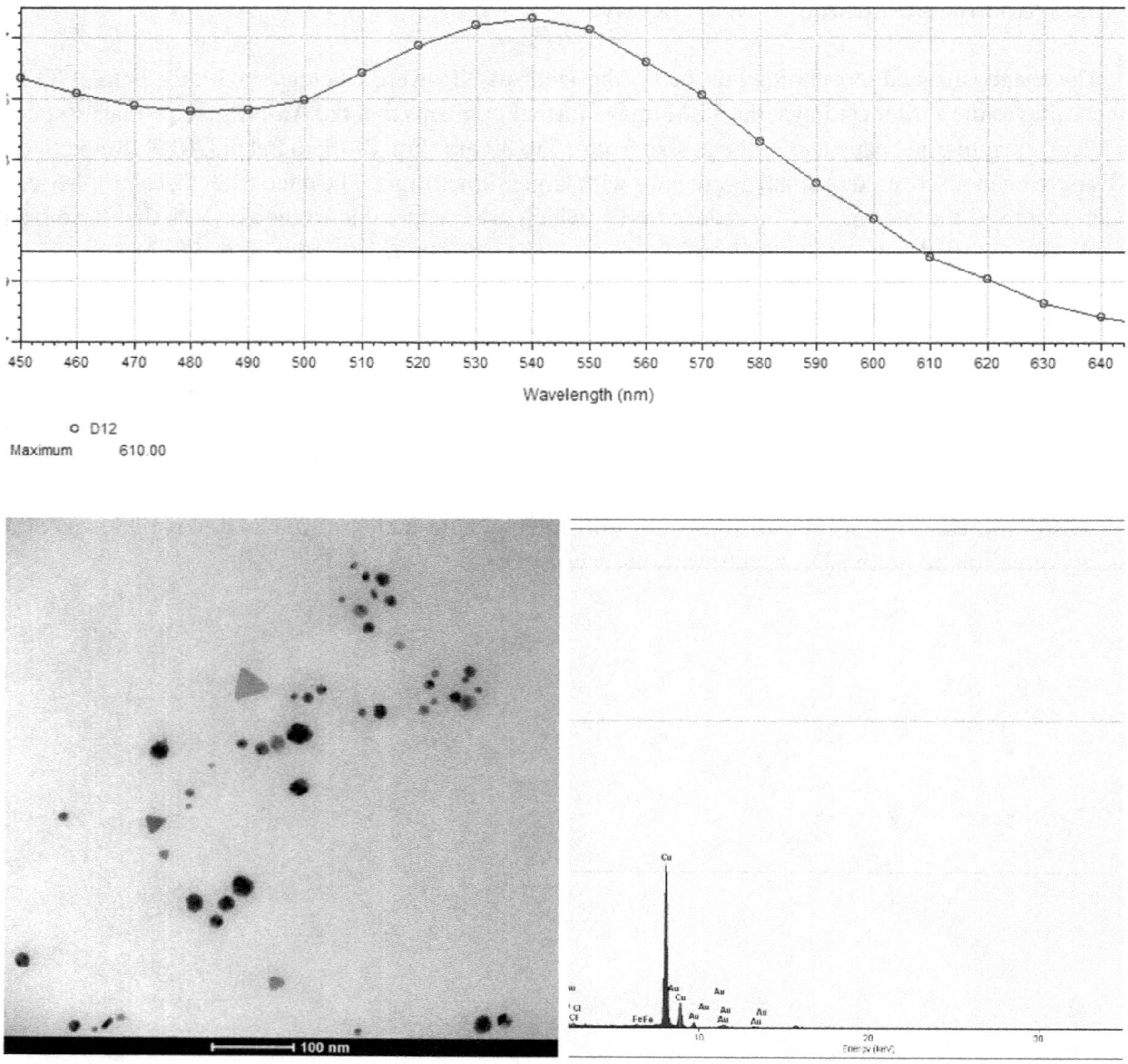

3.3 Antibacterial Activity

The Au-NPs biosynthesized from *E. globulus* were tested for their antibacterial activity against common human pathogens such as *Staphylococcus aureus* (NCIM 5021) and *Pseudomonas aeruginosa* (NCIM 5029) using the agar well diffusion method. The Au-NPs showed moderate antibacterial activity 20 and 18 mm zone of inhibition respectively against common bacterial pathogens (Table 2). However, leaf extract lacked significant antibacterial activity against bacteria. Muthiah et al. also demonstrated similar results for Au-NPs prepared from *E. Concoction* (Kiran et al. 2020; Muthiah et al. 2020). The effectiveness of NPs established as an ordinary antibacterial agent.

Table 2. Antibacterial activity of Au-NPs of prepared from Eucalyptus globulus leaf extract

Sr. No.	Name of organism	Diameter of zone of inhibition (mm)	
		Au-NPs	Leaf extract
1	*Staphylococcus aureus* (NCIM 5021)	20	14
2	*Pseudomonas aeruginosa* (NCIM 5029)	18	12

4. DISCUSSION

The synthesis of NPs using plant extracts, particularly from various *Eucalyptus* species, demonstrates an eco-friendly and cost-effective approach that aligns with the principles of green chemistry. This discussion examines the findings from several studies focusing on the synthesis of silver (Ag-NPs), zinc oxide (ZnO-NPs) and gold nanoparticles (Au-NPs) using *Eucalyptus* extracts.

Ag-NPs Synthesis using *Eucalyptus* sp.

The study by Mohammed (2015) successfully utilized aqueous extracts of *Eucalyptus camaldulensis* leaves as reducing and stabilizing agents for synthesizing Ag-NPs. The formation of Ag-NPs was confirmed by the characteristic color change and a surface plasmon resonance (SPR) peak at 450 nm. The particle size analysis showed varying sizes: 28 nm (SEM), 16 nm (X-Ray) and 68 nm (DLS). The presence of flavonoids, polyphenols and proteins in the extract played a crucial role in reducing and capping the NPs, resulting in stable, monodisperse and negatively charged colloidal nanosilver particles. This method underscores the efficiency and simplicity of using plant extracts for NP synthesis, providing a viable alternative to conventional chemical methods.

Ya-yuan Mo et al. (2015) focused on synthesizing Ag-NPs using extracts from *Eucalyptus* leaves and *Larrea tridentate*. The method proved efficient, cost-effective and environmentally friendly, producing functionalized NPs with spherical morphology. The synthesized Ag-NPs demonstrated potential for diverse applications due to their high-quality and stable characteristics. This approach reinforces the feasibility of using plant extracts for large-scale and sustainable NP synthesis.

The synthesis of Ag-NPs using *Eucalyptus globulus* leaf extracts, as reported by Balčiūnaitienė et al. (2022), highlighted the eco-friendly and cost-effective nature of this approach. The Ag-NPs produced were approximately 17.5 ± 5.89 nm in size, influenced by the flavonoids, phenolic compounds, hydroxycinnamic acid and other phytochemicals in the extracts. These compounds not only facilitated the reduction and stabilization of Ag-NPs but also imparted antioxidant properties and antibacterial potency. This dual functionality enhances the potential applications of Ag-NPs in various fields, including biomedicine and environmental remediation.

Multiple studies have reported the successful synthesis of Ag-NPs using *Eucalyptus* extracts. For instance, *Eucalyptus camaldulensis* leaves were found to effectively reduce and stabilize Ag-NPs, indicated by a color change and a SPR peak at 450 nm. The NPs had an average size of 28 nm, 16 nm and

68 nm depending on the analysis method used (Alghoraibi et al. 2020). Similarly, *Eucalyptus globulus* leaf extracts resulted in Ag-NPs with an average size of 17.5 nm, exhibiting significant antibacterial and antioxidant activities (Balčiūnaitienė et al. 2022). These findings suggest that *Eucalyptus* extracts can produce Ag-NPs with uniform size distribution and potent biological activities.

Green Synthesis of ZnO-NPs using *Eucalyptus* sp.

Barzinjy & Azeez (2020) demonstrated the rapid and cost-effective biosynthesis of ZnO-NPs using *Eucalyptus globulus* leaf extract. The phytochemicals in the extract acted as reducing, capping and stabilizing agents, producing hexagonal ZnO-NPs with an average diameter of 35 nm and high crystallinity. These ZnO-NPs exhibited thermal stability and potential adsorption sites for dyes and heavy metal ions, indicating their applicability in fields such as environmental purification and catalysis. This study highlights the versatility of *Eucalyptus* extracts in synthesizing different types of NPs with desirable properties for specific applications.

The biosynthesis of ZnO-NPs using *Eucalyptus globulus* leaf extract and zinc nitrate hexahydrate salt was also explored. Barzinjy & Azeez (2020) reported the formation of thermally stable hexagonal ZnO-NPs with an average size of approximately 35 nm. This indicates that *Eucalyptus* extracts can be used to produce ZnO-NPs with stable thermal properties, which are crucial for various industrial applications.

Au-NPs synthesis using *Eucalyptus* sp.

The synthesis of Au-NPs using *Eucalyptus* bud concoction, as explored by Muthiah et al. (2020), involved the bio-reduction of chloroauric acid. The resulting Au-NPs varied in shape from spherical to cubical and ranged in size from 30 to 90 nm. Although these NPs did not exhibit significant antibacterial activity, they showed promising anti-proliferative and cytotoxic effects on HEp-2 and Vero cell lines. These findings suggest potential applications in nanomedicine, particularly in cancer therapy. This study emphasizes the importance of exploring various plant parts and their extracts for synthesizing NPs with specific biological activities.

Kiran et al. (2020) highlighted the green synthesis of Au-NPs using *Eucalyptus tereticornis* leaf extract. The Au-NPs were spherical, stable, well-dispersed and exhibited significant antibacterial activity against specific strains. Additionally, the biosynthesized Au-NPs displayed antioxidant activity and cytotoxicity towards MCF-7 cell lines, suggesting their potential in pharmaceutical and nanomedicine applications. This study underscores the therapeutic potential of NPs synthesized using *Eucalyptus* extracts.

Eucalyptus extracts have also been effective in the synthesis of Au-NPs. Muthiah et al. (2020) demonstrated that a concoction of *Eucalyptus* buds could reduce gold ions, resulting in Au-NPs with inhibitory effects on cancer cell lines. Additionally, *Eucalyptus tereticornis* leaf extract was used to synthesize Au-NPs, which showed significant antibacterial activity and potential applications in nanomedicine (Kiran et al. 2020). These studies highlight the versatility of *Eucalyptus* extracts in producing NPs with therapeutic potential.

The discussed studies collectively highlight the efficacy and environmental benefits of using *Eucalyptus* extracts for NP synthesis. The varying sizes, shapes and functionalities of the NPs synthesized demonstrate the versatility of these natural reducing and stabilizing agents. The antioxidant, antibacterial and potential therapeutic properties of the synthesized NPs underscore their wide-ranging applications in biomedicine, environmental remediation and catalysis. These green synthesis methods offer a sustainable alternative to conventional NP synthesis, aligning with the growing need for eco-friendly and cost-effective technologies.

The synthesis of NPs using *Eucalyptus* extracts has been extensively studied, demonstrating the plant's efficacy in producing various types of NPs with significant biological and catalytic activities. This discussion section will delve into the key findings from several studies, highlighting the potential of *Eucalyptus* extracts in nanotechnology applications.

Rabeea et al. (2021) investigated the microwave-assisted synthesis of Ag-NPs using *Eucalyptus* leaves, which exhibited excellent catalytic activity in dye degradation. This suggests that *Eucalyptus*-based NPs could be employed in environmental remediation processes, such as the degradation of toxic dyes in wastewater.

The discussed studies collectively underscore the potential of *Eucalyptus* extracts in the green synthesis of various NPs. The ability to produce NPs with specific properties, such as antibacterial, antioxidant and catalytic activities, places *Eucalyptus* extracts as a valuable resource in nanotechnology. Further research could explore the optimization of synthesis parameters and the scalability of these processes for industrial applications. The versatility, cost-effectiveness and environmental friendliness of using *Eucalyptus* extracts make them an attractive option for sustainable NP synthesis.

5. CONCLUSIONS

Eucalyptus extracts have proven to be effective in the green synthesis of various NPs, including silver, zinc oxide and gold NPs. These NP exhibit multiple beneficial properties, such as antibacterial, antioxidant, catalytic and anticancer activities. The synthesis processes using *Eucalyptus* extracts are generally simple, cost-effective and scalable, making them suitable for potential applications in medical, pharmaceutical and agricultural fields. The success of these methods highlights the potential of *Eucalyptus* extracts as sustainable alternatives to conventional chemical synthesis methods, promoting environmental sustainability and advancing nanotechnology applications.

6. FUTURE PROSPECTS

Nanotechnology is an innovative and rapidly developing field in which nanomaterials are used in a variety of industries: In healthcare, nanomaterials are used for diagnostics and targeted drug delivery. In the computing and information sector, they are improving software, hardware and network systems for processing, collecting and distributing data, as well as AI, machine learning, robotics and sensors. In bioprocessing, nanodevices are used in biosensors, pathogen detection, food monitoring and smart packaging. In agriculture, nanotechnology is important in production, treatment, storage, packaging, product transport, conversion and pesticides. In the food industry, nanotechnology improves the productivity, expiration date, packaging and resistance to poultry, meat and product antibacterial drugs. The use of antimicrobial and anti-caking properties also improves packaging safety.

In construction, nanomaterials are integrated into cement and combined with polymer additives to improve stability and strength. In textiles, NPs create breathable, smooth fabrics that are stain-resistant, wrinkle-resistant and conductive. In the automotive industry, nanotechnology improves car parts, engines and coatings, making them more durable, thermally stable and lightweight. In healthcare, it plays a key role in diagnostics, treatment, disease prevention and the development of targeted drugs and diagnostic tools for cancer detection.

Nanomaterials are also used in cosmetics, especially for sun-protection products. In environmental and industrial applications, nanotechnology is increasingly being used in raw materials processing, waste treatment, energy storage and pollution control and in the oil and gas industry to improve oil recovery, refining and transportation. Renewable energy sectors, such as solar and hydrogen production, are benefiting from nanotechnological advancements. The wood industry uses nanomaterials for biodegradable products, preservatives and wood composites. Finally, in the chemical industry, nanotechnology is applied to surfactants used in cleaning, pharmaceuticals and food production. Applications of nanotechnology in various sectors are illustrated in Fig. 3.

Figure 3. Graphical representation of nanotechnology applications in various sectors.

REFERENCES

Abdelkhalek, A., Salem, M. Z., Kordy, A. M., Salem, A. Z., & Behiry, S. I. (2020). Antiviral, antifungal, and insecticidal activities of *Eucalyptus* bark extract: HPLC analysis of polyphenolic compounds. *Microbial Pathogenesis*, 147, 104383. DOI: 10.1016/j.micpath.2020.104383 PMID: 32659315

Abuzeid, H. M., Julien, C. M., Zhu, L., & Hashem, A. M. (2023). Green synthesis of nanoparticles and their energy storage, environmental, and biomedical applications. *Crystals*, 13(11), 1576. DOI: 10.3390/cryst13111576

Akhtar, M. S., Panwar, J., & Yun, Y. S. (2013). Biogenic synthesis of metallic nanoparticles by plant extracts. *ACS Sustainable Chemistry & Engineering*, 1(6), 591–602. DOI: 10.1021/sc300118u

Alghoraibi, I., Soukkarieh, C., Zein, R., Alahmad, A., Walter, J. G., & Daghestani, M. (2020). Aqueous extract of *Eucalyptus camaldulensis* leaves as reducing and capping agent in biosynthesis of silver nanoparticles. *Inorganic and Nano-Metal Chemistry*, 50(10), 895–902. DOI: 10.1080/24701556.2020.1728315

Alghoraibia, I., Soukkarieh, C., Zein, R., Alahmad, A., Walter, J. G., & Daghestani, M. (2019). Aqueous extract of *Eucalyptus camaldulensis* leaves as reducing and capping agent in green synthesis of silver nanoparticles. *Current Nanomaterials*, 4(1). Advance online publication. DOI: 10.2174/2405461504666190625121807

Ali, K., Ahmed, B., Dwivedi, S., Saquib, Q., Al-Khedhairy, A. A., & Musarrat, J. (2015). Microwave accelerated green synthesis of stable silver nanoparticles with *Eucalyptus globulus* leaf extract and their antibacterial and antibiofilm activity on clinical isolates. *PLoS One*, 10(7), e0131178. DOI: 10.1371/journal.pone.0131178 PMID: 26132199

Álvarez-Chimal, R., & Arenas-Alatorre, J. Á. (2023). Green synthesis of nanoparticles. A biological approach. In *Green Chemistry for Environmental Sustainability*. IntechOpen.

Bahrulolum, H., Nooraei, S., Javanshir, N., Tarrahimofrad, H., Mirbagheri, V. S., Easton, A. J., & Ahmadian, G. (2021). Green synthesis of metal nanoparticles using microorganisms and their application in the agrifood sector. *Journal of Nanobiotechnology*, 19(1), 1–26. DOI: 10.1186/s12951-021-00834-3 PMID: 33771172

Balamurugan, M., & Saravanan, S. (2017). Green synthesis of silver nanoparticles by using *Eucalyptus globulus* leaf extract. Journal of The Institution of Engineers (India): Series A, 98, 461-467.

Balamurugan, M., Saravanan, S., & Soga, T. (2014). Synthesis of iron oxide nanoparticles by using *Eucalyptus globulus* plant extract. *E-Journal of Surface Science and Nanotechnology*, 12(0), 363–367. DOI: 10.1380/ejssnt.2014.363

Balčiūnaitienė, A., Liaudanskas, M., Puzerytė, V., Viškelis, J., Janulis, V., Viškelis, P., & Jankauskaitė, V. (2022). *Eucalyptus globulus* and *Salvia officinalis* extracts mediated green synthesis of silver nanoparticles and their application as an antioxidant and antimicrobial agent. *Plants*, 11(8), 1085. DOI: 10.3390/plants11081085 PMID: 35448813

Barzinjy, A. A., & Azeez, H. H. (2020). Green synthesis and characterization of zinc oxide nanoparticles using *Eucalyptus globulus* Labill. leaf extract and zinc nitrate hexahydrate salt. *SN Applied Sciences*, 2(5), 991. DOI: 10.1007/s42452-020-2813-1

Bashir, T., & Qureshi, M. Z. (2015). Phytosynthesis of silver nanoparticles using *E. Camaldulensis* leaf extract and their characterization. *Journal of the Chilean Chemical Society*, 60(1), 2861–2863. DOI: 10.4067/S0717-97072015000100019

Bayda, S., Adeel, M., Tuccinardi, T., Cordani, M., & Rizzolio, F. (2019). The history of nanoscience and nanotechnology: From chemical–physical applications to nanomedicine. *Molecules (Basel, Switzerland)*, 25(1), 112. DOI: 10.3390/molecules25010112 PMID: 31892180

Damera, D. P., Manimaran, R., Krishna Venuganti, V. V., & Nag, A. (2020). Green synthesis of full-color fluorescent carbon nanoparticles from *Eucalyptus* twigs for sensing the synthetic food colorant and bioimaging. *ACS Omega*, 5(31), 19905–19918. DOI: 10.1021/acsomega.0c03148 PMID: 32803087

Das, R. K., Pachapur, V. L., Lonappan, L., Naghdi, M., Pulicharla, R., Maiti, S., Cledon, M., Dalila, L. M. A., Sarma, S. J., & Brar, S. K. (2017). Biological synthesis of metallic nanoparticles: Plants, animals and microbial aspects. *Nanotechnology for Environmental Engineering*, 2(1), 1–21. DOI: 10.1007/s41204-017-0029-4

Dhand, C., Dwivedi, N., Loh, X. J., Ying, A. N. J., Verma, N. K., Beuerman, R. W., & Ramakrishna, S. (2015). Methods and strategies for the synthesis of diverse nanoparticles and their applications: A comprehensive overview. *RSC Advances*, 5(127), 105003–105037. DOI: 10.1039/C5RA19388E

Dubey, M., Bhadauria, S., & Kushwah, B. S. (2009). Green synthesis of nanosilver particles from extract of *Eucalyptus hybrida* (safeda) leaf. *Digest Journal of Nanomaterials and Biostructures*, 4(3), 537–543.

Dudhane, A. A., Waghmode, S. R., Bhosale, M. A., & Mhaindarkar, V. P. (2017). *Caesalpinia pulcherrima* mediated green synthesis of silver nanoparticles: Evaluation of their antimicrobial and catalytic activity. *International Journal of Nanoparticles*, 9(3), 153–165. DOI: 10.1504/IJNP.2017.089195

Dzimitrowicz, A., Berent, S., Motyka, A., Jamroz, P., Kurcbach, K., Sledz, W., & Pohl, P. (2019). Comparison of the characteristics of gold nanoparticles synthesized using aqueous plant extracts and natural plant essential oils of *Eucalyptus globulus* and *Rosmarinus officinalis*.*Arabian Journal of Chemistry*, 12(8), 4795–4805. DOI: 10.1016/j.arabjc.2016.09.007

El-Naggar, N. E. A., Shiha, A. M., Mahrous, H., & Mohammed, A. A. (2022). Green synthesis of chitosan nanoparticles, optimization, characterization and antibacterial efficacy against multi drug resistant biofilm-forming *Acinetobacter baumannii*.*Scientific Reports*, 12(1), 19869. DOI: 10.1038/s41598-022-24303-5 PMID: 36400832

El-Rahman, A. F. A., & Mohammad, T. G. M. (2013). Green synthesis of silver nanoparticle using *Eucalyptus globulus* leaf extract and its antibacterial activity. *Journal of Applied Sciences Research*, 9(13), 6437–6440.

Faiz, N., Sivaswamy, V., Veeraraghavan, V. P., Kumar, R. (2023). Green Synthesis of Gold Nanoparticles Using *Eucalyptus* and *Piper Longum* and Its Subsequent Antiinflammatory Activity Evaluation.

Gardea-Torresdey, J. L., Tiemann, K. J., Gamez, G., Dokken, K., Tehuacanero, S., & Jose-Yacaman, M. (1999). Gold nanoparticles obtained by bio-precipitation from gold (III) solutions. *Journal of Nanoparticle Research*, 1(3), 397–404. DOI: 10.1023/A:1010008915465

Ghosh, P., Han, G., De, M., Kim, C. K., & Rotello, V. M. (2008). Gold nanoparticles in delivery applications. *Advanced Drug Delivery Reviews*, 60(11), 1307–1315. DOI: 10.1016/j.addr.2008.03.016 PMID: 18555555

Ghosh, S., Patil, S., Ahire, M., Kitture, R., Gurav, D. D., Jabgunde, A. M., Kale, S., Pardesi, K., Shinde, V., Bellare, J., Dhavale, D. D., & Chopade, B. A. (2012). *Gnidia glauca* flower extract mediated synthesis of gold nanoparticles and evaluation of its chemocatalytic potential. *Journal of Nanobiotechnology*, 10(1), 1–9. DOI: 10.1186/1477-3155-10-17 PMID: 22548753

Gohil, K., Dhakhda, S., Patel, V., Rathod, A., & Singh, P. K. (2024). Green Synthesis of Silver Nanoparticles using Various Plant Extracts and Evaluation of their Antimicrobial Activities. *Biosciences Biotechnology Research Asia*, 21(2), 765–777. DOI: 10.13005/bbra/3263

González-Burgos, E., Liaudanskas, M., Viškelis, J., Žvikas, V., Janulis, V., & Gómez-Serranillos, M. P. (2018). Antioxidant activity, neuroprotective properties and bioactive constituents analysis of varying polarity extracts from *Eucalyptus globulus* leaves. *Yao Wu Shi Pin Fen Xi*, 26(4), 1293–1302. PMID: 30249328

Gullón, P., Gullón, B., Astray, G., Munekata, P. E., Pateiro, M., & Lorenzo, J. M. (2020). Value-added compound recovery from invasive forest for biofunctional applications: Eucalyptus species as a case study. *Molecules (Basel, Switzerland)*, 25(18), 4227. DOI: 10.3390/molecules25184227 PMID: 32942656

Hassan, A. K., Hashim, L. Q., Rezooqi, A. M., & Hashim, M. F. (2024). Biosynthesis of Fe/Pd Bimetallic Nanoparticles and Used for Removal of Synthetic Oily Wastewater. *Journal of Petroleum Research and Studies*, 14(2), 121–137. DOI: 10.52716/jprs.v14i2.867

Ibraheem, I. J., Hassan, K. T., Ali, H. H., & Obaid, A. S. (2020). One-step biosynthesis of gold nanoparticles using natural reductive extracts. *Nano Biomedicine and Engineering*, 12(4), 331–337. DOI: 10.5101/nbe.v12i4.p331-337

Iravani, S. (2011). Green synthesis of metal nanoparticles using plants. *Green Chemistry*, 13(10), 2638–2650. DOI: 10.1039/c1gc15386b

Iravani, S., Korbekandi, H., Mirmohammadi, S. V., & Zolfaghari, B. (2014). Synthesis of silver nanoparticles: Chemical, physical and biological methods. *Research in Pharmaceutical Sciences*, 9(6), 385–406. PMID: 26339255

Jeevanandam, J., Chan, Y. S., & Ku, Y. H. (2018). Aqueous *Eucalyptus globulus* leaf extract-mediated biosynthesis of MgO nanorods. *Applied Biological Chemistry*, 61(2), 197–208. DOI: 10.1007/s13765-018-0347-7

Jha, A. K., Prasad, K., Prasad, K., & Kulkarni, A. R. (2009). Plant system: Nature's nanofactory. *Colloids and Surfaces. B, Biointerfaces*, 73(2), 219–223. DOI: 10.1016/j.colsurfb.2009.05.018 PMID: 19539452

Joudeh, N., & Linke, D. (2022). Nanoparticle classification, physicochemical properties, characterization, and applications: A comprehensive review for biologists. *Journal of Nanobiotechnology*, 20(1), 262. DOI: 10.1186/s12951-022-01477-8 PMID: 35672712

Katta, V. K. M., Dubey, R. S., & Saravanan, S. (2024). Green Synthesis of Iron Nanoparticles using *Eucalyptus* Leaf Extracts and Characterization for Photocatalytic Studies. *International Journal of Nanoscience and Nanotechnology*, 20(1), 11–23.

Kiran, M. S., & Betageri, V. S., CR, R. K., Vinay, S. P., & Latha, M. S. (2020). Synthesis of gold nanoparticles using novel *Eucalyptus tereticornis* and their in-vitro antibacterial, antioxidant and anticancer studies. *Advances in Natural Sciences: Nanoscience and Nanotechnology*, 11(4), 045012.

Kumar, J. A., Krithiga, T., Manigandan, S., Sathish, S., Renita, A. A., Prakash, P., Prasad, B. S. N., Kumar, T. R. P., Rajasimman, M., Hosseini-Bandegharaei, A., Prabu, D., & Crispin, S. (2021). A focus to green synthesis of metal/metal based oxide nanoparticles: Various mechanisms and applications towards ecological approach. *Journal of Cleaner Production*, 324, 129–198. DOI: 10.1016/j.jclepro.2021.129198

Majeed, H., Iftikhar, T., Nadeem, M. A., & Nazir, M. A. (2024). Green synthesis of *Eucalyptus globulus* zinc nanoparticles and its use in antimicrobial insect repellent paint formulation in bulk industrial production. *Heliyon*, 10(2), e24467. DOI: 10.1016/j.heliyon.2024.e24467 PMID: 38288019

Mo, Y., Tang, Y., Wang, S., Lin, J., Zhang, H., & Luo, D. (2015). Green synthesis of silver nanoparticles using *Eucalyptus* leaf extract. *Materials Letters*, 144, 165–167. DOI: 10.1016/j.matlet.2015.01.004

Mohammed, A. E. (2015). Green synthesis, antimicrobial and cytotoxic effects of silver nanoparticles mediated by *Eucalyptus camaldulensis* leaf extract. *Asian Pacific Journal of Tropical Biomedicine*, 5(5), 382–386. DOI: 10.1016/S2221-1691(15)30373-7

Moond, M., Singh, S., Sangwan, S., Devi, R., & Beniwal, R. (2022). Green synthesis and applications of silver nanoparticles: A systematic review. *AATCC Journal of Research*, 9(6), 272–285. DOI: 10.1177/24723444221119847

Muthiah, B., Muthukrishnan, L., Anita Lett, J., Sagadevan, S., Kesavan, S., Vennila, S., & Ahmad, N. (2020). *Eucalyptus concoction* mediated synthesis of gold nanoparticles and its bioactive role explored via antimicrobial and cytotoxic studies. *Journal of Nanoscience and Nanotechnology*, 20(10), 6326–6333. DOI: 10.1166/jnn.2020.17897 PMID: 32384982

Naganthran, A., Verasoundarapandian, G., Khalid, F. E., Masarudin, M. J., Zulkharnain, A., Nawawi, N. M., Karim, M., Che Abdullah, C. A., & Ahmad, S. A. (2022). Synthesis, characterization and biomedical application of silver nanoparticles. *Materials (Basel)*, 15(2), 427. DOI: 10.3390/ma15020427 PMID: 35057145

Nam, N. H., Luong, N. H. (2019). Nanoparticles: synthesis and applications. Materials for Biomedical Engineering. 211–40.

Nayab, D. E., & Akhtar, S. (2023). Green synthesized silver nanoparticles from *Eucalyptus* leaves can enhance shelf life of banana without penetrating in pulp. *PLoS One*, 18(3), e0281675. DOI: 10.1371/journal.pone.0281675 PMID: 36888584

Nikam, A., Pagar, T., Ghotekar, S., Pagar, K., & Pansambal, S. (2019). A review on plant extract mediated green synthesis of zirconia nanoparticles and their miscellaneous applications. *Journal of Chemical Reviews*, 1(3), 154–163. DOI: 10.33945/SAMI/JCR.2019.3.1

Pinto, R. J., Lucas, J. M., Morais, M. P., Santos, S. A., Silvestre, A. J., Marques, P. A., & Freire, C. S. (2017). Demystifying the morphology and size control on the biosynthesis of gold nanoparticles using *Eucalyptus globulus* bark extract. *Industrial Crops and Products*, 105, 83–92. DOI: 10.1016/j.indcrop.2017.05.003

Poinern, G. E. J., Chapman, P., Shah, M., & Fawcett, D. (2013). Green biosynthesis of silver nanocubes using the leaf extracts from *Eucalyptus macrocarpa*. Nano Bulletin, 2(1).

Pourmortazavi, S. M., Taghdiri, M., Makari, V., & Rahimi-Nasrabadi, M. (2015). Procedure optimization for green synthesis of silver nanoparticles by aqueous extract of *Eucalyptus oleosa.Spectrochimica Acta. Part A: Molecular and Biomolecular Spectroscopy*, 136, 1249–1254. DOI: 10.1016/j.saa.2014.10.010 PMID: 25456666

Pourmortazavi, S. M., Taghdiri, M., Makari, V., Rahimi-Nasrabadi, M., & Batooli, H. (2017). Reducing power of *Eucalyptus oleosa* leaf extracts and green synthesis of gold nanoparticles using the extract. *International Journal of Food Properties*, 20(5), 1097–1103. DOI: 10.1080/10942912.2016.1203334

Rabeea, M. A., Owaid, M. N., & Muslim, R. F. (2021). Synthesis and characterization of silver nanoparticles by natural organic compounds extracted from *Eucalyptus* leaves and their role in the catalytic degradation of methylene blue dye. *Songklanakarin Journal of Science and Technology*, 43(1).

Rafiee, B., Ghani, S., Sadeghi, D., & Ahsani, M. (2018). Green synthesis of zinc oxide nanoparticles using *eucalyptus mellidora* leaf extract and evaluation of its antimicrobial effects. *Majallah-i Danishgah-i Ulum-i Pizishki-i Babul*, 20(10), 28–35.

Rahimi, H. R., & Doostmohammadi, M. (2019). Nanoparticle synthesis, applications, and toxicity. Applications of Nanobiotechnology, 10.

Rahmatyian, M., & Dadmehr, M. (2020). Biological synthesis of silver nanoparticle based on *Eucalyptus obliqua* and *Mentha spicata* plant extracts and evaluation of their physical properties. *Crop Biotechnology*, 10(31), 61–70.

Ramezani, N., Ehsanfar, Z., Shamsa, F., Amin, G., Shahverdi, H. R., Esfahani, H. R. M., Shamsaie, A., Bazaz, R. D., & Shahverdi, A. R. (2008). Screening of medicinal plant methanol extracts for the synthesis of gold nanoparticles by their reducing potential. *Zeitschrift für Naturforschung. Section B. A Journal of Chemical Sciences*, 63(7), 903–908. DOI: 10.1515/znb-2008-0715

Ramrakhiani, L., & Ghosh, S. (2018). Metallic nanoparticle synthesised by biological route: Safer candidate for diverse applications. *IET Nanobiotechnology / IET*, 12(4), 392–404. DOI: 10.1049/iet-nbt.2017.0076 PMID: 29768220

Rocha, V., Ferreira-Santos, P., Aguiar, C., Neves, I. C., & Tavares, T. (2024). Valorization of plant by-products in the biosynthesis of silver nanoparticles with antimicrobial and catalytic properties. *Environmental Science and Pollution Research International*, 31(9), 14191–14207. DOI: 10.1007/s11356-024-32180-w PMID: 38278998

Sadiq, M. U., Shah, A., Haleem, A., Shah, S. M., & Shah, I. (2023). *Eucalyptus globulus* mediated green synthesis of environmentally benign metal based nanostructures: A Review. *Nanomaterials (Basel, Switzerland)*, 13(13), 2019. DOI: 10.3390/nano13132019 PMID: 37446535

Sankpal, P. S., Bandal, O. B., Bailgekar, A. V., Bilur, P., & Chavanpatil, S. (2020). Screening of in-vitro anthelmintic activity of silver nanoparticles of *Eucalyptus globulus* against *Pheretima posthuma.Global Journal for Research Analysis*, 9(2), 53–55.

Shah, M., Poinern, G. E. J., Sharma, S. B., & Fawcett, D. (2015). Biogenic synthesis of gold and silver nanoparticles using the leaf extract from *Eucalyptus macrocarpa.International Journal of Sciences*, 4(6), 27–33.

Shahid, M., Ijaz, N., Shahid, B., Tufail, T., Ain, H. B. U., Hussain, M., Basharat, S., Ikram, A., & Al Jbawi, E. (2023). *Eucalyptus globulus* Labill. Mediated synthesis of ZnO nanoparticles, their Optimization and characterization. *Cogent Food & Agriculture*, 10(1), 2293332. DOI: 10.1080/23311932.2023.2293332

Shala, A. Y., & Gururani, M. A. (2021). Phytochemical properties and diverse beneficial roles of *Eucalyptus globulus* Labill.: A Review. *Horticulturae*, 7(11), 450. DOI: 10.3390/horticulturae7110450

Siddique, S., & Chow, J. C. (2020). Application of nanomaterials in biomedical imaging and cancer therapy. *Nanomaterials (Basel, Switzerland)*, 10(9), 1700. DOI: 10.3390/nano10091700 PMID: 32872399

Sila, M. J., Nyambura, M. I., Abong'o, D. A., Mwaura, F. B., & Iwuoha, E. (2019). Biosynthesis of silver nanoparticles from *Eucalyptus corymbia* leaf extract at optimized conditions. *Nano Hybrids and Composites*, 25, 32–45. DOI: 10.4028/www.scientific.net/NHC.25.32

Siripireddy, B., & Mandal, B. K. (2017). Facile green synthesis of zinc oxide nanoparticles by *Eucalyptus globulus* and their photocatalytic and antioxidant activity. *Advanced Powder Technology*, 28(3), 785–797. DOI: 10.1016/j.apt.2016.11.026

Sunder, S. S. (1993, October). The ecological, economic and social effects of *Eucalyptus*. In Proceedings of the regional expert consultation on eucalyptus (Vol. 1). FAO Rome Italy.

Thakkar, K. N., Mhatre, S. S., & Parikh, R. Y. (2010). Biological synthesis of metallic nanoparticles. *Nanomedicine; Nanotechnology, Biology, and Medicine*, 6(2), 257–262. DOI: 10.1016/j.nano.2009.07.002 PMID: 19616126

Thakur, K., Bala, I., Rajeshwer, D. M., & Bhatt, A. K. (2017). Evaluation of effectiveness of biologically synthesized silver nanoparticles of *Eucalyptus globules* leaf extract against pathogenic and acne-inducing bacteria. *Journal of Nanomedicine & Nanotechnology*, 8(3), 443. DOI: 10.4172/2157-7439.1000443

Thesing, A., Nascimento, J. E. R. D., Jacob, R. G., & Ferreira, J. (2018, February). *Eucalyptus* oil-mediated synthesis of gold nanoparticles. *Journal of Chemistry and Chemical Engineering. New York.*, 12(2), 52–59.

Tsilo, P. H., Basson, A. K., Ntombela, Z. G., Dlamini, N. G., & Pullabhotla, R. V. S. R. (2024). Green synthesis and characterisation of silver nanoparticles utilising a bioflocculant obtained from *Pichia kudriavzevii* isolated from kombucha tea SCOBY. Advances in Materials and Processing Technologies, 1-17.

Vecchio, M. G., Loganes, C., & Minto, C. (2016). Beneficial and healthy properties of *Eucalyptus* plants: A great potential use. *The Open Agriculture Journal*, 10(1), 52–57. DOI: 10.2174/1874331501610010052

Verma, A. K., & Kumar, P. (2022). On recent developments in biosynthesis and application of Au and Ag nanoparticles from biological systems. *Journal of Nanotechnology*, 2022(1), 5560244. DOI: 10.1155/2022/5560244

Vitta, Y., Figueroa, M., Calderon, M., & Ciangherotti, C. (2020). Synthesis of iron nanoparticles from aqueous extract of *Eucalyptus robusta* Sm and evaluation of antioxidant and antimicrobial activity. *Materials Science for Energy Technologies*, 3, 97–103. DOI: 10.1016/j.mset.2019.10.014

Waghmode, S. R., Dudhane, A. A., & Mhaindarkar, V. P. (2021). *Syzygium cumini* mediated green synthesis of silver nanoparticles for reduction of 4-nitrophenol and assessment of its antibacterial activity. In *Nanocomposite Materials for Biomedical and Energy Storage Applications*. IntechOpen.

Wang, T., Jin, X., Chen, Z., Megharaj, M., & Naidu, R. (2014). Green synthesis of Fe nanoparticles using *Eucalyptus* leaf extracts for treatment of eutrophic wastewater. *The Science of the Total Environment*, 466, 210–213. DOI: 10.1016/j.scitotenv.2013.07.022 PMID: 23895784

Weng, X., Guo, M., Luo, F., & Chen, Z. (2017). One-step green synthesis of bimetallic Fe/Ni nanoparticles by *Eucalyptus* leaf extract: Biomolecules identification, characterization and catalytic activity. *Chemical Engineering Journal*, 308, 904–911. DOI: 10.1016/j.cej.2016.09.134

Zhang, X. F., Liu, Z. G., Shen, W., & Gurunathan, S. (2016). Silver nanoparticles: Synthesis, characterization, properties, applications, and therapeutic approaches. *International Journal of Molecular Sciences*, 17(9), 1534. DOI: 10.3390/ijms17091534 PMID: 27649147

Zhao, R., Xiang, J., Wang, B., Chen, L., & Tan, S. (2022). Recent advances in the development of noble metal NPs for cancer therapy. *Bioinorganic Chemistry and Applications*, 2022(1), 2444516. DOI: 10.1155/2022/2444516 PMID: 35126483

KEY TERMS AND DEFINITIONS

Eucalyptus globulus: *Eucalyptus globulus* is an aromatic tree in the Myrtle Family (Myrtaceae) that typically grows to a height of 150-180 feet with a diameter of 4-7 feet.

Au-NPs: Gold nanoparticles are nanoscale clusters of neutral gold atoms formed through the reduction of gold salts in suitable agents.

Ag-NPs: Silver nanoparticles are nanoparticles of silver that range in size from 1 nm to 100 nm and are produced by reducing silver salts in suitable agents.

TEM: Transmission electron microscopy is a microscopy technique in which an electron beam is sent through a specimen to produce an image.

Antibacterial activity: Anything that kills bacteria, inhibits their growth, or prevents their reproduction.

Chapter 3

Green Synthesis Methods for Metallic Zinc Oxide Nanoparticles: A Sustainable Angle

Aishwarya Jain
Independent Researcher, India

Kiran Bhise
MCE's Allana College of Pharmacy, Pune, India

ABSTRACT

This abstract provides a comprehensive guide on plant-mediated biocompatible metallic zinc oxide nanoparticles (ZnO NPs), emphasizing green synthesis methods' sustainability in nanoparticle production. It explores their characterization, applications in healthcare, agriculture, and environmental cleanup, and discusses challenges and opportunities within the scientific landscape. Green synthesis harnesses natural resources like plants and microorganisms to produce stable nanoparticles, offering non-toxic, cost-effective alternatives aligned with sustainability goals. The chapter highlights the role of plant-derived compounds, particularly polyphenols, in nanoparticle synthesis, and evaluates advanced characterization techniques and potential integration with artificial intelligence in nanobiotechnology applications. Ethical considerations for stakeholders navigating this evolving field are also addressed.

INTRODUCTION

What is Green Synthesis?

Green synthesis refers to a methodology that focuses on the environmentally friendly production of materials, particularly nanoparticles, by utilizing naturally occurring materials or organisms while minimizing hazardous by-products. This process employs eco-friendly techniques, relying on natural resources such as plants and microorganisms instead of toxic solvents or harmful chemicals, thus contrasting sharply with traditional synthesis methods that often harm the environment. Among the benefits

DOI: 10.4018/979-8-3693-6240-2.ch003

of green synthesis are its cost-effectiveness, simplicity, lower energy requirements, and reduced toxic waste generation, while also enabling the production of nanoparticles with tailored properties for various applications including environmental remediation and medical uses. The approach can be categorized into two primary methods: the bottom-up approach, which involves assembling nanoparticles from simpler molecules, and the top-down approach, which breaks down larger materials into nanoparticles. As research advances, the versatility of green synthesis is being further unveiled, positioning it as a pivotal development in ensuring a more sustainable future across various sectors such as nanotechnology, environmental science, and medicine (Silva et al., 2022).

How Does Green Synthesis Compare to Traditional Synthesis in Terms of Efficiency?

Green synthesis offers a more efficient and environmentally friendly approach to the production of nanoparticles compared to traditional synthesis methods. It has been shown to reduce production costs by up to 40% and energy consumption by approximately 30%, while also increasing production output by around 50%(Huston et al., 2021).

This sustainable method leverages natural resources, such as plant extracts and microorganisms, which are less toxic and more readily available than the hazardous chemicals often used in conventional synthesis. As a result, green synthesis not only decreases pollution but also significantly improves environmental and human health. Moreover, green methods are scalable and cost-effective, allowing for large-scale production of metal and metal oxide nanoparticles, which are often produced under mild, ambient conditions. Overall, the advantages of green synthesis make it a superior alternative to traditional techniques, aligning with modern goals of sustainability and reduced environmental impact (Banjara et al., 2024).

Green synthesis refers to material fabrication, including that of NP, through an environment-friendly process using plants, microorganisms, and biowastes for reduction and stabilization agents instead of the harmful chemicals used in conventional modes of synthesis. This method of synthesis belongs to the category of green chemistry and, as such, strives to minimize environmental impact, promoting sustainability (Singh et al., 2018). Some of the key principles of green synthesis are that renewable feedstock, for example, plant extracts and agricultural residues, should always ensure a sustainable supply of raw materials. Additionally, toxic chemicals should be avoided in green synthesis methods; this decreases environmental pollution together with health hazards to a great level. Most green synthesis methods are energy-efficient, since the process usually occurs under conditions of ambient temperature and pressure, saving energy compared to the conventional methods that require high-temperature and high-pressure conditions. The by-products resulting from green synthesis also are usually nontoxic and biodegradable, further reducing environmental impact. Besides, the low cost of producing nanoparticles, which is enabled by the use of readily available natural materials and simpler processes, presents an economical alternative as well (Sharma et al., 2016).

So far, various kinds of nanoparticles, like Ag, Au, Cu, TiO_2, and Fe_3O_4, have already been synthesized using this green synthesis. For instance, AgNP synthesized using *Azadirachta indica* neem leaf extract showed high antibacterial activity, while AuNP synthesized using *Pelargonium graveolens* extract depicted antioxidant and anticancer activities (Ahmed et al., 2016; Iravani et al., 2014). However, among these, the nanoparticles of zinc oxide hold particular importance due to their wide applications in medicine, electronics, cosmetics, and environmental remediation according to the literature review done by (Khan & Al-Thabaiti, 2019)

Much attention is focused on green synthesis because of the unique properties and wide range of applications of ZnO nanoparticles. ZnO NP synthesized by green methods are more biocompatible with a minimum level of cytotoxicity and hence are highly suitable for biomedical applications, including drug delivery and wound-healing areas. Singh et al., 2018. In addition, this green synthesis of ZnO NP goes hand in glove with the efforts of the whole world in making processes in nanotechnology greener and therefore contributes to the protection of an ecologically clean and safe environment that enables the development of technology. This idea is further discussed by Mittal et al. (2013). The emphasis on ZnO NP tends to be related not only to their huge applications but also to the promotion of a sustainable approach in nanomaterial production, which is cost-effective and less hazardous.

There is a growing interest in green synthesis methods driven by several factors:

1. **Environmental Concerns:** The increasing awareness of the detrimental effects of chemical pollutants and the need to reduce the carbon footprint has led researchers to explore eco-friendly alternatives. Traditional methods often involve harmful solvents and by-products that can contribute to environmental degradation and pose health risks.
2. **Regulatory Pressures:** Governments and regulatory bodies worldwide are implementing stricter environmental regulations, pushing industries to adopt greener manufacturing practices.

India:

Environment Protection Act, 1986: Sets standards for emissions and handling of hazardous substances, promoting safer, less toxic manufacturing methods.

National Guidelines for Nanotechnology Research: Encourages sustainable and safe nanotechnology practices, including green synthesis.

Global:

EU REACH Regulation: Requires companies to manage risks associated with chemicals and promotes safer alternatives, including green synthesis of nanoparticles.

U.S. **EPA (TSCA):** Mandates reporting on the environmental and health impacts of nanomaterials, encouraging the use of eco-friendly production methods.

China's Safety Regulations on Hazardous Chemicals: Imposes strict controls on chemical use and emissions, promoting safer synthesis techniques.

ISO Standards: Provides guidelines for safe and sustainable nanomaterial production.

3. **Sustainability Goals:** Industries are striving to meet sustainability targets and corporate social responsibility (CSR) commitments. Green synthesis aligns with these goals by reducing the environmental impact of production processes.
4. **Advances in Biotechnology:** The progress in biotechnology and nanotechnology has enabled the development of efficient green synthesis methods. Researchers are leveraging biological systems, such as plant extracts and microorganisms, to synthesize nanoparticles in a controlled and reproducible manner.

Biotechnological interventions have significantly advanced the green synthesis of zinc oxide (ZnO) nanoparticles by using biological systems to create environmentally friendly and sustainable methods. Key Interventions:

1. **Plant Extracts:** Rich in bioactive compounds (e.g., polyphenols, flavonoids), plant extracts like Aloe vera and neem act as natural reducing and stabilizing agents in ZnO nanoparticle synthesis.
2. **Microbial Synthesis:** Bacteria, fungi, and algae (e.g., *Pseudomonas aeruginosa*, *Fusarium oxysporum*) use enzymes and biomolecules to reduce zinc salts into ZnO nanoparticles, allowing controlled size and shape formation.
3. **Enzyme-Mediated Synthesis:** Enzymes like nitrate reductase offer specific and efficient catalytic processes for ZnO NP synthesis under mild conditions.
4. **Phytochemicals:** Compounds like saponins and tannins from plants serve as reducing and capping agents, stabilizing nanoparticles and enhancing their properties.
5. **Genetic Engineering:** Modified microbes are engineered to enhance their nanoparticle synthesis capabilities, improving yield and consistency.
6. **Bioreactors:** Provide controlled environments for scalable and reproducible synthesis of ZnO nanoparticles using microorganisms or plant cells.

Impact:

These methods reduce the use of toxic chemicals, lower energy consumption, and enable sustainable and scalable production of ZnO nanoparticles, which are more biocompatible and functional for applications in medicine, electronics, cosmetics, and environmental remediation.

5. **Economic Benefits:** Green synthesis can reduce production costs by utilizing low-cost and abundant natural resources. It also minimizes the need for expensive waste treatment processes associated with traditional synthesis methods.
6. **Public Perception and Market Demand:** There is a growing demand from consumers for eco-friendly and sustainable products. Companies adopting green synthesis methods can enhance their brand image and meet the expectations of environmentally conscious customers.

Exploration of Green Synthesis Methods for ZnO NP

Green synthesis methods for producing zinc oxide NP (ZnO NP) utilize natural resources such as plant extracts, microorganisms, biowastes, and other biological agents as reducing and stabilizing agents, avoiding hazardous chemicals and offering sustainable alternatives to traditional synthesis techniques (Kuriakose & Philip, 2014). Plant extracts, rich in bioactive compounds like flavonoids and tannins, are effective in reducing zinc salts to ZnO NP and preventing agglomeration. For instance, leaf extracts of *Azadirachta indica* (neem) mixed with zinc acetate can synthesize ZnO NP under mild conditions, making the process simple, cost-effective, and environmentally friendly (Sangeetha et al., 2011).

Microbial synthesis involves using bacteria, fungi, or algae to produce ZnO NP through metabolic processes. For example, *Bacillus subtilis* can synthesize ZnO NP by enzymatically reducing zinc ions. This method produces well-defined nanoparticles and utilizes renewable biological resources (He et al., 2011).

Biowastes, such as fruit peels and vegetable waste, contain organic compounds that can reduce and stabilize ZnO NP. For example, banana peel extract can synthesize ZnO NP at room temperature, promoting waste valorization and supporting circular economy principles (Singh et al., 2018). Other biological agents, including proteins and polysaccharides, can also facilitate green synthesis by acting as capping and reducing agents (Madhumitha & Elango, 2018).

Importance of Environmentally Conscious Approaches in NP Synthesis

Traditional synthesis methods for NP often involve significant chemical pollution, energy consumption, waste generation, and health risks. Common reagents used in these processes, such as sodium borohydride, hydrazine, and various organic solvents, are toxic and can lead to soil and water contamination if improperly handled or disposed of, adversely affecting ecosystems and human health (Khan et al., 2019). Additionally, the use of metal salts as precursors can result in heavy metal contamination, with metals bioaccumulating in the food chain and posing risks to wildlife and humans, including neurological and developmental disorders (Park et al., 2014).

The energy consumption associated with traditional nanoparticle synthesis is also significant. Many conventional methods require high temperatures and pressures, demanding substantial energy inputs and contributing to the depletion of non-renewable energy resources and increased greenhouse gas emissions, which exacerbate climate change (Jeevanandam et al., 2018). Furthermore, the need for specialized, energy-intensive equipment escalates the environmental footprint of these processes (Mittal et al., 2020).

Waste generation is another major concern. Synthesis processes produce chemical waste, including unreacted precursors, solvents, and by-products, which are challenging and costly to manage, with improper disposal leading to environmental contamination (Sharma et al., 2015). Inefficient synthesis methods also produce a significant amount of unusable nanoparticles, contributing to waste and resource inefficiency (Baker et al., 2017).

Health risks associated with traditional synthesis methods include occupational hazards for workers exposed to hazardous chemicals, leading to potential health issues such as respiratory problems, skin irritation, and long-term effects like cancer (Patil et al., 2018). Additionally, the release of toxic substances into the environment can cause public health issues through contaminated water, air, and soil, affecting communities and ecosystems (Gupta & Xie, 2018).

ADVANTAGES OF GREEN SYNTHESIS

Environmental Friendliness

Green synthesis of ZnO nanoparticles (ZnO NP) stands out for its environmental friendliness, employing non-toxic precursors like plant extracts, biowaste, and microorganisms. These materials are biodegradable and far less harmful to the environment than traditional chemical reagents, reducing chemical pollutants and adhering to sustainability principles. By utilizing renewable biological resources, green synthesis conserves non-renewable resources and promotes a circular economy. Moreover, this method minimizes harm to aquatic and terrestrial ecosystems, as the absence of harmful chemicals ensures effluents are less likely to damage biodiversity and ecosystem health. Biodegradable by-products further mitigate long-term environmental impact and simplify disposal.

Cost-Effectiveness

Green synthesis offers significant cost advantages primarily due to its reliance on abundant natural resources like plant extracts, which are typically more affordable than synthetic chemicals. Utilizing agricultural and biowastes as raw materials further enhances cost-effectiveness through waste valourization. Energy savings are realized as green synthesis reactions often proceed at ambient conditions, contrasting sharply with energy-intensive traditional methods. Simplified processes requiring less equipment and fewer steps contribute to reduced operational expenses. Moreover, the use of non-toxic, biodegradable waste simplifies waste management and lowers disposal costs, enhancing overall economic viability. Products synthesized via green methods can also leverage their eco-friendly credentials to attract environmentally conscious consumers, potentially commanding premium market prices.

Reduced Use of Hazardous Chemicals

Green synthesis employs safer alternatives to hazardous chemicals, using natural reducing agents found in plants, such as polyphenols, flavonoids, and terpenoids, which are safer than traditional chemical reducers like hydrazine or sodium borohydride. Biological molecules, such as proteins, polysaccharides, and enzymes, act as non-toxic stabilizers. This method improves occupational safety by reducing workers' exposure to toxic chemicals and lowers public health risks by minimizing the release of toxic substances into the environment. Using safer chemicals also facilitates compliance with environmental and health regulations, reducing the risk of legal issues and associated costs.

Potential for Scalability

The scalability of green synthesis for ZnO NP is facilitated by the abundant and sustainable supply of natural resources such as plant materials and agricultural waste. These resources can be sourced in large quantities without complex infrastructure needs, thanks to the straightforward and efficient methods involved. Green synthesis methods, which involve extracting active compounds from plants and utilizing them in nanoparticle production, offer versatility in raw material selection and synthesis conditions. This adaptability allows for the customization of NP tailored for various applications in medicine, electronics, and environmental sectors. Integration of green synthesis with existing manufacturing processes further streamlines scalability, minimizing capital investment in new infrastructure. Ongoing advancements in green chemistry and nanotechnology continue to improve efficiency, supporting the wider adoption of these sustainable synthesis methods.

Controlled Size and Shape Attributes Achievable Through Green Synthesis Methods

Green synthesis methods can effectively control the size and shape of nanoparticles (NP) by manipulating various synthesis parameters. These techniques are crucial for tailoring the properties of NP to specific applications.

1. Selection of Biological Reducing Agents:

Plant Extracts: Different plant extracts contain varying types and concentrations of bioactive compounds, such as flavonoids, phenolics, and alkaloids, which can influence the reduction rate and stabilization of NP. For instance, extracts from leaves, fruits, or flowers can lead to different nanoparticle sizes and shapes due to their unique biochemical compositions. Different plant extracts contain various bioactive compounds that play a crucial role in the green synthesis of zinc oxide (ZnO) nanoparticles by acting as reducing, capping, and stabilizing agents. These compounds can influence the size, shape, and overall properties of the synthesized nanoparticles.

Examples of Plant Compounds Necessary for Zinc Oxide NP Synthesis

1. **Flavonoids:** Quercetin and Kaempferol are common flavonoids found in many plant extracts, such as those from *Camellia sinensis* (green tea) and *Azadirachta indica* (neem). These compounds act as reducing agents that facilitate the conversion of zinc ions into ZnO NP while providing stability to prevent agglomeration.
2. **Phenolic Compounds:** Gallic Acid and Caffeic Acid are phenolic compounds present in plant extracts like *Punica granatum* (pomegranate) and *Moringa oleifera*. These phenolics have strong antioxidant properties that help reduce zinc salts to ZnO nanoparticles and stabilize them by capping their surface, thus influencing their size and shape.
3. **Alkaloids:** Berberine and Sanguinarine, found in plants such as Berberis species and Argemone mexicana, respectively, are alkaloids that act as both reducing agents and stabilizers. Their presence helps in the formation of ZnO NP by reducing zinc ions and stabilizing the nanoparticles, preventing aggregation.
4. **Terpenoids:** Limonene and Menthol are terpenoids found in citrus fruits (*Citrus limon*) and mint (*Mentha spicata*), respectively. These compounds assist in reducing zinc ions and help stabilize ZnO NP, often resulting in smaller, more uniform particles due to their strong capping capabilities.
5. **Tannins:** Ellagic Acid and Catechins are tannins commonly found in plant extracts such as *Terminalia chebula* (haritaki) and *Camellia sinensis* (green tea). These tannins serve as powerful antioxidants, reducing zinc ions to ZnO NP while preventing their aggregation through surface capping.
6. **Polysaccharides:** Pectin and Cellulose, found in fruits like apples (*Malus domestica*) and vegetables like okra (*Abelmoschus esculentus*), are polysaccharides that provide a protective layer around the nanoparticles, enhancing stability and preventing clumping during the synthesis process.
7. **Proteins and Amino Acids:** Amino Acids like cysteine and lysine, found in various plant proteins (e.g., from *Vigna radiata* (mung bean) or *Pisum sativum* (pea)), help reduce zinc ions and act as stabilizing agents by binding to the nanoparticle surface and controlling particle size and morphology.
8. **Saponins:** Saponins present in plants like *Sapindus mukorossi* (soapnut) and Glycyrrhiza glabra (liquorice) are effective in reducing metal ions and stabilizing NP. They provide a dual role by both reducing zinc ions to ZnO and creating a capping layer around the NP, ensuring stability.

The various bioactive compounds in plant extracts, such as flavonoids, phenolics, alkaloids, terpenoids, tannins, polysaccharides, proteins, and saponins, are crucial for the green synthesis of ZnO NP. These compounds not only reduce zinc ions to NP but also stabilize them, influencing their size, shape, and properties, making them suitable for a wide range of applications.

Microorganisms: The use of different bacteria, fungi, or algae can produce NP of distinct sizes and shapes. The metabolic pathways and enzymatic activities of these organisms affect the nucleation and growth processes of NP.

2. Concentration of Precursors and Biological Agents:

    Precursor Concentration: Adjusting the concentration of metal salts used as precursors can control the nucleation rate and particle growth. Higher precursor concentrations typically lead to larger nanoparticles due to the increased availability of metal ions for growth.

    Biological Agent Concentration: The concentration of plant extracts or microbial biomass also influences NP size and shape. Higher concentrations provide more bioactive molecules, which can result in more extensive capping and stabilization, thereby controlling particle growth and preventing aggregation.

3. pH of the Reaction Mixture:

    pH Adjustment: For the green synthesis of zinc oxide (ZnO) NP, an alkaline pH in the range of 8 to 12 is typically optimal. This pH range enhances the reduction of zinc ions and promotes the formation of smaller, more uniform ZnO NP by improving the ionization of bioactive compounds and stabilizing the particles effectively.

4. Temperature of the Reaction:

    Temperature Control: For the green synthesis of zinc oxide (ZnO) NP, maintaining a reaction temperature between 60 to 90°C is generally ideal. This temperature range provides sufficient energy for efficient reduction of zinc ions and controlled nucleation, promoting the formation of smaller and more uniform ZnO NP.

5. Reaction Time:

    Duration of Synthesis: The length of the reaction time determines the extent of NP growth. Shorter reaction times favour the formation of smaller NP, while longer times allow for continued growth and potentially larger particles.

6. Use of Capping Agents:

    Natural Capping Agents: Biological molecules, such as proteins, polysaccharides, and polyphenols, act as capping agents that stabilize nanoparticles and prevent aggregation. These molecules can bind to specific crystal facets, influencing the shape of the NP. For instance, proteins can create specific binding interactions that lead to anisotropic growth, resulting in shaped nanoparticles like rods or triangles.

Importance of Size and Shape in Nanoparticle Applications

The size and shape of nanoparticles are critical parameters that significantly influence their properties and performance in various applications. Here's why these attributes are important:

1. Medical Applications:

    Drug Delivery: Shapes of Zinc Oxide (ZnO) NP and Their Applications:

1. **Spherical ZnO NP:** Widely used in sunscreens, cosmetics, and coatings due to their excellent UV-blocking properties and high surface area.
2. **Rod-shaped (Nanorods) ZnO NP:** Utilized in drug delivery systems for enhanced cellular uptake, in biosensors for improved sensitivity, and in photocatalysis for environmental remediation due to their high aspect ratio.
3. **Needle-Like (Nanoneedles) ZnO NP:** Used in antibacterial agents and wound healing due to their sharp surface, which can disrupt bacterial cell membranes.
4. **Flower-Like (Nanoflowers) ZnO NP:** Employed in gas sensors and dye-sensitized solar cells (DSSCs) because of their large surface area and high photocatalytic activity.
5. **Sheet-Like (Nanosheets) ZnO NP:** Used in flexible electronics and transparent conductive films due to their excellent conductivity and flexibility.

Each shape of ZnO NP is chosen based on specific properties that enhance their effectiveness in various applications, from healthcare to environmental and electronic technologies.

Diagnostics and Imaging: The optical and magnetic properties of nanoparticles are size- and shape-dependent. Zinc oxide NP, for example, exhibit unique plasmonic properties that are useful in imaging and diagnostic applications. Specific shapes, like nanorods, have distinct optical properties that enhance contrast in imaging techniques.

2. Catalysis:
   Surface Area and Active Sites: The catalytic activity of NP is highly dependent on their surface area and the number of active sites. Smaller nanoparticles with high surface-to-volume ratios provide more active sites for reactions. Additionally, certain shapes expose specific crystal facets that can enhance catalytic performance. For example, Zinc oxide NP with a cubic shape can exhibit higher catalytic efficiency due to the exposure of more active {100} facets.
3. Environmental Applications:
   Pollutant Removal: The efficiency of NP in adsorbing or degrading pollutants is influenced by their size and shape. Smaller nanoparticles provide a larger surface area for interaction with pollutants. Shapes like nanoflakes or nanowires can offer enhanced surface reactivity and better pollutant capture.
4. Electronic and Optical Applications:
   Quantum Effects: The electronic properties of NP, such as quantum confinement, depend on their size. Smaller nanoparticles can exhibit unique electronic properties, making them suitable for applications in quantum dots and light-emitting devices.
   Surface Plasmon Resonance (SPR): The optical properties of metal nanoparticles, like Zinc oxide NP, are shape-dependent due to SPR. Nanorods, for instance, have two SPR peaks corresponding to their longitudinal and transverse axes, which can be tuned for specific optical applications like sensors and photothermal therapy.
5. Mechanical and Structural Applications:
   Strength and Durability: The mechanical properties of nanocomposites are influenced by the size and shape of the incorporated NP. For example, nanofibers or nanotubes can significantly enhance the strength and durability of composite materials due to their high aspect ratios and superior mechanical properties.
   Unique physicochemical properties of ZnO NP:

Zinc oxide (ZnO) NP are renowned for their unique optical and electrical properties, making them highly valuable for a wide range of technological applications.

Optical Properties

1. Wide Bandgap:

ZnO NP possess a wide bandgap of approximately 3.37 eV at room temperature. This characteristic is advantageous for optoelectronic applications such as light-emitting diodes (LEDs) and laser diodes. The wide bandgap enables efficient emission and absorption of ultraviolet (UV) light, making ZnO nanoparticles ideal for UV-related applications such as sunscreen formulations and UV-blocking coatings.

2. UV Absorption:

Due to their wide bandgap and high exciton binding energy, ZnO NP exhibit strong absorption of UV light. This property is crucial in applications where efficient UV absorption is needed, such as in protective coatings and biomedical applications.

2. Photoluminescence:

ZnO NP display photoluminescence properties, emitting light in both the UV and visible regions. This emission is valuable in bioimaging, sensor technologies, and optoelectronic devices. The visible emission typically ranges from blue to green, attributed to defect states within the ZnO crystal lattice, such as oxygen vacancies and zinc interstitials.

Exciton Emission:

The high exciton binding energy (~60 meV) of ZnO NP facilitates efficient exciton emission at room temperature. This characteristic is beneficial in UV lasers and light-emitting devices, where strong interactions between excitons and photons are desirable.

3. Transparent Conductors:

ZnO NP are transparent to visible light, which makes them suitable for use in transparent conductive films. These films find applications in touch screens, flat-panel displays, and solar cells, where high transparency and conductivity are essential.

Electrical Properties

1. High Electron Mobility:

ZnO NP exhibit high electron mobility, enhancing their electrical conductivity. This property is crucial for electronic applications such as field-effect transistors (FETs) and thin-film transistors (TFTs). Doping ZnO with elements like aluminium, gallium, or indium can further enhance their conductivity by increasing the carrier concentration.

2. Piezoelectricity:

ZnO NP possess inherent piezoelectric properties, meaning they can generate an electric charge in response to mechanical stress. This property is utilized in sensors, actuators, and energy-harvesting devices. At the nanoscale, ZnO NP exhibit enhanced piezoelectric effects, making them highly sensitive and efficient in converting mechanical energy into electrical energy.

3. Semiconductor Behaviour:

ZnO NP naturally behave as n-type semiconductors due to intrinsic defects such as oxygen vacancies and zinc interstitials. This semiconductor behaviour enables their use in various electronic devices including diodes, transistors, and sensors. Moreover, ZnO NP exhibit photoconductivity, where their electrical conductivity increases upon exposure to light. This property is exploited in photodetectors and photovoltaic devices for converting light energy into electrical signals.

Chemical Properties

1. Chemical Stability:

Resistance to Degradation: ZnO NP are chemically stable under a wide range of environmental conditions, including exposure to air, moisture, and various chemicals. This stability makes them suitable for long-term applications in coatings, sensors, and catalysts.

Antimicrobial Activity: ZnO NP possess intrinsic antimicrobial properties, attributed to the generation of reactive oxygen species (ROS) and the release of zinc ions. These properties are exploited in medical applications, such as antimicrobial coatings, wound dressings, and in the treatment of bacterial infections.

2. Catalytic Activity:

Photocatalysis: ZnO NP are effective photocatalysts due to their ability to generate electron-hole pairs upon exposure to UV light. These electron-hole pairs can initiate various chemical reactions, such as the degradation of organic pollutants and the production of hydrogen through water splitting. ZnO's photocatalytic activity is utilized in environmental remediation and clean energy applications.

Gas Sensing: The surface chemistry of ZnO NP allows them to interact with gas molecules, making them sensitive gas sensors. They can detect gases such as hydrogen, methane, and carbon monoxide at low concentrations, which is useful in environmental monitoring and industrial safety.

3. Surface Functionalization:

Versatile Surface Chemistry: The surface of ZnO NP can be easily modified with various functional groups, enhancing their compatibility with different materials and biological systems. Surface functionalization is used to improve the dispersion of ZnO NP in composites, enhance their biocompatibility for medical applications, and tailor their properties for specific uses in catalysis and sensing.

ZnO NP exhibit a unique set of physicochemical properties that make them highly valuable for a wide range of applications. Their optical properties, including a wide bandgap and strong UV absorption, make them ideal for optoelectronic devices and UV-blocking materials. High electron mobility and piezoelectricity contribute to their effectiveness in electronic and energy-harvesting devices. Chemical stability, antimicrobial activity, and catalytic efficiency further expand their utility in medical, environmental, and industrial applications.

Diverse applications of ZnO NP:

Biomedical Applications

1. Drug Delivery:

Targeted Drug Delivery: ZnO NP can be functionalized with various ligands, such as antibodies or peptides, to target specific cells or tissues. This targeted approach ensures that the drug is delivered directly to the site of action, enhancing therapeutic efficacy and reducing side effects.

Controlled Release: ZnO NP can be engineered to provide controlled and sustained release of drugs. This is achieved through mechanisms such as pH-sensitive degradation, where the nanoparticles release their payload in response to the acidic environment of a tumour or infected tissue.

Anticancer Activity: ZnO NP themselves have shown potential anticancer activity. They can induce apoptosis (programmed cell death) in cancer cells through the generation of reactive oxygen species (ROS) and the disruption of cellular processes.

2. Bioimaging:

Fluorescence Imaging: ZnO NP exhibit strong photoluminescence, making them useful as fluorescent probes in bioimaging. Their ability to emit light in the UV and visible ranges allows for high-resolution imaging of biological structures.

Multimodal Imaging: ZnO NP can be combined with other imaging modalities, such as magnetic resonance imaging (MRI) or computed tomography (CT), to provide comprehensive diagnostic information. For example, doping ZnO nanoparticles with magnetic materials can create contrast agents for MRI.

Biosensing: ZnO NP can be used in biosensors to detect biomolecules such as glucose, cholesterol, and DNA. Their high surface area and biocompatibility allow for the sensitive and selective detection of these molecules.

3. Antimicrobial Applications:

Wound Dressings: ZnO NP have intrinsic antimicrobial properties that are effective against a wide range of pathogens, including bacteria, fungi, and viruses. They can be incorporated into wound dressings to prevent infections and promote healing.

Coatings and Textiles: ZnO NP can be used to create antimicrobial coatings for medical devices, surfaces, and textiles, reducing the risk of healthcare-associated infections.

Agricultural Applications

1. Crop Protection:

Pesticidal Activity:

ZnO NP have been shown to control Aphids (specifically *Aphis gossypii*), a common pest affecting cotton crops. The NP act as an effective alternative to traditional chemical pesticides by disrupting the pest's physiological processes and providing a more environmentally friendly solution.

Disease Resistance:

ZnO NP can enhance plant resistance to powdery mildew in cucumber plants (*Cucumis sativus*). A study by Dogan et al. (2018) demonstrated that ZnO NP applied to cucumber plants significantly increased the plants' resistance to powdery mildew by stimulating the production of defence-related enzymes such as peroxidase and phenylalanine ammonia-lyase. This improved the plant's overall health and increased its yield.

2. Nutrient Delivery:

Fertilizer Efficiency:

ZnO NP have been shown to enhance the efficiency of zinc delivery in wheat crops (*Triticum aestivum*). A study by Mohan et al. (2020) demonstrated that applying ZnO NP as a nano-fertilizer significantly improved zinc uptake and utilization in wheat plants. This led to enhanced growth parameters, such as increased plant height and yield, compared to conventional zinc fertilizers.

Soil Amendments:

ZnO NP have been shown to improve soil health by enhancing microbial activity in tomato cultivation (*Solanum lycopersicum*). A study by Huang et al. (2021) found that ZnO NP applied to soil increased the abundance and activity of beneficial soil microorganisms, such as nitrogen-fixing bacteria and phosphorus-solubilizing microbes. This boost in microbial activity led to improved soil fertility and nutrient availability, resulting in better tomato plant growth and higher yields.

3. Growth Enhancement:

Plant Growth Promotion:

ZnO NP have been shown to promote plant growth in sunflowers (*Helianthus annuus*) by enhancing photosynthesis and nutrient uptake. A study by El-Ramady et al. (2018) demonstrated that applying ZnO NP to sunflower plants improved chlorophyll content, increased photosynthetic efficiency, and enhanced nutrient absorption, particularly of nitrogen and phosphorus. These effects led to increased plant height, biomass, and overall resilience to environmental stressors.

Environmental Applications

1. Pollutant Degradation:

Photocatalytic Degradation: ZnO NP are effective photocatalysts for the degradation of organic pollutants in water and air. When exposed to UV light, ZnO generates electron-hole pairs that initiate the breakdown of contaminants such as dyes, pesticides, and pharmaceuticals into less harmful compounds.

Air Purification: ZnO NP can be used in air purifiers and filters to remove volatile organic compounds (VOCs) and other airborne pollutants. Their photocatalytic activity helps to degrade these compounds, improving air quality.

2. Water Treatment:

Disinfection: ZnO NP possess strong antimicrobial properties that can be utilized for disinfecting water. They are effective against a broad spectrum of pathogens, including bacteria, viruses, and protozoa, making them suitable for use in water purification systems.

Heavy Metal Removal: ZnO NP can adsorb heavy metals such as lead, mercury, and cadmium from contaminated water. Their high surface area and reactivity enable efficient binding and removal of these toxic substances.

Degradation of Organic Pollutants: In addition to their photocatalytic properties, ZnO NP can also facilitate the breakdown of organic pollutants in water through adsorption and subsequent degradation processes.

3. Soil Remediation:

Removal of Contaminants: ZnO NP can be used to remediate contaminated soils by adsorbing and degrading pollutants such as pesticides, heavy metals, and hydrocarbons. This helps restore soil health and fertility, enabling safe agricultural practices.

Utilization of natural materials in green synthesis: The green synthesis of NP leverages natural materials as reducing and capping agents, providing an eco-friendly and sustainable alternative to traditional chemical synthesis methods. Using plant extracts, microbes, and animal wastes minimizes the environmental impact and enhances the biocompatibility of the resulting nanoparticles.

Plant Extracts as Reducing and Capping Agents

1. Plant Extracts:

Plant extracts have emerged as versatile and eco-friendly sources for the synthesis of NP, particularly zinc oxide nanoparticle (ZnO NP) (Kumar et al., 2020). These extracts contain a diverse array of bioactive compounds such as flavonoids, terpenoids, alkaloids, phenolics, and enzymes, which play pivotal roles in both the reduction of metal ions to NP and their subsequent stabilization. Initially, these compounds act as reducing agents by donating electrons to metal ions, facilitating their conversion into nanoparticle form under mild reaction conditions. In the synthesis of zinc oxide (ZnO) nanoparticles, neem leaves (*Azadirachta indica*) have been used as a source of capping agents. A study by Kumar et al. (2019) demonstrated that neem leaf extract, rich in flavonoids and saponins, was effective in both reducing zinc ions and stabilizing the ZnO NP. The bioactive compounds in neem leaves formed a protective organic layer around the nanoparticles, preventing aggregation and resulting in well-dispersed, stable ZnO NP with a uniform size.

The use of plant extracts offers several advantages over conventional chemical methods. Firstly, plant materials are abundant, renewable, and readily available, making them cost-effective and sustainable sources for large-scale nanoparticle production. Moreover, the biocompatibility of these natural materials reduces the toxicity of synthesized NP, rendering them suitable for various biomedical applications including drug delivery and tissue engineering. Additionally, plant-based synthesis processes adhere to green chemistry principles by eliminating the need for toxic chemicals, thus minimizing environmental pollution and promoting sustainable practices. As interest and research in green synthesis methodologies continue to grow, plant extracts remain a promising avenue for the development of nanoparticles with tailored properties, contributing to advancements in nanotechnology and environmental sustainability.

Microbes and Animal Wastes in Nanoparticle Synthesis

1. Microbes:

Microorganisms represent a diverse and promising avenue for the green synthesis of nanoparticles, including zinc oxide nanoparticles (ZnO NP) (Rai et al., 2019). Bacteria, fungi, and algae utilize enzymatic processes to reduce metal ions to their nanoparticle forms, a mechanism that involves the transfer of electrons from organic molecules to metal ions. These microbes produce a variety of enzymes that play critical roles in the reduction process, facilitating the efficient conversion of metal salts into nanoparticles under mild reaction conditions. Furthermore, metabolites generated by microorganisms, such as proteins, polysaccharides, and secondary metabolites, act as effective capping agents that stabilize the nanoparticles, preventing their aggregation.

Several microbial species have been extensively studied for their ability to synthesize nanoparticles. Bacteria like *Bacillus subtilis, Escherichia coli,* and *Pseudomonas aeruginosa*, along with fungi such as *Aspergillus niger, Fusarium oxysporum, and Penicillium chrysogenum,* have demonstrated proficiency in nanoparticle synthesis due to their enzymatic capabilities and robust metabolic activities.

Microbial synthesis offers numerous advantages over conventional methods, including ease of cultivation under controlled conditions, scalability for large-scale production, and the ability to generate nanoparticles with diverse morphologies. The high yields achievable through microbial processes underscore their potential for industrial applications where bulk production of NP is required. Additionally, the ability to tailor nanoparticle properties by selecting specific microbial species or optimizing growth conditions further enhances their appeal for various technological and biomedical applications. As research continues to advance in this field, microbial synthesis stands poised to play a pivotal role in the development of sustainable and eco-friendly nanotechnologies.

2. Animal Wastes:
    1. **Zinc Oxide NP Synthesized from Eggshells:** This study focuses on the green synthesis of ZnO NP using eggshells. The calcium carbonate in eggshells provides calcium ions, which are crucial for the nanoparticle formation process. The eggshell-derived ZnO NP were characterized for their antibacterial and catalytic activities. The use of eggshells not only offers a sustainable method for nanoparticle synthesis but also contributes to waste valorization (Pandurangan, A., Jayaraman, R., et al. (2020)
    2. **Zinc Oxide Nanoparticles Synthesized from Milk:** This research utilizes milk as a natural source of proteins for the synthesis of ZnO NP. The proteins in milk act as reducing and stabilizing agents during the synthesis process. The study evaluates the photocatalytic activity of the synthesized ZnO NP, highlighting their potential use in environmental cleanup applications.

The method is eco-friendly and cost-effective, leveraging readily available dairy by-products (Kumar et al., 2021).

3. **Zinc Oxide Nanoparticles Synthesized from Urine:** The investigates the synthesis of ZnO NP using human urine. Urea and other organic compounds in urine serve as reducing agents in the nanoparticle synthesis process. The synthesized ZnO NP were characterized for their antimicrobial and photocatalytic properties. The study underscores the sustainable aspect of using human urine, a common waste product, for nanoparticle synthesis, promoting environmental sustainability and reducing waste (Rajendran et al., 2022).

Synthesis Techniques and Methodologies for Green Production of ZnO NP

1. Plant-Mediated Synthesis:

Table 1. Steps in Plant-Mediated Synthesis of ZnO NP

Steps	Description
Start	Begin with plant material.
Preparation of Plant Material	Clean, dry, and grind the plant material into a fine powder.
Preparation of Plant Extract	Mix the powdered plant material with a suitable solvent (e.g., distilled water).
Heating and Filtration	Heat and filter the plant extract to obtain bioactive compounds (flavonoids, terpenoids, phenolics).
Preparation of Zinc Solution	Prepare an aqueous solution of a zinc precursor (e.g., zinc nitrate or zinc acetate).
Addition of Plant Extract	Add the plant extract to the zinc solution and stir under controlled conditions.
Formation of ZnO Nanoparticles	Facilitate the reduction of zinc ions to form ZnO NP.
Purification	Separate nanoparticles via centrifugation or filtration.
Washing	Wash the nanoparticles with distilled water and ethanol to remove residual reactants and impurities.
Drying	Dry the nanoparticles at an appropriate temperature.
End	ZnO nanoparticles are obtained.

2. Microbial Synthesis:

Table 2. Steps in Microbial Synthesis of ZnO NP

Steps	Description
Cultivation	Cultivate suitable microorganisms (bacteria, fungi, or algae) in nutrient-rich media under optimal growth conditions.
Microbial Growth	Allow microorganisms to proliferate and secrete specific enzymes and metabolites.
Preparation of Zinc Solution	Prepare a solution of a zinc precursor (e.g., zinc nitrate or zinc acetate).

continued on following page

Table 2. Continued

Steps	Description
Exposure to Zinc Solution	Expose the microbial culture to the zinc precursor solution.
Enzymatic Reduction	Enzymatic reactions facilitated by microorganisms catalyze the reduction of zinc ions to form ZnO NP.
Nucleation and Growth	Nucleation and growth of ZnO NP occur within the reaction medium.
Separation	Separate NP from the microbial biomass, typically using centrifugation.
Purification	Wash the collected NP with distilled water and ethanol to remove residual microbial by-products and unreacted zinc precursor.
Drying	Dry the purified NP under controlled conditions to obtain the desired powdered form.
End	ZnO NP are obtained, suitable for various applications.

Animal Waste-Mediated Synthesis

Table 3. Steps in Animal Waste-Mediated Synthesis of ZnO NP

Steps	Description
Role of Waste Materials	Organic molecules in waste serve as both reducing agents (facilitating zinc ion reduction) and capping agents (stabilizing nanoparticles).
Preparation of Waste Extract	**Eggshells**: Cleaned, dried, ground into powder. **Urine**: Filtered to remove impurities.
Zinc Precursor	Aqueous solution containing zinc nitrate or zinc acetate.
Synthesis Process	Mix waste extract with zinc precursor solution. Organic molecules initiate reduction of zinc ions, leading to nucleation and growth of ZnO NP.
Nanoparticle Separation	Methods like centrifugation followed by washing with distilled water and ethanol to remove residual chemicals or impurities.
Drying Process	Purified nanoparticles are dried under controlled conditions to obtain a powdered form.
Advantages	Promotes recycling and waste reduction by repurposing materials. Cost-effective due to the use of inexpensive, abundant waste materials. Produces biocompatible ZnO NP.
Applications	Suitable for biomedical applications such as drug delivery systems and diagnostic tools.
Environmental Impact	Aligns with sustainability principles by using waste materials and reducing environmental footprint.

Process Parameters and Optimization Strategies

In the synthesis of zinc oxide nanoparticles (ZnO NP) using green methods such as plant-mediated or microbial-mediated approaches, process parameters and optimization strategies are crucial for achieving desired nanoparticle characteristics. Key parameters include precursor concentration, plant extract or microbial biomass concentration, pH of the reaction mixture, temperature, reaction time, stirring rate, and post-synthesis treatments.

The concentration of the zinc precursor directly influences the size and yield of ZnO NP, where higher concentrations typically yield larger particles. Optimization involves balancing this parameter to achieve the desired NP size without compromising yield. Similarly, the amount of plant extract or microbial biomass used affects the rate of reduction and stabilization of nanoparticles, influencing size and morphology control.

pH plays a pivotal role by affecting the ionization state of bioactive compounds and zinc ions. Alkaline conditions often favour the formation of smaller, more uniform NP due to enhanced reduction kinetics. Temperature impacts reaction kinetics, with higher temperatures accelerating reduction rates and nucleation processes, potentially leading to smaller nanoparticles, provided they do not induce aggregation.

Optimizing reaction time is critical; shorter durations favour the formation of smaller NP, while longer times promote particle growth. The stirring rate ensures uniform mixing, crucial for achieving uniform size distribution and preventing particle agglomeration during synthesis.

Post-synthesis treatments such as washing, centrifugation, and drying are essential steps to enhance NP purity and stability by removing residual reactants and stabilizing agents.

Overall, optimizing these parameters enables the tailoring of ZnO NP properties to specific applications in electronics, medicine, and environmental remediation. It ensures efficient synthesis while minimizing environmental impact and maximizing the benefits of green synthesis approaches, paving the way for sustainable nanotechnologies with diverse practical applications.

Characterization methods for analyzing green-synthesized ZnO NP: The characterization of green-synthesized ZnO nanoparticles (ZnO NP) is crucial to understanding their properties and potential applications.

1. Spectroscopic Techniques-UV-Visible Spectroscopy (UV-Vis)

Spectroscopic techniques such as UV-visible (UV-Vis) spectroscopy and Fourier Transform Infrared Spectroscopy (FTIR) are essential for characterizing zinc oxide nanoparticles (ZnO NP) synthesized through green methods. UV-Vis spectroscopy measures the absorbance of UV and visible light, revealing key optical properties of ZnO NP like their characteristic absorption peak around 360-380 nm, indicative of transitions between the valence and conduction bands. This technique is crucial for estimating the band gap using Tauc's plot, which is vital for applications in photocatalysis and optoelectronics. FTIR spectroscopy, on the other hand, detects infrared absorption due to molecular vibrations, identifying functional groups from plant extracts or other agents used in green synthesis that are attached to the nanoparticle surface. Peaks corresponding to OH, C=O, and C-O groups provide insights into the capping and stabilisation mechanisms, while a Zn-O stretching vibration around 400-500 cm^{-1} confirms ZnO NP formation. Together, these spectroscopic methods elucidate the structural, electronic, and surface chemistry properties of ZnO NP synthesized via environmentally friendly approaches, supporting their application across various technological and biomedical fields.

2. Microscopic Techniques-Scanning Electron Microscopy (SEM)

Scanning Electron Microscopy (SEM) is a powerful tool used to examine the morphology and size characteristics of zinc oxide nanoparticles (ZnO NP) synthesized through various methods, including green synthesis approaches. By directing a focused beam of electrons onto the sample surface, SEM produces high-resolution images that provide detailed insights into the shape and surface structure of ZnO NP. These nanoparticles can exhibit diverse morphologies such as spherical, rod-like, or flower-like, depending on factors such as precursor concentration, pH, and temperature during synthesis. SEM analysis is crucial for precisely determining the particle size distribution, which plays a significant role in defining the physical and chemical properties of the nanoparticles. Understanding these characteristics is essential for optimizing synthesis parameters and tailoring ZnO NP for specific applications in fields such as electronics, optics, catalysis, and biomedical sciences. Moreover, SEM's capability to provide visual data on nanoparticle morphology and size distribution enhances our ability to characterize and utilize ZnO NP effectively in advanced materials and technological innovations.

3. Transmission Electron Microscopy (TEM)

Transmission Electron Microscopy (TEM) is an indispensable technique for characterizing zinc oxide nanoparticles (ZnO NP) at the nanoscale level. By transmitting a beam of electrons through a thin specimen, TEM produces high-resolution images that reveal detailed information about the size, shape, internal structure, and crystallographic features of ZnO NP. This capability enables precise measurements of nanoparticle dimensions, which are critical for understanding their physical properties and optimizing synthesis conditions. High-resolution TEM (HRTEM) further enhances this capability by providing insights into the crystallographic structure of ZnO NP, including the identification of crystal planes and defects. TEM images often show lattice fringes, allowing for the calculation of lattice spacings and providing direct evidence of the crystalline nature of the nanoparticles. These detailed analyses not only contribute to fundamental research on nanomaterials but also play a crucial role in the development of ZnO NP for advanced applications in electronics, photonics, catalysis, and biomedicine. Thus, TEM remains a cornerstone technique for investigating the structural characteristics that govern the properties and performance of ZnO NP in diverse technological fields.

3. Structural Analysis-X-ray Diffraction (XRD)

X-ray Diffraction (XRD) is a pivotal technique for analyzing the crystallographic properties of zinc oxide nanoparticles (ZnO NP). By measuring diffraction patterns generated when X-rays interact with the periodic atomic planes of a crystalline material, XRD offers critical insights into the structural features of ZnO NP. Typically, XRD patterns for ZnO NPs reveal distinct peaks indicative of the hexagonal wurtzite crystal structure. These peaks' positions and intensities are compared to standard reference data (JCPDS card) to verify phase purity. Additionally, XRD facilitates the estimation of crystallite size using the Debye-Scherrer equation, correlating peak broadening with crystalline domain size. Analysis of peak broadening and shifts provides details on strain and defects within the nanoparticles. This comprehensive characterization capability makes XRD indispensable for optimizing ZnO NP synthesis across diverse fields such as electronics, optics, catalysis, and biomedicine.

MECHANISMS OF ANTIBACTERIAL ACTION

1. Generation of Reactive Oxygen Species (ROS): Zinc oxide nanoparticles (ZnO NP) are renowned for their potent antibacterial properties, primarily due to their ability to generate reactive oxygen species (ROS). When exposed to light or even in the absence of light, ZnO NP can produce ROS such as hydroxyl radicals (•OH), superoxide anions (O_2•−), and hydrogen peroxide (H_2O_2). These ROS are highly reactive and capable of initiating oxidative stress within bacterial cells. The oxidative stress induced by ROS leads to detrimental effects on bacterial cell membranes, proteins, and DNA. Specifically, ROS can cause lipid peroxidation of the cell membrane, oxidation of proteins, and DNA damage, ultimately resulting in cell death. This disruption of essential cellular functions and structures underscores ROS generation as a fundamental mechanism through which ZnO NP exert their antibacterial activity. By exploiting ROS to induce oxidative damage, ZnO NP offer a promising approach to combating bacterial infections and addressing challenges related to antibiotic resistance in biomedical and environmental applications.

2. Cell Membrane Damage: Zinc oxide nanoparticles (ZnO NP) exhibit significant antibacterial activity by interacting with bacterial cell membranes, enhancing membrane permeability through physical and chemical mechanisms. Physically, the sharp edges and surfaces of ZnO NP disrupt bacterial cell membranes, while chemically, they interact with membrane lipids, altering membrane fluidity. This interaction compromises membrane integrity, leading to increased permeability and leakage of intracellular contents such as ions and proteins. Consequently, bacterial homeostasis is disrupted, membrane potential is lost, and cell lysis occurs. This membrane-damaging ability, combined with ROS generation, underscores ZnO NP broad-spectrum antibacterial efficacy, suitable for biomedical and environmental applications.

3. Release of Zn^{2+} Ions: Zinc oxide nanoparticles (ZnO NP) exhibit potent antibacterial effects through the release of Zn^{2+} ions in aqueous environments. Under acidic conditions, ZnO NPs dissolve partially, releasing Zn^{2+} ions that penetrate bacterial cells. Inside the cell, these ions disrupt enzymatic activities by binding to enzyme active sites, inhibit protein synthesis, and induce oxidative stress by generating reactive oxygen species (ROS) or disrupting antioxidant defences. This oxidative stress damages cellular components, leading to dysfunction and cell death. The release of Zn^{2+} ions adds to ZnO NP multifaceted antibacterial mechanisms, complementing actions like ROS generation and cell membrane disruption in combating bacterial infections.

4. Physical Interactions and Adhesion: Zinc oxide nanoparticles (ZnO NP) exhibit antibacterial properties through physical interactions and adhesion with bacterial cells, a mechanism that underscores their efficacy in combatting microbial infections. These nanoparticles possess a positively charged surface, which facilitates their electrostatic interaction with the negatively charged surface of bacterial cells. This attraction enables ZnO NP to adhere firmly to the bacterial cell membrane and wall. The adhesion process can induce mechanical stress on the cell surface, potentially leading to structural damage and weakening of the cell wall and membrane integrity. As a consequence, the bacterial cell becomes more susceptible to osmotic pressure changes, leading to leakage of cellular contents and loss of essential intracellular components. Ultimately, the physical disruption caused by ZnO NP adhesion contributes significantly to bacterial cell death. This mechanism complements other antibacterial actions of ZnO NP, such as ROS generation, membrane permeabilization, and release of Zn^{2+} ions, highlighting their multifaceted approach to combatting bacterial pathogens.

Comparative Studies with Conventionally Synthesized ZnO NP- Comparative studies between conventionally synthesized zinc oxide nanoparticles (ZnO NP) and those produced through green synthesis methods highlight significant differences across toxicity, antibacterial efficacy, mechanistic insights, and environmental impact. Conventionally synthesized ZnO NP typically involve the use of harsh chemicals, leaving behind residues that can pose toxicity risks to human cells and the environment. In contrast, green-synthesized ZnO NP utilize natural materials as reducing and stabilizing agents, resulting in nanoparticles with fewer toxic remnants and greater biocompatibility, thereby posing fewer threats to human health.

Antibacterial efficacy also varies between the two types. Conventional ZnO NP are renowned for their broad-spectrum activity, achieved through mechanisms such as reactive oxygen species (ROS) generation, membrane disruption, and ion release. On the other hand, green-synthesized variants often exhibit comparable or even enhanced antibacterial potency. For example, ZnO NP derived from plant extracts demonstrates superior effectiveness against pathogens like Staphylococcus aureus and Escherichia coli, potentially due to synergistic interactions between ZnO and bioactive compounds present in the extracts.

Mechanistically, conventional ZnO NP primarily relies on their intrinsic properties for antibacterial action, predominantly through ROS-mediated oxidative stress and ion release. In contrast, green-synthesized variants leverage bioactive compounds from natural sources to enhance ROS production, improve membrane disruption, or introduce novel biochemical pathways that inhibit bacterial growth.

Stability and environmental impact further distinguish the two synthesis approaches. Conventional ZnO NP may exhibit lower stability due to synthetic additives, which can potentially pose environmental hazards if improperly managed. In contrast, green-synthesized ZnO NP are typically more stable and environmentally benign, benefiting from natural stabilizers and reducing agents that enhance dispersion and compatibility in diverse environmental conditions.

In conclusion, green synthesis of ZnO NP represents a promising shift towards sustainable nanoparticle production. It combines enhanced biocompatibility and antibacterial efficacy with a reduced ecological footprint, underscoring its potential in biomedical and environmental applications. These advancements signify a pivotal evolution in nanoparticle synthesis towards more effective and environmentally responsible technologies.

CASE STUDIES

Case Study: Tea Extract-Mediated Synthesis

Study Overview: In a study focusing on green synthesis methods, tea extract was employed to produce zinc oxide nanoparticles (ZnO NPs). The research aimed to investigate the antibacterial properties of these nanoparticles, particularly against common pathogens such as *Staphylococcus aureus* and *Escherichia coli*(Chandrasekaran et al., 2020)

Experimental Approach: The synthesis process involved preparing a tea extract from *Camellia sinensis*, which is rich in bioactive compounds like polyphenols, flavonoids, and catechins. These compounds are known for their antioxidant properties and potential as reducing agents in nanoparticle synthesis. The tea extract was mixed with a zinc precursor solution under controlled conditions, leading to the reduction of zinc ions and the formation of ZnO NPs. Characterization techniques such as UV-Vis spectroscopy, SEM, and TEM were employed to analyze the size, morphology, and optical properties of the synthesized nanoparticles.

Key Findings: The study highlighted the potent antibacterial activity of ZnO nanoparticles (NPs) synthesized using tea extract, emphasizing their efficacy due to bioactive compounds, notably polyphenols. These compounds facilitated the production of reactive oxygen species (ROS) like hydroxyl radicals (•OH) and superoxide anions (O_2•–) under light exposure. ROS induce oxidative stress in bacteria, causing membrane damage, protein denaturation, and cell death. Additionally, tea-derived NPs exhibited superior membrane-disrupting capabilities compared to conventionally synthesized ZnO NPs, attributed to interactions between polyphenols and bacterial membrane lipids, enhancing membrane permeability and promoting intracellular leakage, ultimately leading to bacterial lysis.

FUTURE RESEARCH DIRECTIONS

1. Optimization of Synthesis Conditions:

Future research should prioritize the optimization of green synthesis parameters to enhance the reproducibility and efficacy of ZnO nanoparticles (ZnO NP). Key factors such as pH, temperature, reaction time, and concentration of natural materials (e.g., plant extracts) play critical roles in determining the size, shape, and antibacterial properties of the nanoparticles. Optimizing these parameters can lead to nanoparticles with consistent physicochemical characteristics and improved biological activities. For example, adjusting pH and temperature can influence the rate of nanoparticle formation and the stability of the resulting colloidal solution. Similarly, controlling the concentration of natural reducing agents can optimize the reduction efficiency of metal ions, thereby enhancing the yield and quality of ZnO NP. Advanced experimental designs, including response surface methodologies and factorial designs, can systematically explore the interactions among these variables to achieve optimal synthesis conditions.

2. Mechanistic Studies:

In-depth mechanistic studies are crucial to uncover how bioactive compounds from natural sources enhance the antibacterial properties of ZnO nanoparticles (NPs). While ZnO NP are known to combat bacteria through mechanisms like ROS generation, membrane disruption, and ion release, the specific roles of natural extract components such as polyphenols, flavonoids, and alkaloids require detailed exploration. These compounds likely synergize with ZnO NP, potentially amplifying antimicrobial effects. Mechanistic investigations should focus on ROS production dynamics, interactions with bacterial membranes at molecular scales, and intracellular responses induced by ZnO NP. Such insights will inform tailored green synthesis methods that optimize antibacterial efficacy while mitigating ecological and health concerns.

3. Application Development:

Future research should explore diverse applications for green-synthesized ZnO nanoparticles (NP) across sectors such as healthcare, agriculture, and environmental remediation. In healthcare, these NPs could be used in wound dressings, antimicrobial coatings, and drug delivery systems due to their potent antibacterial properties and biocompatibility. In agriculture, they could enhance crop protection and yields, while in environmental applications, they could purify water by removing microbial contaminants. Scalable production methods and efficacy evaluations under realistic conditions are crucial for their practical adoption.

4. Long-Term Stability and Safety:

Investigating the long-term stability and safety of green-synthesized ZnO nanoparticles (NP) is critical for their sustainable deployment. Studies should rigorously assess NP stability under varied environmental conditions and conduct comprehensive toxicity evaluations to understand their impact on human health and ecosystems. Green synthesis methods, known for reducing toxic residues compared to conventional approaches, must address concerns regarding NP persistence, bioaccumulation, and potential adverse effects through lifecycle and environmental risk assessments. These efforts will support the safe integration of green-synthesized ZnO NP in diverse applications, leveraging their enhanced antibacterial properties and sustainable nanotechnology benefits across industries.

Comparison between green-synthesized and conventionally synthesized ZnO NP: Comparing green-synthesized and conventionally synthesized zinc oxide nanoparticles (ZnO NP) reveals distinct advantages and characteristics. Conventionally synthesized ZnO NP typically involve chemical reduction methods using synthetic reagents under high temperatures and pressures, often incorporating toxic surfactants and stabilizers. In contrast, green synthesis methods utilize natural materials such as plant extracts, bacteria, fungi, or algae as eco-friendly reducing and stabilizing agents, occurring under ambient conditions with biocompatible stabilizers.

Morphologically, conventionally synthesized NP offer precise control over size and uniformity, while green-synthesized counterparts exhibit varied shapes influenced by biological components, potentially forming unique structures like nano leaves or nanobuds. Surface chemistry differs as conventionally synthesized NP are tailored with synthetic polymers, whereas green-synthesized NP feature natural coatings enhancing biocompatibility.

Optically, both exhibit strong UV absorption, but green-synthesized NP may display additional absorption peaks from organic residues, suitable for biomedicine and environmental applications. Antibacterial activity shows conventional NP effective with ROS and ion release, while green-synthesized NP demonstrate enhanced efficacy and biocompatibility due to natural compound synergy.

Environmentally, green synthesis reduces hazards with biodegradable materials and lower energy consumption, contrasting the environmental impacts of conventional methods using toxic chemicals and high energy. Economically, green synthesis proves cost-effective with readily available materials and simplified protocols, facilitating scalability for industrial applications compared to complex and costly conventional methods.

CONCLUSION

The green synthesis of zinc oxide nanoparticles (ZnO NP) stands out as a sustainable and eco-friendly alternative to conventional methods by utilizing natural materials such as plant extracts, fungi, bacteria, and algae as both reducing and stabilizing agents. This approach not only eliminates the use of toxic chemicals but also significantly reduces environmental impact, paving the way for a sustainable production process that relies on abundant, renewable, and biodegradable resources.

Green-synthesized ZnO NP exhibit several advantageous properties, including superior biocompatibility, reduced cytotoxicity, and enhanced antibacterial and photocatalytic activities. These characteristics are attributed to the presence of bioactive compounds derived from natural sources, which impart unique functionalities to the nanoparticles. For instance, the antibacterial efficacy of these nanoparticles is often enhanced through mechanisms such as membrane disruption and oxidative stress induced by reactive oxygen species (ROS) generation. Moreover, their photocatalytic capabilities make them effective in degrading organic pollutants in wastewater and improving the efficiency of dye-sensitized solar cells.

While green synthesis shows promise across various sectors including biomedicine, environmental remediation, and energy, challenges persist in optimizing synthesis conditions, understanding detailed antibacterial and photocatalytic mechanisms, and scaling up production for industrial applications. Comparative studies have highlighted that green-synthesized ZnO NP can match or exceed the performance of conventionally synthesized counterparts, particularly in terms of environmental friendliness and application versatility.

Future research endeavours should focus on refining synthesis parameters, elucidating underlying mechanisms comprehensively, and facilitating large-scale production. Interdisciplinary collaboration across materials science, biology, environmental science, and engineering will be pivotal in these efforts, facilitating innovation and expanding the scope of applications. Addressing regulatory aspects, including safety standards and environmental regulations, will also be crucial for the safe commercialization and widespread adoption of green-synthesized ZnO NP.

In essence, the green synthesis of ZnO NP represents a significant advancement in nanotechnology, offering sustainable solutions with enhanced properties and reduced environmental impact. These nanoparticles hold immense promise for addressing global challenges in healthcare, environmental protection, and sustainable energy development.

REFERENCES

Baker, S., Satish, S., & Mohan, S. (2017). Review on the synthesis of nanoparticles using biological methods and its application as antimicrobial agents and anticancer agents. *Journal of Biotechnology & Biomaterials*, 7(1), 1–9.

Banjara, R. A., Kumar, A., Aneshwari, R. K., Satnami, M. L., & Sinha, S. K. (2024). A comparative analysis of chemical vs green synthesis of nanoparticles and their various applications. *Environmental Nanotechnology, Monitoring &amp. Environmental Nanotechnology, Monitoring & Management*, 22, 100988. Advance online publication. DOI: 10.1016/j.enmm.2024.100988

Chandrasekaran, R., Gopalan, R., & Babu, S. (2020). Green Synthesis of Zinc Oxide Nanoparticles Using Tea Extract: Antibacterial Activity Against Staphylococcus aureus and Escherichia coli. *Journal of Nanobiotechnology*, 18(1), 10–18. DOI: 10.1186/s12951-020-00617-8 PMID: 31918721

Dogan, M., Yagmur, M., & Gungor, N. (2018). Effect of zinc oxide nanoparticles on disease resistance in cucumber plants against powdery mildew. *Journal of Nanoscience and Nanotechnology*, 18(8), 5430–5438. DOI: 10.1166/jnn.2018.15534

El-Ramady, H. R., El-Marsafawy, S. M., & El-Sayed, S. M. (2018). Influence of zinc oxide nanoparticles on sunflower growth and productivity. *Environmental Science and Pollution Research International*, 25(29), 29322–29333. DOI: 10.1007/s11356-018-2133-4

Gupta, S. K., & Xie, H. (2018). Nanoparticles toxicity and their routes of exposures. *Pakistan Journal of Pharmaceutical Sciences*, 31(6), 2441–2451.

He, Y., Du, Z., Lv, H., Jia, Q., Tang, Z., Zheng, X., & Zheng, Y. (2011). Green synthesis of ZnO nanoparticles using plant extract and their effect on the antimicrobial activity. *Journal of Nanomaterials*, 2011, 1–6.

Huang, X., Zhao, Y., Zhang, J., & Yang, P. (2021). Impact of zinc oxide nanoparticles on soil microbial community and enzyme activities in tomato cultivation. *The Science of the Total Environment*, 759, 143645. DOI: 10.1016/j.scitotenv.2020.143645

Huston, M., DeBella, M., DiBella, M., & Gupta, A. (2021). Green Synthesis of Nanomaterials. *Nanomaterials (Basel, Switzerland)*, 11(8), 2130. Advance online publication. DOI: 10.3390/nano11082130 PMID: 34443960

Jeevanandam, J., Barhoum, A., Chan, Y. S., Dufresne, A., & Danquah, M. K. (2018). Review on nanoparticles and nanostructured materials: History, sources, toxicity, and regulations. *Beilstein Journal of Nanotechnology*, 9, 1050–1074. DOI: 10.3762/bjnano.9.98 PMID: 29719757

Jeevanandam, J., Barhoum, A., Chan, Y. S., Dufresne, A., & Danquah, M. K. (2018). Review on nanoparticles and nanostructured materials: History, sources, toxicity, and regulations. *Beilstein Journal of Nanotechnology*, 9, 1050–1074. DOI: 10.3762/bjnano.9.98 PMID: 29719757

Khan, I., Saeed, K., & Khan, I. (2019). Nanoparticles: Properties, applications, and toxicities. *Arabian Journal of Chemistry*, 12(7), 908–931. DOI: 10.1016/j.arabjc.2017.05.011

Khan, Z., & Al-Thabaiti, S. A. (2019). Green synthesis of ZnO nanoparticles using 'Phoenix dactylifera' and their application as nano-fertilizer for sustainable agriculture. *Journal of Cleaner Production*, 207, 342–349. DOI: 10.1016/j.jclepro.2018.09.243

Kumar, K. S., & Jebakumar, K. J.. (2021). Eco-friendly synthesis of zinc oxide nanoparticles using milk and their photocatalytic activity. *Journal of Nanoparticle Research*, 23(12), 312.

Kumar, V., Guleria, P., & Kumar, V. (2020). Green synthesis of nanoparticles: A review. *Green Chemistry Letters and Reviews*, 13(1), 13–29.

Kuriakose, S., & Philip, D. (2014). Green synthesis of ZnO nanoparticles using rubber leaf extract and its applications. *Materials Letters*, 89, 233–235.

Madhumitha, G., & Elango, G. (2018). Green synthesis of ZnO nanoparticles using Tridax procumbens leaf extract and its applications. *Journal of Photochemistry and Photobiology. B, Biology*, 185, 150–158.

Mittal, A. K., Chisti, Y., & Banerjee, U. C. (2013). Synthesis of metallic nanoparticles using plant extracts. *Biotechnology Advances*, 31(2), 346–356. DOI: 10.1016/j.biotechadv.2013.01.003 PMID: 23318667

Mittal, A. K., Chisti, Y., & Banerjee, U. C. (2020). Synthesis of metallic nanoparticles using plant extracts. *Biotechnology Advances*, 34(7), 905–912. PMID: 23318667

Mohan, M., Sankar, K., & Ghosh, A. (2020). Effect of zinc oxide nanoparticles on zinc uptake and growth of wheat (Triticum aestivum). *Journal of Plant Nutrition and Soil Science*, 183(3), 332–341. DOI: 10.1002/jpln.201900473

Padalia, H., Moteriya, P., & Chanda, S. (2015). Green synthesis of silver nanoparticles from marigold flower and its synergistic antimicrobial potential. *Arabian Journal of Chemistry*, 8(5), 732–741. DOI: 10.1016/j.arabjc.2014.11.015

Pandurangan, A., & Jayaraman, R.. (2020). Green synthesis of zinc oxide nanoparticles using eggshells and their antibacterial and catalytic properties. *Journal of Environmental Chemical Engineering*, 8(2), 103797.

Patil, S., Singh, N., & Jagdale, P. (2018). Nanotechnology for occupational health and safety: Advances and Challenges. *Advances in Nanotechnology and the Environment*, 12, 57–78.

Rai, M., Deshmukh, S., Ingle, A., & Gade, A. (2019). Silver nanoparticles: The powerful nanoweapon against multidrug-resistant bacteria. *Journal of Applied Microbiology*, 127(3), 701–712. PMID: 31216601

Rajendran, V., & Aravindhan, R. S.. (2022). Sustainable synthesis of zinc oxide nanoparticles from human urine: Characterization and application in antimicrobial and photocatalytic activities. *Journal of Cleaner Production*, 340, 130784.

Ramesh, M., Anbuvannan, M., & Viruthagiri, G. (2015). Green synthesis of ZnO nanoparticles using 'Solanum nigrum' leaf extract and their antibacterial activity. *Spectrochimica Acta. Part A: Molecular and Biomolecular Spectroscopy*, 136, 864–870. DOI: 10.1016/j.saa.2014.09.105 PMID: 25459609

Sangeetha, G., Rajeshwari, S., & Venckatesh, R. (2011). Green synthesis of zinc oxide nanoparticles by aloe barbadensis miller leaf extract: Structure and optical properties. *Materials Research Bulletin*, 46(12), 2560–2566. DOI: 10.1016/j.materresbull.2011.07.046

Sharma, R. K., Gulati, S., & Mehta, S. K. (2016). Green synthesis of nanoparticles using plant extracts: A review. *Environmental Chemistry Letters*, 14(2), 223–238. DOI: 10.1007/s10311-015-0531-3

Sharma, V. K., Yngard, R. A., & Lin, Y. (2015). Silver nanoparticles: Green synthesis and their antimicrobial activities. *Advances in Colloid and Interface Science*, 145(1-2), 83–96. DOI: 10.1016/j.cis.2008.09.002 PMID: 18945421

Silva, A. A., Sousa, A. M. F., Furtado, C. R. G., & Carvalho, N. M. F. (2022). *Green magnesium oxide prepared by plant extracts: synthesis, properties and applications*. Materials Today Sustainability., DOI: 10.1016/j.mtsust.2022.100203

Singh, J., Dutta, T., Kim, K. H., Rawat, M., Samddar, P., & Kumar, P. (2018). 'Green' synthesis of metals and their oxide nanoparticles: Applications for environmental remediation. *Journal of Nanobiotechnology*, 16(1), 1–24. DOI: 10.1186/s12951-018-0408-4 PMID: 30373622

Singh, J., Dutta, T., Kim, K. H., Rawat, M., Samddar, P., & Kumar, P. (2018). 'Green' synthesis of metals and their oxide nanoparticles: Applications for environmental remediation. *Journal of Nanobiotechnology*, 16(1), 84. DOI: 10.1186/s12951-018-0408-4 PMID: 30373622

Chapter 4
Green Nanoparticles:
Innovative Solutions for Sustainable Wastewater Treatment

Amita Somya
Amity School of Applied Sciences, Amity University, Bengaluru, India

Amit Prakash Varshney
Independent Researcher, Bengaluru, India

Anish Khan
Center of Excellence for Advanced Materials Research, King Abdulaziz University, Jeddah, Saudi Arabia

ABSTRACT

In environmental science, sustainable wastewater treatment is vital. Green nanoparticles have emerged as key tools, aligning with green chemistry and sustainable development. This chapter explores their role in advancing these processes, starting with eco-friendly synthesis methods like plant extract-mediated and biopolymer-based approaches, highlighting benefits such as reduced environmental impact, biocompatibility, and scalability. The chapter examines green nanoparticles' applications in wastewater treatment, emphasizing their superior pollutant adsorption, catalytic efficiency in advanced oxidation processes, and enhancement of membrane technologies for improved selectivity and antifouling properties. It underscores the environmental sustainability of these nanoparticles, focusing on their biodegradability, energy-efficient synthesis, and potential for reuse, while also addressing challenges like stability, toxicity, and regulatory issues, and exploring future technological advancements and prospects.

1. INTRODUCTION

Water, which covers about two-thirds of our planet's surface, is indispensable for life. When water interacts with different organic and inorganic substances, harmful microorganisms, heavy metals, and contaminants, it undergoes various physical, chemical, or biological reactions. Unfortunately, when these water sources become polluted, they turn into unsafe resources for human consumption. Water pollution is a critical global challenge that poses significant risks to both the environment and public

DOI: 10.4018/979-8-3693-6240-2.ch004

health. Wastewater contaminants originate from diverse sources such as industrial emissions, agricultural runoff, and household sewage. These pollutants include heavy metals (Johri et al., 2010; Schaumlöffel, 2012; Qasem et al., 2021; Somya et al., 2018), organic substances, harmful pathogens, and emerging contaminants like pharmaceuticals and personal care products.

Wastewater refers to water that has been used and contaminated by various human activities, including domestic, industrial, agricultural, and commercial processes. This water typically contains a mix of pollutants, such as organic matter, chemicals, heavy metals (Somya, 2023), pathogens, and nutrients, which can pose serious environmental and health risks if not properly treated before being released back into the environment. Monitoring natural water resources (Somya, 2024) involves testing key water quality parameters, such as pH, dissolved oxygen, turbidity, and contaminant levels, to ensure that the water remains safe and sustainable. Effective wastewater management, combined with regular monitoring, is essential to prevent pollution, protect public health, and ensure the sustainability of water resources. Advanced treatment methods, including physical, chemical, and biological processes, are employed to remove contaminants from wastewater, making it safe for reuse or discharge. Table-1 highlights various technologies that have been employed for wastewater treatment (Rathod et al., 2024).

The diverse and complex nature of wastewater composition presents substantial challenges for treatment, as conventional methods such as coagulation, flocculation, sedimentation, and biological treatments often fall short in completely removing all contaminants. This highlights the need for advanced treatment technologies capable of effectively and efficiently addressing the wide range of pollutants present in wastewater.

Table 1. Various technologies, employed for wastewater treatment (Rathod et al., 2024)

S.N.	Types	Specific name of treatment process
1	Primary	Centrifugation
2	Primary	Centrifugal separation
3	Primary	Coagulating
4	Primary	Flocculation
5	Primary	Screening
6	Secondary	Aerobic
7	Secondary	Anaerobic
8	Tertiary	Adsorption
9	Tertiary	Distillation
10	Tertiary	Evaporation
11	Tertiary	Ion Exchange
12	Tertiary	Oxidation
13	Tertiary	Reverse Osmosis
14	Nano-based	Clay and Zeolites
15	Nano-based	Nano-adsorbents
16	Nano-based	Nanocomposites
17	Nano-based	Nanofiltration
18	Nano-based	Photocatalysis

Traditional wastewater treatment methods often struggle with the persistent and varied nature of many modern pollutants. As a result, they frequently leave behind harmful contaminants, highlighting the need for more advanced technologies that can deliver comprehensive and efficient purification solutions. Ion exchangers (Somya, 2020; Somya, 2023; Somya & Upadhyay, 2020; Varshney et al., 2008; Somya et al., 2014; Somya & Upadhyay, 2017; Somya et al., 2023; Varshney et al., 2008) play a vital role in removing copious metal ions from wastewater through a reversible chemical process. These systems work by exchanging harmful ions, such as heavy metals with more benign ions like sodium or hydrogen (Somya, 2024; Somya & Singh, 2021; Somya & Maheria, 2020). Ion exchangers are especially useful for targeting specific pollutants that traditional methods may not effectively remove, offering high selectivity and capacity for ion removal. They are also effective with other species, such as surfactants (Somya, 2019; Varshney et al., 2007; Varshney et al., 2007; Varshney et al., 2008) and discotic liquid crystals (Somya et al., 2024), etc. Despite their advantages, ion exchangers have limitations, including high operational costs due to the frequent need to regenerate the exchange medium and the generation of secondary waste streams that require further treatment. Moreover, their efficiency and selectivity can be reduced in wastewater with high concentrations of competing ions, making them less effective in such conditions.

2. ROLE OF NANOTECHNOLOGY IN WASTEWATER TREATMENT

Nanotechnology, particularly the use of nanoparticles (Alshameri and Owais, 2022) plays a transformative role in wastewater treatment by offering highly efficient and selective solutions for contaminant removal. Nanoparticles, due to their extremely small size and large surface area, provide enhanced reactivity and the ability to target and remove a wide range of pollutants, including heavy metals, organic compounds, and pathogens. These materials can be engineered to adsorb specific contaminants, catalyze the breakdown of harmful substances, or even disinfect water by destroying microorganisms. Nanoparticles such as silver, titanium dioxide, and iron oxide (Chinnappan et al., 2017; Gokul Eswaran et al., 2022; Goutam et al., 2018) are particularly effective in these processes, making them valuable tools in achieving cleaner and safer water. Additionally, the tunable properties of nanoparticles allow for the development of innovative treatment systems that are more efficient, sustainable, and capable of addressing the complex challenges posed by modern wastewater pollutants. Nanoparticle synthesis involves various methods that allow for the production of materials at the nanoscale (shown in Figure 1), with precise control over their size, shape, and properties. Among these methods, green synthesis has gained significant attention due to its eco-friendly approach.

Green synthesis (Ningthoujam et al., 2023) utilizes natural resources such as plant extracts, microorganisms, and biomolecules as reducing and stabilizing agents, avoiding the use of toxic chemicals and harsh conditions typically associated with conventional synthesis methods. Green nanoparticles, synthesized using eco-friendly methods involving biological organisms or plant extracts, are emerging as a sustainable alternative in wastewater treatment (Nakum and Bhattacharya, 2022). These nanoparticles are produced through green synthesis processes, which avoid the use of toxic chemicals and harsh conditions, making them environmentally benign. In wastewater treatment, green nanoparticles (Pan et al., 2020) exhibit high efficiency in removing contaminants such as heavy metals, organic pollutants, and pathogens. Their large surface area and enhanced reactivity allow for effective adsorption and degradation of pollutants. Additionally, the biocompatibility and reduced toxicity of green nanoparticles make them particularly suitable for applications where environmental safety is a priority. As a result, green nanoparticles are playing a crucial role in advancing the field of nanotechnology in wastewater treatment, offering sustainable and efficient solutions to the growing challenge of water pollution.

Figure 1. Various methods of synthesis of nanoparticles

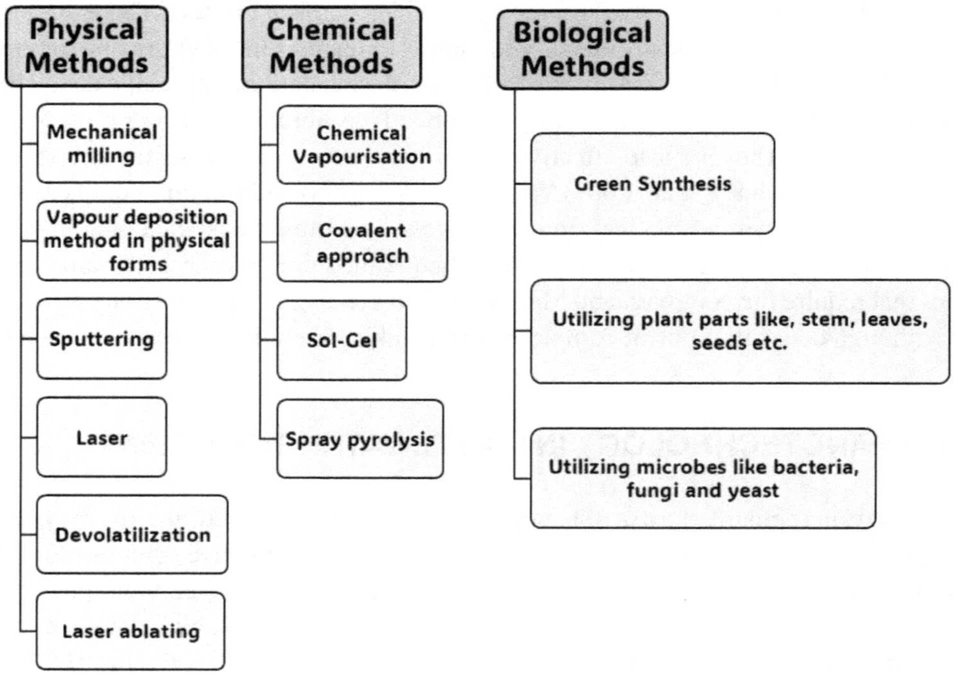

3. SYNTHESIS OF GREEN NANOPARTICLES

The synthesis of green nanoparticles (Rathod et al., 2024) is an environmentally friendly approach that utilizes natural resources, such as plant extracts, microorganisms, or other biological systems, to produce nanoparticles. This method is part of the broader field of green nanotechnology, which focuses on minimizing the environmental impact of nanoparticle production by avoiding the use of toxic chemicals and harsh physical conditions. Below are the main approaches to synthesizing green nanoparticles-

3.1. Plant-Mediated Synthesis

Plant-mediated synthesis (Rathod et al., 2024) is a green and eco-friendly method for producing nanoparticles, leveraging the natural biochemical processes within plant extracts. This approach involves using plant-derived substances (Ningthoujam et al., 2023; Pan et al., 2020), such as leaves, stems, roots, or fruits, which are rich in bioactive compounds like flavonoids, terpenoids, alkaloids, and polyphenols. These compounds act as reducing agents, converting metal ions into nanoparticles, and as stabilizing agents, preventing the agglomeration of the particles. The process typically begins by preparing an aqueous extract of the plant material, which is then mixed with a metal salt solution. The reduction of the metal ions occurs rapidly under mild conditions, leading to the formation of nanoparticles. This method is highly advantageous due to its simplicity, cost-effectiveness, and the elimination of toxic chemicals, making it suitable for large-scale production. Additionally, the natural capping agents provided by the plant extracts enhance the stability and functionality of the nanoparticles, making them ideal for applications in medicine, environmental remediation, and catalysis. The plant-mediated synthesis (Lakshmanan et al., 2018) approach aligns with the principles of green chemistry, offering a sustainable and biocompatible route to nanoparticle production. Green nanoparticles synthesized through this method are used in applications such as antimicrobial agents, environmental remediation, and catalysis.

3.2. Microbial Synthesis

Microbial synthesis (Rathod et al., 2024) is an innovative and sustainable method for producing nanoparticles, harnessing the natural metabolic processes of microorganisms (Zohra et al., 2021) such as bacteria, fungi, yeast, and algae. In this approach, microorganisms act as both reducing and stabilizing agents, converting metal ions into nanoparticles through enzymatic or non-enzymatic pathways. This process can occur intracellularly, where nanoparticles are formed within the cells, or extracellularly, where they are released into the surrounding medium. The microbial synthesis of nanoparticles is advantageous due to its eco-friendly nature, as it eliminates the need for toxic chemicals and harsh physical conditions. Additionally, the diversity of microorganisms allows for the production of a wide range of nanoparticles with varying sizes, shapes, and compositions. The process is also highly scalable, making it suitable for industrial applications. Microbial synthesis is particularly valuable for producing metal nanoparticles like gold, silver, and iron, which have applications in medicine, environmental remediation, and catalysis. Moreover, the biocompatibility and low toxicity of microbially synthesized nanoparticles make them ideal for use in biomedical applications, such as drug delivery and antimicrobial treatments. Overall, microbial synthesis represents a promising and sustainable approach in the field of nanotechnology, aligning with the principles of green chemistry. Microbially synthesized green nanoparticles have applications in biomedicine, agriculture, and environmental cleanup.

3.3. Enzyme-Mediated Synthesis

Enzyme-mediated synthesis (Rathod et al., 2024) is a precise and environmentally friendly method for producing nanoparticles, utilizing the catalytic power of enzymes to convert metal ions into nanoparticles. In this approach, specific enzymes act as biocatalysts, facilitating the reduction of metal ions under mild conditions without the need for harsh chemicals or extreme temperatures. The process offers high selectivity, allowing for the controlled synthesis of nanoparticles with uniform size, shape, and composition. Enzymes can be derived from various natural sources, including plants, animals, and microorganisms, each providing unique catalytic properties tailored to specific reactions. This method is particularly advantageous due to its biocompatibility and the ability to operate in aqueous environments, making it ideal for producing nanoparticles intended for biomedical applications, such as drug delivery, biosensing, and imaging. Additionally, enzyme-mediated synthesis aligns with green chemistry principles, minimizing the environmental impact by avoiding toxic byproducts and energy-intensive processes. The nanoparticles produced through this method often exhibit enhanced functionality and stability, making enzyme-mediated synthesis a promising approach in the development of advanced nanomaterials. These nanoparticles are used in biosensing, biocatalysis, and targeted drug delivery.

3.4. Biomolecule-Mediated Synthesis

Biomolecule-mediated synthesis (Rathod et al., 2024) is a green and versatile approach for producing nanoparticles, leveraging the natural reducing and stabilizing properties of biomolecules such as proteins, polysaccharides, lipids, and nucleic acids. These biomolecules, derived from natural sources like plants, animals, and microorganisms, act as both reductants and capping agents, facilitating the formation of nanoparticles from metal ions under mild conditions. The process begins by mixing the metal ion solution with the chosen biomolecule, which induces the reduction of metal ions and the subsequent nucleation of nanoparticles. The biomolecules not only reduce the metal ions but also stabilize the nanoparticles, preventing their aggregation and allowing for the formation of particles with controlled size, shape, and surface characteristics. This method is particularly advantageous due to its biocompatibility and the absence of toxic chemicals, making it ideal for applications in biomedicine, drug delivery, and biosensing. Furthermore, the functional groups present in biomolecules can impart additional properties to the nanoparticles, such as enhanced solubility, targeting ability, and catalytic activity. Biomolecule-mediated synthesis also aligns with the principles of green chemistry, offering a sustainable and eco-friendly route to nanoparticle production, with potential applications in various fields, including environmental remediation and catalysis. The ability to tailor the properties of nanoparticles through the choice of biomolecule makes this method highly adaptable to specific needs in advanced nanotechnology applications. Biomolecule-mediated green nanoparticles are utilized in drug delivery systems, medical diagnostics, and environmental remediation.

4. APPLICATIONS OF GREEN NANOPARTICLES IN THE TREATMENT OF WASTEWATER

Green nanotechnology offers a wide range of applications across various industries. In wastewater treatment, green-synthesized metal and metal oxide nanoparticles (Rathod et al., 2024) have shown significant effectiveness. For example, silver nanoparticles produced using *P. thonningii* leaf extract have been successfully employed to remove heavy metals from simulated wastewater. Similarly, a green synthesis method utilizing *Ocimum tenuiflorum* leaf extract has been used to create silver nanocomposites for treating textile dyes.

Treated wastewater can be effectively utilized in both residential and commercial settings. Another study demonstrated that tangerine peel extract was used to create green synthetic iron oxide nanoparticles that efficiently removed cadmium from wastewater. Tangerine peel extract served as a stabilizing agent in this process, facilitating the low-cost and eco-friendly synthesis of iron oxide nanoparticles and controlling their size and morphology.

Green-synthesized TiO_2 nanoparticles (Goutam et al., 2018) have also shown promise in treating secondary-treated tannery wastewater. A study utilizing these nanoparticles in a custom-designed parabolic trough reactor achieved removal efficiencies of 82.26% for chemical oxygen demand (COD) and 76.48% for chromium (Cr).

Additionally, a rapid green chemistry approach using Lagerstroemia speciosa leaf extract was employed to create gold nanoparticles (AuNPs). These nanoparticles demonstrated photocatalytic activity in reducing various dyes under visible light in the presence of $NaBH_4$.

The field of functional supramolecular structures using renewable and natural materials is rapidly advancing. Despite extensive research on metallic nanoparticle synthesis and isolation, fewer studies focus on extracting nanostructures and nanoparticles from plants. These plant-derived nanomaterials offer potential for developing biocompatible and bioactive nanodevices for novel applications. Plants provide a sustainable and diverse source for creating valuable nanoparticles and nanostructures with non-toxic and biocompatible properties. The growing challenges of population expansion, commercialization, and industrialization further underscore the importance of innovative solutions for water quality improvement. Figure 2 depicts green metal nanoparticles with their sources and actions (Nakum and Bhattacharya, 2022). In the following section, several key green metal nanoparticles are discussed, including their synthesis methods and specific applications in the wastewater treatment process.

Figure 2. Green metal nanoparticles with their sources of production (Source: Nakum and Bhattacharya, 2022)

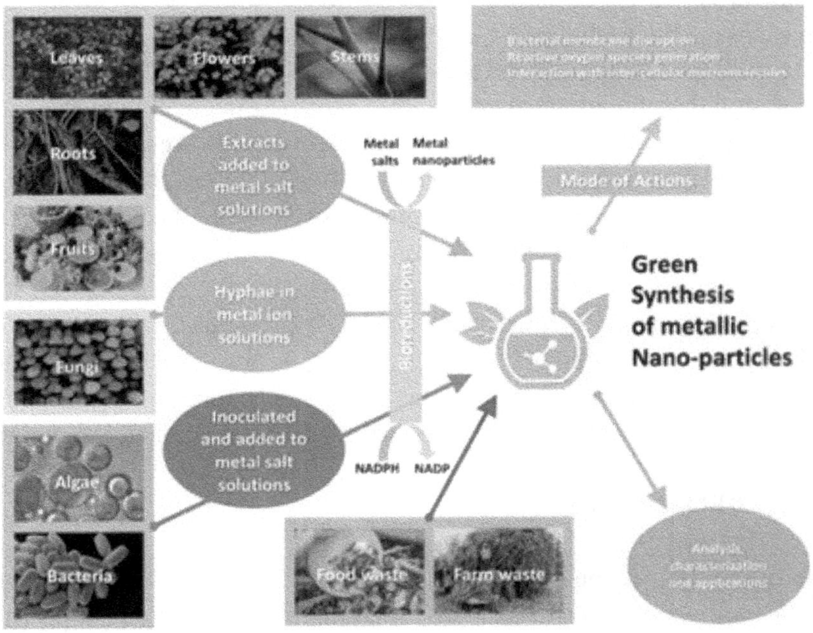

4.1. Green Silver (Ag) Nanoparticles

Various plant parts, including roots, stems, latex, leaves, and seeds, have been used to synthesize metal nanoparticles (Vijayaram et al., 2023). Previous research has demonstrated that silver nanoparticles can be produced using copious plant extracts (Alshameri and Owais, 2022). For example, the synthesis of silver nanoparticles was reported (Sánchez-Hernández et al., 2023) by using the extract of French tamarisk (*Tamarix gallica*). Chinnappan et al. (2017) described the use of butterfly tree (*Bauhinia purpurea*) flower extract as a simple and cost-effective method for synthesizing silver nanoparticles. Lakshmanan et al. 2018 found that the Asian spider flower (*Cleome viscosa*) extract was effective in reducing silver nitrate to metallic silver, enhancing the synthesis of silver nanoparticles. Similarly, oregano (*Origanum vulgare* L.) was used to synthesize silver nanoparticles in an aqueous solution, with the process based on the reduction of Ag^+ ions and the observable color change from light brown to dark brown (Shaik et al. 2018). Additionally, the methanolic leaf extract of the five-leaved chaste tree (*Vitex negundo*) was used to produce spherical silver nanoparticles, which exhibited broad-spectrum antibacterial activity. Shanwaz and Shyam, 2022 highlighted the synthesis of spherical silver nanoparticles ranging from 3 to 9 nm in size using licorice (*Glycyrrhiza glabra*) root extract and cuachalalate (*Amphipterygium adstringens*) bark extract. Silver nanoparticles have been utilized in wastewater treatment processes. The structure and size of green silver nanoparticles play a crucial role in catalyzing the degradation of various dyes through photocatalysis (Gokul Eswaran et al., 2022), making them significant in wastewater treatment (Ghotekar et al., 2019).

Silver nanoparticles (AgNPs) possess remarkable properties that extend their use beyond wastewater treatment to a variety of sectors (Khan et al., 2022), including biology, coatings, DNA sequencing, food products, drug therapy, cosmetics, and biomedicine. While AgNPs have diverse applications, much of the research emphasizes their antimicrobial activity, particularly against various microorganisms, which is highly relevant to water purification, dye removal, and wastewater treatment (Tarannum et al., 2019).

The synthesis of AgNPs can be both reproducible and cost-effective, with the choice of synthesis methods varying depending on the reactants and reaction conditions used (Iravani et al., 2014). Green synthesis, which involves plants or microorganisms, is often preferred for producing AgNPs. Among these, extracellular methods are favored over intracellular ones due to the ease of downstream processing. Biological synthesis methods are more environmentally friendly and cost-effective compared to traditional physical or chemical methods (Xiao et al., 2020).

Characterizing AgNPs typically involves analyzing their properties, with one of the most straightforward methods being the observation of color changes (Pryshchepa et al., 2020) during synthesis. The degradation of toxic chemicals in aqueous solutions using AgNPs occurs in two main ways: chemical reduction using catalytic agents, and catalytic degradation (Khan et al., 2022) under induced light.

4.2. Green Iron (Fe) Nanoparticles

Green synthesis of zero-valent iron nanoparticles has been achieved using extracts from various plants, including thyme (*Thymus vulgaris*), Damask rose (*Rosa damascena*), and stinging nettle (*Urtica dioica*) (Wen et al., 2022). The plant-extracted iron nanoparticles exhibit antioxidant properties, which have been assessed using methods such as the DPPH radical scavenging assay, the Folin-Ciocalteu method, and the ferric reducing antioxidant power (FRAP) assay (Cui et al., 2023). Phytochemical components, such as polyphenols, flavonoids, and amino acids, play a crucial role in the synthesis of these nanoparticles, contributing to their effective antioxidant activity (Ningthoujam et al., 2023).

The green synthesis method for producing zero-valent iron nanoparticles is significant for treating a variety of pollutants, including brominated organic compounds, pesticides, azo dyes, alkaline-earth metals, malachite green, nitrate, monochlorobenzene, antibiotics, and for converting metals such as chromium, cobalt, and copper (Ningthoujam et al., 2023; Zhang et al., 2022; Du et al., 2022). For example, zero-valent iron nanoparticles synthesized using the aqueous leaf extract of rose apple (*Syzygium jambos* L. Alston) have been effective in removing hexavalent chromium. The efficiency of chromium removal (Pan et al., 2020) is influenced by factors such as nanoparticle dosage, temperature, and pH, with optimal pH and nanomaterial concentration being critical.

Similarly, magnetite iron nanoparticles synthesized from coconut husk extract have shown the ability to absorb small amounts of calcium and cadmium (Wang et al., 2020). The synthesis of zero-valent iron nanoparticles, confirmed by color changes in the solution, involves the use of various leaf extracts at different ratios and temperatures. Phytochemicals present in the extracts act as key agents in reducing and stabilizing the nanoparticles during synthesis (Lakkaboyana et al., 2021). Characterization of these zero-valent iron nanoparticles is typically performed using methods such as SEM, TEM, and zeta potential analysis, which confirm their shape, size, and stability.

Zero-valent iron nanoparticles have proven to be effective in the removal of lead from contaminated environments. The efficiency of this process is influenced by both the duration of exposure and the concentration of the nanoparticles. Specifically, a lower concentration of lead requires less time for removal, as the nanoparticles can rapidly neutralize smaller amounts of the contaminant. Conversely, when dealing with higher concentrations of lead, a greater quantity of nanoparticles is required (Eljamal et al., 2020), and the removal process takes longer due to the increased volume of contaminant that needs to be treated. This relationship highlights the importance of optimizing both nanoparticle dosage and treatment duration to achieve effective lead removal in various environmental settings. Additionally, factors such as pH, temperature, and the presence of other contaminants can further influence the performance of zero-valent iron nanoparticles in lead remediation efforts, making it crucial to tailor the treatment conditions to specific site requirements.

4.3. Green Selenium (Se) Nanoparticles

Green selenium nanoparticles offer an environmentally safe, cost-effective, and non-toxic method for producing these nanoparticles on a large scale (Zohra et al., 2021). The plant-mediated approach to synthesizing selenium nanoparticles is particularly advantageous because it not only avoids the use of harmful chemicals but also utilizes the natural phytochemical components found in plants. These components, including polyphenols, flavones, carboxylic acids, aldehydes, amides, and ketones, play a critical role in reducing metal ions during the synthesis process, leading to the formation of selenium nanoparticles (Zohra et al., 2021; Rajeshkumar et al., 2018; Shahabadi et al., 2021).

Different plant parts, such as leaves, fruits, seeds, fruit peels, and roots, have been successfully used to produce selenium nanoparticles in various shapes and sizes (Pyrzynska & Sentkowska, 2021; Shahbaz et al., 2022). The specific characteristics of the nanoparticles, including their size and shape, are influenced by the type of plant material used and the conditions under which synthesis occurs. These characteristics are crucial because they determine the effectiveness of the nanoparticles in various applications.

For example, selenium nanoparticles synthesized through green methods have been shown to effectively remove heavy metals from contaminated solutions. The ability to remove heavy metals such as zinc, copper, and nickel from soil highlights their potential in environmental remediation. Furthermore, studies have demonstrated that selenium nanoparticles can also be used to remove elemental mercury from both soil and air (Wang et al., 2018; Zohra et al., 2021), making them a versatile tool for addressing different types of environmental contamination. This broad applicability, combined with the sustainability and safety of the green synthesis process, underscores the significance of selenium nanoparticles in environmental and industrial applications.

4.4. Green Copper (Cu) Nanoparticles

Green copper nanoparticles (Cu-NPs) have gained attention for their diverse applications in enhancing water quality. One notable application is their ability to adsorb and remove organic pollutants and suspended solids from water. For example, the CuNPs/DA nanocomposite has proven effective in lowering chemical oxygen demand (COD) and biochemical oxygen demand in fishpond water samples (Hanh et al., 2022). In addition, Cu-NPs possess strong antibacterial properties, making them effective against harmful bacteria like *Aeromonas hydrophila* and *Edwardsiella ictaluri*, which pose a threat to aquatic ecosystems (Hanh et al., 2022). In agriculture, copper is vital for controlling fungal diseases and fostering plant growth (Mohammadi Dehcheshmeh et al., 2024), with copper sulfate being a commonly used derivative.

Moreover, Cu-NPs synthesized through eco-friendly methods, such as using aqueous extracts from the *Diaplazium esulentum* fern, have shown impressive efficiency in removing industrial pollutants like Methyl Violet 6B and Methylene Blue (Sinha et al., 2022), achieving over 92% removal within 150 minutes. These nanoparticles also exhibit potent antibacterial activity against both gram-positive and gram-negative bacteria, along with antifungal and anticancer properties, demonstrating their versatility.

Copper nanoclusters (Cu-NCs), which have core sizes smaller than 2 nm, possess unique fluorescence characteristics, making them suitable for monitoring water quality by detecting various analytes such as Hg^{2+}, NO^{2-}, S^{2-}, and I^-. Despite challenges like low quantum yield and oxidation (Lin et al., 2021), these properties offer significant potential. When combined with other transition metal nanoparticles like silver (Ag) and zinc (Zn), Cu-NPs can significantly enhance disinfection efficiency in drinking water applications, particularly by improving bacterial control and demonstrating synergistic effects with Zn-NPs (Alherek & Basu, 2023).

In membrane technology, Cu-doped graphene oxide (GO) nanohybrids integrated into polyphenylsulfone (PPSU) membranes have shown effectiveness in selectively removing organic dyes such as bromothymol blue, congo red, and methylene blue, achieving high dye rejection rates and demonstrating strong antibacterial activity against various bacterial strains (Shukla et al., 2023). Additionally, copper-based nanomaterials have been extensively researched for their role in desalination, effectively reducing salinity in seawater to produce potable water, addressing global freshwater shortages. The environmental benefits of Cu-NPs are further emphasized by their ability to inhibit the growth of bacterial pathogens like *Pseudomonas aeruginosa* and *Staphylococcus aureus*, as well as their antioxidant and antidiabetic properties, making them valuable in water treatment applications (Shanmugam et al., 2023). These varied applications highlight the significant potential of copper nanoparticles in improving water quality across multiple sectors.

4.5. Green Chitosan Nanoparticles

4.5.1. Role of Chitosan as Adsorbent

In 1859, Professor C. Rouget first synthesized chitosan, a biodegradable polysaccharide derived from the deacetylation of chitin. Chitosan's structure, comprising β-1, 4-linked amino-2-deoxy-D-glucose units, features three functional groups—amine, primary, and secondary hydroxyl groups—that facilitate the adsorption of ionic or polar impurities on its surface. Chitosan is valued for its unique properties, including viscosity, solubility in various solvents, polyelectrolytic nature, ability to form polyoxysalts, mucoadhesive behavior, and capacity to form metal chelates and films. These characteristics make it suitable for a wide range of applications in industries such as food, agriculture, pharmaceuticals, and cosmetics.

Chitosan's primary appeal lies in its biodegradability, biocompatibility, and cost-effectiveness, making it particularly useful for biomedical and bioremediation purposes (Kaur and Dhillon, 2014). Its non-toxic, biocompatible, and biodegradable nature has led to its increased use in fields such as biomedicine, food processing, pesticide production, and water treatment. Chitosan's excellent adsorptive properties enable it to effectively remove hazardous contaminants from wastewater, including metal ions and organic pollutants like dyes, pesticides, and pharmaceutical residues.

Chitosan's structural characteristics, along with its amine and hydroxyl functional groups, offer several advantages over other adsorbents. It is abundant and economical, as it is often derived from natural resources and considered waste. Being non-toxic and easily biodegradable, chitosan is environmentally friendly. Its superior adsorptive capacity allows it to remove a variety of contaminants from wastewater. However, despite its many benefits, chitosan's widespread use as an adsorbent is limited due to drawbacks such as low mechanical strength, reduced adsorption capacity, poor stability in acidic environments, and minimal specificity. To enhance its stability and sensitivity, chitosan can be modified physically by altering its form or chemically through crosslinking and grafting (Olivera et al., 2016).

4.5.2. Role of Chitosan Nanoparticles in Removal of Dyes

Chitosan nanoparticles are characterized by a relatively large surface area and a high adsorption capacity, which makes them highly effective as adsorbent materials for decomposing various dyes found in textile wastewater. These properties are further enhanced when the surface of chitosan nanoparticles is modified with functional groups or grafted with suitable compounds, improving both their adsorption efficiency and selectivity. One notable advantage of magnetic chitosan nanoparticles is their ease of separation using a magnet, which also contributes to better recyclability.

In a study conducted by Tanhaei et al. in 2015, chitosan/Al_2O_3/magnetic iron oxide nanoparticles were synthesized and applied for the removal of methyl orange dye from wastewater. Additionally, the researchers also explored the use of simple magnetic chitosan, which proved to be an effective adsorbent for dye removal, as demonstrated by the experimental results with methyl orange.

Another significant development in this area is the creation of glutaraldehyde-crosslinked magnetic chitosan nanocomposites, which are produced through a reduction precipitation process. These nanocomposites have been applied for the removal of Acid Red 2 dye, achieving an impressive adsorption capacity of 90.06 mg g−1. The synthesis of these nanocomposites is environmentally friendly, as it does not involve the use of hazardous chemicals or extreme heat. The adsorption mechanism is based on electrostatic interactions between the anionic Acid Red 2 dye and the positively charged surface of the magnetic chitosan nanocomposite, resulting from the protonation of the amine groups (Kadam and Lee, 2015).

4.5.3. Role of Chitosan Nanoparticles in Removing Pesticides

Crop productivity faces significant challenges due to pests, weeds, and harmful substances that accumulate in the soil. To address these issues, pesticides are commonly employed. However, the use of pesticides introduces another set of problems, as they can leach into nearby water bodies, leading to rapid algal growth and posing serious risks to aquatic ecosystems. Due to their toxic nature, pesticides (Tudi et al., 2021) must be effectively removed from the environment, and one promising approach involves the use of adsorbents.

Chitosan nanoparticles have emerged as a novel solution for pesticide removal, though this application remains relatively unexplored by researchers. In a study conducted by Dehaghi et al. in 2014, chitosan was combined with zinc oxide nanoparticles to enhance its pesticide removal capabilities. This innovative approach leverages the unique properties of chitosan, such as its high surface area and adsorption capacity, to effectively capture and remove harmful pesticides from the environment. The integration of zinc oxide further improves the efficiency of this process, offering a potentially powerful method for mitigating the environmental impact of pesticides while preserving crop productivity.

4.5.4. Role of Chitosan-Infused Nanomaterials in Removal of Organic Pollutants

Chitosan-infused nanomaterials have shown remarkable effectiveness in eliminating a variety of organic pollutants from the environment. Numerous studies have demonstrated the potential of nanochitosan for the removal of polyaromatic hydrocarbons (PAHs) under controlled conditions (pH 5, 30°C), as reported by Tahvildari and Mojahedi in 2016. These studies highlight the significant promise of chitosan-based nanomaterials in addressing organic pollution.

One notable advancement is the development of magnetic chitosan nanoparticles, which were synthesized using a one-step in situ co-precipitation technique. This process resulted in the formation of quasi-spherical nanoparticles with an average size of 10 nm. These magnetic chitosan nanoparticles have been particularly effective in removing humic acid from aqueous solutions at neutral pH (7) and a temperature of 25°C, as documented by Dong et al. in 2014. The reclaimable nature of these nanoparticles, which allows them to be reused over an extended period, further enhances their practicality and cost-effectiveness.

The adsorption capacity of chitosan-based nanoparticles for organic pollutants has garnered significant attention due to their efficient sorbent properties, economic viability, and environmentally friendly nature. Their biodegradability makes them especially valuable, as they provide a sustainable solution for pollution control while minimizing adverse effects on the environment and its inhabitants. The combination of these attributes positions chitosan-based nanomaterials as a highly promising approach for the long-term management of organic pollutants.

5. ADVANTAGES OF GREEN NANOPARTICLES

5.1. Eco-Friendly and Sustainable

Biomolecule-mediated synthesis is aligned with green chemistry principles, as it avoids the use of toxic chemicals and harsh conditions, making it environmentally friendly (Rathod et al., 2024). The biomolecules used are typically derived from renewable natural sources, further enhancing the sustainability of the process.

5.2. Biocompatibility

The nanoparticles produced through biomolecule-mediated synthesis are highly biocompatible, making them suitable for biomedical applications (Shahabadi et al., 2021) such as drug delivery, biosensing, and medical imaging. The natural origin of the biomolecules reduces the risk of toxicity, making these nanoparticles safer for use in living organisms.

5.3. Stability and Functionality

Biomolecules act as capping agents, stabilizing the nanoparticles and preventing aggregation. This leads to the formation of nanoparticles with controlled size and shape (Rathod et al., 2024). The presence of functional groups in biomolecules can impart additional properties to the nanoparticles, such as enhanced solubility, targeting capabilities, and catalytic activity (Vijayaram et al., 2023).

5.4. Mild Reaction Conditions

The synthesis process occurs under mild conditions, such as ambient temperature and neutral pH, which is beneficial for preserving the structural integrity of both the biomolecules and the nanoparticles (Rathod et al., 2024).

5.5. Versatility

Biomolecule-mediated synthesis is versatile, allowing the production of a wide range of nanoparticles, including metals, metal oxides, and hybrid materials (Somya et al., 2023; Pyrzynska and Sentkowska, 2021) with customizable properties based on the choice of biomolecule.

6. LIMITATIONS OF GREEN NANOPARTICLES

6.1. Complexity and Variability

The complexity of natural biomolecules can lead to variability in the synthesis process, resulting in nanoparticles with inconsistent size, shape, or composition. The exact mechanisms of reduction and stabilization are often not fully understood, which can make it challenging to achieve reproducible results (Rathod et al., 2024).

6.2. Scalability Issues

While the method is promising on a laboratory scale, scaling up the process for industrial production can be challenging due to the need for large quantities of biomolecules and the potential variability in natural sources (Rathod et al., 2024).

6.3. Purity and Yield

The presence of other components in the biomolecule extracts can affect the purity of the synthesized nanoparticles, requiring additional purification steps. The yield of nanoparticles might be lower compared to chemical methods, which can limit the feasibility of this approach for large-scale production (Rathod et al., 2024).

6.4. Storage and Stability

Nanoparticles synthesized using biomolecules may have limited shelf life due to potential degradation of the biomolecular capping agents over time. Maintaining the stability of these nanoparticles during storage can be challenging, particularly in the absence of preservatives or stabilizing agents (Rathod et al., 2024).

6.5. Cost Considerations

The extraction and purification of biomolecules can be costly, and the process may require sophisticated equipment, which can increase the overall cost of nanoparticle production. While biomolecule-mediated synthesis is eco-friendly, it may not always be the most cost-effective option, especially for large-scale industrial applications (Rathod et al., 2024).

7. FUTURE SCOPE

The future scope of green nanoparticles in wastewater treatment is highly promising, with the potential to revolutionize how we manage and purify water resources. As environmental concerns grow and the demand for clean water intensifies, green nanoparticles, synthesized through eco-friendly methods using plant extracts, microorganisms, or other natural agents, offer a sustainable alternative to traditional treatment methods. These nanoparticles can be engineered to target specific contaminants, such as heavy metals, organic pollutants, and pathogens, providing highly efficient removal processes with minimal environmental impact. The scalability of green nanoparticle synthesis, combined with their biocompatibility and low toxicity, positions them as key components in advanced wastewater treatment technologies. Future developments may focus on enhancing the selectivity, stability, and reusability of these nanoparticles, making them more cost-effective and widely applicable. Additionally, integrating green nanoparticles into existing treatment systems could lead to more comprehensive solutions for addressing complex and emerging contaminants, ultimately contributing to the global effort to ensure access to clean and safe water.

8. CONCLUSION

This chapter has provided a comprehensive overview of the role of green nanoparticles in wastewater treatment, highlighting their synthesis, advantages, limitations, and diverse applications. The introduction of green synthesis methods, including plant-mediated, microbial, enzyme-mediated, and biomolecule-mediated approaches, underscores the shift towards more sustainable and environmentally friendly nanoparticle production. These methods leverage natural resources to create nanoparticles that are both effective and less harmful to the environment.

The advantages of green nanoparticles, such as their high efficiency, reduced toxicity, and enhanced stability, make them valuable tools in wastewater treatment. They offer significant improvements over traditional methods by targeting a broad spectrum of contaminants, including heavy metals, organic pollutants, and pathogens, while minimizing environmental impact. However, challenges remain, such as scalability issues, variability in nanoparticle properties, and cost considerations.

The chapter has also explored the specific contributions of metal nanoparticles and chitosan-based materials in wastewater treatment. Metal nanoparticles, like those of silver, gold, and iron, provide powerful contaminant removal capabilities due to their high reactivity and surface area. Chitosan and its derivatives complement these effects with their biocompatibility, biodegradability, and high adsorption capacity, leading to enhanced treatment efficiency and sustainability.

Looking forward, the future of green nanoparticles in wastewater treatment is promising. Continued advancements in synthesis techniques, combined with innovations in material design and functionalization, are likely to lead to even more effective and eco-friendly solutions. The integration of green nanoparticles with existing treatment technologies and the exploration of new applications in environmental and industrial contexts will be crucial for addressing the evolving challenges of water pollution. Embracing these innovations not only offers solutions for cleaner and safer water but also supports broader goals of environmental sustainability and public health.

ACKNOWLEDGEMENTS

The author is grateful and thankful to the Chancellor, Vice-Chancellor, Pro-Vice Chancellor, Director, Amity University, Bengaluru for the above project work. Thanks are also due to Miss. Anushi Varshney and Mr. Anirudh for supporting to compile this work.

REFERENCES

Alherek, M., & Basu, O. D. (2023). Impact of low levels of silver, zinc and copper nanoparticles on bacterial removal and potential synergy in water treatment applications. In Journal of Chemical Technology and Biotechnology (Vol. 98, Issue 5, pp. 1137–1146). Wiley. DOI: 10.1002/jctb.7318

Alshameri, A. W., & Owais, M. (2022). Antibacterial and cytotoxic potency of the plant-mediated synthesis of metallic nanoparticles Ag NPs and ZnO NPs: A review. In *OpenNano* (Vol. 8, p. 100077). Elsevier BV., DOI: 10.1016/j.onano.2022.100077

Chinnappan, S., Kandasamy, S., Arumugam, S., Seralathan, K.-K., Thangaswamy, S., & Muthusamy, G. (2017). Biomimetic synthesis of silver nanoparticles using flower extract of Bauhinia purpurea and its antibacterial activity against clinical pathogens. In Environmental Science and Pollution Research (Vol. 25, Issue 1, pp. 963–969). Springer Science and Business Media LLC. DOI: 10.1007/s11356-017-0841-1

Cui, W., Wang, Y., Sun, Z., Cui, C., Li, H., Luo, K., & Cheng, A. (2023). Effects of steam explosion on phenolic compounds and dietary fiber of grape pomace. In *LWT* (Vol. 173, p. 114350). Elsevier BV., DOI: 10.1016/j.lwt.2022.114350

Dong, C., Chen, W., Liu, C., Liu, Y., & Liu, H. (2014). Synthesis of Magnetic Chitosan Nanoparticle and its Adsorption Property for Humic Acid from Aqueous Solution. *Colloids and Surfaces. A, Physicochemical and Engineering Aspects*, 446, 179–189. DOI: 10.1016/j.colsurfa.2014.01.069

Du, C., Xu, N., Yao, Z., Bai, X., Gao, Y., Peng, L., Gu, B., & Zhao, J. (2022). Mechanistic insights into sulfate and phosphate-mediated hexavalent chromium removal by tea polyphenols wrapped nano-zero-valent iron. In *Science of The Total Environment* (Vol. 850, p. 157996). Elsevier BV., DOI: 10.1016/j.scitotenv.2022.157996

Eljamal, O., Thompson, I. P., Maamoun, I., Shubair, T., Eljamal, K., Lueangwattanapong, K., & Sugihara, Y. (2020). Investigating the design parameters for a permeable reactive barrier consisting of nanoscale zero-valent iron and bimetallic iron/copper for phosphate removal. In *Journal of Molecular Liquids* (Vol. 299, p. 112144). Elsevier BV., DOI: 10.1016/j.molliq.2019.112144

Ghotekar, S., Pansambal, S., Pawar, S. P., Pagar, T., Oza, R., & Bangale, S. (2019). Biological activities of biogenically synthesized fluorescent silver nanoparticles using Acanthospermum hispidum leaves extract. In SN Applied Sciences (Vol. 1, Issue 11). Springer Science and Business Media LLC. https://doi.org/DOI: 10.1007/s42452-019-1389-0

Gokul Eswaran, S., Shahid Afridi, P., & Vasimalai, N. (2022). Effective Multi Toxic Dyes Degradation Using Bio-Fabricated Silver Nanoparticles as a Green Catalyst. In Applied Biochemistry and Biotechnology (Vol. 195, Issue 6, pp. 3872–3887). Springer Science and Business Media LLC. DOI: 10.1007/s12010-022-03902-y

Goutam, S. P., Saxena, G., Singh, V., Yadav, A. K., Bharagava, R. N., & Thapa, K. B. (2018). Green synthesis of TiO2 nanoparticles using leaf extract of Jatropha curcas L. for photocatalytic degradation of tannery wastewater. In *Chemical Engineering Journal* (Vol. 336, pp. 386–396). Elsevier BV., DOI: 10.1016/j.cej.2017.12.029

Hanh, T. T., & Chi, L. N. T., & Duy, N. N. (2022). Preparation of Copper Nanoparticles/Diatomite Nanocomposite for Improvement in Water Quality of Fishponds. In S. Hussain (Ed.), Journal of Chemistry (Vol. 2022, pp. 1–9). Hindawi Limited. DOI: 10.1155/2022/3921631

Iravani, S., Korbekandi, H., Mirmohammadi, S., & Zolfaghari, B. (2014). Synthesis of silver nanoparticles: Chemical, physical and biological methods. *Research in Pharmaceutical Sciences*, 9, 385–406. PMID: 26339255

Johri, N., Jacquillet, G., & Unwin, R. (2010). Heavy metal poisoning: The effects of cadmium on the kidney. *Biometals*, 23, 783–792.

Kadam, A. A., & Lee, D. S. (2015). Glutaraldehyde Cross-Linked Magnetic Chitosan Nanocomposites: Reduction Precipitation Synthesis, Characterization, and Application for Removal of Hazardous Textile Dyes. *Bioresource Technology*, 193, 563–567. DOI: 10.1016/j.biortech.2015.06.148 PMID: 26166462

Kaur, S., & Dhillon, G. S. (2014). The Versatile Biopolymer Chitosan: Potential Sources, Evaluation of Extraction Methods and Applications. *Critical Reviews in Microbiology*, 40(2), 155–175. DOI: 10.3109/1040841X.2013.770385 PMID: 23488873

Khan, S. A., Jain, M., Pandey, A., Pant, K. K., Ziora, Z. M., Blaskovich, M. A. T., Shetti, N. P., & Aminabhavi, T. M. (2022). Leveraging the potential of silver nanoparticles-based materials towards sustainable water treatment. In *Journal of Environmental Management* (Vol. 319, p. 115675). Elsevier BV., DOI: 10.1016/j.jenvman.2022.115675

Lakkaboyana, S. K., Khantong, S., Asmel, N. K., Obaidullah, S., Kumar, V., Kannan, K., Venkateswarlu, K., Yuzir, A., & Wan Yaacob, W. Z. (2021). Indonesian Kaolin supported nZVI (IK-nZVI) used for the an efficient removal of Pb(II) from aqueous solutions: Kinetics, thermodynamics and mechanism. In Journal of Environmental Chemical Engineering (Vol. 9, Issue 6, p. 106483). Elsevier BV. DOI: 10.1016/j.jece.2021.106483

Lakshmanan, G., Sathiyaseelan, A., Kalaichelvan, P. T., & Murugesan, K. (2018). Plant-mediated synthesis of silver nanoparticles using fruit extract of Cleome viscosa L.: Assessment of their antibacterial and anticancer activity. In Karbala International Journal of Modern Science (Vol. 4, Issue 1, pp. 61–68). University of Kerbala. DOI: 10.1016/j.kijoms.2017.10.007

Lin, Y.-S., Lin, Y.-F., Nain, A., Huang, Y.-F., & Chang, H.-T. (2021). A critical review of copper nanoclusters for monitoring of water quality. In *Sensors and Actuators Reports* (Vol. 3, p. 100026). Elsevier BV., DOI: 10.1016/j.snr.2021.100026

Mohammadi Dehcheshmeh, I., Poursattar Marjani, A., Sadegh, F., & Ebrahim Soltani, M. (2024). Copper Application and Copper Nanoparticles in Chemistry. In Various Uses of Copper Material [Working Title]. IntechOpen. DOI: 10.5772/intechopen.1004068

Moradi Dehaghi, S., Rahmanifar, B., Moradi, A. M., & Azar, P. A. (2014). Removal of Permethrin Pesticide from Water by Chitosan-Zinc Oxide Nanoparticles Composite as an Adsorbent. *Journal of Saudi Chemical Society*, 18(4), 348–355. DOI: 10.1016/j.jscs.2014.01.004

Nakum, J., & Bhattacharya, D. (2022). Various Green Nanomaterials Used for Wastewater and Soil Treatment: A Mini-Review. In Frontiers in Environmental Science (Vol. 9). Frontiers Media SA. DOI: 10.3389/fenvs.2021.724814

Ningthoujam, R., Sahoo, B., Ghosh, P., Shivani, A., Ganguli, P., & Chaudhuri, S. (2023). Green production of zero-valent iron nanoparticles using pomegranate peel extracts and its use in lindane degradation. In Nanotechnology for Environmental Engineering (Vol. 8, Issue 2, pp. 581–589). Springer Science and Business Media LLC. DOI: 10.1007/s41204-023-00313-0

Olivera, S., Muralidhara, H. B., Venkatesh, K., Guna, V. K., Gopalakrishna, K., & Kumar, K. Y. (2016). Potential Applications of Cellulose and Chitosan Nanoparticles/composites in Wastewater Treatment: A Review. *Carbohydrate Polymers*, 153, 600–618. DOI: 10.1016/j.carbpol.2016.08.017 PMID: 27561533

Pan, Z., Lin, Y., Sarkar, B., Owens, G., & Chen, Z. (2020). Green synthesis of iron nanoparticles using red peanut skin extract: Synthesis mechanism, characterization and effect of conditions on chromium removal. In *Journal of Colloid and Interface Science* (Vol. 558, pp. 106–114). Elsevier BV., DOI: 10.1016/j.jcis.2019.09.106

Pryshchepa, O., Pomastowski, P., & Buszewski, B. (2020). Silver nanoparticles: Synthesis, investigation techniques, and properties. In *Advances in Colloid and Interface Science* (Vol. 284, p. 102246). Elsevier BV., DOI: 10.1016/j.cis.2020.102246

Pyrzynska, K., & Sentkowska, A. (2021). Biosynthesis of selenium nanoparticles using plant extracts. In Journal of Nanostructure in Chemistry (Vol. 12, Issue 4, pp. 467–480). Springer Science and Business Media LLC. DOI: 10.1007/s40097-021-00435-4

Qasem, N. A. A., Mohammed, R. H., & Lawal, D. U. (2021). Removal of heavy metal ions from wastewater: a comprehensive and critical review. In npj Clean Water (Vol. 4, Issue 1). Springer Science and Business Media LLC. DOI: 10.1038/s41545-021-00127-0

Rajeshkumar, S., Veena, P., & Santhiyaa, R. V. (2018). Synthesis and Characterization of Selenium Nanoparticles Using Natural Resources and Its Applications. In Nanotechnology in the Life Sciences (pp. 63–79). Springer International Publishing. DOI: 10.1007/978-3-319-99570-0_4

Rathod, S., Preetam, S., Pandey, C., & Bera, S. P. (2024). Exploring synthesis and applications of green nanoparticles and the role of nanotechnology in wastewater treatment. In *Biotechnology Reports* (Vol. 41, p. e00830). Elsevier BV., DOI: 10.1016/j.btre.2024.e00830

Sánchez-Hernández, E., González-García, V., Correa-Guimarães, A., Casanova-Gascón, J., Martín-Gil, J., & Martín-Ramos, P. (2023). Phytochemical Profile and Activity against Fusarium Species of Tamarix gallica Bark Aqueous Ammonia Extract. In Agronomy (Vol. 13, Issue 2, p. 496). MDPI AG. DOI: 10.3390/agronomy13020496

Schaumlöffel, D. (2012). Nickel species: Analysis and toxic effects. In Journal of Trace Elements in Medicine and Biology (Vol. 26, Issue 1, pp. 1–6). Elsevier BV. DOI: 10.1016/j.jtemb.2012.01.002

Shahabadi, N., Zendehcheshm, S., & Khademi, F. (2021). Selenium nanoparticles: Synthesis, in-vitro cytotoxicity, antioxidant activity and interaction studies with ct-DNA and HSA, HHb and Cyt c serum proteins. In *Biotechnology Reports* (Vol. 30, p. e00615). Elsevier BV., DOI: 10.1016/j.btre.2021.e00615

Shahbaz, M., Akram, A., Raja, N. I., Mukhtar, T., Mashwani, Z. U. R., Mehak, A., Fatima, N., Sarwar, S., Haq, E. U., & Yousaf, T. (2022). Green synthesis and characterization of selenium nanoparticles and its application in plant disease management: A review. In Pakistan Journal of Phytopathology (Vol. 34, Issue 1, pp. 189–102). Pakistan Phytopathological Society. DOI: 10.33866/phytopathol.034.01.0739

Shaik, M., Khan, M., Kuniyil, M., Al-Warthan, A., Alkhathlan, H., Siddiqui, M., Shaik, J., Ahamed, A., Mahmood, A., Khan, M., & Adil, S. (2018). Plant-Extract-Assisted Green Synthesis of Silver Nanoparticles Using Origanum vulgare L. Extract and Their Microbicidal Activities. In Sustainability (Vol. 10, Issue 4, p. 913). MDPI AG. DOI: 10.3390/su10040913

Shanmugam, K., Subha, V., & Thulasimuthu, E. (2023). Environmental Aspects of Copper Nanoparticles Against Green Algae Water and Bactericidal Action From Ethanolic Root Extract of Asparagus Racemosus. MDPI AG. DOI: 10.20944/preprints202306.1501.v1

Shanwaz, M. M., & Shyam, P. (2022). Synthesis of Silver Nanoparticles from Vitex negundo Plant by Green Method and their Bactericidal Effects. In Letters in Applied NanoBioScience (Vol. 12, Issue 2, p. 59). AMG Transcend Association. DOI: 10.33263/LIANBS122.059

Shukla, A. K., Alam, J., Mishra, U., & Alhoshan, M. (2023). Effect of Cu-doped GO nanoparticles on polyphenylsulfone nanocomposite membrane surface and its application for the removal of organic pollutants and antibacterial analysis. In *Materials Today Communications* (Vol. 35, p. 106326). Elsevier BV., DOI: 10.1016/j.mtcomm.2023.106326

Sinha, T., Adhikari, P. P., & Bhandari, V. M. (2022). Sustainable Fabrication of Copper Nanoparticles: A Potent and Affordable Candidate for Water Treatment, Water Disinfection, Antioxidant Activity and Theranostic Agent. In ChemistrySelect (Vol. 7, Issue 15). Wiley. DOI: 10.1002/slct.202103552

Somya, A. (2019). Surfactant based hybrid ion exchangers. *Research Journal of Chemistry and Environment*, 23(3), 96.

Somya, A. (2020). Hybrid Ion Exchangers. In Hybrid Nanomaterials - Flexible Electronics Materials. IntechOpen. DOI: 10.5772/intechopen.92116

Somya, A. (2023). Metal Phosphates: Their Role as Ion Exchangers in Water Purification. In Metal Phosphates and Phosphonates (pp. 341–356). Springer International Publishing. DOI: 10.1007/978-3-031-27062-8_19

Somya, A. (2024). Hybrid cation exchangers: A boon for analytical chemistry. *Journal of Applied Optics*, 263–270. Retrieved from https://appliedopticsjournal.net/index.php/JAO/article/view/111

Somya, A. (2024). An investigation study on water quality of few lakes of Bengaluru. In *Industry 4.0 with Modern Technology* (pp. 200–204). CRC Press., DOI: 10.1201/9781003450924-37

Somya, A., Khan, A., Hussein, M. A., Al-Romaizan, A. N., Malash, A. M., Topare, N. S., & Asiri, A. M. (2023). Biological and environmental studies on pectin-zirconium (IV) phosphate, a biopolymer doped hybrid cation exchanger. In *International Journal of Biological Macromolecules* (Vol. 250, p. 125803). Elsevier BV., DOI: 10.1016/j.ijbiomac.2023.125803

Somya, A., & Maheri, A. K. (2020). Studies on polyoxyethylene octyl phenyl ether supported thorium(IV) phosphates: A new cation exchange material. *Bulletin of Materials Science*, 43(1), 163. DOI: 10.1007/s12034-020-02153-z

Somya, A., Rafiquee, M. Z. A., & Upadhyaya, V. (2018). Role of surfactants in the adsorption studies of n-butyl acetate-cerium (IV) phosphate for some alkaline earths and heavy metal ions. *Research Journal of Chemistry and Environment*, 22(6), 45.

Somya, A., Kumar, S., & Swamynathan, K. (2024). An exploration of AFLC-14-doped ceric phosphate, a novel liquid crystal hybrid ion exchanger: synthesis, ion exchange studies and analytical insights. In *Materials Research Innovations* (pp. 1–9). Informa UK Limited., DOI: 10.1080/14328917.2024.2396655

Somya, A., Rafiquee, M. Z. A., Varshney, K. G., & Al-Lohedan, H. A. (2014). Synthesis, characterization and analytical applications of N-dodecyl pyridinium chloride-cerium (IV) phosphate fibrous ion exchanger: selective for Hg(II) and its binary separations. In Desalination and Water Treatment (Vol. 56, Issue 3, pp. 638–645). Elsevier BV. DOI: 10.1080/19443994.2014.940388

Somya, A., & Singh, M. (2022). Studies on Sodium Lauryl Sulphate-supported Thorium (IV) Phosphate: A New Surfactant-supported Cation Exchange Resin, Useful in Water Purification. *Current Analytical Chemistry, 18*(3), 383–390). https://doi.org/DOI: 10.2174/1573411017666210804102748

Somya, A., & Upadhayaya, V. (2020). Synthesis and ion exchange characterization of n-butyl acetate based tin (IV) phosphate: A new intercalated hybrid ion exchange material. *Research Journal of Chemistry and Environment*, 24(7), 107.

Somya, A., & Upadhyaya, V. (2017). Synthesis, Characterization and Adsorption behaviour of N-dodecyl pyridinium chloride based Tin (IV) phosphate: A new surfactant based hybrid ion exchanger. *Research Journal of Chemistry and Environment*, 21(1), 18.

Tahvildari, K., & Mojahedi, H. (2016). Trends. *Life Sciences*, 5, 2319.

Tanhaei, B., Ayati, A., Lahtinen, M., & Sillanpää, M. (2015). Preparation and Characterization of a Novel chitosan/Al2O3/magnetite Nanoparticles Composite Adsorbent for Kinetic, Thermodynamic and Isotherm Studies of Methyl Orange Adsorption. *Chemical Engineering Journal*, 259, 1–10. DOI: 10.1016/j.cej.2014.07.109

Tarannum, N., Divya, D., & Gautam, Y. K. (2019). Facile green synthesis and applications of silver nanoparticles: a state-of-the-art review. In RSC Advances (Vol. 9, Issue 60, pp. 34926–34948). Royal Society of Chemistry (RSC). https://doi.org/DOI: 10.1039/C9RA04164H

Tudi, M., Daniel Ruan, H., Wang, L., Lyu, J., Sadler, R., Connell, D., Chu, C., & Phung, D. T. (2021). Agriculture Development, Pesticide Application and Its Impact on the Environment. In International Journal of Environmental Research and Public Health (Vol. 18, Issue 3, p. 1112). MDPI AG. DOI: 10.3390/ijerph18031112

Varshney, K. G., Rafiquee, M. Z. A., & Somya, A. (2007). Effect of surfactants on the adsorption behaviour of cerium (IV) phosphate, cation exchanger for alkaline earths and heavy metal ions. In Colloids and Surfaces A: Physicochemical and Engineering Aspects (Vol. 301, Issues 1–3, pp. 69–72). Elsevier BV. DOI: 10.1016/j.colsurfa.2006.12.025

Varshney, K. G., Rafiquee, M. Z. A., & Somya, A. (2007). Effect of surfactants on the adsorption behaviour of tin(IV) phosphate, cation exchanger for alkaline earths and heavy metal ions. In Colloids and Surfaces A: Physicochemical and Engineering Aspects (Vol. 301, Issues 1–3, pp. 224–228). Elsevier BV. DOI: 10.1016/j.colsurfa.2006.12.061

Varshney, K. G., Rafiquee, M. Z. A., & Somya, A. (2008). Triton X-100 based cerium(IV) phosphate as a new Hg(II) selective, surfactant based fibrous ion exchanger: Synthesis, characterization and adsorption behaviour. In Colloids and Surfaces A: Physicochemical and Engineering Aspects (Vol. 317, Issues 1–3, pp. 400–405). Elsevier BV. DOI: 10.1016/j.colsurfa.2007.11.012

Vijayaram, S., Razafindralambo, H., Sun, Y.-Z., Vasantharaj, S., Ghafarifarsani, H., Hoseinifar, S. H., & Raeeszadeh, M. (2023). Applications of Green Synthesized Metal Nanoparticles — a Review. In Biological Trace Element Research (Vol. 202, Issue 1, pp. 360–386). Springer Science and Business Media LLC. DOI: 10.1007/s12011-023-03645-9

Wang, H., Wang, Z., Yue, R., Gao, F., Ren, R., Wei, J., Wang, X., & Kong, Z. (2020). Functional group-rich hyperbranched magnetic material for simultaneous efficient removal of heavy metal ions from aqueous solution. In *Journal of Hazardous Materials* (Vol. 384, p. 121288). Elsevier BV., DOI: 10.1016/j.jhazmat.2019.121288

Wang, X., Zhang, D., Qian, H., Liang, Y., Pan, X., & Gadd, G. M. (2018). Interactions between biogenic selenium nanoparticles and goethite colloids and consequence for remediation of elemental mercury contaminated groundwater. In *Science of The Total Environment* (Vol. 613–614, pp. 672–678). Elsevier BV., DOI: 10.1016/j.scitotenv.2017.09.113

Wen, J., Fu, W., Ding, S., Zhang, Y., & Wang, W. (2022). Pyrogallic acid modified nanoscale zero-valent iron efficiently removed Cr(VI) by improving adsorption and electron selectivity. In Chemical Engineering Journal (Vol. 443, p. 136510). Elsevier BV. DOI: 10.1016/j.cej.2022.136510

Xiao, C., Li, H., Zhao, Y., Zhang, X., & Wang, X. (2020). Green synthesis of iron nanoparticle by tea extract (polyphenols) and its selective removal of cationic dyes. In *Journal of Environmental Management* (Vol. 275, p. 111262). Elsevier BV., DOI: 10.1016/j.jenvman.2020.111262

Zhang, J., Niu, Y., Zhou, Y., Ju, S., & Gu, Y. (2022). Green preparation of nano-zero-valent iron-copper bimetals for nitrate removal: Characterization, reduction reaction pathway, and mechanisms. In Advanced Powder Technology (Vol. 33, Issue 11, p. 103807). Elsevier BV. DOI: 10.1016/j.apt.2022.103807

Zohra, E., Ikram, M., Omar, A. A., Hussain, M., Satti, S. H., Raja, N. I., Mashwani, Z.-U.-R., & Ehsan, M. (2021). Potential applications of biogenic selenium nanoparticles in alleviating biotic and abiotic stresses in plants: A comprehensive insight on the mechanistic approach and future perspectives. In Green Processing and Synthesis (Vol. 10, Issue 1, pp. 456–475). Walter de Gruyter GmbH. DOI: 10.1515/gps-2021-0047

Zohra, E., Ikram, M., Omar, A. A., Hussain, M., Satti, S. H., Raja, N. I., Mashwani, Z.-U.-R., & Ehsan, M. (2021). Potential applications of biogenic selenium nanoparticles in alleviating biotic and abiotic stresses in plants: A comprehensive insight on the mechanistic approach and future perspectives. In Green Processing and Synthesis (Vol. 10, Issue 1, pp. 456–475). Walter de Gruyter GmbH. https://doi.org/DOI: 10.1515/gps-2021-0047

Chapter 5
Biopolymeric Nanoparticles: A Novel Ingredient for Application in the Food Industry

Rimpi Foujdar
https://orcid.org/0009-0003-3730-5115
Haldia Institute of Technology, India

Swati Ray
Haldia Institute of Technology, India

ABSTRACT

Polymer-based nano-delivery systems do not conform to the toxicity standards and efficacy needed to act inside cells. Biopolymeric nanoparticles have biodegradability, compatibility with biological systems, and negligible toxicity, creating an edge over toxic polymers or metal nanoparticles. Besides being nanoparticles, they are naturally antibacterial and eco-friendly. Food-based matrices such as proteins and polysaccharides and complexes such as protein-polysaccharide complex and lipid-biopolymer hybrids are typically used to manufacture these nanoparticles. Manufacturing methods like nanoprecipitation, ionic gelation, microemulsion, and coacervation are discussed in this context. They are proven effective as food additives, vehicles for nutrient retention, packaging matrices, and functional food formulations. For wider application of this tool, testing in different systems to develop safe, effective, and degradable nano-delivery vehicles must be undertaken.

1. INTRODUCTION

Technological innovation is one of the greatest tools for advancement in the field of nanoscience and technology. Nanotechnology is a highly promising area that has gained huge acceptance among peer scientists since it offers a wide range of applications addressing almost all fields of science (Sahoo *et al.*, 2021). The nanoparticles with excellent catalytic, electronic, magnetic, and optical attributes can be reinforced into different matrices and employed to perform a wide range of functions (Onyancha *et al.*, 2021; Welch *et al.*, 2021; Basu *et al.*, 2018). Biopolymer-based protein and polysaccharide matrices are biodegradable, non-toxic, and compatible with any biological entity (Sundar *et al.*, 2010). Due to their natural origin, biocompatibility, and eco-friendly characteristics, they have successfully captivated the

DOI: 10.4018/979-8-3693-6240-2.ch005

market as a biomedicine-based application tool in drug delivery or antimicrobial agents (Vodyashkin *et al.*, 2021). It also holds the potential to be used in various food applications. (Kaur *et al.*, 2023) pointed out the conjugation capabilities of food additives like food color, flavors, anticaking agents, etc. with edible biopolymers to enhance their gradual release and stability. These biopolymers form a degradable protein film constructed using alginate, cellulose, and chitosan which preserves the food products making them shelf-stable. Biopolymers find application as a food packaging material that has improved barrier properties, gas, moisture, and organic solvent permeability, temperature resistivity, and preservation capacity as compared to normal polymer-based packaging materials. The petroleum-based packaging materials are equally harmful to the consumer and environment which can be replaced by these bio-based polymers and their incorporation into packaging as nanoparticles (Basavegowda and Baek, 2020). Nano-encapsulation using protein or polysaccharide-based nanoparticles for nutritional retention and vehicle for food delivery reduces the non-compatibility factor of polymer-based carriers in biological systems. However, the interaction study of these proteins and polysaccharides in the living system needs thorough scientific evaluation (Sadeghi *et al.*, 2016). Some of the recent advancements in developing functional foods have been reported. (Foujdar *et al.*, 2023) reviews all possibilities of utilizing nanoparticles in the formulation of functional foods. An oat-based functional beverage was developed in two different flavors chocolate and vanilla which was admixed with iron nanoparticles (< 300 nm) for improvement of toddler health. It was found that this formulation has improved the level of reactivity and absorptivity of iron compared to normal health drinks (Kirdar, 2015).

Conventionally, they can be fabricated through emulsification, desolvation, coaceveration, or electrospray drying. More advanced techniques are being implemented for maximum utilization of this mechanism covering a large spectrum of associated fields (Sundar *et al.*, 2010).

1.1 Evolution of Biopolymeric Nanoparticles

Taniguchi of Japan first coined the term 'Nanotechnology' when he described particles of nanometer size which were found to be advantageous over the use of microspheres (Taniguchi, 1983; Gao and Xu, 2009). The application of these nanoparticles principally gained an appreciation for use in non-biological systems since there was an issue of stability in storage or interaction with bio-fluids. This limitation led to the foundation of bio-based nanoparticles which proved itself as an alternative to liposome technology in drug delivery systems (Oppenheim, 1981). It improves the effectiveness of drug targeting by fitting itself into the drug delivery colloids and intensifies its uptake in the cell without any toxic side effects of the free drug. For holistic treatment, biopolymeric nanoparticles can be easily circulated into different sites of the body (Sundar *et al.*, 2010). These nanoparticles can be made of natural biomolecules or synthetic polymers. The factors that determine the choice of matrix are nanoparticle size, conformation, charge, permeability, biodegradability, biocompatibility, cytotoxicity, desirable drug loading and release capacity, etc. (Kreuter, 1995).

1.2 Types of Biopolymers

The huge environmental impact that plastic waste generates is primarily due to a lack of skill to manage hazardous waste, infrastructural inadequacy for recovery or recycling, absence of awareness among people, and shortage of alternatives available (Kibria *et al.*, 2023). Biopolymers can be an option that is naturally biodegradable and when embedded with nanoparticles can be used in the packaging and delivery of drugs in biotic systems. Depending on the chemical structure, a wide variety of biopolymers can be manufactured utilizing constituents of living organisms. Utilizing advanced technologies, biopolymers can also be synthesized through fermentation of an enzymatic process (Hassan *et al.*, 2019).

Based on the source of their synthesis, biopolymers can be classified into two major categories with some sub-categories that can be illustrated in Figure 1.

Figure 1. Two major biopolymer categories based on their origin

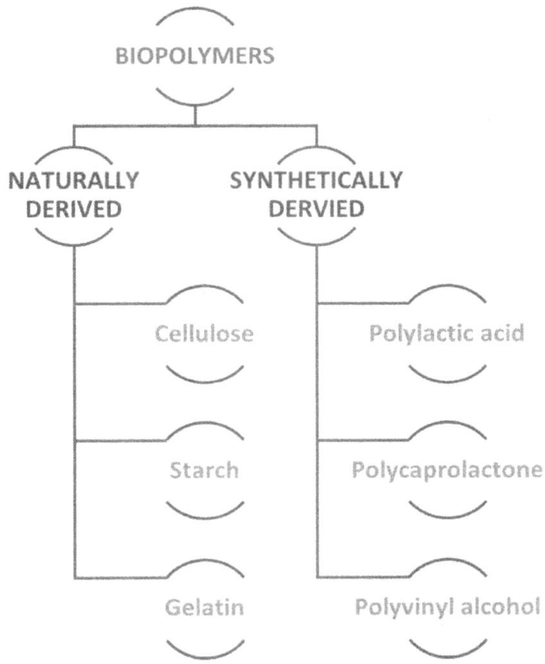

Among the naturally derived materials carbohydrates, lipids, non-polymeric giant lipids, macrocycles, proteins, nucleic acid, etc. are frequently used (Stryer *et al.*, 2002). The polysaccharides are mainly obtained from cellulose and starch; whereas the proteins are sourced from collagen, gelatin, wheat gluten, and casein. Some combination molecules of saccharides, nucleic acid, and amino acids in repeating units with chemically derived side chains are also employed. Microorganisms and GMO-derived molecules like polyhydroxyalkanoates and polylactic acid can be obtained by traditional extraction process (Baranwal *et al.*, 2022). Polysaccharide-based molecules have a specific charge which is the key factor for compatibility with the drug delivery system. The classification with examples can be shown in Figure 2.

Figure 2. Classifications of polysaccharide-based biopolymers based on their charge (Baranwal et al., 2022)

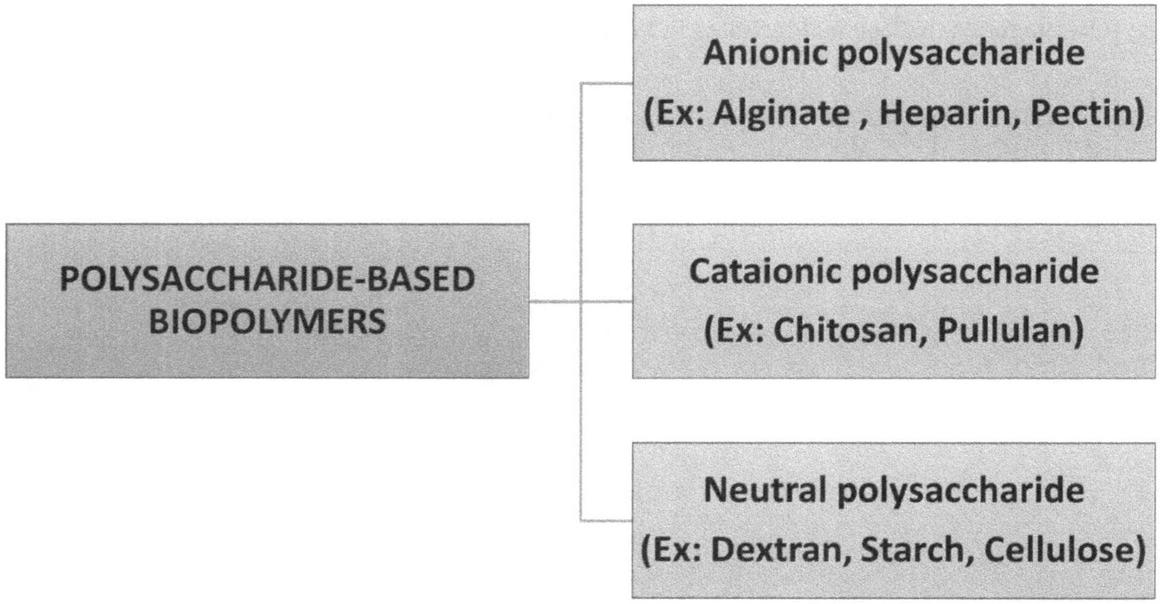

The major plant sources for base material production are obtained from grains (Karmanov *et al.*, 2021), roots (Borah *et al.*, 2017), starchy vegetables (Perotti *et al.*, 2013), fruits (Redondo-Gómez *et al.*, 2020), etc. Cash crops like cotton (Aguilar *et al.*, 2019) oils from oil seeds, and vegetable oils (Sharma and Kundu, 2006) are also used for their synthesis. Cattle are the main animal source and marine animals such as fish, coral sponges, lobster, prawn, and shrimps are the origin of protein-based products. Many microbes such as bacteria, algae, and fungi produce polyhydroxyalkanoates as a secondary metabolite which accumulates in the body of these microbes as transparent granules (Muhamad *et al.*, 2017). This component behaves like a thermoplastic from which biopolymers can be manufactured (Mojaveryazdi *et al*, 2014). Lignin from agricultural wastes, waste papers, and green and woody wastes can also be valorized to obtain biosynthetic polymers (Flaris and Singh, 2009; Hatakeyama and Hatakeyama, 2009).

1.3 Significance of Biopolymeric Nanoparticles and Regulatory Aspect

The conventional use of chemical synthetic polymers for food and pharmaceutical applications can have severe environmental impact and concerning human health. The use of biopolymers for entrapping nanoparticles reduces carbon footprint, helps to manage municipal solid wastes, and minimizes dependency on fossil fuel-based resources (Baranwal *et al.*, 2022). The natural characteristics of these matrices such as biodegradability, biocompatibility, non-immunogenicity, and non-toxicity make them an obvious choice as an alternative to synthetic carriers (Yetisgin *et al.*, 2020). The components such as drugs, antimicrobials, antibiotics, bioactive components, natural extracts, or essential oils (Kučuk *et al.*, 2022) can be incorporated into the biopolymeric nanoparticles either by surface absorption or by incorporation during the manufacturing process. Factors like size, molecular weight, component solubility, and mutual interaction of nanoparticles and the component determine the efficiency of loading in nanoparticles. The components are released from the nanoparticles by desorption, diffusion, or erosion process (Harish *et al.*, 2022; Lu *et al.*, 2011). Nanoparticle incorporation helps in better absorption of the component, protection from degradation, and extends circulation time and targeted delivery due to good solubility and delayed-release (Guo *et al.*, 2019). Biopolymers are macromolecules of covalently bonded repeating units that are considered generally recognized as safe (GRAS). Besides being a carrier, it also finds its application in tissue engineering and wound healing with reduced reported side effects. They are recyclable, cost-effective, and stable for a long time in biological fluids in storage conditions (Muir and Burdick, 2021; Biswas *et al.*, 2022). The use of conjugate complexes and composites in biopolymeric nanoparticle manufacturing has been reported to enhance its existing characteristics with added superior mechanical properties (Bealer *et al.*, 2020).

Though biopolymeric nanoparticles are used successfully in different applications in the health and food sector, persisting health effects and safety concerns regarding their use fall under regulatory issues (Larsson *et al.*, 2019). A proper risk management system is necessary to handle this since there is very limited clinical study reported confirming their safety. However, some manufacturers developing such products have already invested in thorough pre-clinical, clinical, and phase 1 studies which are helping in developing a pool of data supporting their safety (Miao *et al.*, 2018). The silver lining is the approval from the Food and Drug Administration (FDA) for a few biopolymeric nanoparticle-based formulations, especially for cancer treatment agents (Hong *et al.*, 2020). One such FDA-approved commercial anticancer formulation is Abraxane. The hydrophobic drug paclitaxel has been incorporated into an albumin-based biopolymeric nanoparticle which is now used for the treatment of advanced-stage metastatic breast, pancreatic, and lung cancer. These 135 nm nanoparticles are taken as intra-cutaneous injection and are proven to eliminate cancer cells more effectively than their micellar counterpart (Gawde *et al.*, 2018; Yuan *et al.*, 2020). The potential of these products is enormous, but since the end consumer is concerned with human health, more serious and effective scientific investigation is the need of the hour to overcome the barriers related to its safety and stability in biological systems.

This chapter provides detailed information about the evolution of biopolymeric nanoparticles as a carrier in food systems utilizing different biopolymers as matrices. It elaborates on the methodologies for the development of such nanoparticles embedded in biopolymers which can be applied in various health sectors related to food and pharmaceutical industries. It also discusses the toxicological assessment techniques for effective evaluation of their safety and health concerns.

2. EXTRACTION OF BIOPOLYMERS

The main source of biopolymers is biological-based products such as microbes, plants, and animals. Natural polysaccharides such as chitosan, pectin, starch, and cellulose are widely used to produce biocompatible nanoparticles. Similarly, protein is another extensively used biopolymer in the nanoformulation for the food and pharmaceutical industry. The extraction of these biopolymers mainly depends on the type of biopolymers and its sources. Chitosan can form strong fibers and is considered one of the widely used natural biopolymers as a coating agent in food preservation (Senebirathna, 2022). Crustacean shells and the cell walls of fungi are the main sources of chitosan. Commercially, it is extracted from waste byproducts of marine crustaceans such as krill, shrimp, and crab shells as well as from fish scales

(Aranaz *et al,* .2009). The extraction of chitosan is mainly done in 3 steps (Kumari and Rath, 2014). The initial step is demineralization which is carried out using acid treatment. Mostly diluted hydrochloric acid is used. In demineralization, the calcium carbonate is decomposed in water-soluble calcium salts. De-proteinization or removal of protein is the second step which is followed by de-colorization or pigment removal (Kumari and Rath, 2014). Chemical or enzymatic treatment is used to perform both steps. Starch is another widely used biopolymer in the food industry and can be extracted by using two methods: dry and wet extraction methods. Based on the findings of Hernandez-Carmona *et al.* (2017), it has been revealed that the extraction of starch from green plantain peels is one way to minimize the agro-waste into value-added product. Wet grinding methods are used to extract the starch from the peel. Another biopolymer, protein is known for its unique techno-functional properties which contribute to the appearance, texture, and stability of food materials (Chakravarti *et al.*, 2018). Extraction of protein can be effectively done by using various non-thermal techniques that include chemical extraction, aqueous two-phase system, sub-critical water extraction (SWE), enzymatic-assisted extraction (EAE); novel assisting cell disruption techniques such as microwave-assisted extraction, ultrasonic-assisted extraction, pulsed electric field, and high voltage electric field (Franca-Oliveira *et al.*, 2021). Based on the review of research papers, it has been concluded that a high-quality protein with a high extraction yield can be obtained by using these modern extraction techniques.

3. FABRICATION METHOD FOR SYNTHESIS OF BIOPOLYMERIC NANOPARTICLES

Nanoparticle fabrication can be undertaken by several methods. More advanced technologies are being developed each day. Some of the main methods are discussed here-

3.1 Coacervation

The coacervation method is a natural and cost-effective process; widely used for the fabrication of biopolymeric nanoparticles. It is a method in which phase separation occurs by the variation in environmental factors (such as solubility, temperature, pH, and ionic strength) under properly maintained conditions. In this system, one phase contains a high concentration of colloid, known as the coacervate phase, and the other phase where a very minute quantity of colloid is present called as equilibrium phase as shown in Figure 3 (Nairm, 1995). Depending upon the types of biopolymeric systems involved in phase separation; they are classified as simple and complex coacervation methods (Timilsena *et al.*, 2019).

In simple coacervation; a single macromolecule (e.g. biopolymers: Proteins or polysaccharides) is used to prepare the solution and by addition of low-affinity solvent or salt also known as a coacervate agent into it; the reaction mixture is turned into thick liquid solution to create two phases spontaneously-colloid rich coacervate and equilibrium phase.

Figure 3. Schematic diagram of the coacervation method to produce biopolymeric nanoparticles

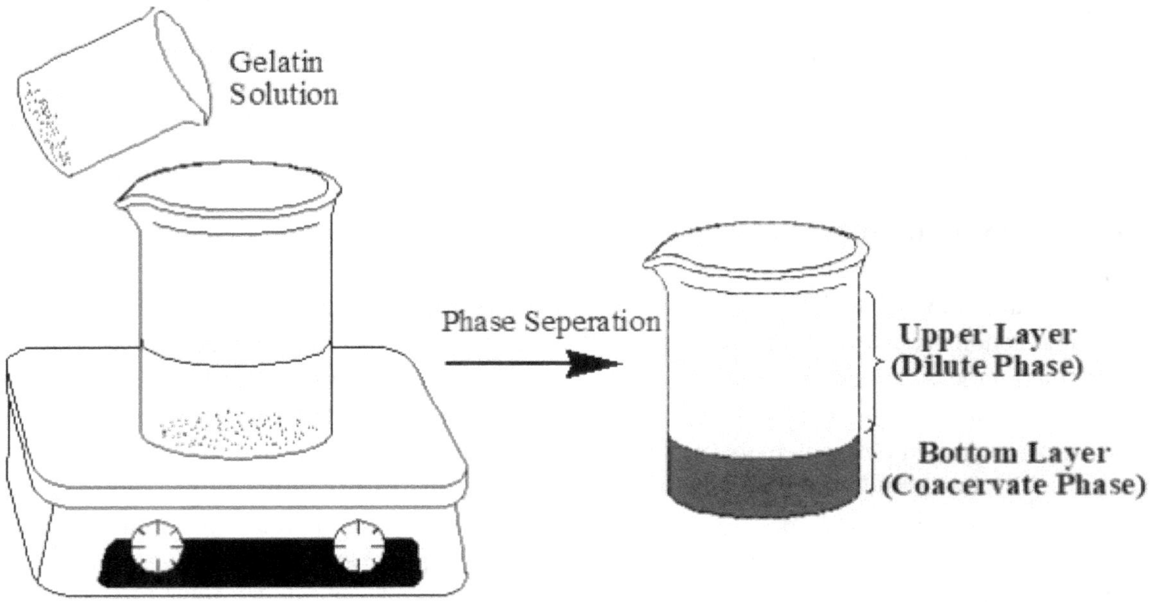

During this process, the structure of coacervate is formed in the nanoscale range due to dehydration. For instance, gelatin nanoparticles are prepared by titration with ethanol into their solution (Mohanty *et al.*, 2005). Phase separation occurs in the form of two layer- dense phase coacervates (gelatin nanoparticles) which are settled down into the bottom and diluted phase which is formed into the upper layer (Mohanty *et al.*, 2005). Another example of a simple coacervation process is the formation of chitosan nanoparticles using sodium sulfate solution (Tavares *et al.*, 2012). The coacervate phases are prepared by simply stirring the sodium sulfate and chitosan solution at a fixed concentration and volume (Tavares *et al.*, 2012).

On the contrary, complex coacervation involves ionic interaction between two or more opposite-charge biopolymers (usually polysaccharides and proteins) that leads to phase separation and coacervate formation (Newton, 1991; Madan, 1978). This method is mostly suitable for encapsulation of bioactive compounds within biopolymeric nanoparticles. The main mechanism that is associated with the formation of complex coacervates is the electrostatic interaction between the charged molecules. The hydrophobic interaction within dipolar proteins and simultaneous conformational changes in nucleic acid results from intermolecular hydrogen bonding and Van der Waals forces that contribute significantly to the formation of complex coacervates (Xiao *et al.*, 2014; Schmitt and Turgeon, 2011; Turgeon *et al.*, 2007). Bovine serum albumin and gum acacia complex nanoparticles have been synthesized previously using the complex coacervation method (Hedayati *et al.*, 2012). The complex nanoparticles were prepared by regulating various factors such as ionic strength, temperature, pH, agitation speed, protein-to-polysaccharide ratio, concentration of total biopolymer, and glutaraldehyde concentration. Most results suggested a negligible effect of total biopolymer concentration. The most influencing parameters which affect the average particle size were pH, ionic strength, and temperature. At optimum conditions (pH=4.9, ionic strength= 0, and temperature= 4°C), the smallest size of obtained nanoparticles was 108 nm. Using a similar methodology, one study was carried out to fabricate starch nanoparticles. It showed the formation of starch nanoparticles by using two types of solution of pre-modified starch: positively charged with an amino group and negatively charged with a carboxyl group. They were mixed with various molar ratios under constant stirring at room temperature for 15 min. Nanoparticles are recovered by centrifugation and lyophilization process. The optimal molar ratio was 1:3. At this optimal condition the obtained nanoparticles have 138.9±0.03 nm particle size (Barthold *et al.*, 2016).

3.2 Ionic Gelation Method

Gelation is a method by which a biopolymer solution changes from liquid to solid as a result of a crosslinking reaction occurring within the solution. This is an ancient method and recently has been re-utilized for the synthesis of biopolymeric nanoparticles. Ionic gelation occurs mainly due to electrostatic interaction between biopolymers and cross-linkers. Here reaction occurs between cationic biopolymer (usually chitosan) and anionic polymers (usually tripolyphosphate: TPP) (Calvo *et al.*, 1997). Studies revealed that TPP has limited sites of interaction for ionic gelation of chitosan nanoparticles because chitosan is a polycationic biopolymers (Avadi *et al.*, 2010). Another limiting factor of using TPP is its chemical nature. So, the use of biocompatible and biodegradable polymers with more interaction sites of negative charge is usually preferred for efficient gelation with polycationic biopolymers (Avadi *et al.*, 2010). Arabic gum, sodium alginate, and carrageenan are some examples of anionic biopolymers (Zhang *et al.*, 2014; Avadi *et al.*, 2010; Shinde and Nagarsenker, 2009). A schematic diagram has been represented to provide a clear view of the biomolecule-assisted ionic gelation method (Figure 4). The ionic gelation method has been used by Rahaiee *et al.* (2023) for the preparation of chitosan-alginate nanoparticles to protect total saffron extract Produced chitosan-alginate nanoparticles showed 218 -251 nm particle size and 27.5-53.5 mV zeta potential. The same methodology was also used for the formulation of betel leaf extract-based alginate nanoparticles using calcium chloride (Saryanti *et al.*, 2020). Alginate Nanoparticles were developed by the slow addition of calcium chloride solution into the sodium alginate solution under continuous stirring at room temperature (Mohamed and Laraba-Djebari, 2016). Synthesized nanoparticles were spherical with the particle size range from 173-297 nm.

Figure 4. Schematic diagram of ionic gelation method to produce biopolymeric nanoparticles

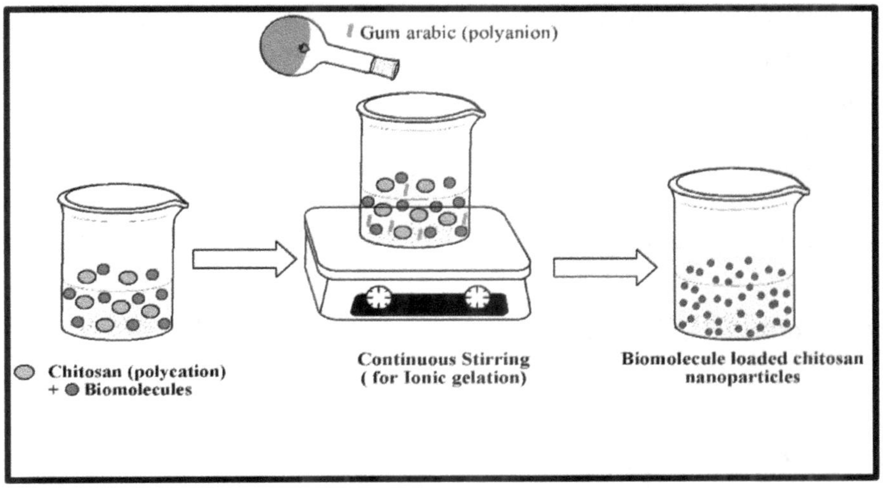

3.3 Microemulsion

Microemulsions are transparent, thermodynamically stable, isotropic colloidal dispersion solutions that mainly contain three components: a polar, a nonpolar, and a surfactant phase. In some cases, cosurfactant or cosolvent might be required. Microemulsions are considered as super solvents due to their ability to get soluble in both water and oil phases (Bansal *et al.*, 2008, Asgari *et al.*, 2019). The structural arrangement of microemulsion exists in two forms that is oil-in-water and water-in-oil. In the biopolymeric microemulsion method, the water-in-oil microemulsion is mainly prepared (Asgari *et al.*, 2019). A nonpolar organic solvent containing surfactant is prepared to make a structural arrangement as reversed micelles (hydrophilic head towards the inward side and the hydrophobic tails outwards side to the non-polar solvent). Aqueous biopolymer solution comes in contact with the oil phase and quickly diffuses into the hydrophilic areas within the reverse micelles system. After homogenization, the solution is produced in the form of a water-in-oil microemulsion. Ultrasonicator or homogenizer is used for efficient homogenization (Figure 5).

Microemulsion gel is also used to prepare solidified biopolymeric nanoparticles (Asgari *et al.* 2019). Gelation can be done by either heating, cooling, or electrolyzing. Heat assisted gelation method is mainly achieved by using biopolymers which tend to get converted to gel when high heat is introduced as in globular proteins. During the heating process the protein aggregates due to a change in intermolecular bonding and forms a nano-sized three-dimensional network. After aggregation, the mixture solution is cooled down immediately and heat-induced nanoparticles will be cleaned, isolated, and dried properly to get nano-sized particles. Whey protein isolate nanoparticles were prepared using this gelation method (Spotti *et al.*, 2024). The water-based system was prepared using 12% Whey protein isolate (WPI). WPI is encapsulated within an emulsion and heat is introduced to form gel which leads to the formation of nano-sized particles. The study revealed that the size and morphology of the nanoparticles are greatly influenced by the type of surfactants used in the preparation of the emulsion system. The study also concluded that polyglycerol polyricinoleate (PGPR) showed good stability and produce controlled-sized nanoparticles (300 nm) upon heating as compared to other surfactants (Sorbitan monooleate 80 (Span 80), 3:1 mixture of span80 and polyoxyethylene sorbitan monooleate) (Spotti *et al.*, 2024).

Another microemulsion-based method for the development of biopolymeric nanoparticles is biopolymer precipitation within the microemulsion system. The method is mainly carried out by the addition of precipitating agent in a biopolymer-based emulsion system. For instance, starch nanoparticles can be obtained by the addition of ethanol in span 60-based reverse micelles (Chin *et al.* 2014). They also suggested that ethanol served as an efficient co-solvent to produce the precipitated nanoparticles. Using this precipitation method curcumin is encapsulated within starch nanoparticles (Chin *et al.* 2014).

Figure 5. Schematic diagram of ionic gelation method to produce biopolymeric nanoparticles

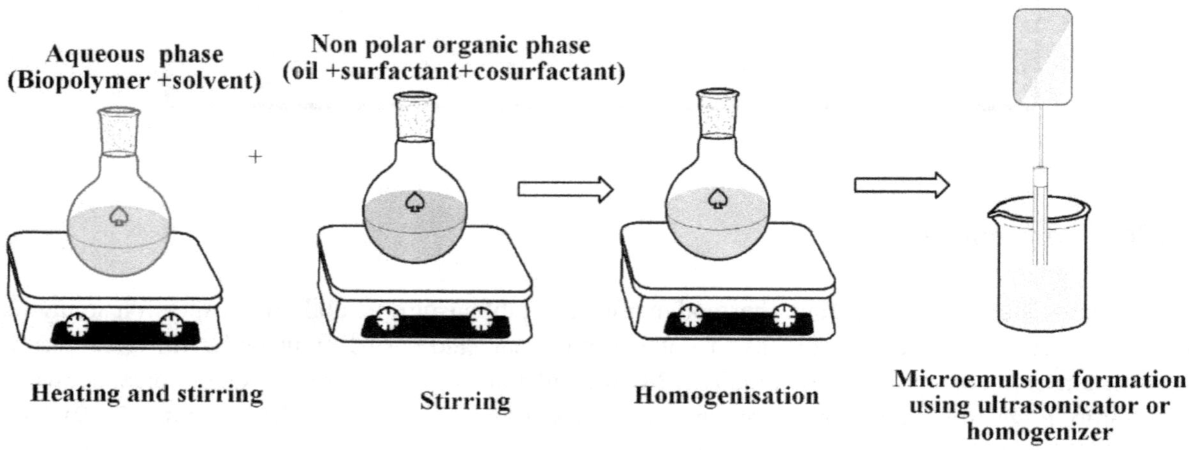

3.4 Anti-Solvent Precipitation Method

Anti-solvent method (solvent displacement/solvent de-solvation method) is a suitable and industrial method for the generation of biopolymeric (protein and polysaccharides) nanoparticles (Chacon *et al.*, 2023; Wang *et al.*, 2012; Langer *et al.*, 2003). This method has been used to produce nanoparticles from polysaccharides such as starch, chitosan, pectin, and maltodextrin, proteins such as zein, gelatin, etc. The pharmaceutical industry is already using this method for the preparation of their nanoparticle-based products. The method involved different production phases: supersaturation, nucleation, particle growth, and coagulation (Joye and McClements, 2013). In this method, the nanoparticles are primarily obtained by the addition of a low solubility-based solvent to the solvent in which the biopolymer is dissolved (Joye and McClements, 2014). This induces supersaturation that provides a driving force to form solute nano-precipitates. The main mechanism of nanoparticle production here is the disparity of molecular interaction between the solute, solvent, and the anti-solvent. For specific solvent or anti-solvent combinations, the interaction between solute-solute can become strong enough to overwhelm other interactions, and the entropy of mixing can cause precipitation in nano-size. It is necessary to ensure that the formed nanoparticles should have strong repulsion among them to avoid aggregation. Water or organic solvent (methanol, ethanol, and acetone) is generally used as a solvent/ anti-solvent for the formation of biopolymeric nanoparticles (Foujdar et al., 2023; Chacon *et al.*, 2023; Nishimoto-Sauceda *et al.*, 2022; Davidov-Pardo *et al.*, 2015; Montes *et al.*, 2011). The selection of solvent or anti-solvent mainly relies on the concept that -"two liquid systems must be miscible throughout the concentration range used in the nanoparticles production".

In recent research, it has been revealed that phenolic compounds can be synthesized in nano forms by using this anti-solvent precipitation method. Curcumin nanoparticles have been also developed using this technique. The anti-solvent method is generally considered a low-cost technique for the formation of biopolymeric nanoparticles and is widely used in the food industry because there is no need for any specific equipment or complex operating conditions and can be easily scaled up. Table 1 describes suitable fabrication methods and their application in food systems.

Table 1. An overview of biopolymeric nanoparticles, preparation method, and its application in food systems

Types of biopolymeric nanoparticles	Method used for nanoparticle production	Method in detail	Particle size (nm)	Encapsulation efficiency (%)	Application	References
Bovine serum albumin and gum arabic	Complex coacervation	Fabrication was carried out through agitation. The effect of several parameters such as ratio of protein to polysaccharide (pr: ps), pH, ionic strength, temperature (T), overall biopolymer concentration, agitation speed, and glutaraldehyde concentration were observed.	108	-	Used as a nano-delivery system	Hedayati *et al.*, 2012
Zein and Chitosan	Complex Coacervation	Complex nanoparticles of zein-chitosan were fabricated in ethanol-water (70%) solution using atmospheric cold plasma treatment.	1191-370	82.7%	Efficient encapsulation of resveratrol. Can be used to develop novel functional food fields.	Cheng *et al.*, 2020
Zein and Gum Arabic	Complex Coacervation	Zein-gum Arabic complex nanoparticles were developed using complex coacervation and ultrasonic methods.	152.17	74.2%	Encapsulation of resveratrol in complex nanoparticles has promising applications in functional beverages and the food industry.	Liu *et al.*, 2022

continued on following page

Table 1. Continued

Types of biopolymeric nanoparticles	Method used for nanoparticle production	Method in detail	Particle size (nm)	Encapsulation efficiency (%)	Application	References
Pectin and zein	Complex Coacervation	Coacervates are formed using polyanion carbohydrate and zein nanoparticles in an ethanol solution.	-	-	-	Kaushik *et al.*, 2018
Starch	Coacervation	Positive and negative charge-based water-soluble starch were developed. Starch nanoparticles formed via coacervation as well as protein were encapsulated within nanoparticles.	140-350	43	Exhibited a good biocompatibility with A549 cell (lung cancer cell) Can be used in nutraceutical formulation.	Barthold *et al.*, 2016
Zein and alginate	Antisolvent precipitation method	Zein (core) and alginate (shell) nanoparticles was developed using antisolvent precipitation method	160	-	Exhibited good stability against pH and ionic strength variation. Can be useful to encapsulate the micronutrients.	Hu and Mc Clements, 2015
Hyaluronic acid, chitosan, and zein	Antisolvent precipitation	A core-shell nanoparticles were developed using a combined method of antisolvent precipitation and layer-by-layer deposition method. At first, the antisolvent precipitation method was used to load the curcumin in the core of zein and hydraulic acid-based nanoparticles (hydrophobic core) and then piperine was loaded at the outer layer of nanoparticles and finally all were encapsulated with chitosan via electrostatic deposition method.	599	90.2 for curcumin and 86.4 for piperine	Nanoparticles exhibited good stability against light, thermal fluctuation up to 2 months	Chen *et al.*, 2019

continued on following page

Table 1. Continued

Types of biopolymeric nanoparticles	Method used for nanoparticle production	Method in detail	Particle size (nm)	Encapsulation efficiency (%)	Application	References
Lysine, Zein	Antisolvent Precipitation	A conjugated nanoparticles were developed using zein, lysine and pomegranate peel purified ellagitannin.	174.33	-	Conjugated nanoparticles exhibited a good antioxidant and antimicrobial properties. In vitro cell study showed that nanoconjugates were nontoxic to normal *Vero* cell line	Foujdar et al., 2023
Zein, sodium alginate, and sodium caseinate	ionic gelation method	An aqueous solution having curcumin, sodium caseinate, zein, and sodium alginate-based nanocomplex was prepared by maintaining a strong alkaline condition (12 pH) and then acidic condition (pH 4).	202	98	Enhance the stability, dispersibility, and bioactivity of hydrophobic nutraceuticals	Li *et al.*, 2022
Chitosan and soya lecithin	Ionic gelation method	Lecithin-ethanol solution as an organic phase was prepared and added into chitosan solution (1% v/v). For electrostatic interaction, the mixture solution was kept under stirring for 30 min. The temperature of nanosuspension was slightly increased (40 °C) and stirred for 4 hrs to completely evaporate.	117.1	92	Encapsulation of curcumin. Within developed biopolymeric nanoparticles. Exhibited potential antioxidant activity. As a nanocarrier in the development of functional food.	Pathak *et al.*, 2015
Pectin	Complexation	Nisin was loaded into pectin nanoparticles via methoxylation.	-	-	Showed good antibacterial activity	Krivorotova *et al.*, 2016
Chitosan	Self-assembly	Different formulation was developed on the combination of cinnamaldehyde, methyl eugenol, and estragol	-	-	Effective to kill fungal strain (*Aspergillus flavus*) As a green food preservative	Kumar *et al.*, 2020

continued on following page

Table 1. Continued

Types of biopolymeric nanoparticles	Method used for nanoparticle production	Method in detail	Particle size (nm)	Encapsulation efficiency (%)	Application	References
Whey Protein	Microemulsion-cold gelation	Initially an aqueous phase was prepared using aqueous extract assisted date palm extract powder, glucono-δ-lactone, and thermally treated whey protein isolate, calcium chloride. Then micro emulsified phase was developed using sunflower oil and sorbitan monooleate	230	-	Improves the stability of nutraceutical compounds of date palm pit extract.	Sadeghi *et al.* 2014
Starch	Microemulsion Method	Starch nanoparticles were developed by precipitation in the oil-in-water microemulsion method.	83-109	-	Can be used as a nutraceutical carrier as well as the development of biodegradable packaging material	Chin *et al.* 2014

4. APPLICATION OF BIOPOLYMERIC NANOPARTICLES IN THE FOOD INDUSTRY

Based on the unique molecular and physicochemical properties of biopolymeric nanoparticles, it shows a wide range of applications in the food industry, including preservatives, additives formulation, encapsulation, and biodegradable packaging. The following section discusses the same.

4.1 As Preservatives

Antimicrobial properties against various microorganisms are necessary to increase the shelf life of food products to ensure that the food is safe for consumption. It is well known that metallic nanoparticles showed antimicrobial properties (Karagoz *et al.*, 2021, Liu *et al.*, 2021; Singh *et al.*, 2021). However, the limitation of using metallic nanoparticles is that they are highly toxic to bacteria as well as to human cells (Kuang *et al.*, 2022). Biopolymer nanoparticles exhibit antibacterial activity against a wide range of microorganisms including bacteria and fungi (Foujdar *et al.*, 2023; Zohri *et al.*, 2010; Landriscina *et al.*, 2015). The mechanism of action may vary with the type of nanoparticles. Chitosan-based nanoparticles are one of the known biopolymeric nanoparticles that exhibit good antibacterial activity (Chandrasekaran *et al.*, 2020). The antibacterial activity of chitosan nanoparticles is probably due to interaction with cell walls or membranes (Tsai and Suo, 1999). The most widely accepted mechanism for antimicrobial action is the electrostatic interaction between negatively charged bacterial cell membranes and positively charged amino groups of glucosamine (Raafat *et al.*, 2008; Shahidi *et al.*, 1999; Sudarshan *et al.*, 1992).

Due to this interaction, the cell surface changes significantly leading to cell membrane permeability which causes osmotic imbalance and leakage of intracellular components resulting in cell death. The results suggested that chitosan nanoparticles are more effective against pathogenic microorganisms (*K. pneumoniae, E coli, S. aureus, and P. aeruginosa*) as compared to chitin or chitosan (Ma *et al.*, 2017). Chitosan nanoparticles also exhibited promising antimicrobial activity against food pathogenic bacteria (*Bacillus sp.* and *Pseudomonas sp.*) and suggested that it can be used as a potential non-toxic preservative in food (Alarfaj, 2019).

Some studies indicated that a biopolymeric nanoparticle combined with essential oil showed potent antimicrobial activity. For instance, cellulosic nanoparticles, whey protein, and chitosan combined with cinnamaldehyde-based edible film were applied to the outer layer of chestnut. The storage study (16 days at 25°C) suggested that the edible film is effective against *S. aureus, E. coli,* and chestnut fungus and it also helps to reduce the mold growth and calcification index of chestnut (Huang *et al.*, 2020). In another study nano-encapsulation of cumin essential oil with chitosan was done. Slow release of cumin essential oil showed antibacterial activity against E coli and L. Monocytogenes. This study proposes the application of Biopolymeric-based nano-capsules as effective food preservatives (Amiri *et al.*, 2020). Studies have also shown that they are capable of performing the antioxidant activity as well (Maraveas *et al.*, 2021; Zafar *et al*, 2021; Elmowafy *et al.*, 2023)

4.2 As Nano-Vehicles for Delivery of Bioactive Compounds and Nutrients

Bioactive compounds such as plant phenolic compounds are widely known due to their numerous health-beneficial properties such as antioxidant, anti-inflammatory, anti-microbial, anti-cancer, anti-allergenic, anti-thrombotic and anti-atherosclerosis properties (Nishimoto-Sauceda *et al.*, 2022). Despite having excellent properties, its frequent use in the food industry is limited due to its poor solubility in water, fast release, and unstable chemical nature under various environmental factors such as light, heat, ionic strength, and pH. This led to a change in chemical structure and ultimately compromised its specific biological activity. It also shows poor bioavailability during gastrointestinal digestion. To overcome these drawbacks, encapsulation within biopolymeric nanoparticles is a novel way to improve stability as well as bioavailability. Many works are available which prove that encapsulation with biopolymeric nanoparticles showed enhanced stability and activity. For instance, the nanosized saffron extract was encapsulated in chitosan-alginate nanoparticles to improve the stability and delivery of the bioactive compounds (Rahaiee *et al.*, 2023). Investigations revealed that encapsulation helps in controlled release of bioactive compounds and can protect the bioactivity such as antioxidant and anticancer properties during storage conditions. Biopolymers such as chitosan and gum arabic-based nanoparticles can be used for loading or encapsulation of insulin (Avadi *et al.*, 2010). Similarly, chitosan-based cumin essential oil nano-encapsulates were synthesized using the ionic gelation method (Zhang *et al.*, 2024). The results suggested that a 1:3 ratio of chitosan/cumin essential oil-based nano-encapsulates showed the highest encapsulation efficiency (52%) with spherical particle size (584.67 nm). Pectin can be effectively used for nano-encapsulation of bioactive compounds. Pectin-based nano-complexes showed improved bio-accessibility as well as bioavailability (Rosales and Fabi, 2023).

4.3 In Biodegradable Packaging

Effective food packaging is one of the most important criteria to ensure the quality, shelf life, and safety of foods. The main factors that affect these parameters are respiration, chemical reaction, and microbial spoilage which are mainly associated with improper packaging and storage conditions. The synthesis of biodegradable packaging materials with active compounds can help to overcome these problems (Khezerlou *et al.* 2021 a & b). Table 2 enlists some of the biopolymers used for preparing nanoparticle-doped biodegradable packaging material.

Table 2. Nanoparticle-embedded biopolymers for food packaging

Name of the Biopolymer	Nanoparticle entrapped in packaging	Type of food packaged	References
Alginate	Nisin	Beef	Cutter and Siragusa, 1996
Carboxymethyl cellulose (CMC), Arabic gum (AG) & gelatin (GL)	Garlic extracts (GE) and TiO2 nanoparticles (TiO2-NPs)	Fresh Nile tilapia fish fillets	Youssef *et al.*,2021
Carrageenan	Chlortetracycline and oxytetracycline	Poultry	Meyer *et al*, 1959
Chitosan–beeswax	Pollen grains	Le Conte pear	Sultan *et al.*, 2021
Chitosan–beeswax	Acetic and propionic acid	Meat	Ouattara *et al.*, 2000
Cellulose	Pediocin	Meat	Ming *et al.*, 1997
Poly-(butylene adipate-co-terephthalate) (PBAT)	Copper oxide nanoparticles	Packaging of food	Hasanin & Youssef, 2022
Polyurethane, chitin, and resin composites	Zinc Oxide-doped-Silicon dioxide nanoparticles	Recycled packaging	Moustafa *et al.*,2022
Polyvinyl alcohol and chitosan	Essential oil of cinnamon and Titanium dioxide nanoparticles	Fresh chicken breast fillets	Youssef *et al.*, 2023
Starch and derivatives	Potassium sorbate	Strawberry	Garcia *et al.*, 1998

There is an increased amount of research that is undertaken every day to improve the effectiveness of this technology. More novel materials are tried out along with different combinations of nanoparticles. However, the chances of migration of nanoparticles that may generate toxicity cannot be overlooked in this regard.

5. TOXICOLOGICAL ASSESSMENT OF BIOPOLYMERIC NANOPARTICLES

Since polymeric nanoparticles are already proven to be toxic by several studies (Murali *et al.*, 2024; Sharma *et al.*, 2021; Suriya Prabha *et al.*, 2019), the concept of Biopolymeric nanoparticles came into place. The target was to practice more eco-friendly methodologies with minimum use of chemicals. Since many nanoparticles are of microbial origin, their toxicity is much less as compared to synthetic nanomaterials (Saraswat *et al.*, 2023). So far, biopolymeric material like cellulose has been reported to cause lung disease mesothelioma in some textile workers due to the accumulation of particulates in their cell bodies (Catalán and Norppa, 2017). Therefore, the biopolymeric nanomaterials must be evaluated based on their cytotoxic, and genotoxic behavior, immunological response, oxidative stress, presence of endotoxins etc. Though most of them were found to be reportedly non-toxic, their high cost of manufacturing and energy consumption are their limitations for production and distribution on a larger scale (Saraswat *et al.*, 2023). During their manufacturing, the companies should conform to the regulations of that country regarding the use of bio-based nanomaterials concerning its safer use, scaling up, and testing (Catalán and Norppa, 2017). Stability is another issue that must be evaluated before they are applied to products for the delivery of important biomolecules (Putro *et al.*, 2017).

Since they are biological materials, some immune responses may be generated (Zolnik *et al.*, 2010) after their intake in sensitive individuals. However, no such cases have been reported so far by any studies.

CONCLUSION

Traditionally, nanoparticles can be developed from polymers and biological materials such as proteins and polysaccharides. The toxicity of nanoparticles using polymers showed an increase in the number of incidences when introduced into the biological systems. It was found that the bio-based or Biopolymeric nanoparticles are better compatible with biological systems providing eco-friendly degradable characteristics with zero or negligible toxicity. There are different methods available for the fabrication of these particles- some follow traditional methods. At the same time, others are developed to overcome the limitations of the traditional methods practiced thereof. No matter what the source is, each produced nanoparticle is evaluated in terms of size, morphology, stability in biological system, structure, cytotoxicity, and compatibility. Advanced technologies such as zeta analyzer, dynamic light scattering analyzer, scanning electron microscopy, transmission electron microscopy, atomic force microscopy, x-ray diffraction, Fourier transform infrared spectroscopy, etc. are used for proper analysis and evaluation of the nanoparticles generated. If expectations are met after analysis, they can be used as a vector for the transmission of the bioactive molecules, therapeutics, vaccines, etc. to specific target sites like tumor cells or cancerous stem cells. Though it seems such a lucrative option, it is not free from its challenges. Enhancement of evenly distributed stable nanoparticles is necessary for their regular applications in drug delivery systems. Since they are still in their developing stage, more research data and in vivo tests are needed to back up their utilization in medicine, food, and pharmacy. Scientists, biologists, and chemical engineers need to collaborate to successfully apply this tool in clinical drug and therapeutic delivery systems to improve their pharmacokinetics and bio-distribution in biological systems.

ACKNOWLEDGMENT

Swati Ray served as the corresponding author on this chapter.

REFERENCES

Aguilar, N. M., Arteaga-Cardona, F., de Anda Reyes, M. E., Gervacio-Arciniega, J. J., & Salazar-Kuri, U. (2019). Magnetic bioplastics based on isolated cellulose from cotton and sugarcane bagasse. *Materials Chemistry and Physics*, 238, 121921. Advance online publication. DOI: 10.1016/j.matchemphys.2019.121921

Alarfaj, A. A. (2019). Antibacterial effect of chitosan nanoparticles against food spoilage bacteria. *Journal of Pure & Applied Microbiology*, 13(2), 1273–1278. DOI: 10.22207/JPAM.13.2.70

Amiri, N., Afsharmanesh, M., Salarmoini, M., Meimandipour, A., Hosseini, S., & Ebrahimnejad, H. (2020). Effects of nanoencapsulated cumin essential oil as an alternative to the antibiotic growth promoter in broiler diets. *Journal of Applied Poultry Research*, 29(4), 875–885. DOI: 10.1016/j.japr.2020.08.004

Aranaz, I., Mengíbar, M., Harris, R., Paños, I., Miralles, B., Acosta, N., & Heras, Á. (2009). Functional characterization of chitin and chitosan. *Current Chemical Biology*, 3(2), 203–230.

Avadi, M. R., Sadeghi, A. M. M., Mohammadpour, N., Abedin, S., Atyabi, F., Dinarvand, R., & Rafiee-Tehrani, M. (2010). Preparation and characterization of insulin nanoparticles using chitosan and Arabic gum with ionic gelation method. *Nanomedicine; Nanotechnology, Biology, and Medicine*, 6(1), 58–63. DOI: 10.1016/j.nano.2009.04.007 PMID: 19447202

Bansal, T., Mustafa, G., Khan, Z. I., Ahmad, F. J., Khar, R. K., & Talegaonkar, S. (2008). Solid self-nanoemulsifying delivery systems as a platform technology for formulation of poorly soluble drugs. *Critical Reviews™ in Therapeutic Drug Carrier Systems, 25*(1), 63-116.

Baranwal, J., Barse, B., Fais, A., Delogu, G. L., & Kumar, A. (2022). Biopolymer: A Sustainable Material for Food and Medical Applications. *Polymers*, 14(5), 983. DOI: 10.3390/polym14050983 PMID: 35267803

Barthold, S., Kletting, S., Taffner, J., de Souza Carvalho-Wodarz, C., Lepeltier, E., Loretz, B., & Lehr, C. M. (2016). Preparation of nanosized coacervates of positive and negative starch derivatives intended for pulmonary delivery of proteins. *Journal of Materials Chemistry. B*, 4(13), 2377–2386. DOI: 10.1039/C6TB00178E PMID: 32263233

Basavegowda, N., & Baek, K. (2020). Advances in Functional Biopolymer-Based Nanocomposites for Active Food Packaging Applications. *Polymers*, 13(23), 4198. DOI: 10.3390/polym13234198 PMID: 34883701

Basu, A., Ray, S., Chowdhury, S., Sarkar, A., Mandal, D. P., Bhattacharjee, S., & Kundu, S. (2018). Evaluating the antimicrobial, apoptotic, and cancer cell gene delivery properties of protein- capped gold nanoparticles synthesized from the edible mycorrhizal fungus Tricholomacrassum. *Nanoscale Research Letters*, 13(1), 1–16. DOI: 10.1186/s11671-018-2561-y PMID: 29299709

Bealer, E. J., Onissema-Karimu, S., Rivera-Galletti, A., Francis, M., Wilkowski, J., Salas-de la Cruz, D., & Hu, X. (2020). Protein–Polysaccharide Composite Materials: Fabrication and Applications. *Polymers*, 12(2), 464. DOI: 10.3390/polym12020464 PMID: 32079322

Biswas, M. C., Jony, B., Nandy, P. K., Chowdhury, R. A., Halder, S., Kumar, D., Ramakrishna, S., Hassan, M., Ahsan, M. A., Hoque, M. E., & Imam, M. A. (2022). Recent Advancement of Biopolymers and Their Potential Biomedical Applications. *Journal of Polymers and the Environment*, 30(1), 51–74. DOI: 10.1007/s10924-021-02199-y

Borah, P. P., Das, P., & Badwaik, L. S. (2017). Ultrasound treated potato peel and sweet lime pomace-based biopolymer film development. *Ultrasonics Sonochemistry*, 36, 11–19. DOI: 10.1016/j.ultsonch.2016.11.010 PMID: 28069189

Calvo, P., Remunan-Lopez, C., Vila-Jato, J. L., & Alonso, M. J. (1997). Novel hydrophilic chitosan-polyethylene oxide nanoparticles as protein carriers. *Journal of Applied Polymer Science*, 63(1), 125–132. DOI: 10.1002/(SICI)1097-4628(19970103)63:1<125::AID-APP13>3.0.CO;2-4

Catalán, J., & Norppa, H. (2017). Safety aspects of bio-based nanomaterials. *Bioengineering (Basel, Switzerland)*, 4(4), 94. DOI: 10.3390/bioengineering4040094 PMID: 29194367

Chacon, W. D. C., Verruck, S., Monteiro, A. R., & Valencia, G. A. (2023). The mechanism, biopolymers and active compounds for the production of nanoparticles by anti-solvent precipitation: A review. *Food Research International*, 168, 112728. DOI: 10.1016/j.foodres.2023.112728 PMID: 37120194

Chakrabarti, S., Guha, S., & Majumder, K. (2018). Food-derived bioactive peptides in human health: Challenges and opportunities. *Nutrients*, 10(11), 1738. DOI: 10.3390/nu10111738 PMID: 30424533

Chandrasekaran, M., Kim, K. D., & Chun, S. C. (2020). Antibacterial activity of chitosan nanoparticles: A review. *Processes (Basel, Switzerland)*, 8(9), 1173. DOI: 10.3390/pr8091173

Chen, G., Dong, S., Chen, Y., Gao, Y., Zhang, Z., Li, S., & Chen, Y. (2020). Complex coacervation of zein-chitosan via atmospheric cold plasma treatment: Improvement of encapsulation efficiency and dispersion stability. *Food Hydrocolloids*, 107, 105943. DOI: 10.1016/j.foodhyd.2020.105943

Chen, S., McClements, D. J., Jian, L., Han, Y., Dai, L., Mao, L., & Gao, Y. (2019). Core–shell biopolymer nanoparticles for co-delivery of curcumin and piperine: Sequential electrostatic deposition of hyaluronic acid and chitosan shells on the zein core. *ACS Applied Materials & Interfaces*, 11(41), 38103–38115. DOI: 10.1021/acsami.9b11782 PMID: 31509373

Chin, S. F., Azman, A., & Pang, S. C. (2014). Size controlled synthesis of starch nanoparticles by a microemulsion method. *Journal of Nanomaterials*, 2014(1), 763736. DOI: 10.1155/2014/763736

Cutter, C. N., & Siragusa, G. R. (1996). Reduction of *Brochothrix thermosphacta* on beef surfaces following immobilization of nisin in calcium alginate gels. *Letters in Applied Microbiology*, 23(1), 9–12. DOI: 10.1111/j.1472-765X.1996.tb00018.x PMID: 8679147

Davidov-Pardo, G., Joye, I. J., & McClements, D. J. (2015). Encapsulation of resveratrol in biopolymer particles produced using liquid antisolvent precipitation. Part 1: Preparation and characterization. *Food Hydrocolloids*, 45, 309–316. DOI: 10.1016/j.foodhyd.2014.11.023

Elmowafy, M., Shalaby, K., Elkomy, M. H., Alsaidan, O. A. M., Gomaa, H. A., Abdelgawad, M. A., & Mostafa, E. M. (2023). Polymeric Nanoparticles for Delivery of Natural Bioactive Agents: Recent Advances and Challenges. *Polymers*, 15(5), 1123. DOI: 10.3390/polym15051123 PMID: 36904364

Flaris, V., & Singh, G. (2009). Recent developments in biopolymers. *Journal of Vinyl and Additive Technology*, 15(1), 11. DOI: 10.1002/vnl.20171

Foujdar, R., Chopra, H., Janghu, S., & Bera, M. B. (2023). Fabrication of "Lysine modified Zein- Punica granatum peels' purified ellagitannin nanoconjugates (Ly-MZ-PE NCs)" and in- vitro bio-accessibility, antiproliferative activity, and biosafety studies. *Food Bioscience*, 55, 102943. DOI: 10.1016/j.fbio.2023.102943

Foujdar, R., Kumar, S., Chauhan, A. K., & Bera, M. B. (2023). Role of Nanomaterials for Food Processing and Preservation. In *Nanotechnology Horizons in Food Process Engineering* (pp. 47–83). Apple Academic Press. DOI: 10.1201/9781003305378_4

Franca-Oliveira, G., Fornari, T., & Hernandez-Ledesma, B. (2021). A review on the extraction and processing of natural source-derived proteins through eco-innovative approaches. *Processes (Basel, Switzerland)*, 9(9), 1626. DOI: 10.3390/pr9091626

Gao, J. H., & Xu, B. (2009). Applications of nanomaterials inside cell. *Nano Today*, 4(1), 37–51. DOI: 10.1016/j.nantod.2008.10.009

García, M. A., Martino, M. N., & Zaritzky, N. E. (1998). Plasticized starch-based coatings to improve strawberry (Fragaria× ananassa) quality and stability. *Journal of Agricultural and Food Chemistry*, 46(9), 3758–3767. DOI: 10.1021/jf980014c

Gawde, K. A., Sau, S., Tatiparti, K., Kashaw, S. K., Mehrmohammadi, M., Azmi, A. S., & Iyer, A. K. (2018). Paclitaxel and di-fluorinated curcumin loaded in albumin nanoparticles for targeted synergistic combination therapy of ovarian and cervical cancers. *Colloids and Surfaces. B, Biointerfaces*, 167, 8–19. DOI: 10.1016/j.colsurfb.2018.03.046 PMID: 29625422

Guo, S., Fu, D., Utupova, A., Sun, D., Zhou, M., Jin, Z., & Zhao, K. (2019). Applications of polymer-based nanoparticles in vaccine field. *Nanotechnology Reviews*, 8(1), 143–155. DOI: 10.1515/ntrev-2019-0014

Harish, V., Tewari, D., Gaur, M., Yadav, A. B., Swaroop, S., Bechelany, M., & Barhoum, A. (2022). Review on Nanoparticles and Nanostructured Materials: Bioimaging, Biosensing, Drug Delivery, Tissue Engineering, Antimicrobial, and Agro-Food Applications. *Nanomaterials (Basel, Switzerland)*, 12(3), 457. DOI: 10.3390/nano12030457 PMID: 35159802

Hasanin, M. S., & Youssef, A. M. (2022). Ecofriendly bioactive film doped CuO nanoparticles based biopolymers and reinforced by enzymatically modified nanocellulose fibers for active packaging applications. *Food Packaging and Shelf Life*, 34, 100979. DOI: 10.1016/j.fpsl.2022.100979

Hassan, M., Bai, J., & Dou, D. (2019). Biopolymers; Definition, Classification and Applications. *Egyptian Journal of Chemistry*, 62(9), 1725–1737. DOI: 10.21608/ejchem.2019.6967.1580

Hatakeyama, H., & Hatakeyama, T. (2009). Biopolymers. In *Lignin Structure, Properties, and Applications* (pp. 1–63). Springer.

Hedayati, R., Jahanshahi, M., & Attar, H. (2012). Fabrication and characterization of albumin- acacia nanoparticles based on complex coacervation as potent nanocarrier. *Journal of Chemical Technology and Biotechnology*, 87(10), 1401–1408. DOI: 10.1002/jctb.3758

Hernandez-Carmona, F., Morales-Matos, Y., Lambis-Miranda, H., & Pasqualino, J. (2017). Starch extraction potential from plantain peel wastes. *Journal of Environmental Chemical Engineering*, 5(5), 4980–4985. DOI: 10.1016/j.jece.2017.09.034

Hong, S., Choi, D. W., Kim, H. N., Park, C. G., Lee, W., & Park, H. H. (2020). Protein-Based Nanoparticles as Drug Delivery Systems. *Pharmaceutics*, 12(7), 604. DOI: 10.3390/pharmaceutics12070604 PMID: 32610448

Hu, K., & McClements, D. J. (2015). Fabrication of biopolymer nanoparticles by antisolvent precipitation and electrostatic deposition: Zein-alginate core/shell nanoparticles. *Food Hydrocolloids*, 44, 101–108. DOI: 10.1016/j.foodhyd.2014.09.015

Huang, Y., Gu, C., He, S., Zhu, D., Liu, X., & Chen, Z. (2020). Development and characterization of an edible chitosan–whey protein nano composite film for chestnut (Castanea mollissima Bl.) preservation. *Journal of Food Science*, 85(7), 2114–2123. DOI: 10.1111/1750-3841.15174 PMID: 32519374

Joye, I. J., & McClements, D. J. (2013). Production of nanoparticles by anti-solvent precipitation for use in food systems. *Trends in Food Science & Technology*, 34(2), 109–123. DOI: 10.1016/j.tifs.2013.10.002

Karagoz, S., Kiremitler, N. B., Sarp, G., Pekdemir, S., Salem, S., Goksu, A. G., Onses, M. S., Sozdutmaz, I., Sahmetlioglu, E., Ozkara, E. S., Ceylan, A., & Yilmaz, E. (2021). Antibacterial, antiviral, and self-cleaning mats with sensing capabilities based on electrospun nanofibers decorated with ZnO nanorods and Ag nanoparticles for protective clothing applications. *ACS Applied Materials & Interfaces*, 13(4), 5678–5690. DOI: 10.1021/acsami.0c15606 PMID: 33492946

Karmanov, A. P., Kanarsky, A. V., Kocheva, L. S., Belyy, V. A., Semenov, E. I., Rachkova, N. G., Bogdanovich, N. I., & Pokryshkin, S. A. (2021). Chemical structure and polymer properties of wheat and cabbage lignins–Valuable biopolymers for biomedical applications. *Polymer*, 220, 123571. DOI: 10.1016/j.polymer.2021.123571

Kaur, G., Thakur, B., Gill, R., Kaur, R., & Kaur, S. (2023). Biopolymeric conjugation with food additives. *Physical Sciences Reviews*. Advance online publication. DOI: 10.1515/psr-2022-0191

Kaushik, P., Rawat, K., Aswal, V. K., Kohlbrecher, J., & Bohidar, H. B. (2018). Mixing ratio dependent complex coacervation versus bicontinuous gelation of pectin with in situ formed zein nanoparticles. *Soft Matter*, 14(31), 6463–6475. DOI: 10.1039/C8SM00809D PMID: 30051132

Khezerlou, A., Tavassoli, M., Alizadeh Sani, M., Mohammadi, K., Ehsani, A., & McClements, D. J. (2021a). Application of nanotechnology to improve the performance of biodegradable biopolymer-based packaging materials. *Polymers*, 13(24), 4399. DOI: 10.3390/polym13244399 PMID: 34960949

Khezerlou, A., Zolfaghari, H., Banihashemi, S. A., Forghani, S., & Ehsani, A. (2021b). Plant gums as the functional compounds for edible films and coatings in the food industry: A review. *Polymers for Advanced Technologies*, 32(6), 2306–2326. DOI: 10.1002/pat.5293

Kibria, M. G., Masuk, N. I., Safayet, R., Nguyen, H. Q., & Mourshed, M. (2023). Plastic waste: Challenges and opportunities to mitigate pollution and effective management. *International Journal of Environmental Research*, 17(1), 20. DOI: 10.1007/s41742-023-00507-z PMID: 36711426

Kirdar, S. S. (2015). Current and future applications of nanotechnology in the food industry. In Conference paper, ISITES2015, Valencia-Spain.

Kreuter, J. (1995). Nanoparticulate systems in drug delivery and targeting. *Journal of Drug Targeting*, 3(3), 171–173. DOI: 10.3109/10611869509015940 PMID: 8705247

Krivorotova, T., Cirkovas, A., Maciulyte, S., Staneviciene, R., Budriene, S., Serviene, E., & Sereikaite, J. (2016). Nisin-loaded pectin nanoparticles for food preservation. *Food Hydrocolloids*, 54, 49–56. DOI: 10.1016/j.foodhyd.2015.09.015

Kuang, X., Wang, Z., Luo, Z., He, Z., Liang, L., Gao, Q. H., Li, Y., Xia, K., Xie, Z., Chang, R., Wang, Y., Liu, Y., Zhao, S., Su, J., Wang, Y., Situ, W., Chen, M., Zhao, Y., Chen, X., & Liu, H. (2022). Ag nanoparticles enhance immune checkpoint blockade efficacy by promoting of immune surveillance in melanoma. *Journal of Colloid and Interface Science*, 616, 189–200. DOI: 10.1016/j.jcis.2022.02.050 PMID: 35203032

Kučuk, N., Primožič, M., Knez, Ž., & Leitgeb, M. (2022). Sustainable Biodegradable Biopolymer- Based Nanoparticles for Healthcare Applications. *International Journal of Molecular Sciences*, 24(4), 3188. DOI: 10.3390/ijms24043188 PMID: 36834596

Kumar, A., Gupta, V., Singh, P. P., Kujur, A., & Prakash, B. (2020). Fabrication of volatile compounds loaded-chitosan biopolymer nanoparticles: Optimization, characterization and assessment against Aspergillus flavus and aflatoxin B1 contamination. *International Journal of Biological Macromolecules*, 165, 1507–1518. DOI: 10.1016/j.ijbiomac.2020.09.257 PMID: 33038402

Kumari, S., & Rath, P. K. (2014). Extraction and characterization of chitin and chitosan from (Labeo rohit) fish scales. *Procedia Materials Science*, 6, 482–489. DOI: 10.1016/j.mspro.2014.07.062

Landriscina, A., Rosen, J., & Friedman, A. J. (2015). Biodegradable chitosan nanoparticles in drug delivery for infectious disease. *Nanomedicine (London)*, 10(10), 1609–1619. DOI: 10.2217/nnm.15.7 PMID: 26008195

Langer, K., Balthasar, S., Vogel, V., Dinauer, N., Von Briesen, H., & Schubert, D. (2003). Optimization of the preparation process for human serum albumin (HSA) nanoparticles. *International Journal of Pharmaceutics*, 257(1-2), 169–180. DOI: 10.1016/S0378-5173(03)00134-0 PMID: 12711172

Larsson, S., Jansson, M., & Boholm, Å. (2019). Expert stakeholders' perception of nanotechnology: Risk, benefit, knowledge, and regulation. *Journal of Nanoparticle Research*, 21(3), 1–17. DOI: 10.1007/s11051-019-4498-1

Li, Z., Fu, Y., McClements, D. J., & Li, T. (2022). Impact of alginate block type on the structure and physicochemical properties of curcumin-loaded complex biopolymer nanoparticles. *Lebensmittel-Wissenschaft + Technologie*, 162, 113435. DOI: 10.1016/j.lwt.2022.113435

Liu, F., Cheng, X., Xiao, L., Wang, Q., Yan, K., Su, Z., Wang, Y., Ma, C., & Wang, Y. (2021). Inside-outside Ag nanoparticles-loaded polylactic acid electrospun fiber for long-term antibacterial and bone regeneration. *International Journal of Biological Macromolecules*, 167, 1338–1348. DOI: 10.1016/j.ijbiomac.2020.11.088 PMID: 33232699

Liu, Y., Liang, Q., Liu, X., Raza, H., Ma, H., & Ren, X. (2022). Treatment with ultrasound improves the encapsulation efficiency of resveratrol in zein-gum Arabic complex coacervates. *Lebensmittel-Wissenschaft + Technologie*, 153, 112331. DOI: 10.1016/j.lwt.2021.112331

Lu, X. Y., Wu, D. C., Li, Z. J., & Chen, G. Q. (2011). Polymer nanoparticles. *Progress in Molecular Biology and Translational Science*, 104, 299–323. DOI: 10.1016/B978-0-12-416020-0.00007-3 PMID: 22093222

Madan, P. L. (1978). Microencapsulation I. Phase separation or coacervation. *Drug Development and Industrial Pharmacy*, 4(1), 95–116. DOI: 10.3109/03639047809055641

Maraveas, C., Bayer, I. S., & Bartzanas, T. (2021). Recent Advances in Antioxidant Polymers: From Sustainable and Natural Monomers to Synthesis and Applications. *Polymers*, 13(15), 2465. Advance online publication. DOI: 10.3390/polym13152465 PMID: 34372069

Meyer, R. C., Winter, A. R., & Weiser, H. H. (1959). Edible protective coatings for extending the shelf life of poultry. *Food Technology*, 13, 146–148.

Miao, T., Wang, J., Zeng, Y., Liu, G., & Chen, X. (2018). Polysaccharide-based controlled release systems for therapeutics delivery and tissue engineering: From bench to bedside. *Advancement of Science*, 5(4), 1700513. DOI: 10.1002/advs.201700513 PMID: 29721408

Ming, X., Weber, G. H., Ayres, J. W., & Sandine, W. E. (1997). Bacteriocins applied to food packaging materials to inhibit Listeria monocytogenes on meats. *Journal of Food Science*, 62(2), 413–415. DOI: 10.1111/j.1365-2621.1997.tb04015.x

Mohamed, F. A. N., & Laraba-Djebari, F. (2016). Development and characterization of a new carrier for vaccine delivery based on calcium-alginate nanoparticles: Safe immunoprotective approach against scorpion envenoming. *Vaccine*, 34(24), 2692–2699. DOI: 10.1016/j.vaccine.2016.04.035 PMID: 27109567

Mohanty, B., Aswal, V. K., Kohlbrecher, J., & Bohidar, H. B. (2005). Synthesis of gelatin nanoparticles via simple coacervation. *Journal of Surface Science and Technology*, 21(3/4), 149.

Mojaveryazdi, F. S., Muhamad, I. I., Rezania, S., & Benham, H. (2014). Importance of glucose and Pseudomonas in producing degradable plastics. *Jurnal Teknologi*, 69(5). Advance online publication. DOI: 10.11113/jt.v69.3194

Montes, A., Gordillo, M. D., Pereyra, C., & De La Ossa, E. M. (2011). Co-precipitation of amoxicillin and ethyl cellulose microparticles by supercritical antisolvent process. *The Journal of Supercritical Fluids*, 60, 75–80. DOI: 10.1016/j.supflu.2011.05.002

Moustafa, H., Darwish, N. A., & Youssef, A. M. (2022). Rational formulations of sustainable polyurethane/chitin/rosin composites reinforced with ZnO-doped-SiO2 nanoparticles for green packaging applications. *Food Chemistry*, 371, 131193. DOI: 10.1016/j.foodchem.2021.131193 PMID: 34649200

Muhamad, I. I., Sabbagh, F., & Abdul Karim, N. (2017). Plant Secondary Metabolites. In *Polyhydroxyalkanoates: A Valuable Secondary Metabolite Produced in Microorganisms and Plants* (1st ed., Vol. 3, pp. 205–234). Apple Academic Press.

Muir, V. G., & Burdick, J. A. (2020). Chemically modified biopolymers for the formation of biomedical hydrogels. *Chemical Reviews*, 121(18), 10908–10949. DOI: 10.1021/acs.chemrev.0c00923 PMID: 33356174

Murali, N., Rainu, S. K., Sharma, A., Siddhanta, S., Singh, N., & Betal, S. (2024). Remotely Controlled Surface Charge Modulation of Magnetoelectric Nanogenerators for Swift and Efficient Drug Delivery. *ACS Omega*, 9(26), 28937–28950. DOI: 10.1021/acsomega.4c03825 PMID: 38973906

Nairm, J. G. (1995). 3 Coacervation-phase separation technology. []). Academic Press.]. *Advances in Pharmaceutical Sciences*, 7, 93–219. DOI: 10.1016/S0065-3136(06)80005-1

Newton, D. W. (1991). Coacervation: Principles and applications. In *Polymers for controlled drug delivery* (pp. 67-81). CRC Press.

Nishimoto-Sauceda, D., Romero-Robles, L. E., & Antunes-Ricardo, M. (2022). Biopolymer nanoparticles: A strategy to enhance stability, bioavailability, and biological effects of phenolic compounds as functional ingredients. *Journal of the Science of Food and Agriculture*, 102(1), 41–52. DOI: 10.1002/jsfa.11512 PMID: 34460939

Onyancha, R. B., Aigbe, U. O., Ukhurebor, K. E., & Muchiri, P. W. (2021). Facile synthesis and applications of carbon nanotubes in heavy-metal remediation and biomedical fields: A comprehensive review. *Journal of Molecular Structure*, 1238, 130462. DOI: 10.1016/j.molstruc.2021.130462

Oppenheim, R. C. (1981). Solid colloidal drug delivery systems: Nanoparticles. *International Journal of Pharmaceutics*, 8(3), 217–234. DOI: 10.1016/0378-5173(81)90100-9

Ouattara, B., Simard, R. E., Piette, G., Bégin, A., & Holley, R. A. (2000). Inhibition of surface spoilage bacteria in processed meats by application of antimicrobial films prepared with chitosan. *International Journal of Food Microbiology*, 62(1-2), 139–148. DOI: 10.1016/S0168-1605(00)00407-4 PMID: 11139014

Pathak, L., Kanwal, A., & Agrawal, Y. (2015). Curcumin loaded self assembled lipid-biopolymer nanoparticles for functional food applications. *Journal of Food Science and Technology*, 52(10), 6143–6156. DOI: 10.1007/s13197-015-1742-2 PMID: 26396362

Perotti, G. F., Tronto, J., Bizeto, M. A., Izumi, C., Temperini, M. L., Lugão, A. B., Parra, D. F., & Constantino, V. R. (2014). Biopolymer-clay nanocomposites: Cassava starch and synthetic clay cast films. *Journal of the Brazilian Chemical Society*, 25, 320–330. DOI: 10.5935/0103-5053.20130300

Prabha, A. S., Dorothy, R., Jancirani, S., Rajendran, S., Singh, G., & Kumaran, S. S. (2020). Recent advances in the study of toxicity of polymer-based nanomaterials. *Nanotoxicity*, 143-165. https://doi.org/ DOI: 10.1016/B978-0-12-819943-5.00007-5

Putro, J. N., Santoso, S. P., Ismadji, S., & Ju, Y. H. (2017). Investigation of heavy metal adsorption in binary system by nanocrystalline cellulose–bentonite nanocomposite: Improvement on extended Langmuir isotherm model. *Microporous and Mesoporous Materials*, 246, 166–177. DOI: 10.1016/j.micromeso.2017.03.032

Raafat, D., Von Bargen, K., Haas, A., & Sahl, H. G. (2008). Insights into the mode of action of chitosan as an antibacterial compound. *Applied and Environmental Microbiology*, 74(12), 3764–3773. DOI: 10.1128/AEM.00453-08 PMID: 18456858

Rahaiee, S., Shojaosadati, S. A., & Hashemi, M. (2023). An efficient ionic gelation based nano- delivery system to improve the stability and controlled release of saffron extracts. *Biocatalysis and Agricultural Biotechnology*, 52, 102831. DOI: 10.1016/j.bcab.2023.102831

Redondo-Gómez, C., Rodríguez Quesada, M., Vallejo Astúa, S., Murillo Zamora, J. P., Lopretti, M., & Vega-Baudrit, J. R. (2020). Biorefinery of biomass of agro-industrial banana waste to obtain high-value biopolymers. *Molecules (Basel, Switzerland)*, 25(17), 3829. DOI: 10.3390/molecules25173829 PMID: 32842473

Rosales, T. K. O., & Fabi, J. P. (2023). Pectin-based nanoencapsulation strategy to improve the bioavailability of bioactive compounds. *International Journal of Biological Macromolecules*, 229, 11–21. DOI: 10.1016/j.ijbiomac.2022.12.292 PMID: 36586647

Sadeghi, R., Mehryar, L., Karimi, M., & Kokini, J. (2016). Nanocapsule formation by individual biopolymer nanoparticles. In: Nanoencapsulation Technologies for the Food and Nutraceutical Industries, 404-446. https://doi.org/DOI: 10.1016/B978-0-12-809436-5.00011-2

Sadeghi, S., Madadlou, A., & Yarmand, M. (2014). Microemulsification–cold gelation of whey proteins for nanoencapsulation of date palm pit extract. *Food Hydrocolloids*, 35, 590–596. DOI: 10.1016/j.foodhyd.2013.07.021

Sahoo, M., Vishwakarma, S., Panigrahi, C., & Kumar, J. (2021). Nanotechnology: Current applications and future scope in food. *Food Frontiers*, 2(1), 3–22. DOI: 10.1002/fft2.58

Saraswat, P., Singh, S., Prasad, M., Misra, R., Rajput, V. D., & Ranjan, R. (2023). Applications of bio-based nanomaterials in environment and agriculture: A review on recent progresses. *Hybrid Advances*, 4, 100097. DOI: 10.1016/j.hybadv.2023.100097

Saryanti, D., Nugraheni, D., & Astuti, N. S. (2020). Preparation and characterization of Betel Leaves (Piper betle Linn) extract nanoparticle with ionic gelation method. *Journal of Tropical Pharmacy and Chemistry*, 5(1), 15–20. DOI: 10.25026/jtpc.v5i1.224

Schmitt, C., & Turgeon, S. L. (2011). Protein/polysaccharide complexes and coacervates in food systems. *Advances in Colloid and Interface Science*, 167(1-2), 63–70. DOI: 10.1016/j.cis.2010.10.001 PMID: 21056401

Senevirathna, S., Pitawala, J., Asela, U., & Attanayake, A. (2022). Biopolymers: Structure, properties, extraction methods and applications. *Sri Lankan Journal of Applied Sciences*, 1(01), 18–30.

Shahidi, F., Arachchi, J. K. V., & Jeon, Y. J. (1999). Food applications of chitin and chitosans. *Trends in Food Science & Technology*, 10(2), 37–51. DOI: 10.1016/S0924-2244(99)00017-5

Sharma, S., Parveen, R., & Chatterji, B. P. (2021). Toxicology of Nanoparticles in Drug Delivery. *Current Pathobiology Reports*, 9(4), 133–144. DOI: 10.1007/s40139-021-00227-z PMID: 34840918

Sharma, V., & Kundu, P. P. (2006). Addition polymers from natural oils—A review. *Progress in Polymer Science*, 31(11), 983–1008. DOI: 10.1016/j.progpolymsci.2006.09.003

Shinde, U. A., & Nagarsenker, M. S. (2009). Characterization of gelatin-sodium alginate complex coacervation system. *Indian Journal of Pharmaceutical Sciences*, 71(3), 313. DOI: 10.4103/0250-474X.56033 PMID: 20490302

Singh, T. A., Sharma, A., Tejwan, N., Ghosh, N., Das, J., & Sil, P. C. (2021). A state of the art review on the synthesis, antibacterial, antioxidant, antidiabetic and tissue regeneration activities of zinc oxide nanoparticles. *Advances in Colloid and Interface Science*, 295, 102495. DOI: 10.1016/j.cis.2021.102495 PMID: 34375877

Spotti, M. L., Acosta, C. A., Perez, A. A., Spotti, M. J., Quintero-Cerón, J. P., Carrara, C. R., & Fioramonti, S. A. (2024). Whey protein nanoparticles obtained by water in oil emulsification followed by heating: Influence of surfactant type on the stability, rheology and interfacial properties. *Food Hydrocolloids*, 153, 109912. DOI: 10.1016/j.foodhyd.2024.109912

Stryer, L., Berg, J. M., & Tymoczko, J. L. (2002). *Biochemistry and Study Guide* (5th ed.). W.H. Freeman.

Sudarshan, N. R., Hoover, D. G., & Knorr, D. (1992). Antibacterial action of chitosan. *Food Biotechnology*, 6(3), 257–272. DOI: 10.1080/08905439209549838

Sultan, M., Hafez, O. M., Saleh, M. A., & Youssef, A. M. (2021). Smart edible coating films based on chitosan and beeswax–pollen grains for the postharvest preservation of Le Conte pear. *RSC Advances*, 11(16), 9572–9585. DOI: 10.1039/D0RA10671B PMID: 35423446

Sundar, S., Kundu, J., & Kundu, S. C. (2010). Biopolymeric nanoparticles. *Science and Technology of Advanced Materials*, 11(1), 014104. Advance online publication. DOI: 10.1088/1468-6996/11/1/014104 PMID: 27877319

Taniguchi, N. (1983). Current status in, and future trends of, ultraprecision machining and ultrafine materials processing. *CIRP Annals*, 32(2), 573–582. DOI: 10.1016/S0007-8506(07)60185-1

Tavares, I. S., Caroni, A. L. P. F., Neto, A. D., Pereira, M. R., & Fonseca, J. L. C. (2012). Surface charging and dimensions of chitosan coacervated nanoparticles. *Colloids and Surfaces. B, Biointerfaces*, 90, 254–258. DOI: 10.1016/j.colsurfb.2011.10.025 PMID: 22078924

Timilsena, Y. P., Akanbi, T. O., Khalid, N., Adhikari, B., & Barrow, C. J. (2019). Complex coacervation: Principles, mechanisms and applications in microencapsulation. *International Journal of Biological Macromolecules*, 121, 1276–1286. DOI: 10.1016/j.ijbiomac.2018.10.144 PMID: 30352231

Tsai, G. J., & Su, W. H. (1999). Antibacterial activity of shrimp chitosan against Escherichia coli. *Journal of Food Protection*, 62(3), 239–243. DOI: 10.4315/0362-028X-62.3.239 PMID: 10090242

Turgeon, S. L., Schmitt, C., & Sanchez, C. (2007). Protein–polysaccharide complexes and coacervates. *Current Opinion in Colloid & Interface Science*, 12(4-5), 166–178. DOI: 10.1016/j.cocis.2007.07.007

Vodyashkin, A. A., Kezimana, P., Vetcher, A. A., & Stanishevskiy, Y. M. (2021). Biopolymeric Nanoparticles–Multifunctional Materials of the Future. *Polymers*, 14(11), 2287. DOI: 10.3390/polym14112287 PMID: 35683959

Wang, H., Han, S., Sun, J., Fan, T., Tian, C., & Wu, Y. (2012). Amphiphilic dextran derivatives nanoparticles for the delivery of mitoxantrone. *Journal of Applied Polymer Science*, 126(S1), E35–E43. DOI: 10.1002/app.36534

Welch, E. C., Powell, J. M., Clevinger, T. B., Fairman, A. E., & Shukla, A. (2021). Advances in biosensors and diagnostic technologies using nanostructures and nanomaterials. *Advanced Functional Materials*, 31(44), 2104126. DOI: 10.1002/adfm.202104126

Xiao, Z., Liu, W., Zhu, G., Zhou, R., & Niu, Y. (2014). A review of the preparation and application of flavour and essential oils microcapsules based on complex coacervation technology. *Journal of the Science of Food and Agriculture*, 94(8), 1482–1494. DOI: 10.1002/jsfa.6491 PMID: 24282124

Yetisgin, A. A., Cetinel, S., Zuvin, M., Kosar, A., & Kutlu, O. (2020). Therapeutic Nanoparticles and Their Targeted Delivery Applications. *Molecules (Basel, Switzerland)*, 25(9), 2193. DOI: 10.3390/molecules25092193 PMID: 32397080

Youssef, A. M., El-Sayed, H. S., El-Sayed, S. M., Fouly, M., & El-Aziz, M. A. (2023). Novel bionanocomposites based on cinnamon nanoemulsion and TiO2-NPs for preserving fresh chicken breast fillets. *Food and Bioprocess Technology*, 16(2), 356–367. DOI: 10.1007/s11947-022-02934-w

Youssef, A. M., El-Sayed, H. S., Islam, E. N., & El-Sayed, S. M. (2021). Preparation and characterization of novel bionanocomposites based on garlic extract for preserving fresh Nile tilapia fish fillets. *RSC Advances*, 11(37), 22571–22584. DOI: 10.1039/D1RA03819B PMID: 35480459

Yuan, H., Guo, H., Luan, X., He, M., Li, F., Burnett, J., Truchan, N., & Sun, D. (2020). Albumin nanoparticle of paclitaxel (Abraxane) decreases while taxol increases breast cancer stem cells in treatment of triple negative breast cancer. *Molecular Pharmaceutics*, 17(7), 2275–2286. DOI: 10.1021/acs.molpharmaceut.9b01221 PMID: 32485107

Zafar, A., Alruwaili, N. K., Imam, S. S., Alsaidan, O. A., Ahmed, M. M., Yasir, M., Warsi, M. H., Alquraini, A., Ghoneim, M. M., & Alshehri, S. (2021). Development and Optimization of Hybrid Polymeric Nanoparticles of Apigenin: Physicochemical Characterization, Antioxidant Activity and Cytotoxicity Evaluation. *Sensors (Basel)*, 22(4), 1364. DOI: 10.3390/s22041364 PMID: 35214260

Zhang, M., Li, M., Zhang, D., Yu, Y., Zhu, K., Zang, X., & Liu, D. (2024). Preparation and Investigation of Sustained-Release Nanocapsules Containing Cumin Essential Oil for Their Bacteriostatic Properties. *Foods*, 13(6), 947. DOI: 10.3390/foods13060947 PMID: 38540937

Zhang, S., Zhang, Z., & Vardhanabhuti, B. (2014). Effect of charge density of polysaccharides on self-assembled intragastric gelation of whey protein/polysaccharide under simulated gastric conditions. *Food & Function*, 5(8), 1829–1838. DOI: 10.1039/C4FO00019F PMID: 24920131

Zohri, M., Alavidjeh, M. S., Haririan, I., Ardestani, M. S., Ebrahimi, S. E. S., Sani, H. T., & Sadjadi, S. K. (2010). A comparative study between the antibacterial effect of nisin and nisin-loaded chitosan/alginate nanoparticles on the growth of Staphylococcus aureus in raw and pasteurized milk samples. *Probiotics and Antimicrobial Proteins*, 2(4), 258–266. DOI: 10.1007/s12602-010-9047-2 PMID: 26781321

Zolnik, B. S., González-Fernández, Á., Sadrieh, N., & Dobrovolskaia, M. A. (2010). Nanoparticles and the Immune System. *Endocrinology*, 151(2), 458–465. DOI: 10.1210/en.2009-1082 PMID: 20016026

KEY TERMS AND DEFINITIONS

Bioactive components: Bioactive compounds are chemicals found in small amounts in plants, foods, and animal products that may promote good health. They are also known as nutraceuticals.

Biodegradable: Any substance or object capable of being decomposed by bacteria or other living organisms and thereby avoiding pollution.

Biocompatible: Any substance which is not harmful or toxic to living tissue.

Biomaterials: A biomaterial is defined as any substance, whether natural or synthetic, that can be used in the body to treat, augment, or replace tissues, organs, or bodily functions. It includes materials used in the production of medical devices and can be categorized based on their intended function and interaction with the body.

Biopolymers: Biopolymers are natural polymers produced by the cells of living organisms. Like other polymers, biopolymers consist of monomeric units that are covalently bonded in chains to form larger molecules.

Cytotoxicity: Cytotoxicity is the quality of being toxic to cells.

Drug delivery: Drug delivery is a process or method to deliver a drug in human or animal to achieve its therapeutic effect. Selection of a drug delivery method has significant effect on the efficacy of the drug.

In –vivo: In the body.

Nanoparticles: A nanoparticle is a small particle that ranges between 1 to 100 nanometres in size. Undetectable by the human eye, nanoparticles can exhibit significantly different physical and chemical properties to their larger material counterparts.

Therapeutics: It means treatment and care of a patient for the purpose of both preventing and combating disease or alleviating pain or injury. The term comes from the Greek therapeutikos, which means "inclined to serve."

Toxicity: The quality of being toxic or poisonous.

Chapter 6
Plant-Mediated Synthesis of Biocompatible Metal Nanoparticles for Sustainable Agriculture

Hammad Ali Khan
https://orcid.org/0009-0009-0055-2423
Government College University, Faisalabad, Pakistan

Abdul Malik
University of Agriculture, Faisalabad, Pakistan

Qasim Waheed
Kohat University of Science and Technology, Pakistan

Muhammad Shahjahan
Government College University, Faisalabad, Pakistan

Behram Khan Ajat Khel
https://orcid.org/0009-0005-9978-8196
Abdul Wali Khan University, Mardan, Pakistan

Samrin Gul
https://orcid.org/0000-0002-6025-9538
University of Sargodha, Pakistan

Muhammad Huzaifa Mahmood
https://orcid.org/0000-0001-7303-1500
University of Turin, Italy

Ayesha Hameed
https://orcid.org/0000-0003-0904-1661
University of Gujrat, Pakistan

Aamir Raza
https://orcid.org/0009-0001-1867-2660
University of Agriculture, Faisalabad, Pakistan

Samreen Anjum
University of Engineering and Technology, Lahore, Pakistan

ABSTRACT

This chapter explores the use of plant-mediated production of biocompatible metal nanoparticles (MNPs) for sustainable agriculture. It highlights the benefits of this green synthesis method, including its environmental friendliness, affordability, and renewable resource use. The chapter discusses the processes by which phytochemicals in plant extracts act as reducing and capping agents, enabling the creation and stabilization of MNPs with specific features. It also discusses the characterization methodologies used to evaluate the dimensions, morphology, electrical charge, and compatibility with living organisms of the synthesized MNPs. The chapter also discusses the potential uses of MNPs in sustainable agriculture, such

DOI: 10.4018/979-8-3693-6240-2.ch006

as targeted nutrient delivery, plant development promotion, and environmentally friendly pest control. The chapter explores the use of IoT technology for immediate monitoring and accurate implementation of plant-mediated MNPs in agriculture, highlighting the financial implications, cost reductions, and market opportunities for environmentally friendly solutions.

1. INTRODUCTION

Agriculture, the foundation of civilization, faces unprecedented difficulties in the 21st century. Global population growth, expected to reach 9.7 billion by 2025, requires a comparable increase in food supply (Siegel & Siegel, 2015). Traditional agriculture, which uses pesticides and chemical fertilizers, has damaged the ecosystem. Agriculture is threatened by soil erosion, water pollution, and biodiversity loss. About 33% of Earth's soil has been degraded by erosion, salinization, acidification, and chemical pollution, according to the FAO (Gomiero, 2016). These issues make sustainable agricultural methods increasingly important because they boost output and safeguard the environment. Sustainable farming meets current food demands and preserves future food needs. It includes socially responsible, environmentally friendly, and profitable activities. Crop rotation, organic farming, integrated pest management, and renewable resources are becoming more popular worldwide.

Sustainable agriculture requires reducing chemical inputs and alternative strategies for improving soil fertility, managing pests, and protecting eco-friendly but effective crops. Nanotechnology, especially MNP-based nanotechnology, looks promising within this paradigm. Conventional MNP production uses toxic chemicals and complicated processes that increase cost and environmental impact. This chapter explores how plants can create biocompatible MNPs for sustainable agriculture using natural reducing and capping agents in plant extracts instead of standard synthesis methods.

After discussing phytochemicals as reducing and capping agents, the chapter will discuss how plant extracts assist MNP production and stabilization. This study presents plant species and extracts used to synthesize agricultural MNPs. Minnesota-made nanoparticles contain gold, iron, silver, and zinc oxide. The chapter will also discuss the characterization methods used to evaluate these metal nanoparticles' size, shape, electrostatic properties, and compatibility with living things. Many sustainable development applications of plant-mediated MNPs will be discussed in this chapter. They will be important for delivering nutrients to plants, promoting their growth, and controlling pests and diseases sustainably. The goal is to raise awareness of how green MNPs can help sustainable agriculture. This will provide insight for nanotechnology, agricultural, and environmental science professionals, scholars, and scientists. The chapter will emphasize using plants to make biocompatible MNPs as a sustainable alternative to present approaches. It will show how this strategy could revolutionize agriculture and solve environmental issues.

Figure 1 illustrates nanoparticles' multifunctional uses in agriculture. Nano fertilizers improve soil health, seed germination, plant growth, and agricultural output. Nanosensors measure soil heavy metal levels, improving soil management. Nanoremediators reduce pesticides, heavy metals, and persistent organic pollutants sustainably. Nanopesticides address plant pests, illnesses, and health. Nanomaterials boost plant tolerance to salinity stress, minimize heavy metal bioaccumulation, heat stress, and drought stress.

Figure 1. Overview of nanoparticle applications in agriculture.

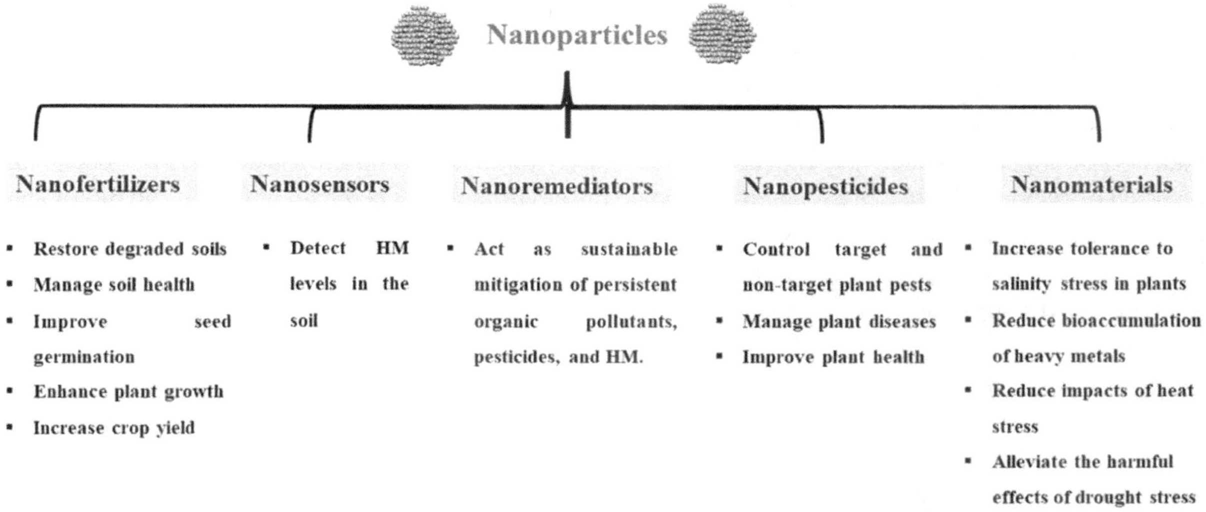

2. CHALLENGES OF CONVENTIONAL METHODS FOR MNP SYNTHESIS

2.1. Limitations of Traditional Methods

Due to their unique features and wide variety of applications, MNPs have been widely investigated and synthesized using standard physical and chemical processes (Shnoudeh et al., 2019). Traditional methods have limitations, leading the search for greener ones.

High production costs plague conventional MNP synthesis. SPuttering, ball milling, and laser ablation demand specialized equipment and energy. High-powered lasers ablate metals into nanoparticles. It needs expensive instruments and upkeep. Chemical reduction, sol-gel, and microemulsion require pure chemicals and solvents, making them pricey. Production expenses rise due to precise reaction conditions like temperature and pressure. Nanoparticle stabilization and cleaning need extra steps and resources, raising costs.

Traditional MNP synthesis and its environmental effects are contentious. Common organic solvents, reducing agents, and chemical synthesis produce hazardous waste. Reactive reducing agents like NaBH4 and N2H4 can contaminate and poison. Water and soil pollution from drug disposal endangers ecosystems and humans. Without chemicals, physical processes can harm the ecosystem. Laser ablation and ball milling are energy-intensive and carbon-intensive. These methods require surfactants and stabilizing chemicals to control nanoparticle size and structure, which can harm the environment if not adequately managed.

Complexity in nature makes conventional synthesis difficult. Nanoparticle production requires precise control of reaction parameters like temperature, pressure, pH, and reaction time in many physical and chemical processes. These sophisticated techniques are hard to scale and imitate, making bulk MNP manufacture problematic. In the sol-gel process, “sol,” a liquid, becomes “gel,”. Hydrolysis and condensation must be managed for this change. Any variation from ideal conditions can impact nanoparticle size, structure, and purity. Due to the difficulty of maintaining stable conditions over larger numbers, extending lab methods to industrial manufacturing is difficult.

Traditional MNP synthesis uses hazardous chemicals and byproducts that harm researchers and workers. Reducing agents, solvents, and nanoparticle dust can cause respiratory, skin, and general health issues when inhaled or touched. Nanoparticle inhalation irritates lungs. Reactive substance management and storage are complicated and expensive due to safety regulations and infrastructure. Poor chemical disposal or accidental releases endanger health and the environment. Chemical, physical, and biological approaches for metal nanoparticle synthesis are shown in Figure 2. Chemical synthesis produces toxins. The pricey Complex equipment and high-energy photons are used in Physical Method. Biosynthesis is difficult yet employs natural materials. These methods must be efficient and sustainable.

Figure 2. Conventional nanoparticle synthesis methods include chemical synthesis, physical processes, and biological synthesis

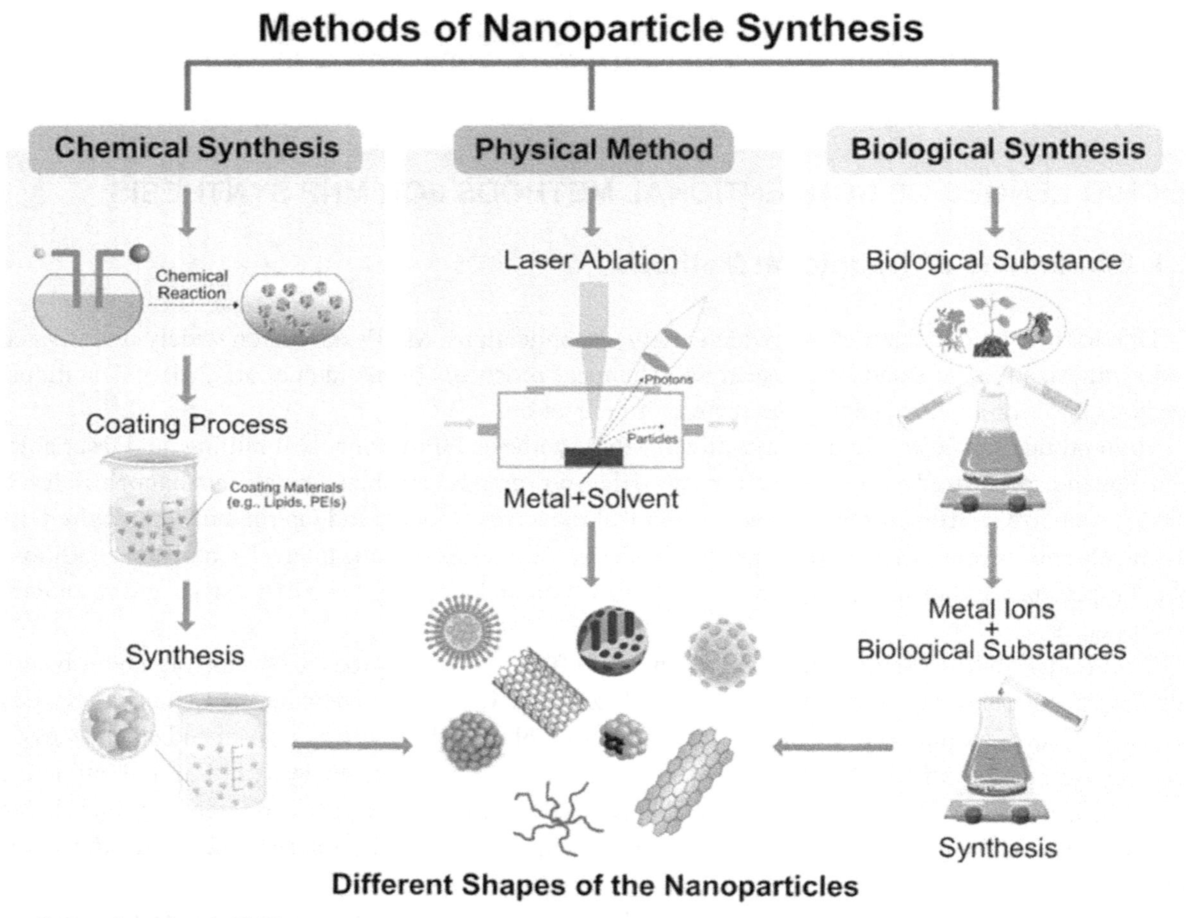

(Source, Jeyaraj et al., 2019)

Due to potential environmental and health risks, traditional MNPs are rigorously reviewed by regulators. Manufacturers, who spend a lot, need help to meet ecological standards including hazardous waste disposal and emissions management. The European Union's Registration, Evaluation, Authorization, and Restriction of Chemicals (REACH) regulatory framework sets strict criteria for chemical manufacturing and handling, including nanoparticles, to ensure safety and environmental preservation. Even though conventional physical and chemical methods for synthesizing MNPs have advanced nanotechnology, their high cost, potential ecological damage, complex procedures, safety concerns, and regulatory compliance issues highlight the need for more environmentally sustainable and eco-friendly alternatives. Plant-mediated technologies, particularly green synthesis methods, may overcome these constraints by using natural resources and lowering environmental impact. Figure 3 shows manmade, accidental, and natural nanomaterials. Natural technologies like viral and green synthesis are sustainable but difficult to govern, while artificial nanomaterials are dangerous. Improper handling can pollute and poison.

Figure 3. Different sources of nanomaterials and their implications on the environment

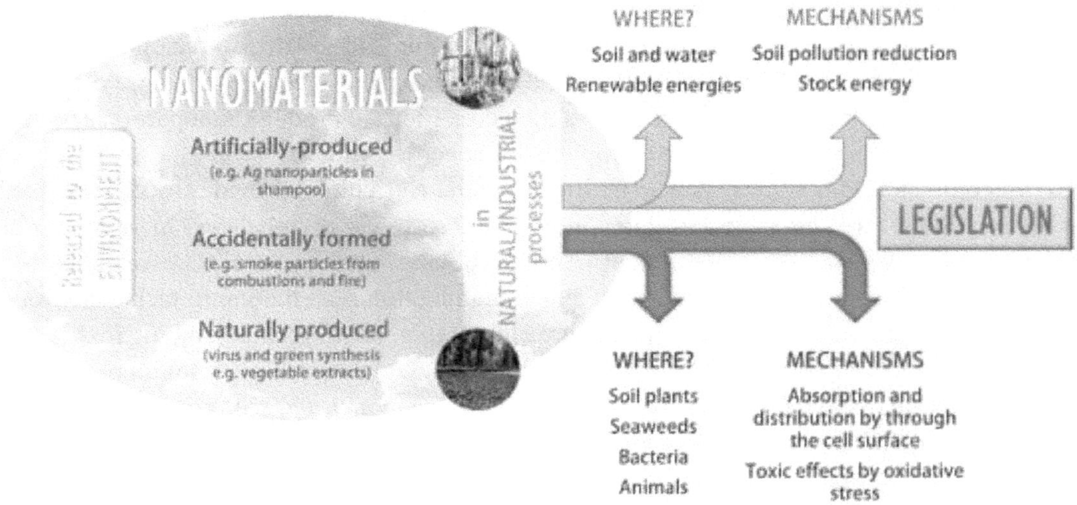

(Source, Martínez et al., 2020)

2.2. Green Synthesis as a Sustainable Alternative

Green synthesis (biological synthesis) makes MNPs from plant extracts, microbes, and enzymes (Jeevanandam et al., 2022). Green chemistry promotes safe, sustainable, and eco-friendly substances and procedures. Green synthesis seeks economical, appealing, and environmentally friendly nanoparticle production. A series of essential ideas distinguishes green MNP synthesis from conventional approaches.

Plants, algae, fungi, and bacteria are abundant and sustainable green synthesis resources. Environmentally utilizing these abundant natural resources reduces the usage of pricey, scarce chemicals. Green synthesis requires non-toxic reducing and stabilizing molecules. Plant extracts contain polyphenols, flavonoids, and terpenoids. These compounds reduce and stabilize metal ions into nanoparticles, preventing buildup. We develop green synthesis technologies to reduce environmental impact. They produce little waste and have no negative consequences at ambient temperature and pressure. Water solutions lessen the environmental impact of organic solvents. Energy efficiency specifies green synthesis at ambient temperature and pressure. Certain physical and chemical processes demand much energy. Green synthesis uses natural ingredients and simplified procedures to lower MNP manufacturing costs. This makes it appropriate for industrial and large-scale production.

Using green synthesis, organisms produce metallic nanoparticles from metal ions. The standard plant-mediated synthesis technique is discussed here. Cleaning and grinding leaves, roots, petals, and fruit yields powder. Heating the powder with distilled water separates bioactive components. When mixed with Zn^{2+}, Ag^{+}, and Au^{3+} metal nanoparticles in plant extract, polyphenols, alkaloids, and proteins are reduced. Nanoparticles of metal ions shrink phytochemicals. UV-Vis, TEM, and XRD show that these plant components cover and stabilize tiny particles, preventing aggregation and controlling their shape. Nanoparticle size, shape, and characteristics differ. The medical and environmental uses of nanoparticles match agriculture.

Green synthesis provides numerous advantages over conventional methods. Reducing hazardous waste and chemical pollution reduces pollution and health risks. The technique is simple to utilize for industrial manufacturing and requires no qualifications. Environmentally friendly nanoparticle production technologies are superior for drug distribution and cancer treatment since they are more biocompatible. Green synthesis adjusts biological supply and reaction conditions to produce nanoparticles with various characteristics. This adaptability facilitates application-specific synthesis.

MNPs' ecological technologies open up multiple sectors. Agriculture provides food, fiber, medicine, and other requirements. These bio-fertilizers, pesticides, and growth enhancers increase agricultural productivity and safeguard the environment. Antimicrobials, imaging, and pharmaceutical delivery systems use their biocompatibility—pollution and contamination remediation to tackle environmental issues. Sustainable environmental solutions require eliminating, treating, or containing toxic substances in waste treatment, pollution control, and water purification. Green synthesis is an innovative and sustainable MNP alternative. Traditional practices harm health, economy, and ecology. Naturally, biological systems solve them. This eco-friendly method represents green chemistry and allows nanotechnology to grow safely and sustainably in many industries.

2.3 Plant-Mediated Synthesis Economic Analysis

Plant-mediated metal nanoparticle (MNP) synthesis is cheaper than physical and chemical approaches. Laser ablation and chemical reduction involve expensive equipment, energy-intensive procedures, and toxic chemicals, which raises production costs (Shnoudeh et al., 2019). Green synthesis uses abundant natural resources, including plant biomaterials, which are cheap and eco-friendly (Jeevanandam et al., 2022).

Plant-mediated approaches can synthesize silver nanoparticles (AgNPs) for up to 50% less than conventional methods due to reduced chemical reagents and energy usage (Shafey, 2020). Green synthesis also simplifies waste disposal, saving industrial production costs (Jeyaraj et al., 2019). These economic advantages make plant-mediated synthesis a viable large-scale agricultural alternative.

2.4 Economic Analysis of Plant-Mediated Synthesis

Plant-mediated synthesis of metal nanoparticles (MNPs) presents a more cost-effective alternative to traditional physical and chemical methods. Traditional synthesis methods, such as laser ablation and chemical reduction, require expensive equipment, energy-intensive processes, and the use of hazardous chemicals, which significantly increase production costs (Shnoudeh et al., 2019). In contrast, green synthesis using plant extracts utilizes abundant natural resources, such as plant biomaterials, which are low-cost and eco-friendly (Jeevanandam et al., 2022).

A comparative analysis shows that the cost of synthesizing silver nanoparticles (AgNPs) using plant-mediated methods can be up to 50% lower than conventional methods, mainly due to the reduced need for chemical reagents and energy consumption (Shafey, 2020). Additionally, the simplified process of green synthesis reduces the need for complex waste disposal measures, which further contributes to cost savings in industrial-scale production (Jeyaraj et al., 2019). These economic advantages make plant-mediated synthesis a feasible alternative for large-scale agricultural applications.

3. PLANT-MEDIATED SYNTHESIS OF MNPS

Biological or green synthesis is the method of manufacture of metal nanoparticles (MNPs) employing plant extracts. This approach generates a plant extract solution with bioactive chemicals, including flavonoids, terpenoids, and polyphenols, by combining several sections of a plant's leaves, fruit, roots, stems, flowers, and seeds. These molecules lower metal ions to create nanoparticles, which are subsequently blended with a metal salt solution. As reducing agents, the bioactive chemicals in the extract start nucleation and gather nanoparticles. The hue shift in the solution denotes the effective nanoparticle development. This environmentally friendly methodology presents a viable substitute for more traditional chemical and physical synthesis techniques. Figure 4 shows the method of biological or green synthesis, that is, the synthesis of metal nanoparticles (MNPs) using plant extracts.

Figure 4. Illustration of plant-mediated synthesis of metal nanoparticles (MNPs) showing the use of plant extracts

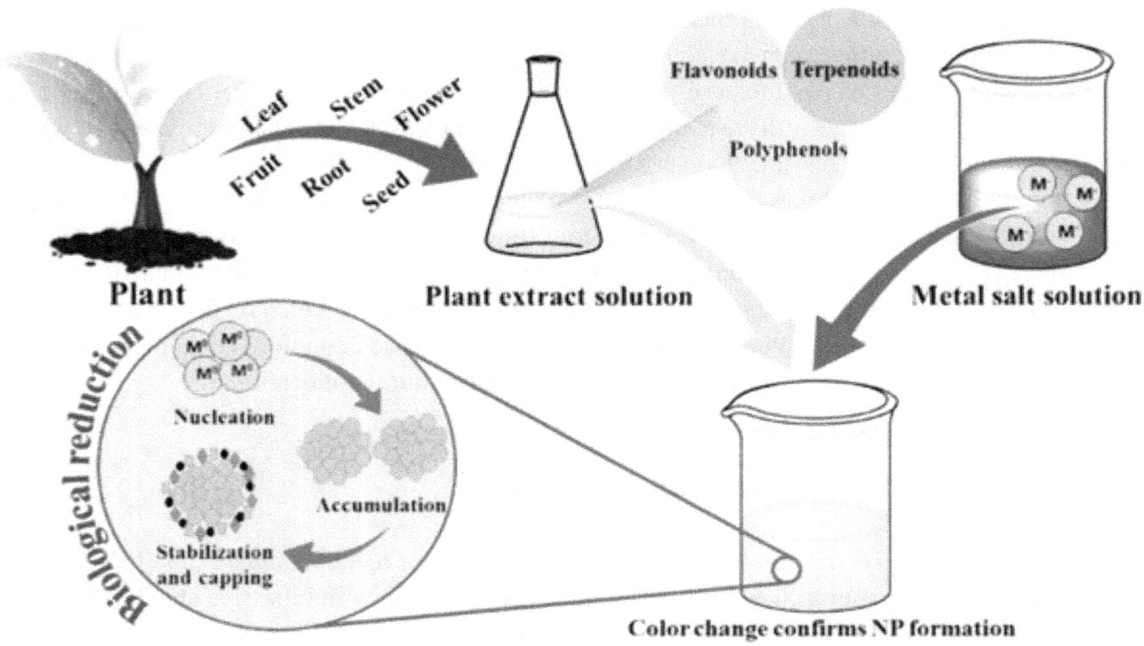

(Source, Dikshit et al., 2021)

3.1. Mechanisms of Plant Extracts in MNP Synthesis

Green synthesis offers a possible and ecologically friendly approach to producing MNPs from plant extracts (Shafey, 2020). Plant extracts are synthesized with both capping and reducing agents. They assist in developing and improving MNPs by means of complex systems combining several bioactive compounds. Plant extracts include polyphenols, tannins, saponins, alkaloids, terpenoids, flavonoids, and proteins, among other bioactive compounds (El-Seedi et al., 2019). Crucially, in the process of lowering metal ions, these compounds feature hydroxyl, carbonyl, and amine groups. The bioactive compounds in plant extracts coupled with a solution with metal ions (like Ag^+, Au^{3+}, and Zn^{2+}) transform the metal ions into metallic nanoparticles. By means of the electron-donating power of the phenolic groups present in polyphenols, silver ions (Ag^+) are converted into metallic silver (Ag^0) during the synthesis of silver nanoparticles (AgNPs).

The general reaction can be represented as:

$Ag^+ + \text{Phytochemical} \rightarrow Ag^0 + \text{Oxidized Phytochemical}$

Figure 5 illustrates the green synthesis of metal nanoparticles (MNPs) using plant materials. The process begins with the extraction of plant materials, which contain various bioactive compounds, to obtain the plant extract. This extract is then mixed with a metal salt solution. Through processes like heating, stirring, filtration, or calcination, the metal ions in the solution are reduced to form nanoparticles. This method results in the formation of nanoparticles that can be applied in various fields such as medicine, agriculture, food packaging, environment, textiles, and the dye industry, emphasizing their eco-friendly nature (Ahmed et al., 2023).

Figure 5. Green synthesis of metal nanoparticles using plant extracts, highlighting the process

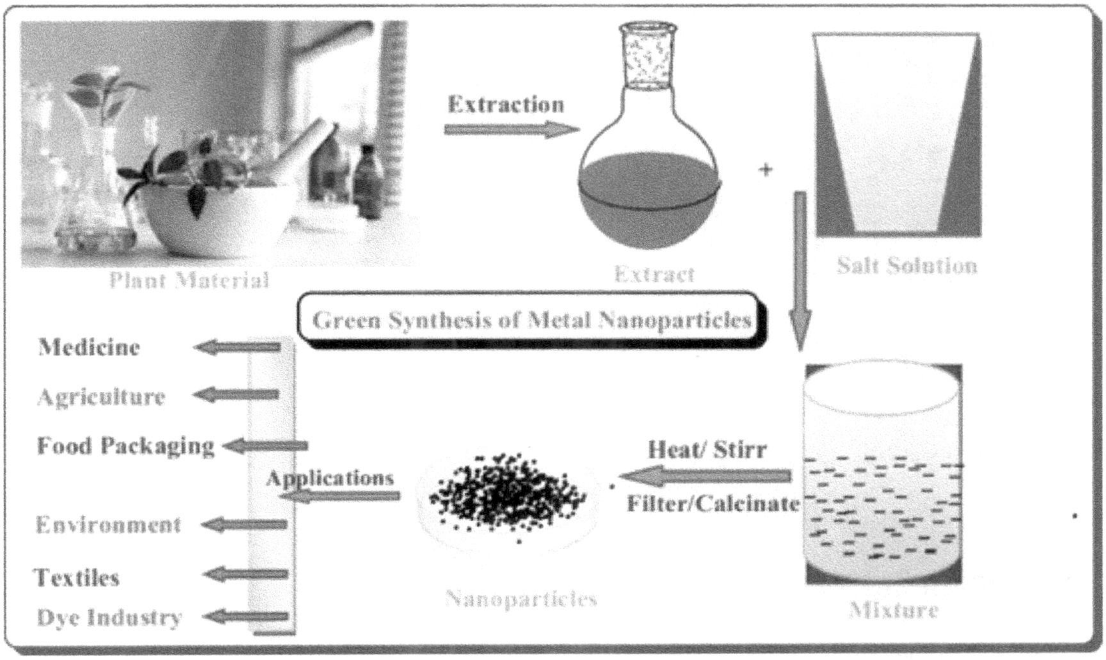

(Source, Ahmed et al., 2023)

Figure 6 provides a comprehensive overview of the plant-mediated synthesis and characterization of metal nanoparticles. It starts with the selection of plant components and microbial inoculum to prepare the plant extract. This extract is then combined with a metal salt solution, initiating the nanoparticle synthesis. The biological reduction process involves several stages: nucleation, accumulation, stability, and capping of nanoparticles. The resulting nanoparticles are then characterized using various techniques such as X-ray Diffraction (XRD), Fourier Transform Infrared Spectroscopy (FTIR), Energy Dispersive X-ray Analysis (EDX), Scanning Electron Microscopy (SEM), Transmission Electron Microscopy (TEM), Thermogravimetric Analysis (TGA), and UV-Visible Spectroscopy. The synthesized nanoparticles find applications in diverse areas like agriculture, medicine, food packaging, and wastewater remediation, showcasing the versatility and eco-friendly potential of this method (Dikshit et al., 2021).

Figure 6. Comprehensive process of plant-mediated synthesis and characterization of metal nanoparticles

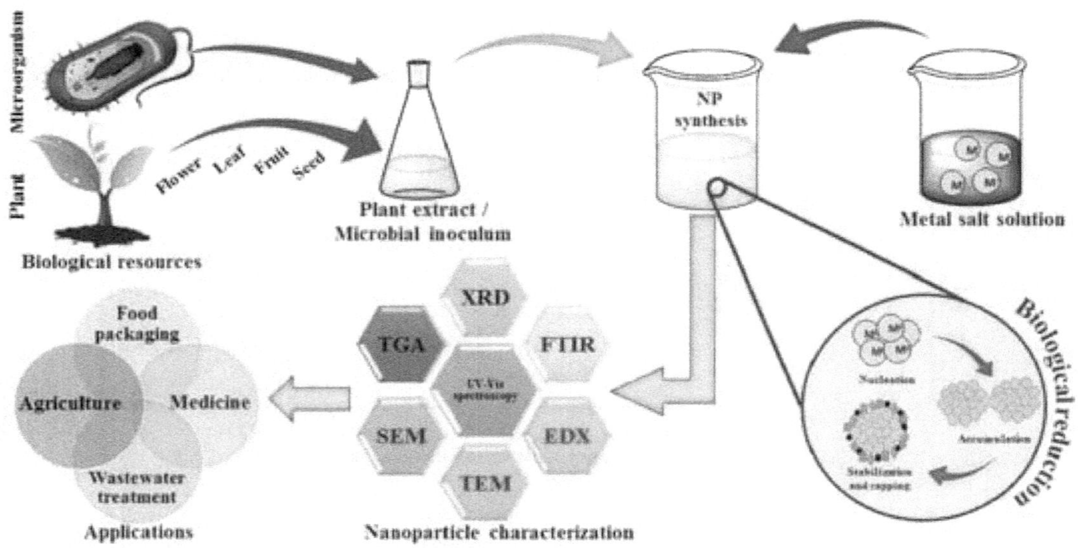

(Source, Dikshit et al., 2021)

3.2. Influencing Factors

Metal nanoparticle (MNP) size, shape, and surface characteristics are influenced by various factors, including plant species, components, extraction methods, and synthesis environment (Berta et al., 2021). Plant species like green tea and neem, rich in polyphenols and flavonoids, can form MNPs that cap and reduce, creating smaller, more stable nanoparticles. Protein-rich plants like soybean and wheat can impact MNP structure and longevity through complex metal ion interactions. Bioactive compounds in plant leaves, roots, stems, flowers, and fruits also affect MNP production and characteristics. The extraction method used, such as water or organic solvents, plays a crucial role in MNP production. The synthesis reaction environment, including pH, temperature, and metal ion concentration, also affects MNP size, shape, and surface characteristics (Shafaey, 2020). By intentionally choosing and optimizing these elements, scientists can customize nanoparticle size, configuration, and surface qualities for environmental restoration, medicine, and agriculture. Understanding these factors is essential for creating eco-friendly, sustainable technologies that maximize their potential.

4. CHARACTERIZATION TECHNIQUES FOR MNPs

4.1. Analytical Techniques

MNPs are defined, and their size, shape, surface characteristics, and other salient features are obtained using many analytical techniques. Understanding the physical, chemical, and structural aspects of MNPs that directly affect their performance in many uses enables one to offer three prominent methods in this subject. Several analytical methods are needed to understand MNPs and improve their efficiency for various applications. UV-Vis spectroscopy, TEM, DLS, XRD, FTIR, SEM, zeta potential testing, and EDS provide nanoparticle size, shape, surface characteristics, crystalline structure, and elemental composition (Catalano et al., 2021). These methods allow researchers to modify the synthesis process to manufacture MNPs with specified qualities, boosting their value in environmental restoration, industry, and medicine.

4.1.1. Ultra-Violet Visible (UV-Vis) Spectroscopy

UV-Vis spectroscopy uses nanoparticles to absorb UV or visible light. Surface plasmon resonance (SPR) from conduction electron vibrations on metal nanoparticles vibrating with approaching light gives them distinctive optical properties. A colloidal solution of MNPs is prepared for UV-Vis spectroscopy, and its light absorption spectra are quantified over multiple wavelengths. The surface plasmon resonance peak wavelength indicates nanoparticle size and organization, which determines peak absorption. The surface plasmon resonance (SPR) spectrum for silver nanoparticles (AgNPs) is 400–450 nm. A red shift in this peak indicates particle size change or aggregation. UV-Vis spectroscopy may measure particle size because smaller particles absorb light at shorter wavelengths and larger particles at longer wavelengths. SPR peak variations can indicate nanoparticle aggregation or dispersion.

Figure 7 shows the UV-Vis absorption spectra of silver (Ag), gold (Au), copper (Cu), and platinum (Pt) nanoparticles. Each nanoparticle has a unique SPR peak at a wavelength, revealing its optical properties. Ag has absorbance maxima at 452 nm, Au at 567, Cu at 613, and Pt at 368. The platinum nanoparticle absorption spectrum is shown inset at a wider wavelength range. Knowing the size, distribution, and probable aggregation of nanoparticles in a colloidal fluid is crucial.

Figure 7. UV-Vis absorption spectra of silver, gold, copper, and platinum nanoparticles, showing distinct SPR peaks

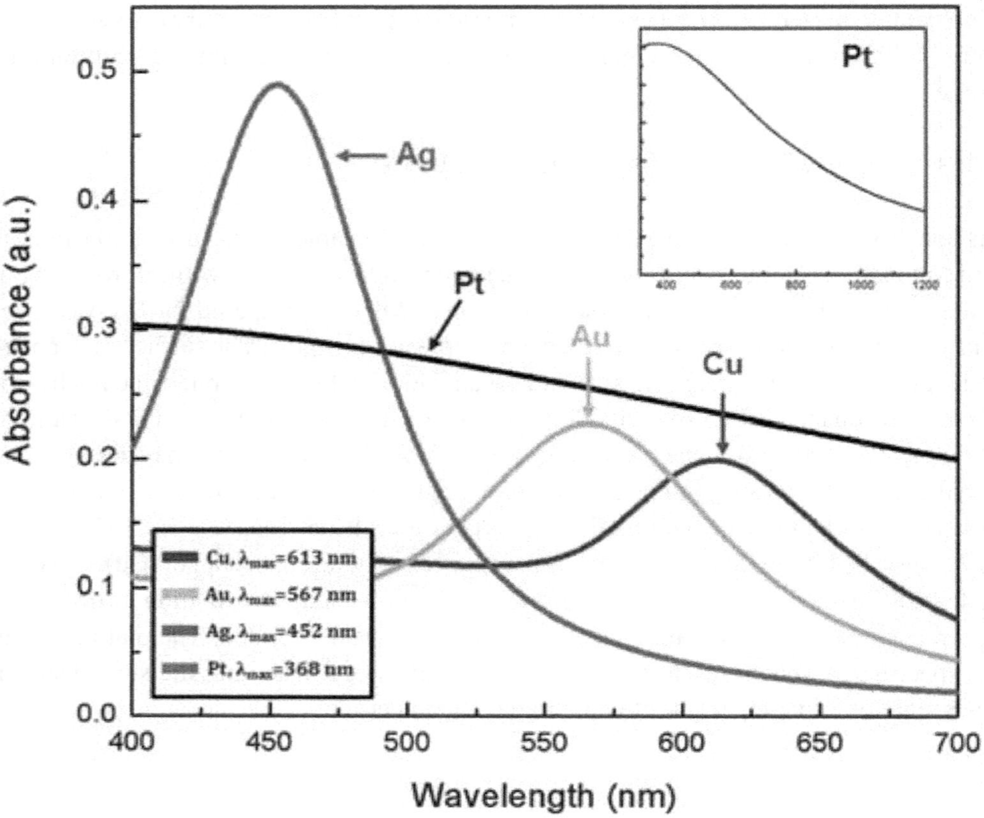

(Source, Shin et al., 2010)

4.1.2. Transmission Electron Microscopy (TEM)

TEM uses an electron beam and a fragile material to provide an image of nanoparticle size, form, and structure. A small amount of solution with nanoparticles can be placed on a copper grid covered with carbon film for transmission electron microscopy (TEM). The TEM instrument receives the dry grid. Detectors pick up electrons the sample lets through to produce images. High-resolution transmission electron microscopy (TEM) pictures of nanoparticles allow exact measurement of their size and form. HRTEM can show nanoparticle defects and their crystalline shape. TEM pictures of gold nanoparticles (AuNPs) may show spherical particles with tightly packed sizes, usually 10–20 nm.

4.1.3. Dynamic Light Scattering (DLS)

Directly connected to their size, DLS underlines Brownian motion random movement of nanoparticles in a liquid media. The method consists of examining the variations in the intensity of scattered light generated by moving nanoparticles. Dynamic light scattering (DLS) is performed with a cuvette to hold a colloidal state of nanoparticles. The DLS apparatus then features an internal cuvette. A laser beam whose scattered light is measured at a predetermined angle targets the sample. We find the hydrodynamic size of the nanoparticles by use of changes in the intensity of the scattered light. The size distribution of nanoparticles in a liquid solution can be obtained by dynamic light scattering (DLS). It can enable the identification of changes in particle size resulting from clusters or environmental influences. This is seen from the dynamic light scattering (DLS) investigation of zinc oxide nanoparticles (ZnONPs), which can expose a hydrodynamic diameter between 30 and 50 nm. This measurement suggests that in a liquid solution, the particles are consistently spaced.

4.1.4. X-ray Diffraction (XRD)

One can find the crystalline arrangement of nanoparticles by analyzing the diffraction patterns produced when X-rays traverse the sample. The patterns generate details on phase composition and crystal structure. Inside the X-ray diffraction (XRD) equipment, a powdered sample of nanoparticles is positioned, and the X-rays are exactly focused toward the sample. Different angles of sight reveal the diffraction-oriented ray, so the resulting diffraction pattern is studied to find the crystal structure and phase degree of purity. X-ray diffraction (XRD) is one approach that may be able to determine the phase composition and crystal structure of nanoparticles. Applying the Scherrer equation and considering the spreading of diffraction peaks helps one to shrink the crystallite. X-ray diffraction (XRD) can help iron oxide nanoparticles (Fe^3O^4) confirm their spinel crystal structure and project the size of the individual crystals.

4.1.5. Fourier Transform Infra-Red (FTIR) Spectroscopy

FTIR spectroscopy measures infrared radiation absorption using nanoparticles, therefore revealing information on their chemical composition and surface functional groups. Using the FTIR technique, they mix a minute quantity of nanoparticle sample with potassium bromide (KBr), and compression produces a pellet. After that, this pellet fits the FTIR spectrometer and records the absorption spectra. FTIR spectroscopy allows one to find the specific functional groups and chemical bonds existent on the surface of nanoparticles. It also can detect any changes in surface chemistry resulting from capping or functionalizing agents. For AgNPs covered in plant extract, for instance, the FTIR spectra might reveal distinct peaks corresponding to hydroxyl, carboxyl, and amine groups.

4.1.6. Scanning Electron Microscopy (SEM)

By means of a concentrated stream of electrons, a scanning electron microscope (SEM) generates finely detailed images exposing the form and surface composition of nanoparticles. Under the SEM approach, a minute quantity of dried nanoparticle sample is positioned on a conductive substrate and covered with a thin film of gold or carbon. Images are obtained by secondary or backscattering electron detection after the specimen is under the scanning electron microscope (SEM). The Scanning Electron Microscope (SEM) approach allows one to study the surface texture and structure of nanoparticles. It provides a perceptive study of their size and form, which balances the TEM (Transmission Electron Microscope) data obtained. SEM pictures clearly reveal the constant size distribution and spherical morphology of titanium dioxide nanoparticles (TiO^2).

4.1.7. Zeta Potential Measurements

Zeta potential quantifies the electric charge present on a liquid suspension surface of nanoparticles. This charge influences the inclination and stability of the nanoparticles in the direction of aggregation. The electrophoretic mobility of the particles under an electric field decides it. In the zeta potential measuring process, a zeta potential analyzer receives a colloidal solution of nanoparticles. One generates an electric field, and particle velocity can be observed. One learns about the zeta potential using electrophoretic mobility. Zeta potential helps to define the stability of nanoparticles in a liquid solution. Strong stability corresponds either with a positive or negative absolute zeta potential value. On the other side, low numbers indicate a propensity for the nanoparticles to cluster together. Zeta potential studies on gold nanoparticles (AuNPs) covered with citrate ions, for instance, allow one to obtain a negative value. The negative value of the electrical repulsion between the nanoparticles shows their stability.

4.1.8. Energy-dispersive X-ray Spectroscopy (EDS)

Elemental composition study of nanoparticles is usually obtained using EDS coupled with SEM or TEM. The method is the identification under electron impact of the specific X-rays generated by the elements in the sample. Under a scanning electron microscope (SEM) or transmission electron microscope (TEM), the EDS technique positions the nanoparticle sample under a detector suited for energy-dispersive X-ray spectroscopy. The produced X-rays are analyzed following scanning to ascertain the elemental composition. By offering both qualitative and quantitative information on the composition of nanoparticles, energy-dispersive X-ray spectroscopy (EDS) helps to identify secondary components or pollutants within the sample. EDS studies of silver nanoparticles (AgNPs) help to verify the existence of silver and highlight any production-related contaminants.

4.2. Physicochemical Properties

Understanding the physicochemical properties of MNPs facilitates one to grasp their behavior, stability, and application value in many different fields. The next section clarifies the analytical techniques that provide comprehensive data on the qualities of interest dimensions, shape, electrostatic potential, chemical constitution, and crystal lattice arrangement. In this sense, we clarify how each technique enhances awareness of particular physical features. Combining different analytical methods guarantees a thorough understanding of MNP physicochemical characteristics. Researchers can maximize the synthesis process and personalize nanoparticles for particular tasks by means of size, shape, surface charge, chemical composition, crystalline structure, and optical properties (Mourdikoudis et al., 2018). These realizations will enable the development of efficient, stable, and useful nanoparticles relevant to various spheres, including environmental research, medicine, and agriculture.

4.2.1. Size and Shape

TEM and SEM

Transmission electron microscopy (TEM) and scanning electron microscopy (SEM) provide high-resolution pictures that enable the exact size of nanoparticles to be estimated. Approaching the nanoscale range, transmission electron microscopy (TEM) is rather successful in precisely measuring the sizes of individual nanoparticles. Transmission electron microscopy (TEM) and scanning electron microscopy (SEM) can give complete information on the structure and morphology of the nanoparticles regardless of their geometries: spherical, rod-shaped, cubic, or more complex. For uses like catalysis or drug distribution where the shape influences functionality, this is critically crucial. For instance, whereas transmission electron microscopy (TEM) photographs of gold nanoparticles can show spherical particles with diameters ranging from 10 to 20 nanometers, scanning electron microscopy (SEM) can provide information regarding the surface roughness and shape of bigger nanoparticles.

Figure 8 TEM pictures show different-sized and shaped nanoparticles. With a scale bar of 200 nm, it presents a homogeneous array of nanoparticles showing somewhat constant size and dispersion. With a scale bar of 100 nm, this figure shows a closer view of nanoparticles with a more finely detailed shape. (c) Shows a single nanoparticle with a surrounding 100 nm scale bar among smaller particles. With a 20 nm scale bar, it offers a dramatically magnified perspective of two nanoparticles, stressing nanoscale-level intricate structures. These pictures show how well TEM can provide high-resolution visualizations, which are essential for examining nanoparticle size and form.

Figure 8. TEM images illustrating various nanoparticles, showcasing their size and shape at different magnifications

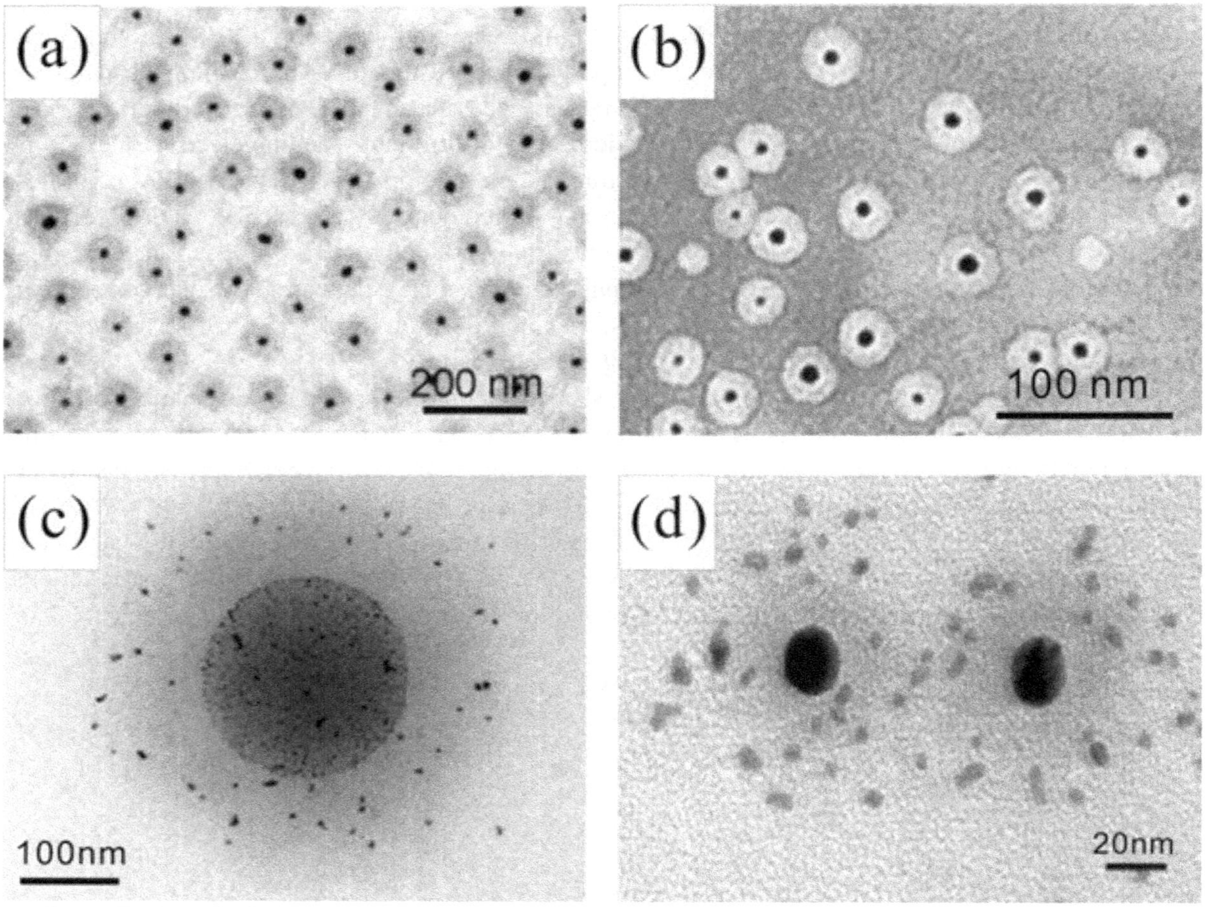

(Source, Tan & Lee 2017)

DLS

DLS determines the hydrodynamic radius of nanoparticles in a liquid media, therefore offering details on average particle size and dispersion. This knowledge helps one assess the aggregating state of nanoparticles in a colloidal fluid. For example, a DLS investigation on zinc oxide nanoparticles would reveal a hydrodynamic diameter between 30 and 50 nm, implying an equal dispersion of the particles in the fluid.

4.2.2. Surface Charge and Stability

Zeta potential tests give the surface charge of nanoparticles by means of their colloidal stability, so guiding knowledge of this information. Positive or negative big absolute zeta potential values show the existence of repulsive solid forces among particles, so indicating strong stability and resistance to cluster development. Moreover, the zeta potential shows the consistency of nanoparticles in suspension. Stable nanoparticles usually have zeta potential values that are either higher than +30 mV or lower than -30 mV. For gold nanoparticles covered in citrate ions, for instance, a zeta potential of -40 mV shows outstanding stability brought about by electrostatic repulsion.

4.2.3. Chemical Composition and Functional Groups

FAIR

FTIR spectroscopy is the technique utilized to identify and investigate the functional groups and chemical bonds present on the surface of nanoparticles. One needs this knowledge in surface chemistry as well as any interactions with other materials or biological systems. FTIR can identify changes in surface chemistry by means of functionalization or capping agents, so offering useful data on the effectiveness of surface alterations. For example, the FTIR spectra of silver nanoparticles covered in plant extract could reveal various peaks corresponding to hydroxyl, carboxyl, and amine groups. These peaks indicate a good method of coating.

EDS

EDS offers both qualitative and quantitative information about the composition of nanoparticles. One should verify the existence of specific metals and identify other elements or pollutants. Energy-dispersive X-ray spectroscopy (EDS) is used to evaluate the purity of the produced nanoparticles so that they satisfy specified application criteria. By means of EDS analysis of silver nanoparticles, one can verify the existence of silver and identify any generated contaminants during manufacturing.

4.2.4. Crystalline Structure and Phase Composition

XRD

X-ray diffraction (XRD) can determine the crystalline structure of nanoparticles by means of their diffraction patterns. This book offers particular details on atom spatial arrangement inside the nanoparticles and phase composition. By enlarging diffraction peaks in the XRD series, the Scherrer equation enables one to approximate the size of crystallites. By means of X-ray diffraction (XRD) investigations on iron oxide nanoparticles, one could validate their spinel crystal structure and offer an approximative size of the individual crystals.

HRTEM

HRTEM enables careful investigation of the crystalline arrangement by allowing the viewing of atomic-level details to help find defects like lattice faults, dislocations, and grain boundaries. By disclosing their crystallographic orientation and structure, HRTEM offers important fresh viewpoints on the stability and development mechanisms of individual nanoparticles. Gold nanoparticle HRTEM images provide complete information on the atomic arrangement and expose any structural defects in the particles.

4.2.5. Optical Properties

SPR Phenomenon

UV-Vis spectroscopy detects the light absorption of nanoparticles, whose size, form, and chemical composition affect their surface plasmon resonance (SPR) phenomena. The strength and placement of the SPR peak provide an interesting study of these features. Variations in particle size, shape, or aggregation state shown by changes in the SPR peak help to define the homogeneity and stability of the nanoparticles by means of useful information. Usually showing a peak in the 400–450 nm range, surface plasmon resonance (SPR) of silver nanoparticles corresponds with UV-Vis spectroscopy of them. A redshift, that is, a movement toward longer wavelengths, indicates either growing particle size or the aggregation process.

5. APPLICATIONS OF PLANT-MEDIATED MNPS IN AGRICULTURE

The synthesis of metal nanoparticles (MNPs) using plant-mediated methods presents numerous applications in agriculture, leveraging the environmentally friendly and sustainable nature of green synthesis. This approach utilizes the inherent reducing and capping agents in plant extracts to produce biocompatible MNPs, which can significantly enhance agricultural practices. Figure 9 outlines the various methods for synthesizing nanoparticles, categorized into physical, chemical, and biological methods. It highlights the focus on biological methods, mainly green synthesis using plants and microorganisms. Physical Methods Include techniques such as electrochemical processes, ultra-sonication, laser ablation, irradiation, evaporation, and condensation—Chemical Methods Encompass methods like etching, mechanical milling, laser ablation, sputtering, and thermal decomposition. Biological Methods (Green Synthesis) Utilize plants (leaves, roots, flowers, fruits) and microorganisms (fungi, bacteria, algae, yeast) for synthesizing nanoparticles. Plant-mediated synthesis is particularly emphasized due to its sustainable approach, where different parts of plants (leaves, roots, flowers, fruits) are used to produce nanoparticles.

Figure 9. Methods for nanoparticle synthesis, highlighting physical, chemical, and biological (green synthesis) approaches

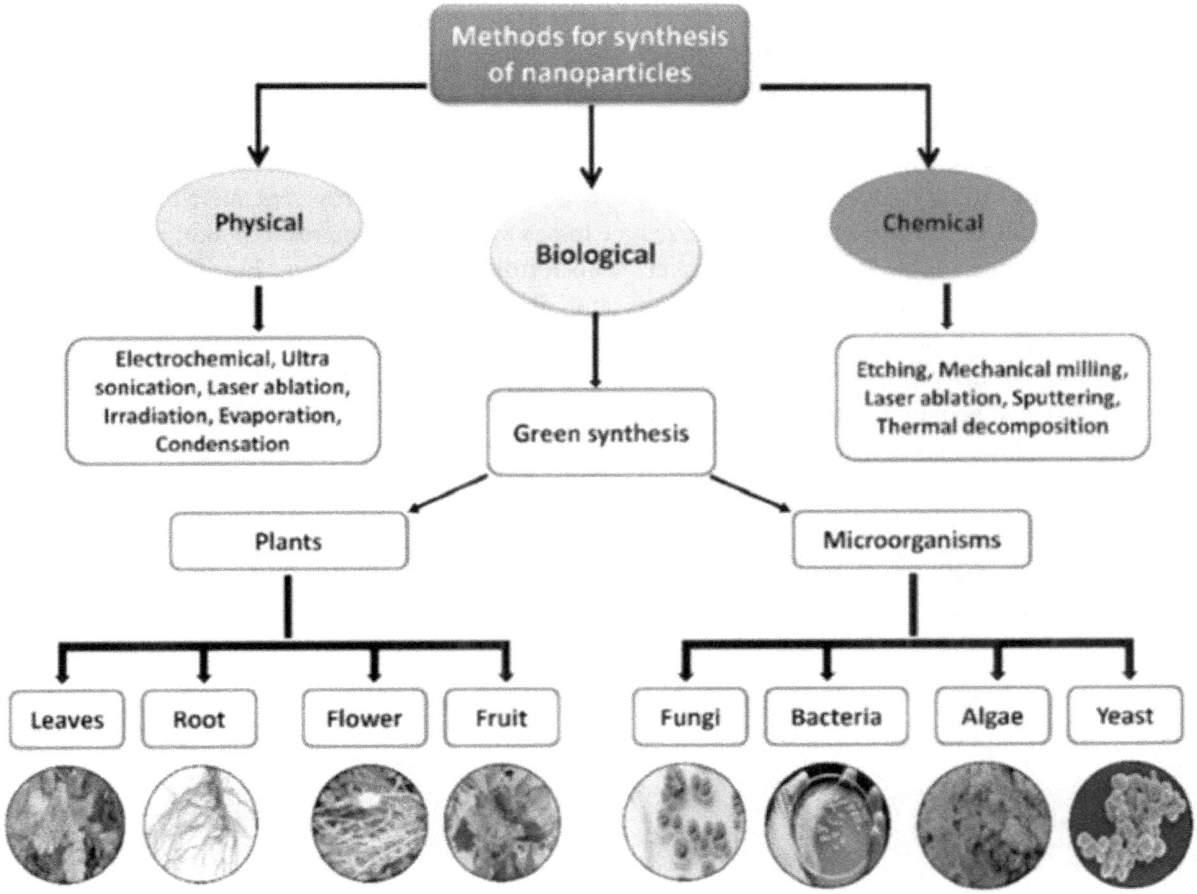

(Source, Rani et al., 2023)

5.1. Targeted Nutrient Delivery

Using MNPs in agriculture has opened more opportunities to raise agricultural efficiency and output; one quite exciting use is targeted nutrient delivery (Le Wee et al., 2022). Conventional methods of fertilizer application usually result in significant losses brought about by low absorption efficiency, volatilization, and leaching. This increases costs and helps to contaminate the surroundings. Sustainable means to improve the accuracy and efficacy of nutrient transfer to plants are plant-mediated metal nanoparticles (MNPs). Plant-mediated MNPs could fundamentally change agricultural nutrient delivery. MNP-based fertilizers offer a sustainable and innovative solution to existing farming by enhancing nutrient usage efficiency, minimizing environmental consequences, and raising crop production. Future research should give top importance to solving the issues linked to their application to guarantee the safe and efficient integration of this possible technology into agricultural techniques.

Figure 10 shows the major role nanotechnology plays in transforming agricultural methods through the use of nanofertilizers. These sophisticated fertilizers improve the efficiency and sustainability of nitrogen delivery to plants, therefore optimizing resource consumption and lowering environmental effects. The figure emphasizes numerous important features of nano fertilizers, including their capacity to give stable suspension, boost surface area coverage, and distribute nutrients more precisely, thus improving the efficiency of nutrient administration. Designed to control several biotic and abiotic challenges like nutrient deficits, salinity, water shortages, drought, heat, light, and pest pressures, nano fertilizers minimize nutrient movement to non-target areas, reduce losses by means of improved photolysis protection, and boost pest control, so enabling Moreover, nano fertilizers promote nano-bio interaction, which guarantees greater absorption, regulated release, and targeted supply of nutrients, thereby enhancing the physicochemical characteristics of soil and plants. By encouraging effective resource management, ensuring food security, and improving public acceptability, these developments greatly help to foster sustainability and economic progress. In line with the objectives of sustainable agriculture, the use of nano fertilizers thus results in enhanced crop yields, better control of environmental pressures, and general improvement of the quality of the ecosystem.

Figure 10. Significant nanotechnology-based agriculture nano fertilizers result in more sustainable and effective farming methods.

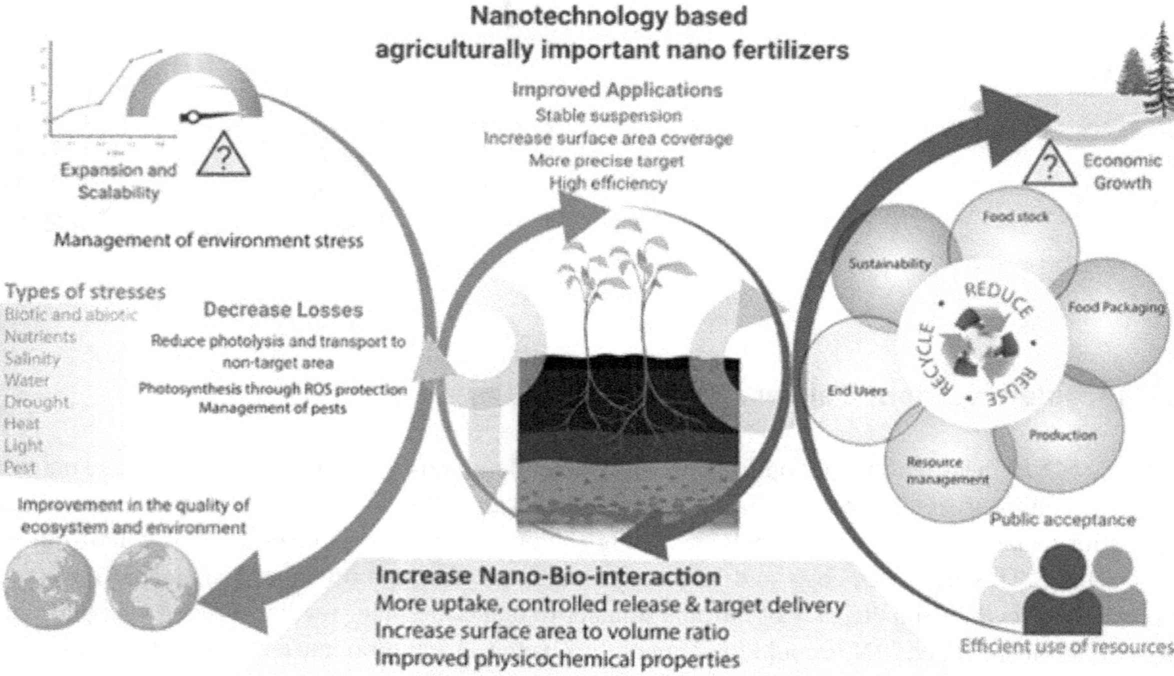

(Source, Mittal et al., 2020)

5.1.1. Mechanisms of Targeted Nutrient Delivery

Targeting nutrients with MNPs is the method of conjugating or encapsulating important nutrients with nanoparticles and subsequently transporting them to the plant system (Manna & Bandyopadhyay, 2019). The special features of MNPs' small size, great surface area, and capacity to be modified with different ligands help to enable effective absorption and distribution inside plant tissues. Enclosed either inside or connected to MNPs are nutrients like nitrogen (N), phosphorous (P), potassium (K), and micronutrients, including zinc and iron. This encapsulation ensures the long availability of nutrients for plants, therefore preventing early degradation or loss of them. MNPs can be regulated to release nutrients in a controlled way in response to particular environmental stimuli such as pH, temperature, or root exudates. This guarantees that nutrients will be accessible to plants at their most need, thereby maximizing their utilization and lowering waste. Given their small size, MNPs pass through plant cell walls and membranes more precisely than other nutrient transporters can. Functionalizing MNPs improves their attraction to plant root surfaces by means of biocompatible ligands, hence improving their absorption and movement inside the plant.

5.1.2. Advantages of Targeted Nutrient Delivery with MNPs

MNPs could help plants to make far better use of nutrients. Although controlled release systems guarantee a constant supply of nutrients that meet plant needs, lowering the necessity for regular fertilizer and encapsulation helps to limit nutrient waste. MNP-based fertilizers reduce the risk of water and soil contamination by controlling nutrient losses resulting from leaching and volatilization. This promotes the implementation of more environmentally friendly farming techniques and shields ecosystems against the negative consequences of excessively high fertilizer consumption. For farmers, MNP-based fertilizers are financially reasonable even if their initial expenses are higher. Long-term gains from them include reduced fertilizer application rates, increased agricultural yields, and less environmental cleanup expenses. Customizing MNPs enables one to satisfy the particular dietary needs of various crops. This customization guarantees ideal nutrient balance for every crop, thereby improving development and production.

5.1.3. Examples of MNPs in Targeted Nutrient Delivery

Important for plant growth, zinc also stimulates the synthesis of proteins and activates enzymes. Thanks in part to zinc oxide nanoparticles (ZnONPs), wheat and maize, among other crops, have shown improved absorption of zinc (Al Jabri et al., 2022). Studies have shown that by providing sufficient zinc, ZnONPs can enhance general biomass (Chen et al., 2023), raise chlorophyll levels, and improve root development. Synthesis of chlorophyll and plant electron transport depends on iron. Effective iron transfer from iron oxide nanoparticles to plants helps to lower the frequency of iron shortage chlorosis. Fe^3O^4 nanoparticles (NPs) increase iron absorption in crops such as soybean and spinach, so encouraging better and more strong growth. Silicon builds plant cell walls and increases stress tolerance. Silicon's distribution to plants can be enhanced by silicon dioxide nanoparticles (SiO^2NPs), therefore enhancing their resistance to both biotic and abiotic stressors. Studies on rice and cucumbers indicate that SiO^2NPs can increase agricultural output, encourage plant development, and strengthen resistance against diseases and pests.

5.2. Plant Growth Promotion

Plant-mediated MNPs present a sustainable and effective approach to increasing plant health and productivity by numerous mechanisms; MNPs applied in agriculture have shown considerable potential in boosting plant development and raising agricultural output (Kanwar et al., 2019). These benefits span improved meal absorption, hormonal function stimulation, environmental pressure resistance, and illness avoidance. Plant-mediated metal nanoparticles (MNPs) offer great possibilities for promoting plant development and increasing agricultural output. Micronutrient nanoparticles (MNPs) present a sustainable and efficient solution to contemporary agriculture by enhancing the absorption of nutrients, inducing hormonal activity, and shielding crops from illnesses and environmental problems. By means of the safe and efficient integration of MNPs into agricultural systems, addressing the issues of their use would help to enable environmental sustainability and world food security.

5.2.1. Mechanisms of Plant Growth Promotion by MNPs

Metal nanoparticles (MNPs) can help plants to absorb vital nutrients. Their small size and great surface area enable them to enter plant tissues more successfully, therefore delivering nutrients exactly to their required locations (Su et al., 2019). Boosting the absorption of zinc, a necessary element for activating enzymes and synthesis of proteins, zinc oxide nanoparticles (ZnONPs) thereby support plant development and growth. MNPs can affect the hormonal equilibrium in plants by boosting the synthesis of growth-promoting hormones like auxins, gibberellins, and cytokinins. Studies have revealed that the presence of silver nanoparticles (AgNPs) can influence the amounts of these hormones, so either more root elongation or shoot development is promoted. Metal nanoparticles (MNPs) have antibacterial characteristics that help plants fight bacterial, fungal, and viral infections. This preventive measure improves the general state of plants and lessens the occurrence of disease. Copper nanoparticles (CuNPs) have significant antifungal characteristics that help shield crops, including tomatoes and cucumbers, against fungal invasion. Metal nanoparticles (MNPs) enable plants to resist environmental conditions, including lack of water, too high salt levels, and the presence of hazardous heavy metals. They accomplish this by regulating the expression of genes connected to stress and boosting antioxidant action. TiO^2NPs and titanium oxide nanoparticles increase agricultural photosynthetic efficiency and drought resilience, including in wheat and maize.

5.2.2. Examples of MNPs in Plant Growth Promotion

Particularly in terms of increased plant height, root length, and biomass, studies on maize (Zea mays) revealed that ZnONPs greatly influenced the development of the plants. Improved the growth of the nanoparticles (Srivastav et al., 2021), which is required for the synthesis of chlorophyll and the activation of enzymes by means of better absorption of zinc. By raising the zinc availability in the soil, zinc oxide nanoparticles (ZnONPs) ensure that plants obtain sufficient zinc for this essential mineral. Studies on soybean (Glycine max) have shown that AgNPs generated higher root and shoot development by varying the levels of growth-promoting hormones (Jampílek & Kráľová, 2019). Moreover, nanoparticles protect against root rot brought on by soil bacteria. AgNP silver nanoparticles interact with plant hormonal systems to increase auxin and cytokinin levels. These hormones are absolutely crucial for both root and shoot development. Fe3O4NPs raised general biomass, root growth, and chlorophyll content in spinach

(Spinacia oleracea). The synthesis of chlorophyll and the electron flow during photosynthesis depends on iron's enhanced accessibility created by the nanoparticles; it provides a continuous supply of iron, which supports vigorous plant development and helps avoid iron deficiency chlorosis (Li et al., 2024).

Copper nanoparticles (CuNPs) were proven to boost the development and output of cucumber (Cucumis sativus) by shielding the plants against fungal diseases and so increasing nutrient absorption (Jampílek & Kráľová, 2022). By means of their impacts on solid antifungal characteristics of copper nanoparticles (CuNPs), illness occurrence is reduced, and better development is encouraged. Using titanium oxide nanoparticles (TiO^2NPs), improved photosynthetic efficiency, and wheat (Triticum aestivum) drought resilience were obtained. The nanoparticles enhanced light absorption and hence lowered oxidative stress caused by dry surroundings. TiO^2NPs assist plants in more effectively responding to environmental needs by raising their photosynthetic efficiency.

5.2.3. Advantages of Using MNPs for Plant Growth Promotion

Crop productivity and quality could be improved significantly by using MNPs. Improved nutrient uptake and improved resistance to diseases create better and more plentiful plants. Higher yields and improved nutritional value are revealed by studies on crops produced under ZnONPs and Fe3O4NPs. Reducing reliance on pesticides and artificial fertilizers will enable MNPs to promote acceptance of more ecologically friendly farming methods. This decline helps to slow down environmental harm and lowers farmer input expenses. CuNPs and AgNPs, for instance, have antibacterial qualities that aid in lowering the requirement for chemical bactericides and fungicides. Micronutrient powders (MNPs) assist in protecting the environment by improving nutrient use efficiency and reducing dependency on chemical inputs. They help to lower the environmental consequences of farming methods, preserve soil fertility, and hence lower water contamination. For example, the controlled release of MNPs lowers nutrient leakage into aquatic bodies, therefore preventing eutrophication.

5.3. Disease and Pest Control

MNPs from plant extracts can control agricultural diseases and pests sustainably (Maity et al., 2022). Antibacterial, antifungal, and pesticidal properties distinguish plant-derived MNPs. These traits protect crops against pests and illnesses. Silver nanoparticles (AgNPs) made from neem (Azadirachta indica) leaves have strong antibacterial and antifungal effects against Pseudomonas syringae, Xanthomonas campestris, Fusarium oxysporum, and Aspergillus niger. Tomato plants treated with neem-derived silver nanoparticles (AgNPs) had less bacterial spot disease, boosting plant health and crop yield. Antifungal copper nanoparticles (CuNPs) from aloe vera leaves prevent Phytophthora infestans and Alternaria solani, which cause late and early blight in potatoes and tomatoes, respectively. Copper nanoparticles (CuNPs) reduced late blight in potato plants, boosting plant health and tuber production.

Zinc oxide nanoparticles (ZnONPs) from Holy Basil (Ocimum sanctum) leaves are antibacterial and antifungal. Nanoparticles treat infectious diseases like Ralstonia solanacearum and Botrytis cinerea. ZnONPs also reduce whiteflies and aphids, decreasing the spread of viral diseases. ZnONPs reduced gray mold on grapevines, enhancing fruit quality and yield. Iron oxide nanoparticles (Fe3O4NPs) made from Moringa oleifera leaves have strong antibacterial and antifungal properties against a wide range of pathogens, including Xanthomonas oryzae and Magnaporthe oryzae. They reduce the severity of bacterial leaf blight and rice blasts in rice plants, improving plant development and agricultural output.

Turmeric plant roots have been utilized to produce gold nanoparticles (AuNPs), which are antibacterial against Agrobacterium tumefaciens and antifungal against Sclerotinia sclerotiorum. AuNPs stimulate the plant's immune system, making it more disease-resistant. Turmeric-derived gold nanoparticles reduced white mold on soybeans, improving plant vigor and yield.

Plant-based, it's sustainable and environmentally friendly. They use renewable and natural resources, decreasing pesticide use and environmental impact. They are a safer pesticide alternative since they reduce crop residues and human and non-target animal risks. MNPs protect crops and reduce pesticide use by targeting a wide range of hazardous bacteria and organisms. MNPs also increase plant health and growth, boosting crop quality and farm output.

Still, several challenges must be resolved to employ plant-mediated metal nanoparticles safely and successfully in agriculture. Long-term effects on soil fertility, beneficial microorganisms, and unwanted species need further study. Clear restrictions are needed to apply these technologies safely. Scalable and affordable production methods must be developed simultaneously for widespread use. Overall, plant-mediated MNPs are a good and environmentally friendly way to control agricultural pests and diseases. Their antimicrobial properties and eco-friendly production make them a preferable pesticide alternative. Further research and development will help solve current issues and fully utilize MNPs to increase crop protection and productivity, boosting sustainable agriculture worldwide.

5.4 Case Studies and Real-world Applications

Several case studies have demonstrated the effectiveness of plant-mediated MNPs in improving agricultural practices. For instance, the use of zinc oxide nanoparticles (ZnONPs) synthesized from green tea extracts has been shown to increase wheat yield by 20%, improve root growth, and enhance chlorophyll content, all of which contribute to higher crop productivity (Chen et al., 2023). Similarly, silver nanoparticles (AgNPs) synthesized using neem extracts have been used successfully in disease management, reducing bacterial and fungal infections in tomato plants by 40% (Jampílek & Kráľová, 2019).

These real-world applications highlight the practical benefits of MNPs in agriculture, particularly in improving plant growth and disease resistance while minimizing environmental harm. The scalability of these solutions remains a challenge, but ongoing research is working to optimize the production processes for industrial use.

6. BIOCOMPATIBILITY AND ENVIRONMENTAL CONSIDERATIONS

6.1. Importance of Biocompatibility

MNP biocompatibility is crucial for agricultural use. MNPs can be used without harming the environment or living things. We term this biocompatibility. Biocompatibility ensures that MNPs improve crop output and health without harming plants, soil, beneficial microbes, animals, or consumers. In this session, we discuss biocompatibility and pertinent concerns for employing MNPs (metal nanoparticles) in agriculture. Biocompatibility is crucial to MNPs' agricultural use. To employ MNPs in sustainable agriculture, they must be safe for plants, soil ecosystems, non-target species, and humans. Researchers can generate MNPs with agricultural benefits and lower environmental and health concerns by optimizing synthesis, doses, and application. Nanotechnology in agriculture will boost food production and sustainability if biocompatibility is studied and improved (Batool et al., 2010).

Figure 11 shows the direct and indirect effects of metal and metallic oxide-engineered nanoparticles in soils. Nanoparticles in pesticides, insecticides, herbicides, fungicides, and fertilizers are direct routes. Cosmetics, clothing, plastic additives, pigments, semiconductors, and antibacterial agents use them indirectly. These nanoparticles alter soil hydraulic conductivity, porosity, texture, bulk density, water content, and aggregation. Cation exchange capacity, soil pH, redox potential, electrical conductivity, and nutrient availability are also affected. To protect the environment and living things, nanoparticle biocompatibility must be considered because these changes in soil parameters can dramatically impact soil health and quality.

Figure 11. Impact of metal and metallic oxide engineered nanoparticles on soil health and quality

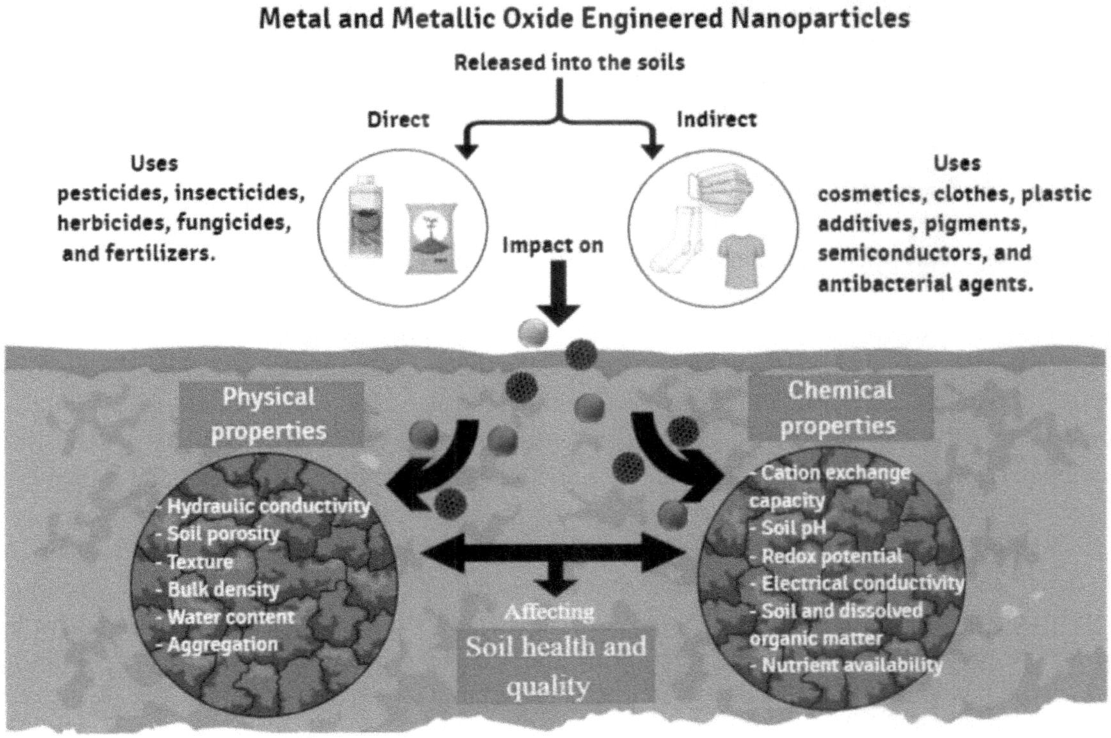

(Source, Suazo-Hernández et al., 2023)

6.1.1. Key Aspects of Biocompatibility in Agriculture

Biocompatible metal nanoparticles (MNPs) must not harm crops they benefit. This protects plant growth, development, and physiology from nanoparticles. Toxins diminish growth, chlorosis, necrosis, and germination. MNPs should improve nutrition absorption and usage without imbalances or shortages. Zinc oxide nanoparticles (ZnONPs) may boost zinc availability without affecting other element absorption. The soil microbiome maintains soil structure, nutrient cycling, and plant health. Biocompatible metal nanoparticles must not harm nitrogen-fixing bacteria, mycorrhizal fungi, or decomposers. MNPS shouldn't affect soil pH, texture, or water-holding ability to harm plants. They should mix into the soil without harming it (Adeyemi et al., 2021).

Biocompatibility is a substance's compatibility with non-use species like other fauna, natural predators (ladybugs, spiders), and pollinators. These important species should be protected by MNP design. Avoiding hazardous residues requires MNPs to decay or become non-toxic, which stops terrestrial-aquatic interactions. Crop-based biocompatible MNPs should not affect people. This involves preventing edible plant tissues from accumulating human-harmful nanoparticles. To ensure the safe use of agricultural MNPs in food production systems, the FAO and EPA created safety criteria (Oriola et al., 2021).

Neem-derived AgNPs are antibacterial and treat plant ailments. According to studies, silver nanoparticles (AgNPs) in suitable concentrations do not damage soil microbes or plant growth. The slow decline diminishes accumulation and environmental presence. Zinc oxide nanoparticles (ZnONPs) boost maize and wheat zinc absorption and nutrition. Zinc oxide nanoparticles (ZnONPs) boost plant growth and yield without phytotoxicity. In authorized quantities, they rarely influence soil microbes. Fe^3O^4NPs restore rice and spinach iron. Fe3O4NP iron promotes plant growth. They retain soil microbes and fungi and sustain the ground (Adeyemi & Fawole, 2023).

6.1.2. Strategies to Enhance Biocompatibility

Crucially, the appropriate metal nanoparticle (MNp) concentration must be exactly determined. Although a tiny quantity would not provide the expected advantages, an excess can be hazardous. Dose-response studies are essential in determining suitable and efficient application rates that ensure both safety and efficacy. Functionalizing MNPs with biocompatible coatings such as polysaccharides or proteins helps one to lower possible toxicity and improve one's lifetime. Furthermore, these coatings enhance the interactions among MNPs, plant surfaces, and soil particles. Extensive environmental impact studies (EIAs) should be conducted to identify potential dangers and provide strategies of action to reduce them before MNPs are generally applied in agriculture. These tests ought to encompass their circumstances as well as investigations on soil condition, water cleanliness, and effects on non-targeted species. Using plant-mediated synthesis methods for MNP manufacturing guarantees that the nanoparticles are ecologically benign from the beginning. These techniques reduce agents by means of reduction and encapsulation using natural molecules that minimize their impact on the environment, minimizing the possibility of introducing dangerous substances into it (Adeyemi et al., 2021).

6.2. Environmental Implications

Agricultural MNPs have environmental pros and cons. MNPs must be thoroughly evaluated to ensure sustainable and safe agricultural practices, even if they increase crop output, fertilizer distribution, and pest control. The expected ecological implications of rural MNP use on soil fertility, water systems, non-target species, and ecosystem stability are explored here. Agriculture benefits from MNPs' increased agricultural output, disease management, and nutrition. To ensure safe and sustainable farming systems, we must effectively control the environmental impacts of these operations. Scientists and legislators can limit hazards by designing nanotechnology agricultural plans and understanding how MNPs affect soil quality, water systems, non-target species, and ecosystem stability. The constant study, environmental monitoring, and sustainable MNP manufacturing and application technologies will be needed to balance the pros and negatives of MNPs in agriculture (Batool et al., 2010).

Figure 12 summarizes MNP applications in agriculture and their environmental impacts: A: Displays emulsions, filled microgels, micelles, colloidosomes, liposomes, nanosomes, and biopolymer nanoparticle-based emulsions. B displays nanoparticle translocation criteria via plant structures based on size, charge, and surface makeup. C: Shows how plants absorb nanoparticles used as nano-pesticides and nano-fertilizers utilizing folious, branch feeding, trunk injection, and root uptake. D: Summarize nanoparticles' benefits and harms for plants based on their physicochemical properties. The benefits include increased light absorption, root and shoot growth, nutrient transfer, and fruit production. Genotoxicity, plant form changes, and water and nutrient absorption are negative effects.

Long-term environmental and health effects of MNPs in agriculture are currently being studied, including soil health, water quality, and biodiversity. Although plant-mediated MNPs are biocompatible, prolonged exposure to high quantities of nanoparticles like silver (AgNPs) and zinc oxide nanoparticles (ZnONPs) may disturb soil microbial populations and nutrient cycling (Gomiero, 2016). Runoff of MNPs can also poison aquatic organisms. Suazo-Hernández et al. (2023) found that titanium dioxide nanoparticles (TiO2NPs) can cause oxidative stress in fish and invertebrates, potentially harming aquatic ecosystems. Preliminary studies imply negligible health concerns from eating crops produced in nanoparticle-enriched soil when MNPs are utilized within specified limits (Adeyemi et al., 2021).

Figure 12. Comprehensive overview of the application, uptake routes, and environmental implications of MNPs in agriculture

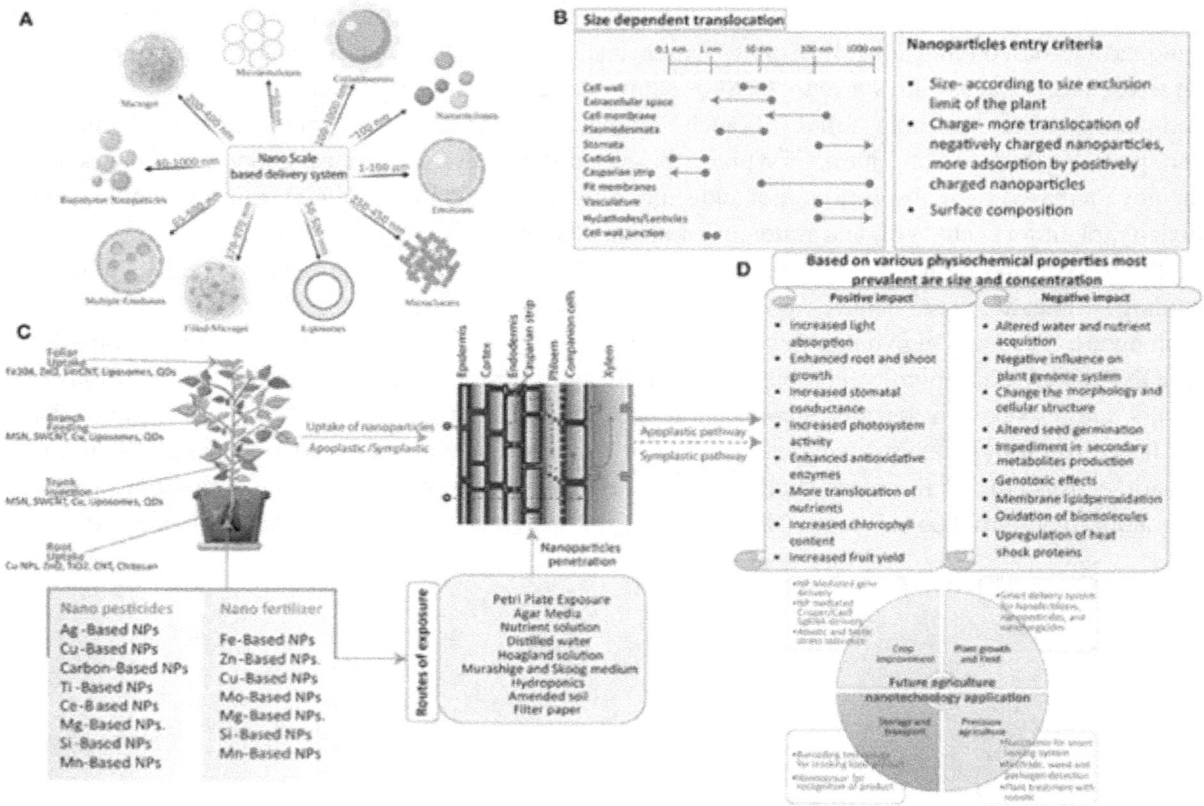

(Source, Mittal et al., 2020)

6.2.1. Impact on Soil Health

Necessary for the process of nutrient cycling, breakdown of organic matter, and the well-being of plants, MNPs could affect the diversity, amount, and functionality of soil microbial communities. Studies have shown that increased amounts of silver nanoparticles (AgNPs) can prevent the spread of beneficial soil bacteria and fungi, hence possibly upsetting the soil microbiome. Interaction between MNPs and soil enzymes helps to control their activity and thereby affects soil biochemical processes. Studies have indicated, for example, that zinc oxide nanoparticles (ZnONPs) can alter the activity of soil enzymes such as urease and phosphatase, which are vital for nitrogen and phosphatase cycling. The pH, texture, and structure of the soil can be changed by the addition of metal nanoparticles (MNPs), therefore influencing water retention, aeration, and root development. Copper nanoparticles (CuNPs) have been shown to affect aggregate stability and soil pH, therefore influencing both plant development and soil erosion (Oriola et al., 2021).

6.2.2. Impact on Water Systems

Run-off and leaching let MNPs applied in agricultural areas into water systems and maybe poll surface and groundwater. Agricultural runoff has been discovered to include silver nanoparticles, which raises issues regarding their probable influence on aquatic habitats and sources of drinking water. MNPs present in water can be absorbed by aquatic life; they build up in their bodies and may be harmful as they move up the food chain. Studies reveal that titanium dioxide nanoparticles (TiO2NPs) can build up in aquatic life, including fish and invertebrates, generating oxidative stress and other detrimental effects (Adeyemi & Fawole, 2023).

6.2.3. Impact on Non-Target Organisms

MNPs can endanger pollinators like bees and butterflies as well as beneficial insects like ladybugs and spiders either directly by touch or by swallowing polluted nectar and pollen. The behavior and survivability of honeybees, which are pretty important for providing industry pollination services, have been demonstrated to be affected by silver nanoparticles. MNPs can travel the food chain, so their effect on higher trophic levels might differ. This covers birds and other animals consuming polluted prey or vegetation. Studies have shown plants to be able to absorb gold nanoparticles (AuNPs), which can subsequently be passed on to herbivores and their predators. This has sparked concerns about possible long-term environmental consequences (Adeyemi et al., 2021).

6.2.4. Ecosystem Stability

MNPs added to agricultural systems might upset the balance of species interactions, causing possible loss of biodiversity and changing ecosystem services. Like AgNPs, antimicrobial nanoparticles, when widely used, can reduce the diversity of microorganisms in soil and water, therefore influencing the process of nutrient cycling and the capacity of ecosystems to recover from disruptions. The combined and cumulative effects of several MNPs, as well as other agricultural inputs like pesticides and fertilizers, could provide synergistic effects on the surroundings. The synergy between MNPs and conventional agrochemicals could magnify their toxicity, therefore causing unexpected impacts on the environment (Batool et al., 2010).

6.2.5. Strategies for Mitigating Environmental Impacts

By designing MNPs with formulations that enable controlled release, one can help to decrease the number of MNPs emitted into the environment, hence lowering the possible leaching and runoff potential. Encapsulating MNPs in biodegradable polymers can attain their controlled release, therefore enhancing their effectiveness and reducing their environmental effects. By using plants capable of absorbing and storing MNPs from the soil, contaminated sites can be restored, and environmental contamination can be slowed down. By lowering their mobility and bioavailability, hyperaccumulator plants could help to clean metal nanoparticle-polluted soils. By using plant-mediated synthesis methods for MNPs, one less depends on dangerous chemicals and produces nanoparticles that may be less detrimental to the environment. By using plant extracts, the synthesis of MNPs helps produce environmentally safe and harmonic nanoparticles in line with natural ecosystems. Comprehensive ecological monitoring systems will help us monitor MNP occurrence and its effects on agricultural surroundings. After that, this data could help direct risk analyses and create workable plans of action. Following exact rules and guidelines for the safe use of MNPs in agriculture guarantees their benefits and helps to lower possible environmental risks (Adeyemi & Fawole, 2023).

7. FUTURE DIRECTIONS

Fast-growing plant-mediated metal nanoparticle (MNP) production for agriculture presents research opportunities to improve agricultural practices. Future studies will increase MNPs' ecological sustainability, safety, and efficacy, benefiting crop development and protection. Synthesis approach optimization has huge potential for research. MNP synthesis must be constant to ensure homogeneity and replicability across plant species and extraction methods. Researchers should prioritize optimizing plant components, extraction conditions, metal ion concentration, and reaction time. These reduced technologies must be industrialized to be profitable. Green chemistry also enables plant extract-based synthesis that is environmentally benign. Studying several plant species and developing abundant, affordable, and effective plant-derived reducing and capping agents can manufacture MNPs with minimal environmental impact.

Understanding how functionalizing MNPs are enhanced is another important field of study. Understanding the complex principles of MNP growth using plant extracts would help explain how phytochemicals and metal ions create nanoparticles. Advanced analytical approaches, including high-resolution microscopy, spectroscopy, and molecular modeling, can assist in developing this understanding. Improvements to MNP surfaces are needed to improve their agricultural applications. MNP stability, bioavailability, and plant-specific capacity will improve with research on functionalizing them with biocompatible chemicals, ligands, or polymers. MNPs that release pesticides or nutrients under control in response to environmental cues are needed. Environmental and toxicological research will determine MNP safety in agriculture. MNPs' long-term environmental and toxicological effects on agriculture require comprehensive risk assessments. MNP fate, mobility, and modification in terrestrial and aquatic environments should be studied. After assessing soil health, microbial ecosystems, non-target species, and food safety, legal norms and public approval follow. Developing biocompatible and biodegradable MNPs can also lower their environmental toxicity and persistence. MNP breakdown routes and byproducts in the region, biodegradable nanoparticle manufacturing, and functionalization are important issues of study.

Intelligent nano-pesticides and nano-fertils are another option. Creatively designed nano fertilizers can deliver nutrients to crops efficiently and controllably, eliminating nutrient losses. The study should prioritize developing MNPs that selectively release nutrients based on plant demands or environmental circumstances. Use stimuli-responsive materials that change behavior in response to pH, temperature, or moisture. Nano-pesticides can lessen environmental effects even if they control insects perfectly. Researchers could design more efficient and environmentally friendly pest management methods by building MNPs that largely target pests and pathogens without harming beneficial organisms. Plant-mediated metal nanoparticle (MNP) synthesis has great potential in agriculture, transforming crop output and protection. Researchers can safely and sustainably employ MNPs by optimizing synthesis, understanding mechanical aspects, conducting environmental assessments, and developing intelligent agricultural nanotechnologies. These projects will improve agricultural, food security, sustainability, and technology applications.

7.1 Practical Integration Guidelines

Integrating MNPs into traditional farming practices requires careful consideration of dosage, application methods, and crop-specific needs. Studies suggest that using MNP-based fertilizers at low concentrations can significantly improve nutrient uptake and plant growth without causing phytotoxicity (Li et al., 2024). For instance, the use of Fe3O4 nanoparticles in soybean cultivation has resulted in a 15% increase in iron content and overall biomass (Srivastav et al., 2021).

Depending on the specific crop and environmental conditions, farmers can integrate MNPs through soil amendments, foliar sprays, or seed treatments. Educational programs and training sessions are essential to ensure the safe and effective use of these technologies. Best practices for integrating MNPs include regular soil testing, adherence to recommended application rates, and monitoring for any adverse effects on crop health.

7.2 Public Perception and Acceptance

Public perception plays a critical role in the adoption of nanotechnology in agriculture. While MNPs offer numerous benefits, including enhanced crop yields and reduced environmental impact, there is a general lack of public awareness about the safety and sustainability of these technologies (Batool et al., 2018). Public concerns often revolve around potential risks to human health and the environment, which can hinder widespread acceptance.

To address these concerns, stakeholders must engage in transparent communication, providing clear evidence of the safety and benefits of MNPs. Public education campaigns, workshops, and collaborations with agricultural communities can help demystify nanotechnology and foster trust. Acceptance is likely to increase as more research validates the safety of plant-mediated MNPs in food production systems (Oriola et al., 2021).

8. CONCLUSION

This chapter investigates the technique of generating biocompatible MNPs from plants and their possible uses for sustainable agriculture. The proclamation emphasizes the increasing need for sustainable solutions and their benefits from green synthesis. Green synthesis generates MNPs at an environmentally friendly and reasonably reasonable cost by means of plant extracts as reducing and capping agents. The chapter looks at how polyphenols, flavonoids, and proteins bioactive compounds could ensure their necessary size, shape, and surface properties for use in agriculture and help to manufacture MNPs. Many factors influence MNP's size, shape, and surface properties: plant species selected, specific plant parts employed, extraction methods applied, and synthesis conditions used. Characterizing techniques include UV-Vis spectroscopy, TEM, SEM, DLS, XRD, FTIR, and zeta potential studies, which offer a thorough knowledge of MNP physicochemical properties.

In agriculture, plant-mediated MNPs find application in nutrient delivery to particular sites, plant development promotion, and pest and disease management. Illustrations show how well MNPs made from plant extracts could increase nutrient absorption, activate hormonal activities, fight against diseases, and foster resilience to stress, thus producing more abundant crops. With an eye on strategies to increase biocompatibility and minimize environmental damage, the book explores the significance of biocompatibility and ecological elements. Future directions of research could be improving synthesis techniques, learning fundamental mechanisms, investigating environmental and toxicological concerns, and producing enhanced nano fertilizers and nano-pesticides with cognitive capacity. In essence, modern agriculture may benefit much from the sustainable approach of producing biocompatible MNPs from plants. It offers a workable substitute for the traditional MNP synthesis. Surmounting current challenges and using MNPs ' complete capabilities for a more ecologically friendly and strong agricultural future will depend on further research and development.

REFERENCES

Adeyemi, J. O., & Fawole, O. A. (2023). Metal-based nanoparticles in food packaging and coating technologies: A review. *Biomolecules*, 13(7), 1092. DOI: 10.3390/biom13071092 PMID: 37509128

Adeyemi, J. O., Oriola, A. O., Onwudiwe, D. C., & Oyedeji, A. O. (2022). Plant extracts mediated metal-based nanoparticles: Synthesis and biological applications. *Biomolecules*, 12(5), 627. DOI: 10.3390/biom12050627 PMID: 35625555

Ahmed, I., Mir, F. A., & Banday, J. A. (2023). Synthesis of metal and metal oxide nanoparticles using plant extracts characterization and applications. *BioNanoScience*, 13(4), 1541–1557. DOI: 10.1007/s12668-023-01194-y

Al Jabri, H., Saleem, M. H., Rizwan, M., Hussain, I., Usman, K., & Alsafran, M. (2022). Zinc oxide nanoparticles and their biosynthesis: Overview. *Life (Chicago, Ill.)*, 12(4), 594. DOI: 10.3390/life12040594 PMID: 35455085

Batool, S., Batool, F., Zia, A., & Saeed, R. (2018). The struggle for women rights: A study of the emergence of feminism in Pakistan,(1947 to 2010). *Journal of the Punjab University Historical Society*, 31(1), 211–221.

Berta, L., Coman, N.-A., Rusu, A., & Tanase, C. (2021). A review on plant-mediated synthesis of bimetallic nanoparticles, characterization, and their biological applications. *Materials (Basel)*, 14(24), 7677. DOI: 10.3390/ma14247677 PMID: 34947271

Catalano, P. N., Chaudhary, R. G., Desimone, M. F., & Santo-Orihuela, P. L. (2021). A survey on analytical methods for the characterization of green synthesized nanomaterials. *Current Pharmaceutical Biotechnology*, 22(6), 823–847. DOI: 10.2174/1389201022666210104122349 PMID: 33397235

Chen, F., Li, Y., Zia-ur-Rehman, M., Hussain, S. M., Qayyum, M. F., Rizwan, M., Alharby, H. F., Alabdallah, N. M., Alharbi, B. M., & Ali, S. (2023). Combined effects of zinc oxide nanoparticles and melatonin on wheat growth, chlorophyll contents, cadmium (Cd), and zinc uptake under Cd stress. *The Science of the Total Environment*, 864, 161061. DOI: 10.1016/j.scitotenv.2022.161061 PMID: 36565889

Dikshit, P. K., Kumar, J., Das, A. K., Sadhu, S., Sharma, S., Singh, S., Gupta, P., & Kim, B. S. (2021). Green synthesis of metallic nanoparticles: Applications and limitations. *Catalysts*, 11(8), 902. DOI: 10.3390/catal11080902

Dikshit, P. K., Kumar, J., Das, A. K., Sadhu, S., Sharma, S., Singh, S., Gupta, P., & Kim, B. S. (2021). Green synthesis of metallic nanoparticles: Applications and limitations. *Catalysts*, 11(8), 902. DOI: 10.3390/catal11080902

El-Seedi, H. R., El-Shabasy, R. M., Khalifa, S. A., Saeed, A., Shah, A., Shah, R., Iftikhar, F. J., Abdel-Daim, M. M., Omri, A., Hajrahand, N. H., Sabir, J. S. M., Zou, X., Halabi, M. F., Sarhan, W., & Guo, W. (2019). Metal nanoparticles fabricated by green chemistry using natural extracts: Biosynthesis, mechanisms, and applications. *RSC Advances*, 9(42), 24539–24559. DOI: 10.1039/C9RA02225B PMID: 35527869

Gomiero, T. (2016). Soil degradation, land scarcity, and food security: Reviewing a complex challenge. *Sustainability (Basel)*, 8(3), 281. DOI: 10.3390/su8030281

Jampílek, J., & Kráľová, K. (2019). Beneficial effects of metal-and metalloid-based nanoparticles on crop production. Nanotechnology for Agriculture: Advances for Sustainable Agriculture, 161-219.

Jampílek, J., & Kráľová, K. (2022). Impact of copper-based nanoparticles on economically important plants. In *Copper nanostructures: Next-generation of agrochemicals for sustainable agroecosystems* (pp. 293–339). Elsevier. DOI: 10.1016/B978-0-12-823833-2.00030-1

Jeevanandam, J., Kiew, S. F., Boakye-Ansah, S., Lau, S. Y., Barhoum, A., Danquah, M. K., & Rodrigues, J. (2022). Green approaches for the synthesis of metal and metal oxide nanoparticles using microbial and plant extracts. *Nanoscale*, 14(7), 2534–2571. DOI: 10.1039/D1NR08144F PMID: 35133391

Jeyaraj, M., Gurunathan, S., Qasim, M., Kang, M. H., & Kim, J. H. (2019). A comprehensive review on the synthesis, characterization, and biomedical application of platinum nanoparticles. *Nanomaterials (Basel, Switzerland)*, 9(12), 1719. DOI: 10.3390/nano9121719 PMID: 31810256

Kanwar, M. K., Sun, S., Chu, X., & Zhou, J. (2019). Impacts of metal and metal oxide nanoparticles on plant growth and productivity. Nanomaterials and plant potential, 379-392.

Le Wee, J., Law, M. C., Chan, Y. S., Choy, S. Y., & Tiong, A. N. T. (2022). The potential of Fe-based magnetic nanomaterials for the agriculture sector. *ChemistrySelect*, 7(17), e202104603. DOI: 10.1002/slct.202104603

Li, P., Xia, Y., Song, K., & Liu, D. (2024). The Impact of Nanomaterials on Photosynthesis and Antioxidant Mechanisms in Gramineae Plants: Research Progress and Future Prospects. *Plants*, 13(7), 984. DOI: 10.3390/plants13070984 PMID: 38611512

Maity, D., Gupta, U., & Saha, S. (2022). Biosynthesized metal oxide nanoparticles for sustainable agriculture: Next-generation nanotechnology for crop production, protection and management. *Nanoscale*, 14(38), 13950–13989. DOI: 10.1039/D2NR03944C PMID: 36124943

Manna, I., & Bandyopadhyay, M. (2019). A review on the biotechnological aspects of utilizing engineered nanoparticles as delivery systems in plants. *Plant Gene*, 17, 100167. DOI: 10.1016/j.plgene.2018.100167

Martínez, G., Merinero, M., Pérez-Aranda, M., Pérez-Soriano, E. M., Ortiz, T., Villamor, E., Begines, B., & Alcudia, A. (2020). Environmental impact of nanoparticles' application as an emerging technology: A review. *Materials (Basel)*, 14(1), 166. DOI: 10.3390/ma14010166 PMID: 33396469

Mittal, D., Kaur, G., Singh, P., Yadav, K., & Ali, S. A. (2020). Nanoparticle-based sustainable agriculture and food science: Recent advances and future outlook. *Frontiers in Nanotechnology*, 2, 579954. DOI: 10.3389/fnano.2020.579954

Mourdikoudis, S., Pallares, R. M., & Thanh, N. T. (2018). Characterization techniques for nanoparticles: Comparison and complementarity upon studying nanoparticle properties. *Nanoscale*, 10(27), 12871–12934. DOI: 10.1039/C8NR02278J PMID: 29926865

Oriola, A. O., Onwudiwe, D. C., & Oyedeji, A. O. (2021). Study on the synthesis and applications of metal nanoparticles in agriculture.

Rani, N., Singh, P., Kumar, S., Kumar, P., Bhankar, V., & Kumar, K. (2023). Plant-mediated synthesis of nanoparticles and their applications: A review. *Materials Research Bulletin*, 163, 112233. DOI: 10.1016/j.materresbull.2023.112233

Shafey, A. M. E. (2020). Green synthesis of metal and metal oxide nanoparticles from plant leaf extracts and their applications: A review. *Green Processing and Synthesis*, 9(1), 304–339. DOI: 10.1515/gps-2020-0031

Shin, D. O., Jeong, J. R., Han, T. H., Koo, C. M., Park, H. J., Lim, Y. T., & Kim, S. O. (2010). A plasmonic biosensor array by block copolymer lithography. *Journal of Materials Chemistry*, 20(34), 7241–7247. DOI: 10.1039/c0jm01319f

Shnoudeh, A. J., Hamad, I., Abdo, R. W., Qadumii, L., Jaber, A. Y., Surchi, H. S., & Alkelany, S. Z. (2019). Synthesis, characterization, and applications of metal nanoparticles. In *Biomaterials and bionanotechnology* (pp. 527–612). Elsevier. DOI: 10.1016/B978-0-12-814427-5.00015-9

Siegel, F. R., & Siegel, F. R. (2015). Population assessments: 2013–2050–2100: Growth, stability, contraction. Countering 21st Century Social-Environmental Threats to Growing Global Populations, 1-8.

Srivastav, A., Ganjewala, D., Singhal, R. K., Rajput, V. D., Minkina, T., Voloshina, M., Srivastava, S., & Shrivastava, M. (2021). Effect of ZnO nanoparticles on growth and biochemical responses of wheat and maize. *Plants*, 10(12), 2556. DOI: 10.3390/plants10122556 PMID: 34961025

Su, Y., Ashworth, V., Kim, C., Adeleye, A. S., Rolshausen, P., Roper, C., White, J., & Jassby, D. (2019). Delivery, uptake, fate, and transport of engineered nanoparticles in plants: A critical review and data analysis. *Environmental Science. Nano*, 6(8), 2311–2331. DOI: 10.1039/C9EN00461K

Suazo-Hernández, J., Arancibia-Miranda, N., Mlih, R., Cáceres-Jensen, L., Bolan, N., & Mora, M. D. L. L. (2023). Impact on some soil physical and chemical properties caused by metal and metallic oxide engineered nanoparticles: A review. *Nanomaterials (Basel, Switzerland)*, 13(3), 572. DOI: 10.3390/nano13030572 PMID: 36770533

Tan, N. P. B., & Lee, C. H. (2017). Environment-friendly approach in the synthesis of metal/polymeric nanocomposite particles and their catalytic activities on the reduction of p-nitrophenol to p-aminophenol. In *Green chemical processing and synthesis*. IntechOpen.

Chapter 7

Effect of Nano-Fertilizer on Crop Plant Growth and Its Effect on Rhizospheric Microorganism and Its Application:

A Comprehensive Review

Santosh Kumar Sethi
https://orcid.org/0000-0003-0197-0570
School of Biotechnology, Gangadhar Meher University, Sambalpur, India

Smita Rani Dalei
School of Biotechnology, Gangadhar Meher University, Sambalpur, India

Augustine Okpani Oko
Department of Biotechnology, Ebonyi State University, Abakaliki, Nigeria

Asish Kumar
Department of Biotechnology, Guru Ghasidas University, India

Arati Kindo
School of Biotechnology, Gangadhar Meher University, Sambalpur, India

ABSTRACT

Nanofertilizers represent a groundbreaking advancement in agricultural techniques, poised to transform how nutrients are delivered and managed in the interaction between plants and soil. Unique characteristics of nanoparticles facilitate improved nutrient absorption, precise delivery, and enhanced availability, potentially revolutionizing conventional fertilization methods. Numerous studies demonstrate significant enhancements in growth metrics, seed yield, and overall plant vigor following nanofertilizer application. Furthermore, plants treated with nanofertilizers exhibit heightened resilience to both biological and environmental stressors. Comparative assessments against traditional fertilizers underscore the superior efficacy of nanofertilizers while also highlighting economic and environmental considerations. The regulatory landscape is evolving in response to the rapid advancement of nanotechnology in agriculture,

DOI: 10.4018/979-8-3693-6240-2.ch007

with ongoing research aiming to address existing gaps and shape the future trajectory of nanofertilizer utilization.

1. INTRODUCTION

The overuse of pesticides and fertilizers in agriculture to boost yields has proven to be excessive, as a significant portion goes unused, causing harm to both the environment and human health. Consequently, farmers face the major challenge of shifting to nanopesticides and nanofertilizers, aiming to decrease reliance on mineral fertilizers, improve yields, and promote sustainable agricultural development (Bayda et al. 2019). Nanotechnology has significant potential because of their distinct physical, chemical, and biological characteristics, nanoparticles (NPs) have the unique ability to alter common chemicals and devices (Viswanathan et al. 2019). Nanoparticles (NPs) are defined as natural and manufactured materials having at least one dimension between 1 and 100 nm (Shang et al. 2019). Traditional agriculture is facing previously unheard-of difficulties as a result of the fast-expanding population and environmental deterioration (Sang et al. 2019). For many years, fertilizers and insecticides have been essential to supplying the demands of agricultural production (Guo et al. 2018). On the other hand, overuse of these pesticides also hinders the development of sustainable farming practices (Raliya et al. 2017; Singh et al. 2021). According to Sanzari et al. (2019), extending the use of nanotechnology can result in innovative ways to improve sustainable agriculture and satisfy food demands. According to current research, nanotechnology can successfully handle a number of agricultural difficulties, such as misuse of fertilizers and pesticides and plant stress caused by extreme weather (Demirer et al. 2021; Giraldo et al. 2019; Kamle et al. 2020; Kumar et al. 2019; Lahiani et al. 2013). In addition, application of nanotechnology includes monitoring of plant growth status, fast and easy genome modification and transgene expression in intact plant cells (Kwak et al. 2019; Kwak et al. 2017). Nanofertilizers, their ability to deliver nutrients directly to the target area can reduce nutrient runoff, a major environmental problem with traditional fertilizers. The spread of nanofertilizers in agriculture promises many benefits, such as higher yields, better plant health and better ecological results. But like with everything new in technology, it's critical to comprehend how it will affect the ecosystem as a whole as crops. In this regard, the purpose of this review is to explain the application and importance of nanofertilizers, focusing on its effect on the well-being of the plant (Figure 1). The transition from traditional to nanotechnology-enhanced agricultural practices is not simply a matter of changing the type of fertilizer. This is an evolution in how we understand and approach crop nutrition. The nuances of this transition are diverse. While conventional fertilizers often release nutrients either too quickly, leading to waste, or too slowly, inhibiting plant growth, nanofertilizers, with their controlled release mechanisms, try to strike the right balance. Their size allows a closer interaction with plant cells, which promotes better uptake of nutrients and thus better plant health (Kumar et al. 2015). Plant welfare is not limited to growth and yield. This includes the plant ability to deal with stress, its nutritional profile and overall vigor. Because of their exact effect, nanofertilizers may handle these areas of plant protection more precisely than traditional fertilizers. Although the prospect sounds promising, it is equally important to consider all unintended consequences. The nanoscale world behaves differently, and while certain effects are beneficial, they can have unknown consequences for plant health and the surrounding environment. Hence, this review aims to investigate areas that require more investigation in addition to highlighting the apparent advantages of nano-fertilizer on various crop plant growths and its effect on specific rhizospheric microorganisms.

Figure 1. Agricultural use of nanofertilizer

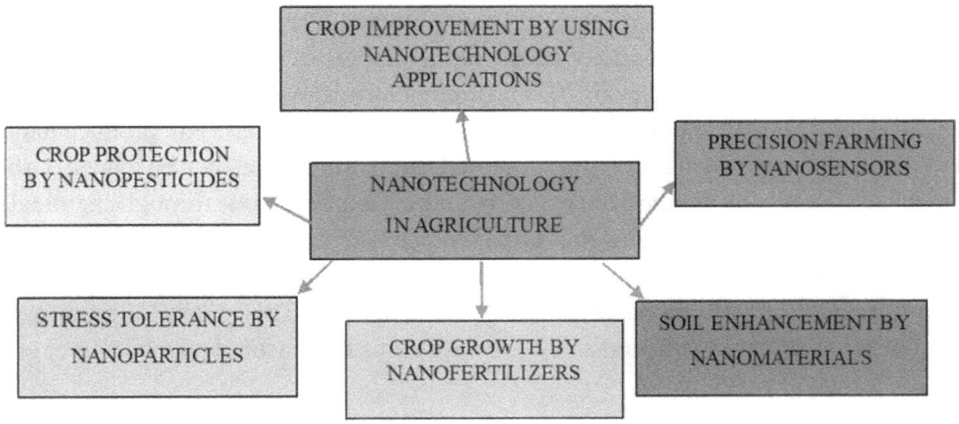

1.1 Nanofertilizers

The most recent innovation that promises to alter the way we nurture is the advent of nanofertilizers. However, its critical to figure out what nanofertilizers are and how they're categorized in order to evaluate their potential and address the issues they raise. In essence there aren't many differences between standard fertilizer and nanofertilizer. Both supply vital nutrients to plants that they need for healthy development and overall function. The difference is in scale and delivery mechanism. Nanofertilizers are basically fertilizers designed at the nanoscale, typically 1-100 nanometers in at least one dimension (Kumar et al., 2021). This nanoscale restructuring significantly increases surface area to volume, increases reactivity, and enables unique interactions with plant tissues that cannot be achieved with larger traditional fertilizer particles. The formulation of nanofertilizers is where science and nature intertwine. They are rich in different trace elements, potassium, phosphorus, nitrogen, and other essential nutrients. On the other hand, these nutrients are coated, encapsulated, or embedded in nanoparticles, which can be metal, polymer, or lipid based. The nanoscale delivery system provides a more controlled release of nutrients, allowing plants to take them up more efficiently, reducing waste and potential environmental pollution (Manjunatha et al. 2019). It is also extremely important to note that fertilizers might alter their chemical and biological characteristics in addition to their physical characteristics when they are smaller in size. Because nanoparticles are more easily absorbed by plant tissues and cells, nutrients may be delivered precisely to the areas of the cell where they are most required (Harish et al. 2022). Diving deeper into the world of nanofertilizers, we discover a diverse landscape. This diversity is critical for maximizing the agricultural advantages of nanotechnology, as well as for feeding various crops and soil types (Fig. 2). Organic nanofertilizers: They are made to supply nutrients inside an organic matrix and are produced from biological ingredients like proteins or lipids. For example, lipid-based nanoparticles can be loaded with nutrients, ensuring a slow and steady release. Inorganic nanofertilizers: These consist of inorganic nanoparticles such as silica, zinc oxide or even gold nanoparticles. They are often used to deliver certain micronutrients to plants, taking advantage of their unique properties such as high stability or specific interactions with plant tissues (Avellan et al. 2021). Biologically derived nanofertilizers: These are synthesizing using biological processes. For example, certain fungi and bacteria can produce nanoparticles when provided with metal ions. Such biologically derived nanoparticles can then be loaded with nutrients and used as fertilizer. Chemically synthesized nanofertilizers: These are made through chemical reactions and processes, often in laboratories. Precision is an advantage here, as chemist can control the size, shape and properties of nanoparticles with high precision (Tao et al. 2021). Multipurpose nanofertilizers: These represents the pinnacle of agricultural nanotechnology development. They not only supply nutrients but also have the ability to deter pests and enhance soil health by using sensor elements (Fierer et al. 2021).

Figure 2. Application of nanofertilizer and its prospects

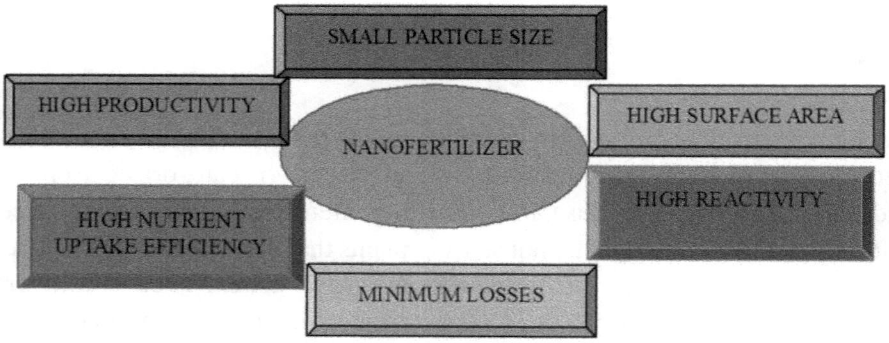

1.2 Synthesis of Nanofertilizer

Nanofertilizer nanoparticles can be synthesized by various physical methods, such as physical fragmentation, Vapor condensation, sol-gel process, interferometric lithography, microemulsion deposition, which require high pressure and temperature and high energy consumption (Singh et al. 2021). Chemical methods are usually dangerous and toxic, because the chemical materials still pollute the environment and can be dangerous for the person handling them. The biological method is therefore much more favorable, because it is essentially less toxic, environmentally friendly and more energy efficient (Fig. 3). The main reaction in nanoparticles biosynthesis is reductional oxidation. Plant phytochemicals or microbial enzymes present in microbes have some reducing or antioxidant properties that resulted in the decrease of metal compounds to the corresponding nanoparticles (Singh et al. 2016; Singh et al. 2019) (Table-1; Fig-1-4).

The biological synthesis of nanoparticles used in nanofertilizers is mainly carried out from plant extracts isolated from different parts, such as fruits, stems, roots and flowers (Table 1). A significant portion of the production of different metals and metal oxide nanoparticles is played by certain microorganisms (Fig. 4), namely bacteria, fungus, and yeast. They may naturally possess the capacity to create nanoparticles both intracellularly and extracellularly. Due to its ease of use, ease of industrialization, and ability to manufacture vast amounts of nanoparticles using a straightforward upstream procedure, the extracellular method has shown to be more economical than the intracellular route (Fig-3 &4).

Table 1. Uses of different metal nanoparticles synthesized using various plants parts and products

PLANT USED	PLANT PART USED	NANOPARTICLES	PLANT METABOLITE INVOLVED	REFERENCES
Acalypha indica (Copperleaf)	Leave extract(10g)	Ag (Silver), Au (Gold) (20-30 nm)	Quercetin, plant pigment	Krishnaraj et al. 2010
Aloe vera	Leave extract (35g)	In_2O_3 (Indium Oxide) (5-50nm)	Biomolecules	Maensiri et al. 2008
Alternanthera sessilis (Dwarf copperleaf)	Whole plant	Ag (Silver) (435 nm)	Amine, carboxyl group	Niraimathi et al., 2013
Andrographis paniculata (Creat or Green chiretta)	Leave extract (Shade dried leaves)	Ag (Silver) (423 nm peak from uv-visible spectroscopy)	Alkaloids, flavonoids	Suriyakalaa et al. 2013
Carica papaya (Papaya)	Fruits (25g)	Ag (Silver) (15 nm from FWHM)	Hydroxyl flavones, catechin	Jain et al. 2009
Cassia fistula (Golden shower)	Stem bark extract (60 mg of bark powder)	Au (Gold) (190-900 nm)	Hydroxyl group	Daisy and Supriya, 2012
Cinnamon zeylanicum (Cinnamon)	Leaves (Bark extract) (2.5g)	Ag (Silver)	Water soluble organics	Sathishkumar et al. 2009
Citrullus colocynthis(Bitter cucumber)	Calli extract (0.5mg L-1 IAA)	Ag (Silver) (31 nm by AFM)	Polyphenols	Satyavani et al. 2011
Citrus sinensis (Orange)	Peel extract (4 g)	Ag (Silver) (35 and 10 nm size)	Water soluble compounds	Kaviya et al. 2011

continued on following page

Table 1. Continued

PLANT USED	PLANT PART USED	NANOPARTICLES	PLANT METABOLITE INVOLVED	REFERENCES
Dillenia indica (Elephant apple)	Fruit extract	Ag (Silver) (421nm)	Biomolecules	Singh et al. 2013
Dioscorea bulbifera (Air potato)	Tuber extract (5 g)	Ag (Silver) (540nm)	Diosgenin, ascorbic acid	Ghosh et al. 2012

2. MECHANISM OF MICROBIAL SYNTHESIS OF NANOPARTICLES

Ions and the microbial cell wall are essential for the synthesis of nanoparticles in this kind of process. An active site for the binding of metal ions is created by the different proteins and polysaccharides found in the microbial cell wall. The technique of bioreduction, which reduces the trapped ions to elementary atoms, starts with the electrons being transferred from NADH via NADH-dependent reductase, an embedded electron carrier in the plasma membrane. In addition, nuclei grow and form nanoparticles that aggregate them in the cytoplasm or cell wall. Proteins, peptides, and acids like cysteine, tyrosine, and tryptophan keep these nanoparticles stable within the cell. Intracellularly synthesized nanoparticles require ultrasonication to obtain purified nanoparticles. The process of extracellular synthesis is accountable for the reduction of metal ions into metal nanoparticles. These particles are produced by the nitrate reductase enzyme, which catalyzes the nitrogen cycle's conversion of nitrate to nitrite.

2.1 Effect of Nanofertilizers on Plant

The effect of nanofertilizers on plants depends on several factors, such as nanoscale properties, soil, environment, use of nanoparticles and plant species. So far, the foliar mode of entry can give the best response as nutrients can be easily utilized by nanosized pores present in plasmodesmata of leaves (Mahil et al. 2019; Iqbal, 2019). In addition, nanomaterials are transported through root hairs, after symplastic and apoplastic movements, to reach the xylem and then into leaves and stems (Mittal et al. 2020). However, there are some reports that show the accumulation of nanoparticles on the surface of the primary root system, leading to a reduction in the size of the pore wall, root water transport and water transport (Asli and Neumann, 2009; Martínez -Fernández et al. 2016). Similarly, phosphorus nanofertilizer showed sustained release of phosphorus for up to 40-50 days and nutritional and nutrient efficiency of Glycine max (Soybean) (Liu and Lal, 2014). Squash (Cucurbita pepo) showed additional physiological and biochemical responses under foliar application of potassium nanofertilizer (Gerdini, 2016). Apart from macronutrient-based nanofertilizers, micronutrient-based nanofertilizers have also demonstrated noteworthy advantages for plants. For instance, it was discovered that using zinc oxide nanoparticles enhanced the sprouting of peanut (*Arachis hypogaea*) seeds (Seleiman et al. 2020). In peanut plants, the application of iron nanoparticles also appeared to accelerate root growth (Rui et al. 2016). Furthermore, by raising α-amylase activity, silver nanoparticles have demonstrated a strong potential to speed up rice (Oryza sativa) seed germination (Mahil and Kumar, 2019). It has also been observed that rice treated with silica-based nanofertilizers can reduce heavy metal levels (Wang et al. 2016). Furthermore, it has been revealed that plant resilience to biotic stress is increased by nanoparticles. Important plant diseases including Phytophthora, Fusarium, Colletotrichum, Alternaria, etc. have been demonstrated to be effectively combated by metal and oxide nanomaterials (Malandrakis et al. 2019; Shenashen et al. 2017; Vanti et al. 2019; Imada et al. 2016). Satti et al. (2021) reported a reduction in the incidence of spot disease caused by Bipolaris sorokiniana in wheat. Also, application of silver nanoparticles (5μg/ml) reduced the black spot disease of Arabidopsis thaliana caused by Alternaria brassicicola by 70-80% (Kumari et al. 2020) (Table 2, Figure 3 and 4).

Table 2. Impact of nanofertilizers on crop plants and their rhizosphere structural and functional micro-organisms

Sl. No.	Nanomaterial	Plant name	Effects on microorganisms	Effect on microbial function	Results	Reference
1	Mettalic silver	*Cucumis sativus (Cucumber)*	Increased growth promoting bacteria activity Plant growth promoting Rhizobacteria (PGPR)	Improved carbon, nitrogen, and other biogeochemical cycle	An increase in the length of roots and shoots as well as biochemical indicators like proline, protein, and antioxidants	Nawaz and Bano, 2020
2	ZnO NPs (Zinc oxide)	*Phoenix dactylifera (Date palm)*	Number of fungal and bacterial cultivable heterotrophic colony forming units reduced significantly	Reduction in carbon and nitrogen mineralization efficiency On soil microbes	Decrease in dissolved organic carbon and mineral nitrogen in date palm plant (122%)	Rashid et al. 2017
3	Iron oxide (FeO)	*Zea mays (Maize)*	An increase in *Bradyrhizobium* and ammonia oxidizing bacteria activity	Improved soil nitrification, microbial activity	Improved plant growth and yield in maize	He et al. 2016
4	Silver nanoparticles	*Cucumis sativa (cucumber)*	Increased actinobacteria and planctomyces population	Improved nitrogen fixation efficiency	Improvement in phenotypic characteristics	Zhang et al. 2020
5	Pristine and sulfidized Zn0 NPs	*Glycine max (Soybean)*	Significant effects on bacterial communities' composition and structure	Drastic impact on carbon and nitrogen metabolism	Over exposure to zinc may have an impact on the development and growth of soybeans	Chen et al. 2023
6	Zinc oxide (ZnO)	*Lactuca sativa (Lettuce)*	Increased abundance of cyanobacteria, bacteria, and protozoa	Enhancement of organic matter decomposition and nitrogen fixing	Fresh biomass and net photosynthetic rate both increased by 6.2%	Xu et al. 2018
7	Cu and Zn NPs	*Raphanus sativus (Radish)*	Reduced *Azotobacter* genus abundance in the soil	Decrease in catalyze and dehydrogenase activities	Decrease in germination and roots length in Radish plant	Kolesnikov et al. 2019

continued on following page

Table 2. Continued

Sl. No.	Nanomaterial	Plant name	Effects on microorganisms	Effect on microbial function	Results	Reference
8	Silica	*Zea mays (Maize)*	Phosphate solubilizing and nitrogen fixing bacteria were more abundant, but silicate solubilizing	Change in microbial biomass carbon and nitrogen content	Increased germination and absorption of silica	Ragaraj et al. 2014
9	High dose of ZnO NPs	*Medicago sativa (Alfalfa)*	Reduction in the quantity of bacteroids and in the diversity and relative abundance of soil microorganisms	Decreased nitrogen-fixing ability	Decrease in root nodules and plant biomass	Sun et al. 2022
10	Copper oxide (CuO) 10-1000 mg/ kg direct soil application	*Carya illinoinensis (pecan tree)*	Decreased abundance of Acidobacteria and Actinobacteria	No significant effect on nitrogen, sulphur or halogen biogeochemical cycles but slight reduction in nitrification and denitrification was noted	No harmful effect on plants	Salas-Leiva et al. 2021
11	Copper oxide (CuO) 50 mg/ kg, direct application to soil	*Triticum aestivum (Wheat)*	Population of *Sphingo bacterium, Devosia, Pseudomonas, Rhizobium, Pseudo xanthomonas, Shinella, Dyadobacter* and *Pantoea* increased	Improved nitrogen fixation and reduced denitrification process	Improved photosynthetic rate in plants	Guan et al. 2020
12	Titanium Oxide (TiO_2) 1-1000 mg/kg, direct application to soil	*Triticum aestivum (Wheat)*	Increased abundance of actinobacteria and plactomyces	Nitrogen fixation efficiency improved	Improved fresh biomass weight, shoot, root weight and chlorophyll content	Moll et al. 2017
13	TiO_2 50 µg/ml, foliar spray	*Brassica napus (Rapeseed)*	Improved performance of *B. amylo liquefaciens*	Improved adhesion of beneficial bacteria to roots	Enhanced plant growth and yield	Palmqvist et al. 2015
14	Zinc oxide (ZnO) 0.3-0.9 g/L, foliar spray	*Glycine max (Soybean)*	Increased diversity of *Bradyrhizobium japonicum, Pseudomonas putida, Azospirillum lipoferum*	Increased nitrogen fixation	Improved plant height, nodulation, grain yield and grain weight	Seyedsharifi and Khorramdel, 2015
15	Zinc oxide (ZnO) 1-400 mg/kg, direct application to soil	*Glycine max (Soybean)*	Enhanced rhizobial activity	Improve nitrogen cycling	Seedling growth is promoted	Yusefi-Tanha et al. 2020
16	Molybdenum (Mo) 10 mg/L, seed treatment	*Cicer arietinum (Chick pea)*	Increase in the abundance of the members of *Rhizobium*	NA	Improved germination, peroxidise activity and reduce lipid peroxidation	Shcherbakova et al. 2017
17	Zeolite 50 ppm, foliar spray	*Z. mays (Maize)*	Increased relative abundance of *Bacillus* sp.	Improved degradation of organic matter	25% increase in fresh plant biomass	Khati et al. 2018

continued on following page

Table 2. Continued

Sl. No.	Nanomaterial	Plant name	Effects on microorganisms	Effect on microbial function	Results	Reference
18	Carbon black 0.1-1000 mg/kg	*G. max (Soybean)*	Increased abundance of organic matter decomposing bacterial community while lowered abundance of methanogenic, chitinolytic and nitrate reducing bacteria	Improved xylanolytic, cellulolytic activity and aerobic ammonia oxidation	NA	Ge et al. 2018
19	Zinc oxide (ZnO) 100-4000 mg/kg, direct application to soil	*Arachis hypogaea*	Increased dominance of Lactobacillus	NA	Improved plant growth and yield	Oghenerume et al. 2020

3. MODE OF APPLICATION OF NANOFERTILIZER IN AGRICULTURE

3.1 Foliar Application

The take-up of nanoparticles relies upon plant physiology (Schwab et al. 2016). NPs often enter plants via the xylem and phloem and integrate into the trichome, shame, disgrace, and hydathode (Wang et al. 2016). The trichome base and stomatal width, which can vary from 5 to 20 nm, cannot entirely prevent the entrance of nanoparticles into the plant cell (Fleischer et al. 1999). Relocation within the vascular system is impeded by the Caspian strip (Aubert et al. 2012). Plant germs can be eliminated by combining NFs with nanoparticles. The polymer-walled nanocapsule's synthetic bonds can be broken by plant cell stress proteins. When plants sense a threat from phytopathogens, they release phloem to safeguard against pollution (Ropitaux et al., 2020). Additionally, the accumulation of nanoparticles on the leaflet surface, which might result in stomach obstruction and modify gas exchange (Table 2) (Ma et al. 2010).

3.2 Root Application

NPs enter through epidermis of the root crosses endodermis arrives at the xylem, where they are shipped to the elevated piece of the plant. NPs enter the cell wall through pores when they are somewhere in the range of 3 and 8 nm (Zulfiqar et al. 2019; Rajput et al. 2020; Ali et al. 2021). NPs can likewise enter through the root tip meristem, or at the places of sidelong root arrangement, since there are wounds in the Casparian strip. To enter the epidermal layers of the roots, NPs should infiltrate cell walls and plasma films. From that point they enter the vascular tissue (xylem). The measures of the pores of the cell walls are 3 to 8 nm (Carpita and Gibeaut, 1993), which is a tiny size for NPs to enter, notwithstanding, it has been demonstrated that NPs prompt the development of huge pores in cell walls where they can be incorporated (Navarro et al., 2008). The feeder roots' hairs allow the microelements to enter the plant. Therefore, the organic acids or phenols of the root exudates dissolve the microspheres encapsulating Ca, Mg, Fe, S, or Zn (Pérez-de-Luque, 2017). After the use of composts in soil, a significant part of the supplements is lost because of draining as an outcome of which soil and water are contaminated. Not just this, certain agrosynthetic compounds are answerable for ozone depleting substances and environmental change (Rochette et al. 2018; Wesołowska et al. 2021). Taking everything into account, In a review, twofold layered hydroxide nanocomposites were utilized for the controlled arrival of supplements (Benício et al. 2017). Wang et al. (2021) concentrated on the sluggish arrival of coordinated superabsorbent manure and the water maintenance limit of soils with this compost. They found that the surface cross-connected item had great sluggish delivery property and furthermore excellent soil dampness preservation. Strangely plants can likewise respond to NPs. The breadth of *Zea mays* seedlings root cell wall pores were diminished from 6.6 to 3.0 nm after bentonite and TiO_2 nanoparticles were applied (Asli and Neumann, 2009).

3.3 Mechanism of Action

At the intersection of science and nature, always trying to find effective ways to exploit the abundance of the soil. In this dynamic, nanofertilizers emerge as both evolution and revolution, redefining traditional plant nutrition paradigms. But their effectiveness is based on their underlying mechanisms of action, which redefine the way plants acquire and use essential nutrients. These tiny particles make it easier for nutrients to reach the roots of plants by penetrating their protective layer (Sharma et al. 2023). The majority of the nutrient-rich particles engage directly with plant tissues due to their huge surface area compared to volume, which promotes more effective transport. This productive conveyance framework implies that even at lower application rates, nanofertilizers can accomplish or try and surpass the adequacy of their regular partners (Kah et al. 2018). However, nanofertilizers can be made to deliver these nutrients directly to specific parts of the plant (Fig. 5). This accuracy is achieved through several different mechanisms. Some nanofertilizers are designed to attract certain cellular components to the root system, ensuring that they release nutrients exactly where they are generally required (Saleh et al. 2022). Others may trigger their nutrient release in response to specific environmental or physiological cues, such as changes in soil pH, moisture content, or even the presence of certain enzymes in plant tissue. Such a targeted approach ensures efficient use of nutrients but can also mitigate potential environmental impacts. For example, Phosphorus, if used in excess, can enter water bodies and cause problems such as algal blooms. Nanofertilizers can significantly reduce such environmental risks by ensuring a targeted and efficient delivery of phosphorus to plant roots (Manjunatha et al. 2019) (Fig-3&4).

Figure 3. Image showing the uptake, transport and penetration of nanoparticles in plant cell

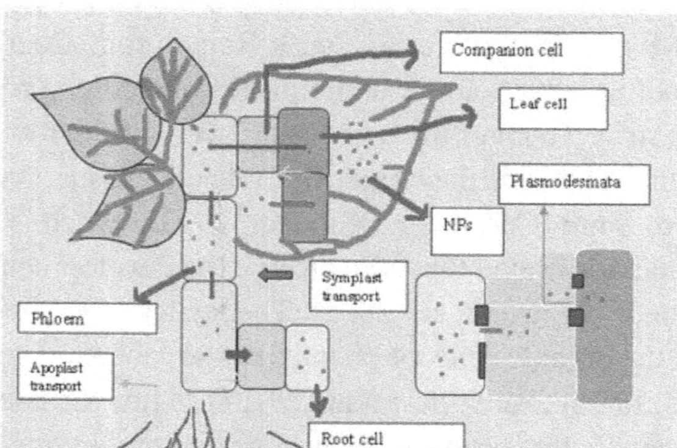

Bioavailability refers to the proportion of a substance that enters the body system and is actively used. It denotes the portion of supplied nutrients that may be efficiently absorbed and used by plants in relation to fertilizers. Nanofertilizers have shown significant increases in nutrient bioavailability for several reasons. Nanoencapsulation can protect phosphorus from such interactions and ensure that the majority remains biological (Manjunatha et al. 2019). Finally, there is the potential for synergy. In addition to nutrients, some nanofertilizers can also provide other compounds that improve the plant's ability to absorb and use these nutrients. Some nanoparticles can, for example, stimulate root growth or improve cell permeability, ensuring a more efficient use of the accompanying nutrients (Ahluwalia et al. 2021).

Figure 4. Mechanism of Nanofertilizer impacts on different plant growth determining factors

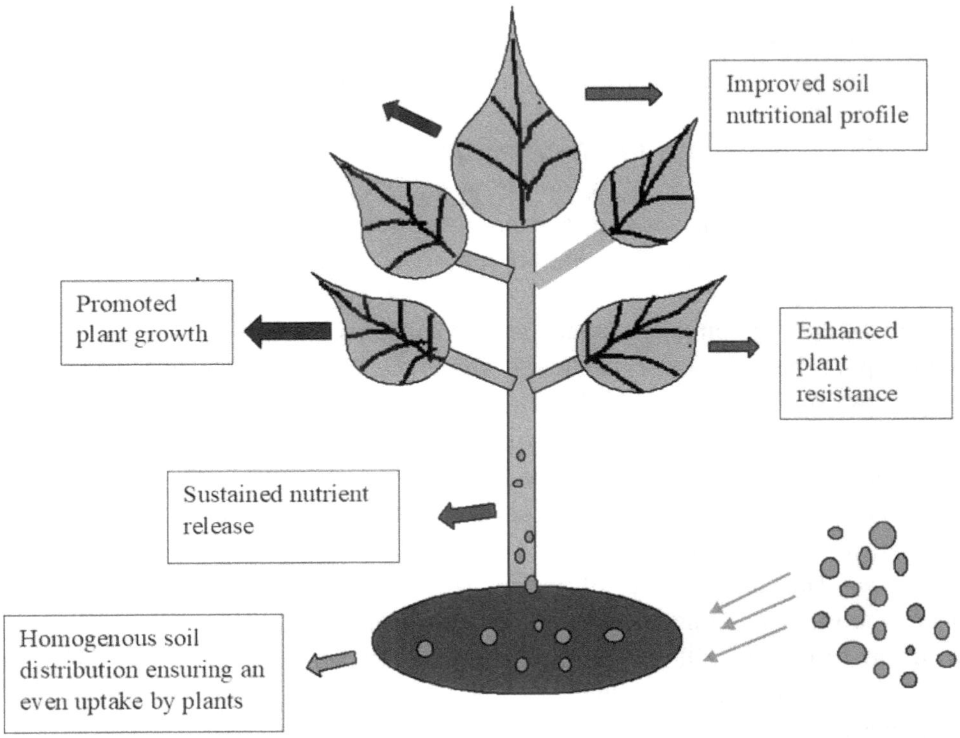

4. IMPACT OF NANOFERTILIZERS WITH PGPRS (PLANT GROWTH PROMOTING RHIZOBACTERIA)

Nanoparticles, when utilized as nanofertilizer, have been demonstrated to upgrade crop development and quality (Fig. 6) (Merghany et al. 2019; Babu et al. 2022). Dissimilar to mass substance manures, which are expected in high dosages, nanofertilizers can be applied in generally more modest amounts. Applying a lower measurement of nanofertilizer can limit the potential for supplement misfortune through draining and volatilization, and in this manner further develops supplement use effectiveness (Raliya et al. 2017). Three elements – natural, extraneous, and method of organization – influence the productivity of nanofertilizers. Nano-detailing strategies, molecule size, and surface covering are instances of inborn factors. While extraneous variables incorporate soil surface, profundity, pH, temperature, natural matter, and microbial action (Zulfiqar et al. 2019). Besides, the system of conveyance through plant roots or leaves (foliar) extensively affects the take-up, conduct, and bioavailability of nanofertilizers (Mahil and Kumar, 2019). Because of their connection with natural materials in the dirt, nanofertilizers might change the dirt surface science, which could affect plants and microorganisms. Then again, microorganisms and their activities might possibly modify how nanoparticles act (Frenk et al. 2013; Zulfiqar et al. 2019; Toksha et al. 2021). In a review directed by Kaur et al. (2022), the impact of titanium dioxide (TiO_2) NPs on the dirt rhizosphere of mung bean crop was assessed. The TiO2 NPs were displayed to invigorate development of soil microflora (N-fixers and alkali oxidizers) as well as increment enzymatic movement for dehydrogenase, phosphatase, protease, urease, and catalase at low focuses (1.0, 2.5, 5.0, and 10.0 mg/L) contrasted with the higher fixation (20 mg/L) (Helal et al. 2023) (Table-1 & 2).

4.1 Effect of NFs on Rhizosphere Microorganisms

The rhizosphere is home to many microorganisms compared to the bulk soil, and is therefore a key niche that influences microbial community structure and plant growth-related functions. Several microbes mediate plant-beneficial processes such as nitrogen fixation, nutrient solubilization, and synthesis of bioactive metabolites that occur in this dynamic niche. Metal ions can form complexes with compounds generated by rhizosphere bacteria and root exudates, which can impact the metals' bioavailability to both plants and microorganisms (Chen et al. 2017). The microbial population and, thus, microbially driven biological processes in the rhizosphere can be directly impacted by the bioavailability, size, and other special chemical characteristics of nanoparticles. Interestingly, microorganisms can significantly promote the biotransformation of metal nanoparticles, which can subsequently change the bioavailability of these ions (Lv et al. 2019). According to this study, nanoparticles have a substantial anti-bacterial effect on plant root surfaces. The application of silver nanoparticles showed a different response between the corn rhizosphere and the whole microbial community, especially a significant change in the bacterial community compared to fungi (Sillen et al. 2015). However, changes in the microbial community of the rhizosphere showed a favorable effect on corn growth, which suggests either a possible suppression of pathogenic bacteria in the rhizomes or an increase in the number of beneficial bacteria for the plant (Sillen et al. 2015). Similarly, CeO_2 nanoparticles (300-2000 mg/L) showed very beneficial effects on strawberry plants, but reduced rhizosphere microbial diversity (Dai et al. 2020). Other studies show that due to slow release due to low solubility, CuO nanoparticles can be a source of copper that is less toxic compared to their ion ($CuSO_4$) (Guan et al. 2020). Additionally, their use may result in the emergence of a nitrogen-fixing specialized microbial population and a decrease in denitrification processes inside the wheat rhizosphere (Guan et al. 2020). Using nanosilica also increased the population of phosphate-solubilizing and nitrogen-fixing bacteria in the maize rhizosphere (Rangaraj et al. 2014). The utilization of CuO and ZnO nanoparticles further developed Cu and Zn collection in Salvia miltiorrhiza roots (Wei et al. 2021). The microbial community in the rhizosphere may alter as a result of exposure to these nanoparticles; the community's makeup changed according on the kind and concentration of the particles. The results of this study demonstrated that exposure to nanoparticles enhanced the growth-promoting (Sphingomonas), biopesticide-producing (Flavisolibacter), superoxide dismutase-producing (Methylobacillus, Humicola and Aminobacter), acidophilic (Arenimonas) and metal-tolerant (Cu, Zn, Pb, Fe) microbes (Thiobacillus and Metarhizium). Nanofertilizers can also alter the beneficial properties of various rhizobacteria, which in turn can improve plant growth. The application of zinc oxide nanofertilizers at a concentration of 8 mg/L improved the production of Bacillus subtilis exopolysaccharides (EPS), providing better aggregation and retention properties to the soil (Dimkpa et al. 2015). Zinc oxide nanofertilizers at a concentration of 18 mg/kg increased siderophore production in Pseudomonas chloraraphis (Dimkpa et al. 2011). A study conducted by Karunakaran et al. (2014) showed the ecotoxicological effects of silica and alumina nanoparticles at concentrations above 1000 mg/L on PGPR such as Bacillus megaterium, *Pseudomonas fluorescens* and *Azotobacter vinelandii*. Copper oxide nanofertilizers at concentrations of 250-500 mg/kg were found to affect Pseudomonas chlororaphis bacterial diversity (Dimkpa et al., 2015). Zinc oxide nanoparticles reduced indoleacetic acid (IAA) but increased siderophore production, while copper oxide nanoparticles reduced indoleacetic acid (IAA) but increased siderophore production, while copper oxide nanoparticles showed IAA and a reverse effect on siderophore production (Dimkpa et al. 2015; Dimkpa, 2014). At greater concentration (2000 mg/kg), molybdenum nanoparticles have been observed to drastically modify the rhizobia-soybean symbiotic relationship by altering soybean root structure (Yang et al. 2020). Application of CeO nanofertilizers at concentrations of 50-100 mg/kg inhibited the nitrogen fixation process of Rhizobium in a symbiotic association with soybean (Priester et al. 2012) (Table-1& 2, Figure 3&4).

Figure 5. Comparison of different fertilizers with nanofertilizers on plant growth

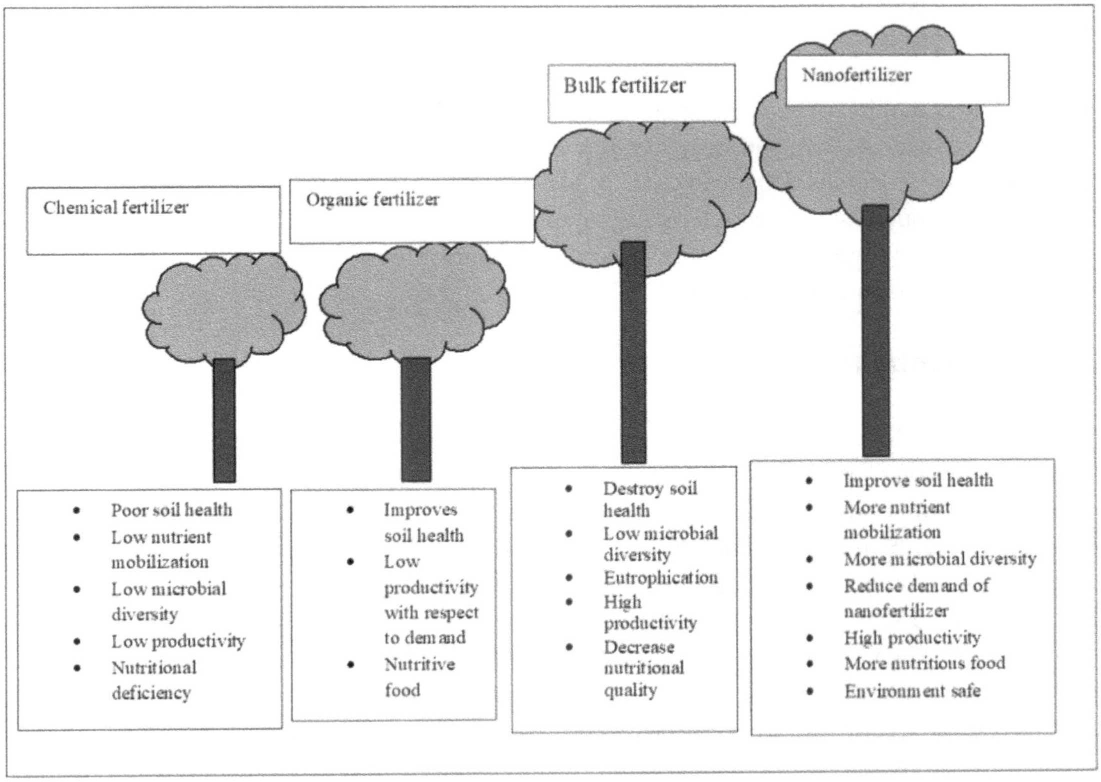

5. CURRENT CHALLENGES AND FUTURE PROSPECTS

As we approach the era of nano-enhanced agriculture, it is critical to comprehend the underlying challenges as well as the endless potential that lied ahed. Nanofertilizers abound in promises of sustainability and efficiency, but there are also some drawbacks. A well-rounded comprehension of the possible obstacles, deficiencies, and future-oriented paths will enhance our readiness for a time when technology and agriculture will seamlessly merge in the future. Any scientific innovation's strength rests in the quality of the supporting research. Indeed, even while huge headway has been made with nanofertilizers, there are as yet clear holes that should be filled. First off, the majority of what is now known about nanofertilizers is empirical. These ideas lack a complete mechanistic understanding, even if they offer a foundation for applications. Understanding a phenomenon's mechanism of occurrence is equally as significant as understanding its existence. This kind of information can help with product improvement, outcome prediction, and safety (Andrew et al. 2019). The long-term effects of nano fertilizers are another major area of uncertainty. The majority of studies have relatively short durations no more than a few crop cycles. Years, if not decades, of understanding are necessary due to the permanent nature of soil and perennial crops. It is critical to evaluate the long-term build-up, interactions with soil bacteria, and possible breakdown mechanisms of nanoparticles. Lastly, a lot of the research that is done nowadays is done in labs, greenhouses, and simulation chambers. Although these offer controlled environments, they frequently fail to capture the intricacy and unpredictability of open field circumstances.

5.1 Environmental and Economic Impacts of Nanofertilizers

The novel nanofertilizers increase the effectiveness of nutrient consumption by utilizing the extraordinary qualities of nanoparticles that include zeolites and nano clays. NFs, in contrast to their traditional equivalents, have a number of benefits, such as variable solubility, trustworthy and effective performance, regulated release mechanisms, and improved targeted action, less ecotoxicity and effortless, secure delivery and disposal processes (Channab et al. 2024). NFs effectively save nutrients that might else go to waste by allowing quick and thorough factory uptake, hence reducing possible environmental impact. Likewise, compared to traditional diseases, their better phrasings allow for the more effective support of sustainable crop development and affair (Channab et al. 2024).

5.2 Environmental Positive Impacts

With some uses, NPs can positively impact their environment. NPs can affect their environmental factors with a portion of their applications (Fig.7). Among the great many potential exercises NPs can have to decrease contamination and upgrade natural wellbeing, for example, advancing creation frameworks with less energy use or assisting during the time spent water and soil remediation, we with having zeroed in on those exercises with high importance and interest in genuine exploration (Martínez et al. 2020; Weijie et al. 2020; Meier et al. 2020). Along these lines, the portion expected to accomplish the impact is brought down as well as the subsequent aftereffects and contamination delivery to the climate and, in this manner, the energy that would be consumed to do later soil medicines. A few models are Zn and Al NPs utilized as covering specialists to create a controlled arrival of supplements and the work of mesoporous silica to safeguard the pesticides dynamic parts against photodegradation (Medina-Pérez et al. 2019).

The Increased need for energy in today's world drives scientists to discover innovative, environmentally friendly methods of producing it. In this sense, nanotechnology has turned into a chance to foster spotless and sustainable power sources. Due to its efficiency, adaptability, and simplicity of use, solar photovoltaic energy generated from solar radiation has emerged as the most intriguing energy source (Diallo et al. 2014). It also has minimal or no adverse effects on natural resources like soils or water. The high production costs and the low absorption capacity of solar energy-only 40% of the sun's energy is converted into electrical energy- are two major problems of solar cells (Echiegu, 2016). Since carbon nanotubes are cheaply synthesized and have an effective energy-gathering and storing capacity, they have been investigated as a potential substitute material for cell architecture. They also resist abrupt shocks and temperature fluctuations (Ong et al. 2010). TiO2 NPs are becoming more and more popular as a possible material for solar energy collection because of their outstanding photocatalytic activity and excellent adsorption (Abdel-Mottaleb et al. 2011). However, the use of carbon nanotubes in windmills has been shown to have a number of benefits, including lightweight blades that can be expanded to capture energy or use as the building blocks for antifreeze coatings, which extend the life of windmills in cold areas. Utilizing certain NPs that lower the temperature and depth required to extract geothermal energy is also preferred (Echiegu, 2016).

5.3 Environmental Consequences of Nanofertilizer Misuse

Environmental elements including the soils phosphate concentration, organic matter content, P^H level, and ionic strength can all have an impact on the chemical properties of nanoparticles (Ameen et al. 2021 and Dimkpa, 2018). The activity of nanoparticles and organic matter in the soil determines their stability in the soil. This contact might alter the surface chemistry, which could have an impact on the bacteria and plants in the soil. However, the behavior of nanoparticles can be influenced by the activity and processes of microorganisms. One of the most crucial things to consider while attempting to comprehend how soil nanoparticles behave is the pH of the soil. Zinc oxide nanoparticles transform into ions more quickly in low pH environments. The nanoparticles tend to adhere to one another at high pH values (Read et al. 2016). Most of the nanomaterials were shown to be harmful to the microbial population at pH values between 4.5 and 7.2. A research of four nanomaterials (zinc oxide nanoparticles, titanium dioxide nanoparticles, cerium dioxide nanoparticles, and magnetite nanoparticles) revealed that higher microbial diversity in saline-alkali soil profited more from zinc oxide nanoparticles than in weakly acidic soil. It has been shown that acidic soil is more sensitive to ZnO nanoparticle impacts on soil microbial enzyme activities than calcareous soil (You et al. 2018). According to Asadishad et al. (2018), at concentrations of 1 to 100 mg/kg, nanoparticles of copper oxide, zinc oxide, and silver are toxic to soil microorganisms; however, nanoparticles of titanium oxide are not harmful at these levels (Asadishad et al. 2018).

6. CONCLUSION

Nanofertilizers represent a revolutionary advance in sustainable agriculture, promising improved nutrient utilization, yield and overall plant well-being. While they offer a number of benefits, from targeted nutrient delivery to plant stress tolerance, they also have potential ecological and health impacts. Combining nanofertilizers with their conventional counterparts highlights their superior efficiency, but also highlights economic and environmental challenges. Going forward, it is important to close research gaps, take advantages of technological advances and create a strong regulatory framework. Collaboration between scientists, policy makers and stakeholders are central to ensuring that nanofertilizers fulfill their potential in shaping a sustainable and food secure future.

In view of the writing analyzed in this survey, the utilization of nanoparticles or nanofertilizers has been displayed to significantly affect plant or yield and soil properties. The effects of exposure to nanoparticles in the agricultural environment can therefore be beneficial or harmful to the health of agro ecosystems and the subsequent consumption of crops. Furthermore, environmental risk assessment of nanoparticles is in its infancy. Hence, further examinations are expected to explore the effects of nanoparticles applied at various kind of portion and ecological circumstances on microbial networks and works, especially in the long term. Despite the increase in the production of nanoparticles used in agriculture, most risk assessment experiments are conducted in vitro, using cells rather than animals as test subjects. Therefore, more research is needed on the effects on soil and human health, as there is uncertainty about the negative consequences of nanoparticle applications.

ACKNOWLEDGEMENT

Gratefully acknowledge to the authority of Gangadhar Meher University for providing financial assistance through sanctioning a minor research project grant to Dr. Santosh Kumar Sethi through Seed Fund Project (Sanctioned order No. 4934/GMU Dt. 02-12-2023)

REFERENCES

Abdel-Mottaleb, M. S. A., Byrne, J. A., & Chakarov, D. (2011). Nanotechnology and solar energy. *International Journal of Photoenergy*, 1, 194146. DOI: 10.1155/2011/194146

Ahluwalia, O., Singh, P. C., & Bhatia, R. (2021). A review on drought stress in plants: Implications, mitigation and the role of plant growth promoting rhizobacteria. *Resources. Environmental Sustainability*, 5, 100032.

Al-Juthery, WA, H., & MN Al-Shami, Q. (2019). The effect of fertigation with nano NPK fertilizers on some parameters of growth and yield of potato (Solanum tuberosum L.). *AL-Qadisiyah. Journal for Agriculture Sciences*, 9(2), 225–232.

Andrew, D. P., Pedersen, P. M., & McEvoy, C. D. (2019). Research methods and design in sport management. *Human Kinetics*. https://books.google.co.in/books?id=8rG8DwAAQBAJ

Asadishad, B., Chahal, S., Akbari, A., Cianciarelli, V., Azodi, M., Ghoshal, S., & Tufenkji, N. (2018). Amendment of agricultural soil with metal nanoparticles: Effects on soil enzyme activity and microbial community composition. *Environmental Science & Technology*, 52(4), 1908–1918. DOI: 10.1021/acs.est.7b05389 PMID: 29356510

Asli, S., & Neumann, P. M. (2009). Colloidal suspensions of clay or titanium dioxide nanoparticles can inhibit leaf growth and transpiration via physical effects on root water transport. *Plant, Cell & Environment*, 32(5), 577–584. DOI: 10.1111/j.1365-3040.2009.01952.x PMID: 19210640

Aubert, T., Burel, A., Esnault, M. A., Cordier, S., Grasset, F., & Cabello-Hurtado, F. (2012). Root uptake and phytotoxicity of nanosized molybdenum octahedral clusters. *Journal of Hazardous Materials*, 219, 111–118. DOI: 10.1016/j.jhazmat.2012.03.058 PMID: 22510541

Avellan, A., Yun, J., Morais, B. P., Clement, E. T., Rodrigues, S. M., & Lowry, G. V. (2021). Critical review: Role of inorganic nanoparticle properties on their foliar uptake and in planta translocation. *Environmental Science & Technology*, 55(20), 13417–13431. DOI: 10.1021/acs.est.1c00178 PMID: 33988374

Babu, S., Singh, R., Yadav, D., Rathore, S. S., Raj, R., Avasthe, R., & Singh, V. K. (2022). Nanofertilizers for agricultural and environmental sustainability. *Chemosphere*, 292, 133451. DOI: 10.1016/j.chemosphere.2021.133451 PMID: 34973251

Bayda, S., Adeel, M., Tuccinardi, T., Cordani, M., & Rizzolio, F. (2019). The history of nanoscience and nanotechnology: From chemical–physical applications to nanomedicine. *Molecules (Basel, Switzerland)*, 25(1), 112. DOI: 10.3390/molecules25010112 PMID: 31892180

Benício, L. P. F., Constantino, V. R. L., Pinto, F. G., Vergütz, L., Tronto, J., & da Costa, L. M. (2017). Layered double hydroxides: New technology in phosphate fertilizers based on nanostructured materials. *ACS Sustainable Chemistry & Engineering*, 5(1), 399–409. DOI: 10.1021/acssuschemeng.6b01784

Burke, D. J., Pietrasiak, N., Situ, S. F., Abenojar, E. C., Porche, M., Kraj, P., & Samia, A. C. S. (2015). Iron oxide and titanium dioxide nanoparticle effects on plant performance and root associated microbes. *International Journal of Molecular Sciences*, 16(10), 23630–23650. DOI: 10.3390/ijms161023630 PMID: 26445042

Carpita, N. C., & Gibeaut, D. M. (1993). Structural models of primary cell walls in flowering plants: Consistency of molecular structure with the physical properties of the walls during growth. *The Plant Journal*, 3(1), 1–30. DOI: 10.1111/j.1365-313X.1993.tb00007.x PMID: 8401598

Channab, B. E., Idrissi, A. E., Ammar, A., Dardari, O., Marrane, S. E., Gharrak, A. E., Akil, A., Es-semlali, Y., & Zahouily, M. (2024). Recent advances in nano-fertilizers: Synthesis, crop yield impact, and economic analysis. *Nanoscale*, 16(9), 4484–4513. DOI: 10.1039/D3NR05012B PMID: 38314867

Chen, C., Guo, L., Chen, Y., Qin, P., & Wei, G. (2023). Pristine and sulfidized zinc oxide nanoparticles alter bacterial communities and metabolite profiles in soybean rhizocompartments. *The Science of the Total Environment*, 855, 158697. DOI: 10.1016/j.scitotenv.2022.158697 PMID: 36099947

Chen, Y. T., Wang, Y., & Yeh, K. C. (2017). Role of root exudates in metal acquisition and tolerance. *Current Opinion in Plant Biology*, 39, 66–72. DOI: 10.1016/j.pbi.2017.06.004 PMID: 28654805

Dai, Y., Chen, F., Yue, L., Li, T., Jiang, Z., Xu, Z., & Xing, B. (2020). Uptake, transport, and transformation of CeO2 nanoparticles by strawberry and their impact on the rhizosphere bacterial community. *ACS Sustainable Chemistry & Engineering*, 8(12), 4792–4800. DOI: 10.1021/acssuschemeng.9b07422

Daisy, P., & Saipriya, K. (2012). Biochemical analysis of Cassia fistula aqueous extract and phytochemically synthesized gold nanoparticles as hypoglycemic treatment for diabetes mellitus. *International Journal of Nanomedicine*, •••, 1189–1202. DOI: 10.2147/IJN.S26650 PMID: 22419867

Dan, Y., Zhang, W., Xue, R., Ma, X., Stephan, C., & Shi, H. (2015). Characterization of gold nanoparticle uptake by tomato plants using enzymatic extraction followed by single-particle inductively coupled plasma–mass spectrometry analysis. *Environmental Science & Technology*, 49(5), 3007–3014. DOI: 10.1021/es506179e PMID: 25633589

Davarpanah, S., Tehranifar, A., Davarynejad, G., Abadía, J., & Khorasani, R. (2016). Effects of foliar applications of zinc and boron nano-fertilizers on pomegranate (Punica granatum cv. Ardestani) fruit yield and quality. *Scientia Horticulturae*, 210, 57–64. DOI: 10.1016/j.scienta.2016.07.003

Demirer, G. S., Silva, T. N., Jackson, C. T., Thomas, J. B. W., Ehrhardt, D., Rhee, S. Y., & Landry, M. P. (2021). Nanotechnology to advance CRISPR–Cas genetic engineering of plants. *Nature Nanotechnology*, 16(3), 243–250. DOI: 10.1038/s41565-021-00854-y PMID: 33712738

Diallo, M. S., Fromer, N. A., & Jhon, M. S. (2014). Nanotechnology for sustainable development: Retrospective and outlook. *Journal of Nanoparticle Research*, 15(11), 2044. DOI: 10.1007/s11051-013-2044-0

Dimkpa, C. O., McLean, J. E., Britt, D. W., & Anderson, A. J. (2015). Nano-CuO and interaction with nano-ZnO or soil bacterium provide evidence for the interference of nanoparticles in metal nutrition of plants. *Ecotoxicology (London, England)*, 24(1), 119–129. DOI: 10.1007/s10646-014-1364-x PMID: 25297564

Drostkar, E., Talebi, R., & Kanouni, H. (2016). Foliar application of Fe, Zn and NPK nano-fertilizers on seed yield and morphological traits in chickpea under rainfed condition. *Journal of Resources and Ecology*, 4(1), 221–228.

Echiegu, E. A. (2016). Nanotechnology as a tool for enhanced renewable energy application in developing countries. *Journal of Fundamentals of Renewable Energy and Applications, 6*(6), 01-09.

Elsheery, N. I., Sunoj, V. S. J., Wen, Y., Zhu, J. J., Muralidharan, G., & Cao, K. F. (2020). Foliar application of nanoparticles mitigates the chilling effect on photosynthesis and photoprotection in sugarcane. *Plant Physiology and Biochemistry*, 149, 50–60. DOI: 10.1016/j.plaphy.2020.01.035 PMID: 32035252

Fierer, N., Wood, S. A., & de Mesquita, C. P. B. (2021). How microbes can, and cannot, be used to assess soil health. *Soil Biology & Biochemistry*, 153, 108111. DOI: 10.1016/j.soilbio.2020.108111

Fleischer, A., O'Neill, M. A., & Ehwald, R. (1999). The pore size of non-gramineous plant cell walls is rapidly decreased by borate ester cross-linking of the pectic polysaccharide rhamnogalacturonan II. *Plant Physiology*, 121(3), 829–838. DOI: 10.1104/pp.121.3.829 PMID: 10557231

Frenk, S., Ben-Moshe, T., Dror, I., Berkowitz, B., & Minz, D. (2013). Effect of metal oxide nanoparticles on microbial community structure and function in two different soil types. *PLoS One*, 8(12), 84441. DOI: 10.1371/journal.pone.0084441 PMID: 24349575

Giannousi, K., Avramidis, I., & Dendrinou-Samara, C. (2013). Synthesis, characterization and evaluation of copper-based nanoparticles as agrochemicals against Phytophthora infestans. *RSC Advances*, 3(44), 21743–21752. DOI: 10.1039/c3ra42118j

Giraldo, J. P., Wu, H., Newkirk, G. M., & Kruss, S. (2019). Nanobiotechnology approaches for engineering smart plant sensors. *Nature Nanotechnology*, 14(6), 541–553. DOI: 10.1038/s41565-019-0470-6 PMID: 31168083

Guan, X., Gao, X., Avellan, A., Spielman-Sun, E., Xu, J., Laughton, S., & Lowry, G. V. (2020). CuO nanoparticles alter the rhizospheric bacterial community and local nitrogen cycling for wheat grown in a calcareous soil. *Environmental Science & Technology*, 54(14), 8699–8709. DOI: 10.1021/acs.est.0c00036 PMID: 32579348

Guerra, F. D., Attia, M. F., Whitehead, D. C., & Alexis, F. (2018). Nanotechnology for environmental remediation: Materials and applications. *Molecules (Basel, Switzerland)*, 23(7), 1760. DOI: 10.3390/molecules23071760 PMID: 30021974

Guo, H., White, J. C., Wang, Z., & Xing, B. (2018). Nano-enabled fertilizers to control the release and use efficiency of nutrients. *Current Opinion in Environmental Science & Health*, 6, 77–83. DOI: 10.1016/j.coesh.2018.07.009

Ha, N. M. C., Nguyen, T. H., Wang, S. L., & Nguyen, A. D. (2019). Preparation of NPK nanofertilizer based on chitosan nanoparticles and its effect on biophysical characteristics and growth of coffee in green house. *Research on Chemical Intermediates*, 45(1), 51–63. DOI: 10.1007/s11164-018-3630-7

Harish, V., Tewari, D., Gaur, M., Yadav, A. B., Swaroop, S., Bechelany, M., & Barhoum, A. (2022). Review on nanoparticles and nanostructured materials: Bioimaging, biosensing, drug delivery, tissue engineering, antimicrobial, and agro-food applications. *Nanomaterials (Basel, Switzerland)*, 12(3), 457. DOI: 10.3390/nano12030457 PMID: 35159802

Imada, K., Sakai, S., Kajihara, H., Tanaka, S., & Ito, S. (2016). Magnesium oxide nanoparticles induce systemic resistance in tomato against bacterial wilt disease. *Plant Pathology*, 65(4), 551–560. DOI: 10.1111/ppa.12443

Iqbal, M. A. (2019). Nano-fertilizers for sustainable crop production under changing climate: A global perspective. *Sustainable crop. Production*, 8, 1–13.

Jain, D., Daima, H. K., Kachhwaha, S., & Kothari, S. L. (2009). Synthesis of plant-mediated silver nanoparticles using papaya fruit extract and evaluation of their anti-microbial activities. *Digest Journal of Nanomaterials and Biostructures*, 4(3), 557–563.

Kah, M., Kookana, R. S., Gogos, A., & Bucheli, T. D. (2018). A critical evaluation of nanopesticides and nanofertilizers against their conventional analogues. *Nature Nanotechnology*, 13(8), 677–684. DOI: 10.1038/s41565-018-0131-1 PMID: 29736032

Kamle, M., Mahato, D. K., Devi, S., Soni, R., Tripathi, V., Mishra, A. K., & Kumar, P. (2020). Nanotechnological interventions for plant health improvement and sustainable agriculture. *3 Biotech, 10*(4), 168.

Karunakaran, G., Suriyaprabha, R., Manivasakan, P., Rajendran, V., & Kannan, N. (2014). Influence of nano and bulk SiO2 and Al2O3 particles on PGPR and soil nutrient contents. *Current Nanoscience*, 10(4), 604–612. DOI: 10.2174/15734137113096660126

Kaur, H., Kalia, A., Sandhu, J. S., Dheri, G. S., Kaur, G., & Pathania, S. (2022). Interaction of TiO2 nanoparticles with soil: Effect on microbiological and chemical traits. *Chemosphere*, 301, 134629. DOI: 10.1016/j.chemosphere.2022.134629 PMID: 35447207

Kolesnikov, S. I., Timoshenko, A. N., Kazeev, K. S., Akimenko, Y. V., & Myasnikova, M. A. (2019). Ecotoxicity of copper, nickel, and zinc nanoparticles assessment on the basis of biological indicators of chernozems. *Eurasian Soil Science*, 52(8), 982–987. DOI: 10.1134/S106422931908009X

Kumari, M., Pandey, S., Mishra, S. K., Giri, V. P., Agarwal, L., Dwivedi, S., & Mishra, A. (2020). Omics-based mechanistic insight into the role of bioengineered nanoparticles for biotic stress amelioration by modulating plant metabolic pathways. *Frontiers in Bioengineering and Biotechnology*, 8, 242. DOI: 10.3389/fbioe.2020.00242 PMID: 32363178

Kwak, S. Y., Lew, T. T. S., Sweeney, C. J., Koman, V. B., Wong, M. H., Bohmert-Tatarev, K., & Strano, M. S. (2019). Chloroplast-selective gene delivery and expression in planta using chitosan-complexed single-walled carbon nanotube carriers. *Nature Nanotechnology*, 14(5), 447–455. DOI: 10.1038/s41565-019-0375-4 PMID: 30804482

Kwak, S. Y., Wong, M. H., Lew, T. T. S., Bisker, G., Lee, M. A., Kaplan, A., & Strano, M. S. (2017). Nanosensor technology applied to living plant systems. *Annual Review of Analytical Chemistry (Palo Alto, Calif.)*, 10(1), 113–140. DOI: 10.1146/annurev-anchem-061516-045310 PMID: 28605605

Lahiani, M. H., Dervishi, E., Chen, J., Nima, Z., Gaume, A., Biris, A. S., & Khodakovskaya, M. V. (2013). Impact of carbon nanotube exposure to seeds of valuable crops. *ACS Applied Materials & Interfaces*, 5(16), 7965–7973. DOI: 10.1021/am402052x PMID: 23834323

Liu, R., & Lal, R. (2014). Synthetic apatite nanoparticles as a phosphorus fertilizer for soybean (Glycine max). *Scientific Reports*, 4(1), 5686. DOI: 10.1038/srep05686 PMID: 25023201

Lucas, W. J., & Lee, J. Y. (2004). Plasmodesmata as a supracellular control network in plants. *Nature Reviews. Molecular Cell Biology*, 5(9), 712–726. DOI: 10.1038/nrm1470 PMID: 15340379

Lv, J., Christie, P., & Zhang, S. (2019). Uptake, translocation, and transformation of metal-based nanoparticles in plants: Recent advances and methodological challenges. *Environmental Science. Nano*, 6(1), 41–59. DOI: 10.1039/C8EN00645H

Ma, X., Geiser-Lee, J., Deng, Y., & Kolmakov, A. (2010). Interactions between engineered nanoparticles (ENPs) and plants: Phytotoxicity, uptake and accumulation. *The Science of the Total Environment*, 408(16), 3053–3061. DOI: 10.1016/j.scitotenv.2010.03.031 PMID: 20435342

Maensiri, S., Laokul, P., Klinkaewnarong, J., Phokha, S., Promarak, V., & Seraphin, S. (2008). Indium oxide (In_2O_3) nanoparticles using Aloe vera plant extract: Synthesis and optical properties. *Journal of Optoelectronics and Advanced Materials*, 10(3), 161–165.

Medina-Pérez, G., Fernández-Luqueño, F., Vazquez-Nuñez, E., López-Valdez, F., Prieto-Mendez, J., Madariaga-Navarrete, A., & Miranda-Arámbula, M. (2019). Remediating polluted soils using nanotechnologies: Environmental benefits and risks. *Polish Journal of Environmental Studies*, 28(3), 1013–1030. DOI: 10.15244/pjoes/87099

Meier, S., Moore, F., Morales, A., González, M. E., Seguel, A., Meriño-Gergichevich, C., & Mejías, J. (2020). Synthesis of calcium borate nanoparticles and its use as a potential foliar fertilizer in lettuce (Lactuca sativa) and zucchini (Cucurbita pepo). *Plant Physiology and Biochemistry*, 151, 673–680. DOI: 10.1016/j.plaphy.2020.04.025 PMID: 32353673

Navarro, E., Piccapietra, F., Wagner, B., Marconi, F., Kaegi, R., Odzak, N., & Behra, R. (2008). Toxicity of silver nanoparticles to Chlamydomonas reinhardtii. *Environmental Science & Technology*, 42(23), 8959–8964. DOI: 10.1021/es801785m PMID: 19192825

Nawaz, S., & Bano, A. (2020). Effects of PGPR (*Pseudomonas* sp.) and Ag-nanoparticles on enzymatic activity and physiology of cucumber. *Recent Patents on Food, Nutrition & Agriculture*, 11(2), 124–136. DOI: 10.2174/2212798410666190716162340 PMID: 31322080

Niraimathi, K. L., Sudha, V., Lavanya, R., & Brindha, P. (2013). Biosynthesis of silver nanoparticles using Alternanthera sessilis (Linn.) extract and their antimicrobial, antioxidant activities. *Colloids and Surfaces. B, Biointerfaces*, 102, 288–291. DOI: 10.1016/j.colsurfb.2012.08.041 PMID: 23006568

Oghenerume, P., Eduok, S., Ita, B., John, O., & Basssy, I. (2020). Impact of zinc oxide nanoparticles amended organic manure on *Arachis hypogaea* growth response and rhizosphere bacterial community. *International Journal of Plant and Soil Science*, 32(5), 24–35. DOI: 10.9734/ijpss/2020/v32i530279

Ong, Y. T., Ahmad, A. L., Zein, S. H. S., & Tan, S. H. (2010). A review on carbon nanotubes in an environmental protection and green engineering perspective. *Brazilian Journal of Chemical Engineering*, 27(2), 227–242. DOI: 10.1590/S0104-66322010000200002

Pérez-de-Luque, A. (2017). Interaction of nanomaterials with plants: What do we need for real applications in agriculture? *Frontiers in Environmental Science*, 5, 12. DOI: 10.3389/fenvs.2017.00012

Preetha, P. S., & Balakrishnan, N. (2017). A review of nano fertilizers and their use and functions in soil. *International Journal of Current Microbiology and Applied Sciences*, 6(12), 3117–3133. DOI: 10.20546/ijcmas.2017.612.364

Priester, J. H., Ge, Y., Mielke, R. E., Horst, A. M., Moritz, S. C., Espinosa, K., & Holden, P. A. (2012). Soybean susceptibility to manufactured nanomaterials with evidence for food quality and soil fertility interruption. *Proceedings of the National Academy of Sciences of the United States of America*, 109(37), E2451–E2456. DOI: 10.1073/pnas.1205431109 PMID: 22908279

Rangaraj, S., Gopalu, K., Rathinam, Y., Periasamy, P., Venkatachalam, R., & Narayanasamy, K. (2014). Effect of silica nanoparticles on microbial biomass and silica availability in maize rhizosphere. *Biotechnology and Applied Biochemistry*, 61(6), 668–675. DOI: 10.1002/bab.1191 PMID: 24329970

Rashid, M. I., Shahzad, T., Shahid, M., Ismail, I. M., Shah, G. M., & Almeelbi, T. (2017). Zinc oxide nanoparticles affect carbon and nitrogen mineralization of Phoenix dactylifera leaf litter in a sandy soil. *Journal of Hazardous Materials*, 324, 298–305. DOI: 10.1016/j.jhazmat.2016.10.063 PMID: 27810328

Read, D. S., Matzke, M., Gweon, H. S., Newbold, L. K., Heggelund, L., Ortiz, M. D., & Svendsen, C. (2016). Soil pH effects on the interactions between dissolved zinc, non-nano-and nano-ZnO with soil bacterial communities. *Environmental Science and Pollution Research International*, 23(5), 4120–4128. DOI: 10.1007/s11356-015-4538-z PMID: 25903189

Rizwan, M., & Ali, S., ur Rehman, M. Z., Malik, S., Adrees, M., Qayyum, M. F. & Ahmad, P. (2019). Effect of foliar applications of silicon and titanium dioxide nanoparticles on growth, oxidative stress, and cadmium accumulation by rice (Oryza sativa). *Acta Physiologiae Plantarum*, 41, 1–12.

Rui, M., Ma, C., Hao, Y., Guo, J., Rui, Y., Tang, X., & Zhu, S. (2016). Iron oxide nanoparticles as a potential iron fertilizer for peanut (Arachis hypogaea). *Frontiers in Plant Science*, 7, 815. DOI: 10.3389/fpls.2016.00815 PMID: 27375665

Saif, S., Tahir, A., & Chen, Y. (2016). Green synthesis of iron nanoparticles and their environmental applications and implications. *Nanomaterials (Basel, Switzerland)*, 6(11), 209. DOI: 10.3390/nano6110209 PMID: 28335338

Satti, S. H., Raja, N. I., Javed, B., Akram, A., Mashwani, Z. U. R., Ahmad, M. S., & Ikram, M. (2021). Titanium dioxide nanoparticles elicited agro-morphological and physicochemical modifications in wheat plants to control Bipolaris sorokiniana. *PLoS One*, 16(2), 0246880. DOI: 10.1371/journal.pone.0246880 PMID: 33571310

Satyavani, K., Gurudeeban, S., Ramanathan, T., & Balasubramanian, T. (2011). Biomedical potential of silver nanoparticles synthesized from calli cells of *Citrullus colocynthis* (L.) Schrad. *Journal of Nanobiotechnology*, 9(1), 43. DOI: 10.1186/1477-3155-9-43 PMID: 21943321

Schwab, F., Zhai, G., Kern, M., Turner, A., Schnoor, J. L., & Wiesner, M. R. (2016). Barriers, pathways and processes for uptake, translocation and accumulation of nanomaterials in plants–Critical review. *Nanotoxicology*, 10(3), 257–278. DOI: 10.3109/17435390.2015.1048326 PMID: 26067571

Seleiman, M. F., Almutairi, K. F., Alotaibi, M., Shami, A., Alhammad, B. A., & Battaglia, M. L. (2020). Nano-fertilization as an emerging fertilization technique: Why can modern agriculture benefit from its use? *Plants*, 10(1), 2. DOI: 10.3390/plants10010002 PMID: 33375026

Seyedsharifi, R., & Khorramdel, S. (2015). Effects of nano-zinc oxide and seed inoculation by plant growth promoting rhizobacteria (PGPR) on yield, yield components and grain filling period of soybean (*Glycine max* l.). *Iranian Journal of Field Crops Research*, 13(4), 738–753.

Shah, V., Collins, D., Walker, V. K., & Shah, S. (2014). The impact of engineered cobalt, iron, nickel and silver nanoparticles on soil bacterial diversity under field conditions. *Environmental Research Letters*, 9(2), 024001. DOI: 10.1088/1748-9326/9/2/024001

Shang, Y., Hasan, M. K., Ahammed, G. J., Li, M., Yin, H., & Zhou, J. (2019). Applications of nanotechnology in plant growth and crop protection: A review. *Molecules (Basel, Switzerland)*, 24(14), 2558. DOI: 10.3390/molecules24142558 PMID: 31337070

Sharma, B., Tiwari, S., Kumawat, K. C., & Cardinale, M. (2023). Nano-biofertilizers as bio-emerging strategies for sustainable agriculture development: Potentiality and their limitations. *The Science of the Total Environment*, 860, 160476. DOI: 10.1016/j.scitotenv.2022.160476 PMID: 36436627

Shcherbakova, E. N., Shcherbakov, A. V., Andronov, E. E., Gonchar, L. N., Kalenskaya, S. M., & Chebotar, V. K. (2017). Combined pre-seed treatment with microbial inoculants and Mo nanoparticles changes composition of root exudates and rhizosphere microbiome structure of chickpea (Cicer arietinum L.) plants. *Symbiosis*, 73(1), 57–69. DOI: 10.1007/s13199-016-0472-1

Shenashen, M., Derbalah, A., Hamza, A., Mohamed, A., & El Safty, S. (2017). Antifungal activity of fabricated mesoporous alumina nanoparticles against root rot disease of tomato caused by Fusarium oxysporium. *Pest Management Science*, 73(6), 1121–1126. DOI: 10.1002/ps.4420 PMID: 27558672

Sillen, W. M., Thijs, S., Abbamondi, G. R., Janssen, J., Weyens, N., White, J. C., & Vangronsveld, J. (2015). Effects of silver nanoparticles on soil microorganisms and maize biomass are linked in the rhizosphere. *Soil Biology & Biochemistry*, 91, 14–22. DOI: 10.1016/j.soilbio.2015.08.019

Singh, H., Sharma, A., Bhardwaj, S. K., Arya, S. K., Bhardwaj, N., & Khatri, M. (2021). Recent advances in the applications of nano-agrochemicals for sustainable agricultural development. *Environmental Science. Processes & Impacts*, 23(2), 213–239. DOI: 10.1039/D0EM00404A PMID: 33447834

Singh, S., Saikia, J. P., & Buragohain, A. K. (2013). A novel 'green'synthesis of colloidal silver nanoparticles (SNP) using *Dillenia indica* fruit extract. *Colloids and Surfaces. B, Biointerfaces*, 102, 83–85. DOI: 10.1016/j.colsurfb.2012.08.012 PMID: 23010109

Singh, S. P., Subhadarshini, S., Dilnashin, H., Keswani, C., Sansinenea, E., & Shende, S. S. (2021). Biological synthesis of nanofertilizer and their effects on crop health. In *Microbial synthesis of nanomaterials* (pp. 201–216). Nova Science publishers.

Slomberg, D. L., & Schoenfisch, M. H. (2012). Silica nanoparticle phytotoxicity to Arabidopsis thaliana. *Environmental Science & Technology*, 46(18), 10247–10254. DOI: 10.1021/es300949f PMID: 22889047

Soliman, A. S., El-Feky, S. A., & Darwish, E. (2015). Alleviation of salt stress on Moringa peregrina using foliar application of nanofertilizers. *Journal of Horticulture and Forestry*, 7(2), 36–47. DOI: 10.5897/JHF2014.0379

Sun, H., Peng, Q., Guo, J., Zhang, H., Bai, J., & Mao, H. (2022). Effects of short-term soil exposure of different doses of ZnO nanoparticles on the soil environment and the growth and nitrogen fixation of alfalfa. *Environmental Pollution*, 309, 119817. DOI: 10.1016/j.envpol.2022.119817 PMID: 35872284

Suriyakalaa, U., Antony, J. J., Suganya, S., Siva, D., Sukirtha, R., Kamalakkannan, S., & Achiraman, S. (2013). Hepatocurative activity of biosynthesized silver nanoparticles fabricated using Andrographis paniculata. *Colloids and Surfaces. B, Biointerfaces*, 102, 189–194. DOI: 10.1016/j.colsurfb.2012.06.039 PMID: 23018020

Toksha, B., Sonawale, V. A. M., Vanarase, A., Bornare, D., Tonde, S., Hazra, C., & Chatterjee, A. (2021). Nanofertilizers: A review on synthesis and impact of their use on crop yield and environment. *Environmental Technology & Innovation*, 24, 101986. DOI: 10.1016/j.eti.2021.101986

Torney, F., Trewyn, B. G., Lin, V. S. Y., & Wang, K. (2007). Mesoporous silica nanoparticles deliver DNA and chemicals into plants. *Nature Nanotechnology*, 2(5), 295–300. DOI: 10.1038/nnano.2007.108 PMID: 18654287

Vanti, G. L., Nargund, V. B., N, B. K., Vanarchi, R., Kurjogi, M., Mulla, S. I., Tubaki, S., & Patil, R. R. (2019). Synthesis of Gossypium hirsutum-derived silver nanoparticles and their antibacterial efficacy against plant pathogens. *Applied Organometallic Chemistry*, 33(1), 4630. DOI: 10.1002/aoc.4630

Viswanathan, V. K., Manoharan, S. R. R., Subramanian, S., & Moon, A. (2019). Nanotechnology in spine surgery: A current update and critical review of the literature. *World Neurosurgery*, 123, 142–155. DOI: 10.1016/j.wneu.2018.11.035 PMID: 30447449

Wang, P., Lombi, E., Zhao, F. J., & Kopittke, P. M. (2016). Nanotechnology: A new opportunity in plant sciences. *Trends in Plant Science*, 21(8), 699–712. DOI: 10.1016/j.tplants.2016.04.005 PMID: 27130471

Wang, S., Wang, F., Gao, S., & Wang, X. (2016). Heavy metal accumulation in different rice cultivars as influenced by foliar application of nano-silicon. *Water, Air, and Soil Pollution*, 227(7), 1–13. DOI: 10.1007/s11270-016-2928-6

Wang, W., Yang, Z., Zhang, A., & Yang, S. (2021). Water retention and fertilizer slow release integrated superabsorbent synthesized from millet straw and applied in agriculture. *Industrial Crops and Products*, 160, 113126. DOI: 10.1016/j.indcrop.2020.113126

Wei, X., Cao, P., Wang, G., Liu, Y., Song, J., & Han, J. (2021). CuO, ZnO, and γ-Fe2O3 nanoparticles modified the underground biomass and rhizosphere microbial community of *Salvia miltiorrhiza* (Bge.) after 165-day exposure. *Ecotoxicology and Environmental Safety*, 217, 112232. DOI: 10.1016/j.ecoenv.2021.112232 PMID: 33864980

Weijie, M., Chongnv, W., Xuming, P., Weixin, J., Yuhang, W., & Benhui, S. (2020). TiO2 nanoparticles and multi-walled carbon nanotubes monitoring and bioremediation potential using ciliates *Pseudocohnilembus persalinus.Ecotoxicology and Environmental Safety*, 187, 109825. DOI: 10.1016/j.ecoenv.2019.109825 PMID: 31677570

Wesołowska, M., Rymarczyk, J., Góra, R., Baranowski, P., Sławiński, C., Klimczyk, M., & Schimmelpfennig, L. (2021). New slow-release fertilizers-economic, legal and practical aspects: A Review. *International Agrophysics*, 35(1), 11–24. DOI: 10.31545/intagr/131184

Wu, H., Tito, N., & Giraldo, J. P. (2017). Anionic cerium oxide nanoparticles protect plant photosynthesis from abiotic stress by scavenging reactive oxygen species. *ACS Nano*, 11(11), 11283–11297. DOI: 10.1021/acsnano.7b05723 PMID: 29099581

Xu, J., Luo, X., Wang, Y., & Feng, Y. (2018). Evaluation of zinc oxide nanoparticles on lettuce (Lactuca sativa L.) growth and soil bacterial community. *Environmental Science and Pollution Research International*, 25(6), 6026–6035. DOI: 10.1007/s11356-017-0953-7 PMID: 29238929

Yang, J., Song, Z., Ma, J., & Han, H. (2020). Toxicity of molybdenum-based nanomaterials on the soybean–rhizobia symbiotic system: Implications for nutrition. *ACS Applied Nano Materials*, 3(6), 5773–5782. DOI: 10.1021/acsanm.0c00943

You, T., Liu, D., Chen, J., Yang, Z., Dou, R., Gao, X., & Wang, L. (2018). Effects of metal oxide nanoparticles on soil enzyme activities and bacterial communities in two different soil types. *Journal of Soils and Sediments*, 18(1), 211–221. DOI: 10.1007/s11368-017-1716-2

Yusefi-Tanha, E., Fallah, S., Rostamnejadi, A., & Pokhrel, L. R. (2020). Zinc oxide nanoparticles (ZnONPs) as a novel nanofertilizer: Influence on seed yield and antioxidant defense system in soil grown soybean (Glycine max cv. Kowsar). *The Science of the Total Environment*, 738, 140240. DOI: 10.1016/j.scitotenv.2020.140240 PMID: 32570083

Zhang, H., Huang, M., Zhang, W., Gardea-Torresdey, J. L., White, J. C., Ji, R., & Zhao, L. (2020). Silver nanoparticles alter soil microbial community compositions and metabolite profiles in unplanted and cucumber-planted soils. *Environmental Science & Technology*, 54(6), 3334–3342. DOI: 10.1021/acs.est.9b07562 PMID: 32088952

Zulfiqar, F., Navarro, M., Ashraf, M., Akram, N. A., & Munné-Bosch, S. (2019). Nanofertilizer use for sustainable agriculture: Advantages and limitations. *Plant Science*, 289, 110270. DOI: 10.1016/j.plantsci.2019.110270 PMID: 31623775

Chapter 8
Harnessing the Therapeutic Potential of Metallic Bionanoparticles:
Integrating Medicinal Plant as Green Factories

Swati Nilkanthrao Hade
Dr. D.Y. Patil ACS College, Pune, India

ABSTRACT

The integration of nanotechnology and biological systems has led to the emerging field of nanobiotechnology, particularly in the development of nanomedicine. This chapter demonstrates that MNPs synthesized from plant extracts, such as silver, gold, copper, and zinc oxide nanoparticles, have significant therapeutic potential. Their antioxidant properties are crucial in mitigating oxidative stress-related diseases, while anticancer activities facilitate targeted drug delivery and enhanced cytotoxic effects against cancer cells. Additionally, antimicrobial properties offer a promising alternative to combat antibiotic-resistant pathogens. Overall, the green synthesis of MNPs represents a promising advancement in nanomedicine, combining traditional medicinal knowledge with modern nanotechnology. This approach has the potential to revolutionize drug delivery systems, reduce the side effects of conventional therapies, and contribute to the development of new, effective treatments for various diseases.

1. INTRODUCTION

The novel field of nanobiotechnology has emerged as a result of the recent fusion of biological systems and nanoscale technologies (Bogunia-Kubik & Sugisaka, 2002). Nanomedicine, which uses nanotechnology to healthcare, medicine, and therapy, offers the possibility to produce new therapeutic agent which involves the exact manipulation of human biological systems at the molecular level utilizing designed nanodevices and nanostructures. This developing discipline focuses on monitoring, fixing, and creating these systems to better medical science and diagnostics. An important issue is ensuring the proper circulation of medications and therapeutic agents throughout the patient's body, enhancing their

DOI: 10.4018/979-8-3693-6240-2.ch008

effectiveness and limiting negative effects. By harnessing the unique features of nanoparticles, nanomedicine strives to produce new treatments that enhance the efficacy and precision of medical interventions.

By integrating nanotechnology into therapeutic development, it is possible to transform treatment modalities and greatly enhance patient outcomes, opening the door to ground-breaking breakthroughs in the treatment of a wide range of illnesses (Kim, 2010; Sanna and Sechi, 2020; Morrow, 2007). Large-scale material utilized in drug administration faces a number of difficulties, including poor solubility and stability, intestinal absorption and bioavailability, sustained release and targeted drug delivery, side effects, and many more. Numerous studies on nanodrug drug delivery systems have concentrated on the creation and development of nanostructures as a means of overcoming all these obstacles. The possibility for more accurate and efficient medical treatments is being advanced by these cutting-edge methods in nanomedicine and nanodrug delivery (Ochubiojo et al., 2012).

Nanoscale drug delivery systems have the potential to enhance the therapeutic efficacy of existing drugs with fewerside effects.It also enables the creation of entirely new categories of therapeutics and inspires a reevaluation of pharmaceutically suboptimal yet biologically active new molecular entities that were once considered unfeasible for development (Doyle, 2013).

The utilization of bioactive compounds from medicinal plants holds tremendous potential for addressing various health concerns. However, challenges such as poor bioavailability and stability have limited their therapeutic efficacy. Plant-mediated synthesis of nanoparticles represents a promising and sustainable approach in nanotechnology, leveraging the natural reducing and stabilizing agents found in plants.

Nanoparticles possess unique properties that allow for their potential use in numerous fields. As a result, substantial research efforts are focused on developing new methods for nanoparticle synthesis. One promising approach is the biological synthesis route, which is efficient, cost-effective, and environmentally friendly. Biological entities like bacteria, fungi, yeast, algae, and plants have demonstrated significant bioreduction capabilities (Vijayaraghava & Ashokkumar, 2017). The components of biological materials are responsible for the reduction process, often initiated by various compounds within the cells. These include phenolic, carbonyl, amine, and amide groups, as well as proteins, pigments, flavonoids, terpenoids, alkaloids, and other reducing agents (Asmathunisha et al., 2013). However, it enables the synthesis of metallic nanoparticles in various sizes and shapes.

2. BIOSYNTHESIS AND MECHANISM OF PLANT BASED METAL NANOPARTICLES (MNPS)

Utilization of plant extract for the preparation of nanoparticles is reasonably uncomplicated and approachable method for the synthesis of metal and metal oxide nanoparticle. Figure 1 illustrates the biosynthesis of metallic nanoparticles using plant extracts. Plant extract considered to be a safe and superior source of the production of metal nanoparticle (Singh et al., 2018). Plants, in particular, are capable of reducing metal ions, offering an economical and efficient means of synthesizing metallic nanoparticles due to their abundance of active metabolites (Puja et al., 2023).

Figure 1. Biological synthesis of metal nanoparticles (MNPs) using plant extract.

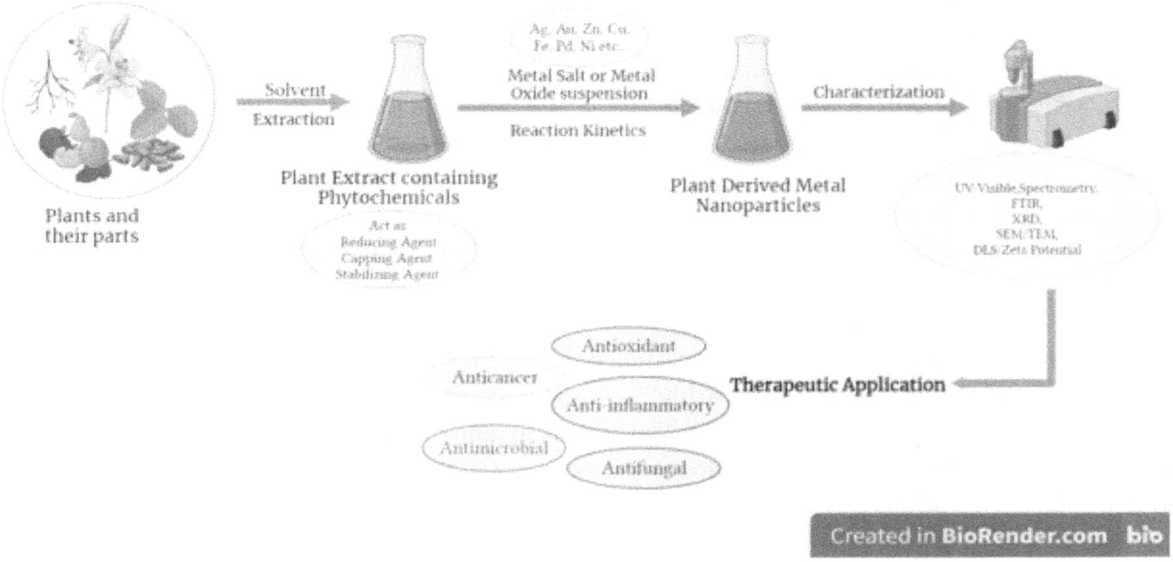

As previously stated, a variety of plant metabolites, including terpenoids, polyphenols, sugars, alkaloids, phenolic acids, and proteins, actively participate in the bioreduction of metal ions, which leads to the synthesis of nanoparticles. Since flavonoids have multiple functional groups that facilitate the production of nanoparticles. Makarov et al. (2014) remarkably states that the electrons or carbonyl groups in some flavonoids have the ability to bind metal ions. Quercetin, for example, is a flavonoid with exceptionally significant chelating activity because it can chelate at three different locations: the C3 and C5 positions for carbonyl and hydroxyl groups, and the C3' and C4' sites for catechol groups. These groups chelate a variety of metal ions, such as Fe_2^+, Fe_3^+, Cu_2^+, Zn_2^+, Fe_2^+, and Co_2^+. Indeed, the existence of such mechanisms might be responsible for flavonoids capacity to adsorb onto the forming nanoparticle's surface.

There are three main steps involved in the mechanism of metal nanoparticle production from plant and plant extracts includes 1) the activation phase, which involves the reduction of metal ions and nucleation of the reduced metal atoms; 2) the growth phase, which involves the spontaneous coaxing of small adjacent nanoparticles into larger particles (direct formation of nanoparticles through heterogeneous nucleation and growth, and additional metal ion reduction; a procedure known as Ostwald ripening), accompanied by an increase in the thermodynamic stability of the nanoparticles; and 3) the process termination phase, which determines the final shape of the nanoparticles.

Diosgenin (steroidal saponin), found in *Dioscorea deltoidea* extract, has the ability to convert Ag^+ (Silver) ions into metallic Ag^0, according to Hussain et al. (2014) Diosgenin's hydroxy groups convert Ag^+ into Ag^0. The comparatively stable Ag_2^+ clusters are created when the neutral atom Ag^0 combines with Ag+. Ag^{2+} clusters dimerize to produce Ag_4^{2+} (a yellow-colored silver sol that stops growing at this stage and is stable for a long period in the presence of a diosgenin, even in the presence of air).

In recent times, nanoparticles have often been utilized alongside natural products to mitigate toxicity concerns. The green chemistry approach to designing drug-loaded nanoparticles is highly recommended as it reduces harmful components in the biosynthesis process. Consequently, employing green nanoparticles for drug delivery can diminish the side effects of medications (Lam et al., 2017).

Presently, more than 40% of pharmaceutical formulations are sourced from natural ingredients, encompassing a range of commercially available medications. These include well-known drugs such as digoxin, chloroquine, quinine, lumefantrine, atovaquone, aspirin, and artemisinin. This reliance on natural sources highlights the significant role that natural compounds play in modern medicine, providing effective treatments for a variety of conditions (Xulu et al., 2022; Woodley et al., 2021).

Considering the current needs, this chapter aims to report the therapeutic prospective of green synthesized metallic nanoparticleand their significant application in drug delivery system and Nanomedicine.

Recently, Puri et al. (2024) discussed how the integration of traditional medicine with modern nanotechnology has led to the emergence of biogenic metallic phytonanoparticles, offering significant health benefits. Derived through green chemistry techniques, phytonanoparticles are produced using plant extracts to synthesize metallic nanoparticles such as copper, silver, iron, zinc, and titanium oxide. These nanoparticles are characterized by their non-toxicity, sustainability, and economic efficiency.

Phytonanoparticles exhibit unique physical, chemical, and biological properties, leading to their diverse applications in healthcare. Their antioxidant properties are crucial for combating oxidative stress, which is linked to various chronic diseases and aging. Additionally, their anti-inflammatory effects help reduce inflammation, a common underlying factor in many health conditions, including arthritis and cardiovascular diseases (Oselusi et al., 2024).

The antimicrobial activity of phytonanoparticles makes them effective against a range of pathogens, offering a natural alternative to synthetic antibiotics. This is particularly valuable in addressing antibiotic resistance. Their anti-cancer properties are another significant health benefit, as phytonanoparticles can induce apoptosis in cancer cells, inhibit their proliferation, and reduce tumor growth (Díaz-Puertas et al., 2023).

Furthermore, phytonanoparticles are being explored for their potential in targeted drug delivery systems. Their small size and large surface area enhance drug solubility and bioavailability, allowing for efficient delivery to specific sites in the body. This targeted approach minimizes side effects and maximizes therapeutic efficacy.

Overall, the development of biogenic metallic phytonanoparticles represents a promising advancement in modern healthcare. By harnessing the benefits of traditional medicine and green chemistry, these nanoparticles offer a sustainable and innovative solution for various health challenges.

3. THERAPEUTIC APPLICATIONS OF MNPS

Biosynthesis is a promising approach for the sustainable production of MNP formulation with potential to revolutionize against a wide array of diseases. This section provides a concise overview of the roles of MNPs in therapeutic and biomedical applications. Additionally, Figure 2 highlights the advantages of green-synthesized nanoparticles.

3.1 Antioxidant Activity

Living cells generate free radicals that cause diseases such as cardiovascular issues, neurodegeneration, inflammation, cancer, and aging by damaging vital biomolecules (Phaniendra et al., 2015; Ge et al., 2022). Conventional chemical antioxidants, and butylated hydroxytoluene (BHT) are used to inhibit degenerative reactions in proceed foods but pose long-term health risks(Barve et al., 2011). Consequently, there is growing interest in nano-antioxidants, which are cost-effective and lack harmful side effects. Researchers have synthesized numerous metal nanoparticles (NPs) using plant extracts, leveraging the antioxidant properties of both medicinal plants and metals like Ag, Au, Cu, Zn, Fe, and Pd. Plants contain bioactive compounds such as polyphenols, flavonoids, terpenoids, tannins, and saponins, which act as reducing, capping, and stabilizing agents in the synthesis of phytochemical-derived metal nanoparticles. Due to their fine-tuned surface area, particle size, and surface activity, Plant derived nanoparticles exhibit powerful antioxidant effects, good biocompatibility, and action stability (Bedlovičová et al., 2020). Numerous studies have examined the free radical scavenging properties of medicinal plants and nanoparticles synthesized from their extracts (Kumar et al., 2010; Ansari et al.,2016). Antioxidants are known to neutralize free radicals by converting superoxide (O_2), hydrogen peroxide (H_2O_2), and hydroxyl radicals into harmless molecules such as oxygen and water (H_2O) (Ahamad, 2019, 2020). The reactive oxygen species (ROS) derived from nanoparticles can damage the cell membranes of microorganisms and cancer cells, leading to their antimicrobial and anticancer effects. The ROS scavenging capacity of phytochemical-derived metal nanoparticles (PDMNPs) can be assessed using various free radical quenching-based assays, including 2,2-diphenyl-1-picrylhydrazyl (DPPH), ferric reducing antioxidant power (FRAP), trolox equivalent antioxidant capacity (TEAC), oxygen radical absorbance capacity (ORAC), total oxy radical scavenging capacity (TOSC), and total radical-trapping antioxidant parameter (TRAP) (Apak et al., 2016). These assays provide a comprehensive evaluation of the antioxidant potential of plant derived metal nanoparticles, highlighting their significance in biomedical applications.

A study by Mukhopadhyay (2018) highlights the antioxidant properties of biofabricated nanoparticles (Pt-NPs and AuAg-NPs) using DPPH and ABTS (3-ethybenzothiazoline-6-sulphonic acid) assay. DPPH, representing lipid radicals with nitrogen radical, exhibits a deep purple color which changes to pale yellow when neutralized by antioxidant compounds (Serpen et al., 2007). Similarly, ABTS is used to measure the antioxidant activity of chain-breaking or hydrogen-donating compounds, changing from bluish-green to colorless upon neutralization. The observed color changes in both assays indicate that the investigated nanoparticles have the ability to transfer electrons or hydrogen atoms to the free radicals, demonstrating their scavenging activity in a nearly linear manner.Their findings demonstrated the superior antioxidant activity of biosynthesized nanoparticles compared to their natural biological sources. The enhanced activity is attributed to the surface-adsorbed antioxidant bioorganic compounds, which help generate an electrostatic field on the active surface of the nanoparticles. This electrostatic field facilitates the transfer of electrons or hydrogen atoms, thereby improving the nanoparticles ability to neutralize free radicals (Seralathan et al., 2014).

Jurkow et al. (2020) reveals that metal nanoparticles (NPs) have varying effects on oakleaf lettuce seedlings antioxidant properties and growth. Foliar exposure to Au, Ag, and Pt nanoparticles did not alter ascorbate peroxidase activity but increased total peroxidase activity with Au-NPs and decreased it with Ag-NPs and Pt-NPs. Glutathione (GSH) levels rose with both concentrations of Au-NPs and Pt-NPs, while L-ascorbic acid content remained unchanged. Ag-NPs and Pt-NPs at 40 ppm enhanced total phenolics content by 17% and 15%, respectively. Carotenoid levels increased with Ag-NPs and Au-NPs at both 20 and 40 ppm, and Pt-NPs at 20 ppm. Seedlings treated with 40 ppm of Ag-NPs and Pt-NPs showed higher total antioxidant capacity and increased chlorophyll a (with Ag-NPs). Pt-NPs at 40 ppm also improved fresh and dry weights of lettuce shoots, indicating a positive effect on growth.Flavonoids, tannins, and/or carbohydrates are reported by Zayed et al. (2020)as the main constituents in *Pimpinella anisum* extract, which are primarily responsible for the reduction and capping of the nanoparticles based on FTIR (Fourier transform infrared spectroscopy) studies and phytochemical results. The green-synthesized silver nanoparticles (AgNPs) exhibited good *in-vitro* antioxidant and antimicrobial activities, indicating their potential for developing new eco-friendly drugs for the biomedical field.Overall, the application of metal NPs influenced various antioxidant activities with specific effects depending on the type and concentration of nanoparticles used (Kalisz et al; 2023).

3.2 Anticancer Activity

Cancer has become one of the most critical health challenges worldwide. Cancer is caused by driver mutations that activate proto-oncogenes and inactivate tumor suppressor genes, leading to hallmark traits such as insensitivity to anti-growth signals, limitless replicative potential, evasion of apoptosis, continuous angiogenesis, self-sufficient growth signals, and tissue invasion and metastasis(Hanahan-etal.,2011). Additionallyit include evading the immune response, reprogramming energy metabolism, tumor-activated inflammation, and genome mutations (Khursheed et al.,2022) Cancer progresses from a pre-cancerous lesion to a malignant tumor due to genetic factors interacting with external stimuli, including biological agents (viruses and bacteria), chemical carcinogens (tobacco smoke, arsenic), and physical carcinogens (ultraviolet and ionizing radiation). Metastasis, the spread of cancer cells, is the leading cause of cancer death (Parsa, 2011).

Accurate diagnosis is essential for effective cancer treatment as each type requires specific therapies. Traditional treatments include chemotherapy, radiotherapy, and surgery. Nanomedicines, particularly metallic nanoparticles (MNPs), offer a non-invasive alternative with anti-tumor potential. MNPs can deliver multiple anti-cancer drugs simultaneously, overcoming challenges of conventional chemotherapy. Functionalized MNPs improve energy deposition control in cancer cells and serve dual roles in diagnosis and therapy, enhancing cancer cell imaging and enabling precise drug delivery (Rahman et al., 2021 &Sharma et al.,2018). The biomedical applications of MNPs significantly enhance cancer treatment through improved targeting, gene silencing, and drug delivery (Evans et al., 2018).

Al-Sheddi et al.(2018) investigated the biosynthesis of silver nanoparticles (AgNPs) using the*Nepeta deflersiana* plant and their anticancer potential against human cervical cancer (HeLa) cells. The biosynthesized AgNPs (ND-AgNPs) induced concentration-dependent cytotoxicity in HeLa cells. They caused oxidative stress, evidenced by increased ROS and lipid peroxidation (LPO) levels and decreased glutathione (GSH) levels. The rise in intracellular ROS led to mitochondrial membrane damage, cell cycle alterations, and ultimately apoptosis and necrosis through SubG1 cell cycle arrest. Their findings suggest that ND-AgNPs have significant anticancer potential and could aid in developing new cancer therapies.

Characterization Studies like FTIR, UV–Vis (UV-Visible Spectroscopy), SEM (Scanning Electron Microscopy), XRD (X-ray diffraction) indicated that proteins stabilized the nanoparticles. AgNPs from *Premnaserratifolia* leaf extract showed anticancer activity in CCl_4^-induced liver cancer, with non-toxic and protective effects in mice, controlling disease progression without lethal effects. These AgNPs present a cost-effective alternative for cancer treatment. (Paul et al.,2015). Similar study performed by Sre et al. (2015) indicated that bioreduction of Ag^+ ions to Ag nanoparticles, likely due to the phytochemical constituents in the plant extract, such as terpenes, flavonols, alkaloids, tannins, and phenols, which act as reducing agents. The synthesized Ag nanoparticles demonstrated potent cytotoxicity against MCF-7 and HEP G2 cancer cell lines, suggesting their potential as chemopreventive agents. Another study also investigated that biologically synthesiseAuNPs *in vitro* using the aqueous flower extract of *Moringa olif-era*showed cytotoxicity against A549 (human lung epithelial adenocarcinoma cells) and did not induced cytotoxic effect against normal healthy peripheral lymphocytes (PLs)(Anand et al., 2015). The green synthesis method using *Aloe barbadensis* leaf extract as a stabilizing and capping agent was employed to prepare ZnO nanoparticles loaded with Doxorubicin and Gemcitabine. This environmentally friendly approach resulted in a system that demonstrated significant cytotoxicity towards the MDA-MB-231 breast cancer cell line, indicating its potential as an effective treatment strategy(Batool et al., 2021). Nevertheless MNPs are particularly advantageous in cancer treatment due to precise control over their shape, size, charge, and surface modification. Additionally, MNPs are more easily taken up by cells compared to non-metallic nanoparticles of the same size, providing a significant advantage for cancer therapy. To advance the clinical application of MNPs for cancer treatment, further research is essential to ensure their effectiveness and safety. Developing mass-scale, cost-effective production methods for MNP functionalized with anticancer drugs and targeting ligands is crucial. Enhancing large-scale nanoparticle synthesis and conducting successful clinical trials will pave the way for nanoparticles with potential therapeutic applications, including personalized cancer therapy. This progress sets the stage for Plant based metallic nanoparticles to become powerful tools in the fight against cancer (Bharadwaj et al., 2021; Bukhari et al., 2021).

3.3 Antimicrobial Activity

For centuries, elemental metals like silver (Ag) and gold (Au) Copper (Cu) have been widely used for their antibacterial properties. Utilizing biological synthesis methods, metal nanoparticles (MNPs) such as silver nanoparticles (AgNPs), gold nanoparticles (AuNPs) and copper and copper oxide (CuNPs) can be produced from plant extracts (Alomar et al., 2024).

The emergence of antibiotic-resistant bacteria has necessitated the development of novel antimicrobial agents. Metal nanoparticles (MNPs), including silver (Ag), gold (Au), copper (Cu), zinc oxide (ZnO), and titanium dioxide (TiO2) nanoparticles, have garnered significant attention due to their potent antimicrobial properties and diverse mechanisms of action (Agarwalla et al., 2024). Recently many researchers explore the effectiveness of MNPs as antimicrobial agents, their modes of action, and their potential applications.

Plant-derived metal nanoparticles exhibit antimicrobial activity due to the release of metal ions. These ions interact with the cell membrane, causing structural damage to membrane by deactivating enzyme functions can lead to cell lysis and death (Kumar et al., 2020). They can induce the production of ROS, such as superoxide anions, hydrogen peroxide, and hydroxyl radicals. These reactive species cause oxidative stress, damaging cellular components like DNA, proteins, and lipids (Rajeshkumar et al., 2017). Plant derived MNPs can bind to bacterial DNA and proteins, inhibiting replication and transcription, which ultimately prevents bacterial growth and proliferation (Mba et al., 2021). Plant based Copper/copper oxide nanoparticles (Cu/CuO NPs) have antibacterial properties that stem from multiple pathways, described by Letchumanan et al. (2021). Bacterial cell lysis is a result of CuO NPs' interaction with bacterial cell walls, which damages the membranes. Cu^{2+} ions are released by these nanoparticles once they are within the bacterial cytoplasm. Oxidative stress is triggered by these ions, which lead the bacteria to generate reactive oxygen species (ROS). Subsequently, the oxidative damage affects essential components of cells including DNA and mitochondria, eventually resulting in cellular death. CuO Nanoparticles have a wide antibacterial activity against both Gram-positive and Gram-negative bacteria can be explained by this mechanism. The bacterial cell walls of both groups are disrupted by the nanoparticles, allowing Cu^{2+} ions to enter and cause the formation of intracellular ROS, which furthers lead to cell death.

According to the study conducted by Sre et al. (2015) the synthesized silver nanoparticles (AgNPs) using aqueous *Erythrina indica* root extract demonstrated strong antibacterial activity against both Gram-positive and Gram-negative bacteria. This activity is likely due to silver's high affinity for reacting with sulfur or phosphorus-containing biomolecules within the cell. Consequently, sulfur-containing proteins in the membrane or inside the cells, along with phosphorus-containing elements like DNA, are preferential binding sites for AgNPs, leading to cell death. Most antibacterial report indicates that the inhibition caused by biosynthesized Ag_2O nanoparticles is primarily due to membrane disruption through the generation of surface oxygen species, which play a crucial role in ultimately leading to the death of pathogens(Pradheesh et al., 2020). There is extensive literature on the antimicrobial activity of metal and metal oxide nanoparticles. Silver nanoparticles are particularly effective, as they decompose the polymer sub-units of microbial cell membranes. Plant-mediated silver nanoparticles disrupt the cell membrane, leading to the destruction of the bacterial protein synthesis mechanism. The silver nanoparticles synthesized using *Citrus sinensis* peel extract showed antibacterial activity against *Staphylococcus aureus*, *Pseudomonas aeruginosa*, and *Escherichia coli* (Bukhari et al., 2021). The current research tested tea tree (Melaleuca alternifolia) oil against oral microbes, including *Porphyro monasgingivalis*, *Enterococcus faecalis*, and *Streptococcus* mutans. It was found that tea tree oil alone was more efficacious than its Zn or Fe nanoparticle counterparts. This may be due to the clumping of nanoparticles, the concentration used, the heterogeneity of phytochemicals in the capping process, and limitations in the number of functional nano-bio interfaces (Mohammed et al., 2024). Recently another research has focused on Ruthenium oxide nanoparticles (RuNPs). In that study, RuNPs were synthesized using *Iris kashmiriana* plant extract and characterized by FTIR, SEM, EDX (Energy dispersive X-ray), and XRD. The particles, synthesized in dimensions of 25–30 nm, demonstrated potent antibacterial activity against *Escherichia coli*, *P. aeruginosa*, *Klebsiella pneumoniae*, and *Streptococcus pneumoniae*, as evidenced by a zone of inhibition for erythromycin (Imtiyaz etal., 2024). Additionally, different phytofabricated metallic nanoparticles in combination exhibited antimicrobial activity against various bacterial strains, including *E. coli*, *Bacillus cereus*, *Salmonella typhi*, and *S. aureus*. They also showed effectiveness against different species of fungi, such as *Candida albicans*, *Penicillium spp.*, and *Aspergillus spp.* These findings underscore the potential utility of different metallic nanoparticles as effective antimicrobial agents (Akila, 2024.; Muhammed, 2024.; Sagar, 2024.; Thirumoorthy, 2024).

Biosynthesis With their broad range of modes of action, metal nanoparticles offer a prospective substitute for traditional antibiotics in terms of strong antibacterial activity. Their uses are widespread in a variety of industries, including as textiles, water treatment, food safety, and medicine. To maximize their utility, reduce any potential toxicity, and get past issues with large-scale production and environmental effect, more research and development is needed in the future. MNPs have great potential as next-generation antimicrobial agents as the problem of antibiotic resistance keeps growing.

Figure 2. Advantages of Green Synthesis of Nanoparticle.

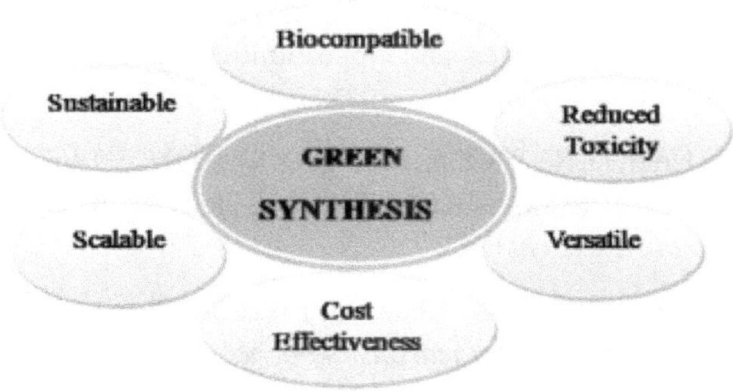

4. CONCLUSION

We highlight the diverse range of medicinal plants used for nanoparticle production through an extensive review of recent literature, highlighting their rich phytochemical profiles and bioreductive potential. Furthermore, we delve into the mechanisms governing the biosynthesis of metallic nanoparticles and explore the factors influencing nanoparticle morphology, size, and stability.

The therapeutic uses of these biofabricated nanoparticles in a variety of medical fields are covered in this chapter's transition. We examine the various therapeutic advantages provided by metallic nanoparticles made from extracts from medicinal plants, including their antibacterial, anticancer, and antioxidant qualities. We also discuss their prospective applications in regenerative medicine, imaging diagnostics, and targeted drug delivery, highlighting their adaptability and practical significance. This project aims to shed light on the revolutionary potential of biosynthesized metallic nanoparticles from medicinal plants in enhancing therapeutic treatments and solving urgent healthcare concerns through a synthesis of cutting-edge scientific insights and future prospects.

REFERENCES

Agarwalla, S., Singh, S. K., Ibrahim, M. A., Noothalapati, H., &Duraiswamy, S. (2024). Antimicrobial Coatings: Current Mechanisms, Challenges, and Opportunities. *Functional Coatings: Innovations and Challenges,* 231-274.

Ahamad, T., Ansari, W. A., Negi, D. S., & Khan, M. F. (2019). NON-ENZYMATIC NATURAL REACTIVE OXYGEN SCAVENGERS (ROS): A REVIEW ON STRUCTURES AND MODE OF ACTION. *Era's. The Journal of Medical Research*, 6(2).

Ahamad, T., Khan, M. A., Ansari, W. A., & Khan, M. F. (2020). Antioxidant and anticancer activities of selected indian medicinal plant viz., *artocarpuslakoocha, kigelia pinnata, and amaranthusviridis. Era's. The Journal of Medical Research*, 7(1), 46–51.

Al-Sheddi, E. S., Farshori, N. N., Al-Oqail, M. M., Al-Massarani, S. M., Saquib, Q., Wahab, R., & Siddiqui, M. A. (2018). Anticancer potential of green synthesized silver nanoparticles using extract of *Nepeta deflersiana*against human cervical cancer cells (HeLA). *Bioinorganic Chemistry and Applications*, 2018(1), 9390784. PMID: 30515193

Alomar, T. S., AlMasoud, N., Awad, M. A., AlOmar, R. S., Merghani, N. M., El-Zaidy, M., & Bhattarai, A. (2024). Designing Green Synthesis-Based Silver Nanoparticles for Antimicrobial Theranostics and Cancer Invasion Prevention. *International Journal of Nanomedicine*, 19, 4451–4464. DOI: 10.2147/IJN.S440847 PMID: 38799694

Ansari, J. A., Khan, A. R., Ahmad, M. K., Fatima, N., Rastogi, N., Khan, M. F., Jilani, H., & Mahdi, A. A. (2016). Phytochemicals, antioxidant and antiproliferative studies of some medicinal plants from Indian sub-continent. *British Journal of Pharmaceutical Research*, 11(6), 1–11. DOI: 10.9734/BJPR/2016/26198

Apak, R., Ozyurek, M., Guclu, K., & Capanoglu, E. (2016). Antioxidant activity/capacity measurement. 3. Reactive oxygen and nitrogen species (ROS/RNS) scavenging assays, oxidative stress biomarkers, and chromatographic/chemometric assays. *Journal of Agricultural and Food Chemistry*, 64(5), 1046–1070. DOI: 10.1021/acs.jafc.5b04744 PMID: 26689748

Arockia John Paul, J., Karunai Selvi, B., & Karmegam, N. (2015). Biosynthesis of silver nanoparticles from Premnaserratifolia L. leaf and its anticancer activity in CCl 4-induced hepato-cancerous Swiss albino mice. *Applied Nanoscience*, 5(8), 937–944. DOI: 10.1007/s13204-014-0397-z

Asmathunisha, N., & Kathiresan, K. (2013). A review on biosynthesis of nanoparticles by marine organisms. *Colloids and Surfaces. B, Biointerfaces*, 103, 283–287. DOI: 10.1016/j.colsurfb.2012.10.030 PMID: 23202242

Barve, M., Mashru, M., Jagtap, C., Patgiri, B. J., & Prajapati, P. K. (2011). Therapeutic potentials of metals in ancient India: A review through Charaka Samhita. *Journal of Ayurveda and Integrative Medicine*, 2(2), 55. DOI: 10.4103/0975-9476.82523 PMID: 21760689

Batool, M., Khurshid, S., Daoush, W. M., Siddique, S. A., & Nadeem, T. (2021). Green synthesis and biomedical applications of ZnO nanoparticles: Role of PEGylated-ZnO nanoparticles as doxorubicin drug carrier against MDA-MB-231 (TNBC) cells line. *Crystals*, 11(4), 344. DOI: 10.3390/cryst11040344

Bedlovičová, Z., Strapáč, I., Baláž, M., & Salayová, A. (2020). A brief overview on antioxidant activity determination of silver nanoparticles. *Molecules (Basel, Switzerland)*, 25(14), 3191. DOI: 10.3390/molecules25143191 PMID: 32668682

Bogunia-Kubik, K., & Sugisaka, M. (2002). From molecular biology to nanotechnology and nanomedicine. *Bio Systems*, 65(2-3), 123–138. DOI: 10.1016/S0303-2647(02)00010-2 PMID: 12069723

Bukhari, A., Ijaz, I., Gilani, E., Nazir, A., Zain, H., Saeed, R., Alarfaji, S. S., Hussain, S., Aftab, R., & Naseer, Y. (2021). Green synthesis of metal and metal oxide nanoparticles using different plants' parts for antimicrobial activity and anticancer activity: A review article. *Coatings*, 11(11), 1374. DOI: 10.3390/coatings11111374

Bukhari, A., Ijaz, I., Gilani, E., Nazir, A., Zain, H., Saeed, R., & Naseer, Y. (2021). Green synthesis of metal and metal oxide nanoparticles using different plants' parts for antimicrobial activity and anticancer activity: A review article. *Coatings*, 11(11), 1374. DOI: 10.3390/coatings11111374

Dauthal, P., & Mukhopadhyay, M. (2018). Antioxidant activity of phytosynthesized biomatrix-loaded noble metallic nanoparticles. *Chinese Journal of Chemical Engineering*, 26(5), 1200–1208. DOI: 10.1016/j.cjche.2017.12.014

Díaz-Puertas, R., Álvarez-Martínez, F. J., Falco, A., Barrajón-Catalán, E., & Mallavia, R. (2023). Phytochemical-based nanomaterials against antibiotic-resistant bacteria: An updated review. *Polymers*, 15(6), 1392. DOI: 10.3390/polym15061392 PMID: 36987172

Doyle, J. D. (2013). Nanotechnology in Anesthesia and Medicine. *Advances in Anesthesia*, 31(1), 181–200. DOI: 10.1016/j.aan.2013.08.010

Hanahan, D., & Weinberg, R. A. (2011). Hallmarks of cancer: the next generation. *cell,* 144(5), 646-674.

Hussain, S., Bashir, O., Khan, Z., & AL-Thabaiti, S. A. (2014). Steroidal saponin based extracellular biosynthesis of AgNPs. *Journal of Molecular Liquids*, 199, 489–494. DOI: 10.1016/j.molliq.2014.09.047

Imtiyaz, A., & Singh, A. (2024). Green synthesized ruthenium oxide nanoparticles mediated through Iris Kashmiriana (Mazar-Graveyard) plant extract and antimicrobial activity. *Journal of Inorganic and Organometallic Polymers and Materials*, 34(5), 1–8. DOI: 10.1007/s10904-023-02968-3

Jurkow, R., Pokluda, R., Sękara, A., & Kalisz, A. (2020). Impact of foliar application of some metal nanoparticles on antioxidant system in oakleaf lettuce seedlings. *BMC Plant Biology*, 20(1), 1–12. DOI: 10.1186/s12870-020-02490-5 PMID: 32576147

Kalisz, A., Kornaś, A., Skoczowski, A., Oliwa, J., Jurkow, R., Gil, J., Sękara, A., Sałata, A., & Caruso, G. (2023). Leaf chlorophyll fluorescence and reflectance of oakleaf lettuce exposed to metal and metal (oid) oxide nanoparticles. *BMC Plant Biology*, 23(1), 329. DOI: 10.1186/s12870-023-04305-9 PMID: 37340375

Katarzyna, B. K., & Masanori, S. (2002). From molecular biology to nanotechnology and nanomedicine. *Bio Systems*, 65(2-3), 123–138. DOI: 10.1016/S0303-2647(02)00010-2 PMID: 12069723

Khursheed, R., Dua, K., Vishwas, S., Gulati, M., Jha, N. K., Aldhafeeri, G. M., Alanazi, F. G., Goh, B. H., Gupta, G., Paudel, K. R., Hansbro, P. M., Chellappan, D. K., & Singh, S. K. (2022). Biomedical applications of metallic nanoparticles in cancer: Current status and future perspectives. *Biomedicine and Pharmacotherapy*, 150, 112951. DOI: 10.1016/j.biopha.2022.112951 PMID: 35447546

Kumar, H., Bhardwaj, K., Kuča, K., Kalia, A., Nepovimova, E., Verma, R., & Kumar, D. (2020). Flower-based green synthesis of metallic nanoparticles: Applications beyond fragrance. *Nanomaterials (Basel, Switzerland)*, 10(4), 766. DOI: 10.3390/nano10040766 PMID: 32316212

Letchumanan, D., Sok, S. P., Ibrahim, S., Nagoor, N. H., & Arshad, N. M. (2021). Plant-based biosynthesis of copper/copper oxide nanoparticles: An update on their applications in biomedicine, mechanisms, and toxicity. *Biomolecules*, 11(4), 564. DOI: 10.3390/biom11040564 PMID: 33921379

Makarov, V. V., Love, A. J., Sinitsyna, O. V., Makarova, S. S., Yaminsky, I. V., Taliansky, M. E., & Kalinina, N. O. (2014). "Green" nanotechnologies: Synthesis of metal nanoparticles using plants. *Acta Naturae (English Ed.)*, 6(1 (20)), 35–44. DOI: 10.32607/20758251-2014-6-1-35-44 PMID: 24772325

Mba, I. E., & Nweze, E. I. (2021). Nanoparticles as therapeutic options for treating multidrug-resistant bacteria: Research progress, challenges, and prospects. *World Journal of Microbiology & Biotechnology*, 37(6), 1–30. DOI: 10.1007/s11274-021-03070-x PMID: 34046779

Mohammed, A. E., Aldahasi, R. M., Rahman, I., Shami, A., Alotaibi, M., BinShabaib, M. S., ALHarthi, S. S., & Aabed, K. (2024). The antimicrobial activity of tea tree oil *(Melaleuca alternifolia)* and its metal nanoparticles in oral bacteria. *PeerJ*, 12, e17241. DOI: 10.7717/peerj.17241 PMID: 38854801

Morrow, K. J.Jr, Bawa, R., & Wei, C. (2007). Recent advances in basic and clinical nanomedicine. *The Medical Clinics of North America*, 91(5), 805–843. DOI: 10.1016/j.mcna.2007.05.009 PMID: 17826104

Muhammed, A., Asere, T. G., & Diriba, T. F. (2024). Photocatalytic and antimicrobial properties of ZnO and Mg-Doped ZnO nanoparticles synthesized using *lupinus albus* leaf extract. *ACS Omega*, 9(2), 2480–2490. DOI: 10.1021/acsomega.3c07093 PMID: 38250416

Oselusi, S. O., Sibuyi, N. R., Meyer, M., & Madiehe, A. M. (2024). Phytonanotherapeutic Applications of Plant Extract-Synthesized Silver Nanoparticles in Wound Healing—A Prospective Overview. *BioNanoScience*, 14(3), 1–21. DOI: 10.1007/s12668-024-01535-5

Parsa, N. (2012). Environmental factors inducing human cancers. *Iranian Journal of Public Health*, 41(11), 1. PMID: 23304670

Phaniendra, A., Jestadi, D. B., & Periyasamy, L. (2015). Free radicals: Properties, sources, targets, and their implication in various diseases. *Indian Journal of Clinical Biochemistry*, 30(1), 11–26. DOI: 10.1007/s12291-014-0446-0 PMID: 25646037

Pradheesh, G., Suresh, S., Suresh, J., & Alexramani, V. (2020). Antimicrobial and Anticancer Activity Studies on Green Synthesized Silver Oxide Nanoparticles from the Medicinal Plant *Cyathea nilgiriensisHolttum.International Journal of Pharmaceutical Investigation*, 10(2), 146–150. DOI: 10.5530/ijpi.2020.2.27

Puja, A. M., Rupa, E. J., Kim, Y. J., & Yang, D. C. (2023). Medicinal plant enriched metal nanoparticles and nanoemulsion for inflammation treatment: A narrative review on current status and future perspective. *Immuno*, 3(2), 182–194. DOI: 10.3390/immuno3020012

Rahman, M., Alam, K., Hafeez, A., Ilyas, R., & Beg, S. (2021). Metallic nanoparticles in drug delivery and cancer treatment. In *Nanoformulation strategies for cancer treatment* (pp. 107–119). Elsevier. DOI: 10.1016/B978-0-12-821095-6.00008-2

Rajeshkumar, S., & Bharath, L. V. (2017). Mechanism of plant-mediated synthesis of silver nanoparticles–a review on biomolecules involved, characterization and antibacterial activity. *Chemico-Biological Interactions*, 273, 219–227. DOI: 10.1016/j.cbi.2017.06.019 PMID: 28647323

Sagar, P. V., Ramadevi, D., Basavaiah, K., & Botsa, S. M. (2024). Green synthesis of silver nanoparticles using aqueous leaf extract of *Saussureaobvallata* for efficient catalytic reduction of nitrophenol, antioxidant, and antibacterial activity. *Water Science and Engineering*, 17(3), 274–282. DOI: 10.1016/j.wse.2023.09.004

Sanna, V., & Sechi, M. (2020). Therapeutic Potential of Targeted Nanoparticles and Perspective on Nanotherapies. *ACS Medicinal Chemistry Letters*, 11(6), 1069–1073. DOI: 10.1021/acsmedchemlett.0c00075 PMID: 32550978

Seralathan, J., Stevenson, P., Subramaniam, S., Raghavan, R., Pemaiah, B., Sivasubramanian, A., & Veerappan, A. (2014). Spectroscopy investigation on chemo-catalytic, free radical scavenging and bactericidal properties of biogenic silver nanoparticles synthesized using Salicornia brachiata aqueous extract. *Spectrochimica Acta. Part A: Molecular and Biomolecular Spectroscopy*, 118, 349–355. DOI: 10.1016/j.saa.2013.08.114 PMID: 24056313

Serpen, A., Capuano, E., Fogliano, V., & Gökmen, V. (2007). A new procedure to measure the antioxidant activity of insoluble food components. *Journal of Agricultural and Food Chemistry*, 55(19), 7676–7681. DOI: 10.1021/jf071291z PMID: 17708649

Sharma, A., Goyal, A. K., & Rath, G. (2018). Recent advances in metal nanoparticles in cancer therapy. *Journal of Drug Targeting*, 26(8), 617–632. DOI: 10.1080/1061186X.2017.1400553 PMID: 29095640

Sre, P. R., Reka, M., Poovazhagi, R., Kumar, M. A., & Murugesan, K. (2015). Antibacterial and cytotoxic effect of biologically synthesized silver nanoparticles using aqueous root extract of *Erythrina indica lam.Spectrochimica Acta. Part A: Molecular and Biomolecular Spectroscopy*, 135, 1137–1144. DOI: 10.1016/j.saa.2014.08.019 PMID: 25189525

Thirumoorthy, G., Balasubramanian, B., George, J. A., Nizam, A., Nagella, P., Srinatha, N., Pappuswamy, M., Alanazi, A. M., Meyyazhagan, A., Rengasamy, K. R. R., & Veerappa Lakshmaiah, V. (2024). Phytofabricated bimetallic synthesis of silver-copper nanoparticles using *Aerva lanata* extract to evaluate their potential cytotoxic and antimicrobial activities. *Scientific Reports*, 14(1), 1270. DOI: 10.1038/s41598-024-51647-x PMID: 38218918

Vanna, S., & Mario, S. (2020). Therapeutic Potential of Targeted Nanoparticles and Perspective on Nanotherapies. *ACS Medicinal Chemistry Letters*, 11(6), 1069–1073. DOI: 10.1021/acsmedchemlett.0c00075 PMID: 32550978

Vijayaraghavan, K., & Ashokkumar, T. (2017). Plant-mediated biosynthesis of metallic nanoparticles: A review of literature, factors affecting synthesis, characterization techniques and applications. *Journal of Environmental Chemical Engineering*, 5(5), 4866–4883. DOI: 10.1016/j.jece.2017.09.026

Woodley, C. M., Amado, P. S., Cristiano, M. L., & O'Neill, P. M. (2021). Artemisinin inspired synthetic endoperoxide drug candidates: Design, synthesis, and mechanism of action studies. *Medicinal Research Reviews*, 41(6), 3062–3095. DOI: 10.1002/med.21849 PMID: 34355414

Xulu, J. H., Ndongwe, T., Ezealisiji, K. M., Tembu, V. J., Mncwangi, N. P., Witika, B. A., & Siwe-Noundou, X. (2022). The use of medicinal plant-derived metallic nanoparticles in theranostics. *Pharmaceutics*, 14(11), 2437. DOI: 10.3390/pharmaceutics14112437 PMID: 36365255

Zayed, M. F., Mahfoze, R. A., El-Kousy, S. M., & Al-Ashkar, E. A. (2020). *In-vitro* antioxidant and antimicrobial activities of metal nanoparticles biosynthesized using optimized *Pimpinella anisum* extract. *Colloids and Surfaces. A, Physicochemical and Engineering Aspects*, 585, 124–167. DOI: 10.1016/j.colsurfa.2019.124167

KEY TERMS AND DEFINITIONS

Nanobiotechnology: Nanobiotechnology is the use of tiny tools and materials to interact with living things for medical and scientific purposes.

Nanodrug: A nanodrug is a medicine that uses tiny particles, called nanoparticles, to deliver drugs more precisely to targeted areas in the body, improving effectiveness and reducing side effects.

Nanoparticle: Nanoparticles are tiny particles, 1-100 nanometers in size, with special properties used in medicine, electronics, and more.

Reactive Oxygen Species (ROS): A type of unstable molecule that contains oxygen and that easily reacts with other molecules in a cell. Reactive oxygen species are free radicals and also called oxygen radicals.

Apoptosis: Apoptosis is a natural process of programmed cell death that occurs in the body to eliminate damaged or unnecessary cells, helping to maintain healthy tissue.

Metastasis: Metastasis is the process by which cancer cells spread from the original tumor to other parts of the body, forming new tumors.

Angiogenesis: Angiogenesis is defined as the growth and development of a vascular network, sprouting from pre-existing vasculature, to support the growth of tissue in an adult organism as opposed to vasculogenesis.

Chapter 9
Plant Biogenic Nanoparticles as Antifungal Agents:
Synthesis Characterization and Mechanistic Insights

Anjali Roy
https://orcid.org/0009-0001-2759-7201
Assam University, Silchar, India

Sharmistha Sharma
https://orcid.org/0009-0006-1101-504X
Assam University, Silchar, India

Joydeb Dey
https://orcid.org/0009-0000-8968-475X
Assam University, Silchar, India

Saurabh Yadav
https://orcid.org/0000-0001-7856-5595
Hemvati Nandan Bahuguna Garhwal University, India

Himanshu Kishore Prasad
https://orcid.org/0000-0003-0215-1323
Assam University, Silchar, India

ABSTRACT

Fungal infections, causing over 1.5 million annual deaths, are a significant global health threat prioritized by the World Health Organization. Plant-based nanoparticle synthesis provides an eco-friendly, rapid, and stable alternative to traditional methods. This chapter reviews the synthesis, characterization, antifungal applications, mechanisms, and clinical potential of biogenic nanoparticles, which interact with fungal cell walls, disrupt cellular processes, and modulate immune responses. Synergistic effects with conventional antifungal drugs and incorporation into new formulations and delivery systems enhance their therapeutic potential. However, clinical translation faces challenges related to toxicity, pharmacokinetics, and regulatory issues. This chapter offers valuable insights into advancing antifungal

DOI: 10.4018/979-8-3693-6240-2.ch009

therapeutics using biogenic nanoparticles to combat fungal infections globally.

1. INTRODUCTION

Fungal infections pose a significant global health burden and affect millions of people worldwide. According to recent estimates, more than 1 billion people suffer from fungal infections, with more than 150 million cases of serious fungal diseases annually (Bongomin et al., 2017; Denning, 2024). These infections are responsible for more than 1.5 million deaths each year, surpassing the mortality rates of tuberculosis and malaria (Rodrigues & Nosanchuk, 2020). The World Health Organization (WHO) has recently highlighted the growing threat of fungal infections in its first list of priority fungal pathogens, published in 2022. This landmark report identified 19 fungal pathogens that pose the greatest threat to public health, categorizing them into three priority groups: critical, high, and medium (WHO, 2022). This report emphasizes the increasing incidence of invasive fungal infections, particularly in immunosuppressed individuals, and the growing concern about antifungal resistance. Among the critical priority pathogens are *Candida auris*, *Aspergillus fumigatus*, and *Cryptococcus neoformans,* which are associated with high mortality rates and are becoming increasingly resistant to available treatments. The global recognition of fungal threats by the WHO underscores the urgency of developing innovative approaches, such as biogenic nanoparticles, to combat these emerging pathogens and address the gaps in current antifungal strategies (Rodrigues et al., 2022).

Plant-based nanoparticle synthesis (PBNP), or phytosynthesis, has garnered attention as an environmentally friendly alternative to traditional physical and chemical methods that require high energy input and can produce toxic by-products (Kulkarni et al., 2023). Plant extracts contain various bioactive components such as alkaloids, tannins, terpenoids, flavonoids, and other phytochemicals that act as reducing and stabilizing agents in nanoparticle synthesis (Sharma et al., 2019). These compounds not only facilitate nanoparticle formation but also enhance the bioactivity of nanoparticles, making them suitable for biomedical applications such as photothermal therapy, biosensing, antioxidants, antimicrobials, and drug delivery. Phytosynthesis offers a more rapid synthesis rate than microbial methods, and the resulting nanoparticles exhibit greater stability. Bioactive compounds in plant extracts contribute to this stability by acting as capping agents and preventing nanoparticle aggregation (Sharma et al., 2019; Patil & Chandrasekaran, 2020; Kulkarni et al., 2023). The biogenesis of nanoparticles involves reduction of metal ions by plant extracts to form stable and well-dispersed nanoparticles. Certain parameters, such as the optical selection of plant species, extraction method, metal ion concentration, and reaction conditions, are key factors in determining the physicochemical properties and antifungal effectiveness of the resulting nanoparticles. Characterization techniques, including ultraviolet-visible (UV-vis) spectroscopy, Fourier transform infrared (FTIR) spectroscopy, X-ray diffraction (XRD), scanning electron microscopy (SEM), and transmission electron microscopy (TEM), offer valuable insights into the size, shape, surface functionality, and crystallinity of the nanoparticles (Gangadoo et al., 2015).

The antifungal mode of action of biogenic nanoparticles is a multifaceted process that involves interaction with the fungal cell wall, disruption of cellular processes, and modulation of immune responses. Furthermore, biogenic nanoparticles modulate the immune response and enhance the clearance of fungal pathogens by immune cells. The potential synergistic effects of biogenic nanoparticles with conventional antifungal medications were investigated, offering new avenues for the development of competent antifungal therapies. Combining biogenic nanoparticles with antifungal medications may enhance their efficacy, reduce the required dosage, and overcome drug resistance mechanisms. Furthermore, biogenic nanoparticles can be incorporated into novel antifungal formulations and delivery systems, such as hydrogels, nanofibers, and nanocomposites, to improve their stability, bioavailability, and targeted distribution to the infected location. Although plant-based syntheses have numerous advantages, some challenges persist. These include controlling the size, shape, and polydispersity of nanoparticles, understanding the precise mechanisms of synthesis, optimizing industrial-scale production, and addressing potential toxicity concerns (Kulkarni et al., 2023). Clinical translation of biogenic nanoparticles as antifungal agents faces several obstacles, including concerns regarding toxicity, pharmacokinetics, and regulatory aspects. Comprehensive toxicological evaluations are necessary to guarantee the health of biogenic nanoparticles for human use (Gangadoo et al., 2015; Jadoun et al., 2021; Sousa et al., 2021). In this chapter, we provide a comprehensive overview of the synthesis, characterization, and antifungal applications of biogenic nanoparticles, with a focus on their mechanism of action and potential for clinical translation (Figure 1).

Figure 1. A comprehensive overview of the role of plant-derived biogenic nanoparticles (PBNPs) in antifungal mechanisms.

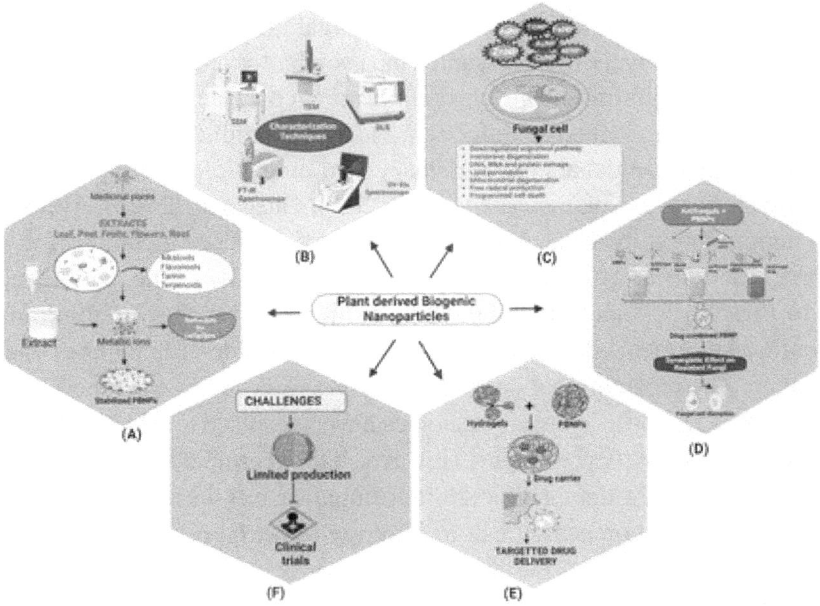

(A) Biosynthesis of PBNPs from different bioactive secondary metabolites of medicinal plant extracts, which react with metallic ions to form reduced and stable PBNPs. (B) Different characterization techniques applied for the validation of the synthesized nanomolecules: Fourier Transform Infrared Spectroscopy (FT-IR), UV-Vis Spectroscopy; Transmission Electron Microscopy (TEM), Scanning Electron Microscopy (SEM), and Dynamic Light Scattering (DLS). (C) Mechanism of action of different metallic nanoparticles, including gold, silver, iron oxide, copper, copper oxide, zinc, zinc oxide, and titanium oxide, against fungal targets. (D) Methods of conjugation of PBNPs with the available synthetic antifungal drugs to create a synergistic effect on resistant fungi. (E) Improvement in drug delivery to different affected organs by the collaboration of nanoparticles with nanocomposites, such as hydrogels and nanofibers. (F) Challenges affecting the application and production of nanoparticles for further pharmacological development (Akhter et al., 2024; Mittal et al., 2013; Ribeiro et al., 2022; Singh et al., 2016).

2. FUNGAL INFECTIONS AND THE NEED FOR NOVEL ANTIFUNGAL AGENTS

Fungal infections represent a significant and escalating global health challenge, resulting in millions of deaths worldwide. Despite this substantial burden, fungal diseases remain neglected in terms of research funding and public health (Rodrigues & Nosanchuk, 2020). Factors contributing to the growing impact of fungal infections include an increased population of immunosuppressed individuals, emergence of antifungal resistance, limited therapeutic options, diagnostic challenges, and inadequate awareness among healthcare providers (Bongomin et al., 2017). In response to these challenges, the World Health Organization released its first list of priority fungal pathogens in 2022, categorizing 19 fungal entities to guide research and public health efforts (WHO, 2022). Those with compromised immune systems or other diseases such as cancer are more vulnerable to fungal infections. Superficial infections such as candidiasis, dermatophytosis, onychomycosis, and Tinea versicolor are caused by various fungi. Subcutaneous infections, including sporotrichosis, chromoblastomycosis, and eumycetoma, arise from fungi that penetrate injured skin. Serious internal infections, such as histoplasmosis, coccidioidomycosis, and blastomycosis, can provoke bronchial allergies. Urinary tract fungal infections, predominantly caused by *Candida spp.*, are prevalent among women (Fang et al., 2023; Gnat et al., 2021). Although pharmaceuticals derived from chemically synthesized molecules have shown efficacy in treating fungal diseases, they often lead to adverse reactions and are not economically feasible. Over time, pathogens build biofilms, become resistant to these drugs, and, in some cases, exhibit resistance to multiple medications. The effectiveness of traditional drugs is decreasing, resulting in a considerable number of deaths attributed to multidrug resistance. The recent global outbreak of SARS-CoV-2 has worsened resistance problems and weakened the immune response. Various strategies have been used to address these challenges, such as combining drugs or adjusting their dosages. However, this treatment has increased the incidence of serious side effects. Consequently, demand for new antifungal agents with various mechanisms of action has increased, leading to the emergence of nanotechnology (Hoenigl et al., 2022).

Nanotechnology has revolutionized the healthcare sector by garnering significant attention for its antimicrobial properties because of its stability, minuteness, solubility, and adaptability, which offer a platform for investigating innovative approaches to address the limitations associated with synthetic drugs. However, nanoparticles produced using conventional physicochemical methods may pose risks, underscoring the importance of biosynthesis from natural sources. Ongoing research on this groundbreaking methodology has sparked interest in the combination of nanoparticles with conventional drugs to mitigate resistance to pathogens, minimize dosage, and improve the effectiveness of novel drug delivery agents (Nath & Banerjee, 2013).

Biogenic nanoparticles produced using biological entities are different from conventional nanoparticles in several aspects: they are less toxic, cost-effective, and less stable than conventional nanoparticles. The stability of biogenic nanoparticles can be improved by using capping agents such as plant extracts. Specific applications of biogenic nanoparticles include drug delivery, biosensing, and antimicrobial agents

(Sharma et al., 2019; Patil & Chandrasekaran, 2020; Kulkarni et al., 2023). Unlike traditional antifungals, which often target a single aspect of fungal survival, making resistance development straightforward, biogenic nanoparticles exert their antifungal effects by acting on multiple cellular targets simultaneously (Padmavathi & Bhadramaraju, 2024). Additionally, biogenic nanoparticles can synergistically enhance the effectiveness of existing antifungals. Biofilms, a collection of microorganisms encased in a protective matrix, pose a significant challenge in the treatment of fungal infections because of their increased resistance to antimicrobials and host immune responses. Biogenic nanoparticles have the potential to disrupt and eradicate biofilms, offering a significant advantage over traditional antifungals (Padmavathi & Bhadramaraju, 2024). This is achieved by damaging the extracellular polymeric substance (EPS) matrix that holds the biofilm together, compromising its stability and integrity (Pappas et al., 2018). A major drawback of traditional antifungal drugs is their cytotoxicity, which often leads to adverse side-effects. Biogenic nanoparticles, particularly those synthesized using green methods, exhibit lower cytotoxicity and higher biocompatibility than chemically synthesized nanoparticles and conventional drugs. This advantage makes them a safer and more tolerable option for antifungal therapies. Thus, biogenic nanoparticles are a promising alternative to traditional antifungals, offering enhanced efficacy, reduced resistance development, biofilm disruption capabilities, and improved safety profiles (Padmavathi & Bhadramaraju, 2024).

3. MEDICINAL PLANTS AS AN ORIGIN OF BIOLOGICALLY ACTIVE SUBSTANCES FOR THE GREEN SYNTHESIS OF NANOPARTICLES

Plants serve as an accessible and sustainable source for nanoparticle biosynthesis, providing a greener, cost-effective alternative to traditional chemical and physical methods, often termed "green synthesis" or "phytosynthesis." The effectiveness of plants in nanoparticle synthesis is due to the presence of various phytochemicals in their parts, including alkaloids, flavonoids, phenols, tannins, carbohydrates, steroids, saponins, terpenoids, ketones, aldehydes, amides, and carboxylic acids. These phytochemicals function as reducing and capping agents during the biosynthesis process. Reducing agents donate electrons to metal ions, reducing them to their elemental form, which then acts as nucleation points for nanoparticle growth. Capping agents adhere to the surface of the nanoparticles, preventing further growth and aggregation while stabilizing them in a colloidal solution. Specific phytochemicals and their roles can vary depending on the plant species and the type of nanoparticle (Patil & Chandrasekaran 2020; Kulkarni et al., 2023) (Table 1). A diverse array of plant species has shown remarkable potential for the green synthesis of nanoparticles, offering an eco-friendly alternative to traditional chemical methods.

Table 1. Antifungal potential of plant-derived Biogenic Nanoparticles (PBNP) from various medicinal plant extracts and their prospective mechanism of action against fungal pathogens.

Nanoparticles synthesized	Medicinal plant used	Extracted plant part	Fungal target	Characterization techniques	Particle size	Preparative solution with extracts	Antifungal mechanisms	References
Silver	*Zingiber officinale*	Rhizome	*Candida albicans*	UV-Vis Spectroscopy, FTIR, TEM	~12nm	Silver nitrate	Hydrophobic interaction with cell membrane causing cellular wall degradation	(Mohammadi et al., 2019)
	Lamium album	Leaf	*Candida albicans, C. krusei, C. tropicalis, C. dubliniensis, C. glabrata, C. parapsilosis, Cryptococcus neoformans, Aspergillus flavus, A. clavatus, A. fumigatus, Pseudallescheria boydii and Exophiala dermatitidis*	TEM, XRD, DLS, UV-Vis Spectroscopy, FTIR, ZP	~25nm	Silver nitrate	Entry of nanoparticles due to semi permeable membrane expansion	(Zareshahrabadi et al., 2021)
	Hyssopus officinalis and *Salvia officinalis*	Leaf	*Fusarium proliferatum, Aspergillus flavus, A. niger, Candida albicans*	UV-Vis, SEM-EDX, TEM-EDX	~12nm and ~39nm respectively	Silver nitrate	In vitro inhibition of fungal cell wall through antioxidation	(Dėnė et al., 2024)
Copper Oxide	*Bougainvillea*	Flower	*Aspergillus niger*	TEM, XRD, FTIR, UV-Vis Spectroscopy,	~2-15nm	Copper acetate monohydrate	Antifungal activity is higher compared to fluconazole	(Shammout & Awwad, 2021)
Copper	*Citrus medica* Linn.	Fruits	*Fusarium graminearum, F. oxysporum* and *F. culmorum*	UV-Vis Spectroscopy, XRD	~20nm	Copper sulfate	DNA, RNA and protein disintegration by binding of thiol group to nanoparticle	(Shende et al., 2015)
	Celastrus paniculatus	Leaf	*Fusarium oxysporum*	TEM, SEM, FTIR, UV-Vis Spectroscopy, EDS	~2-10nm	Copper sulfate + Sodium hydroxide (pH adjustment)	Spore germination inhibition, protein degradation	(Mali et al., 2020)

continued on following page

Table 1. Continued

Nanoparticles synthesized	Medicinal plant used	Extracted plant part	Fungal target	Characterization techniques	Particle size	Preparative solution with extracts	Antifungal mechanisms	References
Gold	*Acer pentapomicum*	Leaves	*Fusarium oxysporium, Penicillium chrysogenum, Rizopus oryzae* and *Aspergillus niger*	SEM, FTIR, EDX, XRD, UV-Vis Spectroscopy	~25nm	Gold chloride	Growth inhibition by cell membrane decay, decrease in ATP synthesis, Proton pump blocked, ROS accumulation	(Khan et al., 2018)
	Vetiveria zizanioides	Root	*Penicillium sp., Fusarium sp., Aspergillus flavus, A. fumigatus*	ZP, DLS, SEM, UV-Vis Spectroscopy,	~10-35nm	Chloroauric-acid	Minute particle size gives large surface area and entry into the cell stops DNA replication transcription, translation of proteins, causing death	(Swain et al., 2016)
	Cannabis sativa	Leaf	*Penicillium sp., Mucor sp., Fusarium sp., Aspergillus sp., A. flavus, A.fumigatus*	ZP, DLS, SEM, UV-Vis Spectroscopy,	~10-35nm	Chloroauric acid	Cell regulatory pathways hampered	(Swain et al., 2016)
Zinc oxide	*Cinnamomum camphora (L.)* Presl	Leaf	*Alternaria alternata*	FTIR, EDX, SEM, TEM, XRD	~20nm	Zinc acetate	Germination of spores limited, growth of conidiophores inhibited, germ tube and hyphae impaired	(Zhu et al., 2021)
	Lantana aculeata Linn.	Leaf	*Aspergillus flavus, Fusarium oxysporium, Parthenium funiculosum*	UV-Vis Spectroscopy, FTIR, XRD, TEM SEM	~12nm	Zinc nitrate	Higher antifungal activity in comparison to chemically synthesized nanoparticles	(Narendhran & Sivaraj, 2016)
	Salvia officinalis	Leaf	*Candida albicans*	XRD, TEM, UV-Vis Spectroscopy, FTIR, SEM, Differential Scanning Calorimetry (DCS)	~38nm	Ferric nitrate	Downregulation or ergosterol biosynthetic pathway in a concentration-dependent manner	(Abomuti et al., 2021)

3.1 Naturally Occurring Compounds with Antifungal Attributes

The extraction of various bioactive compounds with medicinal properties is an emerging need for the latest antimicrobial drug discovery against drug-resistant pathogens and has been identified in plants from inception as a natural defense mechanism. A wide range of natural substances, including alkaloids, terpenoids, flavonoids, essential oils, tannins, lecithin, and peptides are effective against a spectrum of infectious microbes (Nath & Banerjee, 2013). Although abundant, humans have been successful in isolating only a handful of them (Haruna & Yahaya, 2021).

Alkaloids are a class of amino acids that produce secondary metabolites with nitrogen as a functional group (Castillo et al., 2012). Terpenoids, the most abundant class of secondary metabolites, are produced from isoprene units and biosynthesized from primary metabolites through the mevalonic acid (MVA) and methylerythritol-4-phosphate (MEP) pathways in the cytoplasm and chloroplasts, respectively (Bergman et al., 2019). Polyphenols are another broad class of phenolic compounds with a hydroxyl benzene group, with phenylalanine being the simplest compound to form complex tannins that can be hydrolyzed or condensed, including flavonoids. The biosynthesis of hydrolyzed and condensed tannins occurs via a series of hydrolysis and oxidation reactions, respectively. Flavonoids are synthesized via the phenylpropanoid metabolic pathway to produce 4-coumaroyl-CoA from phenylalanine (Castillo et al., 2012; Liu et al., 2021). One crucial step is to extract these chemicals from various plant species and identify their antifungal capabilities. The antifungal properties of these compounds include inhibition of mycelial growth and germination, fungal pathogen-induced spore formation, and cell membrane disruption (Castillo et al., 2012).

3.2 Advantages of Green Synthesis over Traditional Methods

Nanotechnology, a significant advancement in the current era, has not only advanced in the realm of industry because of its distinctive and captivating physicochemical characteristics, such as notable thermal conductance, metallic reduction, electrical stability, and catalytic potential, but has also paved the way in the healthcare sector for its efficacy against cancerous components and microbial pathogens (Singh et al., 2016). Nanoparticles can be used in various fields. Medicinal applications include anti-inflammatory, antifungal, antibacterial, antiviral, antidiarrheal sedative, antitumor, digestive, antiseptic, antiplasmodial, diuretic, neuroprotective, antioxidant, drug discovery, and anticancer effects in immunosuppressed individuals (Dar et al., 2023; Sasidharan et al., 2010). Furthermore, nanoparticles have applications in nutrition, beauty aids, garments, and ceramics, among other areas, within this extensive spectrum of nanoparticle usage (Akhter et al., 2024; Singh et al., 2016).

Traditional nanoparticle synthesis involves two methods: top-down and bottom-up. The top-down method aims to disrupt large metallic substances into tiny nanoclusters using severe physical forces such as sol-gel, ball milling, photoablation, and pyrolysis (Mittal et al., 2013). The bottom-up approach, also known as "molecular nanotechnology," relies on the assembly of nanoparticles starting from fundamental atoms and molecules congregated into inorganic and organic compounds (Nath & Banerjee, 2013). Chemical biosynthesis of metallic nanoparticles involves certain chemicals, such as stabilizers and reagents. Some common stabilizers used are Tween80, alkyl thiols, anthracene, povidone, and reagents including formic aldehyde, diazane, and sodium tetrahydroborate (Bao et al., 2021). Physical synthesis, owing to severe heat and physical forces, often results in structural imperfections. However, chemical synthesis has toxic effects on the environment owing to its hazardous nature. Both methodologies, along with various adverse consequences, including cytogenicity, hepatotoxicity, oncogenicity, mutagenicity, and other generic fatal toxicities, exhibit a high level of complexity, lack of sustainability, and increased costs, which underscores the need to explore a superior alternative (Abid et al., 2022; Akhter et al., 2024; Bao et al., 2021).

Green nanoparticle synthesis using bioactive compounds, on the other hand, is a safer and sustainable alternative because it is biodegradable with minimal side effects and has a greater impact on fungal toxicity against pathogens (Nath & Banerjee, 2013). The green method involves utilization of plant extraction or microbial isolation to produce nanoparticles, in which both the solidifying agent and reductant are derived from these sources. Complex organic compounds and bioactive secondary metabolites are the main products of the extraction process, contributing not only to the improvement of nanoparticle stability but also to the reduction of metallic ions (Singh et al., 2016). This methodology is preferred over alternative physicochemical methods because of its ability to produce large amounts of nanoparticles via biogenic synthesis and with a specified structure of metallic particles (Mittal et al., 2013). Although both microbial and plant extract alternatives are used for green synthesis, researchers prefer plant extracts because of their eco-friendliness, affordability, time efficiency, low chances of contamination, and labor savings, which are limited to microbial isolates. Microbial isolation is often challenging and costly because of the requirement for extensive experimental procedures, lengthy incubation periods, numerous purification stages with lower yields of viable products, and risk of frequent contamination (Akhter et al., 2024; Bao et al., 2021; Singh et al. 2016).

4. SYNTHESIS AND OPTIMIZATION OF BIOGENIC NANOPARTICLES FOR ANTIFUNGAL APPLICATIONS

4.1 Plant Species and Extraction Methods

With the advancement of latest technologies, several scientific techniques have been developed to extract biogenic nanoparticles from medicinal plant species. These approaches have prompted a broader understanding of the chemical and biological makeup of plant-extracted nanoparticles, thus creating a new avenue for drug discovery (Dar et al., 2023). In addition to the intracellular approach for nanoparticle synthesis, extracellular synthesis using microorganisms is preferred. Intracellular synthesis requires downstream processing, such as extraction through different methods, including maceration, sonification, Soxhlet extraction, and advanced techniques, such as solid-phase, microwave-mediated, and supercritical liquid-based extraction, leading to breakage of the cell wall and purification of these extracts through repeated centrifugation. However, an extracellular approach does not require these critical steps. Both techniques require incubation, metal salt solution, and washing by centrifugation to degenerate cell membranes. In addition, extracellular synthesis allows metallic detoxification through reductants, thereby enhancing the stability of these nanoparticles and preventing their further accumulation. Plant-derived nanoparticles are a simpler alternative to organic compounds, and derived secondary metabolites are used directly as reducing and stabilizing agents for synthesized nanoparticles (Singh et al., 2016) (Table 1).

4.2 Metal Ion Concentration and Reaction Conditions

Based on their molecular characteristics and physicochemical properties, nanoparticles can be lipid-soluble, water-soluble, or in a metallic polymer form. Among these, metallic nanoparticles are the preferred choice in the medical, agricultural, and industrial fields because of their high stability and thermal conductivity. Silver, gold, titanium, copper oxide, zinc oxide, magnesium oxide, iron oxide, cadmium, and alginate are common metals used in nanoparticle synthesis (Akhter et al., 2024;

Dėnė et al., 2024). Three crucial components are necessary to produce metallic nanoparticles that can be obtained from bioactive molecules: a stabilizing agent, reducing agent, and solvent medium. Metal ions are depleted in hot gaseous surroundings to disperse and stabilize them. The high reactivity of these particles allows for their build-up without protection, leading to morbidity. This can be prevented by metal ion reduction using various reductants already present in the biomass-extracted compounds (Akhter et al., 2024; Raveendran et al., 2003). The biosynthesis rate of these particles is determined by the relatively rapid reduction of metallic ions and slower production of nanoparticles, aided by carbonyl and electron groups in nanosubstances, thus increasing the binding of metal ions (Khandel et al., 2018). In addition, size, shape, structure, and other physical characteristics determine the overall composition of these nanoparticles, and their production is determined by adaptation to temperature, pH, and extract concentration (Akhter et al., 2024).

4.3 Characterization Techniques for Physicochemical Properties

Nanoparticles are characterized by various assessments that focus on their physicochemical properties. These characterization methods include transmission electron microscopy (TEM), Fourier transform infrared spectroscopy (FTIR), scanning electron microscopy (SEM), ultraviolet-visible (UV-vis) spectroscopy, X-ray diffraction (XRD), energy-dispersive X-ray spectroscopy (EDX), zeta potential (ZP), dynamic light scattering (DLS), particle size, and selected area electron diffraction (SAED) (Akhter et al., 2024; Mittal et al., 2013). Both SEM and TEM are quantitative spectroscopic imaging methods used for the determination of particle structures and measurements; however, TEM has a comparatively higher resolution (Lee et al., 2020). Infrared spectroscopy is another emerging technique for quantifying nanoparticle characteristics because of its ability to detect nanoparticles at all surface levels. Infrared radiation determines the rotational and vibrational aspects of these molecules, which can be further detected by FTIR (Eid, 2021). Ultraviolet–visible (UV–vis) spectroscopy is one of the best methods for examining the optical properties of nanoparticles, particularly gold and silver. Nanoparticles absorb and scatter light of different wavelengths from the ultraviolet (UV) and visible regions and are quantified. This method not only allows the quantification of optical properties, but also deals with the determination of morphological aspects, as well as the conductivity and resistance of these metal ions (Begum et al., 2018). To analyze the crystalline size and structure of the nanoparticles at the atomic level, a new approach, that is, XRD, was applied. In this technique, the dispersion of X-rays diffused in the nanoparticles is analyzed (Gangadoo et al., 2015; Ingham, 2015; Mittal et al., 2013). Dynamic light scattering depends on the dispersion of light in an aqueous solution, and the correlation between the diffusion constant and Brownian motion of suspended nanoparticles was evaluated (Gangadoo et al., 2015). Zeta potential is a valuable tool for elucidating the electrochemical charge, constancy, and diffusion rate of nanoparticles. Energy-dispersive X-ray spectroscopy (EDX) was used in conjunction with SEM to generate maps of the samples (Akhter et al., 2024).

5. ANTIFUNGAL MECHANISM OF ACTION OF BIOGENIC NANOPARTICLES

Biogenic nanoparticles have emerged as a promising alternative to traditional nanoparticles for the treatment of fungal infections, offering several significant advantages. Several studies have investigated the antifungal mechanism of biogenic nanoparticles, proving their efficacy in fungal diagnosis and drug delivery. However, the precise mechanisms underlying their antifungal activities remain unclear (Nisar et al., 2019). We further explored the potential modes of action of a few green metallic nanoparticles by focusing on their antifungal mechanisms (Figure 2).

Figure 2. Possible mechanisms of action of plant-derived biogenic nanoparticles (PBNPs) trigger programmed cell death in targeted fungal cells.

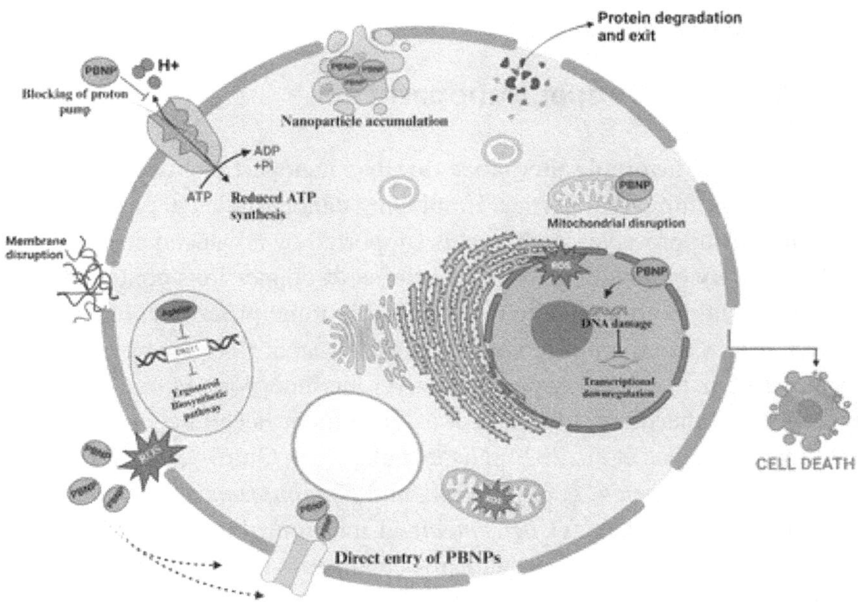

This includes the direct invasion of PBNPs through the cell wall, accumulation to the periphery of cells across the cell membrane, degenerating the cell wall through some unclear mechanisms, disrupting membrane stability by negatively regulating *ERG11* of the ergosterol biosynthetic pathway through silver NPs, production of free radicals causing cell wall rupture, mitochondrial damage, denaturation of DNA, RNA, and protein, and blocking the H+/ATP proton pump by the accumulation of PBNPs. These possible modes of action ultimately lead to programmed cell death (Nisar et al., 2019).

5.1 Antifungal Action of Silver Nanoparticles (AgNps)

AgNPs have been extensively studied owing to their rapid production and potent toxicity against microbial pathogens. Mechanisms include the invasion of AgNPs into the cellular structure of fungi through ergosterol of the cell wall, disruption of membrane surface stability, and exit of internal organelles through pores. The ergosterol biosynthetic pathway encoded by the *ERG11* gene is downregulated, repressing lanosterol demethylase and further disrupting the membrane. Proton pumps are destroyed, affecting the proteins and phospholipids present in the lipid bilayer. The phosphorus and sulfur groups of DNA react with nanoparticles, which affect DNA replication. Reactive oxygen species produced by internal organelles through mitochondrial respiration interact with these nanoparticles and cause oxidative damage, leading to DNA and RNA denaturation, protein modification, lipid peroxidation, cell damage, and reduction in ATP, leading to programmed cell death (Figure 2). Excessive metabolic activity of drugs leads to a surplus of ROS, which cannot be further reduced by enzymes, such as superoxide dismutase or peroxidases. The peroxidation of lipids that form lipid peroxide is transformed into aldehydes, alarming cellular disruptions caused by fungal species. Another reason may be solely due to the release of toxic Ag ions (Nisar et al., 2019). Multiple ongoing studies examining the potential of nanoparticles extracted from several plants with medicinal properties and their antimycotic mechanism have been scrutinized so far (Table 1).

5.2 Antifungal Action of Copper Nanoparticles

Copper nanoparticles are favorable alternatives to silver nanoparticles because of their cost efficiency, enhanced stability, and antifungal properties. Employing nanoparticles in pharmaceuticals for the treatment of various immunosuppressive conditions is supported by advanced clinical trials. The alteration in pH levels and conductivity resulting from the release of copper ions contributes to the degradation of cellular walls, impacting both replication and transcriptional processes. Interactions between metal ions and sulfhydryl groups can disrupt protein synthesis. Recent findings by Quirós et al. suggested an increase in fungal incontinence when copper and silver are incorporated into a complex called electrospun cellulose acetate. Furthermore, free-radical damage induces cell degeneration, culminating in the harnessing of pathogens (Lopes et al., 2024; Nisar et al., 2019; Quirós et al., 2016). Antifungal activity test against *Fusarium graminearum*, *F. oxysporum,* and *F. culmorum* through biogenic green synthesis of copper nanoparticles extracted from *Citrus medica* Linn. contributes to the possibility of modulating antifungal mechanisms. The proposed mechanisms encompass the formation of hollow cavities in the cell membrane, further affecting the stability of the membrane. Effective binding of nanoparticles to the surface of the cell membrane increases its affinity for the thiol groups present in the membrane, leading to the accumulation of free radicals, denaturation of amino acids of proteins and nuclei, and death of pathogens (Shende et al., 2015).

5.3 Antifungal Action of Gold Nanoparticles

The antifungal mechanisms of gold include membrane damage and ATP degeneration. These disrupt the ribosome-bacterial tRNA interaction, hindering protein synthesis. Oxidative phosphorylation, proton pump degeneration, and DNA denaturation occur through base interactions, leading to chemical stimuli. These tiny particles increased the surface area, allowing for higher nanoparticle invasion. The morphology was also found to affect the antifungal mechanism. However, the detailed mechanism of action is not yet fully understood (López-Lorente et al., 2019). These particles penetrate deep into the membrane by creating pits and stabilizing it. Free radicals accumulate in the cell cytoplasm, allowing the destabilization of mitochondrial proteins and causing cell cessation. Antioxidant activity was also elevated in the presence of nanoparticles. Regulatory mechanisms are attributed to smaller particle size, enabling rapid penetration into the cell walls and disruption of internal metabolic processes (Khan et al., 2018).

5.4 Antifungal Action of Zinc Nanoparticles

These semi-insulators are favored because of their ease of production and antifungal properties. FDA approved its use in the food industry, sterile material packaging, and agriculture. Zinc oxides release oxygen species, thereby facilitating antifungal applications. Nanoparticles combat fungal growth by accumulating DNA and RNA. Rupture of the membrane releases the inner cytoplasmic membrane, leading to death. Oxides also aid in the breakdown of toxins from molds. Pathogen hostility increases with smaller particles, which cause cell membrane rupture (Miri et al., 2019; Nisar et al., 2019). The zinc oxide nanoparticles suppressed the growth of the germinal tube and spores, negatively affecting the germination and development of the hyphae. Light-induced activation of ZnO particles generates a higher number of free radicals that are incorporated into the cell, causing severe fatal consequences. Peroxidation of lipids, nucleic acid damage, and protein denaturation are some of the causes that disrupt plasma membrane permeability, causing leakage and exit, leading to cell apoptosis (Zhu et al., 2021).

5.5 Antifungal Action of Other Nanoparticles

Plant-derived platinum nanoparticles (PtNPs) have demonstrated promising antifungal properties, particularly against various *Candida* species that are common causative agents of fungal infections. Research has shown that PtNPs synthesized using plant extracts, such as those from *Xanthium strumarium* leaves, show significant antifungal activity against a range of pathogens, including *Candida albicans, C. tropicalis, C. parapsilosis*, as well as *Aspergillus flaves,* and *A. niger*. The antifungal mechanism of PtNPs, particularly those synthesized using gum kondagogu, appears to involve induction of oxidative stress. This process leads to the formation of reactive oxygen species (ROS), which damage fungal mycelial morphology and membranes, resulting in cell damage and growth inhibition (Jan et al., 2021; Mikhailova, 2020). Zinc oxide nanoparticles have proven effective against *Fusarium sp.* in a concentration-dependent manner, whereas silica silver nanoparticles have shown remarkable potency against *Botrytis cinerea* (Mouhamad et al., 2022). Iron oxide nanoparticles (Fe NPs), particularly those synthesized using leaf extracts of *Platanus orientalis* or mycosynthesized with *Aspergillus terreus*, have shown broad-spectrum antifungal activity against various pathogens, including *Aspergillus niger, Mucor piriformis,* and *Candida albicans* (Sandhu et al., 2024).

6. SYNERGISTIC EFFECTS OF BIOGENIC NANOPARTICLES WITH CONVENTIONAL ANTIFUNGAL DRUGS

Antifungal agents pose significant risks to individuals, especially those with compromised immune systems. Common fungal infections, such as those caused by *Candida* spp. and *Aspergillus fumigatus*, have a mortality rate of up to 50%. Over time, these pathogens can develop resistance to therapeutic drugs, which worsens the situation and underscores the need for further research and development of more effective formulations (Mehta & Pai, 2022; Ribeiro et al., 2022; Yassin et al., 2022). In recent years, antifungal resistance caused by pharmaceutical medications has been observed through several mechanisms. For example, certain mutations or overexpression of genes that encode the ergosterol biosynthesis pathway, such as *ERG5, ERG1, ERG25, ERG11, MDR1,* and *CDR2,* can protect the plasma membrane of fungi from lysis and develop antifungal resistance. The outflow of drugs from the intercellular space through expressional regulation of certain drug carriers may also be the cause. Nanoparticles have come into play to combat health crises and fatal losses. These substances have better outcomes in the development of effective drug control in response to severe damage caused by these infectious agents. Moreover, biogenic nanoparticles are much safer and more sustainable, and thus more effective in the long term (Ahmady et al., 2024; Jia & Sun, 2021).

6.1 Procedure for Combining Nanoparticles with Antifungal Drugs

Incorporation of metallic nanoparticles with antifungal agents can be conducted by multiple processes that revolve around nanoparticle synthesis, addition of conjugating agents, and characterization; first, combining agents can be added to a solution where nanoparticles have been newly prepared. These agents can function as reducers or stabilizers and can be added to a solution in which nanoparticles have been newly prepared. Another alternative includes the separate preparation of both agents and nanoparticles and then combining them in the presence of another reducing agent. A three-step combination is also possible, where synthesis occurs first, followed by functionalization of the nanoparticles, and finally, the addition of combinatory particles. Functionalization is a key step in biological applications to enhance surface chemistry through the addition of an external moiety, thereby improving efficacy and reducing toxicity. The amalgamation of different chemical species into nanoparticles through cell-specific receptor-ligand molecules and covalent bonds display various properties, such as better dispersive ability and stabilized electrostatic repulsion. Biomolecules such as phosphines, phosphonates, thiol groups, amines, nitriles, disulfides, and carboxylic acids participate in the functioning of particles by attaching to their surface molecules, thereby creating a biofilm (Georgilis et al., 2020; Ma et al., 2013; Ribeiro et al., 2022).

6.2 Case Studies on Synergistic Effects of Nanoparticles and Pharmaceutical Drugs

Antifungal drugs can be classified into different categories according to their specific site of activity: azoles and allylamines, which impede the production of ergosterol; polyenes, which engage in physico-chemical interactions with sterols in the fungal membrane; echinocandins, which block the cell wall; and 5-fluorocytosine, which hinders the synthesis of macromolecules (Chen & Sorrell, 2007). Examining the combinatory antifungal effect with incorporation of nanoparticles is determined by certain measures, including colony counting on cultured plates, measuring their optical density, their zone of inhibition, and even by measuring the minimum inhibitory concentrations (MIC). Studying the synergistic role through MIC evaluation is one of the most common practices adopted by researchers (Park et al., 2017; Ribeiro et al., 2022). This new area of research has gained significant interest, and several studies have focused on antifungal components, such as polyenes and azoles, against *Candida* strains. Thus far, most research has focused on the direct combination of metallic nanoparticles with antifungal agents, with a limited number of studies exploring procedures that involve the functionalization of these nanoparticles after biosynthesis. The biogenic synthesis of fluconazole-conjugated AgNPs was evaluated and found to have an improved synergistic outcome in different *Candida* species tested, including *Candida albicans, C. glabrata, C. parapsilosis, C. krusie,* and *C. tropicalis* (Zainab et al., 2022). The biosynthesis of secondary metabolites of AgNPs of *Polyalthia longifolia* combined with amphotericin B showed a synergistic effect on *C. albicans* (Halbandge et al., 2017). Other antifungal compounds evaluated using AgNPs were miconazole and voriconazole (Halbandge et al., 2017; Zainab et al., 2022). Similarly, a study involving *C. albicans* exposed to AgNPs and fluconazole marked the first instance of inhibition of biofilm formation. However, an increase in oxygen radicals was observed because of this combined effect. AgNPswere shown to reduce the fluconazole minimal inhibitory concentration (MIC) (Jia and Sun, 2021). Itraconazole combined with AgNPs effectively increased the sensitivity to various *Candida* strains, thus enhancing therapeutic formulations (Yassin et al., 2022). Furthermore, a cooperative effect with Amphotericin B was induced in *Candida* during pH and oxidative stress, highlighting histatin-5 protein as a potential site-specific ligand for the fungus (Park et al., 2017). The synergistic effects imposed on these strains by different conjugated drugs depend on the type of drug and site-specific activities. They may also have added effects; however, the lack of comprehensive assessments is a limiting factor in the exploration of this emerging technology (Ribeiro et al., 2022). All these studies tend to expand the scope of research possibilities and potential outcomes that need to be explored further.

6.3 Advantages of Conjugated Nanoparticles with Antifungal Drugs

Current research on the application of synthetic drugs combined with nanoparticles has captured the attention of scientific researchers. This study has proven to be effective in addressing serious health conditions such as HIV/AIDS, cancer, heart attack, and infection. The phenomenon of antifungal resistance in various pathogens has been observed to have a synergistic interaction with nanoparticles in drug delivery. However, only a few mechanisms underlying this enhanced advantage have been documented. Studies have shown that the interaction of AgNPs with these drugs increases the degradation caused by the ergosterol biosynthesis pathway because of the disruption or modification of enzymes and genes involved in the regulation of the pathway (Park et al., 2017). A study of miconazole in combination with AgNPs showed increased ROS accumulation in both the membrane and cytoplasmic regions, affecting various organelles. Furthermore, ergosterol present in the plasma membrane was also affected, causing cell rupture (Kumar & Poornachandra, 2015). Research on fluconazole and voriconazole in conjunction with silver nanoparticles has shown a significant reduction in the budding process of fungal strains, and the proton efflux pumps were also affected by the blockage of conjugated drugs (Ribeiro et al., 2022). All these mechanisms, resulting from the synergy of nanoparticles with drugs, are beneficial in the development of modified formulations with maximum efficacy against drug-resistant pathogenic fungi. This emerging field of research not only significantly reduces the required pharmaceutical dosage but also addresses multidrug resistance in fungal pathogens and minimizes toxicity. Additionally, it prevents the emergence of new resistant pathogens. Conjugation enables the formation of autonomous nanoparticles, providing increased stability and effectiveness within the host body. This also leads to a higher recovery rate for patients by broadening the range of antifungal targets in the intracellular system (Ma et al., 2013; Park et al., 2017; Ribeiro et al., 2022).

7. NOVEL ANTIFUNGAL FORMULATIONS AND DELIVERY SYSTEMS INCORPORATING BIOGENIC NANOPARTICLES

Drug delivery systems (DDS) involve formulating, dosing, and managing drugs for human and animal welfare. Biobased nanomaterials are modified for efficient drug delivery of antifungal agents and adjusted for proper binding, availability, reduced dosage, and preserved activity. These modifications address biodegradable carriers for polar and nonpolar molecules at both the cellular and intracellular levels to ensure the site-specific delivery of therapeutic agents (Gopi et al., 2016). Hydrogels are hydrophilic polymers used as carriers for drug delivery. They are incorporated into nanoparticles through physical adsorption, embranglement, and polymer grafting. Natural hydrogels include chitosan, cellulose, gelatin, alginate, pectin, starch, and guar gum, which exhibit properties such as hydrophilicity, biodegradability, pH sensitivity, viscosity, gel formation, and cost efficiency. Cellulose and its derivatives carboxymethylcellulose (CMC), chitosan, and carboxymethyl chitosan (CMCS) are efficient carriers in DDS. Poly (ethylene glycol) and poly (vinyl alcohol) are synthetic biopolymers embodied in metallic nanoparticles, especially silver, owing to their pliability and hydrosolubility (Amiri et al., 2021; Gholamali & Yadollahi, 2021; Gopi et al., 2016). A mixture of different nanocomposites or modifications by placing polymers in the inner portion with nanoparticles as outer coverings can also be embodied in a polymer matrix composite (PMS) (Shariatinia, 2020). Hydrogels incorporated with nanoparticles acted as potent carriers for the successful delivery of drug formulations into the targeted part of the body (Figure 3). Nanofibers have also emerged as promising vehicles to facilitate drug delivery. Their small size and high surface-area-to-volume ratio surpass those of microfibers by a thousand times, thereby promoting specific drug binding to even the smallest target areas and ensuring remarkable stability. Nanofiber synthesis typically involves electrospinning by electrical forces to extrude a solution containing both the drug and polymer into a fiber with dimensions below 1 - 2μm in diameter. Additional advantages such as a lower liquefaction point, amorphous transformation, and lucidity further bolster their reliability. The nanofibers are composed of aqueous, methacrylate, polylactide, or vinyl acetate polymers. Not only do nanofibers shield drugs from degradation, they also regulate controlled and targeted drug release (Sousa et al., 2021).

Figure 3. Illustrative representation of an effective drug delivery system (DDS) by modifying hydrogels for the incorporation of nanoparticles possessing antifungal activity.

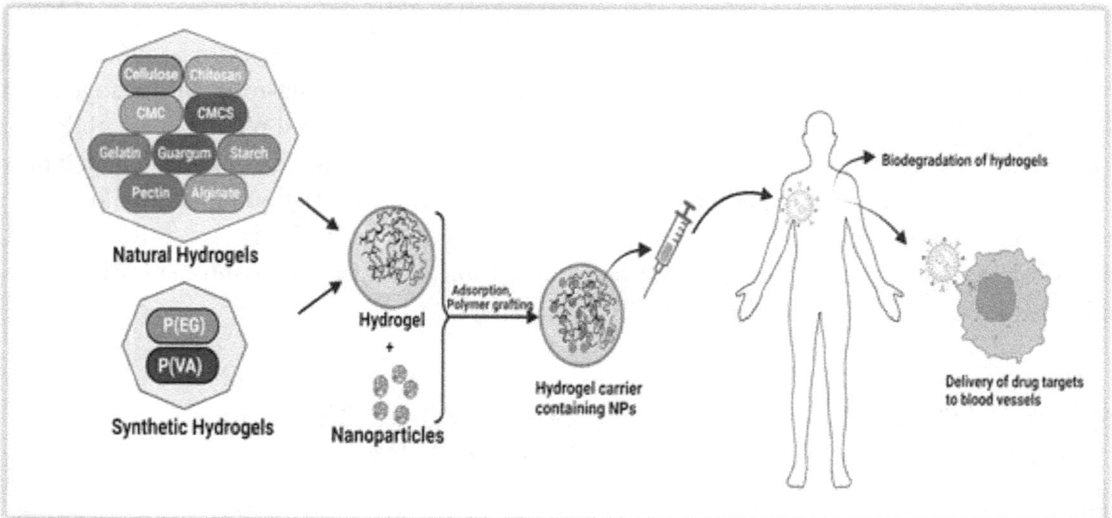

Hydrogels are mostly naturally derived, including cellulose, a derivative of cellulose, carboxymethyl cellulose (CMC), chitosan, a derivative of chitosan, carboxymethyl chitosan (CMCS), pectin, starch, alginate, guargum, and collagen-derived gelatin. Synthetically derived hydrogels are P(EG), that is, polyethylene glycol, and P(VA), that is, polyvinyl alcohol. These hydrogels combine with nanoparticles through adsorption and polymer grafting, making them an efficient carrier for the delivery of the drug into the bloodstream of the diseased body, after which they undergo biodegradation. Drugs can be formulated in the body through oral methods, inhalation, and surgery (Gholamali & Yadollahi, 2021; Gopi et al., 2016).

8. CHALLENGES AND OPPORTUNITIES IN THE CLINICAL TRANSLATION OF BIOGENIC NANOPARTICLES AS ANTIFUNGAL AGENTS

As far as the safety of humanity is concerned, the toxicity assessment of nanoparticles is an indispensable procedure, as the interplay of nanoparticles has become an integral part of human society succumbing to numerous septicities. These assessments deal with *in vivo* estimations based on multiple factors, such as the type of overlaying media for coating, contamination-free base, and type of metal ions utilized, thereby reducing the predilections to improve their adequacy. There are certain limitations, even at the preclinical trial level, which result from a lack of adequate manufacturing of nanomolecules, thus limiting their obtainability in the market and even their acceptance for permission for research at the medical level (Nami et al., 2021). Further development of nanotechnology-based materials should focus on certain improvements to eliminate the constraints that are currently faced to elevate their utilization and provide an enormous scope against standard remedial measures. Enhancing industrial production by optimizing an assemblage of regulatory aspects will not only increase the output of these materials and make them available for further medical research, but it will also increase the likelihood of financing that ultimately benefits humanity (Sousa et al., 2021).

9. FUTURE PERSPECTIVES AND CONCLUDING REMARKS

Plant-based nanoparticles are emerging as promising future treatments for fungal infections, offering enhanced antifungal activity and potential to overcome drug resistance. Current antifungal treatments face limitations due to emerging drug resistance and side effects; however, nanotechnology presents a potential solution. Plant extracts can be used to synthesize biogenic nanoparticles, particularly silver nanoparticles (AgNPs), with potent antifungal properties surpassing those of traditional azoles. These nanoparticles disrupt fungal cell membranes, inhibit cell division, and damage cell walls by affecting N-acetyl glucosamine synthesis. Nanoparticles can also be loaded with antifungal drugs for enhanced efficacy and targeted delivery, as demonstrated by lipase-responsive polycaprolactone nanoparticles loaded with amphotericin B. Combined plant-based nanoparticles with existing antifungal drugs show synergistic effects, allowing for lower drug concentrations and potentially reduced side effects. Future research should focus on translating these promising *in vitro* findings into clinical applications, including exploring diverse plant sources, optimizing nanoparticle characteristics, conducting *in vivo* studies, and developing suitable dosage forms. Although plant-based nanoparticles offer hope for combating drug-resistant fungal infections, further research is crucial for their successful clinical implementation.

Plant-based nanoparticles are emerging as a promising solution for antifungal warfare, addressing concerns regarding pesticide overuse and ecological impacts. This "green nanotechnology" offers sustainable synthesis methods using plant extracts rich in phytochemicals as reducing and stabilizing agents. These nanoparticles demonstrate broad-spectrum antifungal activity against various fungal pathogens, which is crucial for combating fungal resistance. Research is exploring the functionalization of these nanoparticles with specific targeting ligands for more precise delivery. The combination of plant extracts and nanoparticles has synergistic effects, potentially improving antifungal activity and reducing the development of resistance. Future research directions include a deeper understanding of their mechanisms of action, comprehensive toxicological studies to assess safety, and the development of scalable, cost-effective production methods. Although challenges remain, plant-based nanoparticles represent a promising frontier in sustainable agriculture and biomedicine, potentially revolutionizing antifungal treatments with more effective and environmentally friendly strategies.

In conclusion, this book chapter will serve as a helpful guide for researchers, healthcare professionals, and stakeholders interested in biogenic nanoparticles and their antifungal applications. By providing an assessment of the synthesis, characterization, and mechanistic insights into these nanoparticles, along with their viability for clinical translation, the aim of this chapter is to drive advances in the field of antifungal therapeutics and contribute to the worldwide battle against fungal infections.

REFERENCES

Abid, N., Khan, A. M., Shujait, S., Chaudhary, K., Ikram, M., Imran, M., Haider, J., Khan, M., Khan, Q., & Maqbool, M. (2022). Synthesis of nanomaterials using various top-down and bottom-up approaches, influencing factors, advantages, and disadvantages: A review. *Advances in Colloid and Interface Science*, 300, 102597. DOI: 10.1016/j.cis.2021.102597 PMID: 34979471

Abomuti, M. A., Danish, E. Y., Firoz, A., Hasan, N., & Malik, M. A. (2021). Green synthesis of ZnO nanoparticles using Salvia officinalis leaf extract and their photocatalytic and antifungal activities. *Biology (Basel)*, 10(11), 1075. DOI: 10.3390/biology10111075 PMID: 34827068

Ahmady, L., Gothwal, M., Mukkoli, M. M., & Bari, V. K. (2024). Antifungal drug resistance in *Candida*: A special emphasis on amphotericin B. *APMIS*, 132(5), 291–316. DOI: 10.1111/apm.13389 PMID: 38465406

Akhter, M. S., Rahman, M. A., Ripon, R. K., Mubarak, M., Akter, M., Mahbub, S., Al Mamun, F., & Sikder, M. T. (2024). A systematic review on green synthesis of silver nanoparticles using plants extract and their bio-medical applications. *Heliyon*, 10(11), e29766. DOI: 10.1016/j.heliyon.2024.e29766 PMID: 38828360

Amiri, M., Khazaeli, P., Salehabadi, A., & Salavati-Niasari, M. (2021). Hydrogel beads-based nanocomposites in novel drug delivery platforms: Recent trends and developments. *Advances in Colloid and Interface Science*, 288, 102316. DOI: 10.1016/j.cis.2020.102316 PMID: 33387892

Bao, Y., He, J., Song, K., Guo, J., Zhou, X., & Liu, S. (2021). Plant-Extract-Mediated Synthesis of Metal Nanoparticles. *Journal of Chemistry*, 2021(1), 6562687. DOI: 10.1155/2021/6562687

Begum, R., Farooqi, Z. H., Naseem, K., Ali, F., Batool, M., Xiao, J., & Irfan, A. (2018). Applications of UV/Vis Spectroscopy in Characterization and Catalytic Activity of Noble Metal Nanoparticles Fabricated in Responsive Polymer Microgels: A Review. *Critical Reviews in Analytical Chemistry*, 48(6), 503–516. DOI: 10.1080/10408347.2018.1451299 PMID: 29601210

Bergman, M. E., Davis, B., & Phillips, M. A. (2019). Medically Useful Plant Terpenoids: Biosynthesis, Occurrence, and Mechanism of Action. *Molecules (Basel, Switzerland)*, 24(21), 3961. Advance online publication. DOI: 10.3390/molecules24213961 PMID: 31683764

Bongomin, F., Gago, S., Oladele, R. O., & Denning, D. W. (2017). Global and Multi-National Prevalence of Fungal Diseases-Estimate Precision. *Journal of Fungi (Basel, Switzerland)*, 3(4), 57. DOI: 10.3390/jof3040057 PMID: 29371573

Castillo, F., Aguilar, C. N., Hernández, D., Gallegos, G., & Rodríguez, R. (2012). *Antifungal properties of bioactive compounds from plants*. INTECH Open Access Publisher., DOI: 10.5772/26598

Chen, S. C., & Sorrell, T. C. (2007). Antifungal agents. *The Medical Journal of Australia*, 187(7), 404–409. DOI: 10.5694/j.1326-5377.2007.tb01313.x PMID: 17908006

Dar, R. A., Shahnawaz, M., Ahanger, M. A., & Majid, I. U. (2023). Exploring the Diverse Bioactive Compounds from Medicinal Plants: A Review. *The Journal of Phytopharmacology*, 12(3), 189–195. DOI: 10.31254/phyto.2023.12307

Dėnė, L., Chrapačienė, S., Laurinaitytė, G., Rudinskaitė, A., Viškelis, J., Viškelis, P., & Balčiūnaitienė, A. (2024). Green Synthesis of Silver Nanoparticles with *Hyssopus officinalis* and *Salvia officinalis* Extracts, Their Properties, and Antifungal Activity on *Fusarium* spp. *Plants*, 13(12), 1611. DOI: 10.3390/plants13121611 PMID: 38931043

Denning, D. W. (2024). Global incidence and mortality of severe fungal disease. *The Lancet. Infectious Diseases*, 24(7), e428–e438. DOI: 10.1016/S1473-3099(23)00692-8 PMID: 38224705

Eid, M. M. (2021). Characterization of Nanoparticles by FTIR and FTIR-Microscopy. In S. Mallakpour & C. M. Hussain, *Handbook of Consumer Nanoproducts* (pp. 1–30). Springer Singapore. DOI: 10.1007/978-981-15-6453-6_89-1

Fang, W., Wu, J., Cheng, M., Zhu, X., Du, M., Chen, C., Liao, W., Zhi, K., & Pan, W. (2023). Diagnosis of invasive fungal infections: Challenges and recent developments. *Journal of Biomedical Science*, 30(1), 42. DOI: 10.1186/s12929-023-00926-2 PMID: 37337179

. Gangadoo, S., Taylor-Robinson, A., & Chapman, J. (2015). Nanoparticle and biomaterial characterisation techniques. *Materials Technology, 30*(sup5), B44–B56. DOI: 10.1179/1753555714Y.0000000201

Georgilis, E., Abdelghani, M., Pille, J., Aydinlioglu, E., van Hest, J. C. M., Lecommandoux, S., & Garanger, E. (2020). Nanoparticles based on natural, engineered or synthetic proteins and polypeptides for drug delivery applications. *International Journal of Pharmaceutics*, 586, 119537. DOI: 10.1016/j.ijpharm.2020.119537 PMID: 32531450

Gholamali, I., & Yadollahi, M. (2021). Bio-nanocomposite Polymer Hydrogels Containing Nanoparticles for Drug Delivery: A Review. *Regenerative Engineering and Translational Medicine*, 7(2), 129–146. DOI: 10.1007/s40883-021-00207-0

Gnat, S., Łagowski, D., Nowakiewicz, A., & Dyląg, M. (2021). A global view on fungal infections in humans and animals: Opportunistic infections and microsporidioses. *Journal of Applied Microbiology*, 131(5), 2095–2113. DOI: 10.1111/jam.15032 PMID: 33556223

Gopi, S., Amalraj, A., & Thomas, S. (2016). Effective drug delivery system of biopolymers based on nanomaterials and hydrogels-a review. *Drug Designing: Open Access*, 5(129), 2169–0138. DOI: 10.4172/2169-0138.1000129

Halbandge, S. D., Mortale, S. P., & Karuppayil, S. M. (2017). Biofabricated silver nanoparticles synergistically activate amphotericin B against mature biofilm forms of Candida albicans. *The Open Nanomedicine Journal*, 4(1), 1–16. Advance online publication. DOI: 10.2174/1875933501704010001

Haruna, A., & Yahaya, S. M. (2021). Recent Advances in the Chemistry of Bioactive Compounds from Plants and Soil Microbes: A Review. *Chemistry Africa*, 4(2), 231–248. DOI: 10.1007/s42250-020-00213-9

Hoenigl, M., Seidel, D., Sprute, R., Cunha, C., Oliverio, M., Goldman, G. H., Ibrahim, A. S., & Carvalho, A. (2022). COVID-19-associated fungal infections. *Nature Microbiology*, 7(8), 1127–1140. DOI: 10.1038/s41564-022-01172-2 PMID: 35918423

Ingham, B. (2015). X-ray scattering characterisation of nanoparticles. *Crystallography Reviews*, 21(4), 229–303. DOI: 10.1080/0889311X.2015.1024114

Jadoun, S., Arif, R., Jangid, N. K., & Meena, R. K. (2021). Green synthesis of nanoparticles using plant extracts: A review. *Environmental Chemistry Letters*, 19(1), 355–374. DOI: 10.1007/s10311-020-01074-x

Jan, H., Gul, R., Andleeb, A., Ullah, S., Shah, M., Khanum, M., Ullah, I., Hano, C., & Abbasi, B. H. (2021). A detailed review on biosynthesis of platinum nanoparticles (PtNPs), their potential antimicrobial and biomedical applications. *Journal of Saudi Chemical Society*, 25(8), 101297. DOI: 10.1016/j.jscs.2021.101297

Jia, D., & Sun, W. (2021). Silver nanoparticles offer a synergistic effect with fluconazole against fluconazole-resistant Candida albicans by abrogating drug efflux pumps and increasing endogenous ROS. *Infection, Genetics and Evolution*, 93, 104937. DOI: 10.1016/j.meegid.2021.104937 PMID: 34029724

Khan, A. S., Bakht, J., & Syed, F. (2018). Green synthesis of gold nanoparticles using Acer pentapomicum leaves extract its characterization, antibacterial, antifungal and antioxidant bioassay. *Digest Journal of Nanomaterials and Biostructures*, 13(2), 579–589. DOI: 10.3390/biom10060835 PMID: 32486004

Khandel, P., Yadaw, R. K., Soni, D. K., Kanwar, L., & Shahi, S. K. (2018). Biogenesis of metal nanoparticles and their pharmacological applications: Present status and application prospects. *Journal of Nanostructure in Chemistry*, 8(3), 217–254. DOI: 10.1007/s40097-018-0267-4

Kulkarni, D., Sherkar, R., Shirsathe, C., Sonwane, R., Varpe, N., Shelke, S., More, M. P., Pardeshi, S. R., Dhaneshwar, G., Junnuthula, V., & Dyawanapelly, S. (2023). Biofabrication of nanoparticles: Sources, synthesis, and biomedical applications. *Frontiers in Bioengineering and Biotechnology*, 11, 1159193. DOI: 10.3389/fbioe.2023.1159193 PMID: 37200842

Kumar, C. G., & Poornachandra, Y. (2015). Biodirected synthesis of Miconazole-conjugated bacterial silver nanoparticles and their application as antifungal agents and drug delivery vehicles. *Colloids and Surfaces. B, Biointerfaces*, 125, 110–119. DOI: 10.1016/j.colsurfb.2014.11.025 PMID: 25460601

Lee, B., Yoon, S., Lee, J. W., Kim, Y., Chang, J., Yun, J., Ro, J. C., Lee, J.-S., & Lee, J. H. (2020). Statistical Characterization of the Morphologies of Nanoparticles through Machine Learning Based Electron Microscopy Image Analysis. *ACS Nano*, 14(12), 17125–17133. DOI: 10.1021/acsnano.0c06809 PMID: 33231065

Liu, W., Feng, Y., Yu, S., Fan, Z., Li, X., Li, J., & Yin, H. (2021). The Flavonoid Biosynthesis Network in Plants. *International Journal of Molecular Sciences*, 22(23), 12824. DOI: 10.3390/ijms222312824 PMID: 34884627

Lopes, I. S., Soares, J. K. C., de Medeiros, L. S., & Courrol, L. C. (2024). Evaluation of ALA-capped silver, copper, and silver–copper nanoparticles for controlling fungal plant pathogens. *Microbial Pathogenesis*, 191, 106672. DOI: 10.1016/j.micpath.2024.106672 PMID: 38705219

López-Lorente, Á. I., Cárdenas, S., & González-Sánchez, Z. I. (2019). Effect of synthesis, purification and growth determination methods on the antibacterial and antifungal activity of gold nanoparticles. *Materials Science and Engineering C*, 103, 109805. DOI: 10.1016/j.msec.2019.109805 PMID: 31349488

Ma, L., Kohli, M., & Smith, A. (2013). Nanoparticles for Combination Drug Therapy. *ACS Nano*, 7(11), 9518–9525. DOI: 10.1021/nn405674m PMID: 24274814

Mali, S. C., Dhaka, A., Githala, C. K., & Trivedi, R. (2020). Green synthesis of copper nanoparticles using *Celastrus paniculatus* Willd. leaf extract and their photocatalytic and antifungal properties. *Biotechnology Reports (Amsterdam, Netherlands)*, 27, e00518. DOI: 10.1016/j.btre.2020.e00518 PMID: 32923378

Mehta, N., & Pai, R. (2022). Amalgamation of Nanoparticles within Drug Carriers: A Synergistic Approach or a Futile Attempt? *Pharmaceutical Nanotechnology*, 10(5), 354–367. DOI: 10.2174/2211738510666220902150449 PMID: 36056844

Mikhailova, E. O. (2020). Silver Nanoparticles: Mechanism of Action and Probable Bio-Application. *Journal of Functional Biomaterials*, 11(4), 84. DOI: 10.3390/jfb11040084 PMID: 33255874

Miri, A., Mahdinejad, N., Ebrahimy, O., Khatami, M., & Sarani, M. (2019). Zinc oxide nanoparticles: Biosynthesis, characterization, antifungal and cytotoxic activity. *Materials Science and Engineering C*, 104, 109981. DOI: 10.1016/j.msec.2019.109981 PMID: 31500056

Mittal, A. K., Chisti, Y., & Banerjee, U. C. (2013). Synthesis of metallic nanoparticles using plant extracts. *Biotechnology Advances*, 31(2), 346–356. DOI: 10.1016/j.biotechadv.2013.01.003 PMID: 23318667

Mohammadi, M., Shahisaraee, S. A., Tavajjohi, A., Pournoori, N., Muhammadnejad, S., Mohammadi, S. R., Poursalehi, R., & Delavari H, H. (2019). Green synthesis of silver nanoparticles using Zingiber officinale and Thymus vulgaris extracts: Characterisation, cell cytotoxicity, and its antifungal activity against Candida albicans in comparison to fluconazole. *IET Nanobiotechnology / IET*, 13(2), 114–119. DOI: 10.1049/iet-nbt.2018.5146 PMID: 31051440

Mouhamad, R. S., Al Khafaji, K. A., Al-Dharob, M. H., & Al-Abodi, E. E. (2022). Antifungal, antibacterial and anti-yeast activities evaluation of oxides of silver, zinc and titanium nanoparticles. *Chemistry International*, 8(4), 159–166. DOI: 10.5281/zenodo.7317423

Nami, S., Aghebati-Maleki, A., & Aghebati-Maleki, L. (2021). Current applications and prospects of nanoparticles for antifungal drug delivery. *EXCLI Journal*, 20, 562–584. DOI: 10.17179/excli2020-3068 PMID: 33883983

Narendhran, S., & Sivaraj, R. (2016). Biogenic ZnO nanoparticles synthesized using L. aculeata leaf extract and their antifungal activity against plant fungal pathogens. *Bulletin of Materials Science*, 39(1), 1–5. DOI: 10.1007/s12034-015-1136-0

Nath, D., & Banerjee, P. (2013). Green nanotechnology–a new hope for medical biology. *Environmental Toxicology and Pharmacology*, 36(3), 997–1014. DOI: 10.1016/j.etap.2013.09.002 PMID: 24095717

Nisar, P., Ali, N., Rahman, L., Ali, M., & Shinwari, Z. K. (2019). Antimicrobial activities of biologically synthesized metal nanoparticles: An insight into the mechanism of action. *Journal of Biological Inorganic Chemistry*, 24(7), 929–941. DOI: 10.1007/s00775-019-01717-7 PMID: 31515623

Padmavathi, S., & Bhadramaraju, R. (2024). 'A Review on Nanotechnology And Its Applications In The Field Of Pharmacy. *Educational Administration: Theory and Practice*, 30(1), 964–973. DOI: 10.53555/kuey.v30i1.5856

Pappas, P. G., Lionakis, M. S., Arendrup, M. C., Ostrosky-Zeichner, L., & Kullberg, B. J. (2018). Invasive candidiasis. *Nature Reviews. Disease Primers*, 4(1), 18026. DOI: 10.1038/nrdp.2018.26 PMID: 29749387

Park, S. C., Kim, Y. M., Lee, J. K., Kim, N. H., Kim, E. J., Heo, H., Lee, M. Y., Lee, J. R., & Jang, M. K. (2017). Targeting and synergistic action of an antifungal peptide in an antibiotic drug-delivery system. *Journal of Controlled Release*, 256, 46–55. DOI: 10.1016/j.jconrel.2017.04.023 PMID: 28428067

Patil, S., & Chandrasekaran, R. (2020). Biogenic nanoparticles: A comprehensive perspective in synthesis, characterization, application and its challenges. *Journal of Genetic Engineering and Biotechnology*, 18(1), 67. DOI: 10.1186/s43141-020-00081-3 PMID: 33104931

Quirós, J., Gonzalo, S., Jalvo, B., Boltes, K., Perdigón-Melón, J. A., & Rosal, R. (2016). Electrospun cellulose acetate composites containing supported metal nanoparticles for antifungal membranes. *The Science of the Total Environment*, 563, 912–920. DOI: 10.1016/j.scitotenv.2015.10.072 PMID: 26524992

Raveendran, P., Fu, J., & Wallen, S. L. (2003). Completely "Green" Synthesis and Stabilization of Metal Nanoparticles. *Journal of the American Chemical Society*, 125(46), 13940–13941. DOI: 10.1021/ja029267j PMID: 14611213

Ribeiro, A. I., Dias, A. M., & Zille, A. (2022). Synergistic Effects Between Metal Nanoparticles and Commercial Antimicrobial Agents: A Review. *ACS Applied Nano Materials*, 5(3), 3030–3064. DOI: 10.1021/acsanm.1c03891 PMID: 36568315

Rodrigues, M. L., de Oliveira, H. C., & Albuquerque, P. C. (2022). Antifungal development and the urgency of minimizing the impact of fungal diseases on public health. *ACS Bio & Med Chem Au*, 2(6), 506–511. DOI: 10.1021/acsbiomedchemau.2c00055

Rodrigues, M. L., & Nosanchuk, J. D. (2020). Fungal diseases as neglected pathogens: A wake-up call to public health officials. *PLoS Neglected Tropical Diseases*, 14(2), e0007964. DOI: 10.1371/journal.pntd.0007964 PMID: 32078635

Sandhu, Z. A., Raza, M. A., Alqurashi, A., Sajid, S., Ashraf, S., Imtiaz, K., Aman, F., Alessa, A. H., Shamsi, M. B., & Latif, M. (2024). Advances in the Optimization of Fe Nanoparticles: Unlocking Antifungal Properties for Biomedical Applications. *Pharmaceutics*, 16(5), 645. DOI: 10.3390/pharmaceutics16050645 PMID: 38794307

Sasidharan, S., Chen, Y., Saravanan, D., Sundram, K. M., & Yoga Latha, L. (2010). Extraction, Isolation and Characterization of Bioactive Compounds from Plants' Extracts. *African Journal of Traditional, Complementary, and Alternative Medicines*, 8(1), 1–10. DOI: 10.4314/ajtcam.v8i1.60483 PMID: 22238476

Shammout, M., & Awwad, A. (2021). A novel route for the synthesis of copper oxide nanoparticles using Bougainvillea plant flowers extract and antifungal activity evaluation. *Chemistry International*, 7(1), 71–78. DOI: 10.5281/zenodo.4042902

Shariatinia, Z. (2020). Biopolymeric Nanocomposites in Drug Delivery. In Nayak, A. K., & Hasnain, M. S. (Eds.), *Advanced Biopolymeric Systems for Drug Delivery* (pp. 233–290). Springer International Publishing., DOI: 10.1007/978-3-030-46923-8_10

Sharma, D., Kanchi, S., & Bisetty, K. (2019). Biogenic synthesis of nanoparticles: A review. *Arabian Journal of Chemistry*, 12(8), 3576–3600. DOI: 10.1016/j.arabjc.2015.11.002

Shende, S., Ingle, A. P., Gade, A., & Rai, M. (2015). Green synthesis of copper nanoparticles by Citrus medica Linn. (Idilimbu) juice and its antimicrobial activity. *World Journal of Microbiology & Biotechnology*, 31(6), 865–873. DOI: 10.1007/s11274-015-1840-3 PMID: 25761857

Singh, P., Kim, Y. J., Zhang, D., & Yang, D. C. (2016). Biological Synthesis of Nanoparticles from Plants and Microorganisms. *Trends in Biotechnology*, 34(7), 588–599. DOI: 10.1016/j.tibtech.2016.02.006 PMID: 26944794

Sousa, M. G. C., Rezende, T. M. B., & Franco, O. L. (2021). Nanofibers as drug-delivery systems for antimicrobial peptides. *Drug Discovery Today*, 26(8), 2064–2074. DOI: 10.1016/j.drudis.2021.03.008 PMID: 33741497

Swain, S., Barik, S. K., Behera, T., Nayak, S. K., Sahoo, S. K., Mishra, S. S., & Swain, P. (2016). Green Synthesis of Gold Nanoparticles Using Root and Leaf Extracts of Vetiveria zizanioides and Cannabis sativa and its Antifungal Activities. *BioNanoScience*, 6(3), 205–213. DOI: 10.1007/s12668-016-0208-y

World Health Organization. (2022). *WHO fungal priority pathogens list to guide research, development and public health action*. World Health Organization. https://www.who.int/publications/i/item/9789240060241

Yassin, M. T., Mostafa, A. A.-F., Al-Askar, A. A., & Al-Otibi, F. O. (2022). Synergistic antifungal efficiency of biogenic silver nanoparticles with itraconazole against multidrug-resistant candidal strains. *Crystals*, 12(6), 816. DOI: 10.3390/cryst12060816

Zainab, S., Hamid, S., Sahar, S., & Ali, N. (2022). Fluconazole and biogenic silver nanoparticles-based nano-fungicidal system for highly efficient elimination of multi-drug resistant Candida biofilms. *Materials Chemistry and Physics*, 276, 125451. DOI: 10.1016/j.matchemphys.2021.125451

Zareshahrabadi, Z., Karami, F., Taghizadeh, S., Iraji, A., Amani, A. M., Motamedi, M., & Zomorodian, K. (2021). Green synthesis of silver nanoparticles using aqueous extract of Lamium album and their antifungal properties. *Journal of Nano Research*, 67, 55–67. . DOI: 10.4028/www.scientific.net/JNanoR.67.55

Zhu, W., Hu, C., Ren, Y., Lu, Y., Song, Y., Ji, Y., Han, C., & He, J. (2021). Green synthesis of zinc oxide nanoparticles using Cinnamomum camphora (L.) Presl leaf extracts and its antifungal activity. *Journal of Environmental Chemical Engineering*, 9(6), 106659. DOI: 10.1016/j.jece.2021.106659

KEY TERMS AND DEFINITIONS

Biogenic Nanoparticles: The nanomaterials which are synthesized through biological entities, i.e., plants or microbes.

Antifungal Agents: Antifungal or antimycotic agents are synthetic or naturally occurring drugs that help in treating infections caused by fungal agents by suppressing or inhibiting their growth.

Green Synthesis: An eco-friendly and renewable approach by utilizing natural resources and biogenic compounds to produce a wide range of Nanoparticles without using hazardous chemicals.

Medicinal Plants: These are herbs that produce different secondary metabolites that have various medicinal properties and have been used in traditional medicinal practices since ages.

Anti-Fungal Nanoparticles: Nanoparticles developed for use as antifungal agents.

Nanoparticle Mode of Action: Metallic nanoparticles exert toxicity towards infectious microbes by various mechanisms causing death of the fungi.

Antifungal- Nanoparticle conjugation: MNPs in combination with antifungal drugs enhance the synergistic effect on resistant fungi.

Clinical Translation: The process of applying laboratory or pre-clinical findings to clinical practice applications to improve individual and community health.

Chapter 10

Green Synthesis of Nanoparticles and Their Conjugations With Antibiotics for Treating Bacterial Infections

Jyoti Srivastava
https://orcid.org/0009-0008-9019-4182
Ajeenkya D.Y. Patil University, Pune, India

Khushbu Iqbal Siddiqui
Ajeenkya D.Y. Patil University, Pune, India

Nawneet Kurrey
Ajeenkya D.Y. Patil University, Pune, India

Mohammed Shabab
https://orcid.org/0009-0001-2199-6199
Ajeenkya D.Y. Patil University, Pune, India

ABSTRACT

The escalating threat of antibiotic resistance poses a formidable challenge to the global healthcare system, necessitating novel approaches to combat multidrug-resistant bacterial infections. The synthesis of nanoparticle antibiotic conjugates emerges as a promising avenue. Nanoparticle antibiotic conjugates represent a groundbreaking strategy that harnesses the unique properties of nanoparticles and antibiotics to enhance antimicrobial efficacy. This chapter delves into the innovative fusion of nanoparticles and antibiotics through green synthesis techniques, aiming to address the pressing challenges of antibiotic resistance. By exploring these hybrid nanostructures' synthesis, characterization, and therapeutic potential, this chapter sheds light on their significance in revolutionizing infectious disease management. This article highlights potential mechanisms of action of Nanoparticle antibiotic conjugates as next-generation antimicrobial agents and future implications and challenges.

DOI: 10.4018/979-8-3693-6240-2.ch010

1. INTRODUCTION

Nanotechnology delineates the science of examining the materials in the nano range, particularly between 1 to 100 nm which have applications in the food, pharmaceutical, and detergent industries (Andra et al., 2019; Berekaa, 2015; Hedberg et al., 2014). Nanoparticle (NP) synthesis has a long history of 4500 years ago using ceramic-based materials including asbestos nanofibers. The Ancient Egyptians synthesized lead sulfide nanoparticles (PbS NPs) by the chemical process for hair dye (Jeevanandam et al., 2018). The use of nanoparticles for therapeutic purposes began at the beginning of the 20th century by Paul Ehrlich as "magic bullets" to which specific drugs are loaded and used to treat diseases (Kreuter, 2007). Nanoparticles display significant properties like electronic, magnetic, physicochemical, and greater surface area to volume ratio. The NPs are synthesized through physical methods, chemical methods, & green synthesis approach. In the physical approach, the production of nanoparticles can be performed naturally or synthetically, imparting unique properties to them. By physical synthesis approach, nanoparticles were synthesized by the "top-down" method that involves the breaking down of solid materials into nanosized particles by external forces like grinding, milling, sputtering, plasma arcing, spray pyrolysis, thermal/laser ablation etc. (Gour & Jain, 2019). By this method, energy needed to synthesize nanoparticles is supplied by various physical, chemical, and thermal techniques (Amendola et al., 2011). In addition, another traditional method i.e. "bottom-up" involves combining of gas or liquid atoms or molecules. The synthesis of NPs by the "bottom-up" protocol is more advantageous than the "top-down" as in the former method, the size of nanoparticles is controlled precisely (Abid et al., 2022; Jiang et al., 2022).

The other approach to nanoparticle synthesis is the chemical synthesis method includes the sol-gel method, etc. (Panigrahi, Kundu, Ghosh, Nath, & Pal, 2004; Revathy, Sajini, Augustine, & Joseph, 2023). In this method, the key components are the metallic precursors, stabilizing agents & reducing agents. Reducing agents like hydrazine hydrate, ethylene glycol, glucose, ascorbate, $NaBH_4$, etc. are used (Guzmán et al., 2010; Zielińska et al., 2009).

Green synthesis method employs environmentally friendly reducing agents as well in this method reducing agents can be used as end-capping material and dispersants (Devi, Boda, Shah, Parveen, & Wani, 2019). The use of natural reducing agents decreases energy consumption and the generation of toxic pollutants. Due to the lower probability of failure, economic, & ease of characterization, green nanoparticle synthesis is more effective than traditional methods. By green synthesis, nanoparticles are synthesized from microbial biomass including bacteria, fungi, algae, etc., and extracts from plant leaves, flowers (HB, 2022; Kumar et al., 2020), roots, fruits (Demirezen et al., 2019), peelings and seeds (Jagtap & Bapat, 2013; Kumar, Yadav, & Yadav, 2010).

In addition, toxicants released from the physical and chemical synthesis pose serious environmental threats that favor green synthesis of NPs (Abdelghany et al., 2018). Green synthesis of nanoparticles requires the naturally and environmentally friendly materials extracted from plants and microbial biomass. The green synthesis of NPs involves three significant processes: the screening of solvent medium, the selection of reducing agents, & the selection of non-toxic stability agents for nanoparticles. The reducing agents used by green synthesis include plant extracts, fungal biomass, bacterial biomass, polysaccharides, algal biomass, etc, and the metallic solutions based on copper, iron, zinc, silver, gold metals, etc. with varying degrees of biocompatibility (Kurhade et al., 2021; Shah et al., 2015; Yadi et al., 2018). Depending upon the biocompatibility of using various metals, gold, and silver-based nanoparticle synthesis is discussed below:

a) Gold nanoparticles (AuNPs)

The green synthesis of AuNPs for therapeutic applications began in the early 2000s as researchers sought environmentally friendly approach to synthesize these nanoparticles. The extracts from plants such as Aloe vera, tea leaves, and various fruits could effectively reduce gold salts to form nanoparticles. Gold nanoparticles are greatly biocompatible to be used for therapeutic purposes. In addition, surface of AuNPs can be easily modified using different ligands, antibodies, or drugs and is suitable for controlled drug delivery (Ahmed & Ikram, 2016).

In general, as an example, green biosynthesis of AuNPs has been represented in **Figure 1**.

Figure 1. Green synthesis approaches of AuNPs

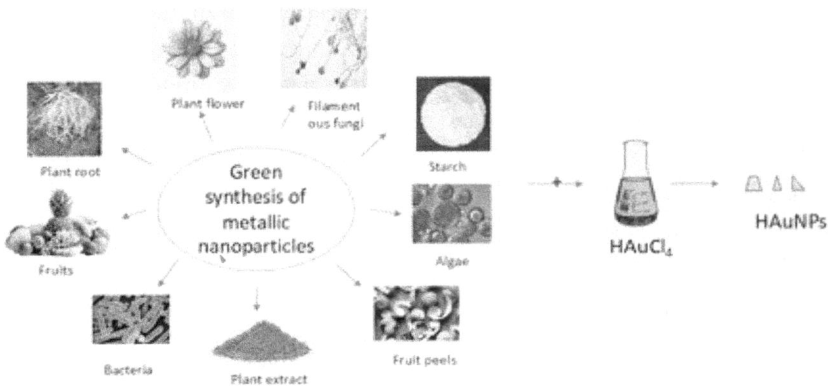

b) Silver nanoparticles (AgNPs)

Green-synthesized AgNPs hold strong antimicrobial properties, presenting them successful in applications like wound dressings, and antibacterial drug delivery. The usage of natural stabilizers & reducing agents in green synthesis allows for exact control over the size and shape of silver nanoparticles which can be used for therapeutic purposes. The green synthesis of AgNPs can be produced using plant extract, microbial biomass, algal biomass etc. as a potential reducing agent producing stable nanoparticles (Srikar et al., 2016).

c) Copper nanoparticles (CuNPs)

Budych *et al* delineated the green synthesis of CuNPs using peppermint extract by the green reduction method and conjugate with rifampicin antibiotic and observed an increase in antimicrobial potential on *Staphylococcus aureus, Escherichia coli, Bacillus pumilis, Pseudomonas fluorescens* (Woźniak-Budych et al., 2017). In addition, Duhalt *et al* conjugated antibodies with copper nanoparticles for targeted bacterial infections of *E . coli* and *B. subtilis* (Ontiveros-Robles et al., 2023).

d) Zinc nanoparticles (ZnNPs)

The synthesis of ZnNPs plays a pivotal role in diverse applications. Tyagi *et al* emphasized that the conjugation of ciprofloxacin antibiotic with ZnNPs has significantly increased the antibacterial activity of 2.9 fold and 2.8 fold against *E. coli* and *Streptococcus spp.* Respectively than ciprofloxacin alone (Tyagi et al., 2020). The study was supported by Selvam *et al* studied the green synthesis of ZnO-NPs using *Hardwickia binata* leaves extract and then conjugated with ciprofloxacin which resulted in significant inhibition of *Staphylococcus aureus* (Manimegalai et al., 2024).

2. NANOPARTICLE-ANTIBIOTIC CONJUGATES

Nanoparticle-antibiotic conjugates shows a cutting-edge approach in the fight against bacterial infections. Combining nanoparticles with antibiotics aims to enhance the efficacy, targeting, and delivery of antimicrobial agents (Vassallo et al., 2020). This synergy leverages the unique properties of nanoparticles to overcome limitations associated with traditional antibiotics, such as poor solubility, rapid degradation, and non-specific distribution (Nazli et al., 2022). This overview explores the synthesis techniques of nanoparticle-antibiotic conjugates, their applications, and the challenges faced in this innovative field. Nanoparticle-antibiotic conjugates are composite materials where antibiotics are covalently or non-covalently attached to nanoparticles (Labruère, Sona, & Turos, 2019). The primary goal of this conjugation is to combine the broad therapeutic benefits of antibiotics with the advantageous properties of nanoparticles, such as enhanced cellular uptake, controlled drug release, and targeted delivery (Kalhapure et al., 2015). This approach addresses several issues in antibiotic therapy, including bacterial resistance, side effects, and limited drug efficacy.

2a. Advantages of Green Nanoparticle-Antibiotic Conjugates

1. **Enhanced Solubility:** NPs can improve the solubility of poorly soluble antibiotics, increasing their bioavailability and effectiveness (Vassallo et al., 2020).
2. **Targeted Delivery:** By attaching antibiotics to nanoparticles that can recognize specific bacterial targets, it is possible to carry the drug directly to the infection site, minimizing systemic side effects (Wang et al., 2020).
3. **Controlled Release:** Nanoparticles can be tailored to release antibiotics in a gradual and targeted manner, optimizing treatment efficacy while reducing the need for repeated doses. (Wang et al., 2022).
4. **Reduced Resistance:** Targeted delivery and controlled release may help in reducing the likelihood of bacterial resistance by maintaining effective drug concentrations at the infection site (Zazo, Colino, & Lanao, 2016).

3. SYNTHESIS TECHNIQUES FOR NANOPARTICLE-ANTIBIOTIC CONJUGATES

The synthesis of nanoparticle-antibiotic conjugates involves various methods to achieve stable and effective conjugation. These approaches can be broadly categorized into covalent and non-covalent approaches.

3a. Covalent Bonding Methods

1. **Direct Coupling:** In direct coupling, the antibiotic is covalently bonded to the nanoparticle surface through chemical reactions involving functional groups. For instance, gold nanoparticles (AuNPs) can be functionalized with thiol groups that react with the antibiotic's functional groups, such as amines or carboxyls (Rajchakit & Sarojini, 2017). This method provides strong and stable conjugation but requires precise control over reaction conditions to avoid altering the antibiotic's activity.
2. **Click Chemistry:** Click chemistry is a versatile method used to achieve highly specific and efficient covalent bonding. The most common click chemistry reaction is the azide-alkyne cycloaddition. In this process, nanoparticles and antibiotics are modified with azide and alkyne groups, respectively, allowing for a highly efficient and selective reaction to form the conjugate (Flauzino et al., 2023). This technique offers robust coupling with minimal by-products and can be applied to a variety of nanoparticles and antibiotics. Reyes et al investigated the conjugation of AgNPs and antimicrobial peptides in which azide ($-N_3$) group proficient in undergoing a click reaction with acetylide-containing peptide during which amino terminal copper and nickel (ATCUN) motif can not react throughout synthesis (Gakiya-Teruya et al., 2020).
3. **Ester or Amide Formation:** This technique involves the formation of ester or amide bonds between functional groups on nanoparticles and antibiotics. For example, carboxyl-functionalized nanoparticles can react with amine groups on antibiotics to form amide bonds (Abdelghafour et al., 2022). This method is commonly used with polymeric nanoparticles and antibiotics that possess suitable functional groups.

3b. Non-Covalent Bonding Methods

1. **Adsorption:** Antibiotics can imbibe onto the surface of nanoparticles through non-covalent interactions, like van der Waals forces / hydrophobic interactions. This method is simpler and often used when the antibiotic's biological activity is sensitive to covalent modifications (Ivanković, Kern, & Rožman, 2021). However, non-covalent interactions may be weaker and less stable, potentially leading to undesired release or loss of the antibiotic.
2. **Electrostatic Assembly:** Nanoparticles with charged surfaces can bind to oppositely charged antibiotics through electrostatic interactions. This method is suitable for antibiotics with ionic or polar groups and can be used to create stable conjugates if the electrostatic forces are strong enough to maintain the association (Oh et al., 2019).
3. **Layer-by-Layer (LbL) Assembly:** This approach involves the sequential deposition of layers of nanoparticles and antibiotics (Zhu & Loh, 2015). By irregular layers of positively and negatively charged substances, it is possible to build up a multilayered structure that can encapsulate or immobilize antibiotics. This method offers particular control over the composition and thickness of coating.

4. CHARACTERIZATION OF NANOPARTICLE-ANTIBIOTIC CONJUGATES

Nanoparticle-antibiotic conjugates represent a sophisticated approach to enhancing antibiotic efficacy and overcoming challenges associated with conventional therapies. Characterizing these conjugates is crucial for understanding their nature, behavior, and possible applications. This comprehensive article explores the various techniques used to characterize nanoparticle-antibiotic conjugates, focusing on their structural, chemical, and biological properties, as well as the insights gained from these analyses and their implications for practical applications.

4a. Importance of Characterization

Characterizing nanoparticle-antibiotic conjugates is crucial for several reasons:

- **Quality Control:** Ensures that the conjugates are synthesized with the desired properties and specifications.
- **Functional Understanding:** Provides insights into how the conjugates interact with biological systems and their effectiveness in delivering antibiotics.
- **Regulatory Compliance:** Meets the requirements for safety and efficacy in clinical and environmental applications.
- **Optimization:** Helps in refining synthesis methods and improving performance for specific applications (Singh et al., 2018; Walemba, 2020).

4b. Structural Characterization

Understanding the structural properties of nanoparticle-antibiotic conjugates is fundamental to optimizing their design and functionality.

a. **TEM:** TEM is a significant tool by which we can visualize the size, shape, & morphology of NPs. It provides high-resolution images of conjugates, allowing researchers to observe the distribution and uniformity of nanoparticles. TEM can also reveal the arrangement of antibiotics on the nanoparticle surface, which is crucial to understanding how fine the conjugates can deliver drug (Abushiba et al., 2019; Krishna et al., 2015). Shaikh et al investigated the morphology of conjugated cefotaxime with gold nanoparticles characterized by TEM and showed an increase in size than the free unconjugated nanoparticles (Shaikh et al., 2017).
b. **SEM:** SEM method offers surface imaging and provides info on the surface topology and morphology of nanoparticle-antibiotic conjugates. While TEM provides internal structure details, SEM is useful for assessing surface features and ensuring that the conjugates maintain their structural integrity after synthesis (Zhuang et al., 2023).
c. **AFM:** AFM delivers topographical pictures of nanoparticle-antibiotic conjugates with nanometer resolution. It can measure surface roughness, particle size, and the mechanical properties of the conjugates. AFM is mainly useful to characterize nanoparticles with complex surface modifications and for assessing the uniformity of conjugated layers (Gutiérrez-Mejía et al., 2024).

d. **XRD:** XRD is useful for determining the crystal-like structure of NPs. It provides information on the phase composition and crystallinity, which can affect the stability and performance of nanoparticle-antibiotic conjugates. XRD helps in confirming the successful attachment of antibiotics to nanoparticles without altering their crystalline structure (Kalita et al., 2016).

4c. Chemical Characterization

Chemical characterization focuses on understanding the chemical composition and bonding within nanoparticle-antibiotic conjugates.

a. **FTIR:** FTIR is utilized to know the functional groups and confirm chemical bonds between nanoparticles and antibiotics. By analyzing the infrared absorption spectra, FTIR can reveal changes in functional groups that occur during the conjugation process. This technique is essential for verifying the presence of specific chemical interactions, such as ester or amide bonds, between the antibiotic and the nanoparticle (Kaur & Kumar, 2019). Samie et al investigated FTIR analysis of conjugated gold nanoparticles with *Pyrenacantha grandiflora* Baill extracts revealing the formation of C–H group and carbonyl (C=O) groups, –OH carbonyl, and C≡C groups (Murei, Pillay, & Samie, 2021).
b. **XPS:** XPS delivers info about the elemental composition & chemical states of the elements present in the nanoparticle-antibiotic conjugates. It is particularly helpful for analyzing the surface chemistry and confirming the successful attachment of antibiotics to the nanoparticles. XPS can also help in assessing the oxidation states and chemical environments of the elements involved (Ran et al., 2022).
c. **NMR Spectroscopy:** NMR spectroscopy is managed to verify the chemical environment of specific nuclei within the antibiotic molecules and their interaction with nanoparticles. While less commonly used for nanoparticle characterization, It gives detailed information about the molecular structure and can help in elucidating the chemical bonding between the nanoparticle and the antibiotic (Agarwal, Nair, Mazumder, & Poluri, 2018; Arachchi et al., 2022).
d. **Elemental Analysis:** The techniques, like Inductively Coupled Plasma Mass Spectrometry (ICP-MS), are active in revealing the elemental composition of nanoparticle-antibiotic conjugates. This analysis provides quantitative information about the concentration of nanoparticles and antibiotics in the conjugate, helping to assess the efficiency of the conjugation process (Lim, Petaroudi, Pearce, Yasakçi, & Unak, 2021). The structural and chemical characterizations of nanoparticle-antibiotic conjugates are shown in **Figure 2.**

Figure 2. Characterization of nanoparticle-antibiotic conjugates

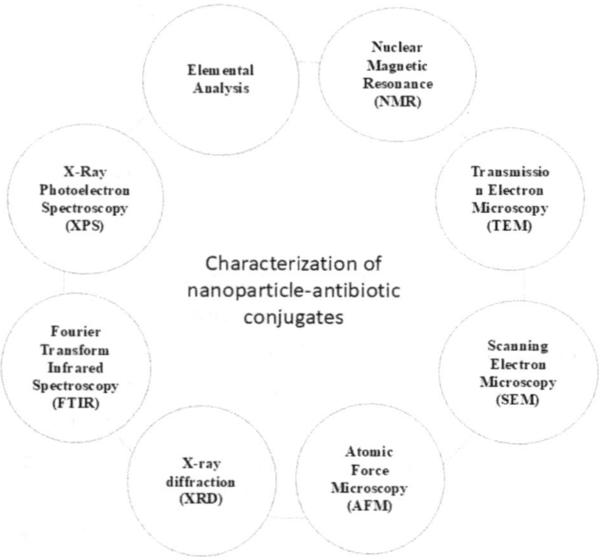

4d. Physical Characterization

Physical characterization focuses on the physical properties of nanoparticle-antibiotic conjugates, which are crucial for their performance in practical applications.

a. **Dynamic Light Scattering (DLS):** DLS deals with size distribution and stability of nanoparticle-antibiotic conjugates in solution. It provides information on the hydrodynamic diameter of the conjugates and their aggregation behavior. DLS is important for assessing the uniformity of particle size and ensuring that the conjugates remain stable in aqueous environments (Hassan et al., 2017; Kaur & Kumar, 2021).
b. **ZP:** ZP measures the surface charge of NPs, which affects their stability and interactions with biological systems. By determining the zeta potential of nanoparticle-antibiotic conjugates, researchers can assess their colloidal stability and predict their behavior in biological fluids (Lowry et al., 2016).
c. **SPR:** SPR is utilized to examine the interactions between nanoparticles and antibiotics in real-time. This technique provides information on binding kinetics, affinity, and specificity. SPR is particularly useful for understanding how well antibiotics are bound to nanoparticles and how this affects their performance (Sparks, Jenkins, & Fratti, 2019).
d. **TA:** TA techniques, like TGA and DSC, give information on the thermal stability and thermal transitions of nanoparticle-antibiotic conjugates. TGA can be used to quantify the amount of antibiotic attached to the nanoparticles, while DSC helps in understanding the thermal behavior and stability of the conjugates (Mishra, Goswami, & Bhatt, 2021).

4e. Biological Characterization

Biological characterization assesses the interaction of nanoparticle-antibiotic conjugates with biological systems and evaluates their effectiveness in practical applications.

a. **In Vitro Antimicrobial Testing:** In vitro tests, such as MIC assays and disk diffusion assays, are performed to evaluate the antimicrobial activity of nanoparticle-antibiotic conjugates. These tests determine the effectiveness of the conjugates in inhibiting bacterial growth compared to free antibiotics (Abdel-Wahab et al., 2021). Shakil et al investigated the significant improvement in MIC using cefotaxime-loaded AuNPs against *E. coli* and *K. pneumoniae* of 1.009 and 2.018 mg/L respectively which were resistant to cefotaxime alone (Shaikh et al., 2017). The improvement in MIC was supported by Saadh *et al* investigating the MIC of ciprofloxacin conjugated with AgNPs was 0.05 ± 0.0012 μg/mL as compared with 0.75 ± 0.0047 μg/mL alone for ciprofloxacin (Saadh, 2022).
b. **Cytotoxicity Assays:** Cytotoxicity assays, such as MTT or LDH assays, assess the biocompatibility of nanoparticle-antibiotic conjugates with mammalian cells. These assays help in determining whether the conjugates are toxic to healthy cells, which is crucial for ensuring safety in clinical applications (Skłodowski et al., 2023). Feng *et al* demonstrated that silver and copper bimetallic nanoparticles have potential antibacterial activity with low cytotoxic effect including ROS generation proving that better antibacterial therapeutic formulations (Hao et al., 2024).
c. **Pharmacokinetics and Biodistribution Studies:** In pharmacokinetics, we know about the absorption, distribution, metabolism, and excretion of nanoparticle-antibiotic conjugates *in vivo*. Biodistribution studies track the distribution of conjugates within the body to evaluate their targeting efficiency and accumulation in specific tissues or organs (Labruère et al., 2019; Martin et al., 2015). Perez *et al* supported improvement in pharmacokinetics and biodistribution kinetics of antibiotic-loaded DNA biomolecule-based Trojan horse-inspired nanostructures and the greater antibacterial efficacies leading to the significant impact on the rise of bacterial infections (Giráldez-Pérez et al., 2022).
d. **Stability Studies:** Stability studies assess the physical and chemical stability of nanoparticle-antibiotic conjugates under various storage conditions. Stability studies include evaluating changes in size, morphology, and chemical composition over time to ensure that the conjugates remain effective throughout their shelf life (Bhatia, Mittal, & Malik, 2021).

4f. Challenges in Characterization

Despite the advances in characterization techniques, several challenges are mentioned below:

- **Complexity of Conjugates:** The complexity of nanoparticle-antibiotic conjugates requires a combination of techniques to obtain a comprehensive understanding of their properties.
- **Dynamic Behavior:** Nanoparticle-antibiotic conjugates may behave differently in biological environments compared to in vitro conditions, necessitating extensive in vivo studies.
- **Reproducibility:** Ensuring consistent results across different batches and experimental conditions is crucial for reliable characterization.

5. MECHANISM OF ACTION OF NANOPARTICLE-ANTIBIOTIC CONJUGATES

Current situation of AMR and also with the toxic impact of antimicrobial medications, the advancement of nanomedicine is substantial for efficient application. It's been observed that the nanoparticles tend to enhance cellular permeability leading to efficient actions against those pathogenic bacteria. According to Dong *et al*, the antimicrobial actions of AgNPs are confirmed to include DNA binding, impacting important proteins and other cell functions, and generation of free radicals (Figure 3). AgNPs were produced using *Bacillus sp.* and conjugated with ciprofloxacin, methicillin, gentamycin, and rifampicin, this made the integration of AgNPs in various products for clinical and other applications. To foster the increase the efficacy, AgNPs-based preparations have been developed. Amongst this, AgNPs-antibiotic conjugates have shown greater efficacy than these components against bacterial infections (Hwang, Hwang, Choi, Kim, & Lee, 2012; Thomas et al., 2020). Upon formation of conjugates, it was found to reduce the MIC of antibiotics effectively. MIC of conjugates displayed decrease, particularly the MIC of methicillin was decreased from 250 µg/mL to 7.8 µg/mL against *S. epidermidis*. In addition, to develop and use the nanoparticle-antibiotic conjugates, the required dose of antibiotic is low as compared to the individual native dose minimizing the toxicity issues in humans (Tao, 2018).

Also, Kotoky *et al* demonstrated the biosynthesis of AuNPs with a reducing agent as extracts of *Adiantum philippens* by microwave irradiation and conjugated with broad-spectrum amoxicillin antibiotic (GNP-Amox) by electrostatic interaction. The conjugate was efficient against Gram-positive and Gram-negative bacteria affecting the protein synthesis and effective on MRSA (Figure 3) (Kalita et al., 2016). In addition, the synthetic peptides named NBC2253 and NBC2254 were synthesized by solid phase synthesis and then conjugated with gold nanoclusters (AuNC) showed significant antibacterial activity, particularly against *E. coli* and MRSA indicating the promising agents in combating the multidrug resistant bacteria (Prada et al., 2019). Also, medical device-based infections are one of the major concerns caused by coagulase-negative Staphylococci which are tough to cure due to antibiotic resistance. The spread of infections was regulated by applications of AuNPs fabricated by *Bacillus sp.* and antibiotic conjugates on the surface of medical devices called as surface engineering agents (Roshmi, Soumya, Jyothis, & Radhakrishnan, 2015).

In some cases, on the surface of the nanoparticle-vancomycin conjugates the cyclic 9-amino acid peptide CARGGLKSC was attached which binds specifically to *Staphylococcus aureus* bacteria posing significant activity (Hussain et al., 2018). The different targets of nanoparticle-antibiotic conjugates on the bacterial constituents are depicted in **Figure 3**.

Figure 3. Different targets of nanoparticle-antibiotic conjugates in bacteria.

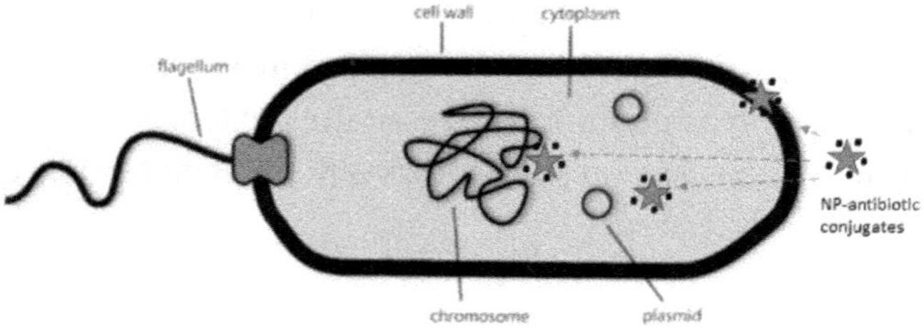

Table 1. Green synthesis of nanoparticles and their conjugations with antibiotics against susceptible microorganisms and their targets.

Metallic nanoparticle	Reducing agent for synthesis of nanoparticles	Antimicrobial agent for conjugation	Target	Susceptible microorganism	Reference
AuNPs	*Fusarium oxysporum*	Tetracycline	30s subunit of ribosome disrupting protein synthesis	*Escherichia coli, Pseudomonas aeruginosa, Bacillus cereus, Staphylococcus aureus*	(Naimi-Shamel, Pourali, & Dolatabadi, 2019)
AuNP-Apt	DNA aptamer	Antimicrobial peptides	Disruption of protein synthesis	*Vibrio vulnificus*	(Lee et al., 2017)
AgNP's	Glucan from mushroom	Ampicillin	Cell wall disruption	*Klebsiella pneumoniae YSI6A*	*(Sen et al., 2013)*
IO-NPs	Flower extract of *Nerium oleander*	Streptomycin	30s subunit of ribosome disrupting protein synthesis	*S. aureus, E. coli*	(Sharma et al., 2024)

6. CONCLUSION

The increased risk of antibiotic resistance poses a formidable challenge to the global healthcare system, necessitating novel approaches to combat multidrug-resistant bacterial infections, which is the need of the hour. The synthesis of nanoparticle-antibiotic conjugates emerges as promising therapeutic formulations for treating bacterial infections effectively. Nanoparticle antibiotic conjugates represent a groundbreaking strategy that harnesses the unique properties of nanoparticles and antibiotics to enhance antimicrobial efficacy. This chapter delves into the innovative fusion of nanoparticles and antibiotics through green synthesis techniques, aiming to address the pressing challenges of antibiotic resistance. The green synthesis of NPs involved the use of reducing agents from natural sources such as plant extracts, fungal biomass, bacterial biomass, polysaccharides, algal biomass, etc. With the reducing agent, the biocompatible metallic solutions are incubated at the specified temperature and the filtered solution is then characterized. The biocompatible nanoparticles are derived from silver, gold, platinum, copper, etc. metals. The synthesized nanoparticles are then characterized by SEM, TEM, XRD, AFM, FTIR, XPS, NMR, etc. Upon successful synthesis of nanoparticles, the antibiotics are conjugated by covalent and non-covalent bonding methods and further characterized. The efficient nanoparticle-antibiotic conjugation is revealed by FTIR, identifying the functional groups, and then studying the in vitro antimicrobial studies. By exploring these hybrid nanostructures' synthesis, characterization, and therapeutic potential, this chapter sheds light on their significance in revolutionizing infectious disease management. This article highlights potential mechanisms of action of Nanoparticle antibiotic conjugates as next-generation antimicrobial agents and future implications and challenges.

7. FUTURE IMPLICATIONS AND CHALLENGES

The synthesis of nanoparticle antibiotic conjugates presents exciting prospects for the development of innovative antimicrobial therapies. However, several challenges, such as regulatory hurdles, scalability issues, and potential toxicity concerns, require to be addressed to facilitate the clinical translation of these nanostructures. Future research directions may focus on optimizing drug delivery mechanisms, exploring combination therapies, and conducting comprehensive preclinical studies to validate the safety and efficacy of nanoparticle antibiotic conjugates in clinical settings.

In addition, the synthesis of nanoparticles of antibiotic conjugates represents a paradigm shift in the field of infectious disease management, offering a sustainable and effective solution to the growing threat of antibiotic resistance. By harnessing the power of nanotechnology and antimicrobial agents, these hybrid nanostructures hold enormous promise for developing the treatment and innovation in this field are crucial to unlock the full therapeutic potential of nanoparticle antibiotic conjugates and pave the way for a future where antibiotic-resistant bacteria can be effectively controlled and treated. Despite the novel ways of efficient delivery of antimicrobial drugs through nanoparticles, there are concerns regarding potential toxicity. Nanoparticles may accumulate unintentionally in various body organs such as the liver, kidneys, spleen, brain, etc. leading to prolonged health defects. Also, prolonged exposure to gold or silver nanoparticles may lead to toxic effects causing oxidative stress, inflammation, and cell damage. The release of AgNPs into the environment through medical wastes, and sewage disrupts the ecosystem by affecting the non-target organisms, beneficial bacteria, plants and aquatic life. The recycling and safe disposal of gold nanoparticles is essential since gold persists in the environment for the long term. Another challenge associated with of using nanoparticle-antibiotic conjugate is developing bacterial resistance against antibiotics. Repeated exposure of bacteria to the antibiotic-nanoparticle conjugate can lead to the emergence of bacterial resistance. Targeting the selective pathogenic bacteria in the microbiome is another major challenge for antibiotic-nanoparticle conjugates. Affecting non-target bacteria could lead to secondary infections and other complications disrupting the natural balance.

The controlled release of antibiotic molecules from the antibiotic-nanoparticle conjugates remains the major technical barrier at the target site of infection. The synergetic release of both nanoparticles and antibiotics is very important for the maximum therapeutic effect with minimum toxicity and side effects. The production of gold or silver-based nanoparticles at a large scale is a major challenge in maintaining efficacy. In addition, silver and gold materials are costly to use for the production of nanoparticles on a large scale. Another major concern of using gold or silver nanoparticles is their stability for long-term storage. During storage, they may get clustered or destroyed which could affect the efficiency of drug delivery. Maintaining stability by retaining the antimicrobial properties is a critical challenge during prolonged storage of nanoparticles for therapeutic applications. The interaction of gold or silver nanoparticles with other therapeutic formulations is another great concern for use along with the existing therapies. So, treating bacterial infections along with the different disorders is quite challenging with using gold or silver nanoparticles. Hence, the *in vitro* and *in vivo* interaction study of various drugs with these nanoparticles must be carried out for future human use.

These challenges can be overcome by studying the properties and characterization of nanoparticles on the various body tissues or cells, interacting with various clinically approved drugs preventing the cytotoxic effects on the human tissues.

REFERENCES

Abdel-Wahab, F., El Menofy, N., El-Batal, A., Mosallam, F., & Abdulall, A. (2021). Enhanced antimicrobial activity of the combination of silver nanoparticles and different β Lactam antibiotics against methicillin resistant Staphylococcus aureus isolates. *Azhar International Journal of Pharmaceutical and Medical Sciences*, 1(1), 22–31. DOI: 10.21608/aijpms.2021.53610.1017

Abdelghafour, M. M., Deák, Á., Szabó, D., Dékány, I., Rovó, L., & Janovák, L. (2022). Use of self-assembled colloidal prodrug nanoparticles for controlled drug delivery of anticancer, antifibrotic and antibacterial mitomycin. *International Journal of Molecular Sciences*, 23(12), 6807. DOI: 10.3390/ijms23126807 PMID: 35743251

Abdelghany, T., Al-Rajhi, A. M., Al Abboud, M. A., Alawlaqi, M., Ganash Magdah, A., Helmy, E. A., & Mabrouk, A. S. (2018). Recent advances in green synthesis of silver nanoparticles and their applications: About future directions. A review. *BioNanoScience*, 8(1), 5–16. DOI: 10.1007/s12668-017-0413-3

Abid, N., Khan, A. M., Shujait, S., Chaudhary, K., Ikram, M., Imran, M., Haider, J., Khan, M., Khan, Q., & Maqbool, M. (2022). Synthesis of nanomaterials using various top-down and bottom-up approaches, influencing factors, advantages, and disadvantages: A review. *Advances in Colloid and Interface Science*, 300, 102597. DOI: 10.1016/j.cis.2021.102597 PMID: 34979471

Abushiba, M., El-Sherbiny, G., Moghannem, S., Kalaba, M., El-Shatoury, E., & Saeed, A. (2019). Enhancement of Antibiotics activity by microbially synthesized silver nanoparticles. *African Journal of Biological Sciences*, 15(1), 137–153. DOI: 10.21608/ajbs.2019.64008

Agarwal, N., Nair, M. S., Mazumder, A., & Poluri, K. M. (2018). *Characterization of nanomaterials using nuclear magnetic resonance spectroscopy Characterization of nanomaterials*. Elsevier.

Ahmed, S., & Ikram, S. (2016). Biosynthesis of gold nanoparticles: A green approach. *Journal of Photochemistry and Photobiology. B, Biology*, 161, 141–153. DOI: 10.1016/j.jphotobiol.2016.04.034 PMID: 27236049

Amendola, V., Meneghetti, M., Granozzi, G., Agnoli, S., Polizzi, S., Riello, P., & Colombatti, M. (2011). Top-down synthesis of multifunctional iron oxide nanoparticles for macrophage labelling and manipulation. *Journal of Materials Chemistry*, 21(11), 3803–3813. DOI: 10.1039/c0jm03863f

Andra, S., Balu, S. K., Jeevanandham, J., Muthalagu, M., Vidyavathy, M., Chan, Y. S., & Danquah, M. K. (2019). Phytosynthesized metal oxide nanoparticles for pharmaceutical applications. *Naunyn-Schmiedeberg's Archives of Pharmacology*, 392(7), 755–771. DOI: 10.1007/s00210-019-01666-7 PMID: 31098696

Arachchi, S. S., Palma, S. P., Sanders, C. I., Xu, H., Ghosh Biswas, R., Soong, R., Simpson, A. J., & Casabianca, L. B. (2022). Binding Between Antibiotics and Polystyrene Nanoparticles Examined by NMR. *ACS Environmental Au*, 3(1), 47–55. DOI: 10.1021/acsenvironau.2c00047 PMID: 36691656

Berekaa, M. M. (2015). Nanotechnology in food industry; advances in food processing, packaging and food safety. *International Journal of Current Microbiology and Applied Sciences*, 4(5), 345–357.

Bhatia, D., Mittal, A., & Malik, D. K. (2021). Antimicrobial potential and in vitro cytotoxicity study of polyvinyl pyrollidone-stabilised silver nanoparticles synthesised from Lysinibacillus boronitolerans. *IET Nanobiotechnology / IET*, 15(4), 427–440. DOI: 10.1049/nbt2.12054 PMID: 34694715

Demirezen, D. A., Yıldız, Y. Ş., Yılmaz, Ş., & Yılmaz, D. D. (2019). Green synthesis and characterization of iron oxide nanoparticles using Ficus carica (common fig) dried fruit extract. *Journal of Bioscience and Bioengineering*, 127(2), 241–245. DOI: 10.1016/j.jbiosc.2018.07.024 PMID: 30348486

Devi, H. S., Boda, M. A., Shah, M. A., Parveen, S., & Wani, A. H. (2019). Green synthesis of iron oxide nanoparticles using Platanus orientalis leaf extract for antifungal activity. *Green Processing and Synthesis*, 8(1), 38–45. DOI: 10.1515/gps-2017-0145

Flauzino, J. M., Nalepa, M. A., Chronopoulos, D. D., Šedajová, V., Panáček, D., Jakubec, P., & Panáček, A. (2023). Click and Detect: Versatile Ampicillin Aptasensor Enabled by Click Chemistry on a Graphene–Alkyne Derivative. *Small*, 19(51), 2207216. DOI: 10.1002/smll.202207216 PMID: 36703534

Gakiya-Teruya, M., Palomino-Marcelo, L., Pierce, S., Angeles-Boza, A. M., Krishna, V., & Rodriguez-Reyes, J. C. F. (2020). Enhanced antimicrobial activity of silver nanoparticles conjugated with synthetic peptide by click chemistry. *Journal of Nanoparticle Research*, 22(4), 90. DOI: 10.1007/s11051-020-04799-6

Giráldez-Pérez, R. M., Grueso, E. M., Jiménez-Aguayo, R., Carbonero, A., González-Bravo, M., Kuliszewska, E., & Prado-Gotor, R. (2022). Use of Nanoparticles to Prevent Resistance to Antibiotics—Synthesis and Characterization of Gold Nanosystems Based on Tetracycline. *Pharmaceutics*, 14(9), 1941. DOI: 10.3390/pharmaceutics14091941 PMID: 36145689

Gour, A., & Jain, N. K. (2019). Advances in green synthesis of nanoparticles. *Artificial Cells, Nanomedicine, and Biotechnology*, 47(1), 844–851. DOI: 10.1080/21691401.2019.1577878 PMID: 30879351

Gutiérrez-Mejía, F. A., Vásquez-López, C., Vargas-Coronado, R. F., Villa-de la Torre, F. E., Arana-Argaez, V. E., Rodríguez-Buenfil, I. M., Gamboa-Angulo, M. M., Torres, Y., & Cauich-Rodríguez, J. V. (2024). Electrophoretic deposition of TiO2 nanotubes and antibiotics on polyurethane coated stainless steel for improved antibacterial response and cell viability. *Materials Today. Communications*, 40, 109428. DOI: 10.1016/j.mtcomm.2024.109428

Guzmán, M. G., Dille, J., & Godet, S. (2009). Synthesis of silver nanoparticles by chemical reduction method and their antibacterial activity. *Int J Chem Biomol Eng*, 2(3), 104–111.

Hao, Z., Wang, M., Cheng, L., Si, M., Feng, Z., & Feng, Z. (2024). Synergistic antibacterial mechanism of silver-copper bimetallic nanoparticles. *Frontiers in Bioengineering and Biotechnology*, 11, 1337543. DOI: 10.3389/fbioe.2023.1337543 PMID: 38260749

Hassan, M. M., Ranzoni, A., Phetsang, W., Blaskovich, M. A., & Cooper, M. A. (2017). Surface ligand density of antibiotic-nanoparticle conjugates enhances target avidity and membrane permeabilization of vancomycin-resistant bacteria. *Bioconjugate Chemistry*, 28(2), 353–361. DOI: 10.1021/acs.bioconjchem.6b00494 PMID: 27959504

HB, S. S. (2022). Comparison of efficacy of nanoparticles synthesized from leaves and flowers of Russelia equisitiformis. *Inorganic and Nano-Metal Chemistry*, 52(1), 75–81. DOI: 10.1080/24701556.2020.1862218

Hedberg, J., Skoglund, S., Karlsson, M.-E., Wold, S., Odnevall Wallinder, I., & Hedberg, Y. (2014). Sequential studies of silver released from silver nanoparticles in aqueous media simulating sweat, laundry detergent solutions and surface water. *Environmental Science & Technology*, 48(13), 7314–7322. DOI: 10.1021/es500234y PMID: 24892700

Hussain, S., Joo, J., Kang, J., Kim, B., Braun, G. B., She, Z.-G., & Teesalu, T. (2018). Antibiotic-loaded nanoparticles targeted to the site of infection enhance antibacterial efficacy. *Nature Biomedical Engineering*, 2(2), 95–103. DOI: 10.1038/s41551-017-0187-5 PMID: 29955439

Hwang, I., Hwang, J. H., Choi, H., Kim, K.-J., & Lee, D. G. (2012). Synergistic effects between silver nanoparticles and antibiotics and the mechanisms involved. *Journal of Medical Microbiology*, 61(12), 1719–1726. DOI: 10.1099/jmm.0.047100-0 PMID: 22956753

Ivanković, K., Kern, M., & Rožman, M. (2021). Modelling of the adsorption of pharmaceutically active compounds on carbon-based nanomaterials. *Journal of Hazardous Materials*, 414, 125554. DOI: 10.1016/j.jhazmat.2021.125554 PMID: 33684815

Jagtap, U. B., & Bapat, V. A. (2013). Green synthesis of silver nanoparticles using Artocarpus heterophyllus Lam. seed extract and its antibacterial activity. *Industrial Crops and Products*, 46, 132–137. DOI: 10.1016/j.indcrop.2013.01.019

Jeevanandam, J., Barhoum, A., Chan, Y. S., Dufresne, A., & Danquah, M. K. (2018). Review on nanoparticles and nanostructured materials: History, sources, toxicity and regulations. *Beilstein Journal of Nanotechnology*, 9(1), 1050–1074. DOI: 10.3762/bjnano.9.98 PMID: 29719757

Jiang, Z., Li, L., Huang, H., He, W., & Ming, W. (2022). Progress in laser ablation and biological synthesis processes:"Top-Down" and "Bottom-Up" approaches for the green synthesis of Au/Ag nanoparticles. *International Journal of Molecular Sciences*, 23(23), 14658. DOI: 10.3390/ijms232314658 PMID: 36498986

Kalhapure, R. S., Suleman, N., Mocktar, C., Seedat, N., & Govender, T. (2015). Nanoengineered drug delivery systems for enhancing antibiotic therapy. *Journal of Pharmaceutical Sciences*, 104(3), 872–905. DOI: 10.1002/jps.24298 PMID: 25546108

Kalita, S., Kandimalla, R., Sharma, K. K., Kataki, A. C., Deka, M., & Kotoky, J. (2016). Amoxicillin functionalized gold nanoparticles reverts MRSA resistance. *Materials Science and Engineering C*, 61, 720–727. DOI: 10.1016/j.msec.2015.12.078 PMID: 26838902

Kaur, A., & Kumar, R. (2019). Enhanced bactericidal efficacy of polymer stabilized silver nanoparticles in conjugation with different classes of antibiotics. *RSC Advances*, 9(2), 1095–1105. DOI: 10.1039/C8RA07980C PMID: 35517620

Kaur, A., & Kumar, R. (2021). Untangling the effect of surfactants as an intermediate at gold nanoparticle-antibiotic interface for enhanced bactericidal effect. *ES Food & Agroforestry*, 7(5), 30–40. DOI: 10.30919/esfaf563

Kreuter, J. (2007). Nanoparticles—A historical perspective. *International Journal of Pharmaceutics*, 331(1), 1–10. DOI: 10.1016/j.ijpharm.2006.10.021 PMID: 17110063

Krishna, G., Kumar, S. S., Pranitha, V., Alha, M., & Charaya, S. (2015). Biogenic synthesis of silver nanoparticles and their synergistic effect with antibiotics: A study against Salmonella sp. *International Journal of Pharmacy and Pharmaceutical Sciences*, 7(11), 84–88.

Kumar, H., Bhardwaj, K., Kuča, K., Kalia, A., Nepovimova, E., Verma, R., & Kumar, D. (2020). Flower-based green synthesis of metallic nanoparticles: Applications beyond fragrance. *Nanomaterials (Basel, Switzerland)*, 10(4), 766. DOI: 10.3390/nano10040766 PMID: 32316212

Kumar, V., Yadav, S. C., & Yadav, S. K. (2010). Syzygium cumini leaf and seed extract mediated biosynthesis of silver nanoparticles and their characterization. *Journal of Chemical Technology and Biotechnology*, 85(10), 1301–1309. DOI: 10.1002/jctb.2427

Kurhade, P., Kodape, S., & Choudhury, R. (2021). Overview on green synthesis of metallic nanoparticles. *Chemicke Zvesti*, 75(10), 5187–5222. DOI: 10.1007/s11696-021-01693-w

Labruère, R., Sona, A., & Turos, E. (2019). Anti–methicillin-resistant Staphylococcus aureus nanoantibiotics. *Frontiers in Pharmacology*, 10, 1121. DOI: 10.3389/fphar.2019.01121 PMID: 31636560

Lee, B., Park, J., Ryu, M., Kim, S., Joo, M., Yeom, J.-H., Kim, S., Park, Y., Lee, K., & Bae, J. (2017). Antimicrobial peptide-loaded gold nanoparticle-DNA aptamer conjugates as highly effective antibacterial therapeutics against Vibrio vulnificus. *Scientific Reports*, 7(1), 13572. DOI: 10.1038/s41598-017-14127-z PMID: 29051620

Lim, J., Petaroudi, M., Pearce, G., Yasakçi, V., & Unak, P. (2021). Effects of Fluorodeoxyglucose Conjugated and Unconjugated Iron Oxide Magnetic Nanoparticles on Macrophages: A Pilot Study. *Brazilian Archives of Biology and Technology*, 64, e21200736. DOI: 10.1590/1678-4324-2021200736

Lowry, G. V., Hill, R. J., Harper, S., Rawle, A. F., Hendren, C. O., Klaessig, F., Nobbmann, U., Sayre, P., & Rumble, J. (2016). Guidance to improve the scientific value of zeta-potential measurements in nanoEHS. *Environmental Science. Nano*, 3(5), 953–965. DOI: 10.1039/C6EN00136J

Manimegalai, P., Selvam, K., Loganathan, S., Kirubakaran, D., Shivakumar, M. S., Govindasamy, M., Rajaji, U., & Bahajjaj, A. A. A. (2024). Green synthesis of zinc oxide (ZnO) nanoparticles using aqueous leaf extract of Hardwickia binata: Their characterizations and biological applications. *Biomass Conversion and Biorefinery*, 14(11), 12559–12574. DOI: 10.1007/s13399-023-04279-6

Martin, C., Low, W., Gupta, A., Amin, M., Radecka, I., Britland, S., Raj, P., & Kenward, K. (2015). Strategies for antimicrobial drug delivery to biofilm. *Current Pharmaceutical Design*, 21(1), 43–66. DOI: 10.2174/1381612820666140905123529 PMID: 25189862

Mishra, A., Goswami, R., & Bhatt, N. (2021). *Polymeric nanomaterials as broad-spectrum antimicrobial compounds Advances in Polymeric Nanomaterials for Biomedical Applications*. Elsevier.

Murei, A., Pillay, K., & Samie, A. (2021). Syntheses, characterization, and antibacterial evaluation of P. grandiflora extracts conjugated with gold nanoparticles. *Journal of Nanotechnology*, 2021(1), 8687627. DOI: 10.1155/2021/8687627

Naimi-Shamel, N., Pourali, P., & Dolatabadi, S. (2019). Green synthesis of gold nanoparticles using Fusarium oxysporum and antibacterial activity of its tetracycline conjugant. *Journal de Mycologie Médicale*, 29(1), 7–13. DOI: 10.1016/j.mycmed.2019.01.005 PMID: 30709721

Nazli, A., He, D. L., Liao, D., Khan, M. Z. I., Huang, C., & He, Y. (2022). Strategies and progresses for enhancing targeted antibiotic delivery. *Advanced Drug Delivery Reviews*, 189, 114502. DOI: 10.1016/j.addr.2022.114502 PMID: 35998828

Oh, S.-H., Jeong, S., Kim, I. S., Shon, H. K., & Jang, A. (2019). Removal behaviors and fouling mechanisms of charged antibiotics and nanoparticles on forward osmosis membrane. *Journal of Environmental Management*, 247, 385–393. DOI: 10.1016/j.jenvman.2019.06.070 PMID: 31254754

Ontiveros-Robles, J. A., Villanueva-Flores, F., Juarez-Moreno, K., Simakov, A., & Vazquez-Duhalt, R. (2023). Antibody-Functionalized Copper Oxide Nanoparticles with Targeted Antibacterial Activity. *ChemistryOpen*, 12(5), e202200241. DOI: 10.1002/open.202200241 PMID: 37226371

Panigrahi, S., Kundu, S., Ghosh, S., Nath, S., & Pal, T. (2004). General method of synthesis for metal nanoparticles. *Journal of Nanoparticle Research*, 6(4), 411–414. DOI: 10.1007/s11051-004-6575-2

Prada, Y., Guzmán, F., Ortíz, C., Cabanzo, R., Torres, R., & Mejía-Ospino, E. (2019). New synthetic peptides conjugated to gold nanoclusters: Antibiotic activity against Escherichia coli O157: H7 and methicillin-resistant Staphylococcus aureus (MRSA). *The Protein Journal*, 38(5), 506–514. DOI: 10.1007/s10930-019-09840-9 PMID: 31119600

Rajchakit, U., & Sarojini, V. (2017). Recent developments in antimicrobial-peptide-conjugated gold nanoparticles. *Bioconjugate Chemistry*, 28(11), 2673–2686. DOI: 10.1021/acs.bioconjchem.7b00368 PMID: 28892365

Ran, F., Li, C., Hao, Z., Zhang, X., Dai, L., Si, C., Shen, Z., Qiu, Z., & Wang, J. (2022). Combined bactericidal process of lignin and silver in a hybrid nanoparticle on E. coli. *Advanced Composites and Hybrid Materials*, 5(3), 1841–1851. DOI: 10.1007/s42114-022-00460-z PMID: 35434523

Revathy, R., Sajini, T., Augustine, C., & Joseph, N. (2023). Iron-based magnetic nanomaterials: Sustainable approaches of synthesis and applications. *Results in Engineering*, 18, 101114. DOI: 10.1016/j.rineng.2023.101114

Roshmi, T., Soumya, K., Jyothis, M., & Radhakrishnan, E. (2015). Effect of biofabricated gold nanoparticle-based antibiotic conjugates on minimum inhibitory concentration of bacterial isolates of clinical origin. *Gold Bulletin*, 48(1-2), 63–71. DOI: 10.1007/s13404-015-0162-4

Saadh, M. J. (2022). Silver nanoparticle-conjugated antibiotics inhibit in vitro growth of Brucella melitensis. *Veterinary World*, 15(7), 1749–1752. DOI: 10.14202/vetworld.2022.1749-1752 PMID: 36185523

Samadi, N., Hosseini, S., Fazeli, A., & Fazeli, M. (2010). Synthesis and antimicrobial effects of silver nanoparticles produced by chemical reduction method. *Daru : Journal of Faculty of Pharmacy, Tehran University of Medical Sciences*, 18(3), 168. PMID: 22615613

Sen, I. K., Mandal, A. K., Chakraborti, S., Dey, B., Chakraborty, R., & Islam, S. S. (2013). Green synthesis of silver nanoparticles using glucan from mushroom and study of antibacterial activity. *International Journal of Biological Macromolecules*, 62, 439–449. DOI: 10.1016/j.ijbiomac.2013.09.019 PMID: 24076192

Shah, M., Fawcett, D., Sharma, S., Tripathy, S. K., & Poinern, G. E. J. (2015). Green synthesis of metallic nanoparticles via biological entities. *Materials (Basel)*, 8(11), 7278–7308. DOI: 10.3390/ma8115377 PMID: 28793638

Shaikh, S., Rizvi, S. M. D., Shakil, S., Hussain, T., Alshammari, T. M., Ahmad, W., Tabrez, S., Al-Qahtani, M. H., & Abuzenadah, A. M. (2017). Synthesis and Characterization of Cefotaxime Conjugated Gold Nanoparticles and Their Use to Target Drug-Resistant CTX-M-Producing Bacterial Pathogens. *Journal of Cellular Biochemistry*, 118(9), 2802–2808. DOI: 10.1002/jcb.25929 PMID: 28181300

Sharma, V., Sharma, J., Kansay, V., Dutta, A., Raj, M., Singh, M., & Kumar, S. (2024). Green synthesis, characterization and drug-loaded iron oxide nanoparticles derived from Nerium oleander flower extract as a nanocarrier for in vitro antibacterial efficacy. *Nano Express*, 5(1), 015014. DOI: 10.1088/2632-959X/ad2997

Singh, P., Pandit, S., Mokkapati, V., Garg, A., Ravikumar, V., & Mijakovic, I. (2018). Gold nanoparticles in diagnostics and therapeutics for human cancer. *International Journal of Molecular Sciences*, 19(7), 1979. DOI: 10.3390/ijms19071979 PMID: 29986450

Skłodowski, K., Chmielewska-Deptuła, S. J., Piktel, E., Wolak, P., Wollny, T., & Bucki, R. (2023). Metallic nanosystems in the development of antimicrobial strategies with high antimicrobial activity and high biocompatibility. *International Journal of Molecular Sciences*, 24(3), 2104. DOI: 10.3390/ijms24032104 PMID: 36768426

Sparks, R. P., Jenkins, J. L., & Fratti, R. (2019). Use of surface plasmon resonance (SPR) to determine binding affinities and kinetic parameters between components important in fusion machinery. *SNAREs. Methods and Protocols*, •••, 199–210.

Srikar, S. K., Giri, D. D., Pal, D. B., Mishra, P. K., & Upadhyay, S. N. (2016). Green synthesis of silver nanoparticles: A review. *Green and Sustainable Chemistry*, 6(01), 34–56. DOI: 10.4236/gsc.2016.61004

Tao, C. (2018). Antimicrobial activity and toxicity of gold nanoparticles: Research progress, challenges and prospects. *Letters in Applied Microbiology*, 67(6), 537–543. DOI: 10.1111/lam.13082 PMID: 30269338

Thomas, R., Jishma, P., Snigdha, S., Soumya, K., Mathew, J., & Radhakrishnan, E. (2020). Enhanced antimicrobial efficacy of biosynthesized silver nanoparticle based antibiotic conjugates. *Inorganic Chemistry Communications*, 117, 107978. DOI: 10.1016/j.inoche.2020.107978

Tyagi, P. K., Gola, D., Tyagi, S., Mishra, A. K., Kumar, A., Chauhan, N., Ahuja, A., & Sirohi, S. (2020). Synthesis of zinc oxide nanoparticles and its conjugation with antibiotic: Antibacterial and morphological characterization. *Environmental Nanotechnology, Monitoring & Management*, 14, 100391. DOI: 10.1016/j.enmm.2020.100391

Vassallo, A., Silletti, M. F., Faraone, I., & Milella, L. (2020). Nanoparticulate antibiotic systems as antibacterial agents and antibiotic delivery platforms to fight infections. *Journal of Nanomaterials*, 2020(1), 6905631. DOI: 10.1155/2020/6905631

Walemba, E. M. (2020). Synthesis and Characterization of Hemocompatible Antibiotic Nanoparticles.

Wang, S., Gao, Y., Jin, Q., & Ji, J. (2020). Emerging antibacterial nanomedicine for enhanced antibiotic therapy. *Biomaterials Science*, 8(24), 6825–6839. DOI: 10.1039/D0BM00974A PMID: 32996490

Wang, Z., Liu, X., Duan, Y., & Huang, Y. (2022). Infection microenvironment-related antibacterial nanotherapeutic strategies. *Biomaterials*, 280, 121249. DOI: 10.1016/j.biomaterials.2021.121249 PMID: 34801252

Woźniak-Budych, M. J., Przysiecka, Ł., Langer, K., Peplińska, B., Jarek, M., Wiesner, M., Nowaczyk, G., & Jurga, S. (2017). Green synthesis of rifampicin-loaded copper nanoparticles with enhanced antimicrobial activity. *Journal of Materials Science. Materials in Medicine*, 28(3), 1–16. DOI: 10.1007/s10856-017-5857-z PMID: 28150115

Yadi, M., Mostafavi, E., Saleh, B., Davaran, S., Aliyeva, I., Khalilov, R., . . . Panahi, Y. (2018). Current developments in green synthesis of metallic nanoparticles using plant extracts: a review. *Artificial cells, nanomedicine, and biotechnology, 46*(sup3), 336-343.

Zazo, H., Colino, C. I., & Lanao, J. M. (2016). Current applications of nanoparticles in infectious diseases. *Journal of Controlled Release*, 224, 86–102. DOI: 10.1016/j.jconrel.2016.01.008 PMID: 26772877

Zhu, X., & Loh, X. J. (2015). Layer-by-layer assemblies for antibacterial applications. *Biomaterials Science*, 3(12), 1505–1518. DOI: 10.1039/C5BM00307E PMID: 26415703

Zhuang, Z., Li, Z., Gong, G., Li, Q., Zhang, Y., Huang, C., & Guo, Z. (2023). Two-dimensional superlattice films of gold nanoparticle-polystyrene composites: A bioactive platform for bone homeostasis maintenance. *Advanced Composites and Hybrid Materials*, 6(5), 166. DOI: 10.1007/s42114-023-00738-w

Zielińska, A., Skwarek, E., Zaleska, A., Gazda, M., & Hupka, J. (2009). Preparation of silver nanoparticles with controlled particle size. *Procedia Chemistry*, 1(2), 1560–1566. DOI: 10.1016/j.proche.2009.11.004

KEY TERMS AND DEFINITIONS

Nanoparticle: A microscopic particle having dimensions in nanometers (nm)

Antibiotic: A chemical or peptide agent which acts against the bacterial population by disrupting the structure or suppressing the growth.

Biocompatible: The condition of being compatible with the living tissues without any cytotoxic effects.

Chapter 11
Development of Advanced Instruments for Spectroscopic Analyses of Nanoparticles/Nanobioparticles

Akshay Kumar
https://orcid.org/0009-0005-0368-6444
Panjab University, Chandigarh, India

Alka Bali
Panjab University, Chandigarh, India

ABSTRACT

Over the past decade, there has been a growing interest in particles with sizes between 1 and 100 nm because of their many uses in fields including drug delivery, cancer diagnosis, and therapy. There are many different chemical compositions of nanobioparticles, ranging from micelles to metal oxide, from synthetic polymers to large biomolecules. Characterizing nanoparticles is crucial after they are synthesized. Size, shape, dispersity, and surface area are the main characteristics of nanoparticles. Numerous methods have been employed to assess the dimensions, crystallographic arrangement, elemental composition, and several other physical attributes of nanoparticles. Multiple techniques can be used to evaluate physical attributes in various cases. Each technique has different strengths and limitations, which complicate the choice of the most suitable method. The common techniques for the characterization of nanoparticles are as follows: XRD, UV-Vis, FTIR, DLS, SEM, TEM, HRTEM, & TGA. This chapter aims to review the various types of techniques for the characterization of nanoparticles.

INTRODUCTION

The study of nanoparticles and nanobioparticles has transformed the progressive developments in science and technology through the application of nanotechnology. These are tiny structures, which are normally sized between 1 and 100 nanometres (Vijayaram et al., 2024). Nanobioparticles specifically use biological components, including proteins, lipids, and nucleic acids, while nanoparticles could be

DOI: 10.4018/979-8-3693-6240-2.ch011

made of wide-ranging materials, comprising of polymers, carbon-based compounds, and metals like gold and silver (Dreaden et al., 20212). There are green-chemistry based methodologies for the synthesis of metal nanoparticles employing plant extracts or plant-derived compounds as reducing agents. Various approaches for synthesizing nanoparticles are listed in Figure 1.

Figure 1. Various approaches for synthesizing of Nanoparticles

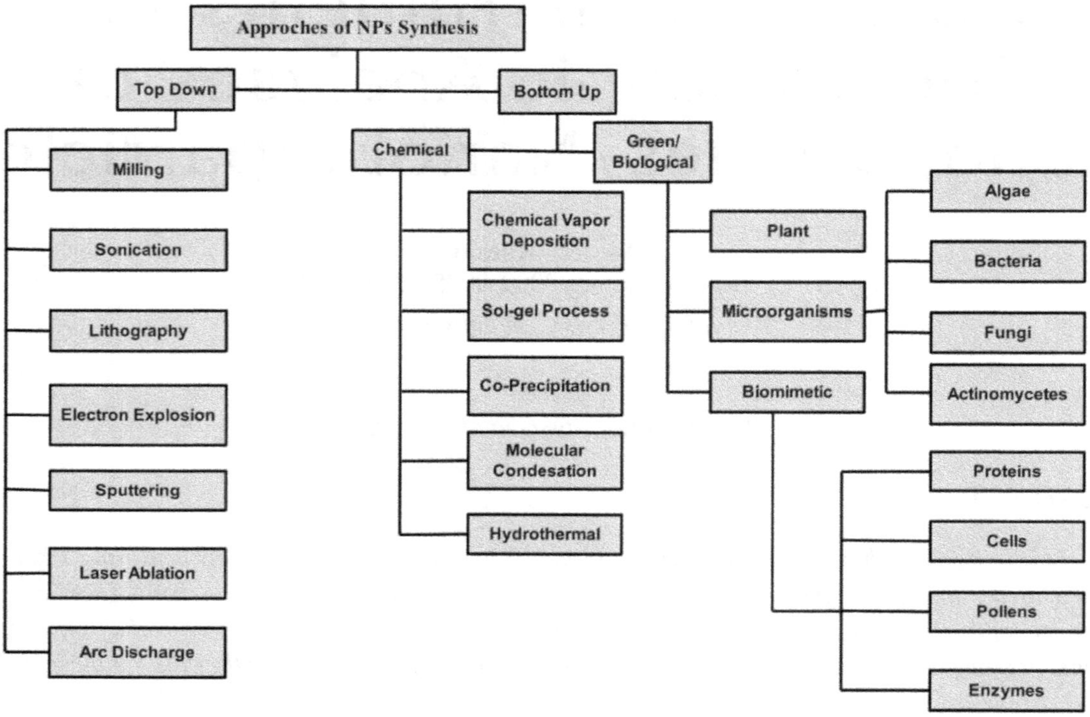

These methods have gained significant attention on account of being eco-friendly, besides their cost-effectiveness, and potential biomedical applications (Mittal et al., 2013). Nanoparticles and nanobioparticles offer several distinct advantages in the pharmaceutical field like enhanced drug delivery, controlled Release, improved pharmacokinetics, biocompatibility, targeting, and specificity (Afzal et al., 2022). Recent advances in nanotechnology have increased its applicability and produced creative solutions in a variety of fields, including electronics, environmental clean-up, medicine, and more. Nanoparticles have transformed drug delivery and diagnostics. In the case of targeted drug delivery, drugs can be delivered to diseased tissues specifically using nanoparticles, which reduces drug side effects. For example, to precisely target tumors, paclitaxel and other chemotherapeutic medicines are encapsulated in lipid-based nanoparticles (Waheed et al., 2024). In diagnostic Imaging iron oxide nanoparticles and quantum dots are used as contrast agents in imaging techniques like fluorescence imaging and MRI, which improve sensitivity and resolution (Avasthi et al., 2020). Conclusively, nanoparticles are an essential component of modern materials science and technology, providing extraordinary potential in the fields of medicine, energy, electronics, and environmental sustainability. Their varied synthesis methods and distinctive size-dependent features continue to stimulate innovation and discovery across a wide range of fields, with the potential to significantly affect global challenges and social demands (Jokerst & Gambhir, 2011).

The characterization of nanoparticles is essential for understanding their physicochemical, and structural characteristics, as the latter are critical for optimizing their pharmaceutical and other applications. Nanoparticles possess very small size and hence, a high ratio of surface area with respect to volume, and this makes them different from their bulkier counterparts with respect to their special features. The first step in characterization of nanoparticles is to ascertain their size and shape. This is usually done with the use of electron microscopy-based methods, most common being Scanning Electron Microscopy (or SEM) and Transmission Electron Microscopy (or TEM) and. While SEM affords information on surface morphology and aggregation states, TEM provides high-resolution pictures that show the size distribution and morphological characteristics of individual nanoparticles. Information about the elemental composition and surface chemistry of nanoparticles involves an investigation of their chemical composition. Techniques like energy dispersive X-ray spectroscopy (EDS) along with X-ray photoelectron spectroscopy (XPS) and are frequently employed. While XPS determines the surface composition and oxidation states of the nanoparticles to help understand their stability and reactivity, EDS offers quantitative elemental analysis. The size of the crystallites, phase purity, and crystalline structure of nanoparticles are all determined by structural analysis utilizing X-ray diffraction (XRD). Applications in catalysis, where nanoparticle crystallinity affects catalytic activity and selectivity, require this knowledge. Furthermore, methods such as Brunauer-Emmett-Teller (BET) and Dynamic Light Scattering (DLS) analysis provide information about the dispersion and accessibility of active sites of nanoparticles by characterizing their size distribution in solution and surface area, respectively (Altammar, 2023). Various techniques for nanoparticles characterization are presented in Figure 2.

Figure 2. Various Techniques of characterization of nanoparticles

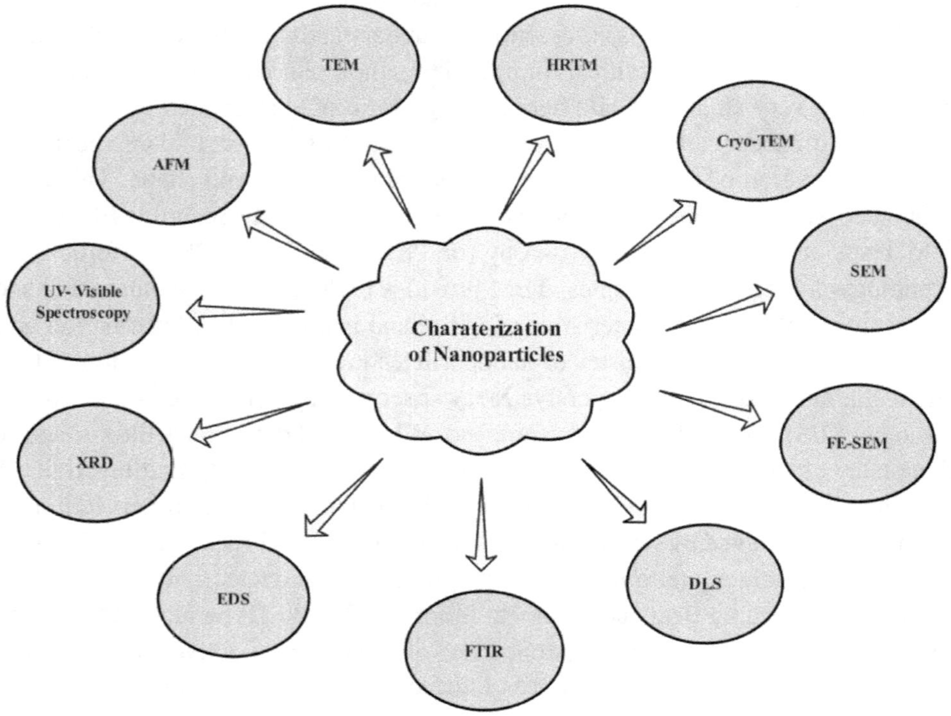

CHARACTERIZATION OF NANOPARTICLES

Characterization of nanoparticles is vital aspect of nanotechnology, essential for understanding at nanoscale level, their physicochemical, and biological characteristics. A comprehensive examination of the nanoparticles is performed relating to their crystal structure, size, shape, chemical composition, etc., the first phase after nanoparticle synthesis. In addition, we can evaluate the surface chemistry to an extent and determine the size distribution, surface charge and area, and the level of aggregation (Mourdikoudis et al., 2018). Various techniques for the characterization of nanoparticles are enlisted in Tables 1 and 2.

Table 1. Summary of the characterization techniques that are used for nanoparticle characterization

Techniques	Main information derived
XRD	Crystal structure, composition, crystalline grain size
DLS	Hydrodynamic size, revealing the presence of aggregates
TEM	Size of nanobioparticles, mono-dispersity in size, particle shape, state of aggregation state, detection and localization/ quantification of nanoparticles in different matrices, growth kinetic studies
Cryo-TEM	Studies involving intricate growth mechanism, aggregation pathways, particularly suited to molecular biology and colloidal chemistry
HRTEM	All information provided by usual TEM analysis, but additionally including crystal structure analysis of distinct nanoparticles. Distinguish mono- and poly-crystalline, and amorphous nanobioparticles.
SEM-HRSEM, T-SEM-EDX	Morphological studies, dispersion of nanobioparticles in cellular matrix and other supports or matrix components, precise study of lateral measurements of nanobioparticles, rapid examination, element composition
Liquid TEM	Depict real time development of nanoparticles, investigation of growth mechanisms, single particulate motion.
HRTEM	All information obtainable from TEM but additionally including crystal structure analysis of distinct nanoparticles. Distinguishing between mono- and polycrystalline from amorphous nanoparticles.
DLS	Hydrodynamic size elucidation, recognition of aggregates
AFM	Nanoparticle size/shape in three-dimensional mode, evaluation of surface covering with nanoparticle morphology
TGA	Composition of stabilizers and their mass

Table 2. Parameters needed to be determined and the corresponding characterization techniques.

Characteristics	Ranges/Parameters	Characterization methods
Particle Size	1 to 100 nm	TEM, DLS, HRTEM, SEM, AFM, DCS
Shape	Various shapes: spherical, rod-like, cuboidal	TEM, HRTEM, EPLS, AFM, FMR, 3D-tomography
Elemental analysis and chemical composition	Detects elements with atomic numbers from 5 (Boron) to 92 (Uranium)	XRD, XPS, SEM-EDX, MFM, LEIS, NMR, ICP-MS, ICP-OES
Size distribution	1 to 100 nm	DLS, ICP-MS
Surface charge	±30 mV	Zeta potential
Concentration	10^8-10^{12} particles per milliliter	UV-Vis, DCS, ICP-MS
Agglomeration state	Larger sizes greater than 1 to 100 nm may indicates agglomeration	Zeta potential, DLS
Structural defects	Point defects, line defects	HRTEM
Detection of NPs	TEM & Cryo-TEM: 1 nm to several hundred nanometers, SEM: 1 nm to several hundred micrometers	TEM, SEM, STEM, Cryo-TEM
Optical properties	UV-Vis (200–800 nm), NIR 12,500 to 4,000 cm^{-1}	UV-Vis, NIR
Density	Vary significantly depending upon composition, size, shape, & surface characteristics	DCS

X-RAY DIFFRACTION

This is a non-destructive technique baesd on the principle of X-ray scattering by the lattice crystal structure of materials, providing valuable understanding of their crystalline structure, phase composition, and crystallite size. This is the most common technique for the characterization of nanobioparticles. Typically, XRD usually provides information relating to the crystalline particle size, lattice parameters, phase nature, and nature of the crystals. It can be used to provide additional details in addition to identification. If the test material is in the form of a mixture, its elemental amounts can be determined from its XRD patterns. Data analysis can further be utilized for determination of degree of crystallinity, the deviation of a specific element from its ideal composition, and its structure. The X-ray beam is partially transmitted on account of the interaction between the atomic planes of the sample; whereas, the remaining portion is absorbed, refracted, scattered, and diffracted. Each element diffracts X-rays differently, depending on the type of atoms and their arrangement (Ingham, 2015). Bragg's law provides a simple rudimentary model to understand the settings required for diffraction and is explained by Equation 1.

$$\lambda = 2d\sin\theta \tag{1}$$

Here "d" is the spatial distance separating the planes. The X-ray diffraction peaks result from scattering of a monochromatic beam of X-rays at a specific angle 2θ from a band of planes. Peak intensities are dependent upon the atomic arrangements within the planes of the lattice (Titus et al., 2018).

Working Principle

Interaction of the X-rays with a crystalline lattice generates constructive and destructive interference, which results in diffraction patterns, and this interaction is the basis for the operation of XRD. When an X-ray beam of monochromatic wavelengths is incident upon a sample, its atomic lattice functions as a diffraction grating in three dimensions, diffracting the X-ray beam at particular angles. A diffractometer is employed to create a diffraction pattern of the crystalline sample solid. The essential components of the diffractometer comprise an X-ray tube, incident beam optics and a detector. At specific angles, the reflected rays afford a peak in the diffractogram. This diffraction pattern can be easily used for identification of a molecule or a mixture of several molecules. Every molecule has a characteristic set of diffraction pattern of peaks which can be used for its identification. Molecules are usually identified by comparison of the diffractogram obtained with a database of previously reported diffraction data (Gopirajah & Anandharamakrishnan, 2019).

Applications

X-ray diffraction provides detailed information about the crystalline structure and chemical compositions of both natural and synthesized materials. It is a practically useful handy analytical tool affording a powerful non-destructive method for the characterization of crystalline materials. High-resolution pictures of biomolecules and biochemical complexes have been obtained *via* X-ray crystallography of their solid phase samples. In fact, with the help of this technique, one can determine the relative positions of the atoms much more precisely compared to AFM. This technique, in conjunction with FTIR and thermal techniques like differential scanning calorimetry (DSC), can resolve several problems confronted with in foods or pharmaceutical formulations (Sharma et al., 2012).

ENERGY-DISPERSIVE X-RAY SPECTROSCOPY (EDS)

EDS is an analytical tool used elemental component analysis and for carrying out X-ray mapping. All atoms have a characteristic electronic configuration. When the sample is bombarded with high-energy electrons (X-rays or g-rays), electrons are ejected out of their inner electronic shells producing a hole or electron vacancy in their inner shells, and subsequently, the higher electronic energy level is relocated to a lower energy level, thus transferring the electrons from higher energy state to the lower energy state. The transfer of electrons emits energy in the form of X-rays; this is detected by the EDS. Measurement of this emitted characteristic X-ray radiation along with its intensity is utilized for estimation of the elemental composition of the sample (Mishra et al., 2017).

Working Principle

Majority of the peaks in the EDS spectrum are attributed to the X-rays, which are generated by distinctive fluorescence light. This technique conforms to the rudimentary principle relating to the periodic table, according to which, every element possesses a distinct lattice arrangement of atoms that generates a particular sequence of peaks on the spectrum of X-ray radiation. Thus, every sample will emit a characteristic X-ray energy pattern in response to a beam of highly energetic particles, such as protons, electrons, or X-rays. With the use of an energy-dispersive spectrometer, we can determine how much energy is emitted by a specimen. Each electronic shell produces its own K-alpha and K-beta transition lines when electrons move from the high energy L and M shells to the lower energy K shells. These lines produce fluorescent photons, which can be detected by a detector. The fluorescence photon energy is equal to the difference in energy between the two transferring energy orbitals. The emitted fluorescent radiation can be represented by equation 2.

$$E = hc/\lambda \quad (2)$$

Investigation of the fluorescent radiation is carried out based upon the energy of a photon. Every atom possesses discrete energy levels and hence, give out characteristic fluorescent radiation, the radiant intensity depending upon the measure of that element comprising the sample (Hodoroaba, 2019).

Applications

EDS is amongst the key tools employed for determining elemental composition and contamination in several systems and is very useful in modern forensic examinations. EDS analysis could be qualitative as well as quantitative, and it maps the spatial organization of the atomic components. It is a non-destructive testing tool for nanoparticle and nanobioparticle characterization (Gul et al., 2020). This is a valuable technique to identify the physicochemical properties and elemental composition of micro- and nanomaterials in the pharmaceutical field (Scimeca et al., 2018).

UV-VIS'S SPECTROSCOPY (UV-Vis)

UV–Visible absorption spectroscopy is the most extensively employed technique for characterization of the electronic structure and for studying the optical characteristics of nanoparticles. It is a non-invasive method of characterization (Philip, 2008). UV-Vis spectroscopy relies on the measurement of absorption of UV and/or visible light by the sample. Every sample has a characteristic electronic distribution due to which, when light is allowed to pass through it, some wavelengths are characteristically absorbed based on the principle of electronic transitions in the molecule, while others are transmitted or reflected. UV-Vis spectrophotometer is an essential instrument in the evaluation of nanoparticles because it can analyze parameters like size, concentration, and stability quickly and non-destructively. By utilizing the unique optical behaviors of nanoparticles, researchers can gain insights into their structure and functionality, advancing both fundamental understanding and practical applications in different fields (Guo et al., 2020).

Working Principle

The fundamental principle behind UV-Vis spectroscopy is to measure how much ultraviolet and visible light a sample absorbs. The UV-Vis spectroscopy focuses on how electromagnetic radiation interacts with matter. Wavelength (λ) or frequency (v) are the properties that define the nature of electromagnetic radiation. The light falling in the ultraviolet or UV region (wavelength range 200–400 nm) and visible region (wavelength range 400–800 nm) is used in UV-Vis spectroscopy. If the energy of electromagnetic radiation matches the difference between the energy of two electronic states of the molecules in the sample, the radiation gets absorbed by the sample when it passes through it (Verma & Mishra, 2018). The intensity of light getting absorbed by a sample is quantitatively related to the sample concentration and the path length traversed by the light in the sample, according to the Beer-Lambert law:

$$A = \varepsilon b C \tag{3}$$

A = absorbance
ε = molar absorptivity
b = length of light path
C = concentration

An absorption spectrum, or a plot of absorbance (A) versus wavelength (λ) or frequency (v), is produced using UV-Vis spectroscopy. This spectrum shows peaks and troughs which corresponds to wavelengths where the sample absorbs light (Li & Hur, 2017).

Applications

UV-Vis spectroscopy has been widely used for the quantitative estimation of nanoparticle concentration in a solution. It is possible to determine the concentration of nanoparticles in a material by measuring the absorbance at specific wavelengths characteristic of the nanoparticle. The size and shape of nanoparticles, especially metallic nanoparticles like gold and silver, influence their absorption spectrum. It is possible to determine the size and size distribution of nanoparticles through UV-Vis spectroscopy (Martínez et al., 2013). When evaluating the colloidal stability of nanoparticle solutions, UV-Vis spectroscopy is a useful tool (Klinavičius et al., 2024). Aggregation or sedimentation of nanoparticles can be indicated by changes in the absorption spectrum over time. Thus, UV-Vis spectroscopy is a versatile and powerful tool useful for nanoparticle characterization (Becheri et al., 2008).

FOURIER TRANSFORM INFRARED SPECTROSCOPY (FTIR)

FTIR, or Fourier transform infrared spectroscopy, is a powerful analytical method that studies the molecular composition and structure of materials by measuring how much infrared light they absorb (Berthomieu & Hienerwadel, 2009). FTIR works on the concept that molecules absorb particular infrared light frequencies that correlate with the vibrational modes of their chemical interactions. This method is useful for identification of structural differences in different types of molecular binding, which can be utilized to decipher the possibility of existence of any type of molecular interactions. Attenuated total reflectance (ATR–FTIR), Transmittance FTIR, and micro-spectroscopy FTIR are popular methods based on FTIR that have been used for characterization of nanoparticles (Mohamed et al., 2017).

Working Principle

The instrument that is used to generate and study the absorption spectrum of a compound is termed as a spectrophotometer. Fourier transform instruments work on the mathematical principle of Fourier transform and thus offer several advantages over conventional instruments. A Fourier transform IR spectrophotometer provides a much more accurate high resolution IR spectrum in a much shorter time compared to a conventional dispersive IR spectrophotometer. A beam of IR irradiation is emitted from a radiation source and a broad spectrum of infrared light is transmitted through the sample. Characteristic wavelengths of IR radiation get absorbed by the sample based on the molecular structures and the functional groups present, corresponding to the stretching and bending vibrational frequencies in the chemical bonds within the molecules. Once the transmitted light has been measured, it is transformed into a spectrum by a Fourier transform algorithm, which shows the absorbance (or transmittance) as a function of wavelength. The IR spectrum retrieved from the FTIR spectrophotometer lies in the mid-IR region, corresponding to a wavelength range of 2.5–15 μm or a wavenumber range of 4000 and 650 cm^{-1} (Mallakpour & Hussain, 2021).

Applications

FTIR spectroscopy is a flexible tool for both qualitative and quantitative analysis that can divulge important information about the molecular structure, chemical bonds and composition of a wide range of materials. The method, on account of its simplicity and diversity, has been employed for a wide range of applications. Further, due to its non-destructive nature, it is invaluable for environmental monitoring, industry, and nanotechnology, advancing a number of scientific and technological domains (Kiefer et al., 2015). FTIR is an extremely useful technique to characterize nanoparticles and it provides useful information relating to the chemical composition, interactions of the particles, and surface functional groups. Interactions of the nanoparticles with polymer matrices are invariably investigated using FTIR spectroscopy. Any changes in IR spectra reveal molecular mobility, bonding interactions, and the extent of dispersion of nanoparticles inside the polymer matrix (Preetha et al., 2013). FTIR is used to distinguish between varying types of nanoparticles or to confirm the presence or absence of specific functional groups. FTIR spectroscopy also provides structural information about nanoparticles, such as phase composition and crystallinity (León et al., 2017). By using sophisticated instruments and techniques such as attenuated total reflection (ATR), FTIR spectroscopy keeps developing as a sensitive and non-destructive way to characterize nanoparticles. This helps to promote advances in nanotechnology and improves our knowledge of how nanoparticles behave in complex structures (Sun et al., 2020).

DYNAMIC LIGHT SCATTERING (DLS)

Dynamic light scattering (DLS) is used for the study of time-based changes in coherent light that is dispersed by a solution containing diffusing particles. The size and diffusivity of the suspended particles are both revealed by the time-correlation function of these scattered intensities. The significance of nanoparticles in science for biomedical applications has increased in recent times. Low toxicity and pronounced stability of the synthesized nanoparticles make them ideal for use as biosensors, bioimaging systems, and targeted drug delivery systems. Nanoparticles can be investigated using microscopy and spectroscopic techniques due to their vast size range (10 nm to 1 μm) and optical properties (Santos & Domingues, 2018).

Working Principle

Dynamic light scattering (DLS) works on the principle of measurement of coherent light which gets scattered over time by objects that scatter light, including big molecules or small particles. The variation of the measurement signals, which can originate from multiple sources, is assessed (Babick, 2020). The fluctuations that DLS analyzes are brought about by the thermal motion of the scattering items and happen at relatively smaller time scales ranging from microseconds to even nanoseconds). DLS has been most commonly employed for the examination of size of particles falling in the sub-micrometer ranges. This technique uses the Brownian motion of the particles, which leads to a permanent rearrangement in space. This affects the interference signal generated by the individual scattering signals and eventually leads to generation of a fluctuating signal by the detector (Falke & Betzel, 2019).

Applications

Dynamic light scattering (DLS) is a extensively employed technique in many different fields since it may give important information about the size and distribution of particles in liquid suspensions (Murdock et al., 2008). Because this technique can quickly and accurately determine the size distribution of nanoparticles in solution, DLS is especially useful in the characterization of nanoparticles. The hydrodynamic size distribution of the nanoparticles in suspension is also revealed by DLS. It has the ability to differentiate between various nanoparticle populations and identify changes in the size distribution brought on by degradation or aggregation. The size distribution's width is quantified by the polydispersity index (PDI), which is measured by DLS (Stetefeld et al., 2016). Ensuring consistency in nanoparticle size is critical for applications such as delivering drugs and nanocomposite materials. DLS can be used to assess stability by monitoring changes in nanoparticle size distribution over time, providing insights into colloidal stability, and the combination of electrophoretic light scattering (ELS) with DLS can be used for determination of zeta potential of nanoparticles. Zeta potential provides information on the stability and surface charge of suspended nanoparticles. DLS is very commonly used in the characterization of nanoparticles across various other fields (Bhattacharjee, 2016).

SCANNING ELECTRON MICROSCOPY (SEM)

The scanning electron microscope (SEM) is amongst the most versatile instruments employed for analyzing the microstructure morphology or topography of materials at high resolution. In contrast to optical microscopes, which utilize visible light, SEM generates images by focusing an electron beam, providing a significantly higher magnification and depth of field. SEM is used to observe the surface of solid objects directly (Zhou et al., 2007). It has a magnification ranging from 10 times to 3 x 10^6 times, a field depth of 4 to 0.4 mm, thus affording a high resolution of 1 to10 nm. Presently, Scanning electron microscopy (SEM) along with transmission electron microscopy (TEM) are the primary techniques utilized for direct imaging as well as for carrying out dimension-based measurements of microstructural and nano-structural entities because of their high resolution and rapid imaging speed. SEM provides detailed surface features, and three-dimensional structures, texture of materials (Sun et al., 2020).

Working Principle

An electron beam subjected to an accelerated voltage of 2 to 30 kV, is a part of a SEM. To produce a 10 nm image of the sample presented on photographic film or onto a display tube, the electron beam is routed through multiple electromagnetic lenses (Inkson, 2016). When the electron beam strikes the sample surface, some electrons are scattered back while others discharge low-energy electrons or secondary electrons (SE). The sample emits electromagnetic radiation at a variety of wavelengths, but essentially, the radiation with wavelength falling in the in visible light range of 400-800 nm (cathodoluminescence) and X-ray range are the most interesting for use. Backscattered electrons (BSE) and the weakly energetic secondary electrons (SE) emitted by the sample are directed towards a scintillator which emits a pulse of light corresponding to the incoming electrons. A photomultiplier is used to amplify the electric signal generated from the emitted light. The signal is steered to the cathode ray tube grid after undergoing the expansion process. In order to accelerate the low energy electrons to make them suitable for emission of visible light when hitting the scintillator, the latter usually has a positive potential of 5 to 10 kV. Using an image processing program on the computer, information can be retrieved by leveraging the reflection of objects (Hidayanti & Harnovan, 2020).

Applications

Scanning Electron Microscopy (SEM) is a crucial tool for characterizing nanoparticles due to its potential to provide high-resolution images and thorough evaluations of size distribution, particle shape, and surface features(Vida-simiti et al., 2004). SEM visualizes the morphology of nanoparticles with additional detail. Nanoparticles can vary in shape, rod-like, including spherical, and irregular shapes (Modena et al., 2019). SEM images can reveal the details of these shapes and provide insights into the particle size distribution within a sample. SEM is also capable of capturing extremely fine surface characteristics like porosity, aggregation behavior, and roughness. SEM is crucial for the comprehensively characterizing the nanoparticles (Buhr et al., 2009). Apart from this, SEM is used for the characterization across various other fields (El et al., 2012).

FIELD EMISSION SCANNING ELECTRON MICROSCOPY (FESEM)

Field Emission Scanning Electron Microscopy (FESEM) offers high-resolution imaging capabilities with remarkable depth of field and magnification. Compared to traditional scanning electron microscopes or SEMs, the field emission scanning electron microscopes (FESEM) use an electron source that emits a highly focused beam of light, usually a cold field emitter or a tungsten tip. This makes FESEM excellent for in-depth surface examination since it allows it to reach much better spatial resolution, typically ranging from a few nanometers to sub-nanometers (Utsunomiya & Ewing, 2003). There are numerous applications of FESEM in both scientific and industrial fields. It is used in materials research to thoroughly examine the topography, microstructure, and surface morphology of materials in nanotechnology. Field Emission Scanning Electron Microscopy (FESEM) offers so many advantages over Scanning Electron Microscopy (SEM), such as that FESEM typically achieves a higher spatial resolution image, produces sharper and clearer images, operates at lower accelerating voltages, and has advanced analytical features that make it a superior option to traditional SEM for applications seeking thorough surface characterization and analysis (Lewczuk & Szyryńska, 2021).

Working Principle

The working principle of FESEM involves applying a strong electric field to the emitter tip which causes electrons to tunnel past the potential barrier at the surface. This leads to the coherent emission of electrons, which can concentrate into a constricted beam employing electromagnetic lenses. This focused electron beam travels across the sample surface; it generates signals that are collected and used to create elemental analysis and high-resolution pictures. These signals include secondary electrons, back-scattered electrons, as well as characteristic X-rays (Mayeen et al., 2018).

Applications

The high-resolution imaging capabilities of field emission scanning electron microscopy (FESEM) enable extensive examination of nanoparticle size, shape, and distribution, as well as surface morphology, making it an invaluable tool in nanoparticle characterization (Nallusamy & Manoj Babu, 2016). It provides clear images. Additionally, FESEM can elucidate the structural uniformity and integrity of nanoparticles synthesized through different methods. FESEM is a crucial tool for many other scientific fields, helping to develop industry, technology, and research due to its comprehensive imaging, elemental analysis, and surface characterization capabilities (Shivani Rasool et al., 2010).

TRANSMISSION ELECTRON MICROSCOPY (TEM)

With conventional light microscopy, it is difficult to resolve the structures at nano-scale. When analyzing fine features with characteristic dimensions below 100 nm, the technique that provides an effective alternative is transmission electron microscopy (TEM). TEM uses an electron beam to probe samples, allowing greater resolution and in-depth analysis of materials at the atomic level. This technique, invented in 1931 (Tang & Yang, 2017), first demonstrated that electron could be accelerated by electromagnetic fields to obtain resolutions substantially higher than those of light-based microscopy. TEM is a highly versatile and efficient technique with several applications in diverse scientific fields. TEM can provide details about morphology, size distribution, and size of the nanoparticles. The only technique for measuring particle size that allows for direct observation and measurement of individual particles is microscopy (Bonnamy & Oberlin, 2016).

Working Principle

The concept behind transmission electron microscopy (TEM) is to examine extremely tiny structures at a resolution much greater than that of optical microscopes by employing an electron beam. A tungsten filament heated to a high temperature of above 2000 ^{0}C may be used or better, a field emission gun (FEG) may be used as typical electron source in TEM that produce an electron beam. Electric fields are used to accelerate these electrons to extremely high speeds. The accelerated electron beam is made to pass through several successive magnetic lenses. These lenses focus and direct the electrons as a focused electron beam to be passed through the sample. The sample is placed on a grid made of materials that electrons can easily pass through, like copper or carbon. As the beam of electrons passes through the sample, there is interaction of the electrons with the nanomaterial atoms. Some electrons get transmitted, while others are scattered or absorbed by the sample. The transmitted and scattered electrons collected at the detector are further processed into an image signal (Buseck, 2018).

Applications

One essential tool for understanding the atomic and molecular structures of various materials is transmission electron microscopy (TEM), which is an essential component in modern scientific study. Transmission electron microscopy has diverse applications across various scientific and industrial fields. Nanoparticle and nanomaterial studies require the use of TEM. It makes it possible to monitor particle size, shape, distribution, and surface morphology, precisely, a critical capability for applications in drug delivery systems, sensors, catalysis, and nanoelectronics (Akbari et al., 2011). It has useful applications in distinguishing between different nanoparticle morphologies, such as rods, wires, spheres, etc., as well as more complex structures. It is also useful in determination of the crystalline structure of nanoparticles. TEM provides a helpful tool for discernment of the internal structure and/or defects in the nanoparticles. Its versatility as well as its ability to offer in-depth analysis and high-resolution imaging make it a crucial tool in a variety of scientific domains (Franken et al., 2017).

HIGH-RESOLUTION TRANSMISSION ELECTRON MICROSCOPY (HRTEM)

High-resolution transmission electron microscopy (HRTEM) is an advanced version of conventional TEM. This technique utilizes a high-energy electron beam which is passed through a thin layer of a sample to achieve excellent spatial resolution, outperforming TEM in a number of crucial areas. While TEM typically achieves resolutions around 0.2 nanometers, HRTEM can characterize nanoparticles in the range of 0.1 or even less, depending on the instrumental specifications and the quality of the sample. HRTEM represents a significant evolution over traditional TEM, offering superior resolution and advanced analytical capabilities (Phung et al., 2003).

Working Principle

HRTEM utilizes a beam of electrons instead of light to image samples. Electrons have much shorter wavelengths as compared to photons used in optical microscopy. The resolution of any imaging system is inversely related to the wavelength of the imaging radiation; this allows HRTEM to reach significantly better resolution. There are variety of interactions of electrons with the atoms comprising the sample as they traverse it, like diffraction and elastic and inelastic scattering. These interactions give essential details about the characteristics, composition, and structure of the substance under investigation. HRTEM can determine the atomic arrangement of a material by utilizing diffraction patterns. The sample crystal lattice diffracts electron waves, which result in interference patterns that can be examined to reconstruct the atoms and defects in spatial distribution (Wang, 1999).

Applications

The characterization of nanoparticles is greatly assisted by High-Resolution Transmission Electron Microscopy (HRTEM), which is indispensable in a number of fields where nanoparticles play a critical role, including catalysis, nanoelectronics, materials science, and biomedical applications. HRTEM can offer broad understanding of the size, shape, structure, composition, and even surface characteristics at the atomic level. Hence, it can be useful for characterization of the nanoparticles (Herrera-Becerra et al., 2008). By generating diffraction patterns and lattice pictures, HRTEM can disclose the crystal structure of nanoparticles. Surfaces of nanoparticles are important for their stability, reactivity, and interactions with their surroundings. High-resolution surface examination of nanoparticles by HRTEM reveals facets, steps, flaws, and surface coatings (Kisielowski et al., 1996).

CRYO-TRANSMISSION ELECTRON MICROSCOPY (Cryo-TEM)

Cryo-transmission electron microscopy (cryo-TEM) is a powerful way for visualization of the nanoparticles. Cryo-TEM permits the samples to be imaged in their native, hydrated states at cryogenic temperatures. By using this preservation method, delicate materials like proteins, viruses, liposomes, and polymers are kept free from structural artifacts and retain their original morphology and functionality. Cryo-TEM can visualize nanoparticles in their dispersed state, providing critical insights into shape, particle size distribution, interactions with surrounding environments, and aggregation behavior (Benjin & Ling, 2020).

Working Principle

It operates on the principle of imaging samples at cryogenic temperatures to preserve the sample's natural state and structure without the need for dehydration, fixation, or staining. Samples are prepared as a thin layered solution to be placed on a grid, which is made up of copper or gold. A very small sample solution is applied to the grid to create a thin film. Then the sample-containing grid is plunged into a cryogen containing liquid nitrogen. Electron beams are generated and focused on the sample (Dalmônico et al., 2019). When electrons are allowed to traverse through the sample, their interactions with the sample atoms results in different types of interactions. Subsequently, the transmitted electrons emerging out through the sample, interact with a detector, which collects signals to form an image. The image is then further processed to enhance resolution and contrast, which provides detailed information about the sample structure (Stewart, 2017).

Applications

The use of cryo-transmission electron microscopy, or cryo-TEM, is essential for the characterization of nanoparticles. It allows nanoparticles to be imaged in their native state and provides high-resolution images of nanoparticles with exceptional clarity (Ballauff & Lu, 2007). The structural investigations of nanoparticles, such as core-shell structures, crystallinity, and assembly configurations, become easier using cryo-TEM. This technique is helpful to elucidate the structure-function relationships, dynamics, and interactions with drugs, ligands, and membrane proteins, which are important for drug discovery and understanding cellular processes. New developments in cryo-TEM equipment, such as better detectors and data gathering mechanisms, improve the sensitivity and resolution of imaging (Crawford et al., 2011).

THERMAL GRAVIMETRIC ANALYSIS (TGA)

Thermogravimetric analysis (TGA) is a quantitative thermo-analytical technique. Thermogravimetric analysis, a type of thermal analysis, involves placing a sample in a controlled environment and measuring its change in weight as a function of time or temperature. TGA is frequently used to gain insight into the composition, moisture content, thermal stability, and decomposition kinetics of materials. As the furnace raises the temperature to 1600 °C, it tracks the mass of a sample, sample masses ranging from 1 mg to several grams. For the purpose of characterizing nanoparticles, thermogravimetric analysis (TGA) is frequently employed since it can yield important details like the thermostability and composition of the nanoparticles (Prime et al., 2009).

Working Principle

The basic concept behind thermogravimetric analysis (TGA) is the ability to analyze a sample's mass change under controlled conditions. The primary use of TGA is in the understanding of specific thermal processes, including vaporization, sublimation, breakdown, adsorption, reduction, and vaporization. A small amount of accurately weighed sample is placed in a sample holder or crucible, which is then inserted into the TGA instrument. The temperature of the crucible is then raised steadily at a pre-defined rate. TGA helps determine the thermal decomposition or oxidation of nanoparticles. The kinetic studies of chemical reactions can be performed under varied conditions using TGA, in addition to the prediction of thermal stability for samples. During this process instrument continuously monitors the mass of the sample with high sensitivity. Finally, data collected from TGA is represented in graphical form as a thermogravimetric curve. This curve plots the percentage weight loss of the sample against temperatures (Loganathan et al., 2017).

Applications

Thermogravimetric analysis is powerful technique for characterizing nanoparticles, offering detailed insights into their composition, thermal stability, and surface properties. The thermal stability of various nanoparticle compositions, coatings, and surface changes can be compared using TGA. This is essential for improving the stability of nanoparticles and improving their synthesis techniques for a diverse range of applications (Saadatkhah et al., 2020). To improve stability or functioning, many nanoparticles are coated with organic compounds (such as stabilizers or surfactants). By calculating the weight loss that occurs during heating, TGA can determine the quantity of organic components present and provide valuable information on the structure and effectiveness of coatings (Mansfield et al., 2014). TGA can also detect and quantify the impurities or contaminants in nanoparticle samples. Summarizing, thermogravimetric analysis affords a flexible and effective technique for identifying the thermal behavior, composition, and stability of nanoparticles (Rami et al., 2020).

ATOMIC FORCE MICROSCOPY (AFM)

AFM is a powerful technique of imaging and characterization, which has transformed the study of samples and materials at the nanoscale. AFM functions differently from conventional optical microscopy in that it scans the surface of a sample using a sharp probe for identifying interaction forces between the sample atoms and the probe tip. Because of its capability of providing accurate measurements at nanoscale levels and high-resolution images, AFM is highly valuable in material research, biology, and nanotechnology. AFM is an important instrument for characterization of nanoparticles as it can provide in-depth information relating to their size, shape, interactions, and surface characteristics. It provides a thorough understanding of nanomaterials (Goldsbury et al., 2009).

Working Principle

The working principle of AFM is based on detection of forces between a sharp probe and the sample surface. An extremely malleable cantilever is linked to the probe, also termed as the AFM tip or AFM sensor. Silicon, or silicon nitride, is used to microfabricate the AFM tips and cantilever, which should be in touch with or close to the surface of interest. This technique accurately measures the nanoscale samples. A wide variety of methods are there for detecting any movement of the cantilever (Leite, 2007), but majority of the AFM instruments employ laser-beam detection. Extremely small forces are produced between the sample surface and the probe when the probe traverses the surface, and these aid the AFM system to record the cantilever deflection. The deflection of the cantilever is known as the "stiffness of the cantilever" which may be computed by Hooke's law. The stiffness is visually recorded and can be pictured on a computer (Vahabi et al., 2013).

Applications

Atomic Force Microscopy (AFM) has diverse applications in different fields because it has the ability to provide precise measurements at nanometer scale and high-resolution imaging. Atomic force microscopy (AFM) provides detailed topographical and mechanical information at the nanoscale, making it highly useful in the characterization of nanoparticles. AFM is able to measure nanoparticle size and shape with great resolution and accuracy (Bellotti et al., 2022). AFM is also used to analyze thin films and surface nanoparticle coatings. It is capable of detecting defects or discontinuities in nanoparticle layers, measuring film thickness, and evaluating surface coverage. AFM can be used in conjunction with other imaging methods for correlating studies, like SEM or TEM (Rao et al., 2007).

SELECTED AREA ELECTRON DIFFRACTION (SAED)

Selected Area Electron Diffraction (SAED) is a powerful technique utilized to find out the crystalline structure of materials at the atomic scale. In this technique, the beam of electrons is directed through a small, selected area of the sample. Interaction of electrons with the atoms of the sample leads to diffraction. This diffraction pattern gives information regarding the arrangement of atoms in the sample, orientation, lattice spacing, and crystal structure. This technique is widely employed for the characterization of nanomaterials, thin films, minerals, and other crystalline substances (Tem, 2008).

Working Principle

Selected Area Electron Diffraction (SAED) operates on the principles of wave diffraction, similar to X-ray diffraction, but SAED utilizes the beam of high-energy electrons instead of X-rays. SAED utilizes a high-energy electron beam, which is typically generated by Transmission Electron Microscope. The electrons are accelerated by using a series of electrostatic lenses to form a coherent and focused beam with energies. The sample for SAED has to be so thin that it does not interfere with the electrons passing through the crystalline structure. Interaction of the electron beam with the crystalline sample produces a diffraction pattern. This pattern consists of a series of bright and dark spots (diffraction spots) that were recorded on a digital detector. The diffraction spots are then analyzed for determination of the crystal structure, defects, and to study orientation relationships within the sample (Smith et al., 2003).

Applications

SAED is widely used in nanotechnology, biology, and geology for various applications, including crystallography, phase identification, strain analysis, and defect characterization. SAED is utilized in the analysis of nanomaterials including nanoparticles and quantum dots to understand their structural properties. For nanoparticle characterization, SAED is a valuable technique because it determines the crystalline structure and phase composition of nanoparticles. Further, it also identifies the presence of different crystallographic phases, polymorphs, or amorphous regions in the nanoparticle. To sum up, SAED is a versatile and efficient technique for characterizing nanoparticles (Lanje et al., 2010).

CONCLUSION

Nanoparticles attracted all areas of research on account of their unique chemical and physical properties. Nanoparticles possess an extremely small size and a high surface area which increases their reactivity. The characterization of nanoparticles is a crucial aspect in assessing their safety and associated potential risks. The characterization of nanoparticles continues to evolve with advancements in analytical techniques and instrumentation. A variety of sophisticated techniques, such as spectroscopy, electron microscopy, surface analysis, and light scattering, help to gain deep insights into nanoparticle's size, surface properties, shape, composition, and behavior at the nanoscale. These techniques thoroughly examine the physical, chemical, and functional properties of nanoparticles, and this is obligatory for improving their synthesis, ensuring their safety, and maximizing their efficiency. Techniques such as electron microscopy (EM) provide crucial insights into the shape, size, and morphology of nanoparticles at high resolution. Techniques like Scanning electron microscopy (SEM) and transmission electron microscopy (TEM) allow the visualization of nanoparticles and the evaluation of their structural integrity and distribution. Spectroscopic methods, including Fourier-transform infrared spectroscopy (FTIR) and Raman spectroscopy, afford crucial details relating to the chemical composition, molecular structure, and surface chemistry of nanoparticles. Dynamic light scattering (DLS) provides information about nanoparticle aggregation behavior, size distribution, and stability in solution. Structural characterization techniques including X-ray diffraction (XRD) provide valuable information concerning the crystal size and crystalline structure of nanoparticles. Therefore, there are various techniques to characterize nanoparticles that provide a detailed understanding of their characteristics and features, enabling the development of innovative applications in healthcare and other fields. In the future advancement of characterization techniques will be essential for realizing the full promise of nanotechnology in resolving global issues and improving quality of life.

REFERENCES

Afzal, O., Altamimi, A. S. A., Nadeem, M. S., Alzarea, S. I., Almalki, W. H., Tariq, A., Mubeen, B., Murtaza, B. N., Iftikhar, S., Riaz, N., & Kazmi, I. (2022). Nanoparticles in drug delivery: From history to therapeutic applications. *Nanomaterials (Basel, Switzerland)*, 12(24), 1–27. DOI: 10.3390/nano12244494 PMID: 36558344

Akbari, B., Tavandashti, M. P., & Zandrahimi, M. (2011). Particle size characterization of nanoparticles-a practicalapproach. *Iranian Journal of Materials Science and Engineering*, 8(2), 48–56.

Altammar, K. A. (2023). A review on nanoparticles: Characteristics, synthesis, applications, and challenges. *Frontiers in Microbiology*, 14(04), 1–20. DOI: 10.3389/fmicb.2023.1155622 PMID: 37180257

Avasthi, A., Caro, C., Pozo-Torres, E., Leal, M. P., & García-Martín, M. L. (2020). Magnetic nanoparticles as MRI contrast agents. *Topics in Current Chemistry*, 378(3), 1–40. DOI: 10.1007/s41061-020-00302-w PMID: 32382832

Babick, F. (2020). Dynamic light scattering (DLS). V. Hodoroaba (Ed.), *Characterization of Nanoparticles* (pp. 137-172). Elsevier Inc. DOI: 10.1016/B978-0-12-814182-3.00010-9

Ballauff, M., & Lu, Y. (2007). "Smart" nanoparticles: Preparation, characterization and applications. *Polymer*, 48(7), 1815–1823. DOI: 10.1016/j.polymer.2007.02.004

Becheri, A., Dürr, M., Lo Nostro, P., & Baglioni, P. (2008). Synthesis and characterization of zinc oxide nanoparticles: Application to textiles as UV-absorbers. *Journal of Nanoparticle Research*, 10(4), 679–689. DOI: 10.1007/s11051-007-9318-3

Bellotti, R., Picotto, G. B., & Ribotta, L. (2022). AFM measurements and tip characterization of nanoparticles with different shapes. *Nanomanufacturing and Metrology*, 5(2), 127–138. DOI: 10.1007/s41871-022-00125-x

Benjin, X., & Ling, L. (2020). Developments, applications, and prospects of cryo-electron microscopy. *Protein Science*, 29(4), 872–882. DOI: 10.1002/pro.3805 PMID: 31854478

Berthomieu, C., & Hienerwadel, R. (2009). Fourier transform infrared (FTIR) spectroscopy. *Photosynthesis Research*, 101(2–3), 157–170. DOI: 10.1007/s11120-009-9439-x PMID: 19513810

Bhattacharjee, S. (2016). DLS and zeta potential - What they are and what they are not. *Journal of Controlled Release*, 235, 337–351. DOI: 10.1016/j.jconrel.2016.06.017 PMID: 27297779

Bonnamy, S., & Oberlin, A. (2016). Materials Science and Engineering of Carbon. M. Inagaki (Ed.), *Transmission Electron Microscopy*. (pp. 45–70). Elsevier. DOI: 10.1016/B978-0-12-805256-3.00004-0

Buhr, E., Senftleben, N., Klein, T., Bergmann, D., Gnieser, D., Frase, C. G., & Bosse, H. (2009). Characterization of nanoparticles by scanning electron microscopy in transmission mode. *Measurement Science & Technology*, 20(8), 1–8. DOI: 10.1088/0957-0233/20/8/084025

Buseck, P. R. (2018). Principles of transmission electron microscopy. P. Buseck (Ed.), *Minerals and Reactions at the Atomic Scale* (pp. 1–36). De Gruyter. DOI: 10.1515/9781501509735-005

Crawford, R., Dogdas, B., Keough, E., Haas, R. M., Wepukhulu, W., Krotzer, S., Burke, P. A., Sepp-Lorenzino, L., Bagchi, A., & Howell, B. J. (2011). Analysis of lipid nanoparticles by cryo-EM for characterizing siRNA delivery vehicles. *International Journal of Pharmaceutics*, 403(1–2), 237–244. DOI: 10.1016/j.ijpharm.2010.10.025 PMID: 20974237

Dalmônico, G. M. L., López, E. O., Longuinho, M. M., Checca, N. R., Farina, M., Ersen, O., Rossi, A. M., & Rossi, A. L. (2019). Insight by Cryo-TEM into the growth and crystallization processes of calcium phosphate nanoparticles in aqueous medium. *Materials Chemistry and Physics*, 237, 1–7. DOI: 10.1016/j.matchemphys.2019.121862

Dreaden, E. C., Alkilany, A. M., Huang, X., Murphy, C. J., & El-Sayed, M. A. (2012). The golden age: Gold nanoparticles for biomedicine. *Chemical Society Reviews*, 41(7), 2740–2779. DOI: 10.1039/C1CS15237H PMID: 22109657

El, S., Koraichi, S., Latrache, H., & Hamadi, F. (2012). Scanning electron microscopy (SEM) and environmental SEM: suitable tools for study of adhesion stage and biofilm formation. V. Kazmiruk (Ed.), *Scanning Electron Microscopy* (pp. 718-725). InTechOpen. DOI: 10.5772/34990

Falke, S., & Betzel, C. (2019). Dynamic Light Scattering (DLS): Principles, perspectives, applications to biological samples. A. Pereira (Ed.), *Radiation in Bioanalysis* (pp. 173-193). Springer. DOI: 10.1007/978-3-030-28247-9

Franken, L. E., Boekema, E. J., & Stuart, M. C. A. (2017). Transmission electron microscopy as a tool for the characterization of soft materials: Application and interpretation. *Advancement of Science*, 4(5), 1–9. DOI: 10.1002/advs.201600476 PMID: 28546914

Goldsbury, C. S., Scheuring, S., & Kreplak, L. (2009). Introduction to atomic force microscopy (AFM) in Biology. *Current Protocols in Protein Science*, 58(1), 1–19. DOI: 10.1002/0471140864.ps1707s58 PMID: 19937721

Gopirajah, R., & Anandharamakrishnan, C. (2019). Characterization methods for nanoparticles. *Food Nanotechnology*, 375, 375–396. DOI: 10.1201/9781315153872-16

Gul, I., Khan, S. M., Mehmood, T., Ahmad, Z., Badshah, H., & Shah, H. (2020). Characterization of cobalt oxide and calcium-aluminum oxide nano-catalyst through scanning electron microscopy, X-ray diffraction, and Energy dispersive X-ray spectroscopy. *Microscopy Research and Technique*, 83(9), 1124–1131. DOI: 10.1002/jemt.23504 PMID: 32394528

Guo, Y., Liu, C., Ye, R., & Duan, Q. (2020). Advances on water quality detection by uv-vis spectroscopy. *Applied Sciences (Basel, Switzerland)*, 10(19), 1–18. DOI: 10.3390/app10196874

Herrera-Becerra, R., Zorrilla, C., Rius, J. L., & Ascencio, J. A. (2008). Electron microscopy characterization of biosynthesized iron oxide nanoparticles. *Applied Physics. A, Materials Science & Processing*, 91(2), 241–246. DOI: 10.1007/s00339-008-4420-7

Hidayanti, F., & Harnovan, A. A. (2020). Application of scanning electron microscopy: A review. *International Journal of Applied Science and Engineering Review*, 1(6), 91–102.

Hodoroaba, V. D. (2019). Energy-dispersive X-ray spectroscopy (EDS). V. Hodoroaba (Ed.), *Characterization of Nanoparticles* (pp. 397-417). Elsevier Inc. DOI: 10.1016/B978-0-12-814182-3.00021-3

Ingham, B. (2015). X-ray scattering characterisation of nanoparticles. *Crystallography Reviews*, 21(4), 229–303. DOI: 10.1080/0889311X.2015.1024114

Inkson, B. J. (2016). Scanning electron microscopy (SEM) and transmission electron microscopy (TEM) for materials characterization. G. Hübschen (Ed.), *Materials Characterization Using Nondestructive Evaluation (NDE) Methods* (pp. 17-43). Elsevier Ltd. DOI: 10.1016/B978-0-08-100040-3.00002-X

Jokerst, J. V., & Gambhir, S. S. (2011). Molecular imaging with theranostic nanoparticles. *Accounts of Chemical Research*, 44(10), 1050–1060. DOI: 10.1021/ar200106e PMID: 21919457

Kiefer, J., Grabow, J., Kurland, H. D., & Müller, F. A. (2015). Characterization of nanoparticles by solvent infrared spectroscopy. *Analytical Chemistry*, 87(24), 12313–12317. DOI: 10.1021/acs.analchem.5b03625 PMID: 26593634

Kisielowski, C., Liliental-Weber, Z., & Weber, E. R. (1996). Quantitative HRTEM: A novel approach towards application oriented basic research. *Brazilian Journal of Physics*, 26(1), 83–94.

Klinavičius, T., Khinevich, N., Tamulevičienė, A., Vidal, L., Tamulevičius, S., & Tamulevičius, T. (2024). Deep learning methods for colloidal silver nanoparticle concentration and size distribution determination from UV-Vis extinction spectra. *The Journal of Physical Chemistry. C, Nanomaterials and Interfaces*, 128(23), 9662–9675. DOI: 10.1021/acs.jpcc.4c02459

Lanje, A. S., Sharma, S. J., Pode, R. B., & Ningthoujam, R. S. (2010). Synthesis and optical characterization of copper oxide nanoparticles *Advances in Applied Science Research 36-401*(2), 36–40.

Leite, F. L. (2007). The Atomic Force spectroscopy as a tool to investigate surface forces: Basic principles and applications. A. Vilas M (Ed.), *Modern Research and Educational Topics in Microscopy*,(pp. 747-755). Formatex.

León, A., Reuquen, P., Garín, C., Segura, R., Vargas, P., Zapata, P., & Orihuela, P. A. (2017). FTIR and raman characterization of TiO_2 nanoparticles coated with polyethylene glycol as carrier for 2-methoxyestradiol. *Applied Sciences (Basel, Switzerland)*, 7(1), 1–9. DOI: 10.3390/app7010049

Lewczuk, B., & Szyryńska, N. (2021). Field-emission scanning electron microscope as a tool for large-area and large-volume ultrastructural studies. *Animals (Basel)*, 11(12), 1–21. DOI: 10.3390/ani11123390 PMID: 34944167

Li, P., & Hur, J. (2017). Utilization of UV-Vis spectroscopy and related data analyses for dissolved organic matter (DOM) studies: A review. *Critical Reviews in Environmental Science and Technology*, 47(3), 131–154. DOI: 10.1080/10643389.2017.1309186

Loganathan, S., Valapa, R. B., Mishra, R. K., Pugazhenthi, G., & Thomas, S. (2017). Thermogravimetric analysis for characterization of nanomaterials. S. Thomas (Ed.), *Thermal and Rheological Measurement Techniques for Nanomaterials Characterization* (pp. 67-108). Elsevier. DOI: 10.1016/B978-0-323-46139-9.00004-9

Mallakpour, S., & Hussain, C. M. (2021). Characterization of nanoparticles by FTIR and FTIR-microscopy handbook of consumer nanoproducts. S. Mallakpour (Ed.), *Handbook of Consumer Nanoproducts* (pp. 978-981). DOI: 10.1007/978-981-15-6453-6

Mansfield, E., Tyner, K. M., Poling, C. M., & Blacklock, J. L. (2014). Determination of nanoparticle surface coatings and nanoparticle purity using microscale thermogravimetric analysis. *Analytical Chemistry*, 86(3), 1478–1484. DOI: 10.1021/ac402888v PMID: 24400715

Martínez, C., J., A. Chequer, N., L. González, J., & Cordova, T. (. (2013). Alternative metodology for gold nanoparticles diameter characterization using PCA technique and UV-VIS spectrophotometry. *Nanoscience and Nanotechnology*, 2(6), 184–189. DOI: 10.5923/j.nn.20120206.06

Mayeen, A., Shaji, L. K., Nair, A. K., & Kalarikkal, N. (2018). Morphological characterization of nanomaterials. S. Bhagyaraj (Ed.), *Characterization of Nanomaterials: Advances and key technologies* (pp. 335-364). Elsevier. DOI: 10.1016/B978-0-08-101973-3.00012-2

Mishra, R. K., Thomas, S., & Zachariah, A. K. (2017). Energy-dispersive X-ray spectroscopy techniques for nanomaterial. S. Thomas (Ed.), *Microscopy methods in nanomaterials characterization* (pp. 383-405). Elsevier. DOI: 10.1016/B978-0-323-46141-2.00012-2

Mittal, A. K., Chisti, Y., & Banerjee, U. C. (2013). Synthesis of metallic nanoparticles using plant extracts. *Biotechnology Advances*, 31(2), 346–356. DOI: 10.1016/j.biotechadv.2013.01.003 PMID: 23318667

Modena, M. M., Rühle, B., Burg, T. P., & Wuttke, S. (2019). Nanoparticle characterization: What to Measure? *Advanced Materials*, 31(32), 1–26. DOI: 10.1002/adma.201901556 PMID: 31148285

Mohamed, M. A., Jaafar, J., Ismail, A. F., Othman, M. H. D., & Rahman, M. A. (2017). Fourier transform infrared (FTIR) spectroscopy. N. Hilal, & A. F. Ismail (Eds.), *Membrane characterization* (pp. 3-29). Elsevier. DOI: 10.1016/B978-0-444-63776-5.00001-2

Mourdikoudis, S., Pallares, R. M., & Thanh, N. T. K. (2018). Characterization techniques for nanoparticles: Comparison and complementarity upon studying nanoparticle properties. *Nanoscale*, 10(27), 12871–12934. DOI: 10.1039/C8NR02278J PMID: 29926865

Murdock, R. C., Braydich-stolle, L., Schrand, A. M., Schlager, J. J., Hussain, S. M., & Al, M. E. T. (2008). Characterization of nanomaterial dispersion in solution prior to in vitro exposure using dynamic light scattering technique. *Toxicological Sciences*, 101(2), 239–253. DOI: 10.1093/toxsci/kfm240 PMID: 17872897

Nallusamy, S., & Manoj Babu, A. (2016). X-Ray differaction and fesem analysis for mixture of hybrid nano particles in heat transfer applications. *Journal of Nano Research*, 37, 58–67. . DOI: 10.4028/www.scientific.net/JNanoR.37.58

Philip, D. (2008). Synthesis and spectroscopic characterization of gold nanoparticles. *Spectrochimica Acta. Part A: Molecular and Biomolecular Spectroscopy*, 71(1), 80–85. DOI: 10.1016/j.saa.2007.11.012 PMID: 18155956

Phung, X., Groza, J., Stach, E. A., Williams, L. N., & Ritchey, S. B. (2003). Surface characterization of metal nanoparticles. *Materials Science and Engineering A*, 359(1–2), 261–268. DOI: 10.1016/S0921-5093(03)00348-4

Preetha, D., Arun, R., Kumari, P., & Aarti, C. (2013). Synthesis and characterization of silver nanoparticles using cannonball leaves and their cytotoxic activity against MCF-7 cell line. *Journal of Nanotechnology*, 2013(1), 1–5.

Prime, R. B., Bair, H. E., & Vyazovkin, S. Riga Alan, & Gallagher, P. K. (2009). Thermogravimetric analysis (TGA). J. D. Menczel & R. B. Prime (Eds.), *Thermal analysis of polymers:Fundamentals and applications* (pp. 241-317). Wily.

Rami, J. M., Patel, C. D., Patel, C. M., & Patel, M. V. (2020). Thermogravimetric analysis (TGA) of some synthesized metal oxide nanoparticles. *Materials Today: Proceedings*, 43, 655–659. DOI: 10.1016/j.matpr.2020.12.554

Rao, A., Schoenenberger, M., Gnecco, E., Glatzel, T., Meyer, E., Brändlin, D., & Scandella, L. (2007). Characterization of nanoparticles using atomic force microscopy. *Journal of Physics: Conference Series*, 61(1), 971–976. DOI: 10.1088/1742-6596/61/1/192

Saadatkhah, N., Carillo Garcia, A., Ackermann, S., Leclerc, P., Latifi, M., Samih, S., Patience, G. S., & Chaouki, J. (2020). Experimental methods in chemical engineering: Thermogravimetric analysis—TGA. *Canadian Journal of Chemical Engineering*, 98(1), 34–43. DOI: 10.1002/cjce.23673

Santos, N. C., & Domingues, M. M. (2018). Application of light scattering techniques to nanoparticle characterization and development. *Frontiers in Chemistry*, 6, 1–12. DOI: 10.3389/fchem.2018.00237 PMID: 29988578

Scimeca, M., Bischetti, S., Lamsira, H. K., Bonfiglio, R., & Bonanno, E. (2018). Energy dispersive X-ray (EDX) microanalysis: A powerful tool in biomedical research and diagnosis. *European Journal of Histochemistry*, 62(1), 89–99. DOI: 10.4081/ejh.2018.2841 PMID: 29569878

Sharma, R., Bisen, D. P., Shukla, U., & Sharma, B. G. (2012). X-ray diffraction: A powerful method of characterizing nanomaterials. *Recent Research in Science and Technology*, 4(8), 77–79.

Shivani Rasool, H. I., Palanisamy, V., Mathisen, C., Schmidt, Ḱ. M., Wong, D. T., & Gimzewski, J. K. (2010). Structural-mechanical characterization and force spectroscopy saliva, using correlative AFM, FESEM, of nanoparticle exosomes in human. *American Chemical Society, 4*(4), 1921–1926.

Smith, J. R., Breakspear, S., & Campbell, S. A. (2003). AFM in Surface Finishing: Part I - An Introduction. *Finishing*, 27(8), 35–40. DOI: 10.1080/00202967.2003.11871484

Stetefeld, J., Mckenna, S. A., & Patel, T. R. (2016). Dynamic light scattering : A practical guide and applications in biomedical sciences. *Biophysical Reviews*, 8(4), 409–427. DOI: 10.1007/s12551-016-0218-6 PMID: 28510011

Stewart, P. L. (2017). Cryo-electron microscopy and cryo-electron tomography of nanoparticles. *Wiley Interdisciplinary Reviews. Nanomedicine and Nanobiotechnology*, 9(2), 1–16. DOI: 10.1002/wnan.1417 PMID: 27339510

Sun, C., Lux, S., Müller, E., Meffert, M., & Gerthsen, D. (2020). Versatile application of a modern scanning electron microscope for materials characterization. *Journal of Materials Science*, 55(28), 13824–13835. DOI: 10.1007/s10853-020-04970-3

Sun, C., Lux, S., Müller, E., Meffert, M., & Gerthsen, D. (2020). Versatile application of a modern scanning electron microscope for materials characterization. *Journal of Materials Science*, 55(28), 13824–13835. DOI: 10.1007/s10853-020-04970-3

Tang, C. Y., & Yang, Z. (2017). Transmission electron microscopy (TEM). N. Hilal, A. F. Ismail (Eds.), *Membrane Characterization* (pp. 145-159). Elsevier. DOI: 10.1016/B978-0-444-63776-5.00008-5

Tem, S. D. (2008). Characterization and simulation technologies of nanomaterial. M. Nayfeh (Ed.), *Fundamentals and Applications of Nano Silicon in Plasmonics and Fullerines* (pp. 153–167). Elsevier. DOI: 10.1016/B978-0-323-48057-4.00007-4

Titus, D., James Jebaseelan Samuel, E., & Roopan, S. M. (2018). Nanoparticle characterization techniques. A. K. Shukla and S. Iravani (Eds.), *Green Synthesis, Characterization and Applications of Nanoparticles* (pp. 303-319). Elsevier. DOI: 10.1016/B978-0-08-102579-6.00012-5

Utsunomiya, S., & Ewing, R. C. (2003). field scanning transmission electron microscopy, scanning transmission electron microscopy-energy dispersive X-ray spectrometry, and energy-filtered transmission. *Environmental Science & Technology*, 37(4), 786–791. https://pubs.acs.org/doi/abs/10.1021/es026053t. DOI: 10.1021/es026053t PMID: 12636280

Vahabi, S., Nazemi Salman, B., & Javanmard, A. (2013). Atomic force microscopy application in biological research: A review. *Iranian Journal of Medical Sciences*, 38(2), 76–83. PMID: 23825885

Verma, G., & Mishra, M. (2018). Development and optimization of UV-Vis spectroscopy - A review. *World Journal of Pharmaceutical Research*, 7(11), 1170–1180. DOI: 10.20959/wjpr201811-12333

Vida-simiti, I., Cluj-napoca, U. T., Chicinas, I., Cluj-napoca, U. T., Batin, G., & Cluj-napoca, U. T. (2004). Applications of Scanning Electron Microscopy (Sem). *Romanian Journal of Physics*, 49, 955–965.

Vijayaram, S., Razafindralambo, H., Sun, Y. Z., Vasantharaj, S., Ghafarifarsani, H., Hoseinifar, S. H., & Raeeszadeh, M. (2024). Applications of green synthesized metal nanoparticles - A review. *Biological Trace Element Research*, 202(1), 360–386. DOI: 10.1007/s12011-023-03645-9 PMID: 37046039

Waheed, I., Ali, A., Tabassum, H., Khatoon, N., Lai, W. F., & Zhou, X. (2024). Lipid-based nanoparticles as drug delivery carriers for cancer therapy. *Frontiers in Oncology*, 14(04), 1–17. DOI: 10.3389/fonc.2024.1296091 PMID: 38660132

Wang, Z. L. (1999). Transmission Electron Microscopy and Spectroscopy of Nanoparticles. *Characterization of Nanophase Materials*, 18, 37–80. DOI: 10.1002/3527600094.ch3

Zhou, W., Apkarian, R., Wang, Z. L., & Joy, D. (2007). Fundamentals of scanning electron microscopy (SEM). W. Zhou, Z. L. Wang (Eds.), *Scanning Microscopy for Nanotechnology: Techniques and applications* (pp. 1–40). DOI: 10.1007/978-0-387-39620-0_1

Chapter 12
Advanced Instrumentation for Characterization of Nanoparticles: A Discourse

Nutan Ankush Napte
Mahatma Phule Krishi Vidyapeeth, Rahuri, India

Sujoy Saha
https://orcid.org/0000-0002-5367-5955
ICAR-National Research Centre for Grapes, Pune, India

ABSTRACT

Nanotechnology will emerge as the most significant technological breakthrough of the twenty-first century. The rapid development of nanotechnology has led to the creation and application of nanoparticles across various fields, including medicine, materials science, agriculture and environmental engineering. It is imperative to characterize these nanoparticles to comprehend their characteristics, actions and possible consequences. Depending on the intended usage, an array of techniques such as SEM, TEM, ZP, DLS, AFM, FTIR, etc., can be employed to examine and describe nanoparticles. Enhancing resolution, sensitivity and accuracy, these approaches are key in providing comprehensive insights into the size, shape, surface characteristics, composition and interactions of nanoparticles. The integration of these techniques allows for the comprehensive characterization of nanoparticles, aiding in the optimization of their design and application. This chapter summarizes the techniques that are commonly used along with their benefits and drawbacks.

INTRODUCTION

Nanotechnology is the study & application of extremely small particles which range between 1 nm to 100 nm. The size of nanoparticles (NPs) is the main aspect that influences their chemical & physical properties (Khan *et al.*, 2019). Unlike their macro state, nanomaterials have distinct physico-chemical properties like optical quality, chemical reactivity, mechanical strength & surface area, which are pivotal

DOI: 10.4018/979-8-3693-6240-2.ch012

for several applications (Baig *et al.*, 2021). These specific qualities of nanomaterials have renewed the interest of researchers in several fields like engineering, chemistry, biology, physics & materials science (Dolez, (2015). NPs can be applied in a broad spectrum of sectors, including biological labelling, medical treatments, industrial applications, everyday products like clothing & cosmetics, optical devices, sensor technology & electronics (Titus *et al.*, 2019). In agriculture, nanotechnology can revolutionize the sector by introducing novel techniques for quick disease detection vis-a-vis targeted control to combat the pathogen. It can play an essential role in increasing potential of vegetation for retaining nutrients, coping with environmental stresses & conservation constraints (Banerjee *et al.*, 2021).

The synthesis of NPs is broadly classified into chemical, physical & biological methods which affect properties & efficiency of NPs (Sajid & Płotka-Wasylka (2020). Physical-chemical techniques are expensive & frequently call for harmful substances. Due to harmful radiation emissions & the production of synthetic reductants & stabilizing agents, these methods have an adverse effect on human & environmental health (Hasan, 2015). In contrast, biological approaches (using plants & microbes) are more effective, cost-efficient & environmentally benign (Bonde *et al.*, 2012). As a result, plant- & microbe-based synthesizing of NPs has gained more popularity than chemical synthesis. Since silver nanoparticles (AgNPs) have antimicrobial properties that fail in their macro state (Sofi *et al.*, 2012), it is evident that their antimicrobial properties rely on their size. To ensure the efficacy of the NPs after their synthesis, characterization must be performed before application.

The increasing use of NPs in research & product development has paved the way for rising a demand to understand the properties & behaviour of NPs. Additionally, there is an increasing awareness of the significance of understanding the effect of NPs on human health, safety & the environment, both in their synthesized & transformed state. Considering all these reasons, the comprehensive characterization of NPs is mandatory (Jose *et al.*, 2017).

In material science, "Characterization" refers to the broad, fundamental processes performed to explore structure & properties of material (Mino *et al.*, 2018). It is vital to evaluate size, form, crystal structure, atomic makeup, surface charge & other features of synthesized nanomaterials (Baer *et al.*, 2013; Raval *et al.*, 2019). Conventional techniques for characterization are being integrated with advanced methods & will be the main emphasis in the present compilation.

CHARACTERIZATION OF NANOPARTICLES

Multiple approaches have been employed to better understand properties of nanomaterials. Various quantitative & qualitative methods are employed for analysis, such as UV-visible spectrometry, scanning electron microscopy (SEM), energy dispersive X-ray spectroscopy (EDS), transmission electron microscopy (TEM), dynamic light scattering (DLS), zeta potential measurement, Fourier transform infrared spectroscopy (FTIR), atomic force microscopy (AFM), scanning tunneling microscopy (STM), scanning probe microscopy (SPM), X-ray photoelectron spectroscopy (XPS) & X-ray diffraction (XRD), among others (Chirayil *et al.*, 2017; Hodoroaba *et al.*, 2019). The fundamental principles & mechanisms of techniques are highlighted in the present discourse.

Figure 1. Different techniques for characterization of nanoparticles

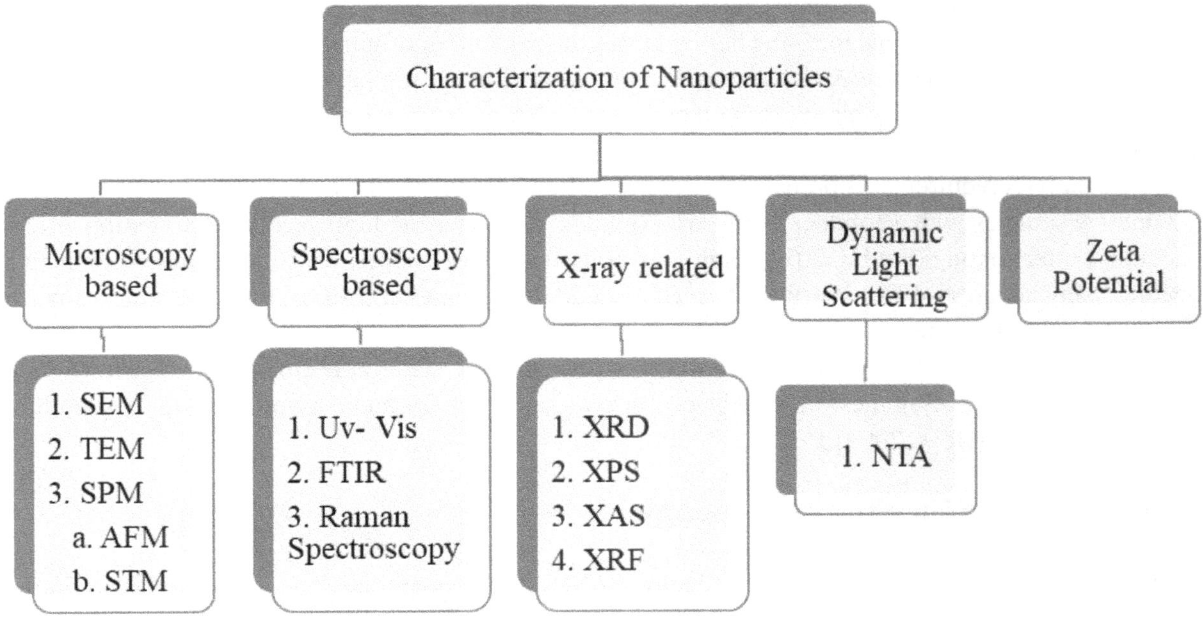

A. MICROSCOPY-BASED NPS CHARACTERIZATION

The microscopy application facilitates reasonable resolution of material observation up to the micron level. Modern technology has advanced to enable the identification of materials with submicron sizes using imaging techniques such as SEM, TEM, AFM & SPM (Tamilselvan *et al.*, 2017). While the principles behind these techniques differ, they all generate highly magnified images.

1. Scanning Electron Microscopy

SEM is versatile, advanced equipment mostly employed to examine material surface phenomena. Unlike light microscopes, which can magnify up to x10,000, it can display incredibly comprehensive three-dimensional (3D) images at higher magnification (up to x3,00,000). SEM can provide qualitative features about the specimen, such as its shape, topography & crystallographic details. Furthermore, compared with TEM, SEM is significantly more user-friendly & allows for faster measurements with cheaper acquisition & maintenance expenses. However, the visuals are black & white because light waves were not employed in their creation. SEM provides an unobstructed image of surface structure of NPs, polymer nanocomposites, nanofibers & nanocoating (Akhtar *et al.*, 2018).

In SEM, an electron gun generates a high-energy electron beam which is focused into a fine spot on the sample using an electromagnetic lens. This beam scans the surface of the sample in a raster pattern. As the beam interacts with the sample, it generates secondary electrons, backscattered electrons & characteristic X-rays. The secondary electrons provide high-resolution, detailed topographical images, while the backscattered electrons reveal compositional contrast by differentiating atomic numbers (Bogner *et al.*, 2007). EDS employs the characteristic X-rays to analyze sample's elemental composition. This version maintains the original meaning but rephrases the content for originality. Finally, detectors gather these signals to create visuals & determine the composition of the sample (Girao *et al.*, 2017). Devi & Joshi (2014) revealed morphology of the biosynthesized AgNPs from endophytic fungi to be more or less spherical with a 10 nm size.

SEM analysis requires well-prepared samples for effective analysis. It is often necessary to coat non-conductive samples with a conductive material (gold or copper) using high vacuum evaporation or low vacuum sputter coating (Echlin, 2011). However, non-conductive samples can also be imaged without a coating using advanced SEM instruments such as Field Emission Gun (FEG) SEM. These instruments operate under conditions like high voltage with low vacuum or low voltage with high vacuum (Fleck & Humbel 2019). When specimens are not properly coated, the electron beam can cause distortion & artifacts in the image. Samples must be placed in a vacuum, which can be a limitation for certain biological specimens (Hoyt, 2009).

2. Transmission Electron Microscopy

TEM is a crucial method for characterization of NPs. It is a quantitative technique to ascertain the distribution, size & form of particles. It is known for its exceptional lateral spatial resolution & its capability to deliver both diffraction patterns & high-resolution images of samples. The whole morphological, crystallographic, atomic structural, electronic structure, coordination number & microanalytical (chemical composition, bonding) information about the materials are obtained using this technique (Goodhew, 2011).

In the TEM, sample mainly interacts with beam of high-energy electron through diffraction instead of absorption. The orientation of atomic planes in relation to electron beam influences diffraction intensity. The electron beam undergoes significant diffraction at specific angles from the incoming beam's axis, while at other angles, most of beam is transmitted. Modern TEMs have specimen holders that allow tilting sample to achieve desired diffraction conditions. By adjusting the aperture to block electrons deflected from the microscope's optical axis, high-contrast images can be produced by allowing only unscattered electrons to pass through. TEM provides several imaging modes, such as Bright Field (BF), Dark Field (DF), High-Resolution TEM (HRTEM) & Selected Area Electron Diffraction (SAED). BF is standard mode, creating images from electrons transmitted through the sample, while DF uses scattered electrons to highlight dispersed areas. HRTEM delivers atomic-level resolution of sample structures & SAED is used to examine crystalline structures & defects (Ma *et al.*, 2006). Shankar *et al.* (2018) reported size distribution histogram of sulphur nanoparticles, which showed a uniform size range of 10–30 nm, with an average diameter of 15.80 ± 7.82 nm.

The practical application of TEM analysis has some limitations as it requires highly skilled workers & is an expensive procedure (both in terms of acquisition & maintenance costs). Sometimes it can be challenging to make ultra-thin samples. NPs must remain stable under high vacuum & beam energy used during TEM characterization, while also maintaining electron transparency. Organic, polymer & hybrid NPs are particularly susceptible to damage from high-energy electron beams. To mitigate this issue, Cryo-electron microscopy (CryoEM) can be utilized as an effective solution (Klaholz, 2019).

Cryogenic electron microscopy is an advanced imaging approach utilised to determine the 3D structure of biological molecules & complexes at almost atomic precision (Hollander & Danino 2015). By quickly freezing the material to a temperature below -150 °C, it is sealed in vitreous ice. Next, the sample is scanned from various perspectives using an electron microscope. As an electron beam interacts with specimen, a collection of two-dimensional (2D) projections is produced which are then used to build a 3D representation of the sample using refined computational methods (Watt *et al.*, 2019). A major advantage of electron microscopes is their capacity to combine high-resolution imaging with spectroscopic methods including EDS, STEM & EELS (Zhu & Durr, 2015).

2.1 Energy Dispersive Spectroscopy

EDS or EDXS is a technique utilized for elemental analysis & determining the proportions of elements at various locations, thereby providing an overall map of a sample (Hodoroaba *et al.*, 2020).

In EDS, when beam with high-energy electrons typically between 10 & 20 keV, strikes the sample, resulting it ionizes the atoms by ejecting electrons from their inner shells. When electrons from higher energy states filled the vacant spaces, the atoms regained stability & emitted energy in the form of element-specific X-rays. Detectors like silicon drift detectors (SDDs), which measure the energy of the X-rays, are used to detect these radiations. Elements in the sample can be identified & quantified since each one emits X-rays at a distinct energy level. The EDS offers rapid data collection, non-destructive examination & spatial element distribution of samples (Herzing *et al.*, 2008).

As long as there are some heavy metal ions present, EDS can be used to estimate composition or quantity of NPs at & adjacent to the surface. The EDS technique confirmed the presence of AgNPs by showing a peak in the silver region at approximately 3 keV on a surface (Narware *et al.*, 2023). EDX have challenges in identifying elements with low atomic numbers (Choel *et al.*, 2005). Elements with atomic numbers below 11 (such as sodium) are undetectable by the Si-Li detector that is shielded via a beryllium window (Goldstein *et al.*, 2017). However, elements with atomic numbers as low as 4 (like beryllium) have been identified in windowless systems. To confirm that the composition of the bimetallic materials is spatially homogeneous, EDX spectra must be obtained by focusing the beam at several positions within the subject (Joshi *et al.,* 2008).

2.2 Scanning Transmission Electron Microscope

STEM integrates features of both TEM & SEM by concentrating fine electron beam that passes across a specimen (Pennycook *et al.*, 2007). It offers many imaging modes & may detail samples' electronic structure & elemental composition up to the smallest possible atomic dimension. Using a field emission gun that emits high-energy electrons (100-300 keV), the specimen interacts with the electrons to produce different signals. To create images & offer compositional data, detectors like Annular Dark Field receive these signals. Equipped with aberration-corrected optics, modern STEM enables imaging & analysis at the atomic scale. The process of ion milling or mechanical thinning is frequently used to create thin specimens for sample processing (Tanaka and Saitoh, 2015).

2.3 Electron Energy Loss Spectroscopy

EELS provides insights into electronic structure, oxidation states & chemical composition at the atomic scale by analyzing the energy of inelastically scattered electrons. It is a quantitative analysis & only applicable to TEM. It can offer comprehensive information on organic & inorganic composition of biological samples with high spatial resolution. Using a high-energy electron beam (usually between 100 & 300 keV), EELS hits a thin sample & these electrons lose energy as they scatter off or travel through the sample (Schneider, 2011). The energy loss is associated with specific excitations in the sample & it is possible to quantify this loss. The resulting EELS spectrum plots the intensity of transmitted electrons versus their energy loss which provides detailed information about the material. EELS offers several advantages over EDS, including more sensitivity to lower atomic number elements, enhanced spatial resolution (down to 1 nm), superior energy resolution (<1 eV) & improved signal-to-noise ratios (Reimer, 2013).

3. Scanning Probe Microscopy

SPM generates 3-D image of surface by mechanical inspection of the surface of particles. AFM & STM are two examples of SPM that have been proven to be efficient for imaging & measuring materials at nanoscale (Bian *et al.*, 2021).

3.1 Atomic Force Microscopy

AFM, a high-resolution form of SPM which is used to measure topographical features at the atomic scale. It is particularly effective for analyzing surface nano-textures across various materials, such as polymer nanocomposites & nanofinished or nanocoated textiles, as well as for measuring nanoscale surface roughness (Kim *et al.*, 2016).

In AFM, a sharp-tipped probe, typically around 10 nm in radius, is attached to end of a cantilever & scans specimen's surface using piezoelectric scanners. A laser is reflected off the cantilever onto a position-sensitive photodiode through an optical lever system, allowing for detection of changes in the interaction between the probe tip & the sample. Images get constructed through a feedback loop between the piezoelectric scanners & the optical detection system while scanning, which maintains a specific operational parameter at a constant level (Sun, 2018).

AFM operates in three distinct scanning modes: contact mode, tapping mode & non-contact mode. In contact mode, the probe tip maintains direct contact with sample surface & a piezoelectric positioning element adjusts the cantilever to alter repulsive forces at the tip. Variations in the cantilever's deflection are then recorded to generate AFM images. The tip is suspended between 50 & 150 Å above the surface of specimen in non contact mode. By scanning above the surface & detecting the attractive forces between specimen & tip, topographical images are produced. Tapping mode operates in ambient air, where the cantilever is vibrated at its resonant frequency typically in the hundreds of kHz using a piezoelectric crystal. The oscillating cantilever moves close enough to the surface to intermittently 'tap' it. As the tip alternates between lifting off & making contact with the surface, at a rate of 50,000 to 500,000 cycles/second, energy loss from tip striking the surface reduces the oscillation amplitude. This reduction is used to map the surface features. Khalid *et al.* (2015) studied copper nanoparticles using tapping mode AFM imaging. An AFM image (2 μm ×2 μm) of 3-D copper NPs with a particle size range of 14 nm to 55 nm.

Some specialized adaptations in AFM measure different types of interactions at the nanoscale, providing a more comprehensive understanding of the properties of a sample. One of the main shortcomings of AFM is that chemical information cannot be easily extracted from AFM images but a specialised form of AFM called Chemical Force Microscopy (CFM) combines chemical sensitivity with high-resolution imaging to solve the problem. The sample surface can be simultaneously mapped topographically & chemically, but precise & consistent functionalization of the AFM tips is required (Paloni *et al.*, 2016). Magnetic Force Microscopy (MFM) measures & visualizes magnetic characteristics at the nanoscale (Kazakova *et al.*, 2019). Friction Force Microscopy (FFM) or Lateral Force Microscopy (LFM), maps & quantifies lateral (horizontal) forces between sample surface & AFM tip (Serry & Schmit 2019).

3.2 Scanning Tunneling Microscope (STM)

STM is a fundamental technology in surface science that offers atomic-scale resolution & comprehensive electronic surface data. It is based on a tunnelling process in quantum mechanics. Electrons are able to cross the vacuum gap between a conductive or semiconductive sample surface & a sharp metallic tip by applying an incredibly small bias voltage between them. The resulting tunnelling current is very sensitive to the distance between the tip & sample, & the microscope can detect changes because of its atomic resolution. For STM, a depth resolution of 0.01 nm & a lateral resolution of 0.1 nm is considered a good resolution. It is used in a range of temperature conditions including vacuum & air, liquid. This method can be difficult as it necessitates clean surfaces & a sharp tip (Chen, 2021).

B. SPECTROSCOPY-BASED CHARACTERIZATION TECHNIQUES

Optical spectroscopy studies the interactions between matter & light across a broad spectrum of wavelengths, ranging from long-wavelength, low-energy radio waves to short-wavelength, high-energy gamma rays (Marques *et al.*, 2021). When light strikes on sample, energy levels of molecules, atoms, or ions in a sample change & resulting in either emission, absorption, or scattering. To find out more information about a sample, a spectrometer is used to assess its transmittance, absorbance & reflectance spectra. A variety of optical spectroscopic techniques for example, ultraviolet-visible spectroscopy (UV-Vis), FTIR & Raman spectroscopy are available for characterizing nanomaterials (Gupta *et al.*, 2024).

1. UV-Vis Spectroscopy

UV-Vis Spectroscopy is a method to ascertain stability & formation of NPs. It provides information concerning the concentration, size, form, & optical characteristics of NPs. The principle relies on the Beer-Lambert law, which asserts that the absorbance of a sample at a specific wavelength is directly proportional to the concentration of the absorbing sample. This can be used to calculate the concentration of NPs in a solution (Akash *et al.*, 2020).

A UV-Vis spectrophotometer consists of several key components, including detector, sample beam, reference beams, monochromator, deuterium lamp for the ultraviolet spectrum & a tungsten lamp for the visible spectrum. The reference beam moves directly from the light source to the detector without interacting with sample, while sample beam interfaces with UV light as its wavelength continuously changes. Absorption happens when the wavelength corresponds to the energy needed for an electron to move to a higher orbital. The ratio of the intensity of the reference beam to that of the sample beam has been recorded by detectors. By identifying the wavelength where the greatest intensity difference occurs, the system determines where the sample absorbs the most ultraviolet light. A significant reduction in the sample beam's intensity relative to the reference beam is noted as the point of maximum UV absorbance. This data is used to prepare the ultraviolet absorbance spectrum (Yang & Zhu, 2023).

When NPs are dispersed in a medium & exposed to UV-Vis light, they display distinctive optical properties attributed to localized surface plasmon resonance. This effect arises as conduction electrons on the surfaces of NPs oscillate in harmony with the incoming light, resulting in specific absorption peaks within the UV-Vis spectrum. The size, form & composition of the NPs can be inferred from the location & form of the plasmon resonance peak. In addition to detecting aggregation or changes in the dispersion state, changes in the UV-Vis spectrum over time can also reveal the stability of NPs in suspension. So, the rapid, non-destructive nature of UV-Vis spectrophotometer makes it apt for regular analysis & quality control in nanoparticle research & applications. Gold NPs synthesized from marine *Enterococcus spp.* have been reported to exhibit an absorption spectrum at 545 nm. Meanwhile, the peaks manifest increases the stability of the NPs with time (Menon *et al.,* 2017).

2. Vibrational Spectroscopy

In vibrational spectroscopy, Raman & infrared (IR) spectroscopies are frequently employed to quantify concentration of particular chemical in a sample & learn more about the interactions between them. They are typically non-destructive (Sathyanarayana, 2015).

2.1 Fourier Transform Infrared Spectroscopy

IR is commonly employed technique for characterizing different types of NPs. By comparing the IR spectra of polymers, composites, organic & inorganic substances to reference spectra, it is possible to investigate & identify materials. The shifts in the unique absorption band pattern indicate a change in composition of material (Bocker *et al.*, 2017).

The components of FTIR spectrometer include a computer, an A/D converter, a detector, an amplifier & a sample cell. A broadband infrared light source emits infrared radiation that enters the sample. A Michelson interferometer generates two paths: one is reflected off a mirror that is stationary while the other is reflected off a movable mirror. After recombining, the two beams create an interference pattern. Certain frequencies in the recombined infrared beam are absorbed by the sample, which lowers their intensity in the reflected or transmitted light. An interferogram is the resultant signal that is detected, recorded & transformed from the time domain to the frequency domain using a mathematical Fourier transform & a spectrum that typically ranges from 4000 to 400 cm^{-1} (Taha *et al.*, 2013). Every kind of chemical bond & functional group absorbs infrared light at certain frequencies, producing a unique absorption spectrum that serves as the molecular fingerprint of the sample (Nan *et al.*, 2021).

Jain *et al.*, (2012) synthesized zinc oxide (ZnO) NPs & were subjected to FTIR analysis, which revealed an intense band around wavenumber 430 cm^{-1}, which seems to be attributed to ZnO vibrations. The bending vibrations of protein amide I & II were represented by bands at wavenumbers 1,625 cm^{-1} & 1,550 cm^{-1}, respectively. It had been observed that the stretching vibrations of amide I superimposed on the side of the hydroxyl group band form the band at wavenumber 3,450 cm^{-1}. Furthermore, the stretching vibrations of amide II & the presence of ambient CO_2 are responsible for the bands at wavenumbers 2,995 cm^{-1} & 2,350 cm^{-1} respectively.

2.2 Raman Spectroscopy

It is non-destructive approach for investigating rotational, vibrational & other low-frequency modes in system. It offers insights into molecular vibrations, which are useful for investigating chemical bonds & molecular interactions as well as for identifying compounds (Mulvaney & Keating 2000). The basic concept of Raman spectroscopy is inelastic scattering of monochromatic light which is typically generated by a laser. The majority of photons that contact with molecules are dispersed elastically (Rayleigh scattering), although a tiny percentage is scattered inelastically. A shift in energy equal to vibrational energy levels of molecules is involved in this inelastic scattering, which sometimes is referred to as Raman scattering. In Stokes Raman scattering, scattered photons produce a longer wavelength (lower energy) & lose energy, whereas in Anti-Stokes Raman scattering, they gain energy & produce a shorter wavelength (higher energy). The vibrational modes of the sample are represented by the peaks where peak position (expressed in wavenumbers, cm^{-3}) & intensity in the Raman spectrum, revealed molecular structure & surroundings (Larkin, 2017).

Raman spectroscopy is widely utilized in field of chemistry because vibrational information about the chemical bonds in molecules is so specific. Consequently, it offers a fingerprint that allows the molecule to be recognised within the 500–2000 cm^{-1} range. Raman gas analyzers serve a broad range of applications. They are employed in medicine, for instance; to monitor anaesthetic & respiratory gas combinations in real-time. Spontaneous Raman spectroscopy is also used for detection of temperature, characterization of materials & ascertain crystallographic orientation of sample in solid-state physics. It is frequently employed to characterize carbon nanotubes (CNTs) (Costa *et al.*, 2008). Key features in Raman spectrum of CNTs include:

G-Band (around 1580 cm^{-1}): Corresponds to in-plane vibration of sp^2-hybridized carbon atoms.
D-Band (around 1350 cm^{-1}): Indicates disorder & defects in carbon lattice.
Radial Breathing Mode (RBM, below 500 cm^{-1}): Provides information about diameter of single-walled carbon nanotubes (SWCNTs).

C. X-RAY-RELATED CHARACTERIZATION TECHNIQUES

Many characterization methods based on X-rays have been developed to analyze the elemental & structural characteristics of a wide range of materials, including low-dimensional nanomaterials. These techniques include XRD, X-ray Absorption scattering (XAS), X-Ray Fluorescence (XRF), X-Ray Photoelectron Spectroscopy (XPS), etc. (Titus *et al.*, 2019).

1. X-Ray Diffraction

It is non-destructive & rapid process, frequently applied as one of initial approaches for characterizing NPs. It is employed in the investigation of the phase composition, crystal structure & further structural characteristics of materials (Epp, 2016). X-ray diffraction depends on the constructive interference between a crystalline sample & monochromatic X-rays. When an X-ray beam interacts with a crystal lattice, it results in general scattering, but most of it is eliminated due to destructive interference. Diffraction happens when the scattered rays are in phase & combine through constructive interference, forming enhanced wavefronts. This process is described by Bragg's law. The intensity of X-rays is measured by detectors that diffract at various angles (2θ) which enables processing of signals with either an electronic devices or computer. Every crystalline substance, such as metal & metal oxide NPs, layered silicate nanoclays & semicrystalline polymers, possesses a distinct atomic structure that manifests a unique diffraction pattern. Graphically each peak represents a group of crystallographic planes in the sample. The structure of particles can be determined by comparing the position & intensity of the diffraction peaks with reference patterns provided by the International Centre for Diffraction Data (ICDD), formerly known as the Joint Committee on Powder Diffraction Standards (JCPDS) (Epp, 2016). It is also possible to determine crystallite size & strain using peak broadening Scherer equation (Bunaciu *et al.*, 2015).

Ramyadevi *et al.*, (2012) displayed the XRD pattern of the copper nanoparticle indicating the crystalline nature. A face-centred cubic structure of copper was represented by the peaks at 43.3165°, 50.4478° & 74.1237°, which correspond to the Miller indices (111), (200) & (220) respectively. Without any impurity phases like copper oxide (CuO), cuprous oxide (Cu_2O), or copper hydroxide ($Cu(OH)_2$), the synthesized NPs had been proven to be phase-pure copper.

XRD requires crystalline samples, as amorphous materials do not produce distinct patterns. Overlapping peaks in complex mixtures complicate phase identification & it is less sensitive to low concentrations of phases. XRD analysis is limited to the surface layer in bulk materials & thin films may face issues related to film thickness & substrate effects. Prolonged X-ray exposure can damage sensitive samples, particularly biological ones & the preferred orientation of crystallites can affect peak intensity, leading to inaccurate results (Kim *et al.*, 2013).

2. X-Ray Photoelectron Spectroscopy

XPS also known electron spectroscopy for chemical analysis (ESCA), is quantitative method for surface analysis. It is utilized to investigate elemental composition, chemical state & electronic state of elements within nanomaterials. This method offers valuable information about surface properties of materials at the nanoscale (Stevie & Donley, 2020).

XPS based on the photoelectric effect, which states that when a material is subjected to X-ray radiation, photoelectrons are emitted from its surface. By measuring & evaluating the kinetic energy of these liberated electrons, the binding energy which represents elements & their chemical states in material is determined. The elements present on the surface of sample can be specified by scanning a broad range of binding energies which specifically provide information regarding the oxidation states & chemical surroundings of element (Aziz & Ismail, 2017). There are some limitations to the types of samples to be analysed requiring ultra-high vacuum (UHV) conditions & it is incapable of identifying helium (Z=2) or hydrogen (Z=1) since they only have valence orbitals instead of core electron orbitals (Krishna & Philip, 2022).

3. X-Ray Absorption Spectroscopy

XAS measures the amount of X-rays absorbed by a sample element when its energy is modified. This enables the assessment of both structural & electrical properties of materials. X-ray Absorption Near Edge Structure (XANES) & Extended X-ray Absorption Fine Structure (EXAFS) are two branches of this technology. XANES covers the range from around 10 eV below to about 50–100 eV above the absorption edge, whereas EXAFS extends from roughly 50–100 eV above the absorption edge to about 1000 eV beyond (Yano & Yachandra, 2009).

XANES described the oxidation state, coordination environment & electronic structure of the absorbing atom, whilst the locations, coordination numbers & variability of surrounding absorbing atoms & adjacent atoms were described by EXAFS. However, the analysis of XANES & EXAFS data can be challenging & usually requires specific expertise & equipment (Yano & Yachandra, 2009). A study conducted by Porcaro *et al.*, (2016) synthesized AgNPs modified with 1-β-thio-D-glucose (TG) & 3-mercapto-1-propansulfonate (3MPS) capping agents. XPS & XAS were employed to examine structural & electrical characteristics of nano-systems.

4. X-Ray Fluorescence

X-ray fluorescence (XRF) is an analytical technique that identifies the elemental composition of materials. It works by detecting the unique "fluorescent" X-rays emitted from a sample when it is exposed to intense X-ray or gamma-ray radiation. This method allows for the non-destructive analysis of various substances & provide valuable information about their elemental make-up. The sample is subjected to radiation from an X-ray source, which prompts ejection of the inner-shell electrons (often the K or L shell) of an atom which is replaced by electrons from outer shells, or higher energy levels. An X-ray photon is generated from the energy difference between lower-energy & higher-energy layers within an atom. When released, this photon carries energy that is characteristic of the elements present in the specimen (Montanha *et al.*, 2020). Paralikar and Rai (2018) synthesized sulphur nanoparticles from medicinal plants, graph obtained from XRF data showed the highest percentage of sulphur (98.33%).

D. DYNAMIC LIGHT SCATTERING

DLS is well-known, non-invasive method for analysing size distribution of small particles in solution. It is additionally referred as Quasi-Elastic Light Scattering (QELS) or Photon Correlation Spectroscopy (PCS) (Falke and Betzel 2019).

The fundamental idea of DLS is to examine variations in the intensity of scattered light of sample over time. The size of particles influences how quickly intensity of scattered light changes because smaller particles are accelerated by solvent molecules & move more quickly when exposed to a laser. These intensity variations can be assessed with the Stokes-Einstein equation to calculate the particle size (radius) & Brownian motion velocity (Sandhu *et al.*, 2018). Narware *et al.*, (2023) measured particle size & size distribution profile of dispersed AgNPs using DLS. They discovered that the average AgNPs particle size was 43.68 nm. Despite a few drawbacks like the requirement for diluted samples & the assumption of spherical particles DLS has remained a widely used technique for describing proteins, polymers & colloids.

1. Nanoparticle Tracking Analysis

NTA is comparable to DLS & employs the free diffusion behaviour of particles in solution to offer comprehensive information about size distribution & concentration of NPs. NTA integrates optical microscopy & image analysis to track the movement of NPs (Tian *et al.*, 2013).

In NTA, the sample chamber is filled with a suspension of NPs that is irradiated by a laser. The emitted dispersed light from the NPs is captured by a camera & microscope. The movement of the NPs has been recorded using the camera throughout time. The software tracks the trajectory of each nanoparticle by analysing the video recordings. The diffusion coefficient is computed using the displacement of each particle over time. The hydrodynamic diameter can be calculated by converting diffusion coefficient through the Stokes-Einstein equation. A size distribution is produced by the software using a basis of the tracked particles (James and Driskell 2013). The number of particles inside the observable volume is counted to determine the particle concentration in the suspension. NTA can analyse data in real-time & requires no sample preparation. However, there are some drawbacks, such as the demand for diluted samples & sensitivity to changes in refractive index (Patois *et al.*, 2012).

E. ZETA POTENTIAL

ZP serves significant role in comprehension & management of colloidal dispersion stability. It offers significant information about the characteristics of particle surface charge & method of interaction in suspension (Bhattacharjee, 2016). Electrophoretic light scattering (ELS) or electroacoustic methods are commonly employed for measuring the ZP. When an electric field is applied, charged particles in a colloidal dispersion migrate towards the electrode with the opposite charge. This speed of movement is determined by the ZP. Doppler Effect of scattered light is correlated with particle velocity, which is a change in frequency driven by the moving particles. The particle velocity determines the electrophoretic mobility & is subsequently converted to ZP using the Henry equation (Hunter, 2013).

Particles with a negative ZP will be drawn to positively charged surfaces & vice versa. Particle stability can be inferred from the magnitude of ZP. The higher magnitude is due to the enhanced electrostatic repulsion indicating increased stability. Higher values (usually greater than ± 30 mV) indicate stable dispersions, whereas lowered ones imply possible aggregation or instability (Clogston & Patri 2011). Measurements of ZP can be influenced by ionic strength of solution, hence careful control of the experimental setup may be necessary. Samples must frequently be diluted for accurate measurements to prevent numerous scattering effects (Kathe *et al.*, 2014).

CHARACTERIZATION CHALLENGES FOR NANOMATERIALS

Nanomaterials exhibit few challenges in their characterization because of their small size, more surface area & intricate nanoscale interactions (Baer *et al.*, 2007), which are as follows:

- Accurate visualisation of nanoscale features is impossible with conventional microscopy methods like optical microscopy due to their lack of resolution. Advanced methods like SEM & TEM are necessary but they are more complex & expensive besides requiring proper sample preparation (Inkson, 2016).

- The characteristics of nanomaterials can be altered by aggregation caused by van der Waals forces, electrostatic interactions & other factors, which may have an impact on measurement reliability.
- Different chemical groups are frequently used to functionalize nanomaterials; these groups might be diverse & challenging to characterise consistently.
- Different levels of crystallinity in nanomaterials can be seen, along with flaws & amorphous areas that can be challenging to identify & examine.
- Accurately detecting & measuring NPs is a challenge for many approaches, particularly when the concentrations are extremely low (Modena *et al.*, 2019).
- Since there are no standards procedures or reference materials for nanomaterial synthesis, different research labs may produce contradictory results.

To address these challenges & unlock the complete potential of nanomaterials across various applications, more research & development of new instruments, standardized procedures & enhanced data analysis techniques will be necessary.

CONCLUSION

Nanotechnology is an innovative field of science & engineering that offers myriads of opportunities for atomic & molecular manipulation of materials. This emerging discipline has sparked advancements in several fields, including electronics, materials science, environmental technology & medicine. Characterising nanomaterials is a complex & challenging task, but extremely important to the advancement of nanotechnology applications. Nanomaterials have special attributes because of their high surface area & nanoscale dimensions, which make it necessary to evaluate their size, shape, composition, surface characteristics & dispersion stability using sophisticated techniques including TEM, SEM, AFM & DLS. Despite phenomenal advancements in this domain, environmental sensitivity, aggregation & the absence of defined standards are major bottlenecks. To address these barriers, characterization approaches must be continuously innovative, reliable & repeatable processes must be created. Removal of these hurdles will enable the realisation of the full potential of nanomaterials thereby paving the path for groundbreaking innovations in a broad spectrum of scientific & industrial domains.

REFERENCES

Akash, M. S. H., Rehman, K., Akash, M. S. H., & Rehman, K. (2020). Ultraviolet-visible (UV-VIS) spectroscopy. *Essentials of pharmaceutical analysis*, 29-56.

Akhtar, K., Khan, S. A., Khan, S. B., & Asiri, A. M. (2018). Scanning Electron Microscopy: Principle and Applications in Nanomaterials Characterization. *Handbook of Materials Characterization*, 113–145. DOI: 10.1007/978-3-319-92955-2_4

Aziz, M., & Ismail, A. F. (2017). X-ray photoelectron spectroscopy (XPS). In *Membrane Characterization* (pp. 81-93), Elsevier.

Baer, D. R., Amonette, J. E., Engelhard, M. H., Gaspar, D. J., Karakoti, A. S., Kuchibhatla, S., Nachimuthu, P., Nurmi, J. T., Qiang, Y., Sarathy, V., Seal, S., Sharma, A., Tratnyek, P. G., & Wang, C. M. (2008). Characterization challenges for nanomaterials. *Surface and Interface Analysis*, 40(3-4), 529–537. DOI: 10.1002/sia.2726

Baer, D. R., Engelhard, M. H., Johnson, G. E., Laskin, J., Lai, J., Mueller, K., Munusamy, P., Thevuthasan, S., Wang, H., Washton, N., Elder, A., Baisch, B. L., Karakoti, A., Kuchibhatla, S., & Moon, D. (2013). Surface characterization of nanomaterials and nanoparticles: Important needs and challenging opportunities. *Journal of Vacuum Science & Technology. A, Vacuum, Surfaces, and Films*, 31(5), 050820. DOI: 10.1116/1.4818423 PMID: 24482557

Baig, N., Kammakakam, I., & Falath, W. (2021). Nanomaterials: A review of synthesis methods, properties, recent progress, and challenges. *Materials Advances*, 2(6), 1821–1871. DOI: 10.1039/D0MA00807A

Banerjee, A., Sarkar, A., Acharya, K., & Chakraborty, N. (2021). Nanotechnology: An Emerging Hope in Crop Improvement. *Letters in Applied NanoBioScience*, 10(4), 2784–2803. DOI: 10.33263/LIANBS104.27842803

Bhattacharjee, S. (2016). DLS and zeta potential–what they are and what they are not? *Journal of Controlled Release*, 235, 337–351. DOI: 10.1016/j.jconrel.2016.06.017 PMID: 27297779

Bian, K., Gerber, C., Heinrich, A. J., Müller, D. J., Scheuring, S., & Jiang, Y. (2021). Scanning probe microscopy. *Nature Reviews. Methods Primers*, 1(1), 36. DOI: 10.1038/s43586-021-00033-2

Bocker, U., Wubshet, S. G., Lindberg, D., & Afseth, N. K. (2017). Fourier-transform infrared spectroscopy for characterization of protein chain reductions in enzymatic reactions. *Analyst*, 142(15), 2812–2818. DOI: 10.1039/C7AN00488E PMID: 28686252

Bogner, A., Jouneau, P. H., Thollet, G., Basset, D., & Gauthier, C. (2007). A history of scanning electron microscopy developments: Towards "wet-STEM" imaging. *Micron (Oxford, England)*, 38(4), 390–401. DOI: 10.1016/j.micron.2006.06.008 PMID: 16990007

Bonde, S. R., Rathod, D. P., Ingle, A. P., Ade, R. B., Gade, A. K., & Rai, M. K. (2012). Murraya koenigii-mediated synthesis of silver nanoparticles and its activity against three human pathogenic bacteria. *Nanoscience Methods*, 1(1), 25–36. DOI: 10.1080/17458080.2010.529172

Bunaciu, A. A., Udriştioiu, E., & Aboul-Enein, H. Y. (2015). X-ray diffraction: Instrumentation and applications. *Critical Reviews in Analytical Chemistry*, 45(4), 289–299. DOI: 10.1080/10408347.2014.949616 PMID: 25831472

Chen, C. J. (2021). *Introduction to Scanning Tunneling Microscopy* (3rd ed., Vol. 69). Oxford University Press. DOI: 10.1093/oso/9780198856559.001.0001

Chirayil, C. J., Abraham, J., Mishra, R. K., George, S. C., & Thomas, S. (2017). Instrumental techniques for the characterization of nanoparticles. In *Thermal and rheological measurement techniques for nanomaterials characterization* (pp. 1–36). Elsevier. DOI: 10.1016/B978-0-323-46139-9.00001-3

Choel, M., Deboudt, K., Osan, J., Flament, P., & Van Grieken, R. (2005). Quantitative determination of low-Z elements in single atmospheric particles on boron substrates by automated scanning electron microscopy– energy-dispersive x-ray spectrometry. *Analytical Chemistry*, 77(17), 5686–5692. DOI: 10.1021/ac050739x PMID: 16131082

Clogston, J. D., & Patri, A. K. (2011). Zeta Potential Measurement. In McNeil, S. (Ed.), *Characterization of Nanoparticles Intended for Drug Delivery. Methods in Molecular Biology, 697*. Humana Press., DOI: 10.1007/978-1-60327-198-1_6

Costa, S., Borowiak-Palen, E., Kruszynska, M., Bachmatiuk, A., & Kalenczuk, R. J. (2008). Characterization of carbon nanotubes by Raman spectroscopy. *Materials Science Poland*, 26(2), 433–441.

Devi, L. S., & Joshi, S. R. (2014). Evaluation of the antimicrobial potency of silver nanoparticles biosynthesized by using an endophytic fungus, Cryptosporiopsis ericae PS4. *Journal of Microbiology (Seoul, Korea)*, 52(8), 667–674. DOI: 10.1007/s12275-014-4113-1 PMID: 24994011

Dolez, P. I. (2015). Nanomaterials definitions, classifications, and applications. In *Nanoengineering* (pp. 3–40). Elsevier. DOI: 10.1016/B978-0-444-62747-6.00001-4

Dufrene, Y. F., Ando, T., Garcia, R., Alsteens, D., Martinez-Martin, D., Engel, A., Gerber, C., & Muller, D. J. (2017). Imaging modes of atomic force microscopy for application in molecular and cell biology. *Nature Nanotechnology*, 12(4), 295–307. DOI: 10.1038/nnano.2017.45 PMID: 28383040

Echlin, P. (2011). *Handbook of sample preparation for scanning electron microscopy and X-ray microanalysis*. Springer Science & Business Media.

Epp, J. (2016). X-ray diffraction (XRD) techniques for materials characterization. In *Materials characterization using nondestructive evaluation (NDE) methods* (pp. 81-124). Woodhead Publishing.

Epp, J. (2016). X-ray diffraction (XRD) techniques for materials characterization. In *Materials characterization using nondestructive evaluation (NDE) methods* (pp. 81-124). Woodhead Publishing.

Falke, S., & Betzel, C. (2019). Dynamic light scattering (DLS) principles, perspectives, applications to biological samples. *Radiation in bioanalysis: Spectroscopic techniques and theoretical methods*, 173-193.

Fleck, R. A., & Humbel, B. M. (Eds.). (2019). *Biological Field Emission Scanning Electron Microscopy, 2 Volume Set*. John Wiley & Sons. DOI: 10.1002/9781118663233

Girao, A. V., Caputo, G., & Ferro, M. C. (2017). Application of Scanning Electron Microscopy–Energy Dispersive X-Ray Spectroscopy (SEM-EDS). *Comprehensive Analytical Chemistry*, 75, 153–168. DOI: 10.1016/bs.coac.2016.10.002

Goldstein, J. I., Newbury, D. E., Michael, J. R., Ritchie, N. W., Scott, J. H. J., & Joy, D. C. (2017). *Scanning electron microscopy and X-ray microanalysis. Springer*, Goodhew, P. (2011). General introduction to transmission electron microscopy (TEM). *Corrected Analytical Transmission Electron Microscopy*, 1-19.

Gupta, P., Rai, N., Verma, A., & Gautam, V. (2024). Microscopy based methods for characterization, drug delivery, and understanding the dynamics of nanoparticles. *Medicinal Research Reviews*, 44(1), 138–168. DOI: 10.1002/med.21981 PMID: 37294298

Hasan, S. A. (2015). A review on nanoparticles: Their synthesis and types. *Research Journal of Recent Sciences*, 4, 9–11.

Herzing, A. A., Watanabe, M., Edwards, J. K., Conte, M., Tang, Z.-R., Hutchings, G. J., & Kiely, C. J. (2008). Energy dispersive X-ray spectroscopy of bimetallic nanoparticles in an aberration corrected scanning transmission electron microscope. *Faraday Discussions*, 138, 337–351. DOI: 10.1039/B706293C PMID: 18447024

Hodoroaba, V. D. (2020). Energy-dispersive X-ray spectroscopy (EDS). Characterization of Nanoparticles, 397–417. DOI: 10.1016/B978-0-12-814182-3.00021-3

Hodoroaba, V.-D. "Energy-dispersive X-ray spectroscopy (EDS)." In *Characterization of Nanoparticles*, pp. 397-417. Elsevier, 2020. DOI: 10.1016/B978-0-12-814182-3.00021-3

Hodoroaba, V. D., Unger, W., & Shard, A. (Eds.). (2019). *Characterization of nanoparticles: measurement processes for nanoparticles*. Elsevier.

Hollander, A., & Danino, D. (2015). Cochleate characterization by cryogenic electron microscopy methods: Cryo-TEM and Cryo-SEM. *Colloids and Surfaces. A, Physicochemical and Engineering Aspects*, 483, 187–192. DOI: 10.1016/j.colsurfa.2015.07.025

Hoyt, M. (2009). Analyses of Nanoparticles in the Environment. *Nanotechnology and the Environment*. CRC Press 99-122 (2009)

Hunter, R. J. (2013). *Zeta potential in colloid science: principles and applications* (Vol. 2). Academic press.

Inkson, B. J. (2016). Scanning electron microscopy (SEM) and transmission electron microscopy (TEM) for materials characterization. In *Materials characterization using nondestructive evaluation (NDE) methods* (pp. 17-43). Woodhead publishing.

Jain, N., Bhargava, A., Tarafdar, J. C., Singh, S. K., & Panwar, J. (2012). A biomimetic approach towards synthesis of zinc oxide nanoparticles. *Applied Microbiology and Biotechnology*, 97(2), 859–869. DOI: 10.1007/s00253-012-3934-2 PMID: 22382164

James, A. E., & Driskell, J. D. (2013). Monitoring gold nanoparticle conjugation and analysis of biomolecular binding with nanoparticle tracking analysis (NTA) and dynamic light scattering (DLS). *Analyst*, 138(4), 1212–1218. DOI: 10.1039/c2an36467k PMID: 23304695

Jose Chirayil, C., Abraham, J., Kumar Mishra, R., George, S. C., & Thomas, S. (2017). Instrumental Techniques for the Characterization of Nanoparticles. Thermal and Rheological Measurement Techniques for Nanomaterials Characterization, 1–36. DOI: 10.1016/B978-0-323-46139-9.00001-3

Joshi, M., Bhattacharyya, A., & Ali, S. W. (2008). Characterization techniques for nanotechnology applications in textiles. Joshi, M. *Indian Journal of Fibre and Textile Research*, 33, 304–317.

Kathe, N., Henriksen, B., & Chauhan, H. (2014). Physicochemical characterization techniques for solid lipid nanoparticles: Principles and limitations. *Drug Development and Industrial Pharmacy*, 40(12), 1565–1575. DOI: 10.3109/03639045.2014.909840 PMID: 24766553

Kazakova, O., Puttock, R., Barton, C., Corte-León, H., Jaafar, M., Neu, V., & Asenjo, A. (2019). Frontiers of magnetic force microscopy. *Journal of Applied Physics*, 125(6), 060901. DOI: 10.1063/1.5050712

Khalid, H., Shamaila, S., Zafar, N., & Shahzadi, S. (2015). Synthesis of copper nanoparticles by chemical reduction method. *SCIENCE International*, 27(4), 3085–3088.

Khan, I., Saeed, K., & Khan, I. (2019). Nanoparticles: Properties, applications and toxicities. *Arabian Journal of Chemistry*, 12(7), 908–931. DOI: 10.1016/j.arabjc.2017.05.011

Kim, S. H., Lee, C. M., & Kafle, K. (2013). Characterization of crystalline cellulose in biomass: Basic principles, applications, and limitations of XRD, NMR, IR, Raman, and SFG. *Korean Journal of Chemical Engineering*, 30(12), 2127–2141. DOI: 10.1007/s11814-013-0162-0

Kim, Y., Kim, W., & Park, J. W. (2016). Principles and applications of force spectroscopy using atomic force microscopy. *Bulletin of the Korean Chemical Society*, 37(12), 1895–1907. DOI: 10.1002/bkcs.11022

Klaholz, B. P. (2019). Deriving and refining atomic models in crystallography and cryo-EM: The latest Phenix tools to facilitate structure analysis. *Acta Crystallographica. Section D, Structural Biology*, 75(10), 878–881. DOI: 10.1107/S2059798319013391 PMID: 31588919

Krishna, D. N. G., & Philip, J. (2022). Review on surface-characterization applications of X-ray photoelectron spectroscopy (XPS): Recent developments and challenges. *Applied Surface Science Advances*, 12, 100332. DOI: 10.1016/j.apsadv.2022.100332

Larkin, P. (2017). *Infrared and Raman spectroscopy: principles and spectral interpretation*. Elsevier.

Ma, H., Shieh, K. J., & Qiao, T. X. (2006). Study of transmission electron microscopy (TEM) and scanning electron microscopy (SEM). *Nature and Science*, 4(3), 14–22.

Marques, A. C., Rojas-Hernandez, R. E., & Almeida, R. M. (2021). Optical spectroscopy methods for the characterization of sol–gel materials. *Journal of Sol-Gel Science and Technology*, 100(1), 1–43. DOI: 10.1007/s10971-021-05592-0

Menon, S., Rajeshkumar, S., & Venkat Kumar, S. (2017). A review on biogenic synthesis of gold nanoparticles, characterization, and its applications. *Resource-Efficient Technologies*, 3(4), 516–527. DOI: 10.1016/j.reffit.2017.08.002

Mino, L., Borfecchia, E., Segura-Ruiz, J., Giannini, C., Martinez-Criado, G., & Lamberti, C. (2018). Materials characterization by synchrotron x-ray microprobes and nanoprobes. *Reviews of Modern Physics*, 90(2), 025007. DOI: 10.1103/RevModPhys.90.025007

Modena, M. M., Rühle, B., Burg, T. P., & Wuttke, S. (2019). Nanoparticle Characterization: What to Measure? *Advanced Materials*, 1901556(32), 1901556. Advance online publication. DOI: 10.1002/adma.201901556 PMID: 31148285

Montanha, G. S., Rodrigues, E. S., Marques, J. P. R., De Almeida, E., Dos Reis, A. R., & Pereira de Carvalho, H. W. (2020). X-ray fluorescence spectroscopy (XRF) applied to plant science: Challenges towards in vivo analysis of plants. *Metallomics*, 12(2), 183–192. DOI: 10.1039/c9mt00237e PMID: 31793600

Mulvaney, S. P., & Keating, C. D. (2000). Raman spectroscopy. *Analytical Chemistry*, 72(12), 145–158. DOI: 10.1021/a10000155 PMID: 10882205

Nan, C., Yue, W., Tao, L., & Yang, X. (2021). Fourier transform infrared nano-spectroscopy: Mechanism and applications. *Applied Spectroscopy Reviews*, 56(7), 531–552. DOI: 10.1080/05704928.2020.1830789

Narware, J., Singh, S. P., Manzar, N., & Kashyap, A. S. (2023). Biogenic synthesis, characterization, and evaluation of synthesized nanoparticles against the pathogenic fungus Alternaria solani. *Frontiers in Microbiology*, 14, 1159251. DOI: 10.3389/fmicb.2023.1159251 PMID: 37138620

Paralikar, P., & Rai, M. (2018). Bio-inspired synthesis of sulphur nanoparticles using leaf extract of four medicinal plants with special reference to their antibacterial activity. *IET Nanobiotechnology / IET*, 12(1), 25–31. DOI: 10.1049/iet-nbt.2017.0079

Patois, E., Capelle, M. A. H., Palais, C., Gurny, R., & Arvinte, T. (2012). Evaluation of nanoparticle tracking analysis (NTA) in the characterization of therapeutic antibodies and seasonal influenza vaccines: Pros and cons. *Journal of Drug Delivery Science and Technology*, 22(5), 427–433. DOI: 10.1016/S1773-2247(12)50069-9

Pennycook, S. J., Lupini, A. R., Varela, M., Borisevich, A., Peng, Y., Oxley, M. P., Benthem, K. V., Chisholm, M. F. (2006). Scanning Transmission Electron Microscopy for Nanostructure Characterization. *Scanning Microscopy for Nanotechnology*, 152–191. DOI: 10.1007/978-0-387-39620-0_6

Poloni, L. N., Zhong, X., Ward, M. D., & Mandal, T. (2016). Best Practices for Real-Time in Situ Atomic Force and Chemical Force Microscopy of Crystals. *Chemistry of Materials*, 29(1), 331–345. DOI: 10.1021/acs.chemmater.6b03082

Porcaro, F., Carlini, L., Ugolini, A., Visaggio, D., Visca, P., Fratoddi, I., Venditti, I., Meneghini, C., Simonelli, L., Marini, C., Olszewski, W., Ramanan, N., Luisetto, I., & Battocchio, C. (2016). Synthesis and Structural Characterization of Silver Nanoparticles Stabilized with 3-Mercapto-1-Propansulfonate and 1-Thioglucose Mixed Thiols for Antibacterial Applications. *Materials (Basel)*, 9(12), 1028. DOI: 10.3390/ma9121028 PMID: 28774148

Ramyadevi, J., Jeyasubramanian, K., Marikani, A., Rajakumar, G., & Rahuman, A. A. (2012). Synthesis and antimicrobial activity of copper nanoparticles. *Materials Letters*, 71, 114–116. DOI: 10.1016/j.matlet.2011.12.055

Raval, N., Maheshwari, R., Kalyane, D., Youngren-Ortiz, S. R., Chougule, M. B., & Tekade, R. K. (2019). Importance of physicochemical characterization of nanoparticles in pharmaceutical product development. In *Basic fundamentals of drug delivery* (pp. 369–400). Academic Press. DOI: 10.1016/B978-0-12-817909-3.00010-8

Reimer, L. (2013). Transmission electron microscopy: physics of image formation and microanalysis. *Springer,* 36

Sajid, M., & Płotka-Wasylka, J. (2020). Nanoparticles: Synthesis, characteristics, and applications in analytical and other sciences. *Microchemical Journal*, 154, 104623. DOI: 10.1016/j.microc.2020.104623

Sandhu, R., Singh, N., Dhankhar, J., Kama, G., & Sharma, R. (2018). Dynamic light scattering (DLS) technique, principle, theoretical considerations and applications. *Nanotechnological and Biochemical Techniques for Assessing the Quality and Safety of Milk and Milk Products*, 135-137.

Sathyanarayana, D. N. (2015). *Vibrational spectroscopy: theory and applications*. New Age International.

Schneider, R. (2011). Electron Energy-Loss Spectroscopy (EELS) and Energy-Filtering Transmission Electron Microscopy (EFTEM). *Surface and Thin Film Analysis: A Compendium of Principles, Instrumentation, and Applications*, 67-91.

Serry, F. M., & Schmit, J. (2019). Characterization and Measurement of Microcomponents with the Atomic Force Microscope (AFM). In *Optical Inspection of Microsystems, Second Edition*, CRC Press, 155-176.

Shankar, S., Pangeni, R., Park, J. W., & Rhim, J. W. (2018). Preparation of sulfur nanoparticles and their antibacterial activity and cytotoxic effect. *Materials Science and Engineering C*, 92, 508–517. DOI: 10.1016/j.msec.2018.07.015 PMID: 30184776

Sofi, W., Gowri, M., Shruthilaya, M., Rayala, S., & Venkatraman, G. (2012). Silver nanoparticles as an antibacterial agent for endodontic infections. *BMC Infectious Diseases*, 12(S1), 60. DOI: 10.1186/1471-2334-12-S1-P60

Stevie, F. A., & Donley, C. L. (2020). Introduction to X-ray photoelectron spectroscopy. *Journal of Vacuum Science & Technology. A, Vacuum, Surfaces, and Films*, 38(6), 063204. DOI: 10.1116/6.0000412

Sun, W. (2018). Principles of atomic force microscopy. *Atomic Force Microscopy in Molecular and Cell Biology*, 1-28.

Taha, M., Hassan, M., Essa, S., & Tartor, Y. (2013). Use of Fourier transform infrared spectroscopy (FTIR) spectroscopy for rapid and accurate identification of yeasts isolated from human and animals. *International Journal of Veterinary Science and Medicine*, 1(1), 15–20. DOI: 10.1016/j.ijvsm.2013.03.001

Tamilselvan, S., Ashokkumar, T., & Govindaraju, K. (2017). Microscopy based studies on the interaction of bio-based silver nanoparticles with Bombyx mori Nuclear Polyhedrosis virus. *Journal of Virological Methods*, 242, 58–66. DOI: 10.1016/j.jviromet.2017.01.001 PMID: 28065747

Tanaka, N., & Saitoh, K. (2015). Basics of STEM. *In Scanning Transmission Electron Microscopy of Nanomaterials: Basics of Imaging And Analysis* (pp. 41-108).

Tian, X., Nejadnik, M. R., Baunsgaard, D., Henriksen, A., Rischel, C., & Jiskoot, W. (2016). A comprehensive evaluation of nanoparticle tracking analysis (NanoSight) for characterization of proteinaceous submicron particles. *Journal of Pharmaceutical Sciences*, 105(11), 3366–3375. DOI: 10.1016/j.xphs.2016.08.009 PMID: 27663383

Titus, D., James Jebaseelan Samuel, E., & Roopan, S. M. (2019). Nanoparticle characterization techniques. Green Synthesis, Characterization and Applications of Nanoparticles, 303–319. DOI: 10.1016/B978-0-08-102579-6.00012-5

Watt, J., Huber, D. L., & Stewart, P. L. (2019). Soft matter and nanomaterials characterization by cryogenic transmission electron microscopy. *MRS Bulletin*, 44(12), 942–948. DOI: 10.1557/mrs.2019.285

Yang, Z., & Zhu, M. (2023). Case Studies: Ultraviolet-Visible (UV-Vis) Spectroscopy. In *Springer Handbook of Advanced Catalyst Characterization* (pp. 265–283). Springer International Publishing. DOI: 10.1007/978-3-031-07125-6_12

Yano, J., & Yachandra, V. K. (2009). X-ray absorption spectroscopy. *Photosynthesis Research*, 102(2-3), 241–254. DOI: 10.1007/s11120-009-9473-8 PMID: 19653117

Zhu, Y., & Durr, H. (2015). The future of electron microscopy. *Physics Today*, 68(4), 32–38. DOI: 10.1063/PT.3.2747

KEY TERMS AND DEFINITIONS

Nanotechnology: Nanotechnology is the study and application of extremely small particles ranging from 1 nm to 100 nm.

Nanoparticles: A nanoparticle is a small particle with sizes ranging from 1 to 100 nanometers.

Characterization Techniques: Characterization techniques are methods that are used to obtain information about the chemical, physical, mechanical, and electrical properties of materials.

Microscopy: Microscopy is used to observe samples and objects that are not visible with the unaided eye with the help of a microscope.

Spectroscopy: Spectroscopy is the study of the absorption and emission of light and other radiation by matter.

Analysis: This refers to the thorough examination of complex entities to understand their nature or determine their essential characteristics.

Properties: Properties are the characteristics that enable us to differentiate one material from another.

Colloidal Stability: The ability of nanoparticles to remain evenly dispersed in a solution without aggregation, influenced by factors like surface charge and particle size.

Morphology: The study of the shape and structure of nanoparticles, which affects their physical and chemical properties.

Particle Size Distribution: A representation of the range of particle sizes in a sample, critical for understanding the behaviour and application of nanoparticles.

Chapter 13 Plant-Based Inorganic Quantum Dots: Synthesis, Characterisation, and Applications

Tejas Chalke
University Institute of Chemical Technology, Kavayitri Bahinabai Chaudhari North Maharashtra University, India

Madhura Bodkhe
https://orcid.org/0009-0002-0066-3405
University Institute of Chemical Technology, Kavayitri Bahinabai Chaudhari North Maharashtra University, India

Sunil Jayant Kulkarni
https://orcid.org/0000-0002-5988-3448
Gharda Institute of Technology, Lavel, India

Ajaygiri K. Goswami
University Institute of Chemical Technology, Kavayitri Bahinabai Chaudhari North Maharashtra University, India

ABSTRACT

The appearance of plant-based inorganic quantum dots (QDs) has attracted a lot of attention in the field of nanotechnology. This is because of the unique optical and electrical properties that these QDs possess. These QDs show potential in a variety of sectors, including biomedical imaging and optoelectronic devices. This growing interest may be traced back to the necessity of finding environmentally safe alternatives to traditional methodological approaches to the synthesis of quantum dots, which frequently include the use of potentially harmful chemicals. Leveraging the inherent abundance, biodegradability, and chemical variety of plants, researchers are digging into sustainable techniques for QD synthesis. The extraction of bioactive chemicals from plant sources is typically required for this process. These compounds serve a dual purpose in the creation of nanocrystals, functioning as both stabilizers and reducing agents.

DOI: 10.4018/979-8-3693-6240-2.ch013

INTRODUCTION

With dimensions ranging from one to ten nanometres, semiconductor nanoparticles (NPs) acknowledged as quantum dots (QDs) are a distinct class of particles (Demir and Castranova, 2017). They received a lot of heed because of their unique characteristics, which span from visual to electrical. They have the potential to drastically change the industrial landscape and are currently extensively employed for a range of cutting-edge uses in medicine biology, agriculture and chemistry (Gupta et al., 2022). Their properties such as physical and chemical have been found to be different from bulk substances (Pan et al., 2013; Danish et al., 2014; Bakar et al., 2022). Numerous applications, such as solar cells, fuel cells, antibacterial activities and photocatalysts, can be enhanced by the strong reactivity and high surface-to-volume ratio of the produced QDs (Bakar et al., 2022). The expanding demand and usage of QDs in various applications provide a risk that these semiconducting QDs will accumulate and leak into the environment, with potentially differing effects on environmental and human health and safety. As a result, it is more important than ever to precisely analyze how exposure to QDs affects biota. According to Wang and Nowack (Wang et al., 2018), most QDs enter the ecosystem by non-direct routes (such dispersion after usage). However, it is evident that these materials can interact with function and biota as active components of trophic food webs; these interactions can be amplified, modified, or both (Gardea-Torresdey et al., 2014). This type of interaction may become important when the materials are applied to agricultural and food production products. Quantum dots can be made utilising a range of methods, including ruthenium-catalysed C60 transformation (Lu et al., 2011), electrochemical techniques like cyclic voltammetry (CV) (Li et al., 2012a), electron beam lithography (Li et al., 2012b), and microwave assisted hydrothermal methods (Tang et al., 2012). The Readying of quantum dots (QDs) using various natural reducing agents, such as extracts of roots, leaves, flowers, fruit, seeds, whole plants, etc., has garnered significant attention recently for greener synthesis approaches (Vanlalveni et al., 2021; Siddiqi et al., 2018; Nasseri et al., 2020; Yan et al., 2023; Centeno et al., 2021; Shivaji et al., 2018; Singh et al., 2018). Native plants are more affordable, environmentally friendly, readily available, and safer than many alternative chemical methods. Materials with nanoscale particle sizes have special physiochemical characteristics because of quantum confinement phenomena. QDs have been found to have distinct electrical and optical characteristics in contrast to their equivalents in bulk, which makes them appropriate for a range of novel applications. Inorganic quantum dots (QDs) derived from plants are a potential new area in biotechnology and nanotechnology that can be used to harness the amazing qualities of semiconductor materials in a friendly and sustainable way. By using plant ingredients and natural processes, these quantum dots can be synthesized in a novel way that deviates from traditional chemical approaches. Plant extracts, such as those from leaves, stems, or roots, are frequently used in plant-based synthesis methods as a source of biocompatible ligands and reducing agents. These bioactive substances, which include proteins, flavonoids, and phenolics, are essential for the nucleation, development, and stabilization of quantum dots, which results in the regulation of their surface chemistry, size and shape. The intrinsic environmental friendliness of plant-based synthesis is one of its main benefits. When compared to conventional synthetic methods, which frequently require dangerous chemicals and high-energy procedures, the environmental effect of producing QDs can be greatly minimized by employing renewable plant resources and green chemistry concepts. Furthermore, the possibility of large-scale production at comparatively cheap cost is provided by the scalability of plant-based synthesis, opening the door for extensive applications across a range of industries. Furthermore, plant-derived materials' biocompatibility offers great promise for biomedical uses. When compared to their chemically manufactured counterparts, plant-based quantum dots show less cytotoxicity, which makes them appropriate for application in biological systems (Nakum and Bhattacharya, 2022).

Since QDs are compatible with living things, they can be used in bioimaging, which allows for high-resolution imaging of cellular structures and functions thanks to the particles' fluorescent characteristics. In addition, it is possible to construct plant-based QDs for targeted medication administration by attaching therapeutic compounds and achieving regulated release profiles through the special surface chemistry of quantum dots. Plant-based quantum dots are promising beyond biomedical uses, with prospective applications in photocatalysis, optoelectronics, and environmental sensing. They are desirable options for light-emitting diodes (LEDs), solar cells, and sensors that identify environmental contaminants due to their customizable electrical and optical characteristics. Researchers hope to fully realize the promise of plant-based quantum dots and spur innovation toward multifunctional and sustainable nanotechnologies by investigating the connections between plant biochemistry and materials science. To ensure the stability and repeatability of plant-based quantum dots for useful applications, difficulties still exist in refining synthesis methods and comprehending the fundamental principles of plant-mediated synthesis. However, the quick development in this area highlights the possibility of using nature's processes to produce cutting-edge materials with never-before-seen capabilities. In order to synthesize Ag nanoparticles with an average particle diameter of 18 ± 0.16 nm, Ndikau et al. successfully devised a one-step green process using Citrullus lanatus fruit rind as a capping and reductant agent (Ndikau et al., 2017). Using cannonball leaf extract, Devaraj et al. presented a green synthesis method for AgNPs. The green produced AgNPs were found to inhibit the MCF-7 breast cancer cell line, according to a cytotoxicity investigation (Devaraj et al., 2013). Using a glucose reduction of aqueous $AgNO_3$ at various amount of concentration of NaOH on molar basis employed as a reaction hastener, Vasireddy et al., (2012) presented a straightforward green synthesis method to produce super-purity AgNPs. By adding gelatine, a biopolymer that is both biocompatible and biodegradable, they were able to control the aggregation of NPs (Vasireddy et al., 2012). Many researchers have published comparable works on the synthesis of the AgNPs utilizing natural reducing agents in the recent past (Sun et al., 2014; Philip et al., 2010; Verma and Mehata, 2016). Figure 1 depicts general overview of inorganic quantum dots. In this review paper, we reviewed the experimental methods used to synthesize and characterize the inorganic QDs. Applications, Properties of QDs are also studied here.

Figure 1. General Overview of Inorganic Quantum dots: Compiled by author

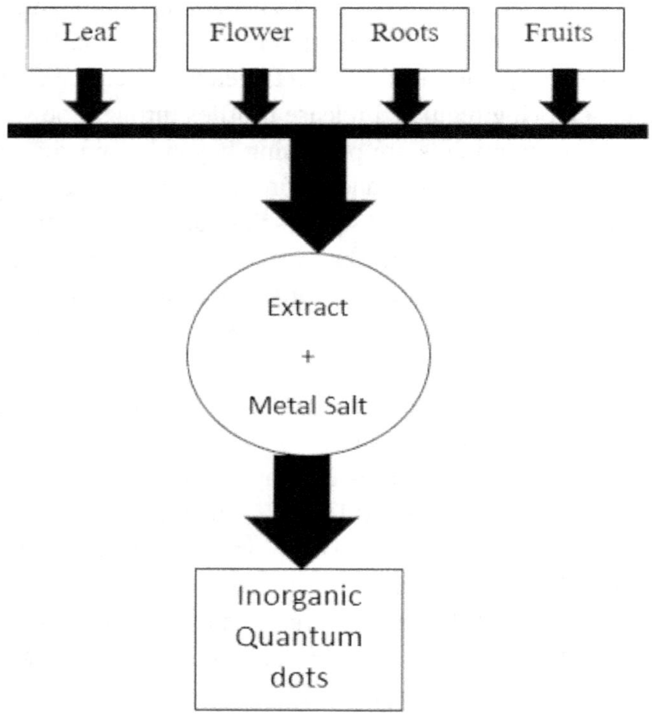

SYNTHESIS

Plant extracts contain bioactive compounds (e.g., polyphenols, flavonoids) that act as reducing and stabilizing agents. Typically involves mixing plant extracts with metal ion precursors (e.g., Cd, Pb, Zn) under controlled conditions (pH, temperature) to form QDs. Green Synthesis Approach includes Environmental Friendliness (Uses environmentally benign and sustainable resources) and Controlled Growth (Enables controlled shape, surface chemistry and size of QDs). Figure 2 depicts the synthesis of inorganic quantum dots. The procedure for making inorganic quantum dots from plant extracts and metal salts involves several key steps. Initially, the plant extract is prepared by collecting the desired plant material, washing it thoroughly, and drying it. The dried material is then ground into a fine powder, which is subsequently boiled in distilled water or an appropriate solvent to extract the bioactive compounds. The resulting extract is filtered to remove any solid residues. Next, a metal salt solution is prepared, usually containing a metal such as cadmium, zinc, or lead. The metal salt is dissolved in distilled water or another suitable solvent. The plant extract is then added to the metal salt solution, typically under constant stirring to ensure uniform mixing. The mixture is heated to a specific temperature, often between 60°C and 100°C, to facilitate the reduction of metal ions by the plant phytochemicals and promote the nucleation and growth of quantum dots. The reaction is allowed to proceed for a certain period, during which the quantum dots form and grow to the desired size. The colour of the solution typically changes as the quantum dots are synthesized, indicating the formation of nanoparticles. After the reaction is complete, the solution is cooled to room temperature. To isolate the quantum dots, the solution is subjected to centrifugation, which separates the quantum dots from any unreacted materials and by-products. The precipitated quantum dots are then washed with a solvent, such as ethanol or distilled water, to remove any impurities. Finally, the purified quantum dots are dried and stored for future use. These quantum dots can be characterized using techniques such as UV-Vis spectroscopy, transmission electron microscopy (TEM), and X-ray diffraction (XRD) to confirm their size, shape, and crystalline structure.

Synthesis of SiQDs

The single-step hydrothermal process that produced the luminous SiQDs was altered from a microwave preparation. 240 mL of deionized (DI) water was used to disperse 11.16 g of trisodium citrate dehydrate in a typical batch of the 300 mL progenitor mixture. The preceding solution was swirled for 20 minutes, and then 60 mL of N-(3 (trimethoxy silyl)propyl) ethylenediamine (DAMO) were introduced. The mixture was then continuously stirred for 40 minutes. The gas of nitrogen was added to the combined mixture to cause the oxygen that was present to evaporate. After that, the precursor solution was put in a stainless-steel sterilizer coated with Kevlar and warmed to 200 °C for 12 hours. The SiQDs underwent purification by dialysis, wherein leftover chemicals (MWCO 1 kDa) were eliminated in 10× DI water for 20 hours, following the typical procedure of cooling to a comfortable temperature. The SiQDs were frozen at the end.

Synthesis of Metal Chalcogenide QDs Originating of Residues From Agriculture Waste

The peels of papayas (*Carica papaya*) were extracted aqueously to create CdS QDs (Bhuvaneswari et al., 2015). Using EDX, prominent peaks for cadmium and sulfide were found. XRD examination showed that the CdS QDs were in their cubic phase. The synthesized CdS QDs varied in terms of size from 2.3 to 3.2 nm on average. Particulate with shape variation were seen in the images from the SEM. Pomelo, the (*Citrus maxima)* peeled extracted from water was investigated as a green biological waste for CdTe QD synthesis (Mahalakhsmi et al., 2019). The 6.6 nm-sized, exceptionally crystalline monodispersed microspheres were as-synthesised. In the spectrum of the FTIR, chemical groups such as carboxyl, hydroxyl, and amino were detected. XRD showed the crystalline form within the phyto-fabricated CdTe QDs. TGA, or thermogravimetric analysis, demonstrated how temperature affects the density of CdTe quantum dots. At 300 °C, a mass change of 29.4% was noted. Through a one-pot responses, the naturally occurring production of CdS QD particles was mediated by the liquid extract of ground watermelon (Citrullus lanatus) peels (Lakshmipathy et al., 2015). There were a lot of carbs in the produced aqueous extract. The FTIR spectrometry was used to further establish this. With a BEG of 2.57 eV, a maximum absorbance peak that emerged at 484 nm verified that the produced CdS particles were in the quantum entanglement zone. Using EDX, robust chemical indications for Cd and S were found to be free of contaminants. XRD patterns demonstrated the CdS QDs' hexa phase. The photosynthesis of CdS QDs was facilitated by the fruit sap of *Opuntia ficus-indica* (Kandasamy et al., 2020). The C–O primary alcohol's flexing vibrations and the aromatic (C–H) and amino (N–H) groups' bend vibrations were detected by FTIR. The BEG was 3.8 eV, and the greatest absorption crest of CdS QDs was seen at 323 nm. The artificially created particles had a crystalline, cubic shape and had no contaminants. The fluorescence from X-rays indicated that S (18.8%) and Cd (80.12%) were present. It was calculated that the average particle size was 9.56 nm.

Synthesis and Characterization of CdSe QDs Suspendable in Water

By carefully adjusting the response time, temperatures, molar ratios of L-Cysteine/Cd/Se, and pH in accordance with earlier published protocols, the dissolved in water CdSe QDs were effectively made. In order to prevent the CdSe QDs from growing too quickly or aggregating, L-cysteine was employed as a stabilizer. It also gave the tiny dots a breathable surface that allowed them to be utilized as in-vivo biolabels. The outcomes of characterizing the optical property of CdSe QDs. In comparison to CdSe (1.70 eV) mass substance that these novel dissolved in water CdSe QDs (0.1 mM) display a clear blue-shift and a wide absorbance spectrum, with a distinctive peak related to the 1S–1S transition at 380 nm (3.27 eV). The impact of nanotechnology was the cause of the change. Additionally, absorption spectroscopy is a useful tool for examining even minute variations in particle dimensions. The CdSe QDs (0.1 mM) luminescence the spectrum, which show a 520 nm-cantered limited fluorescence range and a comparatively significant degree of emission in comparison to other samples. The results of this experiment provide further evidence for the special "quantity containment effects" associated with a high degree of monodispersity in CdSe QDs. The fluorescence spectrum of CdSe QDs changes when they are continuously excited for 60 minutes at 320 nm wavelength (A) and how their intensity curve changes when they stabilize for a few days (three to 29 days) at a temperature lower than 4 C in the fridge . The graphic indicates that even after being exposed to radiation for a day and then left for months at a time, the nanoparticle' sparkle has not diminished. This suggested that CdSe QDs may be used to tasks like protein tagging and biosystem tracking in vivo.

Figure 2. Synthesis Methods: Compiled by the author, Mirzaeifard, Z etal 2014

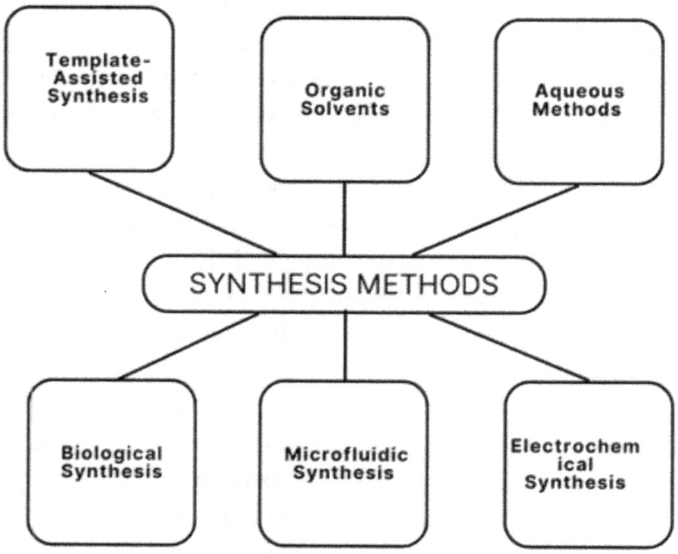

MATERIALS AND PROCEDURES

Material

Samples were pre-treated in iced water for 24 hours, fixed in ethanol-glacial acetic acid (3:1, v/v) for two days at 4°C, and stored at 4°C in 70% ethanol for in situ hybridization of young Allium fistulosum seedlings. The material fixed with ethanol and acetic acid was made in the manner described in. In an alternative method, root meristems were preserved for half an hour in a recently made 4% (w/v) formaldehyde solution that contained phosphate-buffered saline (PBS, pH 7.3). Following this, the meristems were rinsed for 45 minutes in PBS and digested for 25 minutes at 37°C in a solution containing 2.5% cellulase Onozuka R-10, 2.5% pectinase, and 2.5% pectolyase Y-23 (w/v) dissolved in PBS.

Fluorescence in Situ Hybridisation (FISH)

The development of probe unique for the A. fistulosum noncoding satellite gene (Barnes et al., 1985) was conducted as outlined by (Pich et al., 1996). A plasmid VER17 (Yakura et al., 1983) encoding part of the 18S, the 5.8S, most of the 25S and the internal transcribed spacers of Vicia faba 45S rRNA, was employed as a rDNA-specific marker. Through nick translation, either digoxigenin-11-dUTP or biotin-16-dUTP were tagged for in situ hybridization probing. FISH was performed in compliance with (Houben et al., 2001). Using 20 ng of 45S rDNA tagged with digoxigenin and 20 ng of biotin-labelled satellite DNA per slide, in situ hybridization was carried out for the simultaneous probing of rDNA and non-coding or satellite DNA. Using the conventional detection techniques of streptavidin-Alexa 488 or anti-digoxigenin-rhodamine antibody, respectively, at a concentration of 2 µg/ml, hybridization regions of the digoxigenin- or biotin-labelled sensors were detected. Using 20 pM QD 565 streptavidin conjugate (Quantum Dot Corporation,), combination spots of the biotin-labelled sensor were recognized in same fashion. For Alexa 488, Rhodamine, and QD565, the culture times were one hour and two hours, respectively. Either 4 x SSC, 1% BSA, 0.1% Tween 20, or Borate buffer (50 mM boric acid, pH 6.0 or 7.0) were used to create experimental solutions containing QDs and antigens. The tissues were placed in antifade medium containing 10µg/ml DAPI after the last washing procedures and hydration deficit. Retinal patterns have been documented wirelessly using a confocal Laser-Scanning Microscope LSM 510 META (Carl Zeiss Jena GmbH, Jena, Germany) by employing laser line 488 for 543 for rhodamine, Alexa 488, and 364 for QD 565 activation and DAPI.

Immunolabelling of Sectioned Material

After a brief suction penetration, one mm^2 leaf pieces of the plant Zea mays were secured for three hours at ambient temperatures in 50 mM cacodylate buffer (pH 7.2), which included 2.0% (v/v) formaldehyde and 0.5% (v/v) glutaraldehyde. Following fixing, each specimen underwent stepwise dehydration utilizing a freeze substitute unit that is automated that gradually increased the amount of ethanol while simultaneously reducing the room temperature (PLT). The following was actions that were taken: 30%, 40%, and 50% ethanol (v/v) for 1 hour each at 4°C; 75% and 60% ethanol (volume%) for one hour every at -15°C; 90% ethanol (v/v) and twice 100% ethanol (v/v) for one hour each at -35°C. The collected specimens were then incubated in a series of mixes to allow for the penetration of Lowycryl HM20 resin After four hours each of 33% (v/v), 50% (v/v), and 66% (v/v) HM 20 resin in ethanol, 100% (v/v) HM

20 was left longer. Material is placed into gelatine capsules, incubated in new epoxy for three hours, and then polymerized for three days at 35°C with indirect UV light. Sections of the implanted plant material, measuring 0.5 μm in thickness, were sliced with a diamond scalpel and placed on slides at 60°C using an (instrument name should be avoided). These portions were then immersed in PBS + 3% BSA for a twenty-minute period at the ambient temperature (RT) and washed for 3 × 5 minutes in PBS + 1% BSA. Rabbit antibodies were used against chloroplasts and the catalytic zone of the chloroplast H+-ATP synthase (CF1) (21). Following this, an anti-rabbit IgG-biotin conjugate was added (30 min, both diluted in PBS + 1% BSA, RT). The 565-streptavidin conjugation of nano dot was used to identify the supplementary antibodies. Washing came after every stage, apart from obstructing as was already mentioned.

ANALYSIS

Transmission Electron Microscopy Using Transmission

Pipetting a sample of the streptavidin conjugation QD 565 at a dose of 0.2 nM onto a grid covered with formvar was done. To promote the particles' binding to the grids, poly-L-lysine was applied beforehand. Single droplet of 4% uranyl acetate was included after the template had been draining into papers for one minute. After repeating the draining process for fifteen seconds, the grids were allowed to air dry. TEM was used to capture images.

UV-Visible Spectrophotometry

Utilizing a spectrophotometer (instrument) the absorption and emission spectra of nanoparticles made from CdS were examined. Sample' wavelengths of absorption were recorded using conventional quartz 10 mm cuvettes, which have a penetration spectrum ranging from 170–1000 nm. The method states that a 20 cm^{-1} accuracy in wave number recording was achieved. Yet, actual empirical precision was only 80 cm because to random effects and digitization. In a wavelength range of 0 to 1.4, the optically thickness was found with a precision of up to 1% of the length of the optical scale. A machine with a scanner examined the spectrum of light captured by the tape player, which produced the graphic as a JPEG file. After that, a piece of software handled the output file, transforming the spectrum data into a dat-file's quantitative representation. It was used to process the numerical data. Each spectrum was divided into three parts in order to find the length of the wavelength that corresponded to the centres of the absorbance maximum. The Fresnel ratio specified their form. The breadth, location, and max brightness of such elements may all be changed. Such spectral division was performed. Table 1 lists methods for synthesis of inorganic quantum dots.

Table 1. Synthesis Method of Inorganic quantum dots

Synthesis Method	Description	Advantages	Applications	References
Plant-Based	Utilizes Extracts of Plants as stabilizing and reducing agents for QD synthesis.	Sustainable, environmentally friendly.	Biomedical imaging, environmental sensing.	Mirzaeifard & Shariatinia, 2024
Hot-Injection Method	Injection of Predecessor into a hot solvent to control nucleation and growth kinetics of QDs.	High control over size and shape.	Optoelectronics, catalysis.	Kyobe et al, 2020
Aqueous Synthesis	QD synthesis in water-based solutions using water-soluble ligands or capping agents.	Biocompatible, suitable for biomedical applications.	Bioimaging, drug delivery.	Alvand, et al, 2019
Microfluidic Synthesis	Continuous flow reactors to precisely control reaction conditions and QD properties.	Uniform size distribution, scalable.	Optoelectronics, sensors.	Şahin, et al, 2021
Electrochemical Synthesis	Metal ions Electrodeposition onto conductive substrates in the presence of stabilizing agents.	Direct control over QD formation on surfaces.	Solar cells, sensors.	Ahirwar, et al, 2017

PROPERTIES

Basic Properties of Ag-Based QD

Quantum entanglement phenomena is often linked to certain biophysical features of QDs (Cipriano et al., 2020). By altering the electronic spectrum of materials and their energy undergoing an evolution from the tiny level, the effect of quantum imprisonment may be seen.

The most significant characteristic of QDs is the size dependency of their optic band; when something's size in matter approaches tiny scales, electrical excitation moves to a higher energy level, resulting in a distinct energy range. High levels of absorption of light, persistent extended the luminescence (PL), luminescence emission, and tunnelling with a substantial Stokes shift are among the optical characteristics of QDs (Zaini et al., 2020). It should be mentioned that non-cadmium-based QDs generate near-infrared radiation (NIR). This is advantageous for imaging live tissues because of their deep darkness and decaying self-fluorescence in the infrared spectrum, specifically at wavelength between 750 and 940 nm (NIR-I window) and 1000 and 1700 nm (NIR II window). Ag_2S have surfaced as new NIR-II emission quantum dots (QDs) with an energy gap of around 1 eV, excellent catalytic capabilities, and desirable photo- and thermodynamic characteristics. They also exhibit excellent electrical mobility and nearly zero environmental impact (Cui et al., 2015). Nanoscale Ag_2S is a biocompatible semiconductor material that has a high stability, quantity and magnetoresistance. It also has an uneven radiation band. Allowing a light to penetrate the tissue to a high depth and little digestion, light produced in the 700–900 nm range is regarded as an image's aperture (Buz et al., 2019).

Optical Property

Since 2D-QDs emit light throughout a wide assortment of light wavelengths, including the far ultra-violet to the near- inf (NIR), they are quite interesting. (Tang, L et al.,, 2014) These characteristics are influenced by the interactions between dimension (quantum confinement), surface chemical composition, and edges flaws that arise from the various response circumstances, including the synthesis temperature, solvents, working changes, filling factor, duration time, and variations in the ambient pH. (Wang, X. et al., 2016, Benítez-Martínez et al., 2016) The physical properties of 2D-QDs may typically be determined by UV light irradiation using a UV lamp set at 254 or 365 nm. The wavelength of radiation that's most often employed here is 365 nm. 2D-QDs are typically colourless and clear bright yellow (Dong, et al., 2016, Zhao, et al., 2016) or even brown in the light of day,192 yet they release vivid blue (Lin et al., 2013) or green (Li et al., 2016) sometimes crimson White (Luo et al., 2016), or even more hues when exposed to UV light.

The two most important variables in figuring out the PL of the 2D-QDs are their size and surface condition. (Du et al., 2016)While the blue fluorescence of GQDs is governed by intrinsic situations in the highly rigid. The structure the green illumination is most likely the result of fault conditions with oxygen-containing organizations, such as -OH, QO, and -COOH (Arslan, et al., 2015).By using the method of gel electrophoresis, Zhang et al., sorted the crude GQDs into different-sized GQDs with narrow size distributions centred at 5.5, 12.5, and 16 nm in diameter, respectively. The roughness status of GQDs, which varied in a magnitude from 5.5 nm to 16 nm in their study, significantly influenced their PL characteristics. In contrast, the bigger carbon skeleton of GQDs, which varied in dimension from 12.5 to 16 nm, mostly contributed to their PL features. (Zhang, et al., 2016). Various liquid types or pH (Qu et al., 2015) values may adjust the PL of 2D-QDs, leading to varying surface properties and quantum effects. For instance, ZnO QDs generated in methanol, ethanol, and hexanol showed intense green, yellow, and orange fluorescence, respectively, with diameters ranging from 3.3 to 7.7 nm. (Xu, et al., 2018) Our most recent research also found that the size of the that-prepared BNQDs decreased from 4 to 2 nm and the colour of the BNQDs suspension changed from blue to a light green when various solvents, like ethanol, DMF, or NMP, were utilized. (Liu, et al., 2017) When the pH changed from 1 to 14, the GQDs made by the process of a graphite rod showed a noticeable colour change, first going from light yellow to brown and finally red. This was caused by the decay of protons of oxygen-containing functional groups and the accompanying electrostatic charging and doping (Xu et al., 2019). The doping ingredients or the covering agents may also change the colour of the PL. After being modified with 2-a substance called, the PL colour changed from blue to green when exposed to UV light, while the GQDs changed from orange to yellow in the daytime. (Wang, et al.,2016) OLA was used in the synthesis of MoS_2 QDs as an agent to reduce and stabilize them. By controlling the quantity of OLA and the duration of reaction, the resulting MoS_2 QDs displayed tuneable PL from blue to red (Lin et al., 2015). Regarding the effects of doping, it was proposed that whilst boron or fluoro doped may cause a red-shift, hetero-atom dopants like nitrogen or sulphur may cause a blue-shift.182 In light of this, S,N-CO-doped GQDs (S,N-GQDs) were created with three-color emissions, especially blue, red, and green. Each of the three states of S, N, and O, respectively, correspond to the p* orbital of the CQS, CQN, and CQO bindings in the S,N-GQDs when prepared.

Catalytic Property

Owing to their large area of surface and abundance of chemically active sites, 2D-QDs are perfect for use as chemo catalytic (Tokudome et al., 2016) photocatalytic (Wang et al., 2014) electrocatalytic, (Zhao, et al., 2016), enzyme-like contributing1(Peng et al., 2016), or various other substances. It has been shown by both theoretical and experimental evidence that edges sites, as opposed to basal planes, are the primary source of catalytic activity (Zhao, et al., 2016).

Because of this, the 2D-QDs smaller than 10 nm have more exposed edge atoms, a greater specific area, and an enhanced edge-to-volume ratio than the stacked nanosheets of metal .This results in an increase in solvent obligations, such as the enhancement of S edges on Tungsten disulphide or Molybdenum disulphide QDs, (Zhao et al., 2016). Because of their uncompressed condition, that renders them very energetic and insecure, ions may easily join with one another. Strong catalysis characteristics of the 2D-QDs were therefore possible to achieve. (Xu et al., 2015). Furthermore, the 2D-QDs are given additional active sites for enhanced catalysis by molecular doping. (Ke et al., 2016) The element nitrogen may be readily added to the GQD structure via C–N bonds to generate N-GQDs, which is explained by the fact that nitrogen atoms and carbon atoms have comparable atomic sizes and electron structures. (Yeh et al., 2016). The simplified structure illustrates how different sized graphene layers were built in an AB order to form N-GQDs, which were produced by NH3-treatment and further oxidation of GO. The oxygen functions are dispersed across the basal and peripheral planes, while the nitrogen is in the main body of the GQDs (Yeh et al., 2016). The procedure for coordinating among and 20-tetrakis(1-methyl-4-pyridinio) porphyrin and Cd^{2+} may be effectively catalyzed by N-GQDs. (Zhang et al., 2015). The effective doping of the GQDs could provide suggestions and lessons for the preparation of additional 2DQD catalysts. Additionally, some 2D-QDs could be able to take in electrons and function as active sites in photocatalytic processes (Gao et al., 2015).The UCPL feature and wide absorbing in the visible light area of 2D-QDs, such S,N-GQDs, and MoS_2 QDs, make them powerful photocatalysts for the transfer of energy or collecting (Wang et al., 2016).

In addition, 2D-QDs, like GQDs, have peroxidase-like catalytic (Tang et al., 2016) for facilitating the breakdown of H_2O_2 and the production of HO~. Qu's group selectively inhibited the particular oxygenous functional groups on the surface of the GQDs to study the substrate-binding and catalytic sites of the GQDs as a peroxidase, respectively. It was discovered that the carboxylic groups (OQC–O–) served as substrate-binding sites, whilst the ketonic carbonyl groups (–CQO) were the catalytically active sites.

Chiral Property

Chiral QDs generated from semiconductor materials and metals have garnered a lot of attention for asymmetric synthesizing and enantioseparation, photography, selective antibacterial and lethal delivery of medications, and even for forensic and biological tests. (Moloney et al., 2015) Gold's electromagnetic groups' spectrum simplicity, chemical reactivity, and intriguing science make it the dominant material in symmetrical nanoparticles today. (Suzuki et al., 2016). Due to their lesser polarity motion, chiral 2D-QDs haven't been studied very much, despite their through bond edges mutation using L- and D-cysteine, Kotov's group created chiral GQDs in dispersion that had an added twist at the microscopic level (Suzuki et al., 2016). Altering oxidized GQDs with enantiomerically (R) or (S)-2-phenyl-1-propanol may also result in the synthesis of chiral GQDs. It is possible for tiny compounds like pyrene to acquire harmony through the formation of chiral GQDs. It is anticipated that chiral compounds will be employed for embellishment to provide GQDs chirality in the microscopic level serving as a universal technique to produce additional chiral 2D-QDs. Research is expected to be continue Chiral property because of its potential application in both medicine and biology, as well as a variety of benefits, including reduced toxicity and cheaper cost compared to 2D-QDs. (Suzuki et al., 2016).

CHARACTERIZATION TECHNIQUES

The count of the presence of heteroatom doping agents, functional categories, imperfections border arrangements, etc., of 2D-QDs account for all of their superiorities, as shown by numerous characterisation approaches (Cai et al., 2015)– AFM and electron transmission microscopy (TEM), which provide the elevation profiling and lateral-size characteristics of the 2D-QDs, are often employed to confirm the creation of 2D-QDs and further improve the conditions under which they react. Even though 2D-QDs are classified as particles less than 100 nm, the observed particles are often fewer than 10 nm. (Zhang et al., 2016)The average sizes of the BNQDs, BPQDs, WS2 QDs, and MO Ox QDs formed by exfoliation using liquids or/and solvothermal treatments under various reactions are 3.3 nm, 4.9 ~ 1.6 nm, 2.7 ~ 0.8 nm, or 4.4 nm, (Zhao, et al., 2016) accordingly. The size distribution of BPQDs. According to the AFM data, the majority of published 2D-QDs have a thickness of less than 3 nm, which suggests that they are heavily exfoliated (Lin et al., 2016). Given that the majority of 2D components' single-layer thicknesses fall between the 0.5 and 0.7 nm73, the topographic heights of 2D-QDs in the 0.85–2.17 or 0.9–2.1 nm fluctuate represent one to three layers of BN, 1.9 ~ 0.9 nm to four layers of BP, and 0.7 nm to one layer WS_2, (Zhao et al., 2016) accordingly.

Furthermore, the HRTEM image's well-ordered and paralleled lattices imply that the 2D-QDs have excellent crystallinity. The (100) lattice plane of the MoS_2 27 granules the (104) plane from WS_2, the (021) and (020) faces of BP, and the 0.215 nm, 2 0.34 nm, and 0.52 nm, (Zhang et al., 2015) 0.21 nm (Zhao et al., 2016) and 0.27 nm edges of the lattice of the 2D-QDs, as well, relate to these planes. XRD might potentially be used to investigate the 2D-QDs' crystalline structure in more detail. The peaks often widen, shrink, or even vanish after the solvothermal treatment (Dong, et al., 2016), because there is less interlayer activity and the layered bulk materials have less transparency during the exfoliation process (Dong, et al., 2016). The balance of the peaks in the XRD pattern may demonstrate the creation of 2D-QDs.

The two faint XRD spikes at 2y = 32.61 and 58.41 may be ascribed to the (100) and (110) diffractions, however - nonetheless clearly showing the effective exfoliation of the QDs. The distinctive (002) dispersion peak vanishes for MoS_2 QDs. Furthermore, characterizing the crystal's architecture and diameter of 2D-QDs is possible via Raman spectroscopy. Two, furthermore useful data on defect frequency depending on the brightness ratio of D to G peaks at 1589 and 1331 cm alternately might come from the Raman spectra of the GQDs. The substantial drop in this configuration reflected the few-layer structure of BNQDs. Second Likewise, bulk MoS_2 displayed two significant Raman peaks at 407.1 and 380.5 cm1, attributed alternatively to vertical plane (A1g) and in-plane (E12g) oscillations of Mo–S bonds in 2H MoS_2. Although the matching regions of MoS_2 QDs were moved to 406.2 and 380.8 cm respectively, suggesting the A1g relaxing and E1 12g stiffness with decrease layer height, thus verifying the successful synthesis of the MoS_2 QDs with a thickness between 1 and 2. Red shifts in the spectrum of Raman were additionally identified for BNQDs135 and BPQDs relative to the bulk substance of origin. Usually X-ray photoelectron spectroscopy (XPS) and Fourier transitional IR (FTIR) spectrometry are used to investigate the plasma makeup and molecular moods of the 2D-QDs. (Meng et al., 2016). XPS offers a direct means of tracking 2D-QDs' synthesis of the dopant, therefore enabling the ratio of atomic molecules to be calculated. The effective creation of the MoS_2 QDs was confirmed, for instance, by the XPS spectra of of-prepared MoS_2 QDs showing anatomic ratio of Mo: S as 1: 2.09.36. Furthermore, the blurring of the wide scan spectrum of every component elements helps one to identify the functional clusters on the outermost parts of 2D-QDs. (Meng et al., 2016) through matching the matching spot use the multi-peak Gaussian approach. One uncertainty for the first graders is that, while the component with carbon does not constitute included of the layout, the C1s are generally detectable in all the 2D-QDs. This might start from the utilized naturally occurring solvent or could be unavoidably transferred into the identified material via its surrounding atmosphere in the detecting procedure. Usually, FTIR provides data on the functional groups on the 2D-QDs surface. (Qiu et al., 2016).The OH displacement of C-OH, the stretching vibrations of the graphitic areas, and CQO oscillation in the carboxyl group may be respectively ascribed to the peak absorption rates at 1387, 1627, and 1727 cm^1 shown by the FTIR spectra of GQDs (Xu et al., 2014). Whereas the peaks of the N-B-O vibrational show the O atom loaded in the BNQDs in the solvothermal processes process, the peaks at B1390 and 809 cm correspond to the vibrations of the bulk .One prominent point that should be mentioned is that, in a solution in addition for the physical characteristics, additional changes and practicality, 139 which will be addressed in detail in the sections to come, the functional structures on the 2-D QDs, including the oxygen-containing ones, are of importance for their dispersion. Table 2 summarizes synthesis, characterization and applications of plant-based quantum dots.

Table 2. Aspects of quantum dots

Aspect	Details	References
Synthesis	Plant extracts (e.g., from tea, fruits) used as stabilizing and reducing agents for QD synthesis.	Vaseghi, et al 2018
	Green synthesis approach utilizing environmentally benign materials.	
	Control over size and morphology through reaction conditions (e.g., pH, temperature).	
Characterisation	**Structural Analysis**: TEM for size and morphology, XRD for crystal structure, FTIR for functional groups.	Mewada, et al, **2013**
	Optical Properties: UV-Vis spectroscopy for absorption, PL for emission characteristics.	
	Biocompatibility and Stability: Assessing toxicity and stability in biological and environmental contexts.	
Applications	Biomedical Environmental Optoelectronics	Borovaya et al,2021

APPLICATION OF INORGANIC QUANTUM DOTS

Bioimaging Applications

Semiconductor nanoparticles offer distinctive fluorescence attributes that set them apart from widely used organic fluorophores and make them a desirable bioimaging tool. This makes the room for the advancement of effective biomarkers, biosensors, and target cell delivery methods for anticancer drugs. Among QDs' optoelectronic properties include excellent photostability, a wide absorption spectrum, and significant QY of luminescence (Chinnathambi and Shirahta, 2019). There are several benefits that quantum dots offer over organic pigments. They are perfect for multiplex imaging due to their single-light emission and restricted size distribution. Different proteins found in living tissues, such as DNA, collagen, and elastin, emit light across a wide range that includes both visible and ultraviolet light. For imaging deep living tissues, NIR emission is favoured as autofluorescence diminishes. Hence, the inherent brightness of silver-based quantum dots (QDs) at wavelengths ranging from 750–940 nm and 1100–1700 nm has evident advantages in biological applications such identifying intracellular structures, in vivo imaging and cellular imaging for the diagnosis of tumours, among others. To sum up, using nanoprobes for fluorescence imaging has the following advantages: Three things occur: the experiment is made available; high-quality, real-time photographs are obtained; and non-ionizing radiation is used as an excitation source. Ag_2S QDs are suitable for intensive and stable bioimaging because of their brightness and photostability. Compared to commonly utilized heavy metal-based QDs (such as lead or cadmium), Ag_2S QDs have been demonstrated to have less toxicity in both in vitro and in vivo studies. Additionally, Ag_2S QDs' photothermal action gives them their own anticancer properties. The surface of Ag_2S QDs provides a high affinity for electrostatic contact, covalent binding, and successful conjugation of chemotherapeutic medicines. Ag_2S theragnostic nanomaterials were created because Ag_2S QDs can serve as both as an imaging agent and therapeutic image-based diagnostics (Purushothaman and Song, 2021). Non-invasive biomedical imaging methods are becoming more and more common in medical and clinical research since they are significant for both diagnosis and prognosis for a variety of disorders. According to (Rhyner et al., 2006), QDs were linked with antibodies to the prostate-specific membrane antigen with the goal to detect the tumours. The scientists have observed that the formation and maintenance of this antigen-antibody complex at the cancerous growth site forms the basis of targeted therapy for the spread of a human prostate tumour. Borovaya et al., investigated the luminescence properties, penetration, and localization of silver sulphide QDs generated via the "green" technique using fungi in plant cells. The produced Ag_2S QDs' luminescence ranged from 520 to 550 nm, or the green portion of the visible light spectrum. The produced Ag_2S quantum dots had a diameter of 10–25 nm. Silver sulphide QDs, which are mostly found in the nuclei of the epidermal and meristematic cells of the root tips cell, have been shown to pierce Allium cepa root cells (Borovaya et al., 2021).

Phytic Acid Detection

Phytic acid, also called myo-inositol 1,2,3,4,5,6-hexakis dihydrogen phosphate, and phytate anions, are the main sources of phosphorus present in majority of commercial grown grains and some legume species (Reddy et al., 1982). These phosphorus deposits are usually inaccessible to nonruminant animals due to their lack of the enzyme phytase, which is required for their degradation (Zou et al., 2008). When fed grains, nonruminant livestock such as pigs, poultry, and fish enter the digestive system. This results from phytate not being broken down. After excreted, phytate increases manure's phosphorus content and contributes to environmental problems, namely eutrophication (Malline et al., 2003). Additionally, Phytic acid chelates minerals such as calcium, zinc and iron into insoluble complicated materials in the digestive tract. In both human and domestic animal diets, it serves as an antinutrient (Amos et al., 2020). In their paper, Centeno et al., (2021) described the production of graphene quantum dots from ethanolic *Opuntia sp.* extracts by one-step microwave pyrolysis. They then used these dots in the detection of phytic acid sodium salt (PA), dibasic (SPDB), and sodium phosphate monobasic (SPMB); in this case, quenchers were used (Centeno et al., 2021). These GQDTs can eliminate the requirement for inorganic ions in systems that control the on and off fluorescence detecting devices. The results show a phenomenon known as increased fluorescence quenching, which is associated with the formation of J-aggregates via electrostatic interactions between the GQDTs and phosphates or direct structure of (PO_4^{3-}) ions, which are strong proton acceptors. The quenching percentage of phytic acid is 53%, but the quenching percentages of SPDB and SPMB were almost 21% and nearly 26%, respectively. FTIR, UV-Vis, and TEM investigations revealed that graphene quantum dots were made using extract from Opuntia sp., an inexpensive raw material. By producing (PO_4^{3-}) ions, which are strong proton acceptors at pH = 8.8, and improving fluorescence quenching, this produced material can aid in the detection of phosphate. Its limited size distribution of nanostructures and good dispersion make it an easier procedure than those that have been previously reported. Lastly, both static and dynamic quenching mechanics are responsible for the quenching caused by phytic acid. Phytic acid exhibits a quenching value of up to 53%, while monobasic and dibasic potassium phosphate exhibit quenching values of 26 and 21%, respectively.

Immobilization of α- Amylase

Since inorganic quantum dots (QDs) have a unique physicochemical property and a diverse surface chemistry, they have become a viable nanomaterial for the immobilization of enzymes such as alpha amylase. Alpha-amylase is immobilized on QDs by attaching the enzyme molecules to the surface of the QDs, which has various benefits over conventional enzyme immobilization techniques. One important benefit is that quantum dots have a high surface area-to-volume ratio, which means there is plenty of room for enzyme attachment and a high enzyme loading capacity. To achieve improved catalytic efficiency and stability of the immobilized enzyme system, this high loading capacity is essential. Additionally, QDs' surface can be functionalized with a variety of chemical groups to aid in the binding of enzymes, providing exact control over the immobilization procedure and enabling the optimization of both enzyme stability and activity. Furthermore, the high biocompatibility of inorganic quantum dots guarantees that the mounted enzyme's catalytic activity is not significantly affected. Applications in biotechnology and biomedicine, where the immobilized enzyme may come into touch with living systems, benefit greatly from its biocompatibility. In 2023, Karaca Açarı et al. carried out a study in which they synthesized graphene quantum dot (QD) structures using *Hypericum capitatum seeds* (QD-2) and *Hypericum perforatum L. flowers* (QD-1) obtained from an herbalist in the Turkish province of Hatay (Karaca Açarı et al., 2023). Through a variety of characterization techniques, including FTIR, XRD, SEM, and EDX, they were able to demonstrate the creation of QDs and the binding of Amy enzyme. According to their results, 71–81% of the Amy enzyme was successfully immobilized on the QDs by adsorption, suggesting that the material will perform well in biochemical investigations. In QD-1/Amy and QD-2/Amy complexes, some enzymatic capabilities, such as pH and thermal stability, were improved while the ideal pH and temperature stayed the same after immobilization. Furthermore, using green chemistry to manufacture

QDs may make costly enzymes commercially feasible for to be use in a scope of biotechnological applications. However, as mentioned by Noma et al., (2020), disadvantages of this approach include enzyme leakage during adsorption and minor activity losses as a result of mass transfer limits after immobilization. However, the study indicates that QD structures have potential applications in related domains such as enzyme immobilization. Subsequent investigations will focus on the restoration of enzyme activity in response to various environmental stimuli and investigate substitute substances for Amy immobilization.

In-Vitro Catalytic, Antimicrobial and Antioxidant Activities

Rapid population growth and industrialization lead to an exponential increase in wastewater releases containing harmful pollutants such as phthalates, sulphonamides, polychlorinated biphenyls, and synthetic dyes, raising serious environmental and health issues (Chiou et al., 2013; Prasad et al., 2016). Among these, synthetic dyes such as methylene blue (MB) and others are common contaminants found in many industrial wastewaters, including those from the paper, leather, and textile industries. Because MB is not biodegradable, treating the industrial effluent it is associated with can be quite difficult. Adsorption, membrane filtration, and other techniques have been used to treat and decontaminate wastewaters and effluents containing microbiological contaminants (MB), producing large amounts of solid debris in the process. However, these methods require a lot of time, money, and inefficiency, and their cost will only increase due to the additional disposal steps required (Sun et al., 2002). Furthermore, excess reactive oxygen species, or ROS, also known as free radicals, are the cause of many ailments, including cancer and cardiovascular disease. But because of their capacity to scavenge free radicals, antioxidants neutralize these radicals before they have a chance to damage biological targets and induce illnesses of this kind (Kumar et al., 2016). In light of these, nanotechnology employing nanoparticles (NPs) such as copper, copper oxide, silver, platinum, gold, etc., may be utilized as efficient antioxidants, catalysts for the breakdown of the aforementioned organic dyes (resulting in biodegradable products that microbes can easily break down), and antimicrobial/anti-biofilm agents (due to their large surface areas) with potential benefits in terms of cost-effectiveness while remaining effective. In order to create catalysts, antimicrobial/anti-biofilm, and antioxidant agents, researchers have synthesized nanoparticles (NPs) of copper (Cu) using *Thymus vulgaris/Azadirachta Indica* leaf extract, Rubus glaucus/Ziziphus Jujuba/Morus nigra leaf extract for silver (Ag), gold (Au) by Anacardium occidentale/tamarind leaf extract, platinum (Pt) by Fumaria herba extract/Diopyros kaki leaf extract, zinc oxide (ZnO) by Eclipta alba leaf extract, bimetallic Cu–Ag/Cu–Ni by ginger rhizome powder, etc.(Issaabadi et al., 2017; Ismail et al., 2018; Kumar et al., 2016; Nagar and Devra, 2018; Sengan et al., 2019; Singh et al., 2018; Song et al., 2010). For example, using Terminalia catappa leaf extract, Muthulakshmi et al., (2017) performed in situ synthesis of Cu NPs in cellulosic matrix. The resulting NCs of cellulose and Cu NP were found to reveal good antibacterial property towards *Escherichia coli* (Muthulakshmi et al., 2017). Similarly, using *Azadirachta indica* leaf extract, silver nanoparticles were created in situ in a cellulosic matrix and demonstrated strong antibacterial activity against Escherichia coli (Kishanji et al., 2017). In situ production of silver nanoparticles (NPs) and bacterial cellulose nanocrystals (NCs) with significant antibacterial activities against *Escherichia coli, Staphylococcus aureus, and Pseudomonas aeruginosa* was conducted by Wu et al. (Wu et al.,2014). Using polyphenols (mostly mangiferin) from M. indica leaf extract, which function as reducing/capping agents, Rani et al., (2020) synthesized Cu QDs. This work showed that the produced Cu QDs might be used as effective antioxidant, catalyst and antibacterial. In addition to being quite effective as an antimicrobial agent against E. coli and S. aureus, the Cu

QDs that were produced in this way also showed a rapid reduction of MB, indicating that Cu QDs could be effectively used in wastewaters as a catalyst to remove and treat other organic dyes and toxic, non-biodegradable anthropogenic pollutants like MB. This could pave the way for the wastewater treatment in combination with other cutting-edge methods for applications like irrigation of agricultural crops, among other uses (Rani et al., 2020).

Spintronics

Thanks to its potential to reduce energy usage and enhance performance compared to traditional electronics, spintronics has garnered a lot of interest as a potential area for electronics in future. To create new electrical devices, QD-based spintronics aims to harness the electrons' spin degree of freedom in QDs. The spin of an electron can be thought of as its intrinsic rotational momentum. It can have two possible values: “up” or “down.” In order to create innovative and more efficient electrical devices, spintronics aims to utilize the charge and spin degrees of freedom. Larger data densities, faster and more energy-efficient data processing, non-volatile data storage, and many other benefits are expected from spintronics. The full potential can only be realized by efficient spin polarization generation, transit, transfer, manipulation, and detection. These criteria all relate directly to the features of the material as QDs may contain charges because of quantum confinement. Additionally, the dots are used because of their special electronic and optical properties, such as their strong Coulomb interaction, quick switching times and high tunability. Using electrostatics, the confinement potential can be controlled. Unlike traditional electronics, spintronics is unique. The idea of controlling an electron's spin in addition to its charge is the foundation of spintronics. Both charge and spin are present in electrons. Information is stored and processed in spintronics by using the spin of electrons. The field of spintronics holds promise for providing greater functionality and larger data storage capacity than conventional electronics, which solely rely on the charge of electrons for information processing and storage. As a result, new spintronic devices such spin field-effect transistors, spin-polarized light-emitting diodes, and spin filters have been developed, as well as spin-based quantum computing (Cerletti et al., 2005). Graphene QDs and their corresponding spin relaxation times were examined by M. Droth and colleagues in their 2016 study (Droth et al.,2016).

Organic Reductor Agent

Because of their distinctive electrical and optical characteristics, quantum dots (QDs) have become attractive materials for uses in a variety of applications, such as catalysis and optoelectronics. Due to their capacity to participate in redox reactions and donate electrons, quantum dots have garnered increasing heed as organic reducing agents in recent years. In organic synthesis, where the reduction of functional groups is an essential step in the production of complex compounds, these reactions are especially pertinent. When compared to conventional reagents, quantum dots as organic reducing agents have a number of benefits. First off, their precisely adjustable electronic structure permits selective reduction of particular functional groups by providing fine control over the redox potential. Furthermore, quantum dots' selectivity and reactivity in organic transformations can be further increased by simply functionalizing them with organic ligands. Moreover, the small size of quantum dots makes it easier for them to dissolve in organic solvents, allowing for effective reactions in mild circumstances. Quantum dots have been shown in numerous studies to be effective organic reducing agents in a variety of chemical transformations, such as the reduction of unsaturated bonds, carbonyl compounds, and nitro compounds.

Quantum dot-based reagents, for instance, have been used to synthesize fine compounds, agrochemicals, and medicines, demonstrating their potential as flexible instruments in organic synthesis.

ENVIRONMENTAL IMPACT OF INORGANIC PLANT-BASED QUANTUM DOTS

Inorganic plant-based quantum dots (QDs) are newly emerging Category of nanomaterials that offer significant potential in fields such as bioimaging, solar energy conversion, and environmental sensing. Advantages and disadvantages of Inorganic quantum dots over organic quantum dots. These quantum dots are derived from plants, which are abundant and renewable sources, making them an attractive alternative to traditional quantum dots synthesized through more environmentally harmful processes. However, their environmental impact requires careful consideration.

One primary environmental concern associated with inorganic plant-based quantum dots is their potential toxicity. Quantum dots often contain heavy metals like cadmium, lead, or selenium, which can pose significant risks to both ecosystems and human health. When these materials are introduced into the environment, either through disposal or accidental release, they can persist and bioaccumulate, leading to toxic effects in plants, animals, and microorganisms. For instance, studies have shown that cadmium-based QDs can cause oxidative stress in aquatic organisms, affecting their growth and reproductive capabilities.

Moreover, the making and degradation of quantum dots can result in the release of nanoparticles that are not easily removed by conventional water treatment processes. These nanoparticles can infiltrate water supplies and soil, posing long-term contamination risks. The interaction of QDs with environmental factors such as sunlight, water chemistry, and biological organisms cause reactive oxygen species to arise, which further exacerbate their environmental impact by damaging cellular structures in living organisms.

On the positive side, the use of plant-based materials in QD synthesis can mitigate some environmental impacts by reducing reliance on non-renewable resources and decreasing the overall carbon footprint of the production process. Additionally, advancements in the field are focusing on developing more environmentally benign QDs, such as those based on zinc or copper, which are less toxic compared to cadmium or lead-based counterparts. Efforts are also being made to enhance the biodegradability and recyclability of these materials, thereby minimizing their long-term ecological footprint. Table 3 lists advantages and disadvantages of quantum dots.

Table 3. Advantages and disadvantages of Inorganic quantum dots

Quantum dots	Advantages	Disadvantages	References
Organic	Biocompatibility Larger Surface area Precise Targeting Less Hazardous	Less Stable Difficult to obtain high purity	Alaghmandfard et al., 2021
Inorganic	Bio inertness Unique Optical Properties Highly Stable	Less Biocompatible Poor targeting Ability Biological Toxicity	Al-Agel et al., 2017

CONCLUSION

As of 2024, inorganic plant-based quantum dots (QDs) represent a promising advancement in nanotechnology, combining the desirable properties of quantum dots with eco-friendly and sustainable production methods. These quantum dots are synthesized using plant extracts, leveraging the natural Plants' potential to lower metal ions and stabilize nanoparticles. The synthesis process of plant-based quantum dots avoids the use of harsh reaction conditions and toxic chemicals typically associated with conventional methods, thereby reducing environmental impact and aligning with the principles of green chemistry. Plant-based QDs exhibit high biocompatibility, making them appropriate for use in biomedicine, including bioimaging and drug delivery. Their reduced toxicity compared to traditional QDs makes them safer for in vivo applications. These QDs retain the excellent optical properties of conventional quantum dots, such as size-tuneable fluorescence and high quantum yield. They offer potential for use in optoelectronics, including LED technology and solar cells. The use of plant materials, which are abundant and inexpensive, can lower production costs and enhance the scalability of QD synthesis, making the technology more accessible for widespread commercial applications. Ongoing research is focused on streamlining the process of synthesis, improving the stability and performance of plant-based QDs, and expanding their functionalization for diverse applications. Collaborative efforts between academia and industry are crucial for translating laboratory successes into practical, market-ready solutions. while inorganic quantum dots hold immense potential across various applications, they also present significant challenges that must be addressed for their broader adoption. Their potential toxicity is one of the key worries, particularly because of the usage of heavy metals like cadmium, which poses dangers to both human health and the environment. Cadmium is one particularly concerning heavy metal. Furthermore, the synthesis of inorganic quantum dots frequently entails complicated techniques that call for careful control over reaction conditions in order to achieve uniformity in size and characteristics. This might make it challenging to scale up the process for commercial production. Another difficulty is the stability of these quantum dots, which can be susceptible to degradation under specific environmental circumstances, which might result in the loss of their desired optical properties. This presents a problem for the researchers. It is possible that their long-term employment in applications such as biological imaging or photovoltaic systems will be restricted due to their intrinsic instability. In addition, there is a want for synthesis techniques that are less harmful to the environment, as conventional procedures frequently include the utilization of hazardous chemicals and might result in the production of toxic by-products. It is anticipated that the results of research and development on inorganic quantum dots would have a transformative impact on a variety of different industries in the future. At the same time that problems like toxicity, stability, and scalability are being addressed, we can predict a major expansion in the practical applications of quantum dots. Improvements in the biocompatibility and targeting capabilities of quantum dots have the potential to transform diagnosis and therapy in the medical industry. These developments could make it possible to create imaging and medication delivery systems that are extremely precise. In the field of renewable energy, quantum dots are anticipated to play a significant part in improving the efficiency of photovoltaic cells, which might ultimately result in solar energy solutions that are both more economical and more effective. Additionally, their unique optical qualities will likely stimulate advancements in optoelectronics, leading to the creation of next generation displays, sensors, and lighting systems with greater performance and energy efficiency. Environmental applications are also intriguing, with quantum dots potentially being utilized for pollutant monitoring, water purification, and other eco-friendly technology. As green synthesis processes are further perfected, the manufacture of quantum dots could become more sustainable, lowering their environmental footprint and opening new options for environmentally responsible nanotechnology. Overall, the future of inorganic quantum dots is bright, with further study likely to unleash new capabilities and applications that will have a profound impact on technology, medicine, and environmental sustainability.

REFERENCES

Ahirwar, S., Mallick, S., & Bahadur, D. (2017). Electrochemical method to prepare graphene quantum dots and graphene oxide quantum dots. *ACS Omega*, 2(11), 8343–8353. DOI: 10.1021/acsomega.7b01539 PMID: 31457373

Al-Agel, F. A., Suleiman, J., & Khan, S. A. (2017). Studies on silicon quantum dots prepared at different working pressure. *Results in Physics*, 7, 1128–1134. DOI: 10.1016/j.rinp.2017.03.002

Alaghmandfard, A., Sedighi, O., Rezaei, N. T., Abedini, A. A., Khachatourian, A. M., Toprak, M. S., & Seifalian, A. (2021). Recent advances in the modification of carbon-based quantum dots for biomedical applications. *Materials Science and Engineering C*, 120, 111756. DOI: 10.1016/j.msec.2020.111756 PMID: 33545897

Alvand, Z. M., Rajabi, H. R., Mirzaei, A., Masoumiasl, A., & Sadatfaraji, H. (2019). Rapid and green synthesis of cadmium telluride quantum dots with low toxicity based on a plant-mediated approach after microwave and ultrasonic assisted extraction: Synthesis, characterization, biological potentials and comparison study. *Materials Science and Engineering C*, 98, 535–544. DOI: 10.1016/j.msec.2019.01.010 PMID: 30813055

Amos, A., Alvan, A., & Florence, A. (2020). The Anti-nutritional Effect of Phytate on Zinc, Iron and Calcium Bioavailabilities of Some Cereals Staple Foods in Zaria, Nigeria. *European Journal of Nutrition & Food Safety*, 12(1), 1–6. DOI: 10.9734/ejnfs/2020/v12i130178

Arslan, O., Belkoura, L., & Mathur, S. (2015). Swift synthesis, functionalization and phase-transfer studies of ultrastable, visible light emitting oleate@ ZnO quantum dots. *Journal of Materials Chemistry. C, Materials for Optical and Electronic Devices*, 3(45), 11965–11973. DOI: 10.1039/C5TC03377B

Bakar, N. F. A., Tan, H. L., Lim, Y. P., Adrus, N., & Abdullah, J. (2022). Environmental impact of quantum dots. In *Graphene, Nanotubes and Quantum Dots-Based Nanotechnology* (pp. 837-867). Woodhead Publishing.

Barnes, S. R., James, A. M., & Jamieson, G. (1985). The organisation, nucleotide sequence, and chromosomal distribution of a satellite DNA from Allium cepa. *Chromosoma*, 92(3), 185–192. DOI: 10.1007/BF00348692

Benítez-Martínez, S., López-Lorente, Á. I., & Valcárcel, M. (2016). Determination of TiO 2 nanoparticles in sunscreen using N-doped graphene quantum dots as a fluorescent probe. *Mikrochimica Acta*, 183(2), 781–789. DOI: 10.1007/s00604-015-1696-0

Bhuvaneswari, G., & Radjarejesri, S. (2015). Green synthesis and characterization of CdS quantum dots. *International Journal of Chemtech Research*, 8(5), 104–108.

Borovaya, M., Horiunova, I., Plokhovska, S., Pushkarova, N., Blume, Y., & Yemets, A. (2021). Synthesis, properties and bioimaging applications of silver-based quantum dots. *International Journal of Molecular Sciences*, 22(22), 12202. DOI: 10.3390/ijms222212202 PMID: 34830084

Borovaya, M., Horiunova, I., Plokhovska, S., Pushkarova, N., Blume, Y., & Yemets, A. (2021). Synthesis, properties and bioimaging applications of silver-based quantum dots. *International Journal of Molecular Sciences*, 22(22), 12202. DOI: 10.3390/ijms222212202 PMID: 34830084

Buz, P. T., Duman, F. D., Erkisa, M., Demirci, G., Ari, F., Ulukaya, E., & Acar, H. Y. (2019). Development of near-infrared region luminescent N-acetyl-L-cysteine-coated Ag2S quantum dots with differential therapeutic effect. *Nanomedicine (London)*, 14(8), 969–987. DOI: 10.2217/nnm-2018-0214 PMID: 30917096

Cai, Z., Li, F., Wu, P., Ji, L., Zhang, H., Cai, C., & Gervasio, D. F. (2015). Synthesis of nitrogen-doped graphene quantum dots at low temperature for electrochemical sensing trinitrotoluene. *Analytical Chemistry*, 87(23), 11803–11811. DOI: 10.1021/acs.analchem.5b03201 PMID: 26545150

Centeno, L., Romero-García, J., Alvarado-Canché, C., Gallardo-Vega, C., Télles-Padilla, G., Barriga-Castro, E. D., & de León, A. (2021). Green synthesis of graphene quantum dots from Opuntia sp. extract and their application in phytic acid detection. *Sensing and Bio-Sensing Research*, 32, 100412. DOI: 10.1016/j.sbsr.2021.100412

Cerletti, V., Coish, W. A., Gywat, O., & Loss, D. (2005). Recipes for spin-based quantum computing. *Nanotechnology*, 16(4), R27–R49. DOI: 10.1088/0957-4484/16/4/R01

Chinnathambi, S., & Shirahata, N. (2019). Recent advances on fluorescent biomarkers of near-infrared quantum dots for in vitro and in vivo imaging. *Science and Technology of Advanced Materials*, 20(1), 337–355. DOI: 10.1080/14686996.2019.1590731 PMID: 31068983

Chiou, J. R., Lai, B. H., Hsu, K. C., & Chen, D. H. (2013). One-pot green synthesis of silver/iron oxide composite nanoparticles for 4-nitrophenol reduction. *Journal of Hazardous Materials*, 248, 394–400. DOI: 10.1016/j.jhazmat.2013.01.030 PMID: 23416483

Cipriano, L. A., Di Liberto, G., Tosoni, S., & Pacchioni, G. (2020). Quantum confinement in group III–V semiconductor 2D nanostructures. *Nanoscale*, 12(33), 17494–17501. DOI: 10.1039/D0NR03577G PMID: 32808618

Cui, C., Li, X., Liu, J., Hou, Y., Zhao, Y., & Zhong, G. (2015). Synthesis and functions of Ag 2 S nanostructures. *Nanoscale Research Letters*, 10(1), 1–21. DOI: 10.1186/s11671-015-1125-7

Danish, R., Ahmed, F., & Koo, B. H. (2014). Rapid synthesis of high surface area anatase Titanium Oxide quantum dots. *Ceramics International*, 40(8), 12675–12680. DOI: 10.1016/j.ceramint.2014.04.115

Demir, E., & Castranova, V. (2017). Evaluation of the potential genotoxicity of quantum dots. *A Review. Sci. Pages Nanotechnol*, 1, 1–19.

Devaraj, P., Kumari, P., Aarti, C., & Renganathan, A. (2013). Synthesis and characterization of silver nanoparticles using cannonball leaves and their cytotoxic activity against MCF-7 cell line. *Journal of Nanotechnology*, 2013(1), 598328. DOI: 10.1155/2013/598328

Dong, H., Tang, S., Hao, Y., Yu, H., Dai, W., Zhao, G., Cao, Y., Lu, H., Zhang, X., & Ju, H. (2016). Fluorescent MoS2 quantum dots: Ultrasonic preparation, up-conversion and down-conversion bioimaging, and photodynamic therapy. *ACS Applied Materials & Interfaces*, 8(5), 3107–3114. DOI: 10.1021/acsami.5b10459 PMID: 26761391

Droth, M., & Burkard, G. (2016). Spintronics with graphene quantum dots. *physica status solidi (RRL)–. Rapid Research Letters*, 10(1), 75–90.

Du, J., Wang, H., Wang, L., Zhu, S., Song, Y., Yang, B., & Sun, H. (2016). Insight into the effect of functional groups on visible-fluorescence emissions of graphene quantum dots. *Journal of Materials Chemistry. C, Materials for Optical and Electronic Devices*, 4(11), 2235–2242. DOI: 10.1039/C6TC00548A

Gao, W., Wang, M., Ran, C., & Li, L. (2015). Facile one-pot synthesis of MoS 2 quantum dots–graphene–TiO 2 composites for highly enhanced photocatalytic properties. *Chemical Communications*, 51(9), 1709–1712. DOI: 10.1039/C4CC08984G PMID: 25514834

Gardea-Torresdey, J. L., Rico, C. M., & White, J. C. (2014). Trophic transfer, transformation, and impact of engineered nanomaterials in terrestrial environments. *Environmental Science & Technology*, 48(5), 2526–2540. DOI: 10.1021/es4050665 PMID: 24499408

Gupta, A., Mehta, S. K., Kunal, K., Mukhopadhyay, K., & Singh, S. (2022). Quantum dots as promising nanomaterials in agriculture. In *Agricultural Nanobiotechnology* (pp. 243–296). Woodhead Publishing. DOI: 10.1016/B978-0-323-91908-1.00016-X

Higuera-Valenzuela, H. J., Zazueta-Raynaud, A., Heredia-Cancino, J. A., Soto-B, M. A., Cabrera-German, D., Berman-Mendoza, D., ... & Hernadez-Abril, P. A. (2024). Green Synthesis of Silicon Quantum Dots Using Ocimum basilicum var. Purpurascens as an Organic Reductor Agent. *Silicon*, 1-8.

Houben, A., Field, B. L., & Saunders, V. A. (2001). Microdissection and chromosome painting of plant B chromosomes. *Chromosome Painting: Principles, Strategies and Scope*, 115-124.

Ismail, M., Khan, M. I., Khan, S. B., Khan, M. A., Akhtar, K., & Asiri, A. M. (2018). Green synthesis of plant supported CuAg and CuNi bimetallic nanoparticles in the reduction of nitrophenols and organic dyes for water treatment. *Journal of Molecular Liquids*, 260, 78–91. DOI: 10.1016/j.molliq.2018.03.058

Issaabadi, Z., Nasrollahzadeh, M., & Sajadi, S. M. (2017). Green synthesis of the copper nanoparticles supported on bentonite and investigation of its catalytic activity. *Journal of Cleaner Production*, 142, 3584–3591. DOI: 10.1016/j.jclepro.2016.10.109

Kandasamy, K., Venkatesh, M., Khadar, Y. S., & Rajasingh, P. (2020). One-pot green synthesis of CdS quantum dots using Opuntia ficus-indica fruit sap. *Materials Today: Proceedings*, 26, 3503–3506. DOI: 10.1016/j.matpr.2019.06.003

Karaca Açarı, İ., Dik, G., Bakar, B., Ulu, A., Önal, Y., & Ateş, B. (2023). Immobilization of α-amylase onto quantum dots prepared from Hypericum perforatum L. flowers and Hypericum capitatum seeds: Its physicochemical and biochemical characterization. *Topics in Catalysis*, 66(9), 563–576. DOI: 10.1007/s11244-022-01699-y

Ke, C. C., Yang, Y. C., & Tseng, W. L. (2016). Synthesis of Blue-, Green-, Yellow-, and Red-Emitting Graphene-Quantum-Dot-Based Nanomaterials with Excitation-Independent Emission. *Particle & Particle Systems Characterization*, 33(3), 132–139. DOI: 10.1002/ppsc.201500196

Kishanji, M., Mamatha, G., Obi Reddy, K., Varada Rajulu, A., & Madhukar, K. (2017). In situ generation of silver nanoparticles in cellulose matrix using Azadirachta indica leaf extract as a reducing agent. *IJPAC. International Journal of Polymer Analysis and Characterization*, 22(8), 734–740. DOI: 10.1080/1023666X.2017.1369612

Kumar, B., Smita, K., Seqqat, R., Benalcazar, K., Grijalva, M., & Cumbal, L. (2016). In vitro evaluation of silver nanoparticles cytotoxicity on Hepatic cancer (Hep-G2) cell line and their antioxidant activity: Green approach for fabrication and application. *Journal of Photochemistry and Photobiology. B, Biology*, 159, 8–13. DOI: 10.1016/j.jphotobiol.2016.03.011 PMID: 27010841

Kyobe, J. W., Khan, M. D., Kinunda, G., Mubofu, E. B., & Revaprasadu, N. (2020). Synthesis of CdTe quantum dots capped with castor oil using a hot injection solution method. *Materials Science in Semiconductor Processing*, 106, 104780. DOI: 10.1016/j.mssp.2019.104780

Lakshmipathy, R., Sarada, N. C., Chidambaram, K., & Pasha, S. K. (2015). One-step, low-temperature fabrication of CdS quantum dots by watermelon rind: a green approach. *International Journal of Nanomedicine, 10*(sup2), 183-188.

Li, L. L., Ji, J., Fei, R., Wang, C. Z., Lu, Q., Zhang, J. R., Jiang, L.-P., & Zhu, J. J. (2012a). A facile microwave avenue to electrochemiluminescent two-color graphene quantum dots. *Advanced Functional Materials*, 22(14), 2971–2979. DOI: 10.1002/adfm.201200166

Li, R., Liu, Y., Li, Z., Shen, J., Yang, Y., Cui, X., & Yang, G. (2016). Bottom-up fabrication of single-layered nitrogen-doped graphene quantum dots through intermolecular carbonization arrayed in a 2D plane. *Chemistry (Weinheim an der Bergstrasse, Germany)*, 22(1), 272–278. DOI: 10.1002/chem.201503191 PMID: 26593633

Li, Y., Zhao, Y., Cheng, H., Hu, Y., Shi, G., Dai, L., & Qu, L. (2012b). Nitrogen-doped graphene quantum dots with oxygen-rich functional groups. *Journal of the American Chemical Society*, 134(1), 15–18. DOI: 10.1021/ja206030c PMID: 22136359

Lin, H., Wang, C., Wu, J., Xu, Z., Huang, Y., & Zhang, C. (2015). Colloidal synthesis of MoS 2 quantum dots: Size-dependent tunable photoluminescence and bioimaging. *New Journal of Chemistry*, 39(11), 8492–8497. DOI: 10.1039/C5NJ01698C

Lin, L., Xu, Y., Zhang, S., Ross, I. M., Ong, A. C., & Allwood, D. A. (2013). Fabrication and luminescence of monolayered boron nitride quantum dots. *Small (Weinheim an der Bergstrasse, Germany), 10*(1), 60-65.

Lu, J., Yan, M., Ge, L., Ge, S., Wang, S., Yan, J., & Yu, J. (2013). Electrochemiluminescence of blue-luminescent graphene quantum dots and its application in ultrasensitive aptasensor for adenosine triphosphate detection. *Biosensors & Bioelectronics*, 47, 271–277. DOI: 10.1016/j.bios.2013.03.039 PMID: 23587790

Lu, J., Yeo, P. S. E., Gan, C. K., Wu, P., & Loh, K. P. (2011). Transforming C60 molecules into graphene quantum dots. *Nature Nanotechnology*, 6(4), 247–252. DOI: 10.1038/nnano.2011.30 PMID: 21423185

Luo, Z., Qi, G., Chen, K., Zou, M., Yuwen, L., Zhang, X., Huang, W., & Wang, L. (2016). Microwave-Assisted Preparation of White Fluorescent Graphene Quantum Dots as a Novel Phosphor for Enhanced White-Light-Emitting Diodes. *Advanced Functional Materials*, 26(16), 2739–2744. DOI: 10.1002/adfm.201505044

Mahalakhsmi, A., & Baskar, G. (2019). Green synthesis and characterization of cadmium-tellurium quantum dots using pomelo peel aqueous extract. *Journal of Electronic Materials*, 48(9), 5975–5979. DOI: 10.1007/s11664-019-07352-x

Mallin, M. A., & Cahoon, L. B. (2003). Industrialized animal production—A major source of nutrient and microbial pollution to aquatic ecosystems. *Population and Environment*, 24(5), 369–385. DOI: 10.1023/A:1023690824045

Mewada, A., Pandey, S., Shinde, S., Mishra, N., Oza, G., Thakur, M., Sharon, M., & Sharon, M. (2013). Green synthesis of biocompatible carbon dots using aqueous extract of Trapa bispinosa peel. *Materials Science and Engineering C*, 33(5), 2914–2917. DOI: 10.1016/j.msec.2013.03.018 PMID: 23623114

Mirzaeifard, Z., & Shariatinia, Z. (2024). Economical, one-pot, and green synthesis of plant-based carbon quantum dots for efficient visible-light photocatalytic dye degradation and water purification. *Journal of the Taiwan Institute of Chemical Engineers*, 163, 105655. DOI: 10.1016/j.jtice.2024.105655

Mirzaeifard, Z., & Shariatinia, Z. (2024). Economical, one-pot, and green synthesis of plant-based carbon quantum dots for efficient visible-light photocatalytic dye degradation and water purification. *Journal of the Taiwan Institute of Chemical Engineers*, 163, 105655. DOI: 10.1016/j.jtice.2024.105655

Moloney, M. P., Govan, J., Loudon, A., Mukhina, M., & Gun'ko, Y. K. (2015). Preparation of chiral quantum dots. *Nature Protocols*, 10(4), 558–573. DOI: 10.1038/nprot.2015.028 PMID: 25741991

Muthulakshmi, L., Rajini, N., Nellaiah, H., Kathiresan, T., Jawaid, M., & Rajulu, A. V. (2017). Preparation and properties of cellulose nanocomposite films with in situ generated copper nanoparticles using Terminalia catappa leaf extract. *International Journal of Biological Macromolecules*, 95, 1064–1071. DOI: 10.1016/j.ijbiomac.2016.09.114 PMID: 27984140

Nagar, N., & Devra, V. (2018). Green synthesis and characterization of copper nanoparticles using Azadirachta indica leaves. *Materials Chemistry and Physics*, 213, 44–51. DOI: 10.1016/j.matchemphys.2018.04.007

Nakum, J., & Bhattacharya, D. (2022). Various green nanomaterials used for wastewater and soil treatment: A mini-review. *Frontiers in Environmental Science*, 9, 724814. DOI: 10.3389/fenvs.2021.724814

Nasseri, M. A., Keshtkar, H., Kazemnejadi, M., & Allahresani, A. (2020). Phytochemical properties and antioxidant activity of Echinops persicus plant extract: Green synthesis of carbon quantum dots from the plant extract. *SN Applied Sciences*, 2(4), 1–12. DOI: 10.1007/s42452-020-2466-0

Ndikau, M., Noah, N. M., Andala, D. M., & Masika, E. (2017). Green synthesis and characterization of silver nanoparticles using Citrullus lanatus fruit rind extract. *International Journal of Analytical Chemistry*, 2017(1), 8108504. DOI: 10.1155/2017/8108504 PMID: 28316627

Noma, S. A. A., Ulu, A., Acet, Ö., Sanz, R., Sanz-Pérez, E. S., Odabaşı, M., & Ateş, B. (2020). Comparative study of ASNase immobilization on tannic acid-modified magnetic Fe 3 O 4/SBA-15 nanoparticles to enhance stability and reusability. *New Journal of Chemistry*, 44(11), 4440–4451. DOI: 10.1039/D0NJ00127A

Pan, L., Zou, J. J., Wang, S., Huang, Z. F., Yu, A., Wang, L., & Zhang, X. (2013). Quantum dot self-decorated TiO 2 nanosheets. *Chemical Communications*, 49(59), 6593–6595. DOI: 10.1039/c3cc42152j PMID: 23660603

Peng, H., Liu, P., Lin, D., Deng, Y., Lei, Y., Chen, W., Chen, Y., Lin, X., Xia, X., & Liu, A. (2016). Fabrication and multifunctional properties of ultrasmall water-soluble tungsten oxide quantum dots. *Chemical Communications*, 52(61), 9534–9537. DOI: 10.1039/C6CC03245A PMID: 27381501

Philip, D. (2010). Green synthesis of gold and silver nanoparticles using Hibiscus rosa sinensis. *Physica E, Low-Dimensional Systems and Nanostructures*, 42(5), 1417–1424. DOI: 10.1016/j.physe.2009.11.081

Pich, U., Fritsch, R., & Schubert, I. (1996). Closely related Allium species (Alliaceae) share a very similar satellite sequence. *Plant Systematics and Evolution*, 202(3-4), 255–264. DOI: 10.1007/BF00983386

Prasad, P. R., Kanchi, S., & Naidoo, E. B. (2016). In-vitro evaluation of copper nanoparticles cytotoxicity on prostate cancer cell lines and their antioxidant, sensing and catalytic activity: One-pot green approach. *Journal of Photochemistry and Photobiology. B, Biology*, 161, 375–382. DOI: 10.1016/j.jphotobiol.2016.06.008 PMID: 27318296

Purushothaman, B., & Song, J. M. (2021). Ag 2 S quantum dot theragnostics. *Biomaterials Science*, 9(1), 51–69. DOI: 10.1039/D0BM01576H PMID: 33185212

Qiu, J., Li, D., Mou, X., Li, J., Guo, W., Wang, S., Yu, X., Ma, B., Zhang, S., Tang, W., Sang, Y., Gil, P. R., & Liu, H. (2016). Graphene Quantum Dots: Effects of Graphene Quantum Dots on the Self-Renewal and Differentiation of Mesenchymal Stem Cells (Adv. Healthcare Mater. 6/2016). *Advanced Healthcare Materials*, 5(6), 623–623. DOI: 10.1002/adhm.201670026 PMID: 26833812

Qu, D., Sun, Z., Zheng, M., Li, J., Zhang, Y., Zhang, G., & Xie, Z. (2015). Three colors emission from S, N Co-doped graphene quantum dots for visible light H2 production and bioimaging. *Advanced Optical Materials*, 3(3), 360–367.

Rani, H., Singh, S. P., Yadav, T. P., Khan, M. S., Ansari, M. I., & Singh, A. K. (2020). In-vitro catalytic, antimicrobial and antioxidant activities of bioengineered copper quantum dots using Mangifera indica (L.) leaf extract. *Materials Chemistry and Physics*, 239, 122052. DOI: 10.1016/j.matchemphys.2019.122052

Reddy, N. R., Sathe, S. K., & Salunkhe, D. K. (1982). Phytates in legumes and cereals. *Advances in Food Research*, 28, 1–92. DOI: 10.1016/S0065-2628(08)60110-X PMID: 6299067

Rhyner, M. N., Smith, A. M., Gao, X., Mao, H., Yang, L., & Nie, S. (2006). Quantum dots and multifunctional nanoparticles: new contrast agents for tumor imaging.

Şahin, S., Ünlü, C., & Trabzon, L. (2021). Affinity biosensors developed with quantum dots in microfluidic systems. *Emergent Materials*, 4(1), 187–209. DOI: 10.1007/s42247-021-00195-5 PMID: 33718778

Sengan, M., Subramaniyan, S. B., Prakash, S. A., Kamlekar, R., & Veerappan, A. (2019). Effective elimination of biofilm formed with waterborne pathogens using copper nanoparticles. *Microbial Pathogenesis*, 127, 341–346. DOI: 10.1016/j.micpath.2018.12.025 PMID: 30557585

Shivaji, K., Mani, S., Ponmurugan, P., De Castro, C. S., Lloyd Davies, M., Balasubramanian, M. G., & Pitchaimuthu, S. (2018). Green-synthesis-derived CdS quantum dots using tea leaf extract: Antimicrobial, bioimaging, and therapeutic applications in lung cancer cells. *ACS Applied Nano Materials*, 1(4), 1683–1693. DOI: 10.1021/acsanm.8b00147

Siddiqi, K. S., Husen, A., & Rao, R. A. (2018). A review on biosynthesis of silver nanoparticles and their biocidal properties. *Journal of Nanobiotechnology*, 16(1), 1–28. DOI: 10.1186/s12951-018-0334-5 PMID: 29452593

Singh, A. K., Pal, P., Gupta, V., Yadav, T. P., Gupta, V., & Singh, S. P. (2018). Green synthesis, characterization and antimicrobial activity of zinc oxide quantum dots using Eclipta alba. *Materials Chemistry and Physics*, 203, 40–48. DOI: 10.1016/j.matchemphys.2017.09.049

Singh, J., Dutta, T., Kim, K. H., Rawat, M., Samddar, P., & Kumar, P. (2018). 'Green'synthesis of metals and their oxide nanoparticles: Applications for environmental remediation. *Journal of Nanobiotechnology*, 16(1), 1–24. DOI: 10.1186/s12951-018-0408-4 PMID: 30373622

Song, J. Y., Kwon, E. Y., & Kim, B. S. (2010). Biological synthesis of platinum nanoparticles using Diopyros kaki leaf extract. *Bioprocess and Biosystems Engineering*, 33(1), 159–164. DOI: 10.1007/s00449-009-0373-2 PMID: 19701776

Sun, Q., Cai, X., Li, J., Zheng, M., Chen, Z., & Yu, C. P. (2014). Green synthesis of silver nanoparticles using tea leaf extract and evaluation of their stability and antibacterial activity. *Colloids and Surfaces. A, Physicochemical and Engineering Aspects*, 444, 226–231. DOI: 10.1016/j.colsurfa.2013.12.065

Sun, Z., Chen, Y., Ke, Q., Yang, Y., & Yuan, J. (2002). Photocatalytic degradation of a cationic azo dye by TiO2/bentonite nanocomposite. *Journal of Photochemistry and Photobiology A Chemistry*, 149(1-3), 169–174. DOI: 10.1016/S1010-6030(01)00649-9

Suzuki, N., Wang, Y., Elvati, P., Qu, Z. B., Kim, K., Jiang, S., Baumeister, E., Lee, J., Yeom, B., Bahng, J. H., Lee, J., Violi, A., & Kotov, N. A. (2016). Chiral graphene quantum dots. *ACS Nano*, 10(2), 1744–1755. DOI: 10.1021/acsnano.5b06369 PMID: 26743467

Tang, D., Liu, J., Yan, X., & Kang, L. (2016). Graphene oxide derived graphene quantum dots with different photoluminescence properties and peroxidase-like catalytic activity. *RSC Advances*, 6(56), 50609–50617. DOI: 10.1039/C5RA26279H

Tang, L., Ji, R., Cao, X., Lin, J., Jiang, H., Li, X., Teng, K. S., Luk, C. M., Zeng, S., Hao, J., & Lau, S. P. (2012). Deep ultraviolet photoluminescence of water-soluble self-passivated graphene quantum dots. *ACS Nano*, 6(6), 5102–5110. DOI: 10.1021/nn300760g PMID: 22559247

Tang, Y., Su, Y., Yang, N., Zhang, L., & Lv, Y. (2014). Carbon nitride quantum dots: A novel chemiluminescence system for selective detection of free chlorine in water. *Analytical Chemistry*, 86(9), 4528–4535. DOI: 10.1021/ac5005162 PMID: 24655009

Tokudome, Y., Morimoto, T., Tarutani, N., Vaz, P. D., Nunes, C. D., Prevot, V., Stenning, G. B. G., & Takahashi, M. (2016). Layered double hydroxide nanoclusters: Aqueous, concentrated, stable, and catalytically active colloids toward green chemistry. *ACS Nano*, 10(5), 5550–5559. DOI: 10.1021/acsnano.6b02110 PMID: 27124717

Vanlalveni, C., Lallianrawna, S., Biswas, A., Selvaraj, M., Changmai, B., & Rokhum, S. L. (2021). Green synthesis of silver nanoparticles using plant extracts and their antimicrobial activities: A review of recent literature. *RSC Advances*, 11(5), 2804–2837. DOI: 10.1039/D0RA09941D PMID: 35424248

Vaseghi, Z., Nematollahzadeh, A., & Tavakoli, O. (2018). Green methods for the synthesis of metal nanoparticles using biogenic reducing agents: A review. *Reviews in Chemical Engineering*, 34(4), 529–559. DOI: 10.1515/revce-2017-0005

Vasireddy, R., Paul, R., & Mitra, A. K. (2012). Green synthesis of silver nanoparticles and the study of optical properties. *Nanomaterials and Nanotechnology*, 2, 8. DOI: 10.5772/52329

Verma, A., & Mehata, M. S. (2016). Controllable synthesis of silver nanoparticles using Neem leaves and their antimicrobial activity. *Journal of radiation Research and applied sciences, 9*(1), 109-115.

Wang, H., Yuan, X., Wang, H., Chen, X., Wu, Z., Jiang, L., Xiong, W., & Zeng, G. (2016). Facile synthesis of Sb2S3/ultrathin g-C3N4 sheets heterostructures embedded with g-C3N4 quantum dots with enhanced NIR-light photocatalytic performance. *Applied Catalysis B: Environmental*, 193, 36–46. DOI: 10.1016/j.apcatb.2016.03.075

Wang, W., Jimmy, C. Y., Shen, Z., Chan, D. K., & Gu, T. (2014). gC 3 N 4 quantum dots: Direct synthesis, upconversion properties and photocatalytic application. *Chemical Communications*, 50(70), 10148–10150. DOI: 10.1039/C4CC02543A PMID: 25052166

Wang, X., Wang, Y., He, H., Chen, X., Sun, X., Sun, Y., Zhou, G., Xu, H., & Huang, F. (2016). Steering graphene quantum dots in living cells: Lighting up the nucleolus. *Journal of Materials Chemistry. B*, 4(4), 779–784. DOI: 10.1039/C5TB02474A PMID: 32262959

Wang, Y., & Nowack, B. (2018). Environmental risk assessment of engineered nano-SiO2, nano iron oxides, nano-CeO2, nano-Al2O3, and quantum dots. *Environmental Toxicology and Chemistry*, 37(5), 1387–1395. DOI: 10.1002/etc.4080 PMID: 29315795

Wu, J., Zheng, Y., Song, W., Luan, J., Wen, X., Wu, Z., Chen, X., Wang, Q., & Guo, S. (2014). In situ synthesis of silver-nanoparticles/bacterial cellulose composites for slow-released antimicrobial wound dressing. *Carbohydrate Polymers*, 102, 762–771. DOI: 10.1016/j.carbpol.2013.10.093 PMID: 24507345

Xu, Q., Li, W., Ding, L., Yang, W., Xiao, H., & Ong, W. J. (2019). Function-driven engineering of 1D carbon nanotubes and 0D carbon dots: Mechanism, properties and applications. *Nanoscale*, 11(4), 1475–1504. DOI: 10.1039/C8NR08738E PMID: 30620019

Xu, S., Li, D., & Wu, P. (2015). One-pot, facile, and versatile synthesis of monolayer MoS2/WS2 quantum dots as bioimaging probes and efficient electrocatalysts for hydrogen evolution reaction. *Advanced Functional Materials*, 25(7), 1127–1136. DOI: 10.1002/adfm.201403863

Xu, Y., Liu, J., Gao, C., & Wang, E. (2014). Applications of carbon quantum dots in electrochemiluminescence: A mini review. *Electrochemistry Communications*, 48, 151–154. DOI: 10.1016/j.elecom.2014.08.032

Xu, Y., Wang, X., Zhang, W. L., Lv, F., & Guo, S. (2018). Recent progress in two-dimensional inorganic quantum dots. *Chemical Society Reviews*, 47(2), 586–625. DOI: 10.1039/C7CS00500H PMID: 29115355

Yakura, K., & Tanifuji, S. (1983). Molecular cloning and restriction analysis of Eco RI-fragments of Vicia faba rDNA. *Plant & Cell Physiology*, 24(7), 1327–1330.

Yan, H., Li, P., Wen, F., Xu, Q., Guo, Q., & Su, W. (2023). Green synthesis of carbon quantum dots from plant turmeric holds promise as novel photosensitizer for in vitro photodynamic antimicrobial activity. *Journal of Materials Research and Technology*, 22, 17–34. DOI: 10.1016/j.jmrt.2022.11.090

Yang, J. S., Martinez, D. A., & Chiang, W. H. (2017). Synthesis, characterization and applications of graphene quantum dots. *Recent Trends in Nanomaterials: Synthesis and Properties*, 65-120.

Yeh, T. F., Teng, C. Y., Chen, S. J., & Teng, H. (2014). Nitrogen-doped graphene oxide quantum dots as photocatalysts for overall water-splitting under visible light Illumination. *Advanced Materials*, 26(20), 3297–3303. DOI: 10.1002/adma.201305299 PMID: 24677453

Zaini, M. S., Ying Chyi Liew, J., Alang Ahmad, S. A., Mohmad, A. R., & Kamarudin, M. A. (2020). Quantum confinement effect and photoenhancement of photoluminescence of PbS and PbS/MnS quantum dots. *Applied Sciences (Basel, Switzerland)*, 10(18), 6282. DOI: 10.3390/app10186282

Zhang, C., Cui, Y., Song, L., Liu, X., & Hu, Z. (2016). Microwave assisted one-pot synthesis of graphene quantum dots as highly sensitive fluorescent probes for detection of iron ions and pH value. *Talanta*, 150, 54–60. DOI: 10.1016/j.talanta.2015.12.015 PMID: 26838381

Zhang, F., Liu, F., Wang, C., Xin, X., Liu, J., Guo, S., & Zhang, J. (2016). Effect of lateral size of graphene quantum dots on their properties and application. *ACS Applied Materials & Interfaces*, 8(3), 2104–2110. DOI: 10.1021/acsami.5b10602 PMID: 26725374

Zhang, L., Peng, D., Liang, R. P., & Qiu, J. D. (2015). Nitrogen-doped graphene quantum dots as a new catalyst accelerating the coordination reaction between cadmium (II) and 5, 10, 15, 20-tetrakis (1-methyl-4-pyridinio) porphyrin for cadmium (II) sensing. *Analytical Chemistry*, 87(21), 10894–10901. DOI: 10.1021/acs.analchem.5b02450 PMID: 26427520

Zhang, X., Xie, H., Liu, Z., Tan, C., Luo, Z., Li, H., Lin, J., Sun, L., Chen, W., Xu, Z., Xie, L., Huang, W., & Zhang, H. (2015). Black phosphorus quantum dots. *Angewandte Chemie International Edition*, 54(12), 3653–3657. DOI: 10.1002/anie.201409400 PMID: 25649505

Zhao, J., Zhao, L., Lan, C., & Zhao, S. (2016). Graphene quantum dots as effective probes for label-free fluorescence detection of dopamine. *Sensors and Actuators. B, Chemical*, 223, 246–251. DOI: 10.1016/j.snb.2015.09.105

Zhao, X., Ma, X., Sun, J., Li, D., & Yang, X. (2016). Enhanced catalytic activities of surfactant-assisted exfoliated WS2 nanodots for hydrogen evolution. *ACS Nano*, 10(2), 2159–2166. DOI: 10.1021/acsnano.5b06653 PMID: 26785941

Zhu, H., Liu, A., Shan, F., Yang, W., Zhang, W., Li, D., & Liu, J. (2016). One-step synthesis of graphene quantum dots from defective CVD graphene and their application in IGZO UV thin film phototransistor. *Carbon*, 100, 201–207. DOI: 10.1016/j.carbon.2016.01.016

Zou, L. K., Wang, H. N., Pan, X., Tian, G. B., Xie, Z. W., Wu, Q., Chen, H., Xie, T., & Yang, Z. R. (2008). Expression, purification and characterization of a phyA m-phyCs fusion phytase. *Journal of Zhejiang University. Science. B.*, 9(7), 536–545. DOI: 10.1631/jzus.B0720006 PMID: 18600783

KEY TERMS AND DEFINITIONS

Photoluminescence: The release of light from a substance after photon absorption. A crucial characteristic of small particles that makes them advantageous in imaging and display technology.

Quantum Yield: The definition of quantum yield is the ratio of particles absorbed to photons released. A measure of the efficiency of photoluminescence in quantum dots.

Surface Functionalization: Surface functioning is the process of altering a nanoparticle's surfaces to increase its usefulness, stability, and dispersibility. Improves quantum dots' compatibility and performance for certain applications.

Photocatalysis: Photocatalysis may be defined as the process of accelerating a photoreaction when a catalyst is present. Due to their ability to function as photocatalysts, quantum dots have uses in conversion of solar energy and environmental cleanup.

Theranostics: A contemporaneous evaluation and cure of illness employing nanoparticles in tandem with therapy. Quantum dots are potent instruments in customized medicine as they may be designed for precise medication administration and imaging.

Chapter 14

Mapping the Landscape of Biogenic Nanoparticles in Bioinformatics and Nanobiotechnology: AI-Driven Insights

Mrutyunjay Padhiary
https://orcid.org/0000-0002-2236-568X
Assam University, Silchar, India

Dipak Roy
https://orcid.org/0009-0008-7268-6351
Tezpur University, Tezpur, India

Poulami Dey
https://orcid.org/0009-0005-3345-4783
IIT Kharagpur, India

ABSTRACT

The relationship between biogenic nanoparticles, bioinformatics, and nanobiotechnology is investigated in this chapter, with an emphasis on the contribution of artificial intelligence to the advancement of science and technology. There are numerous applications for biological nanoparticles in agriculture, medicine, and environmental cleanup. The chapter emphasizes the significance of thorough investigation, safety concerns, adhering to regulations, and ethical considerations. It also discusses about how AI-powered techniques can be used to make, characterize, and use biogenic nanoparticles, which can lead to new insights based on data, predictive modeling, and personalized methods. Researchers can solve societal demands and promote environmental sustainability and human health by utilizing AI-driven technologies and biogenic nanoparticles through ethical innovation and interdisciplinary collaboration.

DOI: 10.4018/979-8-3693-6240-2.ch014

1. INTRODUCTION

1.1 Background

In the disciplines of bioinformatics and nanobiotechnology, artificial intelligence (AI) has become a potent instrument that is influencing research and innovation. AI algorithms are used in bioinformatics to predict molecular structures, find patterns in genetic sequences, and evaluate large-scale biological datasets (Alam et al., 2024).These AI-driven techniques improve disease diagnosis and therapy by making drug development, personalized medicine, and biomarker identification simpler. Similar to this, AI is essential to the development, characterization, and use of nanomaterials in nanobiotechnology, including biogenic nanoparticles (Ghosh et al., 2019). Through the utilization of AI methodologies, including machine learning, data analytics, and computational modeling, scientists can accelerate the development of innovative nanomaterials with customized characteristics and abilities. AI-driven techniques improve our comprehension of nanoscale phenomena and allow for precise control over the production, assembly, and manipulation of nanoparticles for a variety of applications (Vasoya, 2023).

1.2 Importance of Biogenic Nanoparticles

Biogenic nanoparticles, which are produced naturally by living things (Figure 1), have attracted a lot of interest because of their special qualities and broad range of uses in many different fields. These nanoparticles are perfect for use in environmental remediation, agriculture, medicine, and other fields because of their diverse features, which include size, shape, surface chemistry, and functionality. Because they are non-toxic and biocompatible, biogenic nanoparticles have prospective applications in biomedicine such as targeted medication delivery, imaging, diagnostics, and therapies (Madkour, 2018).These nanoparticles aid in soil remediation, insect control, and crop development in agriculture, supporting food security and sustainable farming methods. Furthermore, biogenic nanoparticles show potential for environmental remediation since they effectively remove pollutants from soil and water, reducing the amount of heavy metals, pesticides, and other pollutant-related environmental contamination (Pramanik & Chakraborty, 2014). Because of their natural synthesis, sustainable characteristics, and wide range of uses, biogenic nanoparticles are significant because they can be used to address social issues and further scientific study.

Figure 1. Schematic of biogenic nanoparticle synthesis

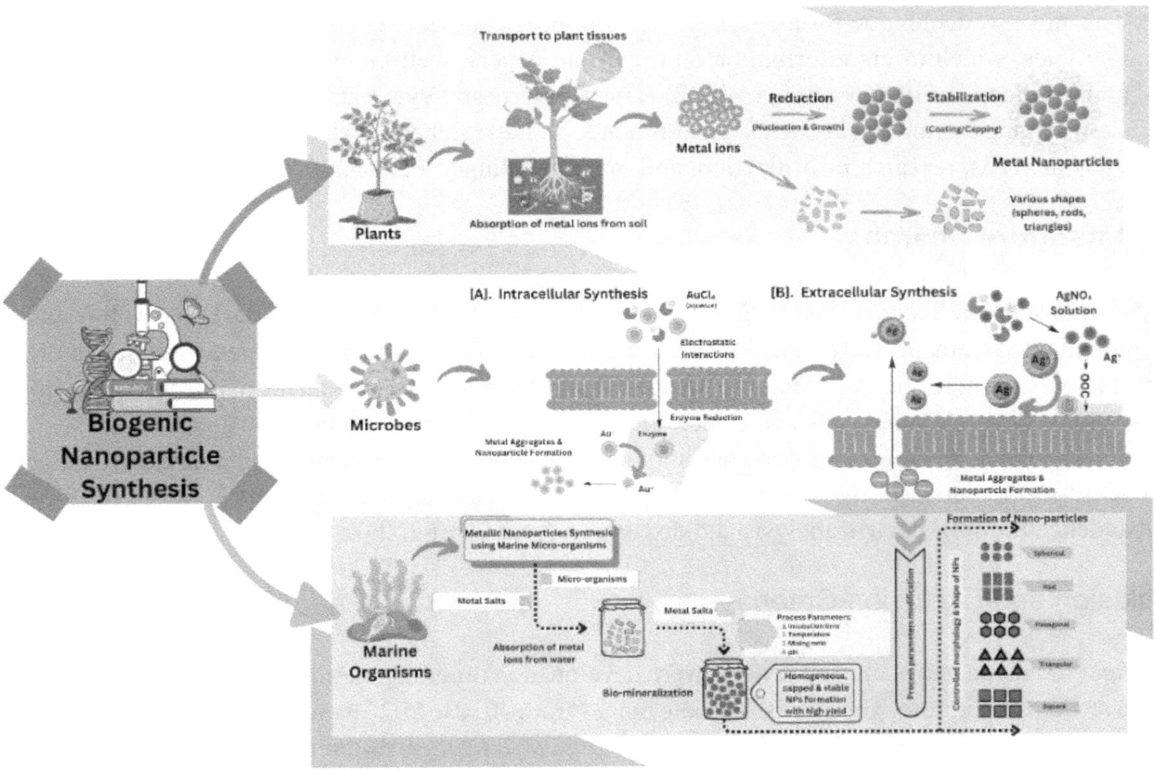

1.3 Purpose and Scope

This chapter aims to give an in-depth investigation of the relationship between biogenic nanoparticles and AI, as well as the uses of these technologies in bioinformatics, nanobiotechnology, and other fields. Our goal is to clarify the exciting prospects of AI-driven approaches to promote advancements in biogenic nanoparticle research and innovation. We will specifically examine the importance of biogenic nanoparticles and emphasize their various uses in environmental cleanup, agriculture, and medicine (Khandel & Shahi, 2018). It will additionally discuss how AI may help overcome obstacles in the characterization, synthesis, and use of nanoparticles, highlighting how it can improve technology and meet societal requirements. This chapter aims to deepen our understanding of this interdisciplinary topic by exploring the interaction between AI and biogenic nanoparticles. Additionally, the study hopes to provide inspiration for future research lines that will pursue the goal of fully utilizing biogenic nanoparticles for societal benefit.

2. BIBLIOMETRIC ANALYSIS

A quantitative technique for assessing scientific publications is conducted with regard to biogenic nanoparticles, which offers information on important papers, well-known writers, new developments, and areas of cooperation (Geng et al., 2022). This supports strategic planning, resource allocation, and decision-making by funding agencies, governments, and researchers. It also helps with knowledge mapping, gap analysis, research prioritization, and interdisciplinary teamwork.

2.1 Literature Search

Google Scholar, Scopus, Web of Science, PubMed, and other academic databases have all been used to implement systematic search tactics (Saad et al., 2021). These databases made a wide range of academic papers, conference proceedings, patents, and gray literature about biogenic nanoparticles available. Keywords, "biogenic nanoparticles," "biologically synthesized nanoparticles," "green synthesis," and terms particular to the fields and applications of nanoparticles (e.g., medicine, agriculture, environmental science) were included in the search queries. Scopus database is used for analysis for uniformity and convenience for bibliometric analysis and discussion (year 2010-2024).

Inclusion and Exclusion Criteria

Clear inclusion and exclusion criteria were developed in order to guarantee the quality and relevance of the retrieved literature. The publishing date ranges from 2010 to 2024 were considered for the inclusion criteria, which also included the preference for the English language, different document categories (such as reviews and research articles), and geographic areas of interest. On the other hand, papers unrelated to biogenic nanoparticles or their uses were excluded, as were duplicates and non-peer-reviewed sources, as well as conference abstracts. Only the Scopus database was used for some of the interpretation, however, other databases were used for the overall conclusions.

Data Extraction

After identifying relevant papers, each source's bibliographic information and metadata were extracted. Publication title, authors, affiliations, year of publication, journal name, abstract, keywords, number of citations, and bibliographic references are among the details included in this study. For all other databases, data extraction was done manually, but for Scopus, automated tools and software were utilized to guarantee accuracy and speed up the process.

2.2 Trends in Research

Emerging Synthesis Methods

The research into new synthesis techniques for biogenic nanoparticles is booming. Conventional methods that make use of microbial fermentation, plant extracts, and biomimetic processes are still being investigated and improved. On the other hand, current patterns suggest that novel approaches including enzymatic synthesis, genetic engineering, and bioinspired manufacturing processes are gaining popularity

(Liu et al., 2023). These methods provide fine control over the size, shape, and surface characteristics of nanoparticles, making it possible to create customized nanoparticles for particular uses.

Multifunctional Nanoparticles

The design of multifunctional biogenic nanoparticles with a variety of characteristics and functions is becoming more and more important. Researchers are investigating methods to augment the adaptability of nanoparticles through the integration of various constituents, including but not limited to medicinal medicines, targeting ligands, imaging probes, and stimuli-responsive moieties (Kang et al., 2017). These multifunctional nanoparticles provide real-time biological process monitoring, simultaneous imaging and therapy, and synergistic therapeutic effects, all of which improve biomedical applications' accuracy and efficacy.

Biomedical Applications

In a number of biomedical applications, such as medication delivery, imaging, diagnostics, and regenerative medicine, biogenic nanoparticles show great potential. This field of study focuses on controlling the release of medicinal substances, bypassing biological barriers, and tailoring the characteristics of nanoparticles to particular biomedical problems (Vinayagasundaram et al., 2021). Furthermore, because of their potential to completely transform healthcare technologies, biogenic nanoparticles are becoming more and more interesting to study in cutting-edge fields including tissue engineering, gene editing, and immunotherapy.

Environmental Remediation

As scientists look for long-term approaches to resource preservation and pollution reduction, the use of biogenic nanoparticles in environmental remediation is becoming more popular. Biogenic nanoparticles have the potential to eliminate pathogens, heavy metals, and organic pollutants from soil and water (Das et al., 2018). Current developments in this field center on refining methods for synthesizing nanoparticles, comprehending how nanoparticles interact with environmental matrices, and creating remediation plans that are scalable for practical use.

Nano-Bio Interface

In the subject of nanobiotechnology, appreciating the interactions between biogenic nanoparticles and biological systems is a major area of scientific interest. This field is seeing trends in the understanding of the relationship between nanoparticles and cells, the study of biological destiny and toxicity mechanisms, and the investigation of immune responses to nanoparticle exposure (Ernst et al., 2021). In order to develop safe and effective nanoparticles for use in biomedical and environmental applications, researchers are working to understand the intricacies of the nano-bio interaction.

Data-Driven Approaches

In biogenic nanoparticle research, there is an increasing tendency toward data-driven techniques due to the widespread availability of data analytics and computational tools. To examine large-scale statistics, forecast the behavior of nanoparticles, and improve synthesis settings, researchers use big data analytics, machine learning algorithms, and computational modeling techniques (Champa-Bujaico et al., 2022). These data-driven methods help scientists find patterns, extract insightful information, and expedite the discovery and development of biogenic nanoparticles for a wide range of uses. The convergence of deep learning, IoT, and sensors has also been extended to the agricultural domain, where AI integration with biogenic nanoparticles can optimize farming and offer precision solutions for sustainable agriculture (Padhiary, 2024c).

The current developments in biogenic nanoparticle research demonstrate a multidisciplinary approach that includes their synthesis, characterization, application, and comprehension of biological system interactions. Through the utilization of modern equipment and interdisciplinary teamwork, scientists want to fully realize the promise of biogenic nanoparticles in addressing urgent societal issues and furthering scientific understanding.

2.3 Key Findings and Insights

Publication Trends

By using bibliometric analysis, it is possible to observe the changing publishing trends and growth dynamics in biogenic nanoparticle research between 2010 and 2024. Emerging research themes, significant publications, and fast growth are all evident in the document numbers, citation rates, and publishing trends during the considered period. One of the most important discoveries is the exponential rise in publications (Figure 2a), which suggests that biogenic nanoparticles are becoming more and more important in scientific research.

Figure 2. Bibliometric analysis: (a) no of publication year wise, (b) geographical distribution, (c) types of publication, and (d) area of research

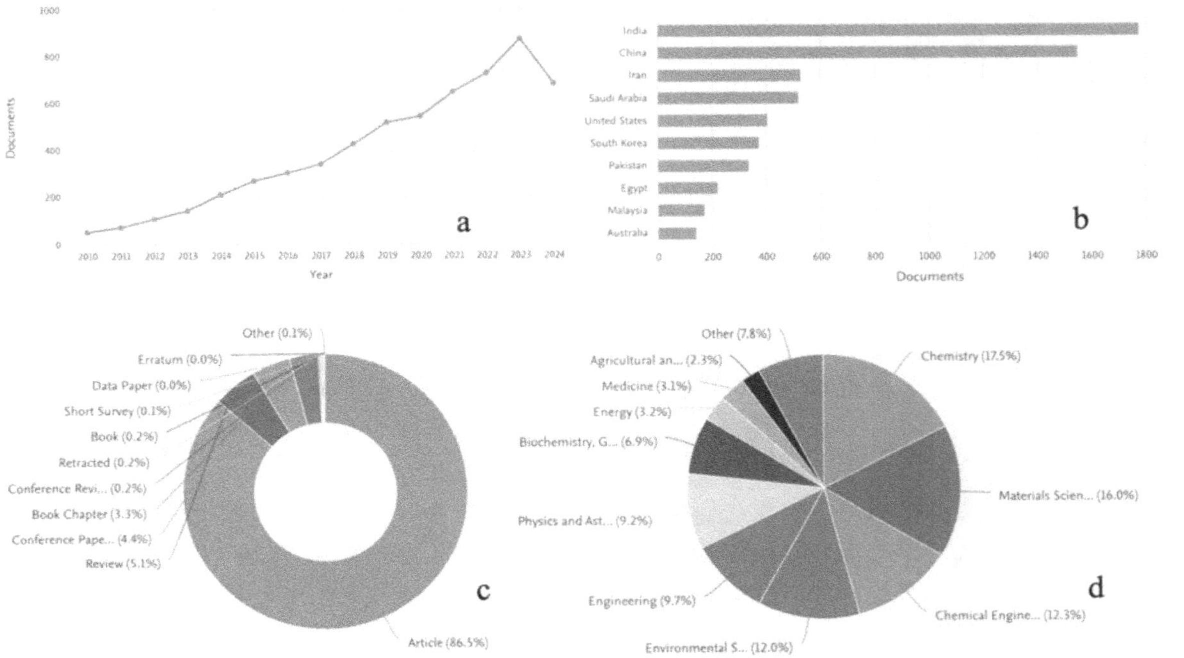

Figure 3. (a) Citation analysis and, (b) prominent authors in the research area

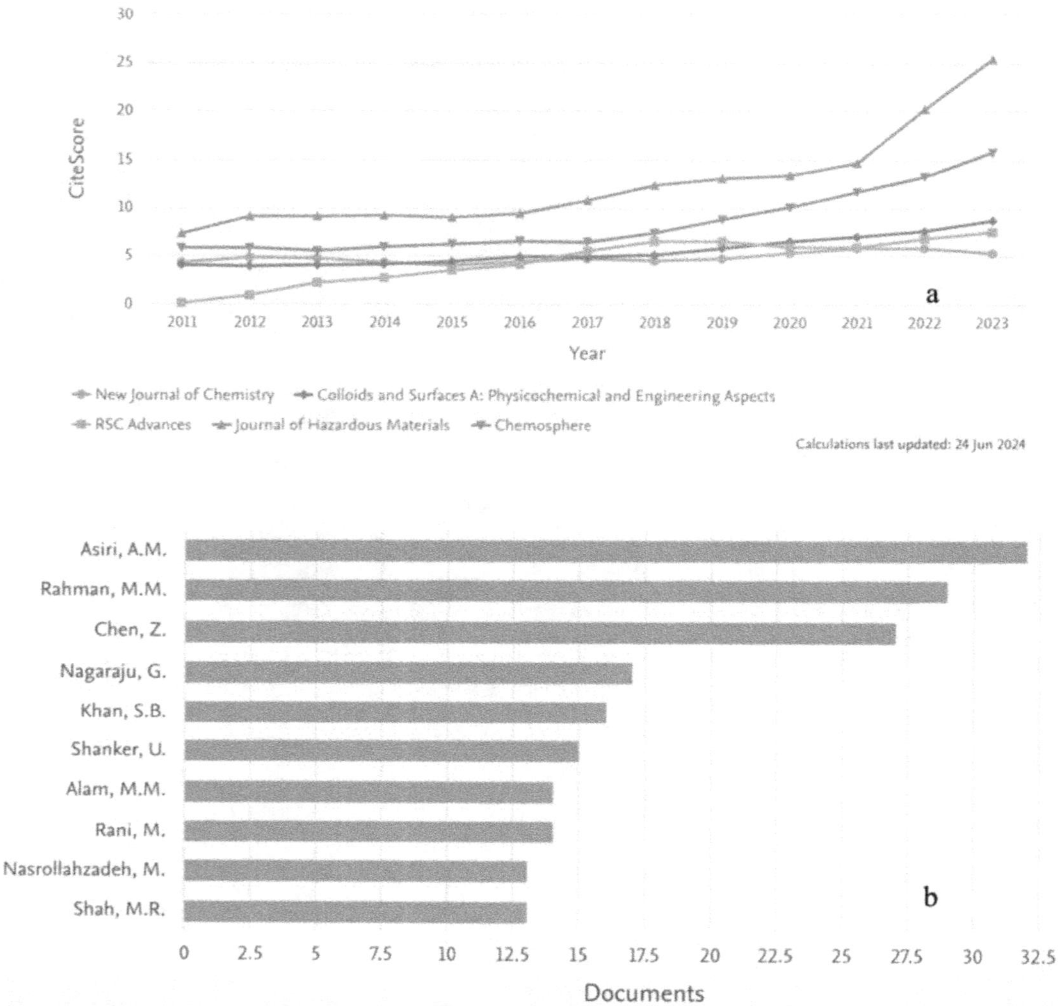

Geographical Distribution

Bibliometric analysis provides insights into the geographic distribution of research on biogenic nanoparticles and shows the contributions of different countries (Figure 2b), regions, and research institutes. Leading research centers, global partnerships, and developing research clusters can all be easily identified by mapping author affiliations and collaboration networks. The dominance of specific countries or organizations in particular fields of study, which suggests local expertise and cooperation networks, is another important discovery. Based on the analysis, it is clear that India is the research leader in this field, with China coming in second.

Authorship Patterns and Citation Analysis

The study reveals the dynamics of scientific collaboration in biogenic nanoparticle research, revealing prolific writers, cooperative groups, and interdisciplinary collaborations (Figure 2c). Key findings include significant researchers, research consortia, and cross-disciplinary collaborations (Figure 2d). Common research themes and topics were identified, including nano-bio interactions, environmental remediation, biomedical applications, and nanoparticle production techniques. Citation analysis showed A. M. Asiri's dominance in authorship (Figure 3b), influential publication, and foundational work, indicating the propagation and acknowledgment of scientific knowledge (Figure 3a).

Interdisciplinary Connections

Bibliometric analysis facilitated the discovery of interdisciplinary collaborations and links in biogenic nanoparticle research. When we look at co-citation networks, keyword co-occurrence matrices, and bibliographic coupling, we can see that there are connections between different fields, big themes, and the sharing of knowledge between many fields (Figure 4). Identifying interdisciplinary research clusters, collaboration networks, and information transfer pathways, all of which point to the integration of knowledge and expertise from several fields, is among the key findings. This analysis revealed significant new patterns, knowledge gaps, and critical findings that provided valuable insights into the dynamics, trends, and implications of biogenic nanoparticle research. Bibliometric analysis can help researchers in the field of biogenic nanoparticles prioritize their work, make informed decisions, and promote scientific innovation.

Figure 4. The keyword network diagram in VOSviewer based on 276 most occurring keywords out of 11754, with a number of occurrences of a keyword being ≥ 10

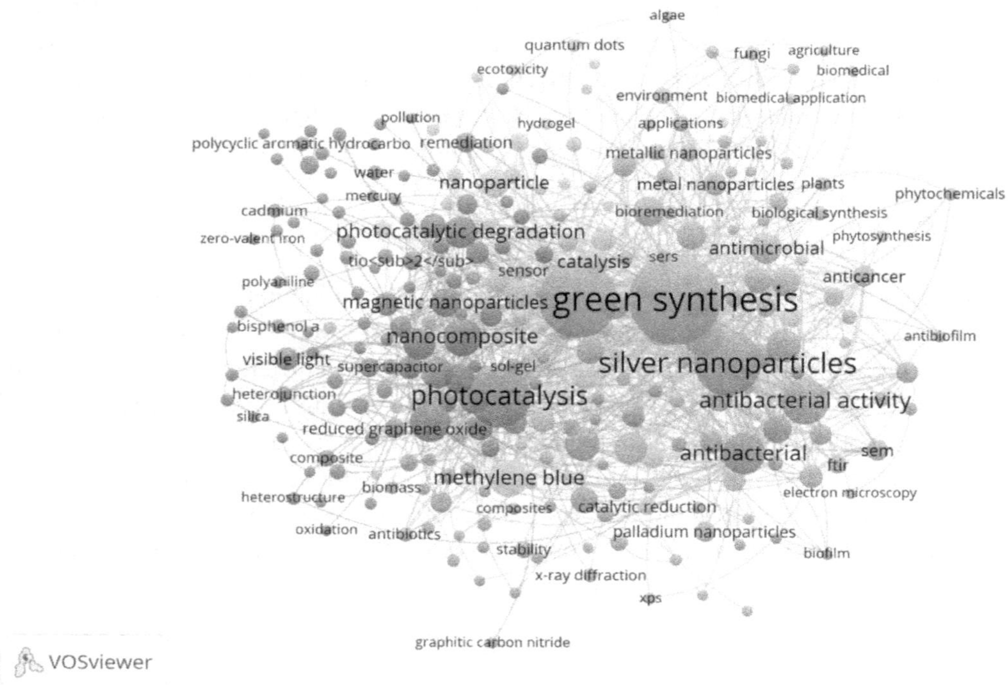

(Based on 5912 publications accessed from Scopus database with search keywords "environmental nanoparticles, biogenic nanoparticles, biologically synthesized nanoparticles, green synthesis, medicine nanoparticles, agriculture nanoparticles" (3rd July 2024).

2.4 Biogenic Nanoparticle Classification

The categorization of biogenic nanoparticles covers the variety of nanoparticles produced from biological sources and offers comprehensive information about their origins, synthesis techniques, properties, uses, and average sizes.

Plant-based nanoparticles, which are generated from plant extracts, are special because they can be used therapeutically and are eco-friendly. Tulsi's stable and homogenous gold nanoparticles are good for antibacterial and anticancer applications. Neem silver nanoparticles are utilized in water purification and wound healing because of their strong antibacterial properties. Aloe vera zinc oxide nanoparticles are excellent for sunscreens and cosmetics because of their photocatalytic and UV-blocking qualities. Green tea's iron oxide nanoparticles, which are employed as medication delivery and MRI contrast agents, show superparamagnetism (Bodade et al., 2023). Eucalyptus copper nanoparticles are useful as catalysts and have antifungal and antioxidant qualities that make them suitable for biological applications. These details are systematically presented in Table 1, which provides a comprehensive overview of plant-based nanoparticles.

Table 1. Plant-based nanoparticles

Types	Source	Synthesis Method	Characteristics	Use	Size (nm)	Reference
Silver nanoparticles	Tulsi (*Ocimum sanctum*)	Green synthesis	Stable, uniform size	Anticancer, antimicrobial	20-50	(Baruah et al., 2021)
Silver nanoparticles	Neem (*Azadirachta indica*)	Bioreduction by plant extract	High antibacterial activity	Wound healing, water purification	10-100	(Mehmood et al., 2016)
Zinc oxide nanoparticles	Aloe vera	Hydrothermal method	Photocatalytic, UV-blocking	Sunscreens, cosmetics	20-60	(Khajeh Mehrizi & Shahi, 2020)
Iron oxide nanoparticles	Green tea (*Camellia sinensis*)	Co-precipitation	Superparamagnetic	MRI contrast agents, drug delivery	5-30	(Periakaruppan et al., 2021)
Copper nanoparticles	Eucalyptus	Microwave-assisted synthesis	Antifungal, antioxidant	Biomedical applications, catalysts	20-50	(Khan & Sharma, 2024), (Hemmati et al., 2020)

Bacteria and fungi use biological processes to make microbial nanoparticles. *Bacillus subtilis* produces silver nanoparticles, which are employed in textiles and biomedical applications due to their antibacterial and antifungal qualities. *Pseudomonas aeruginosa* gold nanoparticles are stable and biocompatible, making them ideal for medication administration and biosensing applications. *Fusarium oxysporum* produces antioxidant and anticancer selenium nanoparticles that are beneficial in medication delivery and nutraceuticals. *Magnetospirillum magnetotacticum* and iron oxide nanoparticles are employed as MRI contrast agents and in hyperthermia therapy due to their superparamagnetism. *Aspergillus fumigatus* zinc oxide nanoparticles possess antibacterial and photocatalytic characteristics. These examples and their attributes are summarized in Table 2, which highlights the versatility of microbial nanoparticles.

Table 2. Microbial nanoparticles

Subtype	Source	Synthesis Method	Characteristics	Use	Size (nm)	Reference
Silver nanoparticles	*Bacillus subtilis*	Bioreduction	Antibacterial, antifungal	Biomedical applications, textiles	20-100	(Kumar et al., 2017)
Gold nanoparticles	*Pseudomonas aeruginosa*	Bioreduction	Stable, biocompatible	Biosensing, drug delivery	10-50	(Vijayakumar et al., 2017)
Selenium nanoparticles	*Fusarium oxysporum*	Enzymatic reduction	Antioxidant, anticancer	Nutraceuticals, drug delivery	6-13	(Kashisaz et al., 2024)
Iron oxide nanoparticles	*Magnetospirillum magnetotacticum*	Magnetosome formation	Superparamagnetic	MRI contrast agents, hyperthermia therapy	30-100	(Mirabello et al., 2016)
Zinc oxide nanoparticles	*Aspergillus fumigatus*	Bioreduction	Antibacterial, photocatalytic	Environmental remediation, sensors	10-70	(Chen et al., 2021)

Marine-based nanoparticles are the products of algae and seaweed, which have special bioactive qualities and harmless to the environment. *Sargassum muticum* makes silver nanoparticles that are antibacterial and safe for the environment, which are used in coatings and biological applications. *Ulva lactuca* gold nanoparticles are stable and biocompatible, making them ideal for imaging and medication delivery. *Gelidium amansii* zinc oxide nanoparticles are perfect for sunscreens and cosmetics because of their photocatalytic and UV-blocking qualities. Iron oxide nanoparticles from *Gracilaria corticata* show superparamagnetism. These particles are used to deliver medicines and create contrast in MRIs. *Kappaphycus alvarezii* copper nanoparticles exhibit antioxidant and antifungal qualities. Detailed information on marine-based nanoparticles is presented in Table 3, showcasing their eco-friendly synthesis and diverse applications.

Table 3. Marine-based nanoparticles

Subtype	Source	Synthesis Method	Characteristics	Use	Size (nm)	Reference
Silver nanoparticles	*Sargassum muticum*	Green synthesis	Antimicrobial, eco-friendly	Biomedical applications, coatings	49-118	(Balaraman et al., 2020)
Gold nanoparticles	*Ulva lactuca*	Green synthesis	Biocompatible, stable	Drug delivery, imaging	10-50	(González-Ballesteros et al., 2019)
Zinc oxide nanoparticles	*Gracilaria edulis*	Hydrothermal method	UV-blocking, photocatalytic	Sunscreens, cosmetics	30-70	(Gowdhami et al., 2019)
Metal oxide nanoparticles	*Gracilaria corticata*	Co-precipitation	Superparamagnetic	MRI contrast agents, drug delivery	5-30	(Naveenkumar et al., 2023)
Copper nanoparticles	*Kappaphycus alvarezii*	Microwave-assisted synthesis	Antifungal, antioxidant	Biomedical applications, catalysts	20-50	(Chandrashekhara et al., 2021)

Bacteria synthesize bacterial nanocellulose nanoparticles, which are then engineered to improve their characteristics for particular uses. *Acetobacter xylinum* produces carboxylated nanocellulose, which is hydrophilic and useful for treating wounds and delivering drugs. Phosphorylated nanocellulose from *Gluconacetobacter hansenii* is used in tissue engineering and bone regeneration because it has a lot of phosphate and is bioactive. *Acetobacter xylinum*'s sulfated nanocellulose is perfect for tissue scaffolds and medicinal devices since it has anti-inflammatory and anticoagulant qualities. *Acetobacter xylinum* produces highly reactive oxidized nanocellulose that can be used in tissue engineering and medication delivery. High-strength and high-crystallinity bacterial cellulose nanocrystals from *Komagataeibacter xylinus* are utilized in biomedical applications and composite materials. These types and their applications are detailed in Table 4, highlighting the innovative uses of bacterial nanocellulose.

Table 4. Bacterial nanocellulose nanoparticles

Subtype	Source	Synthesis Method	Characteristics	Use	Size (nm)	Reference
Carboxylated nanocellulose	*Acetobacter xylinum*	Hydrolysis and carboxylation	Hydrophilic, high surface charge	Drug delivery, wound dressing	50-100	(Aljabali et al., 2020)
Phosphorylated nanocellulose	*Gluconacetobacter hansenii*	Hydrolysis and phosphorylation	High phosphate content, bioactive	Bone regeneration, tissue engineering	50-150	(Jedrzejczak-Krzepkowska et al., 2016)
Sulfated nanocellulose	*Acetobacter xylinum*	Sulfation process	Anticoagulant, anti-inflammatory	Medical devices, tissue scaffolds	50-100	(de Olyveira et al., 2011)
Oxidized nanocellulose	*Acetobacter xylinum*	TEMPO-mediated oxidation	High carboxyl content, reactive	Drug delivery, tissue engineering	50-100	(Rubina et al., 2020)
Bacterial cellulose nanocrystals	*Komagataeibacter xylinus*	Acid hydrolysis	High crystallinity, strong	Composites, biomedical applications	100-200	(Gorgieva & Trček, 2019)

Viral nanoparticles are synthesized from viruses and used in a variety of ways. It is possible for viruses to be uniform and programable. Some examples are plant virus nanoparticles from the cowpea mosaic virus (CPMV), bacteriophage nanoparticles from the M13 bacteriophage, influenza virus nanoparticles from the Influenza A virus, tobacco mosaic virus nanoparticles from the Tobacco mosaic virus, etc. These nanoparticles are perfect for medication delivery and biosensing, and they are utilized in the production of vaccines, gene delivery, and nanocarriers. Comprehensive details of viral nanoparticles are given in Table 5.

Table 5. Viral nanoparticles

Subtype	Source	Synthesis Method	Characteristics	Use	Size (nm)	Reference
Plant virus nanoparticles	Cowpea mosaic virus (CPMV)	Self-assembly of viral capsids	Uniform size, programmable assembly	Vaccines, gene delivery, nanocarriers	30-50	(Beatty & Lewis, 2019)
Bacteriophage nanoparticles	M13 bacteriophage	Filamentous phage display technology	High specificity, bioengineered	Drug delivery, biosensing	20-100	(Ahiwale et al., 2017)
Influenza virus nanoparticles	Influenza A virus	Recombinant virus-like particles	Immunogenic, stable	Vaccine development, diagnostics	80-120	(Nazari et al., 2022)
Tobacco mosaic virus nanoparticles	Tobacco mosaic virus	Virus particle assembly	Robust, high aspect ratio	Nanocarriers, imaging	18x300	(Marín-Caba et al., 2019)
Hepatitis B virus nanoparticles	Hepatitis B virus	Self-assembly of surface antigen	Immunogenic, stable	Vaccine development	22-35	(Pitek et al., 2018)

3. FUTURE CHALLENGES AND OPPORTUNITIES

3.1 Current Challenges

Standardization of Synthesis Methods

The absence of standardized synthesis techniques is one of the main problems in the field of biogenic nanoparticles. A variety of biological processes, such as enzymatic reactions, microbial fermentation, and plant-mediated synthesis, are used to create biogenic nanoparticles (García-Peñas & Sharma, 2023). Reproducibility and scalability are hampered by the inconsistencies in nanoparticle properties that are frequently caused by discrepancies in synthesis conditions, raw ingredients, and biological variables. Standardizing synthesis procedures and refining reaction conditions are crucial for guaranteeing uniform nanoparticle quality and repeatability in various studies and uses.

Characterization and Analysis

Because of their complex compositions and structures, biogenic nanoparticles are difficult to characterize and analyze. The distinct characteristics of biogenic nanoparticles might not be sufficiently captured by conventional characterization methods, including X-ray diffraction (XRD), dynamic light scattering (DLS), and transmission electron microscopy (TEM). Furthermore, biological contaminants and sample variability in nanoparticles make characterization tasks much more challenging. Cryogenic electron microscopy, atomic force microscopy (AFM), and spectroscopic techniques are among the advanced characterization methods that are required to overcome these obstacles and offer complete insights into the structure, shape, and characteristics of nanoparticles (Anbazhagan et al., 2016). Figure 5 illustrates how AI-driven techniques can be employed to characterize biogenic nanoparticles more efficiently and accurately. It showcases the use of machine learning algorithms in combination with imaging and spectroscopic methods.

Figure 5. AI-driven characterization techniques for biogenic nanoparticles

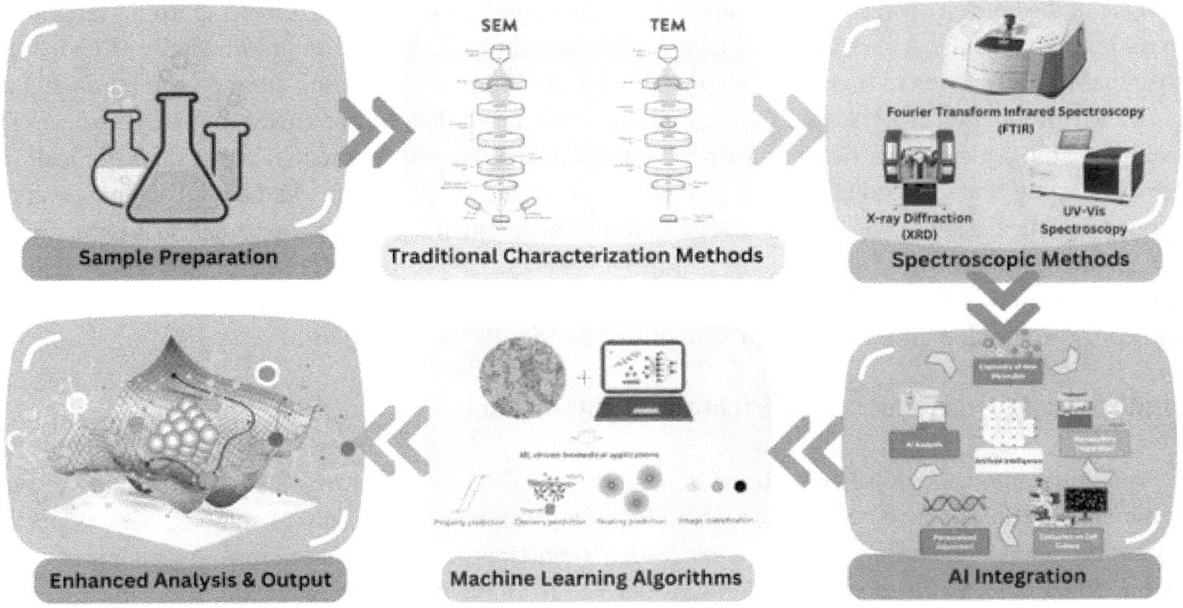

Biocompatibility and Toxicity Assessment

In biomedical applications, determining the toxicity and biocompatibility of biogenic nanoparticles is a crucial task. Because of their natural origin, biogenic nanoparticles are generally regarded as biocompatible; nonetheless, their interaction with biological systems may be affected by differences in synthesis techniques and biological constituents. Furthermore, not much is known about how biogenic nanoparticles may affect organ systems, cellular processes, and ecological health in the long run. To evaluate the safety profile of biogenic nanoparticles and guarantee their safe usage in biomedical and environmental applications, standardized techniques for biocompatibility testing are required. These protocols include in vitro experiments, in vivo studies, and ecological risk assessments (Bernard et al., 2018).

Regulatory Compliance and Safety Standards

The regulatory framework around the use of biogenic nanoparticles is complicated and ever-changing, creating difficulties for regulators, producers, and researchers alike. Depending on their intended application and characteristics, biogenic nanoparticles may fall within the existing regulatory frameworks for nanomaterials, medicines, medical devices, or environmental pollutants. Regulating biogenic nanoparticles has few defined criteria, nevertheless, which makes regulatory compliance erratic and unclear. To guarantee biogenic nanoparticles are used safely and responsibly in a variety of applications, safety standards, risk assessment procedures, and regulatory requirements specific to biogenic nanoparticles must be established (Allan et al., 2021).

Scalability and Commercialization

One of the biggest obstacles still facing biogenic nanoparticle production is moving from a laboratory to an industrial scale. Although biological synthesis techniques provide environmentally friendly and sustainable alternatives to traditional chemical synthesis, commercialization attempts are hampered by issues like production costs, scalability limitations, and batch-to-batch unpredictability. Key concerns for expanding biogenic nanoparticle commercialization and realizing their potential in practical applications are developing scalable and affordable production procedures, optimizing yield and efficiency, and resolving logistical issues (Yadav et al., 2019). In India, the realm of agricultural automation is broadening; yet, the incorporation of AI and biogenic nanoparticles encounters obstacles like cost, scalability, and the assimilation of novel technology. Overcoming these issues necessitates progress in both AI algorithms and hardware implementation to align with global trends (Padhiary, 2024b).

Nanoparticle Production and Environmental Impact

A significant problem is encountered in attaining consistency in the size, shape, and quality of biogenic nanoparticles across various batches. The biological diversity of natural sources, such as plants, microorganisms, and marine organisms, can lead to irregular production, hence impacting their properties and applications. For instance, nanoparticles derived from plant extracts may exhibit diverse physicochemical features owing to variations in cultivation conditions, extraction techniques, and plant species (Joudeh & Linke, 2022). While biogenic nanoparticles are environmentally benign in their synthesis, their post-usage ecological consequences, especially in biomedical applications, require thorough assessment. The prolonged impacts of nanoparticles on soil health, aquatic ecosystems, and human health necessitate additional research.

Ethical and Social Implications

The ethical and societal ramifications of biogenic nanoparticles bring up significant issues related to social justice, health concerns, and environmental damage. It is necessary to address worries about the environmental sustainability of nanoparticle synthesis techniques, possible health risks related to nanoparticle exposure, and fair access to nanoparticle-based technologies. Stakeholder engagement tactics, ethical frameworks, and public discourse are required to guarantee ethical innovation, moral decision-making, and fair sharing of the advantages and disadvantages of biogenic nanoparticles.

3.2 Potential Opportunities

Tailored Nanoparticle Design

Technological developments in biogenic nanoparticle synthesis present prospects for customizing the characteristics and functions of nanoparticles to suit particular uses. Through the optimization of synthesis settings, the integration of bioactive compounds, and surface engineering, scientists can create nanoparticles that exhibit superior stability, biocompatibility, and targeted specificity. Tailored nanoparticle design, which offers precise control over drug administration, imaging, diagnostics, and therapies, makes personalized medicine techniques and better treatment outcomes possible (Sykes et al., 2016).

Multifunctional Nanoparticle Platforms

The development of multifunctional biogenic nanoparticle platforms opens up the possibility of putting together several functions into a single nanoscale carrier. By combining therapeutic compounds, targeting ligands, imaging probes, and stimuli-responsive parts, researchers can make multifunctional nanoparticles that can release drugs under controlled conditions, monitor biological processes in real-time, and be used for both diagnosis and therapy at the same time. Platforms for multifunctional nanoparticles provide adaptability and flexibility in tackling difficult biological problems, such as neurological disorders, infectious diseases, and cancer treatment (Singh et al., 2023).

Smart Nanoparticle Systems

Advances in nanotechnology and materials science have made it possible to create intelligent biogenic nanoparticle systems that can respond to environmental cues and outside stimuli. By adding molecular switches, coatings that are sensitive to stimuli, and responsive polymers, researchers can make smart nanoparticles that react to things like pH, temperature, light, or magnetic fields. These nanoparticles can then undergo controlled release, shape transformation, or surface functionalization. Systems of smart nanoparticles provide options for temporal control over therapeutic interventions, site-specific targeting, and on-demand drug delivery (Karimi et al., 2016).

Precision Medicine Applications

Through the ability to provide targeted and customized therapy based on the unique characteristics of each patient, biologic nanoparticles have the potential to advance precision medicine techniques. By using biocompatible and biodegradable nanoparticle carriers, researchers can get around biological barriers, deliver therapeutic compounds to specific disease sites, and limit effects that aren't meant to happen. Also, progress in proteomics, genetics, and biomarker identification has made it possible to use patient-specific information along with nanoparticle-based treatments to create precision medicine methods that make treatment plans more effective and personalized (Alghamdi et al., 2022).

Environmental Remediation Technologies

By eliminating contaminants from soil, water, and air, biogenic nanoparticles provide environmentally friendly remediation options that are long-lasting. Developments in the synthesis, surface modification, and functionalization of nanoparticles have made it possible to create environmentally sustainable remediation technologies that are effective in eliminating environmental contaminants such as heavy metals, organic pollutants, and microbial infections. It is possible to create biogenic nanoparticles to selectively adsorb, absorb, or degrade contaminants, which will improve soil fertility, water quality, and ecosystem health (Edgar et al., 2024). Recent advancements in AI vision and machine learning are enabling breakthroughs in agricultural applications of biogenic nanoparticles, such as improving soil health monitoring and crop management systems, showcasing AI's potential to optimize environmental remediation processes (Padhiary & Kumar, 2024).

Since nanoparticles facilitate cell-nanoparticle interactions, biomaterial integration, and tissue regeneration, biogenic nanoparticles are essential for biointerface engineering and tissue engineering applications. Developments in nanobiotechnology have made it possible to create biocompatible and bioactive nanoparticle scaffolds for promoting cell adhesion, proliferation, and differentiation. Biogenic nanoparticles can also be added to drug delivery systems, biosensors, and tissue engineering structures to help them better integrate with living cells, support tissue regeneration, and work better as medicines in regenerative medicine (Ye & Peramo, 2014).

4. ROLE OF ARTIFICIAL INTELLIGENCE

4.1 Synthesis Optimization

By utilizing machine learning algorithms and computational modeling approaches, artificial intelligence (AI) plays a crucial role in optimizing the synthesis processes of biogenic nanoparticles. To predict the properties of nanoparticles, determine the optimum way to synthesize them, and maximize reaction yields, researchers can employ AI-driven techniques to analyze massive datasets about synthesis parameters, experimental setups, and nanoparticle properties. Through the use of AI, scientists can find new synthesis techniques more quickly, increase the scalability and repeatability of nanoparticles, and create more sustainable and effective methods for producing nanoparticles (Yao et al., 2023). AI can predict the best biological sources, conditions, and synthesis routes that yield nanoparticles with desired properties. In one study, machine learning was applied to optimize the synthesis of silver nanoparticles from plant extracts, predicting factors like temperature, pH, and extract concentration for uniform size and shape distribution (Nathanael et al., 2023).

4.2 Characterization and Analysis

AI-powered methods provide sophisticated data analytics, image processing, and pattern recognition algorithms, which improve the characterization and study of biogenic nanoparticles. AI algorithms are able to derive useful insights into the structure, shape, and composition of nanoparticles by analyzing complex spectroscopic data, microscope pictures, and multi-dimensional information. Using AI-driven computer models to simulate nanoparticle behavior, predict nanoparticle properties, and optimize experimental parameters can also cut down on the need for expensive and time-consuming experiments. Researchers can overcome the drawbacks of conventional characterization techniques and learn more about the behavior and usefulness of biogenic nanoparticles by incorporating AI into characterization and analysis operations (Banerjee et al., 2022).

AI is now playing a crucial role in automating and improving nanoparticle characterization techniques. By using image recognition algorithms, AI can analyze data from electron microscopy or spectroscopy much faster and more accurately than manual methods, thus aiding in identifying nanoparticles' structural and surface characteristics. AI models can predict the interaction of nanoparticles with biological systems, environmental factors, and even other nanoparticles, aiding in the development of safer and more efficient applications. For instance, AI models have been used to predict the toxicological behavior of silver nanoparticles in aquatic environments, guiding safer production processes (Furxhi et al., 2024).

4.3 Biomedical Applications

AI-driven techniques have revolutionized the use of biogenic nanoparticles in the biomedical field by enabling personalized medicine, drug development, and disease detection. AI algorithms examine genetic profiles, clinical records, and patient data to determine individualized treatment plans based on each patient's unique features. Additionally, AI-driven drug screening technologies accelerate the drug discovery process by identifying possible therapeutic candidates and forecasting drug efficacy. Furthermore, the sensitivity and specificity of biomedical imaging modalities are improved by AI-driven imaging approaches, which make it possible to precisely localize tumors, detect diseases early, and track treatment responses in real-time. Researchers can improve the efficacy, efficiency, and precision of medicines and diagnostics based on biogenic nanoparticles by incorporating AI into biomedical applications (Yip et al., 2024). Besides healthcare uses, biogenic nanoparticles are progressively utilized in agriculture to improve production. Recent developments in precision agriculture have investigated the application of 3D printing technology in intelligent farming. AI-augmented systems, integrated with these advancements, provide greater precision and efficiency in agricultural management, especially in soil assessment nutrient management (Padhiary & Roy, 2024). Figure 6 outlines the utilization of AI-powered biogenic nanoparticles in many domains including healthcare, agriculture, and cosmetics emphasizing the role of AI in facilitating material characteristics and performance analysis.

Figure 6. AI-driven applications of biogenic nanoparticles

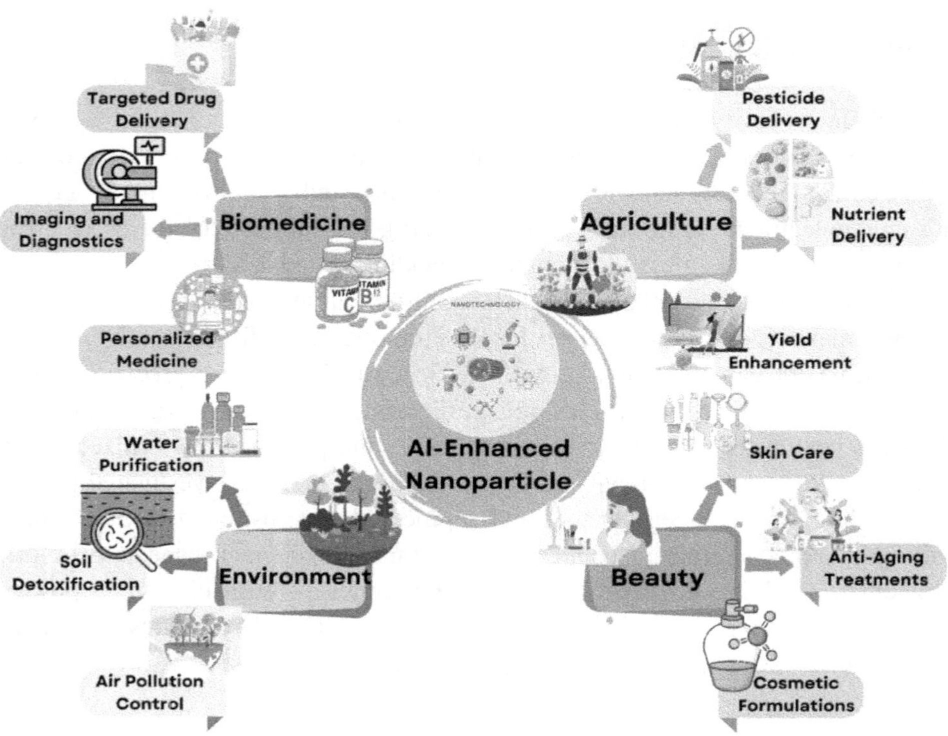

AI has been employed to enhance biogenic nanoparticles in drug delivery systems, especially in the development of tailored cancer medicines. Machine learning techniques forecast the interactions between nanoparticles and specific cell receptors, facilitating effective drug administration. Gold nanoparticles coated with an anti-cancer medication were optimized by the application of AI algorithms. AI-driven methodologies are employed in precision agriculture to enhance crop growth and pest resistance, using zinc oxide nanoparticles sourced from microbial origins (Munir et al., 2024). AI systems assess soil conditions, crop health, and environmental variables in real-time, resulting in enhanced crop yields and more sustainable agricultural methods. In environmental applications, AI models have been applied to predict the efficiency of biogenic nanoparticles in removing pollutants from water and soil. Microbially generated silver nanoparticles have been utilized in water purification, and AI-driven models were developed to forecast nanoparticle behavior across various environmental circumstances, facilitating enhanced pollutant removal.

4.4 Environmental Remediation Technologies

By anticipating environmental fate and transport, creating targeted remediation tactics, and optimizing nanoparticle-based pollution removal procedures, AI-driven methodologies advance environmental remediation technologies. Artificial intelligence systems examine environmental information, pollutant concentrations, and hydrological characteristics to determine the best locations for remediation, forecast pollutant dispersion trends, and enhance remediation tactics instantly. AI-powered modeling tools can also simulate how nanoparticles and pollutants interact, predict how nanoparticles will behave in complex environmental matrices, and improve the design of nanoparticles to remove pollutants more effectively. By incorporating AI into environmental remediation technologies, scientists can create more sustainable, effective, and efficient ways to reduce pollution in the environment and protect the health of ecosystems (Patil et al., 2016).

4.5 Data-Driven Innovation

In the domain of biogenic nanoparticles, AI-driven innovation promotes interdisciplinary cooperation, knowledge exchange, and data-driven decision-making. In order to determine research trends, new opportunities, and knowledge gaps in the field of biogenic nanoparticle research, AI algorithms examine massive databases, scientific literature, and experimental data. Additionally, AI-powered platforms make it easier for academics to collaborate by facilitating data sharing, group analysis, and multidisciplinary research projects. Researchers can quicken the rate of discovery, encourage interdisciplinary collaboration, and convert scientific information into useful real-world applications by utilizing AI-driven innovation (Zhu et al., 2023).

The integration of machine learning and AI vision applications in all-terrain vehicles represents a breakthrough in farm automation, offering new levels of precision and operational efficiency (Padhiary, Saha, et al., 2024). Similarly, Researchers may address important issues, take advantage of new opportunities, and speed up innovation across a range of applications in biomedicine, environmental remediation, and other fields by incorporating AI into biogenic nanoparticle research. Through the use of AI-driven approaches, scholars can propel technological innovation, make significant advancements in the field of biogenic nanoparticles, and effectively, precisely, and sustainably tackle urgent societal issues. AI has the potential to improve agricultural and food processing by integrating 3D printing with advanced technologies such as nanoparticle-based insecticides and precision instruments for food quality evaluation (Padhiary, Barbhuiya, et al., 2024).

5. PREDICTIONS FOR THE FUTURE DIRECTION

5.1 Precision Nanomedicine

It is anticipated that future studies on biogenic nanoparticles will concentrate on precision nanomedicine strategies catered to the unique traits and disease profiles of each patient. Researchers will be able to create biogenic nanoparticles with improved targeting specificity, drug delivery effectiveness, and therapeutic efficacy thanks to developments in genomics, proteomics, and personalized medicine. Additionally, the integration of AI-driven predictive modeling and patient data analytics will make it simpler to develop personalized treatment plans and adaptive therapies based on real-time patient input and biomarker monitoring (Jain, 2016). Figure 7 presents a roadmap highlighting future trends and prospects in the integration of AI with biogenic nanoparticle research. It outlines key advancements expected in areas such as personalized medicine, agriculture, environmental sustainability, and advanced materials, along with potential breakthroughs in nanoparticle synthesis, real-time monitoring, and global scalability.

Figure 7. Future trends and prospects in AI-driven biogenic nanoparticle research

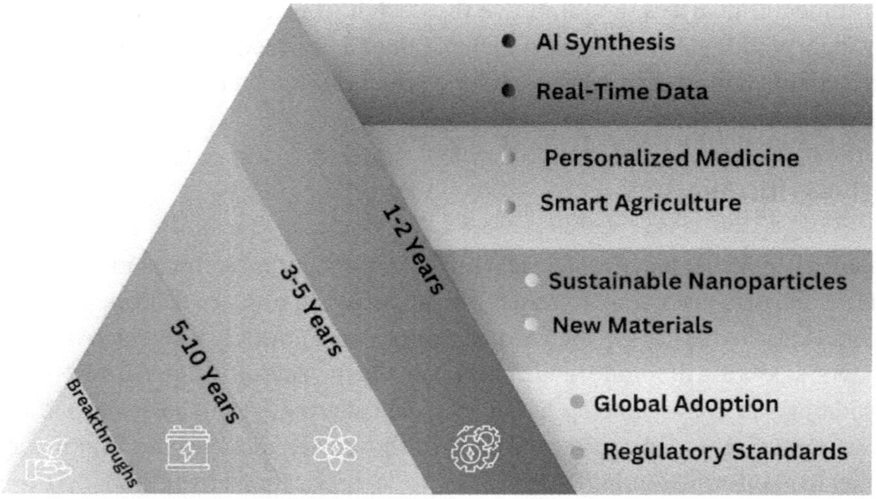

5.2 Biohybrid Nanomaterials

The development of biohybrid nanomaterials, integrating biological components with synthetic nanomaterials, holds promise for expanding the functionality and versatility of biogenic nanoparticles. In the future, scientists will look into adding functional proteins, enzymes, and nucleic acids to biogenic nanoparticle scaffolds to make them more biocompatible, bioactive, and programmable. Biohybrid nanomaterials will also allow biological and synthetic parts to work together more effectively, which will improve therapeutic outcomes and allow biomimetic functions and bio-interface engineering use in regenerative medicine and tissue engineering (Wang et al., 2022).

5.3 Smart Nanoparticle Systems

The development of intelligent biogenic nanoparticle systems that can sense, react, and adapt on their own to changing biological surroundings will be the primary objective of future studies. As materials science, AI, and nanotechnology continue to improve, it will be possible to make nanoparticles that respond to stimuli and can target specific areas, release drugs on their own, and monitor physiological parameters in real-time. Because they offer precise control over medication dosage, release kinetics, and therapeutic interventions customized to each patient's needs, smart nanoparticle systems have the potential to transform drug delivery, diagnostics, and treatments (Chauhan, 2024).

5.4 Environmental Nanotechnology

Biogenic nanoparticles will be used more and more in environmental nanotechnology to solve problems with soil remediation, air filtration, and water purification. Future studies will concentrate on improving remediation systems based on biogenic nanoparticles to eliminate emerging contaminants such as microplastics, medicines, and new pollutants (Rani et al., 2022). Furthermore, researchers will be able to optimize nanoparticle design, synthesis, and deployment techniques for increasing pollutant removal efficiency, minimizing environmental effects, and encouraging sustainable resource management thanks to advancements in AI-driven modeling and predictive analytics (Aithal, 2024).

5.5 Nanobiotechnology Convergence

The fusion of nanotechnology with other emerging fields like biotechnology, nanobiotechnology, and synthetic biology will fuel interdisciplinary research and innovation in biogenic nanoparticles (Lateef et al., 2021). Future research will look into how biogenic nanoparticles can work with cutting-edge biotechnologies such as cell-based therapies, CRISPR-based gene editing, and synthetic biology techniques (Bhatia et al., 2023). Scientists will also be able to look at biological processes and nano-bio interactions at previously unimaginable levels of detail in terms of both space and time when they combine nanotechnology with cutting-edge imaging techniques such as in vivo imaging, single-cell imaging, and super-resolution microscopy.

6. ETHICAL, REGULATORY, AND SOCIETAL CONSIDERATIONS

To promote responsible innovation, fair access, and sustainable development, future biogenic nanoparticle research will place a strong emphasis on ethical, legal, and societal issues. Researchers will work with lawmakers, regulatory bodies, and interested parties to come up with moral standards, safety requirements, and legal frameworks for the moral and responsible use of biogenic nanoparticles in social, environmental, and medical settings (Srivastava, 2023).

6.1 Ethical Considerations

Safety and Risk Assessment

It is crucial to guarantee safety and reduce hazards related to the application of biogenic nanoparticles. To reduce possible harm to human health and the environment, ethical considerations need rigorous safety investigations, which include biocompatibility assessments, toxicity studies, and risk analysis. Researchers must adhere to the ethical principles of beneficence and non-maleficence, which emphasize personal well-being and reduce negative effects (Foret et al., 2023). Figure 8 outlines the ethical and regulatory challenges faced in AI-driven biogenic nanoparticle research, from data privacy to environmental impact, alongside possible mitigation strategies.

Figure 8. Ethical and regulatory considerations in AI-driven nanotechnology

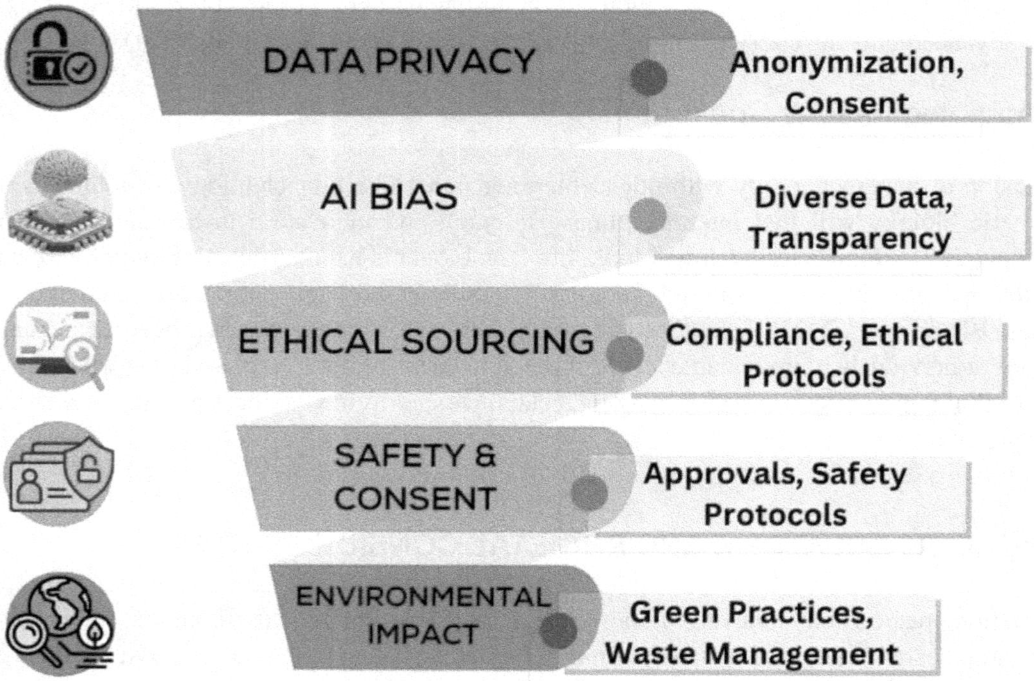

Informed Consent and Human Subjects Research

Obtaining informed permission is crucial for protecting human rights and upholding individual autonomy in research involving human subjects. Ethical considerations require open communication of any potential risks, benefits, and uncertainties associated with participation in studies involving nanoparticles. Under ethical and legal constraints, researchers must provide informed consent, voluntary involvement, and the safeguarding of rights to secrecy and privacy (Spruit et al., 2016).

Environmental Impact and Sustainability

The long-term viability and ecological impact of biogenic nanoparticles over the course of their lifecycle are also ethical considerations. Researchers must assess the environmental effects of creating, using, and discarding nanoparticles while taking into account resource use, energy consumption, and trash production. To reduce ecological harm and enhance long-term environmental health, biodegradable nanoparticle materials, ecologically responsible disposal techniques, and eco-friendly synthesis procedures must be adopted in accordance with ethical ideals of environmental responsibility and sustainability (Dwivedi & Dwivedi, 2022).

Equitable Access and Social Justice

A major area of ethical issues is ensuring social fairness in the dissemination and application of nanoparticle-based technologies, as well as fair access to them. To guarantee that vulnerable people, underprivileged communities, and marginalized groups benefit equally from nanoparticle-based interventions, researchers must address concerns about healthcare disparities, resource inequities, and accessibility hurdles. Proactive measures to lessen gaps, encourage inclusivity, and resolve systemic inequalities in healthcare delivery and access to modern technologies are required under the ethical principles of distributive justice and fairness (Vasudev et al., 2023).

Transparency and Accountability

Every aspect of nanoparticle research and development, including funding sources, conflicts of interest, and research methods, must be transparent and accountable in order to satisfy ethical requirements. To maintain scientific rigor and public trust, researchers must respect the ethical values of integrity, honesty, and transparency by being open and honest about any potential conflicts of interest, financial ties, and study biases. Furthermore, ethical principles need appropriate data sharing, management, and distribution methods to support peer review, scientific cooperation, and the reproducibility of study results (Zarghani et al., 2023).

Ethical Governance and Regulation

For ethical reasons, there needs to be strong governance and regulatory frameworks to make sure that nanoparticle research is done in an ethical way and that legal and moral obligations are met. Institutional review boards (IRBs), ethics committees, and regulatory bodies are essential in reviewing research methods, determining the ethical implications of the findings, and ensuring the welfare and rights of study participants. Biogenic nanoparticles in food processing tackle regulatory and ethical issues in wastewater treatment, providing enhanced filtration and pollutant collection relative to traditional membrane technologies (Padhiary, 2024a). Furthermore, in order to cultivate a culture of ethical awareness, responsibility, and accountability within the scientific community, ethical governance frameworks should support ethical reflection, discussion, and ongoing improvement in nanoparticle research procedures (Hunckler & Levine, 2022).

6.2 Regulatory Frameworks

Like other nanomaterials, biogenic nanoparticles are governed by the same regulatory bodies that oversee biotechnology, nanotechnology, and environmental safety. Depending on their intended use and application, biogenic nanoparticles may be subject to regulations from a number of regulatory bodies, including the Environmental Protection Agency (EPA), the Occupational Safety and Health Administration (OSHA), and the Food and Drug Administration (FDA) in the United States. International agencies that offer regulatory guidance on nanoparticle safety and risk assessment include the European Medicines Agency (EMA), the European Food Safety Authority (EFSA), and the Organisation for Economic Co-operation and Development (OECD) (Allan et al., 2021).

Classification and Definition

Regulatory bodies offer standards for the size, content, intended use, and possible hazards of biogenic nanoparticles in order to classify and define them. Regulatory agencies define nanomaterials differently, but generally speaking, the parameters are centered around size (less than 100 nanometers in at least one dimension), specific surface area, and physicochemical characteristics. Determining the regulatory applicability, evaluating safety hazards, and guaranteeing regulatory compliance all depend on having a clear classification and defining criteria (Rodríguez-Gómez et al., 2023).

Risk Assessment and Safety Evaluation

Biogenic nanoparticles must undergo a thorough risk assessment and safety evaluation in accordance with regulatory frameworks in order to determine their possible effects on the environment and public health. Hazard identification, exposure assessment, toxicity testing, and risk characterization are all included in risk assessment, which evaluates possible health risks, environmental hazards, and safety issues related to nanoparticle exposure. Regulatory bodies set clear guidelines and methods for safety studies, such as animal studies, in vitro assays, and evaluations of nanoparticles' environmental fate, so that scientists can find out how harmful they are, how well they work with living things, and how they affect the environment (More et al., 2021).

Pre-Market Approval and Regulation

Regulation and pre-market approval are required for biogenic nanoparticles used in pharmaceutical or medical applications by organizations like the FDA. Products based on nanoparticles have several regulatory processes dependent on their risk profile, intended application, and classification. Before being approved for sale, nanomedicines such as medications and medical devices based on nanoparticles must pass extensive preclinical and clinical testing to ensure their safety, effectiveness, and quality. To evaluate a product's safety, efficacy, and compliance with regulations, regulatory bodies examine preclinical data, clinical trial findings, manufacturing procedures, and labeling information (Eaglstein, 2014).

Good Manufacturing Practices (GMP)

For the manufacture, distribution, and quality assurance of biogenic nanoparticles and products derived from them, regulatory frameworks place a strong emphasis on adhering to Good Manufacturing Practices (GMP). To guarantee the uniformity, purity, and security of products containing nanoparticles, GMP rules set criteria and specifications for manufacturing facilities, tools, workers, paperwork, and quality control methods. Throughout the course of the product lifecycle, patient safety, regulatory approval, and product quality maintenance depend on compliance with GMP rules (Niazi & Lokesh, 2022).

Post-Market Surveillance and Monitoring

Regulations for biogenic nanoparticles include measures for post-market surveillance and monitoring in order to evaluate long-term safety, track unfavorable events, and identify new hazards. To monitor the safety and efficacy of nanoparticles in real-world situations, regulatory agencies mandate that producers undertake post-market surveillance programs, adverse event reporting systems and pharmacovigilance initiatives. Post-market surveillance data support risk management plans, regulatory decision-making, and changes to product labels to guarantee the continued efficacy and safety of goods containing nanoparticles (Grunwald, 2019).

6.3 Protocols

Ethical Review and Approval

Institutional review boards (IRBs) or ethics committees must provide ethical evaluation and approval prior to researchers using biogenic nanoparticles in a study. The process of ethical assessment guarantees that research protocols follow moral guidelines, uphold human rights, and safeguard the well-being of study participants. In order to receive ethical approval for their research, researchers are required to furnish comprehensive details regarding the study aims, methods, potential dangers, and informed consent processes (Manandhar et al., 2021).

Informed Consent and Participant Protection

For any study involving human beings, including ones using biogenic nanoparticles, informed permission is mandatory. Participants in research must give their free and informed permission after being fully told about the goals, methods, risks, and rewards of the study. Researchers must guarantee confidentiality, privacy, and the protection of participant rights throughout the research process, while also granting participants the opportunity to make educated decisions about their participation (Haik & Bantekas, 2020).

Safety and Risk Assessment

To evaluate potential dangers and limit risks to human health and the environment, researchers working with biogenic nanoparticles must do thorough safety and risk evaluations. To detect potential concerns related to nanoparticle exposure, safety assessments include biocompatibility testing, toxicity investigations, environmental destiny assessments, and risk characterization. It is imperative for researchers to follow safety procedures, utilize suitable personal protective equipment (PPE), and establish risk mitigation strategies to guarantee the secure handling, preservation, and elimination of biogenic nanoparticles (Kumar et al., 2018).

Quality Control and Standardization

Ensuring the reproducibility, dependability, and safety of biogenic nanoparticles in research and application requires maintaining quality control and standardization. To guarantee consistent nanoparticle characteristics and performance, researchers must follow established protocols for the synthesis, characterization, and functionalization of nanoparticles. To preserve the integrity and quality of the product, quality control procedures include stringent analytical technique testing, validation, and calibration; adhering to good laboratory practices (GLP); and putting quality management systems in place (Sarkar et al., 2022).

Regulatory Compliance and Reporting

To maintain regulatory compliance and openness, researchers performing studies with biogenic nanoparticles must adhere to pertinent regulatory criteria and reporting guidelines. Regulatory authorities like the FDA, EPA, and OSHA may have requirements that researchers must follow, depending on how biogenic nanoparticles are meant to be used and applied. To enable peer review, reproducibility, and regulatory approval, researchers must appropriately record and report study findings, methods, and data in compliance with legal requirements, scientific standards, and publication ethics (Van Loo, 2019).

Public Engagement and Communication

It is the duty of researchers to interact with the general public, interested parties, and legislators in order to further openness, discussion, and well-informed choices about the study and use of biogenic nanoparticles. The dissemination of precise and comprehensible information regarding the possible advantages, hazards, and social ramifications of biogenic nanoparticles is a crucial component of successful communication tactics. In order to make sure that studies and uses of biogenic nanoparticles are in line with society's values, requirements, and expectations, researchers must build public confidence, respond to concerns, and ask stakeholders for insights (Fritz et al., 2024).

7. POTENTIAL RISKS AND SOLUTIONS

7.1 Toxicity and Biocompatibility

Risk: Biogenic nanoparticles could be hazardous to biological systems or have negative effects on them, which could have an impact on both the environment and human health. Nanoparticles' size, composition, surface chemistry, and biological interactions are just a few of the variables that affect their toxicity profile (Medici et al., 2021).

Mitigation Strategies:

- Undertake thorough toxicity analyses, encompassing both in vitro and in vivo investigations, to assess the safety and biocompatibility of nanoparticles (Kus- Liśkiewicz et al., 2021).
- Utilize various methods of toxicity testing, such as organ-on-a-chip systems and 3D organoid models, to increase the accuracy of the predictions by imitating physiological conditions (Seo et al., 2022).
- To improve biocompatibility and lessen cytotoxicity, alter the surface characteristics of nanoparticles, such as surface charge, coating, and functionalization (Sanità et al., 2020).
- Use dose optimization techniques to reduce exposure levels and lessen the possibility of harmful impacts on biological systems (Olawale et al., 2021).
- Examine the long-term consequences and possible in vivo bioaccumulation of biogenic nanoparticles to determine their safety over prolonged exposure times (Banu et al., 2021).

7.2 Environmental Impact

Risk: When discharged into the environment, biogenic nanoparticles have the potential to endanger animals, ecological processes, and ecosystems. Ecosystem disturbance, biodiversity loss, and environmental contamination can result from the buildup of nanoparticles in soil, water, and air (Kokilavani et al., 2023).

Mitigation Strategies:

- Evaluate the fate, transport, and ecological impact of biogenic nanoparticles in natural ecosystems by conducting environmental risk assessments (Tran et al., 2024).
- Use green chemistry ideas and eco-friendly synthesis techniques to lessen environmental pollution and ecological imprints (Donato et al., 2022).
- Create materials and coatings using biodegradable nanoparticles to promote natural breakdown and reduce environmental persistence (Abbas et al., 2020).
- Create remediation techniques based on nanoparticles to eliminate environmental contaminants and lessen the harm that toxins do to the ecosystem (Patil et al., 2016).
- In order to track the dispersion, bioavailability, and ecological consequences of nanoparticles in environmental matrices, implement environmental monitoring programs.

7.3 Regulatory Compliance

Risk: Failure to adhere to legal and regulatory rules and guidelines may lead to legal ramifications, regulatory penalties, and delays in the development and launch of new products. Biogenic nanoparticles must adhere to regulatory requirements for quality assurance, safety, and environmental protection.

Mitigation Strategies:

- Keep up with any changes to regulations and guidelines that pertinent regulatory authorities, like the FDA, EPA, and foreign regulatory bodies, have released.
- Early on in the product development process, communicate with regulatory bodies to make sure that expectations and standards are met.
- To ascertain the regulatory classification, process, and compliance requirements that apply to products based on biogenic nanoparticles, conduct comprehensive regulatory assessments.
- To guarantee regulatory compliance and product traceability, and establish strong quality management systems, standard operating procedures (SOPs), and documentation standards.
- Work together with industry partners, legal professionals, and regulatory consultants to negotiate regulatory processes, resolve regulatory issues, and secure prompt approvals for products containing nanoparticles.

7.4 Ethical and Societal Concerns

Risk: Concerns about equity, access, privacy, and public acceptability are among the ethical and societal difficulties surrounding biogenic nanoparticles. The ethical development and deployment of biogenic nanoparticles may be impeded by social resistance, privacy breaches, and the unequal distribution of nanoparticle-based technology.

Mitigation Strategies:

- All phases of nanoparticle research and development, such as study design, participant recruiting, and data management, should take ethical considerations into account.
- Encourage fair access to nanoparticle-based technologies by tackling issues with accessibility, affordability, and healthcare inequalities.
- Interact with stakeholders, neighborhood associations, and legislators to resolve moral issues, establish credibility, and promote public acceptance of biogenic nanoparticles.
- Observe the confidentiality and private rights of research participants while following ethical standards for the gathering, storing, and sharing of data.
- In order to inform moral decision-making and the development of policies, carry out public education and outreach campaigns to increase understanding of the advantages, dangers, and societal ramifications of biogenic nanoparticles. Ask a variety of stakeholders for their opinions and input.

Potential risks associated with biogenic nanoparticles includes toxicity, environmental impact, regulatory compliance, and ethical concerns. By taking proactive steps to reduce risks, researchers and other interested parties can deal with problems and encourage the responsible production and use of biogenic nanoparticles for the good of society and the environment.

8. CONCLUSION

This chapter discusses the synthesis, features, uses, and moral implications of biogenic nanoparticles, as well as their special qualities and possible advantages. It draws attention to the significance of biogenic nanoparticles in a number of areas, such as tissue engineering, medication delivery, environmental remediation, and precision medicine. Biogenic nanoparticle developments, AI-driven approaches, and interdisciplinary cooperation all point to a bright future for bioinformatics and nanobiotechnology. AI-powered methods improve data processing, predictive modeling, and research efficiency, encouraging innovation and scientific advancement in various domains. As AI develops, scientists can use it to solve difficult problems, spur creativity, and fully use biogenic nanoparticles to meet societal requirements and further scientific understanding. The fusion of biogenic nanoparticles with AI-driven technologies heralds a new era of innovation and discovery in bioinformatics and nanobiotechnology. We can harness the revolutionary potential of biogenic nanoparticles to improve human health, safeguard the environment, and improve the quality of life for future generations by embracing interdisciplinary collaboration, ethical values, and responsible innovation.

REFERENCES

Abbas, Q., Yousaf, B., Amina, , Ali, M. U., Munir, M. A. M., El-Naggar, A., Rinklebe, J., & Naushad, M. (2020). Transformation pathways and fate of engineered nanoparticles (ENPs) in distinct interactive environmental compartments: A review. *Environment International*, 138, 105646. DOI: 10.1016/j.envint.2020.105646 PMID: 32179325

Ahiwale, S. S., Bankar, A. V., Tagunde, S., & Kapadnis, B. P. (2017). A Bacteriophage Mediated Gold Nanoparticles Synthesis and Their Anti-biofilm Activity. *Indian Journal of Microbiology*, 57(2), 188–194. DOI: 10.1007/s12088-017-0640-x PMID: 28611496

Aithal, S., & Aithal, P. S. (2024). Predictive Analysis of use of AI-Driven GPTs in Nanomaterials Research Breakthroughs in the 21st Century. [IJAEML]. *International Journal of Applied Engineering and Management Letters*, 8(2), 2. Advance online publication. DOI: 10.47992/IJAEML.2581.7000.0226

Alam, S., Israr, J., & Kumar, A. (2024). Artificial Intelligence and Machine Learning in Bioinformatics. In Singh, V., & Kumar, A. (Eds.), *Advances in Bioinformatics* (pp. 321–345). Springer Nature., DOI: 10.1007/978-981-99-8401-5_16

Alghamdi, M. A., Fallica, A. N., Virzì, N., Kesharwani, P., Pittalà, V., & Greish, K. (2022). The Promise of Nanotechnology in Personalized Medicine. *Journal of Personalized Medicine*, 12(5), 5. Advance online publication. DOI: 10.3390/jpm12050673 PMID: 35629095

Aljabali, A. A. A., Obeid, M. A., Rezigue, M. M., Alqudah, A., Charbe, N. B., Chellappan, D. K., Mishra, V., Pardhi, D. M., Dureja, H., Gupta, G., Prasher, P., Dua, K., Barhoum, A., & Tambuwala, M. M. (2020). Nanocelluloses as a Novel Vehicle for Controlled Drug Delivery. In Barhoum, A. (Ed.), *Handbook of Nanocelluloses: Classification, Properties, Fabrication, and Emerging Applications* (pp. 1–37). Springer International Publishing., DOI: 10.1007/978-3-030-62976-2_36-1

Allan, J., Belz, S., Hoeveler, A., Hugas, M., Okuda, H., Patri, A., Rauscher, H., Silva, P., Slikker, W., Sokull-Kluettgen, B., Tong, W., & Anklam, E. (2021). Regulatory landscape of nanotechnology and nanoplastics from a global perspective. *Regulatory Toxicology and Pharmacology*, 122, 104885. DOI: 10.1016/j.yrtph.2021.104885 PMID: 33617940

Anbazhagan, M., Ramachandran, S., Subramanian, P., Nachimuthu, R., & Kodiveri Muthukaliannan, G. (2016). *Nanomaterials: Classification* (Vol. 23). Biological Synthesis and Characterization., DOI: 10.1007/978-3-319-48009-1_2

Balaraman, P., Balasubramanian, B., Kaliannan, D., Durai, M., Kamyab, H., Park, S., Chelliapan, S., Lee, C. T., Maluventhen, V., & Maruthupandian, A. (2020). Phyco-synthesis of Silver Nanoparticles Mediated from Marine Algae Sargassum myriocystum and Its Potential Biological and Environmental Applications. *Waste and Biomass Valorization*, 11(10), 5255–5271. DOI: 10.1007/s12649-020-01083-5

Banerjee, D., Rajput, D., Banerjee, S., & Saharan, V. A. (2022). Artificial Intelligence and Its Applications in Drug Discovery, Formulation Development, and Healthcare. In Saharan, V. A. (Ed.), *Computer Aided Pharmaceutics and Drug Delivery: An Application Guide for Students and Researchers of Pharmaceutical Sciences* (pp. 309–380). Springer Nature., DOI: 10.1007/978-981-16-5180-9_11

Banu, A. N., Kudesia, N., Raut, A. M., Pakrudheen, I., & Wahengbam, J. (2021). Toxicity, bioaccumulation, and transformation of silver nanoparticles in aqua biota: A review. *Environmental Chemistry Letters*, 19(6), 4275–4296. DOI: 10.1007/s10311-021-01304-w

Baruah, K., Haque, M., Langbang, L., Das, S., Aguan, K., & Singha Roy, A. (2021). *Ocimum sanctum* mediated green synthesis of silver nanoparticles: A biophysical study towards lysozyme binding and anti-bacterial activity. *Journal of Molecular Liquids*, 337, 116422. DOI: 10.1016/j.molliq.2021.116422

Beatty, P. H., & Lewis, J. D. (2019). Cowpea mosaic virus nanoparticles for cancer imaging and therapy. *Advanced Drug Delivery Reviews*, 145, 130–144. DOI: 10.1016/j.addr.2019.04.005 PMID: 31004625

Bernard, M., Jubeli, E., Pungente, M. D., & Yagoubi, N. (2018). Biocompatibility of polymer-based biomaterials and medical devices – regulations, in vitro screening and risk-management. *Biomaterials Science*, 6(8), 2025–2053. DOI: 10.1039/C8BM00518D PMID: 29968869

Bhatia, S., Pooja, , & Yadav, S. K. (2023). CRISPR-Cas for genome editing: Classification, mechanism, designing and applications. *International Journal of Biological Macromolecules*, 238, 124054. DOI: 10.1016/j.ijbiomac.2023.124054 PMID: 36933595

Bodade, R. G., Kumar, R., & Kutty, R. (2023). Crop plant-mediated nanoparticle synthesis and applications. In Bhunia, S., Kumar, B., Singh, P., Oraon, R., & Kim, K.-H. (Eds.), *Nanoparticles in Green Organic Synthesis* (pp. 351–399). Elsevier., DOI: 10.1016/B978-0-323-95921-6.00015-9

Champa-Bujaico, E., García-Díaz, P., & Díez-Pascual, A. M. (2022). *Machine Learning for Property Prediction and Optimization of Polymeric Nanocomposites: A State-of-the-Art.* https://agris.fao.org/search/en/providers/122436/records/647480702d3f560f80b14a2f

Chauhan, S. B. (2024). Current Status of Smart Nanoparticles in Drug Delivery. In Bhardwaj, A. K., Srivastav, A. L., & Rai, S. (Eds.), *Biogenic Wastes-Enabled Nanomaterial Synthesis: Applications in Environmental Sustainability* (pp. 351–372). Springer Nature Switzerland., DOI: 10.1007/978-3-031-59083-2_14

Chen, H., Luo, L., Fan, S., Xiong, Y., Ling, Y., & Peng, S. (2021). Zinc oxide nanoparticles synthesized from Aspergillus terreus induces oxidative stress-mediated apoptosis through modulating apoptotic proteins in human cervical cancer HeLa cells. *The Journal of Pharmacy and Pharmacology*, 73(2), 221–232. DOI: 10.1093/jpp/rgaa043 PMID: 33793807

Chandrashekhara, V.D, Lokesh, S., Kavyashri, V.V, & Prasanna Kumar, M.K, Sanggonda. (2021). Potential Antifungal Activity of Biosynthesized Copper nanoparticles Against Colletotrichum Capsici In Chilly. *International Journal of Biotech Trends and Technology*, 11, 1–7. DOI: 10.14445/22490183/IJBTT-V11I2P601

Das, S., Chakraborty, J., Chatterjee, S., & Kumar, H. (2018). Prospects of biosynthesized nanomaterials for the remediation of organic and inorganic environmental contaminants. *Environmental Science. Nano*, 5(12), 2784–2808. DOI: 10.1039/C8EN00799C

de Olyveira, G. M., Manzine Costa, L. M., Basmaji, P., & Xavier Filho, L. (2011). *Bacterial Nanocellulose for Medicine Regenerative*. DOI: 10.1115/1.4004181

Donato, L., Nasser, I. I., Majdoub, M., & Drioli, E. (2022). Green Chemistry and Molecularly Imprinted Membranes. *Membranes*, 12(5), 5. Advance online publication. DOI: 10.3390/membranes12050472 PMID: 35629798

Dwivedi, S., & Dwivedi, N. (2022). Advances and Challenges of Bionanotechnology in the Sustainable Management of the Environment and Green Economy. In *Bionanotechnology Towards Sustainable Management of Environmental Pollution*. CRC Press. DOI: 10.1201/9781003270959-1

Eaglstein, W. H. (2014). The FDA Approval Process and Drug Development. In Eaglstein, W. H. (Ed.), *The FDA for Doctors* (pp. 39–51). Springer International Publishing., DOI: 10.1007/978-3-319-08362-9_10

Edgar, V.-N., Hermes, P.-H., Denisse, V.-G. J., Andrea, P.-M., Roberto, S.-C. C., Ileana, V.-R., & Fabián, F.-L. (2024). Microorganisms, Plants, and Nanotechnology for Environmental Remediation: A Sustainable Prospect. In *Nano-Bioremediation for Water and Soil Treatment*. Apple Academic Press. DOI: 10.1201/9781003414964-3

Ernst, L. M., Casals, E., Italiani, P., Boraschi, D., & Puntes, V. (2021). The Interactions between Nanoparticles and the Innate Immune System from a Nanotechnologist Perspective. *Nanomaterials (Basel, Switzerland)*, 11(11), 11. Advance online publication. DOI: 10.3390/nano11112991 PMID: 34835755

Foret, F., Chung, D. S., Lavická, J., Přikryl, J., Lee, H., & Drobníková, I. (2023). Proceedings of APCE-CECE-ITP-IUPAC 2022. *Separations, 10*(2), Article 2. DOI: 10.3390/separations10020109

Fritz, L., Baum, C. M., Low, S., & Sovacool, B. K. (2024). Public engagement for inclusive and sustainable governance of climate interventions. *Nature Communications*, 15(1), 4168. DOI: 10.1038/s41467-024-48510-y PMID: 38755215

Furxhi, I., Faccani, L., Zanoni, I., Brigliadori, A., Vespignani, M., & Costa, A. L. (2024). Design rules applied to silver nanoparticles synthesis: A practical example of machine learning application. *Computational and Structural Biotechnology Journal*, 25, 20–33. DOI: 10.1016/j.csbj.2024.02.010 PMID: 38444982

García-Peñas, A., & Sharma, G. (2023). *New Materials for a Circular Economy*. Materials Research Forum LLC. DOI: 10.21741/9781644902639

Geng, Y., Zhu, R., & Maimaituerxun, M. (2022, November 8). *Bibliometric review of carbon neutrality with CiteSpace: Evolution, trends, and framework. | Environmental Science & Pollution Research | EBSCOhost*. DOI: 10.1007/s11356-022-23283-3

Ghosh, S., Girigoswami, K., & Girigoswami, A. (2019). Membrane-Encapsulated Camouflaged Nanomedicines in Drug Delivery. *Nanomedicine (London)*, 14(15), 2067–2082. DOI: 10.2217/nnm-2019-0155 PMID: 31355709

González-Ballesteros, N., Rodríguez-Argüelles, M. C., Prado-López, S., Lastra, M., Grimaldi, M., Cavazza, A., Nasi, L., Salviati, G., & Bigi, F. (2019). Macroalgae to nanoparticles: Study of Ulva lactuca L. role in biosynthesis of gold and silver nanoparticles and of their cytotoxicity on colon cancer cell lines. *Materials Science and Engineering C*, 97, 498–509. DOI: 10.1016/j.msec.2018.12.066 PMID: 30678937

Gorgieva, S., & Trček, J. (2019). Bacterial Cellulose: Production, Modification and Perspectives in Biomedical Applications. *Nanomaterials (Basel, Switzerland)*, 9(10), 10. Advance online publication. DOI: 10.3390/nano9101352 PMID: 31547134

Gowdhami, B., Jaabir, M., Archunan, G., & Suganthy, N. (2019). Anticancer potential of zinc oxide nanoparticles against cervical carcinoma cells synthesized via biogenic route using aqueous extract of Gracilaria edulis. *Materials Science and Engineering C*, 103, 109840.

Grunwald, P. (2019). *Pharmaceutical Biocatalysis: Fundamentals, Enzyme Inhibitors, and Enzymes in Health and Diseases*. CRC Press.

Haik, Y., & Bantekas, I. (2020). Nano-Drug Clinical Trials: Informed Consent and Risk Management through Blockchain. *Pittsburgh Journal of Technology Law and Policy*, 21, 35.

Hemmati, S., Ahany Kamangar, S., Ahmeda, A., Zangeneh, M. M., & Zangeneh, A. (2020). Application of copper nanoparticles containing natural compounds in the treatment of bacterial and fungal diseases. *Applied Organometallic Chemistry*, 34(4), e5465. DOI: 10.1002/aoc.5465

Hunckler, M. D., & Levine, A. D. (2022). Navigating ethical challenges in the development and translation of biomaterials research. *Frontiers in Bioengineering and Biotechnology*, 10, 949280. DOI: 10.3389/fbioe.2022.949280 PMID: 36204464

Jain, K. K. (2016). Chapter Two—Role of Proteomics in the Development of Personalized Medicine. In Donev, R. (Ed.), *Advances in Protein Chemistry and Structural Biology* (Vol. 102, pp. 41–52). Academic Press., DOI: 10.1016/bs.apcsb.2015.09.002

Jedrzejczak-Krzepkowska, M., Kubiak, K., Ludwicka, K., & Bielecki, S. (2016). Bacterial NanoCellulose Synthesis, Recent Findings. In Gama, M., Dourado, F., & Bielecki, S. (Eds.), *Bacterial Nanocellulose* (pp. 19–46). Elsevier., DOI: 10.1016/B978-0-444-63458-0.00002-0

Joudeh, N., & Linke, D. (2022). Nanoparticle classification, physicochemical properties, characterization, and applications: A comprehensive review for biologists. *Journal of Nanobiotechnology*, 20(1), 262. DOI: 10.1186/s12951-022-01477-8 PMID: 35672712

Kang, T., Li, F., Baik, S., Shao, W., Ling, D., & Hyeon, T. (2017). Surface design of magnetic nanoparticles for stimuli-responsive cancer imaging and therapy. *Biomaterials*, 136, 98–114. DOI: 10.1016/j.biomaterials.2017.05.013 PMID: 28525855

Karimi, M., Ghasemi, A., Zangabad, P. S., Rahighi, R., Basri, S. M. M., Mirshekari, H., Amiri, M., Pishabad, Z. S., Aslani, A., Bozorgomid, M., Ghosh, D., Beyzavi, A., Vaseghi, A., Aref, A. R., Haghani, L., Bahrami, S., & Hamblin, M. R. (2016). Smart micro/nanoparticles in stimulus-responsive drug/gene delivery systems. *Chemical Society Reviews*, 45(5), 1457–1501. DOI: 10.1039/C5CS00798D PMID: 26776487

Kashisaz, M., Enayatizamir, N., Fu, P., & Eslahi, M. (2024). Synthesis of nanoparticles using Trichoderma Harzianum, characterization, antifungal activity and impact on Plant Growth promoting Bacteria. *World Journal of Microbiology & Biotechnology*, 40(4), 107. DOI: 10.1007/s11274-024-03920-4 PMID: 38396217

Khajeh Mehrizi, M., & Shahi, Z. (2020). Application of Plant-Based Natural Product to Synthesize Nanomaterial. In Srivastava, M., Srivastava, N., Mishra, P. K., & Gupta, V. K. (Eds.), *Nanomaterials in Biofuels Research* (pp. 53–73). Springer., DOI: 10.1007/978-981-13-9333-4_3

Khan, S. A., & Sharma, R. (2024). Eco-friendly Synthesis of Copper Nanoparticles: An Overview of the Epoch-making Role of Natural Resources, Applications, and Recent Developments. *Current Green Chemistry*, 11(3), 286–295. DOI: 10.2174/0122133461279579231103055412

Khandel, P., & Shahi, S. K. (2018). Mycogenic nanoparticles and their bio-prospective applications: Current status and future challenges. *Journal of Nanostructure in Chemistry*, 8(4), 369–391. DOI: 10.1007/s40097-018-0285-2

Kokilavani, S., Janani, B., Balasurya, S., & Khan, S. S. (2023). Accumulation of Engineered Nanomaterials in Soil, Water, and Air. In Fernandez-Luqueno, F., & Patra, J. K. (Eds.), *Agricultural and Environmental Nanotechnology: Novel Technologies and their Ecological Impact* (pp. 551–582). Springer Nature., DOI: 10.1007/978-981-19-5454-2_21

Kumar, N., Salar, R. K., Kumar, R., Prasad, M., Brar, B., & Nain, V. (2017). Green synthesis of silver nanoparticles and its applications—A review. *Nano Trends: A Journal of Nanotechnology and Its Applications, 19*(3), 1-22.

Kumar, V., Dasgupta, N., & Ranjan, S. (2018). *Nanotoxicology: Toxicity Evaluation, Risk Assessment and Management*. CRC Press. DOI: 10.1201/b21545

Kus-Liśkiewicz, M., Fickers, P., & Ben Tahar, I. (2021). Biocompatibility and Cytotoxicity of Gold Nanoparticles: Recent Advances in Methodologies and Regulations. *International Journal of Molecular Sciences*, 22(20), 20. Advance online publication. DOI: 10.3390/ijms222010952 PMID: 34681612

Lateef, A., Darwesh, O. M., & Matter, I. A. (2021). Microbial Nanobiotechnology: The Melting Pot of Microbiology, Microbial Technology and Nanotechnology. In Lateef, A., Gueguim-Kana, E. B., Dasgupta, N., & Ranjan, S. (Eds.), *Microbial Nanobiotechnology: Principles and Applications* (pp. 1–19). Springer., DOI: 10.1007/978-981-33-4777-9_1

Liu, S., Yu, J.-M., Gan, Y.-C., Qiu, X.-Z., Gao, Z.-C., Wang, H., Chen, S.-X., Xiong, Y., Liu, G.-H., Lin, S.-E., McCarthy, A., John, J. V., Wei, D.-X., & Hou, H.-H. (2023). Biomimetic natural biomaterials for tissue engineering and regenerative medicine: New biosynthesis methods, recent advances, and emerging applications. *Military Medical Research*, 10(1), 16. DOI: 10.1186/s40779-023-00448-w PMID: 36978167

Madkour, L. H. (2018). *Biogenic– Biosynthesis Metallic Nanoparticles (MNPs) for Pharmacological, Biomedical and Environmental Nanobiotechnological Applications*.

Manandhar, S., Sankhe, R., Harish Kumar, B., & Pai, K. S. R. (2021). Clinical protocols for nanopharmacokinetics assessment. In Thorat, N. D., & Kumar, N. (Eds.), *Nano-Pharmacokinetics and Theranostics* (pp. 193–220). Academic Press., DOI: 10.1016/B978-0-323-85050-6.00001-3

Marín-Caba, L., Chariou, P. L., Pesquera, C., Correa-Duarte, M. A., & Steinmetz, N. F. (2019). Tobacco Mosaic Virus-Functionalized Mesoporous Silica Nanoparticles, a Wool-Ball-like Nanostructure for Drug Delivery. *Langmuir*, 35(1), 203–211. DOI: 10.1021/acs.langmuir.8b03337 PMID: 30576145

Medici, S., Peana, M., Pelucelli, A., & Zoroddu, M. A. (2021). An updated overview on metal nanoparticles toxicity. *Seminars in Cancer Biology*, 76, 17–26. DOI: 10.1016/j.semcancer.2021.06.020 PMID: 34182143

Mehmood, A., Murtaza, G., Bhatti, T., Kausar, R., & Ahmed, M. (2016). Biosynthesis, characterization and antimicrobial action of silver nanoparticles from root bark extract of Berberis lycium Royle. *Pakistan Journal of Pharmaceutical Sciences*, 29, 131–136. PMID: 26826826

Mirabello, G., Lenders, J. J. M., & Sommerdijk, N. A. J. M. (2016). Bioinspired synthesis of magnetite nanoparticles. *Chemical Society Reviews*, 45(18), 5085–5106. DOI: 10.1039/C6CS00432F PMID: 27385627

More, S., Bampidis, V., Franz, R., Bragard, C., Halldorsson, T., Hernández-Jerez, A., Hougaard Bennekou, S., Koutsoumanis, K., Lambré, C., Machera, K., Naegeli, H., Nielsen, S., Schlatter, J., & Schrenk, D., Silano (deceased), V., Turck, D., Younes, M., Castenmiller, J., Chaudhry, Q., … Schoonjans, R. (. (2021). Guidance on risk assessment of nanomaterials to be applied in the food and feed chain: Human and animal health. *EFSA Journal*, 19(8). Advance online publication. DOI: 10.2903/j.efsa.2021.6768 PMID: 34377190

Munir, A., Salah, M. A., Ali, M., Ali, B., Saleem, M. H., Samarasinghe, K. G. B. A., De Silva, S. I. S., Ercisli, S., Iqbal, N., & Anas, M. (2024). Advancing Agriculture: Harnessing Smart Nanoparticles for Precision Fertilization. *BioNanoScience*. Advance online publication. DOI: 10.1007/s12668-024-01597-5

Nathanael, K., Cheng, S., Kovalchuk, N. M., Arcucci, R., & Simmons, M. J. H. (2023). Optimization of microfluidic synthesis of silver nanoparticles: A generic approach using machine learning. *Chemical Engineering Research & Design*, 193, 65–74. DOI: 10.1016/j.cherd.2023.03.007

Naveenkumar, S., Kamaraj, C., Ragavendran, C., Vaithiyalingam, M., Sugumar, V., & Marimuthu, K. (2023). Gracilaria corticata red seaweed mediate biosynthesis of silver nanoparticles: Larvicidal, neurotoxicity, molecular docking analysis, and ecofriendly approach. *Biomass Conversion and Biorefinery*. Advance online publication. DOI: 10.1007/s13399-023-04026-x

Nazari, M., Ebrahimzadeh, F., Falaki, M., Hajimomeni, Y., Varma, R. S., & Fallahnia, N. (2022). Nanoparticles in the prevention, diagnosis, and treatment of the influenza virus. *Nanomedicine Research Journal*, 7(1), 19–36. DOI: 10.22034/nmrj.2022.01.002

Niazi, S., & Lokesh, S. (2022). *Biopharmaceutical Manufacturing, Volume 2: Unit processes*. IOP Publishing. https://iopscience.iop.org/book/mono/978-0-7503-3179-1

Olawale, F., Ariatti, M., & Singh, M. (2021). Biogenic Synthesis of Silver-Core Selenium-Shell Nanoparticles Using Ocimum tenuiflorum L.: Response Surface Methodology-Based Optimization and Biological Activity. *Nanomaterials (Basel, Switzerland)*, 11(10), 10. Advance online publication. DOI: 10.3390/nano11102516 PMID: 34684956

Padhiary, M. (2024a). Membrane Technologies for Treating Wastewater in the Food Processing Industry: Practices and Challenges. In *Research Trends in Food Technology and Nutrition* (Vol. 27, pp. 37–62). AkiNik Publications. DOI: 10.22271/ed.book.2817

Padhiary, M. (2024b). Status of Farm Automation, Advances, Trends, and Scope in India. [IJSR]. *International Journal of Science and Research (Raipur, India)*, 13(7), 737–745. DOI: 10.21275/SR24713184513

Padhiary, M. (2024c). The Convergence of Deep Learning, IoT, Sensors, and Farm Machinery in Agriculture. In *Designing Sustainable Internet of Things Solutions for Smart Industries* (pp. 109–142). IGI Global., DOI: 10.4018/979-8-3693-5498-8.ch005

Padhiary, M., Barbhuiya, J. A., Roy, D., & Roy, P. (2024). 3D Printing Applications in Smart Farming and Food Processing. *Smart Agricultural Technology*, 9, 100553. DOI: 10.1016/j.atech.2024.100553

Padhiary, M., & Kumar, R. (2024). Enhancing Agriculture Through AI Vision and Machine Learning: The Evolution of Smart Farming. In Thangam, D. (Ed.), (pp. 295–324). Advances in Computational Intelligence and Robotics. IGI Global., DOI: 10.4018/979-8-3693-5380-6.ch012

Padhiary, M., & Roy, P. (2024). Advancements in Precision Agriculture: Exploring the Role of 3D Printing in Designing All-Terrain Vehicles for Farming Applications. *International Journal of Science and Research (Raipur, India)*, 13(5), 861–868. DOI: 10.21275/SR24511105508

Padhiary, M., Saha, D., Kumar, R., Sethi, L. N., & Kumar, A. (2024). Enhancing Precision Agriculture: A Comprehensive Review of Machine Learning and AI Vision Applications in All-Terrain Vehicle for Farm Automation. *Smart Agricultural Technology*, 8, 100483. DOI: 10.1016/j.atech.2024.100483

Patil, S. S., Shedbalkar, U. U., Truskewycz, A., Chopade, B. A., & Ball, A. S. (2016a). Nanoparticles for environmental clean-up: A review of potential risks and emerging solutions. *Environmental Technology & Innovation*, 5, 10–21. DOI: 10.1016/j.eti.2015.11.001

Periakaruppan, R., Chen, X., Thangaraj, K., Jeyaraj, A., Nguyen, H. H., Yu, Y., Hu, S., Lu, L., & Li, X. (2021). Utilization of tea resources with the production of superparamagnetic biogenic iron oxide nanoparticles and an assessment of their antioxidant activities. *Journal of Cleaner Production*, 278, 123962. DOI: 10.1016/j.jclepro.2020.123962

Pitek, A. S., Hu, H., Shukla, S., & Steinmetz, N. F. (2018). Cancer Theranostic Applications of Albumin-Coated Tobacco Mosaic Virus Nanoparticles. *ACS Applied Materials & Interfaces*, 10(46), 39468–39477. DOI: 10.1021/acsami.8b12499 PMID: 30403330

Pramanik, S., & Chakraborty, P. (2014). Soil productivity analysis under different land uses of Sundarban delta. *Journal of Soil and Water Conservation*, •••, 13.

Rani, M. Ankit, Yadav, J., & Shanker, U. (2022). Sustainable Green Doped Nanomaterials for Emerging Contaminants Removal. In Handbook of Green and Sustainable Nanotechnology: Fundamentals, Developments and Applications (pp. 1-30). Cham: Springer International Publishing.

Rodríguez-Gómez, F. D., Penon, O., Monferrer, D., & Rivera-Gil, P. (2023). Classification system for nanotechnology-enabled health products with both scientific and regulatory application. *Frontiers in Medicine*, 10, 1212949. Advance online publication. DOI: 10.3389/fmed.2023.1212949 PMID: 37601794

Rubina, M. S., Pigaleva, M. A., Naumkin, A. V., & Gromovykh, T. I. (2020). Bacterial Cellulose Film Produced by Gluconacetobacter hansenii as a Source Material for Oxidized Nanofibrillated Cellulose. *Doklady Physical Chemistry*, 493(2), 127–131. DOI: 10.1134/S0012501620080023

Saad, R., Nsour, M., Khader, Y., & Gunaid, M. (2021). Public Health Surveillance Systems in the Eastern Mediterranean Region: Bibliometric Analysis of Scientific Literature (Preprint). *JMIR Public Health and Surveillance*, 7(11), e32639. Advance online publication. DOI: 10.2196/32639 PMID: 34723831

Sanità, G., Carrese, B., & Lamberti, A. (2020). Nanoparticle Surface Functionalization: How to Improve Biocompatibility and Cellular Internalization. *Frontiers in Molecular Biosciences*, 7, 587012. Advance online publication. DOI: 10.3389/fmolb.2020.587012 PMID: 33324678

Sarkar, B., Mahanty, A., Gupta, S. K., Choudhury, A. R., Daware, A., & Bhattacharjee, S. (2022). Nanotechnology: A next-generation tool for sustainable aquaculture. *Aquaculture (Amsterdam, Netherlands)*, 546, 737330. DOI: 10.1016/j.aquaculture.2021.737330

Seo, S., Lee, J. E., Lee, K., & Kim, H. N. (2022). Effects of microenvironmental factors on assessing nanoparticle toxicity. *Environmental Science. Nano*, 9(2), 454–476. DOI: 10.1039/D1EN00925G

Singh, K. R., Natarajan, A., & Pandey, S. S. (2023). Bioinspired Multifunctional Silver Nanoparticles for Optical Sensing Applications: A Sustainable Approach. *ACS Applied Bio Materials*, 6(11), 4549–4571. DOI: 10.1021/acsabm.3c00669 PMID: 37852204

Spruit, S. L., van de Poel, I., & Doorn, N. (2016). Informed Consent in Asymmetrical Relationships: An Investigation into Relational Factors that Influence Room for Reflection. *NanoEthics*, 10(2), 123–138. DOI: 10.1007/s11569-016-0262-5 PMID: 27478516

Srivastava, S. (2023). Ethical issues regarding the use of nanobiotechnology-based products. In Pratap Singh, R., Adetunji, C. O., Singh, R. L., Singh, J., Solanki, P. R., & Singh, K. R. B. (Eds.), *Nanobiotechnology for the Livestock Industry* (pp. 435–473). Elsevier., DOI: 10.1016/B978-0-323-98387-7.00005-7

Sykes, E., Dai, Q., Sarsons, C., Chen, J., Rocheleau, J., Hwang, D., Zheng, G., Cramb, D., Rinker, K., & Chan, W. (2016). Tailoring nanoparticle designs to target cancer based on tumor pathophysiology. *Proceedings of the National Academy of Sciences of the United States of America,* 113. DOI: 10.1073/pnas.1521265113

Tran, T.-K., Nguyen, M.-K., Lin, C., Hoang, T.-D., Nguyen, T.-C., Lone, A. M., Khedulkar, A. P., Gaballah, M. S., Singh, J., Chung, W. J., & Nguyen, D. D. (2024). Review on fate, transport, toxicity and health risk of nanoparticles in natural ecosystems: Emerging challenges in the modern age and solutions toward a sustainable environment. *The Science of the Total Environment*, 912, 169331. DOI: 10.1016/j.scitotenv.2023.169331 PMID: 38103619

Van Loo, R. (2019). Regulatory Monitors: Policing Firms in the Compliance Era. *Columbia Law Review*, 119(2), 369–444.

Vasoya, N. H. (2023). Revolutionizing Nano Materials Processing through IoT-AI Integration: Opportunities and Challenges. *Journal of Materials Science Research and Reviews*, 6(3), 3.

Vasudev, S., Saxena, J., Srivastava, B., & Reddy, P. (2023). *Advancements in Immunology, Nanodrugs and Public health.*

Vijayakumar, S., Vaseeharan, B., Malaikozhundan, B., Gopi, N., Ekambaram, P., Pachaiappan, R., Velusamy, P., Murugan, K., Benelli, G., Suresh Kumar, R., & Suriyanarayanamoorthy, M. (2017). Therapeutic effects of gold nanoparticles synthesized using *Musa paradisiaca* peel extract against multiple antibiotic resistant *Enterococcus faecalis* biofilms and human lung cancer cells (A549). *Microbial Pathogenesis*, 102, 173–183. DOI: 10.1016/j.micpath.2016.11.029 PMID: 27916691

Vinayagasundaram, C., Aruna, V., & Angajala, G. (2021). Review on metal nanoparticles as nanocarriers: Current challenges and perspectives in drug delivery systems. *Emergent Materials*, 5(6), 1593–1615. Advance online publication. DOI: 10.1007/s42247-021-00335-x PMID: 35005431

Wang, C., Zhang, Z., Wang, J., Wang, Q., & Shang, L. (2022). Biohybrid materials: Structure design and biomedical applications. *Materials Today. Bio*, 16, 100352. DOI: 10.1016/j.mtbio.2022.100352 PMID: 35856044

Yadav, S., Lal, S., Yadav, S., Laxman, J., Verma, B., Sushma, M., Choudhary, R., Singh, P., Singh, S., Sharma, V., & Singh, B. (2019). *Use of Nanotechnology in Agri-food Sectors and Apprehensions: An Overview*.

Yao, F., Zhu, P., Chen, J., Li, S., Sun, B., Li, Y., Zou, M., Qi, X., Liang, P., & Chen, Q. (2023). Synthesis of nanoparticles via microfluidic devices and integrated applications. *Mikrochimica Acta*, 190(7), 256. DOI: 10.1007/s00604-023-05838-4 PMID: 37301779

Ye, D., & Peramo, A. (2014). Implementing tissue engineering and regenerative medicine solutions in medical implants. *British Medical Bulletin*, 109(1), 3–18. DOI: 10.1093/bmb/ldt036 PMID: 24357734

Yip, L. X., Wang, J., Xue, Y., Xing, K., Sevencan, C., Ariga, K., & Leong, D. T. (2024). Cell-derived nanomaterials for biomedical applications. *Science and Technology of Advanced Materials*, 25(1), 2315013. DOI: 10.1080/14686996.2024.2315013 PMID: 38476511

Zarghani, M., Nemati-Anaraki, L., Sedghi, S., Chakoli, A. N., & Rowhani-Farid, A. (2023). Iranian researchers' perspective about concept and effect of open science on research publication. *BMC Health Services Research*, 23(1), 437. DOI: 10.1186/s12913-023-09420-9 PMID: 37143102

Zhu, S., Yu, T., Xu, T., Chen, H., Dustdar, S., Gigan, S., Gunduz, D., Hossain, E., Jin, Y., Lin, F., Liu, B., Wan, Z., Zhang, J., Zhao, Z., Zhu, W., Chen, Z., Durrani, T. S., Wang, H., Wu, J., … Pan, Y. (2023). Intelligent Computing: The Latest Advances, Challenges, and Future. *Intelligent Computing, 2*, 0006. DOI: 10.34133/icomputing.0006

KEY TERMS AND DEFINITIONS

Artificial Intelligence (AI): The simulation of human intelligence in machines that can learn, reason, and self-correct to perform tasks autonomously.

Bioinformatics: The application of computational and statistical techniques to manage and analyse biological data, often used in genomics and molecular biology.

Nanobiotechnology: An interdisciplinary field that merges nanotechnology with biological sciences, focusing on applications at the molecular and nanoscale.

Biogenic Nanoparticles: Nanoparticles synthesized using biological organisms, such as plants, bacteria, and fungi, often offering eco-friendly alternatives to chemical synthesis.

Machine Learning (ML): A subset of AI where algorithms automatically improve through experience, commonly used for pattern recognition and predictive analytics in complex datasets.

Nanomedicine: The use of nanoparticles in medical applications, including diagnostics, drug delivery, and therapeutic treatments.

Green Synthesis: The eco-friendly production of nanoparticles using natural resources, such as plant extracts, microbes, and biowastes, minimizing harmful chemicals.

3D Printing: An additive manufacturing process that creates three-dimensional objects by layering material, often used in agriculture for custom tools and devices.

Smart Farming: The application of digital technologies and AI to optimize farming processes, enhancing precision in activities like irrigation, planting, and harvesting.

Precision Agriculture: Farming practices that use technology, such as GPS, sensors, and AI, to improve the efficiency and accuracy of field management.

Nanotoxicology: The study of the toxic effects of nanoparticles on living organisms and the environment, critical for assessing the safety of nanomaterials.

Bibliometric Analysis: A quantitative method for analyzing scientific literature to track research trends, influence, and impact in specific fields.

AI Vision Systems: Systems equipped with AI that process and interpret visual data, widely used in agriculture for real-time plant monitoring and disease detection.

Nanocellulose: Biodegradable nanoparticles derived from plant cellulose, used in applications ranging from packaging to biomedical materials.

Viral Nanoparticles: Nanoparticles derived from viruses, often employed in targeted drug delivery systems and vaccine development.

Predictive Modelling: The use of statistical techniques and AI algorithms to predict future outcomes, such as crop yields or disease outbreaks in agriculture.

Conclusion

In conclusion, *Synthesizing and Characterizing Plant-Mediated Biocompatible Metal Nanoparticles* serves as an indispensable resource for stakeholders across scientific, academic, industrial, and policy-making communities. The book offers a detailed exploration of plant-mediated nanoparticle synthesis, highlighting the advantages of eco-friendly, green synthesis methods over traditional chemical and physical techniques. By examining the key role of plant-derived materials, especially through the lens of Phytonanotechnology, the book showcases the potential of nanoparticles for numerous applications, including environmental sustainability, targeted drug delivery, and nanotechnological innovation.

Chapter 1 sets the foundation by comparing plant-derived nanoparticles with traditional methods, showcasing how green innovations offer safer, more sustainable alternatives. Chapters 2 and 3 focus on specific examples of gold and zinc oxide nanoparticles, using plant extracts to illustrate the potential of these biogenic nanoparticles in various applications. Chapter 4 expands this narrative by presenting green nanoparticles as a solution for sustainable wastewater treatment, addressing one of the most pressing environmental concerns of our time.

The book also ventures into diverse fields of application, as seen in Chapters 5 and 6, which examine the role of biopolymeric nanoparticles in the food industry and agriculture. These chapters demonstrate the far-reaching implications of plant-based nanoparticles, from enhancing food safety to revolutionizing agricultural practices. Chapter 7 continues this focus on agriculture by exploring the impact of nano-fertilizers on crop growth and soil health, underscoring their potential to transform modern farming.

Medical applications are also a key focus, with Chapters 8, 9, and 10 delving into the therapeutic potential of metallic bionanoparticles, their use as antifungal agents, and their conjugation with antibiotics to treat bacterial infections. These chapters highlight the emerging role of plant-mediated nanoparticles in healthcare, offering innovative solutions for drug delivery, disease treatment, and infection control.

Chapters 11 and 12 shift the discussion toward advanced instrumentation and techniques for characterizing nanoparticles, emphasizing the importance of accurate measurements in unlocking the full potential of these materials. The final chapters, 13 and 14, explore futuristic applications such as plant-based inorganic quantum dots and the integration of artificial intelligence in nanoparticle research, pointing toward exciting new frontiers in nanotechnology and bioinformatics.

Throughout the book, readers are introduced to a wide range of spectroscopic and imaging techniques used for characterizing the properties of these biocompatible nanoparticles, providing them with a solid foundation in both the practical and theoretical aspects of nanoparticle science. Additionally, the exploration of artificial intelligence's role in advancing nanoparticle applications broadens the scope of the book, connecting it with cutting-edge technologies in medicine, environmental science, and more.

Ultimately, this book not only guides readers through the intricate processes of green nanoparticle synthesis but also equips them to address the ethical, societal, and technological challenges posed by this rapidly evolving field. With its comprehensive approach, the book paves the way for future research and applications, offering readers the tools to navigate and contribute to the future of nanotechnology.

Compilation of References

Abbas, Q., Yousaf, B., Amina, , Ali, M. U., Munir, M. A. M., El-Naggar, A., Rinklebe, J., & Naushad, M. (2020). Transformation pathways and fate of engineered nanoparticles (ENPs) in distinct interactive environmental compartments: A review. *Environment International*, 138, 105646. DOI: 10.1016/j.envint.2020.105646 PMID: 32179325

Abdelghafour, M. M., Deák, Á., Szabó, D., Dékány, I., Rovó, L., & Janovák, L. (2022). Use of self-assembled colloidal prodrug nanoparticles for controlled drug delivery of anticancer, antifibrotic and antibacterial mitomycin. *International Journal of Molecular Sciences*, 23(12), 6807. DOI: 10.3390/ijms23126807 PMID: 35743251

Abdelghany, T., Al-Rajhi, A. M., Al Abboud, M. A., Alawlaqi, M., Ganash Magdah, A., Helmy, E. A., & Mabrouk, A. S. (2018). Recent advances in green synthesis of silver nanoparticles and their applications: About future directions. A review. *BioNanoScience*, 8(1), 5–16. DOI: 10.1007/s12668-017-0413-3

Abdelkhalek, A., Salem, M. Z., Kordy, A. M., Salem, A. Z., & Behiry, S. I. (2020). Antiviral, antifungal, and insecticidal activities of *Eucalyptus* bark extract: HPLC analysis of polyphenolic compounds. *Microbial Pathogenesis*, 147, 104383. DOI: 10.1016/j.micpath.2020.104383 PMID: 32659315

Abdel-Mottaleb, M. S. A., Byrne, J. A., & Chakarov, D. (2011). Nanotechnology and solar energy. *International Journal of Photoenergy*, 1, 194146. DOI: 10.1155/2011/194146

Abdel-Wahab, F., El Menofy, N., El-Batal, A., Mosallam, F., & Abdulall, A. (2021). Enhanced antimicrobial activity of the combination of silver nanoparticles and different β Lactam antibiotics against methicillin resistant Staphylococcus aureus isolates. *Azhar International Journal of Pharmaceutical and Medical Sciences*, 1(1), 22–31. DOI: 10.21608/aijpms.2021.53610.1017

Abdul Razak, N. A., Othman, N. H., Mat Shayuti, M. S., Jumahat, A., Sapiai, N., & Lau, W. J. (2022). Agricultural and industrial waste-derived mesoporous silica nanoparticles: A review on chemical synthesis route. *Journal of Environmental Chemical Engineering*, 10(2), 107322. DOI: 10.1016/j.jece.2022.107322

Abid, N., Khan, A. M., Shujait, S., Chaudhary, K., Ikram, M., Imran, M., Haider, J., Khan, M., Khan, Q., & Maqbool, M. (2022). Synthesis of nanomaterials using various top-down and bottom-up approaches, influencing factors, advantages, and disadvantages: A review. *Advances in Colloid and Interface Science*, 300, 102597. DOI: 10.1016/j.cis.2021.102597 PMID: 34979471

Abomuti, M. A., Danish, E. Y., Firoz, A., Hasan, N., & Malik, M. A. (2021). Green synthesis of ZnO nanoparticles using Salvia officinalis leaf extract and their photocatalytic and antifungal activities. *Biology (Basel)*, 10(11), 1075. DOI: 10.3390/biology10111075 PMID: 34827068

Aboyewa, J., Sibuyi, N., Meyer, M., & Oguntibeju, O. (2021). Green Synthesis of Metallic Nanoparticles Using Some Selected Medicinal Plants from Southern Africa and Their Biological Applications. *Plants*, 10(9), 1929. DOI: 10.3390/plants10091929 PMID: 34579460

Abushiba, M., El-Sherbiny, G., Moghannem, S., Kalaba, M., El-Shatoury, E., & Saeed, A. (2019). Enhancement of Antibiotics activity by microbially synthesized silver nanoparticles. *African Journal of Biological Sciences*, 15(1), 137–153. DOI: 10.21608/ajbs.2019.64008

Abuzeid, H. M., Julien, C. M., Zhu, L., & Hashem, A. M. (2023). Green Synthesis of Nanoparticles and Their Energy Storage, Environmental, and Biomedical Applications. *Crystals*, 13(11), 1576. DOI: 10.3390/cryst13111576

Adeyemi, J. O., & Fawole, O. A. (2023). Metal-based nanoparticles in food packaging and coating technologies: A review. *Biomolecules*, 13(7), 1092. DOI: 10.3390/biom13071092 PMID: 37509128

Adeyemi, J. O., Oriola, A. O., Onwudiwe, D. C., & Oyedeji, A. O. (2022). Plant extracts mediated metal-based nanoparticles: Synthesis and biological applications. *Biomolecules*, 12(5), 627. DOI: 10.3390/biom12050627 PMID: 35625555

Adrianto, N., Panre, A., Istiqomah, N., Riswan, M., Apriliani, F., & Suharyadi, E. (2022). Localized surface plasmon resonance properties of green synthesized silver nanoparticles. *Nano-Structures & Nano-Objects*, 31, 1–12. DOI: 10.1016/j.nanoso.2022.100895

Afzal, O., Altamimi, A. S. A., Nadeem, M. S., Alzarea, S. I., Almalki, W. H., Tariq, A., Mubeen, B., Murtaza, B. N., Iftikhar, S., Riaz, N., & Kazmi, I. (2022). Nanoparticles in drug delivery: From history to therapeutic applications. *Nanomaterials (Basel, Switzerland)*, 12(24), 1–27. DOI: 10.3390/nano12244494 PMID: 36558344

Agarwalla, S., Singh, S. K., Ibrahim, M. A., Noothalapati, H., &Duraiswamy, S. (2024). Antimicrobial Coatings: Current Mechanisms, Challenges, and Opportunities. *Functional Coatings: Innovations and Challenges,* 231-274.

Agarwal, N., Nair, M. S., Mazumder, A., & Poluri, K. M. (2018). *Characterization of nanomaterials using nuclear magnetic resonance spectroscopy Characterization of nanomaterials*. Elsevier.

Aguilar, N. M., Arteaga-Cardona, F., de Anda Reyes, M. E., Gervacio-Arciniega, J. J., & Salazar-Kuri, U. (2019). Magnetic bioplastics based on isolated cellulose from cotton and sugarcane bagasse. *Materials Chemistry and Physics*, 238, 121921. Advance online publication. DOI: 10.1016/j.matchemphys.2019.121921

Ahamad, T., Ansari, W. A., Negi, D. S., & Khan, M. F. (2019). NON-ENZYMATIC NATURAL REACTIVE OXYGEN SCAVENGERS (ROS): A REVIEW ON STRUCTURES AND MODE OF ACTION. *Era's. The Journal of Medical Research*, 6(2).

Ahamad, T., Khan, M. A., Ansari, W. A., & Khan, M. F. (2020). Antioxidant and anticancer activities of selected indian medicinal plant viz., *artocarpuslakoocha, kigelia pinnata, and amaranthusviridis. Era's. The Journal of Medical Research*, 7(1), 46–51.

Ahirwar, S., Mallick, S., & Bahadur, D. (2017). Electrochemical method to prepare graphene quantum dots and graphene oxide quantum dots. *ACS Omega*, 2(11), 8343–8353. DOI: 10.1021/acsomega.7b01539 PMID: 31457373

Ahiwale, S. S., Bankar, A. V., Tagunde, S., & Kapadnis, B. P. (2017). A Bacteriophage Mediated Gold Nanoparticles Synthesis and Their Anti-biofilm Activity. *Indian Journal of Microbiology*, 57(2), 188–194. DOI: 10.1007/s12088-017-0640-x PMID: 28611496

Ahluwalia, O., Singh, P. C., & Bhatia, R. (2021). A review on drought stress in plants: Implications, mitigation and the role of plant growth promoting rhizobacteria. *Resources. Environmental Sustainability*, 5, 100032.

Ahmad, B., Khan, M. I., Naeem, M. A., Alhodaib, A., Fatima, M., Amami, M., Al-Abbad, E. A., Kausar, A., Alwadai, N., Nazir, A., & Iqbal, M. (2022). Green synthesis of NiO nanoparticles using Aloe vera gel extract and evaluation of antimicrobial activity. *Materials Chemistry and Physics*, 288, 126363. DOI: 10.1016/j.matchemphys.2022.126363

Ahmad, H., Venugopal, K., Rajagopal, K., De Britto, S., Nandini, B., Pushpalatha, H. G., Konappa, N., Udayashankar, A. C., Geetha, N., & Jogaiah, S. (2020). Green Synthesis and Characterization of Zinc Oxide Nanoparticles Using Eucalyptus globules and Their Fungicidal Ability Against Pathogenic Fungi of Apple Orchards. *Biomolecules*, 10(3), 425. DOI: 10.3390/biom10030425 PMID: 32182874

Ahmad, N. (2022). Fozia, D., Jabeen, M., Ul-Haq, Z., Ahmad, I., Wahab, A., … Yahya Khan, M. (2022). Green Fabrication of Silver Nanoparticles using Euphorbia serpens Kunth Aqueous Extract, Their Characterization, and Investigation of Its In Vitro Antioxidative, Antimicrobial, Insecticidal, and Cytotoxic Activities. *BioMed Research International*, •••, 1–11.

Ahmad, N., Sharma, S., Alam, M. K., Singh, V. N., Shamsi, S. F., Mehta, B. R., & Fatma, A. (2010). Rapid synthesis of silver nanoparticles using dried medicinal plant of basil. *Colloids and Surfaces. B, Biointerfaces*, 81(1), 81–86. DOI: 10.1016/j.colsurfb.2010.06.029 PMID: 20656463

Ahmad, T., Bustam, M., Irfan, M., Moniruzzaman, M., Asghar, H., & Bhattacharjee, S. (2019). Mechanistic investigation of phytochemicals involved in green synthesis of gold nanoparticles using aqueous Elaeis guineensis leaves extract: Role of phenolic compounds and flavonoids. *Biotechnology and Applied Biochemistry*, 66(4), 698–708. Advance online publication. DOI: 10.1002/bab.1787 PMID: 31172593

Ahmady, L., Gothwal, M., Mukkoli, M. M., & Bari, V. K. (2024). Antifungal drug resistance in *Candida*: A special emphasis on amphotericin B. *APMIS*, 132(5), 291–316. DOI: 10.1111/apm.13389 PMID: 38465406

Ahmed, I., Mir, F. A., & Banday, J. A. (2023). Synthesis of metal and metal oxide nanoparticles using plant extracts characterization and applications. *BioNanoScience*, 13(4), 1541–1557. DOI: 10.1007/s12668-023-01194-y

Ahmed, S., & Ikram, S. (2016). Biosynthesis of gold nanoparticles: A green approach. *Journal of Photochemistry and Photobiology. B, Biology*, 161, 141–153. DOI: 10.1016/j.jphotobiol.2016.04.034 PMID: 27236049

Aithal, S., & Aithal, P. S. (2024). Predictive Analysis of use of AI-Driven GPTs in Nanomaterials Research Breakthroughs in the 21st Century. [IJAEML]. *International Journal of Applied Engineering and Management Letters*, 8(2), 2. Advance online publication. DOI: 10.47992/IJAEML.2581.7000.0226

Ajaykumar, A. P., Mathew, A., Chandni, A. P., Varma, S. R., Jayaraj, K. N., Sabira, O., Rasheed, V. A., Binitha, V. S., Swaminathan, T. R., Basheer, V. S., Giri, S., & Chatterjee, S. (2023). Green Synthesis of Silver Nanoparticles Using the Leaf Extract of the Medicinal Plant, Uvaria narum and Its Antibacterial, Antiangiogenic, Anticancer and Catalytic Properties. *Antibiotics (Basel, Switzerland)*, 12(3), 564. DOI: 10.3390/antibiotics12030564 PMID: 36978431

Akash, M. S. H., Rehman, K., Akash, M. S. H., & Rehman, K. (2020). Ultraviolet-visible (UV-VIS) spectroscopy. *Essentials of pharmaceutical analysis*, 29-56.

Akbari, B., Tavandashti, M. P., & Zandrahimi, M. (2011). Particle size characterization of nanoparticles-a practicalapproach. *Iranian Journal of Materials Science and Engineering*, 8(2), 48–56.

Akhtar, K., Khan, S. A., Khan, S. B., & Asiri, A. M. (2018). Scanning Electron Microscopy: Principle and Applications in Nanomaterials Characterization. *Handbook of Materials Characterization*, 113–145. DOI: 10.1007/978-3-319-92955-2_4

Akhtar, M. S., Panwar, J., & Yun, Y. S. (2013). Biogenic synthesis of metallic nanoparticles by plant extracts. *ACS Sustainable Chemistry & Engineering*, 1(6), 591–602. DOI: 10.1021/sc300118u

Akhter, M. S., Rahman, M. A., Ripon, R. K., Mubarak, M., Akter, M., Mahbub, S., Al Mamun, F., & Sikder, M. T. (2024). A systematic review on green synthesis of silver nanoparticles using plants extract and their bio-medical applications. *Heliyon*, 10(11), e29766. DOI: 10.1016/j.heliyon.2024.e29766 PMID: 38828360

Al Jabri, H., Saleem, M. H., Rizwan, M., Hussain, I., Usman, K., & Alsafran, M. (2022). Zinc oxide nanoparticles and their biosynthesis: Overview. *Life (Chicago, Ill.)*, 12(4), 594. DOI: 10.3390/life12040594 PMID: 35455085

Al-Agel, F. A., Suleiman, J., & Khan, S. A. (2017). Studies on silicon quantum dots prepared at different working pressure. *Results in Physics*, 7, 1128–1134. DOI: 10.1016/j.rinp.2017.03.002

Alaghmandfard, A., Sedighi, O., Rezaei, N. T., Abedini, A. A., Khachatourian, A. M., Toprak, M. S., & Seifalian, A. (2021). Recent advances in the modification of carbon-based quantum dots for biomedical applications. *Materials Science and Engineering C*, 120, 111756. DOI: 10.1016/j.msec.2020.111756 PMID: 33545897

Alamdari, S. G., Amini, M., Jalilzadeh, N., Baradaran, B., Mohammadzadeh, R., Mokhtarzadeh, A., & Oroojalian, F. (2022). Recent advances in nanoparticle-based photothermal therapy for breast cancer. *Journal of Controlled Release*, 349, 269–303. DOI: 10.1016/j.jconrel.2022.06.050 PMID: 35787915

Alam, S., Israr, J., & Kumar, A. (2024). Artificial Intelligence and Machine Learning in Bioinformatics. In Singh, V., & Kumar, A. (Eds.), *Advances in Bioinformatics* (pp. 321–345). Springer Nature., DOI: 10.1007/978-981-99-8401-5_16

Alarfaj, A. A. (2019). Antibacterial effect of chitosan nanoparticles against food spoilage bacteria. *Journal of Pure & Applied Microbiology*, 13(2), 1273–1278. DOI: 10.22207/JPAM.13.2.70

Albeladi, S., Malik, M., & Al-thabaiti, S. (2020). Facile biofabrication of silver nanoparticles using Salvia officinalis leaf extract and its catalytic activity towards Congo red dye degradation. *Journal of Materials Research and Technology*, 9(5), 10031–10044. DOI: 10.1016/j.jmrt.2020.06.074

Alghamdi, M. A., Fallica, A. N., Virzì, N., Kesharwani, P., Pittalà, V., & Greish, K. (2022). The Promise of Nanotechnology in Personalized Medicine. *Journal of Personalized Medicine*, 12(5), 5. Advance online publication. DOI: 10.3390/jpm12050673 PMID: 35629095

Alghoraibia, I., Soukkarieh, C., Zein, R., Alahmad, A., Walter, J. G., & Daghestani, M. (2019). Aqueous extract of *Eucalyptus camaldulensis* leaves as reducing and capping agent in green synthesis of silver nanoparticles. *Current Nanomaterials*, 4(1). Advance online publication. DOI: 10.2174/24054615046 66190625121807

Alghoraibi, I., Soukkarieh, C., Zein, R., Alahmad, A., Walter, J. G., & Daghestani, M. (2020). Aqueous extract of *Eucalyptus camaldulensis* leaves as reducing and capping agent in biosynthesis of silver nanoparticles. *Inorganic and Nano-Metal Chemistry*, 50(10), 895–902. DOI: 10.1080/24701556.2020.1728315

Alherek, M., & Basu, O. D. (2023). Impact of low levels of silver, zinc and copper nanoparticles on bacterial removal and potential synergy in water treatment applications. In Journal of Chemical Technology & Biotechnology (Vol. 98, Issue 5, pp. 1137–1146). Wiley. DOI: 10.1002/jctb.7318

Ali, K., Ahmed, B., Dwivedi, S., Saquib, Q., Al-Khedhairy, A. A., & Musarrat, J. (2015). Microwave accelerated green synthesis of stable silver nanoparticles with *Eucalyptus globulus* leaf extract and their antibacterial and antibiofilm activity on clinical isolates. *PLoS One*, 10(7), e0131178. DOI: 10.1371/journal.pone.0131178 PMID: 26132199

Ali, N. H., & Mohammed, A. M. (2021). Biosynthesis and characterization of platinum nanoparticles using Iraqi Zahidi dates and evaluation of their biological applications. *Biotechnology Reports (Amsterdam, Netherlands)*, 30, e00635. DOI: 10.1016/j.btre.2021.e00635 PMID: 34094893

Aljabali, A. A. A., Akkam, Y., Al Zoubi, M. S., Al-Batayneh, K. M., Al-Trad, B., Abo Alrob, O., Alkilany, A., Benamara, M., & Evans, D. J. (2018). Synthesis of Gold Nanoparticles Using Leaf Extract of Ziziphus zizyphus and their Antimicrobial Activity. *Nanomaterials (Basel, Switzerland)*, 8(3), 174. DOI: 10.3390/nano8030174 PMID: 29562669

Aljabali, A. A. A., Obeid, M. A., Rezigue, M. M., Alqudah, A., Charbe, N. B., Chellappan, D. K., Mishra, V., Pardhi, D. M., Dureja, H., Gupta, G., Prasher, P., Dua, K., Barhoum, A., & Tambuwala, M. M. (2020). Nanocelluloses as a Novel Vehicle for Controlled Drug Delivery. In Barhoum, A. (Ed.), *Handbook of Nanocelluloses: Classification, Properties, Fabrication, and Emerging Applications* (pp. 1–37). Springer International Publishing., DOI: 10.1007/978-3-030-62976-2_36-1

Al-Juthery, WA, H., & MN Al-Shami, Q. (2019). The effect of fertigation with nano NPK fertilizers on some parameters of growth and yield of potato (Solanum tuberosum L.). *AL-Qadisiyah. Journal for Agriculture Sciences*, 9(2), 225–232.

Allan, J., Belz, S., Hoeveler, A., Hugas, M., Okuda, H., Patri, A., Rauscher, H., Silva, P., Slikker, W., Sokull-Kluettgen, B., Tong, W., & Anklam, E. (2021). Regulatory landscape of nanotechnology and nanoplastics from a global perspective. *Regulatory Toxicology and Pharmacology*, 122, 104885. DOI: 10.1016/j.yrtph.2021.104885 PMID: 33617940

Alomar, T. S., AlMasoud, N., Awad, M. A., AlOmar, R. S., Merghani, N. M., El-Zaidy, M., & Bhattarai, A. (2024). Designing Green Synthesis-Based Silver Nanoparticles for Antimicrobial Theranostics and Cancer Invasion Prevention. *International Journal of Nanomedicine*, 19, 4451–4464. DOI: 10.2147/IJN.S440847 PMID: 38799694

Alsaiari, N. S., Alzahrani, F. M., Amari, A., Osman, H., Harharah, H. N., Elboughdiri, N., & Tahoon, M. A. (2023). Plant and Microbial Approaches as Green Methods for the Synthesis of Nanomaterials: Synthesis, Applications, and Future Perspectives. *Molecules (Basel, Switzerland)*, 28(1), 463. DOI: 10.3390/molecules28010463 PMID: 36615655

Alshameri, A. W., & Owais, M. (2022). Antibacterial and cytotoxic potency of the plant-mediated synthesis of metallic nanoparticles Ag NPs and ZnO NPs: A review. In *OpenNano* (Vol. 8, p. 100077). Elsevier BV., DOI: 10.1016/j.onano.2022.100077

Alshammari, S. O., Mahmoud, S. Y., & Farrag, E. S. (2023). Synthesis of Green Copper Nanoparticles Using Medicinal Plant Krameria sp. Root Extract and Its Applications. *Molecules (Basel, Switzerland)*, 28(12), 4629. DOI: 10.3390/molecules28124629 PMID: 37375184

Al-Sheddi, E. S., Farshori, N. N., Al-Oqail, M. M., Al-Massarani, S. M., Saquib, Q., Wahab, R., & Siddiqui, M. A. (2018). Anticancer potential of green synthesized silver nanoparticles using extract of *Nepeta deflersiana*against human cervical cancer cells (HeLA). *Bioinorganic Chemistry and Applications*, 2018(1), 9390784. PMID: 30515193

Altammar, K. A. (2023). A review on nanoparticles: Characteristics, synthesis, applications, and challenges. *Frontiers in Microbiology*, 14(04), 1–20. DOI: 10.3389/fmicb.2023.1155622 PMID: 37180257

Alvand, Z. M., Rajabi, H. R., Mirzaei, A., Masoumiasl, A., & Sadatfaraji, H. (2019). Rapid and green synthesis of cadmium telluride quantum dots with low toxicity based on a plant-mediated approach after microwave and ultrasonic assisted extraction: Synthesis, characterization, biological potentials and comparison study. *Materials Science and Engineering C*, 98, 535–544. DOI: 10.1016/j.msec.2019.01.010 PMID: 30813055

Álvarez-Chimal, R., & Arenas-Alatorre, J. Á. (2023). Green synthesis of nanoparticles. A biological approach. In *Green Chemistry for Environmental Sustainability*. IntechOpen.

Álvarez-Chimal, R., & Arenas-Alatorre, J. Á. (2023). *Green Synthesis of Nanoparticles: A Biological Approach* (Shah, K. J., Ed.). IntechOpen.

Al-Zahrani, S., Astudillo-Calderón, S., Pintos, B., Pérez-Urria, E., Manzanera, J. A., Martín, L., & Gomez-Garay, A. (2021). Role of Synthetic Plant Extracts on the Production of Silver-Derived Nanoparticles. *Plants*, 10(8), 1671. DOI: 10.3390/plants10081671 PMID: 34451715

Amendola, V., Meneghetti, M., Granozzi, G., Agnoli, S., Polizzi, S., Riello, P., & Colombatti, M. (2011). Top-down synthesis of multifunctional iron oxide nanoparticles for macrophage labelling and manipulation. *Journal of Materials Chemistry*, 21(11), 3803–3813. DOI: 10.1039/c0jm03863f

Amiri, M., Khazaeli, P., Salehabadi, A., & Salavati-Niasari, M. (2021). Hydrogel beads-based nanocomposites in novel drug delivery platforms: Recent trends and developments. *Advances in Colloid and Interface Science*, 288, 102316. DOI: 10.1016/j.cis.2020.102316 PMID: 33387892

Amiri, N., Afsharmanesh, M., Salarmoini, M., Meimandipour, A., Hosseini, S., & Ebrahimnejad, H. (2020). Effects of nanoencapsulated cumin essential oil as an alternative to the antibiotic growth promoter in broiler diets. *Journal of Applied Poultry Research*, 29(4), 875–885. DOI: 10.1016/j.japr.2020.08.004

Amjad, R., Mubeen, B., & Syed, S. (2021). Green Synthesis and Characterization of Copper Nanoparticles Using Fortunella margarita Leaves. *Polymers*, 13(24), 4364. DOI: 10.3390/polym13244364 PMID: 34960915

Amos, A., Alvan, A., & Florence, A. (2020). The Anti-nutritional Effect of Phytate on Zinc, Iron and Calcium Bioavailabilities of Some Cereals Staple Foods in Zaria, Nigeria. *European Journal of Nutrition & Food Safety*, 12(1), 1–6. DOI: 10.9734/ejnfs/2020/v12i130178

Anandan, M., & Gurumallesh Prabu, H. (2018). Dodonaea viscosa Leaf Extract Assisted Synthesis of Gold Nanoparticles: Characterization and Cytotoxicity Against A549 NSCLC Cancer Cells. *Journal of Inorganic and Organometallic Polymers and Materials*, 28(3), 932–941. DOI: 10.1007/s10904-018-0799-6

Anbazhagan, M., Ramachandran, S., Subramanian, P., Nachimuthu, R., & Kodiveri Muthukaliannan, G. (2016). *Nanomaterials: Classification* (Vol. 23). Biological Synthesis and Characterization., DOI: 10.1007/978-3-319-48009-1_2

Anbu, P., Gopinath, S., & Jayanthi, S. (2020). Synthesis of gold nanoparticles using Platycodon grandiflorum extract and its antipathogenic activity under optimal conditions. *Nanomaterials and Nanotechnology*, 10, 1–9. DOI: 10.1177/1847980420961697

Andra, S., Balu, S. K., Jeevanandham, J., Muthalagu, M., Vidyavathy, M., Chan, Y. S., & Danquah, M. K. (2019). Phytosynthesized metal oxide nanoparticles for pharmaceutical applications. *Naunyn-Schmiedeberg's Archives of Pharmacology*, 392(7), 755–771. DOI: 10.1007/s00210-019-01666-7 PMID: 31098696

Andrew, D. P., Pedersen, P. M., & McEvoy, C. D. (2019). Research methods and design in sport management. *Human Kinetics*. https://books.google.co.in/books?id=8rG8DwAAQBAJ

Angel Ezhilarasi, A., Judith Vijaya, J., Kaviyarasu, K., John Kennedy, L., Ramalingam, R. J., & Al-Lohedan, H. A. (2018). Green synthesis of NiO nanoparticles using Aegle marmelos leaf extract for the evaluation of in-vitro cytotoxicity, antibacterial and photocatalytic properties. *Journal of Photochemistry and Photobiology. B, Biology*, 180, 39–50. DOI: 10.1016/j.jphotobiol.2018.01.023 PMID: 29413700

Ansari, J. A., Khan, A. R., Ahmad, M. K., Fatima, N., Rastogi, N., Khan, M. F., Jilani, H., & Mahdi, A. A. (2016). Phytochemicals, antioxidant and antiproliferative studies of some medicinal plants from Indian sub-continent. *British Journal of Pharmaceutical Research*, 11(6), 1–11. DOI: 10.9734/BJPR/2016/26198

Antonio-Pérez, A., Durán-Armenta, L. F., Pérez-Loredo, M. G., & Torres-Huerta, A. L. (2023). Biosynthesis of Copper Nanoparticles with Medicinal Plants Extracts: From Extraction Methods to Applications. *Micromachines*, 14(10), 1882. DOI: 10.3390/mi14101882 PMID: 37893319

Apak, R., Ozyurek, M., Guclu, K., & Capanoglu, E. (2016). Antioxidant activity/capacity measurement. 3. Reactive oxygen and nitrogen species (ROS/RNS) scavenging assays, oxidative stress biomarkers, and chromatographic/chemometric assays. *Journal of Agricultural and Food Chemistry*, 64(5), 1046–1070. DOI: 10.1021/acs.jafc.5b04744 PMID: 26689748

Arachchi, S. S., Palma, S. P., Sanders, C. I., Xu, H., Ghosh Biswas, R., Soong, R., Simpson, A. J., & Casabianca, L. B. (2022). Binding Between Antibiotics and Polystyrene Nanoparticles Examined by NMR. *ACS Environmental Au*, 3(1), 47–55. DOI: 10.1021/acsenvironau.2c00047 PMID: 36691656

Aranaz, I., Mengíbar, M., Harris, R., Paños, I., Miralles, B., Acosta, N., & Heras, Á. (2009). Functional characterization of chitin and chitosan. *Current Chemical Biology*, 3(2), 203–230.

Arik, N., Elcin, E., Tezcaner, A., & Oktem, H. A. (2023). Optimization of whole-cell bacterial bioreporter immobilization on electrospun cellulose acetate (CA) and polycaprolactone (PCL) fibers for arsenic detection. *Environmental Monitoring and Assessment*, 195(6), 666. DOI: 10.1007/s10661-023-11227-4 PMID: 37178337

Arik, N., Horzum, N., & Truong, Y. B. (2022). Development and Characterizations of Engineered Electrospun Bio-Based Polyurethane Containing Essential Oils. *Membranes*, 12(2), 1–16. DOI: 10.3390/membranes12020209 PMID: 35207129

Armendariz, V., Herrera, I., peralta-videa, J. R., Jose-yacaman, M., Troiani, H., Santiago, P., & Gardea-Torresdey, J. L. (2004). Size controlled gold nanoparticle formation by Avena sativa biomass: Use of plants in nanobiotechnology. *Journal of Nanoparticle Research*, 6(4), 377–382. DOI: 10.1007/s11051-004-0741-4

Arockia John Paul, J., Karunai Selvi, B., & Karmegam, N. (2015). Biosynthesis of silver nanoparticles from Premnaserratifolia L. leaf and its anticancer activity in CCl 4-induced hepato-cancerous Swiss albino mice. *Applied Nanoscience*, 5(8), 937–944. DOI: 10.1007/s13204-014-0397-z

Arslan, O., Belkoura, L., & Mathur, S. (2015). Swift synthesis, functionalization and phase-transfer studies of ultrastable, visible light emitting oleate@ ZnO quantum dots. *Journal of Materials Chemistry. C, Materials for Optical and Electronic Devices*, 3(45), 11965–11973. DOI: 10.1039/C5TC03377B

Asadishad, B., Chahal, S., Akbari, A., Cianciarelli, V., Azodi, M., Ghoshal, S., & Tufenkji, N. (2018). Amendment of agricultural soil with metal nanoparticles: Effects on soil enzyme activity and microbial community composition. *Environmental Science & Technology*, 52(4), 1908–1918. DOI: 10.1021/acs.est.7b05389 PMID: 29356510

Asif, M., Yasmin, R., Asif, R., Ambreen, A., Mustafa, M., & Umbreen, S. (2022). Green Synthesis of Silver Nanoparticles (AgNPs), Structural Characterization, and their Antibacterial Potential. *Dose-Response*, 20(2), 15593258221088708. DOI: 10.1177/15593258221088709 PMID: 35592270

Asli, S., & Neumann, P. M. (2009). Colloidal suspensions of clay or titanium dioxide nanoparticles can inhibit leaf growth and transpiration via physical effects on root water transport. *Plant, Cell & Environment*, 32(5), 577–584. DOI: 10.1111/j.1365-3040.2009.01952.x PMID: 19210640

Asmathunisha, N., & Kathiresan, K. (2013). A review on biosynthesis of nanoparticles by marine organisms. *Colloids and Surfaces. B, Biointerfaces*, 103, 283–287. DOI: 10.1016/j.colsurfb.2012.10.030 PMID: 23202242

Aubert, T., Burel, A., Esnault, M. A., Cordier, S., Grasset, F., & Cabello-Hurtado, F. (2012). Root uptake and phytotoxicity of nanosized molybdenum octahedral clusters. *Journal of Hazardous Materials*, 219, 111–118. DOI: 10.1016/j.jhazmat.2012.03.058 PMID: 22510541

Avadi, M. R., Sadeghi, A. M. M., Mohammadpour, N., Abedin, S., Atyabi, F., Dinarvand, R., & Rafiee-Tehrani, M. (2010). Preparation and characterization of insulin nanoparticles using chitosan and Arabic gum with ionic gelation method. *Nanomedicine; Nanotechnology, Biology, and Medicine*, 6(1), 58–63. DOI: 10.1016/j.nano.2009.04.007 PMID: 19447202

Avasthi, A., Caro, C., Pozo-Torres, E., Leal, M. P., & García-Martín, M. L. (2020). Magnetic nanoparticles as MRI contrast agents. *Topics in Current Chemistry*, 378(3), 1–40. DOI: 10.1007/s41061-020-00302-w PMID: 32382832

Avellan, A., Yun, J., Morais, B. P., Clement, E. T., Rodrigues, S. M., & Lowry, G. V. (2021). Critical review: Role of inorganic nanoparticle properties on their foliar uptake and in planta translocation. *Environmental Science & Technology*, 55(20), 13417–13431. DOI: 10.1021/acs.est.1c00178 PMID: 33988374

Aydın Acar, Ç. (2021). Green Synthesis Of Zinc Oxide Nanoparticles Using Aqueous Extract Of Achiella Millefolium L.: In Vitro Anti-Cancer Potential On Lung And Colon Cancer Cells. *Turkish Journal of Health Science and Life*, 4, 40–45.

Aziz, M., & Ismail, A. F. (2017). X-ray photoelectron spectroscopy (XPS). In *Membrane Characterization* (pp. 81-93), Elsevier.

Babick, F. (2020). Dynamic light scattering (DLS). V. Hodoroaba (Ed.), *Characterization of Nanoparticles* (pp. 137-172). Elsevier Inc. DOI: 10.1016/B978-0-12-814182-3.00010-9

Babu, S., Singh, R., Yadav, D., Rathore, S. S., Raj, R., Avasthe, R., & Singh, V. K. (2022). Nanofertilizers for agricultural and environmental sustainability. *Chemosphere*, 292, 133451. DOI: 10.1016/j.chemosphere.2021.133451 PMID: 34973251

Baer, D. R., Amonette, J. E., Engelhard, M. H., Gaspar, D. J., Karakoti, A. S., Kuchibhatla, S., Nachimuthu, P., Nurmi, J. T., Qiang, Y., Sarathy, V., Seal, S., Sharma, A., Tratnyek, P. G., & Wang, C. M. (2008). Characterization challenges for nanomaterials. *Surface and Interface Analysis*, 40(3-4), 529–537. DOI: 10.1002/sia.2726

Baer, D. R., Engelhard, M. H., Johnson, G. E., Laskin, J., Lai, J., Mueller, K., Munusamy, P., Thevuthasan, S., Wang, H., Washton, N., Elder, A., Baisch, B. L., Karakoti, A., Kuchibhatla, S., & Moon, D. (2013). Surface characterization of nanomaterials and nanoparticles: Important needs and challenging opportunities. *Journal of Vacuum Science & Technology. A, Vacuum, Surfaces, and Films*, 31(5), 050820. DOI: 10.1116/1.4818423 PMID: 24482557

Bagheri, A. R., Aramesh, N., Hasnain, M. S., Nayak, A. K., & Varma, R. S. (2023). Greener fabrication of metal nanoparticles using plant materials: A review. *Chemical Physics Impact*, 7, 100255. DOI: 10.1016/j.chphi.2023.100255

Bahrulolum, H., Nooraei, S., Javanshir, N., Tarrahimofrad, H., Mirbagheri, V. S., Easton, A. J., & Ahmadian, G. (2021). Green synthesis of metal nanoparticles using microorganisms and their application in the agrifood sector. *Journal of Nanobiotechnology*, 19(1), 1–26. DOI: 10.1186/s12951-021-00834-3 PMID: 33771172

Baig, N., Kammakakam, I., & Falath, W. (2021). Nanomaterials: A review of synthesis methods, properties, recent progress, and challenges. *Materials Advances*, 2(6), 1821–1871. DOI: 10.1039/D0MA00807A

Bakar, N. F. A., Tan, H. L., Lim, Y. P., Adrus, N., & Abdullah, J. (2022). Environmental impact of quantum dots. In *Graphene, Nanotubes and Quantum Dots-Based Nanotechnology* (pp. 837-867). Woodhead Publishing.

Baker, S., Satish, S., & Mohan, S. (2017). Review on the synthesis of nanoparticles using biological methods and its application as antimicrobial agents and anticancer agents. *Journal of Biotechnology & Biomaterials*, 7(1), 1–9.

Balamurugan, M., & Saravanan, S. (2017). Green synthesis of silver nanoparticles by using *Eucalyptus globulus* leaf extract. Journal of The Institution of Engineers (India): Series A, 98, 461-467.

Balamurugan, M., Saravanan, S., & Soga, T. (2014). Synthesis of iron oxide nanoparticles by using *Eucalyptus globulus* plant extract. *E-Journal of Surface Science and Nanotechnology*, 12(0), 363–367. DOI: 10.1380/ejssnt.2014.363

Balaraman, P., Balasubramanian, B., Kaliannan, D., Durai, M., Kamyab, H., Park, S., Chelliapan, S., Lee, C. T., Maluventhen, V., & Maruthupandian, A. (2020). Phyco-synthesis of Silver Nanoparticles Mediated from Marine Algae Sargassum myriocystum and Its Potential Biological and Environmental Applications. *Waste and Biomass Valorization*, 11(10), 5255–5271. DOI: 10.1007/s12649-020-01083-5

Balasubramanian, S., Kala, S. M. J., & Pushparaj, T. L. (2020). Biogenic synthesis of gold nanoparticles using Jasminum auriculatum leaf extract and their catalytic, antimicrobial and anticancer activities. *Journal of Drug Delivery Science and Technology*, 57, 101620. DOI: 10.1016/j.jddst.2020.101620

Balčiūnaitienė, A., Liaudanskas, M., Puzerytė, V., Viškelis, J., Janulis, V., Viškelis, P., & Jankauskaitė, V. (2022). *Eucalyptus globulus* and *Salvia officinalis* extracts mediated green synthesis of silver nanoparticles and their application as an antioxidant and antimicrobial agent. *Plants*, 11(8), 1085. DOI: 10.3390/plants11081085 PMID: 35448813

Ballauff, M., & Lu, Y. (2007). "Smart" nanoparticles: Preparation, characterization and applications. *Polymer*, 48(7), 1815–1823. DOI: 10.1016/j.polymer.2007.02.004

Banerjee, A., Sarkar, A., Acharya, K., & Chakraborty, N. (2021). Nanotechnology: An Emerging Hope in Crop Improvement. *Letters in Applied NanoBioScience*, 10(4), 2784–2803. DOI: 10.33263/LIANBS104.27842803

Banerjee, D., Rajput, D., Banerjee, S., & Saharan, V. A. (2022). Artificial Intelligence and Its Applications in Drug Discovery, Formulation Development, and Healthcare. In Saharan, V. A. (Ed.), *Computer Aided Pharmaceutics and Drug Delivery: An Application Guide for Students and Researchers of Pharmaceutical Sciences* (pp. 309–380). Springer Nature., DOI: 10.1007/978-981-16-5180-9_11

Banjara, R. A., Kumar, A., Aneshwari, R. K., Satnami, M. L., & Sinha, S. K. (2024). A comparative analysis of chemical vs green synthesis of nanoparticles and their various applications. *Environmental Nanotechnology, Monitoring &amp. Environmental Nanotechnology, Monitoring & Management*, 22, 100988. Advance online publication. DOI: 10.1016/j.enmm.2024.100988

Bankar, A., Joshi, B., Kumar, A. R., & Zinjarde, S. (2010). Banana peel extract mediated novel route for the synthesis of palladium nanoparticles. *Materials Letters*, 64(18), 1951–1953. DOI: 10.1016/j.matlet.2010.06.021

Bansal, T., Mustafa, G., Khan, Z. I., Ahmad, F. J., Khar, R. K., & Talegaonkar, S. (2008). Solid self-nanoemulsifying delivery systems as a platform technology for formulation of poorly soluble drugs. *Critical Reviews™ in Therapeutic Drug Carrier Systems, 25*(1), 63-116.

Banu, A. N., Kudesia, N., Raut, A. M., Pakrudheen, I., & Wahengbam, J. (2021). Toxicity, bioaccumulation, and transformation of silver nanoparticles in aqua biota: A review. *Environmental Chemistry Letters*, 19(6), 4275–4296. DOI: 10.1007/s10311-021-01304-w

Bao, Y., He, J., Song, K., Guo, J., Zhou, X., & Liu, S. (2021). Plant-Extract-Mediated Synthesis of Metal Nanoparticles. *Journal of Chemistry*, 2021, 14. DOI: 10.1155/2021/6562687

Baranwal, J., Barse, B., Fais, A., Delogu, G. L., & Kumar, A. (2022). Biopolymer: A Sustainable Material for Food and Medical Applications. *Polymers*, 14(5), 983. DOI: 10.3390/polym14050983 PMID: 35267803

Barnes, S. R., James, A. M., & Jamieson, G. (1985). The organisation, nucleotide sequence, and chromosomal distribution of a satellite DNA from Allium cepa. *Chromosoma*, 92(3), 185–192. DOI: 10.1007/BF00348692

Barthold, S., Kletting, S., Taffner, J., de Souza Carvalho-Wodarz, C., Lepeltier, E., Loretz, B., & Lehr, C. M. (2016). Preparation of nanosized coacervates of positive and negative starch derivatives intended for pulmonary delivery of proteins. *Journal of Materials Chemistry. B*, 4(13), 2377–2386. DOI: 10.1039/C6TB00178E PMID: 32263233

Baruah, K., Haque, M., Langbang, L., Das, S., Aguan, K., & Singha Roy, A. (2021). *Ocimum sanctum* mediated green synthesis of silver nanoparticles: A biophysical study towards lysozyme binding and anti-bacterial activity. *Journal of Molecular Liquids*, 337, 116422. DOI: 10.1016/j.molliq.2021.116422

Barve, M., Mashru, M., Jagtap, C., Patgiri, B. J., & Prajapati, P. K. (2011). Therapeutic potentials of metals in ancient India: A review through Charaka Samhita. *Journal of Ayurveda and Integrative Medicine*, 2(2), 55. DOI: 10.4103/0975-9476.82523 PMID: 21760689

Barzinjy, A. A., & Azeez, H. H. (2020). Green synthesis and characterization of zinc oxide nanoparticles using *Eucalyptus globulus* Labill. leaf extract and zinc nitrate hexahydrate salt. *SN Applied Sciences*, 2(5), 991. DOI: 10.1007/s42452-020-2813-1

Basavegowda, N., & Baek, K. (2020). Advances in Functional Biopolymer-Based Nanocomposites for Active Food Packaging Applications. *Polymers*, 13(23), 4198. DOI: 10.3390/polym13234198 PMID: 34883701

Bashir, T., & Qureshi, M. Z. (2015). Phytosynthesis of silver nanoparticles using *E. Camaldulensis* leaf extract and their characterization. *Journal of the Chilean Chemical Society*, 60(1), 2861–2863. DOI: 10.4067/S0717-97072015000100019

Basu, A., Ray, S., Chowdhury, S., Sarkar, A., Mandal, D. P., Bhattacharjee, S., & Kundu, S. (2018). Evaluating the antimicrobial, apoptotic, and cancer cell gene delivery properties of protein- capped gold nanoparticles synthesized from the edible mycorrhizal fungus Tricholomacrassum. *Nanoscale Research Letters*, 13(1), 1–16. DOI: 10.1186/s11671-018-2561-y PMID: 29299709

Batool, M., Khurshid, S., Daoush, W. M., Siddique, S. A., & Nadeem, T. (2021). Green synthesis and biomedical applications of ZnO nanoparticles: Role of PEGylated-ZnO nanoparticles as doxorubicin drug carrier against MDA-MB-231 (TNBC) cells line. *Crystals*, 11(4), 344. DOI: 10.3390/cryst11040344

Batool, S., Batool, F., Zia, A., & Saeed, R. (2018). The struggle for women rights: A study of the emergence of feminism in Pakistan,(1947 to 2010). *Journal of the Punjab University Historical Society*, 31(1), 211–221.

Bayda, S., Adeel, M., Tuccinardi, T., Cordani, M., & Rizzolio, F. (2019). The history of nanoscience and nanotechnology: From chemical–physical applications to nanomedicine. *Molecules (Basel, Switzerland)*, 25(1), 112. DOI: 10.3390/molecules25010112 PMID: 31892180

Bealer, E. J., Onissema-Karimu, S., Rivera-Galletti, A., Francis, M., Wilkowski, J., Salas-de la Cruz, D., & Hu, X. (2020). Protein–Polysaccharide Composite Materials: Fabrication and Applications. *Polymers*, 12(2), 464. DOI: 10.3390/polym12020464 PMID: 32079322

Beatty, P. H., & Lewis, J. D. (2019). Cowpea mosaic virus nanoparticles for cancer imaging and therapy. *Advanced Drug Delivery Reviews*, 145, 130–144. DOI: 10.1016/j.addr.2019.04.005 PMID: 31004625

Becheri, A., Dürr, M., Lo Nostro, P., & Baglioni, P. (2008). Synthesis and characterization of zinc oxide nanoparticles: Application to textiles as UV-absorbers. *Journal of Nanoparticle Research*, 10(4), 679–689. DOI: 10.1007/s11051-007-9318-3

Bedlovičová, Z., Strapáč, I., Baláž, M., & Salayová, A. (2020). A brief overview on antioxidant activity determination of silver nanoparticles. *Molecules (Basel, Switzerland)*, 25(14), 3191. DOI: 10.3390/molecules25143191 PMID: 32668682

Begum, R., Farooqi, Z. H., Naseem, K., Ali, F., Batool, M., Xiao, J., & Irfan, A. (2018). Applications of UV/Vis Spectroscopy in Characterization and Catalytic Activity of Noble Metal Nanoparticles Fabricated in Responsive Polymer Microgels: A Review. *Critical Reviews in Analytical Chemistry*, 48(6), 503–516. DOI: 10.1080/10408347.2018.1451299 PMID: 29601210

Bellotti, R., Picotto, G. B., & Ribotta, L. (2022). AFM Measurements and Tip Characterization of Nanoparticles with Different Shapes. *Nanomanufacturing and Metrology*, 5(2), 127–138. DOI: 10.1007/s41871-022-00125-x

Benício, L. P. F., Constantino, V. R. L., Pinto, F. G., Vergütz, L., Tronto, J., & da Costa, L. M. (2017). Layered double hydroxides: New technology in phosphate fertilizers based on nanostructured materials. *ACS Sustainable Chemistry & Engineering*, 5(1), 399–409. DOI: 10.1021/acssuschemeng.6b01784

Benítez-Martínez, S., López-Lorente, Á. I., & Valcárcel, M. (2016). Determination of TiO 2 nanoparticles in sunscreen using N-doped graphene quantum dots as a fluorescent probe. *Mikrochimica Acta*, 183(2), 781–789. DOI: 10.1007/s00604-015-1696-0

Benjin, X., & Ling, L. (2020). Developments, applications, and prospects of cryo-electron microscopy. *Protein Science*, 29(4), 872–882. DOI: 10.1002/pro.3805 PMID: 31854478

Berekaa, M. M. (2015). Nanotechnology in food industry; advances in food processing, packaging and food safety. *International Journal of Current Microbiology and Applied Sciences*, 4(5), 345–357.

Bergman, M. E., Davis, B., & Phillips, M. A. (2019). Medically Useful Plant Terpenoids: Biosynthesis, Occurrence, and Mechanism of Action. *Molecules (Basel, Switzerland)*, 24(21), 3961. Advance online publication. DOI: 10.3390/molecules24213961 PMID: 31683764

Bernard, M., Jubeli, E., Pungente, M. D., & Yagoubi, N. (2018). Biocompatibility of polymer-based biomaterials and medical devices – regulations, in vitro screening and risk-management. *Biomaterials Science*, 6(8), 2025–2053. DOI: 10.1039/C8BM00518D PMID: 29968869

Berta, L., Coman, N.-A., Rusu, A., & Tanase, C. (2021). A review on plant-mediated synthesis of bimetallic nanoparticles, characterization, and their biological applications. *Materials (Basel)*, 14(24), 7677. DOI: 10.3390/ma14247677 PMID: 34947271

Berthomieu, C., & Hienerwadel, R. (2009). Fourier transform infrared (FTIR) spectroscopy. *Photosynthesis Research*, 101(2–3), 157–170. DOI: 10.1007/s11120-009-9439-x PMID: 19513810

Bharali, A., Deka, B., Sahu, B. P., & Laloo, D. (2023). Major challenges and probable scientific solutions toward the large-scale production of plant-based metallic nanoparticles: A systematic review. *Nanotechnology for Environmental Engineering*, 8(4), 933–941. DOI: 10.1007/s41204-023-00347-4

Bhatia, D., Mittal, A., & Malik, D. K. (2021). Antimicrobial potential and in vitro cytotoxicity study of polyvinyl pyrollidone-stabilised silver nanoparticles synthesised from Lysinibacillus boronitolerans. *IET Nanobiotechnology / IET*, 15(4), 427–440. DOI: 10.1049/nbt2.12054 PMID: 34694715

Bhatia, S., Pooja, , & Yadav, S. K. (2023). CRISPR-Cas for genome editing: Classification, mechanism, designing and applications. *International Journal of Biological Macromolecules*, 238, 124054. DOI: 10.1016/j.ijbiomac.2023.124054 PMID: 36933595

Bhattacharjee, S. (2016). DLS and zeta potential - What they are and what they are not. *Journal of Controlled Release*, 235, 337–351. DOI: 10.1016/j.jconrel.2016.06.017 PMID: 27297779

Bhoye, M., Pansambal, S., Basnet, P., Lin, K.-Y. A., Gutierrez-Mercado, K. Y., Pérez-Larios, A., Chauhan, A., Oza, R., & Ghotekar, S. (2023). Eco-Friendly Synthesis of Ni/NiO Nanoparticles Using Gymnema sylvestre Leaves Extract for Antifungal Activity. *Journal of Composites Science*, 7(3), 105. DOI: 10.3390/jcs7030105

Bhuvaneswari, G., & Radjarejesri, S. (2015). Green synthesis and characterization of CdS quantum dots. *International Journal of Chemtech Research*, 8(5), 104–108.

Bian, K., Gerber, C., Heinrich, A. J., Müller, D. J., Scheuring, S., & Jiang, Y. (2021). Scanning probe microscopy. *Nature Reviews. Methods Primers*, 1(1), 36. DOI: 10.1038/s43586-021-00033-2

Biswas, M. C., Jony, B., Nandy, P. K., Chowdhury, R. A., Halder, S., Kumar, D., Ramakrishna, S., Hassan, M., Ahsan, M. A., Hoque, M. E., & Imam, M. A. (2022). Recent Advancement of Biopolymers and Their Potential Biomedical Applications. *Journal of Polymers and the Environment*, 30(1), 51–74. DOI: 10.1007/s10924-021-02199-y

Bloch, K., Pardesi, K., Satriano, C., & Ghosh, S. (2021). Bacteriogenic Platinum Nanoparticles for Application in Nanomedicine. *Frontiers in Chemistry*, 9, 624344. DOI: 10.3389/fchem.2021.624344 PMID: 33763405

Bocker, U., Wubshet, S. G., Lindberg, D., & Afseth, N. K. (2017). Fourier-transform infrared spectroscopy for characterization of protein chain reductions in enzymatic reactions. *Analyst*, 142(15), 2812–2818. DOI: 10.1039/C7AN00488E PMID: 28686252

Bodade, R. G., Kumar, R., & Kutty, R. (2023). Crop plant-mediated nanoparticle synthesis and applications. In Bhunia, S., Kumar, B., Singh, P., Oraon, R., & Kim, K.-H. (Eds.), *Nanoparticles in Green Organic Synthesis* (pp. 351–399). Elsevier., DOI: 10.1016/B978-0-323-95921-6.00015-9

Bogner, A., Jouneau, P. H., Thollet, G., Basset, D., & Gauthier, C. (2007). A history of scanning electron microscopy developments: Towards "wet-STEM" imaging. *Micron (Oxford, England)*, 38(4), 390–401. DOI: 10.1016/j.micron.2006.06.008 PMID: 16990007

Bogunia-Kubik, K., & Sugisaka, M. (2002). From molecular biology to nanotechnology and nanomedicine. *Bio Systems*, 65(2-3), 123–138. DOI: 10.1016/S0303-2647(02)00010-2 PMID: 12069723

Bonde, S. R., Rathod, D. P., Ingle, A. P., Ade, R. B., Gade, A. K., & Rai, M. K. (2012). Murraya koenigii-mediated synthesis of silver nanoparticles and its activity against three human pathogenic bacteria. *Nanoscience Methods*, 1(1), 25–36. DOI: 10.1080/17458080.2010.529172

Bongomin, F., Gago, S., Oladele, R. O., & Denning, D. W. (2017). Global and Multi-National Prevalence of Fungal Diseases-Estimate Precision. *Journal of Fungi (Basel, Switzerland)*, 3(4), 57. DOI: 10.3390/jof3040057 PMID: 29371573

Bonnamy, S., & Oberlin, A. (2016). Materials Science and Engineering of Carbon. M. Inagaki (Ed.), *Transmission Electron Microscopy.* (pp. 45–70). Elsevier. DOI: 10.1016/B978-0-12-805256-3.00004-0

Borah, P. P., Das, P., & Badwaik, L. S. (2017). Ultrasound treated potato peel and sweet lime pomace-based biopolymer film development. *Ultrasonics Sonochemistry*, 36, 11–19. DOI: 10.1016/j.ultsonch.2016.11.010 PMID: 28069189

Borji, H., Ayoub, G., Bilbeisi, R., Nassar, N., & Malaeb, L. (2020). How Effective Are Nanomaterials for the Removal of Heavy Metals from Water and Wastewater? *Water, Air, and Soil Pollution*, 231(7), 330. DOI: 10.1007/s11270-020-04681-0

Borovaya, M., Horiunova, I., Plokhovska, S., Pushkarova, N., Blume, Y., & Yemets, A. (2021). Synthesis, properties and bioimaging applications of silver-based quantum dots. *International Journal of Molecular Sciences*, 22(22), 12202. DOI: 10.3390/ijms222212202 PMID: 34830084

Buarki, F., AbuHassan, H., Al Hannan, F., & Henari, F. Z. (2022). Green Synthesis of Iron Oxide Nanoparticles Using Hibiscus rosa sinensis Flowers and Their Antibacterial Activity. *Journal of Nanotechnology*, 2022, 1–6. DOI: 10.1155/2022/5474645

Buhr, E., Senftleben, N., Klein, T., Bergmann, D., Gnieser, D., Frase, C. G., & Bosse, H. (2009). Characterization of nanoparticles by scanning electron microscopy in transmission mode. *Measurement Science & Technology*, 20(8), 1–8. DOI: 10.1088/0957-0233/20/8/084025

Bukhari, A., Ijaz, I., Gilani, E., Nazir, A., Zain, H., Saeed, R., Alarfaji, S. S., Hussain, S., Aftab, R., & Naseer, Y. (2021). Green synthesis of metal and metal oxide nanoparticles using different plants' parts for antimicrobial activity and anticancer activity: A review article. *Coatings*, 11(11), 1374. DOI: 10.3390/coatings11111374

Bunaciu, A. A., Udriştioiu, E., & Aboul-Enein, H. Y. (2015). X-ray diffraction: Instrumentation and applications. *Critical Reviews in Analytical Chemistry*, 45(4), 289–299. DOI: 10.1080/10408347.2014.949616 PMID: 25831472

Burke, D. J., Pietrasiak, N., Situ, S. F., Abenojar, E. C., Porche, M., Kraj, P., & Samia, A. C. S. (2015). Iron oxide and titanium dioxide nanoparticle effects on plant performance and root associated microbes. *International Journal of Molecular Sciences*, 16(10), 23630–23650. DOI: 10.3390/ijms161023630 PMID: 26445042

Buseck, P. R. (2018). Principles of transmission electron microscopy. P. Buseck (Ed.), *Minerals and Reactions at the Atomic Scale* (pp. 1–36). De Gruyter. DOI: 10.1515/9781501509735-005

Buz, P. T., Duman, F. D., Erkisa, M., Demirci, G., Ari, F., Ulukaya, E., & Acar, H. Y. (2019). Development of near-infrared region luminescent N-acetyl-L-cysteine-coated Ag2S quantum dots with differential therapeutic effect. *Nanomedicine (London)*, 14(8), 969–987. DOI: 10.2217/nnm-2018-0214 PMID: 30917096

Cai, Z., Li, F., Wu, P., Ji, L., Zhang, H., Cai, C., & Gervasio, D. F. (2015). Synthesis of nitrogen-doped graphene quantum dots at low temperature for electrochemical sensing trinitrotoluene. *Analytical Chemistry*, 87(23), 11803–11811. DOI: 10.1021/acs.analchem.5b03201 PMID: 26545150

Calvo, P., Remunan-Lopez, C., Vila-Jato, J. L., & Alonso, M. J. (1997). Novel hydrophilic chitosan-polyethylene oxide nanoparticles as protein carriers. *Journal of Applied Polymer Science*, 63(1), 125–132. DOI: 10.1002/(SICI)1097-4628(19970103)63:1<125::AID-APP13>3.0.CO;2-4

Carpita, N. C., & Gibeaut, D. M. (1993). Structural models of primary cell walls in flowering plants: Consistency of molecular structure with the physical properties of the walls during growth. *The Plant Journal*, 3(1), 1–30. DOI: 10.1111/j.1365-313X.1993.tb00007.x PMID: 8401598

Castillo, F., Aguilar, C. N., Hernández, D., Gallegos, G., & Rodríguez, R. (2012). *Antifungal properties of bioactive compounds from plants*. INTECH Open Access Publisher., DOI: 10.5772/26598

Catalán, J., & Norppa, H. (2017). Safety aspects of bio-based nanomaterials. *Bioengineering (Basel, Switzerland)*, 4(4), 94. DOI: 10.3390/bioengineering4040094 PMID: 29194367

Catalano, P. N., Chaudhary, R. G., Desimone, M. F., & Santo-Orihuela, P. L. (2021). A survey on analytical methods for the characterization of green synthesized nanomaterials. *Current Pharmaceutical Biotechnology*, 22(6), 823–847. DOI: 10.2174/1389201022666210104122349 PMID: 33397235

Centeno, L., Romero-García, J., Alvarado-Canché, C., Gallardo-Vega, C., Télles-Padilla, G., Barriga-Castro, E. D., & de León, A. (2021). Green synthesis of graphene quantum dots from Opuntia sp. extract and their application in phytic acid detection. *Sensing and Bio-Sensing Research*, 32, 100412. DOI: 10.1016/j.sbsr.2021.100412

Cerletti, V., Coish, W. A., Gywat, O., & Loss, D. (2005). Recipes for spin-based quantum computing. *Nanotechnology*, 16(4), R27–R49. DOI: 10.1088/0957-4484/16/4/R01

Chacon, W. D. C., Verruck, S., Monteiro, A. R., & Valencia, G. A. (2023). The mechanism, biopolymers and active compounds for the production of nanoparticles by anti-solvent precipitation: A review. *Food Research International*, 168, 112728. DOI: 10.1016/j.foodres.2023.112728 PMID: 37120194

Chakrabarti, S., Guha, S., & Majumder, K. (2018). Food-derived bioactive peptides in human health: Challenges and opportunities. *Nutrients*, 10(11), 1738. DOI: 10.3390/nu10111738 PMID: 30424533

Champa-Bujaico, E., García-Díaz, P., & Díez-Pascual, A. M. (2022). *Machine Learning for Property Prediction and Optimization of Polymeric Nanocomposites: A State-of-the-Art*. https://agris.fao.org/search/en/providers/122436/records/647480702d3f560f80b14a2f

Chandrakala, V., Aruna, V., & Angajala, G. (2022). Review on metal nanoparticles as nanocarriers: Current challenges and perspectives in drug delivery systems. *Emergent Materials*, 5(6), 1593–1615. DOI: 10.1007/s42247-021-00335-x PMID: 35005431

Chandran, S. P., Chaudhary, M., Pasricha, R., Ahmad, A., & Sastry, M. (2006). Synthesis of Gold Nanotriangles and Silver Nanoparticles Using Aloevera Plant Extract. *Biotechnology Progress*, 22(2), 577–583. DOI: 10.1021/bp0501423 PMID: 16599579

Chandrasekaran, M., Kim, K. D., & Chun, S. C. (2020). Antibacterial activity of chitosan nanoparticles: A review. *Processes (Basel, Switzerland)*, 8(9), 1173. DOI: 10.3390/pr8091173

Chandrasekaran, R., Alagar Yadav, S., & Sivaperumal, S. (2020). Phytosynthesis and Characterization of Copper Oxide Nanoparticles using the Aqueous Extract of Beta vulgaris L and Evaluation of their Antibacterial and Anticancer Activities. *Journal of Cluster Science*, 31(1), 221–230. DOI: 10.1007/s10876-019-01640-6

Chandrasekaran, R., Gopalan, R., & Babu, S. (2020). Green Synthesis of Zinc Oxide Nanoparticles Using Tea Extract: Antibacterial Activity Against Staphylococcus aureus and Escherichia coli. *Journal of Nanobiotechnology*, 18(1), 10–18. DOI: 10.1186/s12951-020-00617-8 PMID: 31918721

Channab, B. E., Idrissi, A. E., Ammar, A., Dardari, O., Marrane, S. E., Gharrak, A. E., Akil, A., Essemlali, Y., & Zahouily, M. (2024). Recent advances in nano-fertilizers: Synthesis, crop yield impact, and economic analysis. *Nanoscale*, 16(9), 4484–4513. DOI: 10.1039/D3NR05012B PMID: 38314867

Chauhan, S. B. (2024). Current Status of Smart Nanoparticles in Drug Delivery. In Bhardwaj, A. K., Srivastav, A. L., & Rai, S. (Eds.), *Biogenic Wastes-Enabled Nanomaterial Synthesis: Applications in Environmental Sustainability* (pp. 351–372). Springer Nature Switzerland., DOI: 10.1007/978-3-031-59083-2_14

Cheirmadurai, K., Biswas, S., Murali, R., & Thanikaivelan, P. (2014). Green synthesis of copper nanoparticles and conducting nanobiocomposites using plant and animal sources. *RSC Advances*, 4(37), 19507–19511. DOI: 10.1039/c4ra01414f

Chen, C. J. (2021). *Introduction to Scanning Tunneling Microscopy* (3rd ed., Vol. 69). Oxford University Press. DOI: 10.1093/oso/9780198856559.001.0001

Chen, C., Guo, L., Chen, Y., Qin, P., & Wei, G. (2023). Pristine and sulfidized zinc oxide nanoparticles alter bacterial communities and metabolite profiles in soybean rhizocompartments. *The Science of the Total Environment*, 855, 158697. DOI: 10.1016/j.scitotenv.2022.158697 PMID: 36099947

Chen, F., Li, Y., Zia-ur-Rehman, M., Hussain, S. M., Qayyum, M. F., Rizwan, M., Alharby, H. F., Al-abdallah, N. M., Alharbi, B. M., & Ali, S. (2023). Combined effects of zinc oxide nanoparticles and melatonin on wheat growth, chlorophyll contents, cadmium (Cd), and zinc uptake under Cd stress. *The Science of the Total Environment*, 864, 161061. DOI: 10.1016/j.scitotenv.2022.161061 PMID: 36565889

Chen, G., Dong, S., Chen, Y., Gao, Y., Zhang, Z., Li, S., & Chen, Y. (2020). Complex coacervation of zein-chitosan via atmospheric cold plasma treatment: Improvement of encapsulation efficiency and dispersion stability. *Food Hydrocolloids*, 107, 105943. DOI: 10.1016/j.foodhyd.2020.105943

Chen, H., Luo, L., Fan, S., Xiong, Y., Ling, Y., & Peng, S. (2021). Zinc oxide nanoparticles synthesized from Aspergillus terreus induces oxidative stress-mediated apoptosis through modulating apoptotic proteins in human cervical cancer HeLa cells. *The Journal of Pharmacy and Pharmacology*, 73(2), 221–232. DOI: 10.1093/jpp/rgaa043 PMID: 33793807

Chen, J., Li, Y., Fang, G., Cao, Z., Shang, Y., Alfarraj, S., Ali Alharbi, S., Li, J., Yang, S., & Duan, X. (2021). Green synthesis, characterization, cytotoxicity, antioxidant, and anti-human ovarian cancer activities of Curcumae kwangsiensis leaf aqueous extract green-synthesized gold nanoparticles. *Arabian Journal of Chemistry*, 14(3), 103000. DOI: 10.1016/j.arabjc.2021.103000

Chen, S. C., & Sorrell, T. C. (2007). Antifungal agents. *The Medical Journal of Australia*, 187(7), 404–409. DOI: 10.5694/j.1326-5377.2007.tb01313.x PMID: 17908006

Chen, S., McClements, D. J., Jian, L., Han, Y., Dai, L., Mao, L., & Gao, Y. (2019). Core–shell biopolymer nanoparticles for co-delivery of curcumin and piperine: Sequential electrostatic deposition of hyaluronic acid and chitosan shells on the zein core. *ACS Applied Materials & Interfaces*, 11(41), 38103–38115. DOI: 10.1021/acsami.9b11782 PMID: 31509373

Chen, Y. T., Wang, Y., & Yeh, K. C. (2017). Role of root exudates in metal acquisition and tolerance. *Current Opinion in Plant Biology*, 39, 66–72. DOI: 10.1016/j.pbi.2017.06.004 PMID: 28654805

Chilly, C. C. I. Potential Antifungal Activity of Biosynthesized Copper nanoparticles Against Colletotrichum Capsici In Chilly.

Chinnappan, S., Kandasamy, S., Arumugam, S., Seralathan, K.-K., Thangaswamy, S., & Muthusamy, G. (2017). Biomimetic synthesis of silver nanoparticles using flower extract of Bauhinia purpurea and its antibacterial activity against clinical pathogens. In Environmental Science and Pollution Research (Vol. 25, Issue 1, pp. 963–969). Springer Science and Business Media LLC. DOI: 10.1007/s11356-017-0841-1

Chinnathambi, S., & Shirahata, N. (2019). Recent advances on fluorescent biomarkers of near-infrared quantum dots for in vitro and in vivo imaging. *Science and Technology of Advanced Materials*, 20(1), 337–355. DOI: 10.1080/14686996.2019.1590731 PMID: 31068983

Chin, S. F., Azman, A., & Pang, S. C. (2014). Size controlled synthesis of starch nanoparticles by a microemulsion method. *Journal of Nanomaterials*, 2014(1), 763736. DOI: 10.1155/2014/763736

Chiou, J. R., Lai, B. H., Hsu, K. C., & Chen, D. H. (2013). One-pot green synthesis of silver/iron oxide composite nanoparticles for 4-nitrophenol reduction. *Journal of Hazardous Materials*, 248, 394–400. DOI: 10.1016/j.jhazmat.2013.01.030 PMID: 23416483

Chirayil, C. J., Abraham, J., Mishra, R. K., George, S. C., & Thomas, S. (2017). Instrumental techniques for the characterization of nanoparticles. In *Thermal and rheological measurement techniques for nanomaterials characterization* (pp. 1–36). Elsevier. DOI: 10.1016/B978-0-323-46139-9.00001-3

Choel, M., Deboudt, K., Osan, J., Flament, P., & Van Grieken, R. (2005). Quantitative determination of low-Z elements in single atmospheric particles on boron substrates by automated scanning electron microscopy– energy-dispersive x-ray spectrometry. *Analytical Chemistry*, 77(17), 5686–5692. DOI: 10.1021/ac050739x PMID: 16131082

Cipriano, L. A., Di Liberto, G., Tosoni, S., & Pacchioni, G. (2020). Quantum confinement in group III–V semiconductor 2D nanostructures. *Nanoscale*, 12(33), 17494–17501. DOI: 10.1039/D0NR03577G PMID: 32808618

Clogston, J. D., & Patri, A. K. (2011). Zeta Potential Measurement. In McNeil, S. (Ed.), *Characterization of Nanoparticles Intended for Drug Delivery. Methods in Molecular Biology, 697*. Humana Press., DOI: 10.1007/978-1-60327-198-1_6

Costa, S., Borowiak-Palen, E., Kruszynska, M., Bachmatiuk, A., & Kalenczuk, R. J. (2008). Characterization of carbon nanotubes by Raman spectroscopy. *Materials Science Poland*, 26(2), 433–441.

Crawford, R., Dogdas, B., Keough, E., Haas, R. M., Wepukhulu, W., Krotzer, S., Burke, P. A., Sepp-Lorenzino, L., Bagchi, A., & Howell, B. J. (2011). Analysis of lipid nanoparticles by cryo-EM for characterizing siRNA delivery vehicles. *International Journal of Pharmaceutics*, 403(1–2), 237–244. DOI: 10.1016/j.ijpharm.2010.10.025 PMID: 20974237

Cui, C., Li, X., Liu, J., Hou, Y., Zhao, Y., & Zhong, G. (2015). Synthesis and functions of Ag 2 S nanostructures. *Nanoscale Research Letters*, 10(1), 1–21. DOI: 10.1186/s11671-015-1125-7

Cui, W., Wang, Y., Sun, Z., Cui, C., Li, H., Luo, K., & Cheng, A. (2023). Effects of steam explosion on phenolic compounds and dietary fiber of grape pomace. In *LWT* (Vol. 173, p. 114350). Elsevier BV., DOI: 10.1016/j.lwt.2022.114350

Cutter, C. N., & Siragusa, G. R. (1996). Reduction of *Brochothrix thermosphacta* on beef surfaces following immobilization of nisin in calcium alginate gels. *Letters in Applied Microbiology*, 23(1), 9–12. DOI: 10.1111/j.1472-765X.1996.tb00018.x PMID: 8679147

Daisy, P., & Saipriya, K. (2012). Biochemical analysis of Cassia fistula aqueous extract and phytochemically synthesized gold nanoparticles as hypoglycemic treatment for diabetes mellitus. *International Journal of Nanomedicine*, •••, 1189–1202. DOI: 10.2147/IJN.S26650 PMID: 22419867

Dai, Y., Chen, F., Yue, L., Li, T., Jiang, Z., Xu, Z., & Xing, B. (2020). Uptake, transport, and transformation of CeO2 nanoparticles by strawberry and their impact on the rhizosphere bacterial community. *ACS Sustainable Chemistry & Engineering*, 8(12), 4792–4800. DOI: 10.1021/acssuschemeng.9b07422

Dalmônico, G. M. L., López, E. O., Longuinho, M. M., Checca, N. R., Farina, M., Ersen, O., Rossi, A. M., & Rossi, A. L. (2019). Insight by Cryo-TEM into the growth and crystallization processes of calcium phosphate nanoparticles in aqueous medium. *Materials Chemistry and Physics*, 237, 1–7. DOI: 10.1016/j.matchemphys.2019.121862

Damera, D. P., Manimaran, R., Krishna Venuganti, V. V., & Nag, A. (2020). Green synthesis of full-color fluorescent carbon nanoparticles from *Eucalyptus* twigs for sensing the synthetic food colorant and bioimaging. *ACS Omega*, 5(31), 19905–19918. DOI: 10.1021/acsomega.0c03148 PMID: 32803087

Danish, R., Ahmed, F., & Koo, B. H. (2014). Rapid synthesis of high surface area anatase Titanium Oxide quantum dots. *Ceramics International*, 40(8), 12675–12680. DOI: 10.1016/j.ceramint.2014.04.115

Dan, Y., Zhang, W., Xue, R., Ma, X., Stephan, C., & Shi, H. (2015). Characterization of gold nanoparticle uptake by tomato plants using enzymatic extraction followed by single-particle inductively coupled plasma–mass spectrometry analysis. *Environmental Science & Technology*, 49(5), 3007–3014. DOI: 10.1021/es506179e PMID: 25633589

Dar, R. A., Shahnawaz, M., Ahanger, M. A., & Majid, I. U. (2023). Exploring the Diverse Bioactive Compounds from Medicinal Plants: A Review. *The Journal of Phytopharmacology*, 12(3), 189–195. DOI: 10.31254/phyto.2023.12307

Darwish, M., Mafla-Gonzalez, C., Kolenovic, B., Deremer, A., Centeno, D., Liu, T., Kim, D.-Y., Cattabiani, T., Drwiega, T. J., Kumar, I., Li, C., & Traba, C. (2022). Rapid Synthesis of Metal Nanoparticles Using Low-Temperature, Low-Pressure Argon Plasma Chemistry and Self-Assembly. *Green Chemistry*, 24(20), 8142–8154. DOI: 10.1039/D2GC02592B PMID: 37153712

Das, R. K., Pachapur, V. L., Lonappan, L., Naghdi, M., Pulicharla, R., Maiti, S., Cledon, M., Dalila, L. M. A., Sarma, S. J., & Brar, S. K. (2017). Biological synthesis of metallic nanoparticles: Plants, animals and microbial aspects. *Nanotechnology for Environmental Engineering*, 2(1), 1–21. DOI: 10.1007/s41204-017-0029-4

Das, S., Chakraborty, J., Chatterjee, S., & Kumar, H. (2018). Prospects of biosynthesized nanomaterials for the remediation of organic and inorganic environmental contaminants. *Environmental Science. Nano*, 5(12), 2784–2808. DOI: 10.1039/C8EN00799C

Dauthal, P., & Mukhopadhyay, M. (2018). Antioxidant activity of phytosynthesized biomatrix-loaded noble metallic nanoparticles. *Chinese Journal of Chemical Engineering*, 26(5), 1200–1208. DOI: 10.1016/j.cjche.2017.12.014

Davarpanah, S., Tehranifar, A., Davarynejad, G., Abadía, J., & Khorasani, R. (2016). Effects of foliar applications of zinc and boron nano-fertilizers on pomegranate (Punica granatum cv. Ardestani) fruit yield and quality. *Scientia Horticulturae*, 210, 57–64. DOI: 10.1016/j.scienta.2016.07.003

Davidov-Pardo, G., Joye, I. J., & McClements, D. J. (2015). Encapsulation of resveratrol in biopolymer particles produced using liquid antisolvent precipitation. Part 1: Preparation and characterization. *Food Hydrocolloids*, 45, 309–316. DOI: 10.1016/j.foodhyd.2014.11.023

de Olyveira, G. M., Manzine Costa, L. M., Basmaji, P., & Xavier Filho, L. (2011). *Bacterial Nanocellulose for Medicine Regenerative*. DOI: 10.1115/1.4004181

Demir, E., & Castranova, V. (2017). Evaluation of the potential genotoxicity of quantum dots. *A Review. Sci. Pages Nanotechnol*, 1, 1–19.

Demirer, G. S., Silva, T. N., Jackson, C. T., Thomas, J. B. W., Ehrhardt, D., Rhee, S. Y., & Landry, M. P. (2021). Nanotechnology to advance CRISPR–Cas genetic engineering of plants. *Nature Nanotechnology*, 16(3), 243–250. DOI: 10.1038/s41565-021-00854-y PMID: 33712738

Demirezen, D. A., Yıldız, Y. Ş., Yılmaz, Ş., & Yılmaz, D. D. (2019). Green synthesis and characterization of iron oxide nanoparticles using Ficus carica (common fig) dried fruit extract. *Journal of Bioscience and Bioengineering*, 127(2), 241–245. DOI: 10.1016/j.jbiosc.2018.07.024 PMID: 30348486

Dėnė, L., Chrapačienė, S., Laurinaitytė, G., Rudinskaitė, A., Viškelis, J., Viškelis, P., & Balčiūnaitienė, A. (2024). Green Synthesis of Silver Nanoparticles with *Hyssopus officinalis* and *Salvia officinalis* Extracts, Their Properties, and Antifungal Activity on *Fusarium* spp. *Plants*, 13(12), 1611. DOI: 10.3390/plants13121611 PMID: 38931043

Denning, D. W. (2024). Global incidence and mortality of severe fungal disease. *The Lancet. Infectious Diseases*, 24(7), e428–e438. DOI: 10.1016/S1473-3099(23)00692-8 PMID: 38224705

Devaraj, P., Kumari, P., Aarti, C., & Renganathan, A. (2013). Synthesis and characterization of silver nanoparticles using cannonball leaves and their cytotoxic activity against MCF-7 cell line. *Journal of Nanotechnology*, 2013(1), 598328. DOI: 10.1155/2013/598328

Devi, H. S., Boda, M. A., Shah, M. A., Parveen, S., & Wani, A. H. (2019). Green synthesis of iron oxide nanoparticles using Platanus orientalis leaf extract for antifungal activity. *Green Processing and Synthesis*, 8(1), 38–45. DOI: 10.1515/gps-2017-0145

Devi, L. S., & Joshi, S. R. (2014). Evaluation of the antimicrobial potency of silver nanoparticles biosynthesized by using an endophytic fungus, Cryptosporiopsis ericae PS4. *Journal of Microbiology (Seoul, Korea)*, 52(8), 667–674. DOI: 10.1007/s12275-014-4113-1 PMID: 24994011

Dhand, C., Dwivedi, N., Loh, X. J., Ying, A. N. J., Verma, N. K., Beuerman, R. W., & Ramakrishna, S. (2015). Methods and strategies for the synthesis of diverse nanoparticles and their applications: A comprehensive overview. *RSC Advances*, 5(127), 105003–105037. DOI: 10.1039/C5RA19388E

Diallo, M. S., Fromer, N. A., & Jhon, M. S. (2014). Nanotechnology for sustainable development: Retrospective and outlook. *Journal of Nanoparticle Research*, 15(11), 2044. DOI: 10.1007/s11051-013-2044-0

Díaz-Puertas, R., Álvarez-Martínez, F. J., Falco, A., Barrajón-Catalán, E., & Mallavia, R. (2023). Phytochemical-based nanomaterials against antibiotic-resistant bacteria: An updated review. *Polymers*, 15(6), 1392. DOI: 10.3390/polym15061392 PMID: 36987172

Dikshit, P. K., Kumar, J., Das, A. K., Sadhu, S., Sharma, S., Singh, S., Gupta, P., & Kim, B. S. (2021). Green Synthesis of Metallic Nanoparticles: Applications and Limitations. *Catalysts*, 11(8), 902. DOI: 10.3390/catal11080902

Dimkpa, C. O., McLean, J. E., Britt, D. W., & Anderson, A. J. (2015). Nano-CuO and interaction with nano-ZnO or soil bacterium provide evidence for the interference of nanoparticles in metal nutrition of plants. *Ecotoxicology (London, England)*, 24(1), 119–129. DOI: 10.1007/s10646-014-1364-x PMID: 25297564

Din, M. I., Nabi, A. G., Rani, A., Aihetasham, A., & Mukhtar, M. (2018). Single step green synthesis of stable nickel and nickel oxide nanoparticles from Calotropis gigantea: Catalytic and antimicrobial potentials. *Environmental Nanotechnology, Monitoring & Management*, 9, 29–36. DOI: 10.1016/j.enmm.2017.11.005

Dobrucka, R. (2018). Antioxidant and Catalytic Activity of Biosynthesized CuO Nanoparticles Using Extract of Galeopsidis herba. *Journal of Inorganic and Organometallic Polymers and Materials*, 28(3), 812–819. DOI: 10.1007/s10904-017-0750-2

Dogan, M., Yagmur, M., & Gungor, N. (2018). Effect of zinc oxide nanoparticles on disease resistance in cucumber plants against powdery mildew. *Journal of Nanoscience and Nanotechnology*, 18(8), 5430–5438. DOI: 10.1166/jnn.2018.15534

Dolez, P. I. (2015). Nanomaterials definitions, classifications, and applications. In *Nanoengineering* (pp. 3–40). Elsevier. DOI: 10.1016/B978-0-444-62747-6.00001-4

Donato, L., Nasser, I. I., Majdoub, M., & Drioli, E. (2022). Green Chemistry and Molecularly Imprinted Membranes. *Membranes*, 12(5), 5. Advance online publication. DOI: 10.3390/membranes12050472 PMID: 35629798

Dong, C., Chen, W., Liu, C., Liu, Y., & Liu, H. (2014). Synthesis of Magnetic Chitosan Nanoparticle and its Adsorption Property for Humic Acid from Aqueous Solution. *Colloids and Surfaces. A, Physicochemical and Engineering Aspects*, 446, 179–189. DOI: 10.1016/j.colsurfa.2014.01.069

Dong, H., Tang, S., Hao, Y., Yu, H., Dai, W., Zhao, G., Cao, Y., Lu, H., Zhang, X., & Ju, H. (2016). Fluorescent MoS2 quantum dots: Ultrasonic preparation, up-conversion and down-conversion bioimaging, and photodynamic therapy. *ACS Applied Materials & Interfaces*, 8(5), 3107–3114. DOI: 10.1021/acsami.5b10459 PMID: 26761391

Doyle, J. D. (2013). Nanotechnology in Anesthesia and Medicine. *Advances in Anesthesia*, 31(1), 181–200. DOI: 10.1016/j.aan.2013.08.010

Dreaden, E. C., Alkilany, A. M., Huang, X., Murphy, C. J., & El-Sayed, M. A. (2012). The golden age: Gold nanoparticles for biomedicine. *Chemical Society Reviews*, 41(7), 2740–2779. DOI: 10.1039/C1CS15237H PMID: 22109657

Drostkar, E., Talebi, R., & Kanouni, H. (2016). Foliar application of Fe, Zn and NPK nano-fertilizers on seed yield and morphological traits in chickpea under rainfed condition. *Journal of Resources and Ecology*, 4(1), 221–228.

Droth, M., & Burkard, G. (2016). Spintronics with graphene quantum dots. *physica status solidi (RRL)–. Rapid Research Letters*, 10(1), 75–90.

Dubey, M., Bhadauria, S., & Kushwah, B. S. (2009). Green synthesis of nanosilver particles from extract of *Eucalyptus hybrida* (safeda) leaf. *Digest Journal of Nanomaterials and Biostructures*, 4(3), 537–543.

Du, C., Xu, N., Yao, Z., Bai, X., Gao, Y., Peng, L., Gu, B., & Zhao, J. (2022). Mechanistic insights into sulfate and phosphate-mediated hexavalent chromium removal by tea polyphenols wrapped nano-zero-valent iron. In *Science of The Total Environment* (Vol. 850, p. 157996). Elsevier BV., DOI: 10.1016/j.scitotenv.2022.157996

Dudhane, A. A., Waghmode, S. R., Bhosale, M. A., & Mhaindarkar, V. P. (2017). *Caesalpinia pulcherrima* mediated green synthesis of silver nanoparticles: Evaluation of their antimicrobial and catalytic activity. *International Journal of Nanoparticles*, 9(3), 153–165. DOI: 10.1504/IJNP.2017.089195

Dufrene, Y. F., Ando, T., Garcia, R., Alsteens, D., Martinez-Martin, D., Engel, A., Gerber, C., & Muller, D. J. (2017). Imaging modes of atomic force microscopy for application in molecular and cell biology. *Nature Nanotechnology*, 12(4), 295–307. DOI: 10.1038/nnano.2017.45 PMID: 28383040

Du, J., Wang, H., Wang, L., Zhu, S., Song, Y., Yang, B., & Sun, H. (2016). Insight into the effect of functional groups on visible-fluorescence emissions of graphene quantum dots. *Journal of Materials Chemistry. C, Materials for Optical and Electronic Devices*, 4(11), 2235–2242. DOI: 10.1039/C6TC00548A

Dulta, K., Ağçeli, G. K., Chauhan, P., & Chauhan, P. K. (2021). Biogenic Production and Characterization of CuO Nanoparticles by Carica papaya Leaves and Its Biocompatibility Applications. *Journal of Inorganic and Organometallic Polymers and Materials*, 31(4), 1846–1857. DOI: 10.1007/s10904-020-01837-7

Dwivedi, S., & Dwivedi, N. (2022). Advances and Challenges of Bionanotechnology in the Sustainable Management of the Environment and Green Economy. In *Bionanotechnology Towards Sustainable Management of Environmental Pollution*. CRC Press. DOI: 10.1201/9781003270959-1

Dzimitrowicz, A., Berent, S., Motyka, A., Jamroz, P., Kurcbach, K., Sledz, W., & Pohl, P. (2019). Comparison of the characteristics of gold nanoparticles synthesized using aqueous plant extracts and natural plant essential oils of *Eucalyptus globulus* and *Rosmarinus officinalis.Arabian Journal of Chemistry*, 12(8), 4795–4805. DOI: 10.1016/j.arabjc.2016.09.007

Eaglstein, W. H. (2014). The FDA Approval Process and Drug Development. In Eaglstein, W. H. (Ed.), *The FDA for Doctors* (pp. 39–51). Springer International Publishing., DOI: 10.1007/978-3-319-08362-9_10

Echiegu, E. A. (2016). Nanotechnology as a tool for enhanced renewable energy application in developing countries. *Journal of Fundamentals of Renewable Energy and Applications, 6*(6), 01-09.

Echlin, P. (2011). *Handbook of sample preparation for scanning electron microscopy and X-ray microanalysis*. Springer Science & Business Media.

Edgar, V.-N., Hermes, P.-H., Denisse, V.-G. J., Andrea, P.-M., Roberto, S.-C. C., Ileana, V.-R., & Fabián, F.-L. (2024). Microorganisms, Plants, and Nanotechnology for Environmental Remediation: A Sustainable Prospect. In *Nano-Bioremediation for Water and Soil Treatment*. Apple Academic Press. DOI: 10.1201/9781003414964-3

Egan Morriss, C., Kimber, R., Powell, N., & Lloyd, J. (2022). Biotechnological synthesis of Pd-based nanoparticle catalysts. *Nanoscale Advances*, 4(3), 654–679. DOI: 10.1039/D1NA00686J PMID: 35224444

Eid, M. M. (2021). Characterization of Nanoparticles by FTIR and FTIR-Microscopy. In S. Mallakpour & C. M. Hussain, *Handbook of Consumer Nanoproducts* (pp. 1–30). Springer Singapore. DOI: 10.1007/978-981-15-6453-6_89-1

el Borady, O., Ayat, M., Shabrawy, M., & Millet, P. (2020). Green synthesis of gold nanoparticles using Parsley leaves extract and their applications as an alternative catalytic, antioxidant, anticancer, and antibacterial agents. *Advanced Powder Technology*, 31(10), 4390–4400. DOI: 10.1016/j.apt.2020.09.017

El, S., Koraichi, S., Latrache, H., & Hamadi, F. (2012). Scanning electron microscopy (SEM) and environmental SEM: suitable tools for study of adhesion stage and biofilm formation. V. Kazmiruk (Ed.), *Scanning Electron Microscopy* (pp. 718-725). InTechOpen. DOI: 10.5772/34990

Eljamal, O., Thompson, I. P., Maamoun, I., Shubair, T., Eljamal, K., Lueangwattanapong, K., & Sugihara, Y. (2020). Investigating the design parameters for a permeable reactive barrier consisting of nanoscale zero-valent iron and bimetallic iron/copper for phosphate removal. In *Journal of Molecular Liquids* (Vol. 299, p. 112144). Elsevier BV., DOI: 10.1016/j.molliq.2019.112144

Elmowafy, M., Shalaby, K., Elkomy, M. H., Alsaidan, O. A. M., Gomaa, H. A., Abdelgawad, M. A., & Mostafa, E. M. (2023). Polymeric Nanoparticles for Delivery of Natural Bioactive Agents: Recent Advances and Challenges. *Polymers*, 15(5), 1123. DOI: 10.3390/polym15051123 PMID: 36904364

El-Naggar, N. E. A., Shiha, A. M., Mahrous, H., & Mohammed, A. A. (2022). Green synthesis of chitosan nanoparticles, optimization, characterization and antibacterial efficacy against multi drug resistant biofilm-forming *Acinetobacter baumannii.Scientific Reports*, 12(1), 19869. DOI: 10.1038/s41598-022-24303-5 PMID: 36400832

El-Rahman, A. F. A., & Mohammad, T. G. M. (2013). Green synthesis of silver nanoparticle using *Eucalyptus globulus* leaf extract and its antibacterial activity. *Journal of Applied Sciences Research*, 9(13), 6437–6440.

El-Ramady, H. R., El-Marsafawy, S. M., & El-Sayed, S. M. (2018). Influence of zinc oxide nanoparticles on sunflower growth and productivity. *Environmental Science and Pollution Research International*, 25(29), 29322–29333. DOI: 10.1007/s11356-018-2133-4

El-Seedi, H. R., El-Shabasy, R. M., Khalifa, S. A. M., Saeed, A., Shah, A., Shah, R., Iftikhar, F. J., Abdel-Daim, M. M., Omri, A., Hajrahand, N. H., Sabir, J. S. M., Zou, X., Halabi, M. F., Sarhan, W., & Guo, W. (2019). Metal nanoparticles fabricated by green chemistry using natural extracts: Biosynthesis, mechanisms, and applications. *RSC Advances*, 9(42), 24539–24559. DOI: 10.1039/C9RA02225B PMID: 35527869

Elsheery, N. I., Sunoj, V. S. J., Wen, Y., Zhu, J. J., Muralidharan, G., & Cao, K. F. (2020). Foliar application of nanoparticles mitigates the chilling effect on photosynthesis and photoprotection in sugarcane. *Plant Physiology and Biochemistry*, 149, 50–60. DOI: 10.1016/j.plaphy.2020.01.035 PMID: 32035252

Epp, J. (2016). X-ray diffraction (XRD) techniques for materials characterization. In *Materials characterization using nondestructive evaluation (NDE) methods* (pp. 81-124). Woodhead Publishing.

Ernst, L. M., Casals, E., Italiani, P., Boraschi, D., & Puntes, V. (2021). The Interactions between Nanoparticles and the Innate Immune System from a Nanotechnologist Perspective. *Nanomaterials (Basel, Switzerland)*, 11(11), 11. Advance online publication. DOI: 10.3390/nano11112991 PMID: 34835755

Fahimmunisha, B. A., Ishwarya, R., AlSalhi, M. S., Devanesan, S., Govindarajan, M., & Vaseeharan, B. (2020). Green fabrication, characterization and antibacterial potential of zinc oxide nanoparticles using Aloe socotrina leaf extract: A novel drug delivery approach. *Journal of Drug Delivery Science and Technology*, 55, 101465. DOI: 10.1016/j.jddst.2019.101465

Fahmy, S. A., Preis, E., Bakowsky, U., & Azzazy, H. M. E.-S. (2020). Platinum Nanoparticles: Green Synthesis and Biomedical Applications. *Molecules (Basel, Switzerland)*, 25(21), 4981. Advance online publication. DOI: 10.3390/molecules25214981 PMID: 33126464

Fahmy, S., Mamdouh, I., Saleh, B., Issa, M., Bakowsky, U., & Azzazy, H. (2021). Green Synthesis of Platinum and Palladium Nanoparticles Using Peganum harmala L. Seed Alkaloids: Biological and Computational Studies. *Nanomaterials (Basel, Switzerland)*, 11(4), 965. DOI: 10.3390/nano11040965 PMID: 33918743

Faiz, N., Sivaswamy, V., Veeraraghavan, V. P., Kumar, R. (2023). Green Synthesis of Gold Nanoparticles Using *Eucalyptus* and *Piper Longum* and Its Subsequent Antiinflammatory Activity Evaluation.

Falke, S., & Betzel, C. (2019). Dynamic light scattering (DLS) principles, perspectives, applications to biological samples. *Radiation in bioanalysis: Spectroscopic techniques and theoretical methods*, 173-193.

Falke, S., & Betzel, C. (2019). Dynamic Light Scattering (DLS): Principles, perspectives, applications to biological samples. A. Pereira (Ed.), *Radiation in Bioanalysis* (pp. 173-193). Springer. DOI: 10.1007/978-3-030-28247-9

Fang, W., Wu, J., Cheng, M., Zhu, X., Du, M., Chen, C., Liao, W., Zhi, K., & Pan, W. (2023). Diagnosis of invasive fungal infections: Challenges and recent developments. *Journal of Biomedical Science*, 30(1), 42. DOI: 10.1186/s12929-023-00926-2 PMID: 37337179

Farooq, A., Khan, U. A., Ali, H., Sathish, M., Naqvi, S. A. H., Iqbal, S., Ali, H., Mubeen, I., Amir, M. B., Mosa, W. F. A., Baazeem, A., Moustafa, M., Alrumman, S., Shati, A., & Negm, S. (2022). Green Chemistry Based Synthesis of Zinc Oxide Nanoparticles Using Plant Derivatives of Calotropis gigantea (Giant Milkweed) and Its Biological Applications against Various Bacterial and Fungal Pathogens. *Microorganisms*, 10(11), 2195. DOI: 10.3390/microorganisms10112195 PMID: 36363787

Fernando, K. M., Gunathilake, C. A., Yalegama, C., Samarakoon, U. K., Fernando, C. A. N., Weerasinghe, G., Pamunuwa, G. K., Soliman, I., Ghulamullah, N., Rajapaksha, S. M., & Fatani, O. (2024). Synthesis of Silver Nanoparticles Using Green Reducing Agent: Ceylon Olive (Elaeocarpus serratus): Characterization and Investigating Their Antimicrobial Properties. *Journal of Composites Science*, 8(2), 43. DOI: 10.3390/jcs8020043

Fierer, N., Wood, S. A., & de Mesquita, C. P. B. (2021). How microbes can, and cannot, be used to assess soil health. *Soil Biology & Biochemistry*, 153, 108111. DOI: 10.1016/j.soilbio.2020.108111

Flaris, V., & Singh, G. (2009). Recent developments in biopolymers. *Journal of Vinyl and Additive Technology*, 15(1), 11. DOI: 10.1002/vnl.20171

Flauzino, J. M., Nalepa, M. A., Chronopoulos, D. D., Šedajová, V., Panáček, D., Jakubec, P., & Panáček, A. (2023). Click and Detect: Versatile Ampicillin Aptasensor Enabled by Click Chemistry on a Graphene–Alkyne Derivative. *Small*, 19(51), 2207216. DOI: 10.1002/smll.202207216 PMID: 36703534

Fleck, R. A., & Humbel, B. M. (Eds.). (2019). *Biological Field Emission Scanning Electron Microscopy, 2 Volume Set*. John Wiley & Sons. DOI: 10.1002/9781118663233

Fleischer, A., O'Neill, M. A., & Ehwald, R. (1999). The pore size of non-gramineous plant cell walls is rapidly decreased by borate ester cross-linking of the pectic polysaccharide rhamnogalacturonan II. *Plant Physiology*, 121(3), 829–838. DOI: 10.1104/pp.121.3.829 PMID: 10557231

Foret, F., Chung, D. S., Lavická, J., Přikryl, J., Lee, H., & Drobníková, I. (2023). Proceedings of APCE-CECE-ITP-IUPAC 2022. *Separations, 10*(2), Article 2. DOI: 10.3390/separations10020109

Foujdar, R., Chopra, H., Janghu, S., & Bera, M. B. (2023). Fabrication of "Lysine modified Zein- Punica granatum peels' purified ellagitannin nanoconjugates (Ly-MZ-PE NCs)" and in- vitro bio-accessibility, antiproliferative activity, and biosafety studies. *Food Bioscience*, 55, 102943. DOI: 10.1016/j.fbio.2023.102943

Foujdar, R., Kumar, S., Chauhan, A. K., & Bera, M. B. (2023). Role of Nanomaterials for Food Processing and Preservation. In *Nanotechnology Horizons in Food Process Engineering* (pp. 47–83). Apple Academic Press. DOI: 10.1201/9781003305378_4

Franca-Oliveira, G., Fornari, T., & Hernandez-Ledesma, B. (2021). A review on the extraction and processing of natural source-derived proteins through eco-innovative approaches. *Processes (Basel, Switzerland)*, 9(9), 1626. DOI: 10.3390/pr9091626

Francis, S., Joseph, S., Koshy, E. P., & Mathew, B. (2018). Microwave assisted green synthesis of silver nanoparticles using leaf extract of elephantopus scaber and its environmental and biological applications. *Artificial Cells, Nanomedicine, and Biotechnology*, 46(4), 795–804. DOI: 10.1080/21691401.2017.1345921 PMID: 28681662

Franken, L. E., Boekema, E. J., & Stuart, M. C. A. (2017). Transmission electron microscopy as a tool for the characterization of soft materials: Application and interpretation. *Advancement of Science*, 4(5), 1–9. DOI: 10.1002/advs.201600476 PMID: 28546914

Frenk, S., Ben-Moshe, T., Dror, I., Berkowitz, B., & Minz, D. (2013). Effect of metal oxide nanoparticles on microbial community structure and function in two different soil types. *PLoS One*, 8(12), 84441. DOI: 10.1371/journal.pone.0084441 PMID: 24349575

Fritz, L., Baum, C. M., Low, S., & Sovacool, B. K. (2024). Public engagement for inclusive and sustainable governance of climate interventions. *Nature Communications*, 15(1), 4168. DOI: 10.1038/s41467-024-48510-y PMID: 38755215

Furxhi, I., Faccani, L., Zanoni, I., Brigliadori, A., Vespignani, M., & Costa, A. L. (2024). Design rules applied to silver nanoparticles synthesis: A practical example of machine learning application. *Computational and Structural Biotechnology Journal*, 25, 20–33. DOI: 10.1016/j.csbj.2024.02.010 PMID: 38444982

Gakiya-Teruya, M., Palomino-Marcelo, L., Pierce, S., Angeles-Boza, A. M., Krishna, V., & Rodriguez-Reyes, J. C. F. (2020). Enhanced antimicrobial activity of silver nanoparticles conjugated with synthetic peptide by click chemistry. *Journal of Nanoparticle Research*, 22(4), 90. DOI: 10.1007/s11051-020-04799-6

Gao, J. H., & Xu, B. (2009). Applications of nanomaterials inside cell. *Nano Today*, 4(1), 37–51. DOI: 10.1016/j.nantod.2008.10.009

Gao, W., Wang, M., Ran, C., & Li, L. (2015). Facile one-pot synthesis of MoS 2 quantum dots–graphene–TiO 2 composites for highly enhanced photocatalytic properties. *Chemical Communications*, 51(9), 1709–1712. DOI: 10.1039/C4CC08984G PMID: 25514834

García, M. A., Martino, M. N., & Zaritzky, N. E. (1998). Plasticized starch-based coatings to improve strawberry (Fragaria× ananassa) quality and stability. *Journal of Agricultural and Food Chemistry*, 46(9), 3758–3767. DOI: 10.1021/jf980014c

García-Peñas, A., & Sharma, G. (2023). *New Materials for a Circular Economy*. Materials Research Forum LLC. DOI: 10.21741/9781644902639

Gardea-Torresdey, J. L., Rico, C. M., & White, J. C. (2014). Trophic transfer, transformation, and impact of engineered nanomaterials in terrestrial environments. *Environmental Science & Technology*, 48(5), 2526–2540. DOI: 10.1021/es4050665 PMID: 24499408

Gardea-Torresdey, J. L., Tiemann, K. J., Gamez, G., Dokken, K., Tehuacanero, S., & Jose-Yacaman, M. (1999). Gold nanoparticles obtained by bio-precipitation from gold (III) solutions. *Journal of Nanoparticle Research*, 1(3), 397–404. DOI: 10.1023/A:1010008915465

Gawde, K. A., Sau, S., Tatiparti, K., Kashaw, S. K., Mehrmohammadi, M., Azmi, A. S., & Iyer, A. K. (2018). Paclitaxel and di-fluorinated curcumin loaded in albumin nanoparticles for targeted synergistic combination therapy of ovarian and cervical cancers. *Colloids and Surfaces. B, Biointerfaces*, 167, 8–19. DOI: 10.1016/j.colsurfb.2018.03.046 PMID: 29625422

Geng, Y., Zhu, R., & Maimaituerxun, M. (2022, November 8). *Bibliometric review of carbon neutrality with CiteSpace: Evolution, trends, and framework. | Environmental Science & Pollution Research | EBSCOhost*. DOI: 10.1007/s11356-022-23283-3

Georgilis, E., Abdelghani, M., Pille, J., Aydinlioglu, E., van Hest, J. C. M., Lecommandoux, S., & Garanger, E. (2020). Nanoparticles based on natural, engineered or synthetic proteins and polypeptides for drug delivery applications. *International Journal of Pharmaceutics*, 586, 119537. DOI: 10.1016/j.ijpharm.2020.119537 PMID: 32531450

Ghasemi, S., Dabirian, S., Kariminejad, F., Koohi, D. E., Nemattalab, M., Majidimoghadam, S., Zamani, E., & Yousefbeyk, F. (2024). Process optimization for green synthesis of silver nanoparticles using Rubus discolor leaves extract and its biological activities against multi-drug resistant bacteria and cancer cells. *Scientific Reports*, 14(1), 4130. DOI: 10.1038/s41598-024-54702-9 PMID: 38374139

Gholamali, I., & Yadollahi, M. (2021). Bio-nanocomposite Polymer Hydrogels Containing Nanoparticles for Drug Delivery: A Review. *Regenerative Engineering and Translational Medicine*, 7(2), 129–146. DOI: 10.1007/s40883-021-00207-0

Ghoohestani, E., Samari, F., Homaei, A., & Yosuefinejad, S. (2024). A facile strategy for preparation of Fe3O4 magnetic nanoparticles using Cordia myxa leaf extract and investigating its adsorption activity in dye removal. *Scientific Reports*, 14(1), 84. DOI: 10.1038/s41598-023-50550-1 PMID: 38168136

Ghosh, P., Han, G., De, M., Kim, C. K., & Rotello, V. M. (2008). Gold nanoparticles in delivery applications. *Advanced Drug Delivery Reviews*, 60(11), 1307–1315. DOI: 10.1016/j.addr.2008.03.016 PMID: 18555555

Ghosh, S., Girigoswami, K., & Girigoswami, A. (2019). Membrane-Encapsulated Camouflaged Nanomedicines in Drug Delivery. *Nanomedicine (London)*, 14(15), 2067–2082. DOI: 10.2217/nnm-2019-0155 PMID: 31355709

Ghosh, S., Patil, S., Ahire, M., Kitture, R., Gurav, D. D., Jabgunde, A. M., Kale, S., Pardesi, K., Shinde, V., Bellare, J., Dhavale, D. D., & Chopade, B. A. (2012). *Gnidia glauca* flower extract mediated synthesis of gold nanoparticles and evaluation of its chemocatalytic potential. *Journal of Nanobiotechnology*, 10(1), 1–9. DOI: 10.1186/1477-3155-10-17 PMID: 22548753

Ghotekar, S., Pansambal, S., Pawar, S. P., Pagar, T., Oza, R., & Bangale, S. (2019). Biological activities of biogenically synthesized fluorescent silver nanoparticles using Acanthospermum hispidum leaves extract. In SN Applied Sciences (Vol. 1, Issue 11). Springer Science and Business Media LLC. https://doi.org/DOI: 10.1007/s42452-019-1389-0

Giannousi, K., Avramidis, I., & Dendrinou-Samara, C. (2013). Synthesis, characterization and evaluation of copper-based nanoparticles as agrochemicals against Phytophthora infestans. *RSC Advances*, 3(44), 21743–21752. DOI: 10.1039/c3ra42118j

Giráldez-Pérez, R. M., Grueso, E. M., Jiménez-Aguayo, R., Carbonero, A., González-Bravo, M., Kuliszewska, E., & Prado-Gotor, R. (2022). Use of Nanoparticles to Prevent Resistance to Antibiotics—Synthesis and Characterization of Gold Nanosystems Based on Tetracycline. *Pharmaceutics*, 14(9), 1941. DOI: 10.3390/pharmaceutics14091941 PMID: 36145689

Giraldo, J. P., Wu, H., Newkirk, G. M., & Kruss, S. (2019). Nanobiotechnology approaches for engineering smart plant sensors. *Nature Nanotechnology*, 14(6), 541–553. DOI: 10.1038/s41565-019-0470-6 PMID: 31168083

Girao, A. V., Caputo, G., & Ferro, M. C. (2017). Application of Scanning Electron Microscopy–Energy Dispersive X-Ray Spectroscopy (SEM-EDS). *Comprehensive Analytical Chemistry*, 75, 153–168. DOI: 10.1016/bs.coac.2016.10.002

Giri, A. K., Jena, B., Biswal, B., Pradhan, A. K., Arakha, M., Acharya, S., & Acharya, L. (2022). Green synthesis and characterization of silver nanoparticles using Eugenia roxburghii DC. extract and activity against biofilm-producing bacteria. *Scientific Reports*, 12(1), 8383. DOI: 10.1038/s41598-022-12484-y PMID: 35589849

Glaubitz, C., Rothen-Rutishauser, B., Lattuada, M., Balog, S., & Petri-Fink, A. (2022). Designing the ultrasonic treatment of nanoparticle-dispersions via machine learning. *Nanoscale*, 14(35), 12940–12950. DOI: 10.1039/D2NR03240F PMID: 36043853

Gnat, S., Łagowski, D., Nowakiewicz, A., & Dyląg, M. (2021). A global view on fungal infections in humans and animals: Opportunistic infections and microsporidioses. *Journal of Applied Microbiology*, 131(5), 2095–2113. DOI: 10.1111/jam.15032 PMID: 33556223

Gobinath, P., Packialakshmi, P., Hatamleh, A. A., Al-Dosary, M. A., Al-Wasel, Y. A., Balasubramani, R., Surendrakumar, R., & Idhayadhulla, A. (2022). Calotropis gigantea Assisted Synthesis of Zinc Oxide Nanoparticle Catalysis: Synthesis of Novel 3-Amino Thymoquinone Connected 1,4-Dihyropyridine Derivatives and Their Cytotoxic Activity. *Journal of Nanomaterials*, 2022(1), 1–10. DOI: 10.1155/2022/9697057

Goh, E. G., Xu, X., & Mccormick, P. (2014). Effect of particle size on the UV absorbance of zinc oxide nanoparticles. *Scripta Materialia, s 78–79*, 49–52.

Gohil, K., Dhakhda, S., Patel, V., Rathod, A., & Singh, P. K. (2024). Green Synthesis of Silver Nanoparticles using Various Plant Extracts and Evaluation of their Antimicrobial Activities. *Biosciences Biotechnology Research Asia*, 21(2), 765–777. DOI: 10.13005/bbra/3263

Gokul Eswaran, S., Shahid Afridi, P., & Vasimalai, N. (2022). Effective Multi Toxic Dyes Degradation Using Bio-Fabricated Silver Nanoparticles as a Green Catalyst. In Applied Biochemistry and Biotechnology (Vol. 195, Issue 6, pp. 3872–3887). Springer Science and Business Media LLC. DOI: 10.1007/s12010-022-03902-y

Goldsbury, C. S., Scheuring, S., & Kreplak, L. (2009). Introduction to atomic force microscopy (AFM) in Biology. *Current Protocols in Protein Science*, 58(1), 1–19. DOI: 10.1002/0471140864.ps1707s58 PMID: 19937721

Goldstein, J. I., Newbury, D. E., Michael, J. R., Ritchie, N. W., Scott, J. H. J., & Joy, D. C. (2017). *Scanning electron microscopy and X-ray microanalysis. Springer*, Goodhew, P. (2011). General introduction to transmission electron microscopy (TEM). *Corrected Analytical Transmission Electron Microscopy*, 1-19.

Gomiero, T. (2016). Soil degradation, land scarcity, and food security: Reviewing a complex challenge. *Sustainability (Basel)*, 8(3), 281. DOI: 10.3390/su8030281

González-Ballesteros, N., Rodríguez-Argüelles, M. C., Prado-López, S., Lastra, M., Grimaldi, M., Cavazza, A., Nasi, L., Salviati, G., & Bigi, F. (2019). Macroalgae to nanoparticles: Study of Ulva lactuca L. role in biosynthesis of gold and silver nanoparticles and of their cytotoxicity on colon cancer cell lines. *Materials Science and Engineering C*, 97, 498–509. DOI: 10.1016/j.msec.2018.12.066 PMID: 30678937

González-Burgos, E., Liaudanskas, M., Viškelis, J., Žvikas, V., Janulis, V., & Gómez-Serranillos, M. P. (2018). Antioxidant activity, neuroprotective properties and bioactive constituents analysis of varying polarity extracts from *Eucalyptus globulus* leaves. *Yao Wu Shi Pin Fen Xi*, 26(4), 1293–1302. PMID: 30249328

Gopirajah, R., & Anandharamakrishnan, C. (2019). Characterization methods for nanoparticles. *Food Nanotechnology*, 375, 375–396. DOI: 10.1201/9781315153872-16

Gopi, S., Amalraj, A., & Thomas, S. (2016). Effective drug delivery system of biopolymers based on nanomaterials and hydrogels-a review. *Drug Designing: Open Access*, 5(129), 2169–0138. DOI: 10.4172/2169-0138.1000129

Gorgieva, S., & Trček, J. (2019). Bacterial Cellulose: Production, Modification and Perspectives in Biomedical Applications. *Nanomaterials (Basel, Switzerland)*, 9(10), 10. Advance online publication. DOI: 10.3390/nano9101352 PMID: 31547134

Gour, A., & Jain, N. K. (2019). Advances in green synthesis of nanoparticles. *Artificial Cells, Nanomedicine, and Biotechnology*, 47(1), 844–851. DOI: 10.1080/21691401.2019.1577878 PMID: 30879351

Goutam, S. P., Saxena, G., Singh, V., Yadav, A. K., Bharagava, R. N., & Thapa, K. B. (2018). Green synthesis of TiO2 nanoparticles using leaf extract of Jatropha curcas L. for photocatalytic degradation of tannery wastewater. In *Chemical Engineering Journal* (Vol. 336, pp. 386–396). Elsevier BV., DOI: 10.1016/j.cej.2017.12.029

Gowdhami, B., Jaabir, M., Archunan, G., & Suganthy, N. (2019). Anticancer potential of zinc oxide nanoparticles against cervical carcinoma cells synthesized via biogenic route using aqueous extract of Gracilaria edulis. *Materials Science and Engineering C*, 103, 109840.

Grunwald, P. (2019). *Pharmaceutical Biocatalysis: Fundamentals, Enzyme Inhibitors, and Enzymes in Health and Diseases*. CRC Press.

Guan, X., Gao, X., Avellan, A., Spielman-Sun, E., Xu, J., Laughton, S., & Lowry, G. V. (2020). CuO nanoparticles alter the rhizospheric bacterial community and local nitrogen cycling for wheat grown in a calcareous soil. *Environmental Science & Technology*, 54(14), 8699–8709. DOI: 10.1021/acs.est.0c00036 PMID: 32579348

Guerra, F. D., Attia, M. F., Whitehead, D. C., & Alexis, F. (2018). Nanotechnology for environmental remediation: Materials and applications. *Molecules (Basel, Switzerland)*, 23(7), 1760. DOI: 10.3390/molecules23071760 PMID: 30021974

Gul, I., Khan, S. M., Mehmood, T., Ahmad, Z., Badshah, H., & Shah, H. (2020). Characterization of cobalt oxide and calcium-aluminum oxide nano-catalyst through scanning electron microscopy, X-ray diffraction, and Energy dispersive X-ray spectroscopy. *Microscopy Research and Technique*, 83(9), 1124–1131. DOI: 10.1002/jemt.23504 PMID: 32394528

Gullón, P., Gullón, B., Astray, G., Munekata, P. E., Pateiro, M., & Lorenzo, J. M. (2020). Value-added compound recovery from invasive forest for biofunctional applications: Eucalyptus species as a case study. *Molecules (Basel, Switzerland)*, 25(18), 4227. DOI: 10.3390/molecules25184227 PMID: 32942656

Guo, H., White, J. C., Wang, Z., & Xing, B. (2018). Nano-enabled fertilizers to control the release and use efficiency of nutrients. *Current Opinion in Environmental Science & Health*, 6, 77–83. DOI: 10.1016/j.coesh.2018.07.009

Guo, M., Li, W., Yang, F., & Liu, H. (2015). Controllable biosynthesis of gold nanoparticles from a Eucommia ulmoides bark aqueous extract. *Spectrochimica Acta. Part A: Molecular and Biomolecular Spectroscopy*, 142, 73–79. DOI: 10.1016/j.saa.2015.01.109 PMID: 25699695

Guo, S., Fu, D., Utupova, A., Sun, D., Zhou, M., Jin, Z., & Zhao, K. (2019). Applications of polymer-based nanoparticles in vaccine field. *Nanotechnology Reviews*, 8(1), 143–155. DOI: 10.1515/ntrev-2019-0014

Guo, Y., Liu, C., Ye, R., & Duan, Q. (2020). Advances on water quality detection by uv-vis spectroscopy. *Applied Sciences (Basel, Switzerland)*, 10(19), 1–18. DOI: 10.3390/app10196874

Gupta, A., Mehta, S. K., Kunal, K., Mukhopadhyay, K., & Singh, S. (2022). Quantum dots as promising nanomaterials in agriculture. In *Agricultural Nanobiotechnology* (pp. 243–296). Woodhead Publishing. DOI: 10.1016/B978-0-323-91908-1.00016-X

Gupta, P., Rai, N., Verma, A., & Gautam, V. (2024). Microscopy based methods for characterization, drug delivery, and understanding the dynamics of nanoparticles. *Medicinal Research Reviews*, 44(1), 138–168. DOI: 10.1002/med.21981 PMID: 37294298

Gupta, S. K., & Xie, H. (2018). Nanoparticles toxicity and their routes of exposures. *Pakistan Journal of Pharmaceutical Sciences*, 31(6), 2441–2451.

Gutiérrez-Mejía, F. A., Vásquez-López, C., Vargas-Coronado, R. F., Villa-de la Torre, F. E., Arana-Argaez, V. E., Rodríguez-Buenfil, I. M., Gamboa-Angulo, M. M., Torres, Y., & Cauich-Rodríguez, J. V. (2024). Electrophoretic deposition of TiO2 nanotubes and antibiotics on polyurethane coated stainless steel for improved antibacterial response and cell viability. *Materials Today. Communications*, 40, 109428. DOI: 10.1016/j.mtcomm.2024.109428

Guzmán, M. G., Dille, J., & Godet, S. (2009). Synthesis of silver nanoparticles by chemical reduction method and their antibacterial activity. *Int J Chem Biomol Eng*, 2(3), 104–111.

Haik, Y., & Bantekas, I. (2020). Nano-Drug Clinical Trials: Informed Consent and Risk Management through Blockchain. *Pittsburgh Journal of Technology Law and Policy*, 21, 35.

Halbandge, S. D., Mortale, S. P., & Karuppayil, S. M. (2017). Biofabricated silver nanoparticles synergistically activate amphotericin B against mature biofilm forms of Candida albicans. *The Open Nanomedicine Journal*, 4(1), 1–16. Advance online publication. DOI: 10.2174/1875933501704010001

Ha, N. M. C., Nguyen, T. H., Wang, S. L., & Nguyen, A. D. (2019). Preparation of NPK nanofertilizer based on chitosan nanoparticles and its effect on biophysical characteristics and growth of coffee in green house. *Research on Chemical Intermediates*, 45(1), 51–63. DOI: 10.1007/s11164-018-3630-7

Hanahan, D., & Weinberg, R. A. (2011). Hallmarks of cancer: the next generation. *cell,* 144(5), 646-674.

Hanh, T. T., & Chi, L. N. T., & Duy, N. N. (2022). Preparation of Copper Nanoparticles/Diatomite Nanocomposite for Improvement in Water Quality of Fishponds. In S. Hussain (Ed.), Journal of Chemistry (Vol. 2022, pp. 1–9). Hindawi Limited. DOI: 10.1155/2022/3921631

Hano, C., & Abbasi, B. H. (2022). Plant-Based Green Synthesis of Nanoparticles: Production, Characterization and Applications. *Biomolecules*, 12(1), 31. DOI: 10.3390/biom12010031 PMID: 35053179

Hao, Z., Wang, M., Cheng, L., Si, M., Feng, Z., & Feng, Z. (2024). Synergistic antibacterial mechanism of silver-copper bimetallic nanoparticles. *Frontiers in Bioengineering and Biotechnology*, 11, 1337543. DOI: 10.3389/fbioe.2023.1337543 PMID: 38260749

Harish, V., Tewari, D., Gaur, M., Yadav, A. B., Swaroop, S., Bechelany, M., & Barhoum, A. (2022). Review on Nanoparticles and Nanostructured Materials: Bioimaging, Biosensing, Drug Delivery, Tissue Engineering, Antimicrobial, and Agro-Food Applications. *Nanomaterials (Basel, Switzerland)*, 12(3), 457. DOI: 10.3390/nano12030457 PMID: 35159802

Haritha, V., Gowri, S., Janarthanan, B., Faiyazuddin, M., Karthikeyan, C., & Sharmila, S. (2022). Biogenic synthesis of nickel oxide nanoparticles using Averrhoa bilimbi and investigation of its antibacterial, antidiabetic and cytotoxic properties. *Inorganic Chemistry Communications*, 144, 109930. DOI: 10.1016/j.inoche.2022.109930

Harris, A. T., & Bali, R. (2008). On the formation and extent of uptake of silver nanoparticles by live plants. *Journal of Nanoparticle Research*, 10(4), 691–695. DOI: 10.1007/s11051-007-9288-5

Haruna, A., & Yahaya, S. M. (2021). Recent Advances in the Chemistry of Bioactive Compounds from Plants and Soil Microbes: A Review. *Chemistry Africa*, 4(2), 231–248. DOI: 10.1007/s42250-020-00213-9

Hasanin, M. S., & Youssef, A. M. (2022). Ecofriendly bioactive film doped CuO nanoparticles based biopolymers and reinforced by enzymatically modified nanocellulose fibers for active packaging applications. *Food Packaging and Shelf Life*, 34, 100979. DOI: 10.1016/j.fpsl.2022.100979

Hasan, S. A. (2015). A review on nanoparticles: Their synthesis and types. *Research Journal of Recent Sciences*, 4, 9–11.

Hassan, A. K., Hashim, L. Q., Rezooqi, A. M., & Hashim, M. F. (2024). Biosynthesis of Fe/Pd Bimetallic Nanoparticles and Used for Removal of Synthetic Oily Wastewater. *Journal of Petroleum Research and Studies*, 14(2), 121–137. DOI: 10.52716/jprs.v14i2.867

Hassan, M. M., Ranzoni, A., Phetsang, W., Blaskovich, M. A., & Cooper, M. A. (2017). Surface ligand density of antibiotic-nanoparticle conjugates enhances target avidity and membrane permeabilization of vancomycin-resistant bacteria. *Bioconjugate Chemistry*, 28(2), 353–361. DOI: 10.1021/acs.bioconjchem.6b00494 PMID: 27959504

Hassan, M., Bai, J., & Dou, D. (2019). Biopolymers; Definition, Classification and Applications. *Egyptian Journal of Chemistry*, 62(9), 1725–1737. DOI: 10.21608/ejchem.2019.6967.1580

Hatakeyama, H., & Hatakeyama, T. (2009). Biopolymers. In *Lignin Structure, Properties, and Applications* (pp. 1–63). Springer.

HB, S. S. (2022). Comparison of efficacy of nanoparticles synthesized from leaves and flowers of Russelia equisitiformis. *Inorganic and Nano-Metal Chemistry*, 52(1), 75–81. DOI: 10.1080/24701556.2020.1862218

Hedayati, R., Jahanshahi, M., & Attar, H. (2012). Fabrication and characterization of albumin- acacia nanoparticles based on complex coacervation as potent nanocarrier. *Journal of Chemical Technology and Biotechnology*, 87(10), 1401–1408. DOI: 10.1002/jctb.3758

Hedberg, J., Skoglund, S., Karlsson, M.-E., Wold, S., Odnevall Wallinder, I., & Hedberg, Y. (2014). Sequential studies of silver released from silver nanoparticles in aqueous media simulating sweat, laundry detergent solutions and surface water. *Environmental Science & Technology*, 48(13), 7314–7322. DOI: 10.1021/es500234y PMID: 24892700

Helan, V., Prince, J. J., Al-Dhabi, N. A., Arasu, M. V., Ayeshamariam, A., Madhumitha, G., Roopan, S. M., & Jayachandran, M. (2016). Neem leaves mediated preparation of NiO nanoparticles and its magnetization, coercivity and antibacterial analysis. *Results in Physics*, 6, 712–718. DOI: 10.1016/j.rinp.2016.10.005

Hemmati, S., Ahany Kamangar, S., Ahmeda, A., Zangeneh, M. M., & Zangeneh, A. (2020). Application of copper nanoparticles containing natural compounds in the treatment of bacterial and fungal diseases. *Applied Organometallic Chemistry*, 34(4), e5465. DOI: 10.1002/aoc.5465

Hernandez-Carmona, F., Morales-Matos, Y., Lambis-Miranda, H., & Pasqualino, J. (2017). Starch extraction potential from plantain peel wastes. *Journal of Environmental Chemical Engineering*, 5(5), 4980–4985. DOI: 10.1016/j.jece.2017.09.034

Herrera-Becerra, R., Zorrilla, C., Rius, J. L., & Ascencio, J. A. (2008). Electron microscopy characterization of biosynthesized iron oxide nanoparticles. *Applied Physics. A, Materials Science & Processing*, 91(2), 241–246. DOI: 10.1007/s00339-008-4420-7

Herzing, A. A., Watanabe, M., Edwards, J. K., Conte, M., Tang, Z.-R., Hutchings, G. J., & Kiely, C. J. (2008). Energy dispersive X-ray spectroscopy of bimetallic nanoparticles in an aberration corrected scanning transmission electron microscope. *Faraday Discussions*, 138, 337–351. DOI: 10.1039/B706293C PMID: 18447024

He, Y., Du, Z., Lv, H., Jia, Q., Tang, Z., Zheng, X., & Zheng, Y. (2011). Green synthesis of ZnO nanoparticles using plant extract and their effect on the antimicrobial activity. *Journal of Nanomaterials*, 2011, 1–6.

Hidayanti, F., & Harnovan, A. A. (2020). Application of scanning electron microscopy: A review. *International Journal of Applied Science and Engineering Review*, 1(6), 91–102.

Higuera-Valenzuela, H. J., Zazueta-Raynaud, A., Heredia-Cancino, J. A., Soto-B, M. A., Cabrera-German, D., Berman-Mendoza, D., ... & Hernadez-Abril, P. A. (2024). Green Synthesis of Silicon Quantum Dots Using Ocimum basilicum var. Purpurascens as an Organic Reductor Agent. *Silicon*, 1-8.

Hodoroaba, V. D. (2019). Energy-dispersive X-ray spectroscopy (EDS). V. Hodoroaba (Ed.), *Characterization of Nanoparticles* (pp. 397-417). Elsevier Inc. DOI: 10.1016/B978-0-12-814182-3.00021-3

Hodoroaba, V. D., Unger, W., & Shard, A. (Eds.). (2019). *Characterization of nanoparticles: measurement processes for nanoparticles*. Elsevier.

Hoenigl, M., Seidel, D., Sprute, R., Cunha, C., Oliverio, M., Goldman, G. H., Ibrahim, A. S., & Carvalho, A. (2022). COVID-19-associated fungal infections. *Nature Microbiology*, 7(8), 1127–1140. DOI: 10.1038/s41564-022-01172-2 PMID: 35918423

Hollander, A., & Danino, D. (2015). Cochleate characterization by cryogenic electron microscopy methods: Cryo-TEM and Cryo-SEM. *Colloids and Surfaces. A, Physicochemical and Engineering Aspects*, 483, 187–192. DOI: 10.1016/j.colsurfa.2015.07.025

Hong, S., Choi, D. W., Kim, H. N., Park, C. G., Lee, W., & Park, H. H. (2020). Protein-Based Nanoparticles as Drug Delivery Systems. *Pharmaceutics*, 12(7), 604. DOI: 10.3390/pharmaceutics12070604 PMID: 32610448

Hossain, A., Abdallah, Y., Ali, M. A., Masum, M. M., Li, B., Sun, G., Meng, Y., Wang, Y., & An, Q. (2019). Lemon-Fruit-Based Green Synthesis of Zinc Oxide Nanoparticles and Titanium Dioxide Nanoparticles against Soft Rot Bacterial Pathogen Dickeya dadantii. *Biomolecules*, 9(12), 863. DOI: 10.3390/biom9120863 PMID: 31835898

Houben, A., Field, B. L., & Saunders, V. A. (2001). Microdissection and chromosome painting of plant B chromosomes. *Chromosome Painting: Principles, Strategies and Scope*, 115-124.

Hoyt, M. (2009). Analyses of Nanoparticles in the Environment. *Nanotechnology and the Environment*. CRC Press 99-122 (2009)

Huang, X., Wu, H., Liao, X., & Shi, B. (2010). One-step, size-controlled synthesis of gold nanoparticles at room temperature using plant tannin. *Green Chemistry -. Green Chemistry*, 12(3), 395–399. DOI: 10.1039/B918176H

Huang, X., Zhao, Y., Zhang, J., & Yang, P. (2021). Impact of zinc oxide nanoparticles on soil microbial community and enzyme activities in tomato cultivation. *The Science of the Total Environment*, 759, 143645. DOI: 10.1016/j.scitotenv.2020.143645

Huang, Y., Gu, C., He, S., Zhu, D., Liu, X., & Chen, Z. (2020). Development and characterization of an edible chitosan–whey protein nano composite film for chestnut (Castanea mollissima Bl.) preservation. *Journal of Food Science*, 85(7), 2114–2123. DOI: 10.1111/1750-3841.15174 PMID: 32519374

Hu, K., & McClements, D. J. (2015). Fabrication of biopolymer nanoparticles by antisolvent precipitation and electrostatic deposition: Zein-alginate core/shell nanoparticles. *Food Hydrocolloids*, 44, 101–108. DOI: 10.1016/j.foodhyd.2014.09.015

Hunckler, M. D., & Levine, A. D. (2022). Navigating ethical challenges in the development and translation of biomaterials research. *Frontiers in Bioengineering and Biotechnology*, 10, 949280. DOI: 10.3389/fbioe.2022.949280 PMID: 36204464

Hunter, R. J. (2013). *Zeta potential in colloid science: principles and applications* (Vol. 2). Academic press.

Hussain, S., Ali Muazzam, M., Ahmed, M., Ahmad, M., Mustafa, Z., Murtaza, S., Ali, J., Ibrar, M., Shahid, M., & Imran, M. (2023). Green synthesis of nickel oxide nanoparticles using Acacia nilotica leaf extracts and investigation of their electrochemical and biological properties. *Journal of Taibah University for Science : JTUSCI*, 17(1), 1–12. DOI: 10.1080/16583655.2023.2170162

Hussain, S., Bashir, O., Khan, Z., & AL-Thabaiti, S. A. (2014). Steroidal saponin based extracellular biosynthesis of AgNPs. *Journal of Molecular Liquids*, 199, 489–494. DOI: 10.1016/j.molliq.2014.09.047

Hussain, S., Joo, J., Kang, J., Kim, B., Braun, G. B., She, Z.-G., & Teesalu, T. (2018). Antibiotic-loaded nanoparticles targeted to the site of infection enhance antibacterial efficacy. *Nature Biomedical Engineering*, 2(2), 95–103. DOI: 10.1038/s41551-017-0187-5 PMID: 29955439

Huston, M., DeBella, M., DiBella, M., & Gupta, A. (2021). Green Synthesis of Nanomaterials. *Nanomaterials (Basel, Switzerland)*, 11(8), 2130. Advance online publication. DOI: 10.3390/nano11082130 PMID: 34443960

Hwang, I., Hwang, J. H., Choi, H., Kim, K.-J., & Lee, D. G. (2012). Synergistic effects between silver nanoparticles and antibiotics and the mechanisms involved. *Journal of Medical Microbiology*, 61(12), 1719–1726. DOI: 10.1099/jmm.0.047100-0 PMID: 22956753

Ibraheem, I. J., Hassan, K. T., Ali, H. H., & Obaid, A. S. (2020). One-step biosynthesis of gold nanoparticles using natural reductive extracts. *Nano Biomedicine and Engineering*, 12(4), 331–337. DOI: 10.5101/nbe.v12i4.p331-337

Ibrahim, F. M., Najeeb, D. A., & ThamerSadeq, H. (2023). Green preparation of Cu nanoparticles of the avocado seed extract as an adsorbent surface. *Materials Science for Energy Technologies*, 6, 130–136. DOI: 10.1016/j.mset.2022.12.006

Ikhioya, I., Onoh, E., Nkele, C., Abor, B., Obitte, B., Maaza, M., & Ezema, F. (2023). The The Green Synthesis of Copper Oxide Nanoparticles Using the Moringa Oleifera Plant and its Subsequent Characterization for Use in Energy Storage Applications. *East European Journal of Physics*, 1(1), 162–172. DOI: 10.26565/2312-4334-2023-1-20

Imada, K., Sakai, S., Kajihara, H., Tanaka, S., & Ito, S. (2016). Magnesium oxide nanoparticles induce systemic resistance in tomato against bacterial wilt disease. *Plant Pathology*, 65(4), 551–560. DOI: 10.1111/ppa.12443

Imtiyaz, A., & Singh, A. (2024). Green synthesized ruthenium oxide nanoparticles mediated through Iris Kashmiriana (Mazar-Graveyard) plant extract and antimicrobial activity. *Journal of Inorganic and Organometallic Polymers and Materials*, 34(5), 1–8. DOI: 10.1007/s10904-023-02968-3

Ingham, B. (2015). X-ray scattering characterisation of nanoparticles. *Crystallography Reviews*, 21(4), 229–303. DOI: 10.1080/0889311X.2015.1024114

Inkson, B. J. (2016). Scanning electron microscopy (SEM) and transmission electron microscopy (TEM) for materials characterization. G. Hübschen (Ed.), *Materials Characterization Using Nondestructive Evaluation (NDE) Methods* (pp. 17-43). Elsevier Ltd. DOI: 10.1016/B978-0-08-100040-3.00002-X

Inkson, B. J. (2016). Scanning electron microscopy (SEM) and transmission electron microscopy (TEM) for materials characterization. In *Materials characterization using nondestructive evaluation (NDE) methods* (pp. 17-43). Woodhead publishing.

Iqbal, M. A. (2019). Nano-fertilizers for sustainable crop production under changing climate: A global perspective. *Sustainable crop. Production*, 8, 1–13.

Iravani, S. (2011). Green synthesis of metal nanoparticles using plants. *Green Chemistry*, 13(10), 2638–2650. DOI: 10.1039/c1gc15386b

Iravani, S., Korbekandi, H., Mirmohammadi, S. V., & Zolfaghari, B. (2014). Synthesis of silver nanoparticles: Chemical, physical and biological methods. *Research in Pharmaceutical Sciences*, 9(6), 385–406. PMID: 26339255

Irum, S., Jabeen, N., Ahmad, K. S., Shafique, S., Khan, T. F., Gul, H., Anwaar, S., Shah, N. I., Mehmood, A., & Hussain, S. Z. (2020). Biogenic iron oxide nanoparticles enhance callogenesis and regeneration pattern of recalcitrant Cicer arietinum L. *PLoS One*, 15(12), 1–19. DOI: 10.1371/journal.pone.0242829 PMID: 33259506

Islam, N. U., Jalil, K., Shahid, M., Muhammad, N., & Rauf, A. (2019). Pistacia integerrima gall extract mediated green synthesis of gold nanoparticles and their biological activities. *Arabian Journal of Chemistry*, 12(8), 2310–2319. DOI: 10.1016/j.arabjc.2015.02.014

Ismail, M., Khan, M. I., Khan, S. B., Khan, M. A., Akhtar, K., & Asiri, A. M. (2018). Green synthesis of plant supported CuAg and CuNi bimetallic nanoparticles in the reduction of nitrophenols and organic dyes for water treatment. *Journal of Molecular Liquids*, 260, 78–91. DOI: 10.1016/j.molliq.2018.03.058

Issaabadi, Z., Nasrollahzadeh, M., & Sajadi, S. M. (2017). Green synthesis of the copper nanoparticles supported on bentonite and investigation of its catalytic activity. *Journal of Cleaner Production*, 142, 3584–3591. DOI: 10.1016/j.jclepro.2016.10.109

Ivanković, K., Kern, M., & Rožman, M. (2021). Modelling of the adsorption of pharmaceutically active compounds on carbon-based nanomaterials. *Journal of Hazardous Materials*, 414, 125554. DOI: 10.1016/j.jhazmat.2021.125554 PMID: 33684815

Jagtap, U. B., & Bapat, V. A. (2013). Green synthesis of silver nanoparticles using Artocarpus heterophyllus Lam. seed extract and its antibacterial activity. *Industrial Crops and Products*, 46, 132–137. DOI: 10.1016/j.indcrop.2013.01.019

Jain, D., Daima, H. K., Kachhwaha, S., & Kothari, S. L. (2009). Synthesis of plant-mediated silver nanoparticles using papaya fruit extract and evaluation of their anti-microbial activities. *Digest Journal of Nanomaterials and Biostructures*, 4(3), 557–563.

Jain, K. K. (2016). Chapter Two—Role of Proteomics in the Development of Personalized Medicine. In Donev, R. (Ed.), *Advances in Protein Chemistry and Structural Biology* (Vol. 102, pp. 41–52). Academic Press., DOI: 10.1016/bs.apcsb.2015.09.002

Jain, N., Bhargava, A., Tarafdar, J. C., Singh, S. K., & Panwar, J. (2012). A biomimetic approach towards synthesis of zinc oxide nanoparticles. *Applied Microbiology and Biotechnology*, 97(2), 859–869. DOI: 10.1007/s00253-012-3934-2 PMID: 22382164

James, A. E., & Driskell, J. D. (2013). Monitoring gold nanoparticle conjugation and analysis of biomolecular binding with nanoparticle tracking analysis (NTA) and dynamic light scattering (DLS). *Analyst*, 138(4), 1212–1218. DOI: 10.1039/c2an36467k PMID: 23304695

Jampílek, J., & Kráľová, K. (2019). Beneficial effects of metal-and metalloid-based nanoparticles on crop production. Nanotechnology for Agriculture: Advances for Sustainable Agriculture, 161-219.

Jampílek, J., & Kráľová, K. (2022). Impact of copper-based nanoparticles on economically important plants. In *Copper nanostructures: Next-generation of agrochemicals for sustainable agroecosystems* (pp. 293–339). Elsevier. DOI: 10.1016/B978-0-12-823833-2.00030-1

Jan, H., Gul, R., Andleeb, A., Ullah, S., Shah, M., Khanum, M., Ullah, I., Hano, C., & Abbasi, B. H. (2021). A detailed review on biosynthesis of platinum nanoparticles (PtNPs), their potential antimicrobial and biomedical applications. *Journal of Saudi Chemical Society*, 25(8), 101297. DOI: 10.1016/j.jscs.2021.101297

Jedrzejczak-Krzepkowska, M., Kubiak, K., Ludwicka, K., & Bielecki, S. (2016). Bacterial NanoCellulose Synthesis, Recent Findings. In Gama, M., Dourado, F., & Bielecki, S. (Eds.), *Bacterial Nanocellulose* (pp. 19–46). Elsevier., DOI: 10.1016/B978-0-444-63458-0.00002-0

Jeevanandam, J., Barhoum, A., Chan, Y. S., Dufresne, A., & Danquah, M. K. (2018). Review on nanoparticles and nanostructured materials: History, sources, toxicity, and regulations. *Beilstein Journal of Nanotechnology*, 9, 1050–1074. DOI: 10.3762/bjnano.9.98 PMID: 29719757

Jeevanandam, J., Chan, Y. S., & Ku, Y. H. (2018). Aqueous *Eucalyptus globulus* leaf extract-mediated biosynthesis of MgO nanorods. *Applied Biological Chemistry*, 61(2), 197–208. DOI: 10.1007/s13765-018-0347-7

Jeevanandam, J., Kiew, S. F., Boakye-Ansah, S., Lau, S. Y., Barhoum, A., Danquah, M. K., & Rodrigues, J. (2022). Green approaches for the synthesis of metal and metal oxide nanoparticles using microbial and plant extracts. *Nanoscale*, 14(7), 2534–2571. DOI: 10.1039/D1NR08144F PMID: 35133391

Jeyaraj, M., Gurunathan, S., Qasim, M., Kang, M. H., & Kim, J. H. (2019). A comprehensive review on the synthesis, characterization, and biomedical application of platinum nanoparticles. *Nanomaterials (Basel, Switzerland)*, 9(12), 1719. DOI: 10.3390/nano9121719 PMID: 31810256

Jha, A. K., Prasad, K., Prasad, K., & Kulkarni, A. R. (2009). Plant system: Nature's nanofactory. *Colloids and Surfaces. B, Biointerfaces*, 73(2), 219–223. DOI: 10.1016/j.colsurfb.2009.05.018 PMID: 19539452

Jia, D., & Sun, W. (2021). Silver nanoparticles offer a synergistic effect with fluconazole against fluconazole-resistant Candida albicans by abrogating drug efflux pumps and increasing endogenous ROS. *Infection, Genetics and Evolution*, 93, 104937. DOI: 10.1016/j.meegid.2021.104937 PMID: 34029724

Jiang, Z., Li, L., Huang, H., He, W., & Ming, W. (2022). Progress in laser ablation and biological synthesis processes:"Top-Down" and "Bottom-Up" approaches for the green synthesis of Au/Ag nanoparticles. *International Journal of Molecular Sciences*, 23(23), 14658. DOI: 10.3390/ijms232314658 PMID: 36498986

Johri, N., Jacquillet, G., & Unwin, R. (2010). Heavy metal poisoning: The effects of cadmium on the kidney. *Biometals*, 23, 783–792.

Jokerst, J. V., & Gambhir, S. S. (2011). Molecular imaging with theranostic nanoparticles. *Accounts of Chemical Research*, 44(10), 1050–1060. DOI: 10.1021/ar200106e PMID: 21919457

Joshi, M., Bhattacharyya, A., & Ali, S. W. (2008). Characterization techniques for nanotechnology applications in textiles. Joshi, M. *Indian Journal of Fibre and Textile Research*, 33, 304–317.

Joudeh, N., & Linke, D. (2022). Nanoparticle classification, physicochemical properties, characterization, and applications: A comprehensive review for biologists. *Journal of Nanobiotechnology*, 20(1), 262. DOI: 10.1186/s12951-022-01477-8 PMID: 35672712

Joye, I. J., & McClements, D. J. (2013). Production of nanoparticles by anti-solvent precipitation for use in food systems. *Trends in Food Science & Technology*, 34(2), 109–123. DOI: 10.1016/j.tifs.2013.10.002

Jurkow, R., Pokluda, R., Sękara, A., & Kalisz, A. (2020). Impact of foliar application of some metal nanoparticles on antioxidant system in oakleaf lettuce seedlings. *BMC Plant Biology*, 20(1), 1–12. DOI: 10.1186/s12870-020-02490-5 PMID: 32576147

Kadam, A. A., & Lee, D. S. (2015). Glutaraldehyde Cross-Linked Magnetic Chitosan Nanocomposites: Reduction Precipitation Synthesis, Characterization, and Application for Removal of Hazardous Textile Dyes. *Bioresource Technology*, 193, 563–567. DOI: 10.1016/j.biortech.2015.06.148 PMID: 26166462

Kah, M., Kookana, R. S., Gogos, A., & Bucheli, T. D. (2018). A critical evaluation of nanopesticides and nanofertilizers against their conventional analogues. *Nature Nanotechnology*, 13(8), 677–684. DOI: 10.1038/s41565-018-0131-1 PMID: 29736032

Kalhapure, R. S., Suleman, N., Mocktar, C., Seedat, N., & Govender, T. (2015). Nanoengineered drug delivery systems for enhancing antibiotic therapy. *Journal of Pharmaceutical Sciences*, 104(3), 872–905. DOI: 10.1002/jps.24298 PMID: 25546108

Kalisz, A., Kornaś, A., Skoczowski, A., Oliwa, J., Jurkow, R., Gil, J., Sękara, A., Sałata, A., & Caruso, G. (2023). Leaf chlorophyll fluorescence and reflectance of oakleaf lettuce exposed to metal and metal (oid) oxide nanoparticles. *BMC Plant Biology*, 23(1), 329. DOI: 10.1186/s12870-023-04305-9 PMID: 37340375

Kalita, S., Kandimalla, R., Sharma, K. K., Kataki, A. C., Deka, M., & Kotoky, J. (2016). Amoxicillin functionalized gold nanoparticles reverts MRSA resistance. *Materials Science and Engineering C*, 61, 720–727. DOI: 10.1016/j.msec.2015.12.078 PMID: 26838902

Kamble, P. (2009). Protein-mediated synthesis of gold nanoparticles. *Materials Science and Engineering B*, 163(2), 93–98. DOI: 10.1016/j.mseb.2009.05.013

Kamçı, H., Celebioglu, H. U., Taş, R., & Köroğlu, E. (2022). Biogenic Synthesis of Zinc Oxide Nanoparticles Using Saponaria officinalis L., Characterisation and Antibacterial Activities. *European Journal of Science and Technology*, 35, 227–234.

Kamle, M., Mahato, D. K., Devi, S., Soni, R., Tripathi, V., Mishra, A. K., & Kumar, P. (2020). Nanotechnological interventions for plant health improvement and sustainable agriculture. *3 Biotech, 10*(4), 168.

Kandasamy, K., Venkatesh, M., Khadar, Y. S., & Rajasingh, P. (2020). One-pot green synthesis of CdS quantum dots using Opuntia ficus-indica fruit sap. *Materials Today: Proceedings*, 26, 3503–3506. DOI: 10.1016/j.matpr.2019.06.003

Kang, T., Li, F., Baik, S., Shao, W., Ling, D., & Hyeon, T. (2017). Surface design of magnetic nanoparticles for stimuli-responsive cancer imaging and therapy. *Biomaterials*, 136, 98–114. DOI: 10.1016/j.biomaterials.2017.05.013 PMID: 28525855

Kanwar, M. K., Sun, S., Chu, X., & Zhou, J. (2019). Impacts of metal and metal oxide nanoparticles on plant growth and productivity. Nanomaterials and plant potential, 379-392.

Karaca Açarı, İ., Dik, G., Bakar, B., Ulu, A., Önal, Y., & Ateş, B. (2023). Immobilization of α-amylase onto quantum dots prepared from Hypericum perforatum L. flowers and Hypericum capitatum seeds: Its physicochemical and biochemical characterization. *Topics in Catalysis*, 66(9), 563–576. DOI: 10.1007/s11244-022-01699-y

Karade, V. C., Dongale, T. D., Sahoo, S. C., Kollu, P., Chougale, A. D., Patil, P. S., & Patil, P. B. (2018). Effect of reaction time on structural and magnetic properties of green-synthesized magnetic nanoparticles. *Journal of Physics and Chemistry of Solids*, 120, 161–166. DOI: 10.1016/j.jpcs.2018.04.040

Karagoz, S., Kiremitler, N. B., Sarp, G., Pekdemir, S., Salem, S., Goksu, A. G., Onses, M. S., Sozdutmaz, I., Sahmetlioglu, E., Ozkara, E. S., Ceylan, A., & Yilmaz, E. (2021). Antibacterial, antiviral, and self-cleaning mats with sensing capabilities based on electrospun nanofibers decorated with ZnO nanorods and Ag nanoparticles for protective clothing applications. *ACS Applied Materials & Interfaces*, 13(4), 5678–5690. DOI: 10.1021/acsami.0c15606 PMID: 33492946

Karimi, M., Ghasemi, A., Zangabad, P. S., Rahighi, R., Basri, S. M. M., Mirshekari, H., Amiri, M., Pishabad, Z. S., Aslani, A., Bozorgomid, M., Ghosh, D., Beyzavi, A., Vaseghi, A., Aref, A. R., Haghani, L., Bahrami, S., & Hamblin, M. R. (2016). Smart micro/nanoparticles in stimulus-responsive drug/gene delivery systems. *Chemical Society Reviews*, 45(5), 1457–1501. DOI: 10.1039/C5CS00798D PMID: 26776487

Karmanov, A. P., Kanarsky, A. V., Kocheva, L. S., Belyy, V. A., Semenov, E. I., Rachkova, N. G., Bogdanovich, N. I., & Pokryshkin, S. A. (2021). Chemical structure and polymer properties of wheat and cabbage lignins–Valuable biopolymers for biomedical applications. *Polymer*, 220, 123571. DOI: 10.1016/j.polymer.2021.123571

Karunakaran, G., Sudha, K. G., Ali, S., & Cho, E.-B. (2023). Biosynthesis of Nanoparticles from Various Biological Sources and Its Biomedical Applications. *Molecules (Basel, Switzerland)*, 28(11), 4527. DOI: 10.3390/molecules28114527 PMID: 37299004

Karunakaran, G., Suriyaprabha, R., Manivasakan, P., Rajendran, V., & Kannan, N. (2014). Influence of nano and bulk SiO2 and Al2O3 particles on PGPR and soil nutrient contents. *Current Nanoscience*, 10(4), 604–612. DOI: 10.2174/1573413711309666012 6

Kashisaz, M., Enayatizamir, N., Fu, P., & Eslahi, M. (2024). Synthesis of nanoparticles using Trichoderma Harzianum, characterization, antifungal activity and impact on Plant Growth promoting Bacteria. *World Journal of Microbiology & Biotechnology*, 40(4), 107. DOI: 10.1007/s11274-024-03920-4 PMID: 38396217

Kasthuri, J., Veerapandian, S., & Rajendiran, N. (2009). Biological synthesis of silver and gold nanoparticles using apiin as reducing agent. *Colloids and Surfaces. B, Biointerfaces*, 68(1), 55–60. DOI: 10.1016/j.colsurfb.2008.09.021 PMID: 18977643

Kathe, N., Henriksen, B., & Chauhan, H. (2014). Physicochemical characterization techniques for solid lipid nanoparticles: Principles and limitations. *Drug Development and Industrial Pharmacy*, 40(12), 1565–1575. DOI: 10.3109/03639045.2014.909840 PMID: 24766553

Katta, V. K. M., Dubey, R. S., & Saravanan, S. (2024). Green Synthesis of Iron Nanoparticles using *Eucalyptus* Leaf Extracts and Characterization for Photocatalytic Studies. *International Journal of Nanoscience and Nanotechnology*, 20(1), 11–23.

Kaur, A., & Kumar, R. (2019). Enhanced bactericidal efficacy of polymer stabilized silver nanoparticles in conjugation with different classes of antibiotics. *RSC Advances*, 9(2), 1095–1105. DOI: 10.1039/C8RA07980C PMID: 35517620

Kaur, A., & Kumar, R. (2021). Untangling the effect of surfactants as an intermediate at gold nanoparticle-antibiotic interface for enhanced bactericidal effect. *ES Food & Agroforestry*, 7(5), 30–40. DOI: 10.30919/esfaf563

Kaur, G., Thakur, B., Gill, R., Kaur, R., & Kaur, S. (2023). Biopolymeric conjugation with food additives. *Physical Sciences Reviews*. Advance online publication. DOI: 10.1515/psr-2022-0191

Kaur, H., Kalia, A., Sandhu, J. S., Dheri, G. S., Kaur, G., & Pathania, S. (2022). Interaction of TiO2 nanoparticles with soil: Effect on microbiological and chemical traits. *Chemosphere*, 301, 134629. DOI: 10.1016/j.chemosphere.2022.134629 PMID: 35447207

Kaur, S., & Dhillon, G. S. (2014). The Versatile Biopolymer Chitosan: Potential Sources, Evaluation of Extraction Methods and Applications. *Critical Reviews in Microbiology*, 40(2), 155–175. DOI: 10.3109/1040841X.2013.770385 PMID: 23488873

Kaushik, P., Rawat, K., Aswal, V. K., Kohlbrecher, J., & Bohidar, H. B. (2018). Mixing ratio dependent complex coacervation versus bicontinuous gelation of pectin with in situ formed zein nanoparticles. *Soft Matter*, 14(31), 6463–6475. DOI: 10.1039/C8SM00809D PMID: 30051132

Kazakova, O., Puttock, R., Barton, C., Corte-León, H., Jaafar, M., Neu, V., & Asenjo, A. (2019). Frontiers of magnetic force microscopy. *Journal of Applied Physics*, 125(6), 060901. DOI: 10.1063/1.5050712

Ke, C. C., Yang, Y. C., & Tseng, W. L. (2016). Synthesis of Blue-, Green-, Yellow-, and Red-Emitting Graphene-Quantum-Dot-Based Nanomaterials with Excitation-Independent Emission. *Particle & Particle Systems Characterization*, 33(3), 132–139. DOI: 10.1002/ppsc.201500196

Khajeh Mehrizi, M., & Shahi, Z. (2020). Application of Plant-Based Natural Product to Synthesize Nanomaterial. In Srivastava, M., Srivastava, N., Mishra, P. K., & Gupta, V. K. (Eds.), *Nanomaterials in Biofuels Research* (pp. 53–73). Springer., DOI: 10.1007/978-981-13-9333-4_3

Khalid, H., Shamaila, S., Zafar, N., & Shahzadi, S. (2015). Synthesis of copper nanoparticles by chemical reduction method. *SCIENCE International*, 27(4), 3085–3088.

Khan, A., Rashid, A., Younas, R., & Chong, R. (2016). A chemical reduction approach to the synthesis of copper nanoparticles. *International Nano Letters*, 6(1), 21–26. DOI: 10.1007/s40089-015-0163-6

Khandel, P., & Shahi, S. K. (2018). Mycogenic nanoparticles and their bio-prospective applications: Current status and future challenges. *Journal of Nanostructure in Chemistry*, 8(4), 369–391. DOI: 10.1007/s40097-018-0285-2

Khandel, P., Yadaw, R. K., Soni, D. K., Kanwar, L., & Shahi, S. K. (2018). Biogenesis of metal nanoparticles and their pharmacological applications: Present status and application prospects. *Journal of Nanostructure in Chemistry*, 8(3), 217–254. DOI: 10.1007/s40097-018-0267-4

Khan, I., Saeed, K., & Khan, I. (2019). Nanoparticles: Properties, applications, and toxicities. *Arabian Journal of Chemistry*, 12(7), 908–931. DOI: 10.1016/j.arabjc.2017.05.011

Khan, M. F., & Khan, M. A. (2023). Plant-Derived Metal Nanoparticles (PDMNPs): Synthesis, Characterization, and Oxidative Stress-Mediated Therapeutic Actions. *Future Pharmacology*, 3(1), 252–295. DOI: 10.3390/futurepharmacol3010018

Khan, S. A., Jain, M., Pandey, A., Pant, K. K., Ziora, Z. M., Blaskovich, M. A. T., Shetti, N. P., & Aminabhavi, T. M. (2022). Leveraging the potential of silver nanoparticles-based materials towards sustainable water treatment. In *Journal of Environmental Management* (Vol. 319, p. 115675). Elsevier BV., DOI: 10.1016/j.jenvman.2022.115675

Khan, S. A., Shahid, S., & Lee, C.-S. (2020). Green Synthesis of Gold and Silver Nanoparticles Using Leaf Extract of Clerodendrum inerme; Characterization, Antimicrobial, and Antioxidant Activities. *Biomolecules*, 10(6), 835. DOI: 10.3390/biom10060835 PMID: 32486004

Khan, S. A., & Sharma, R. (2024). Eco-friendly Synthesis of Copper Nanoparticles: An Overview of the Epoch-making Role of Natural Resources, Applications, and Recent Developments. *Current Green Chemistry*, 11(3), 286–295. DOI: 10.2174/0122133461279579231103055412

Khan, Z., & Al-Thabaiti, S. A. (2019). Green synthesis of ZnO nanoparticles using 'Phoenix dactylifera' and their application as nano-fertilizer for sustainable agriculture. *Journal of Cleaner Production*, 207, 342–349. DOI: 10.1016/j.jclepro.2018.09.243

Khezerlou, A., Tavassoli, M., Alizadeh Sani, M., Mohammadi, K., Ehsani, A., & McClements, D. J. (2021a). Application of nanotechnology to improve the performance of biodegradable biopolymer-based packaging materials. *Polymers*, 13(24), 4399. DOI: 10.3390/polym13244399 PMID: 34960949

Khezerlou, A., Zolfaghari, H., Banihashemi, S. A., Forghani, S., & Ehsani, A. (2021b). Plant gums as the functional compounds for edible films and coatings in the food industry: A review. *Polymers for Advanced Technologies*, 32(6), 2306–2326. DOI: 10.1002/pat.5293

Khodair, Z. T., Ibrahim, N. M., Kadhim, T. J., & Mohammad, A. M. (2022). Synthesis and characterization of nickel oxide (NiO) nanoparticles using an environmentally friendly method, and their biomedical applications. *Chemical Physics Letters*, 797, 139564. DOI: 10.1016/j.cplett.2022.139564

Khursheed, R., Dua, K., Vishwas, S., Gulati, M., Jha, N. K., Aldhafeeri, G. M., Alanazi, F. G., Goh, B. H., Gupta, G., Paudel, K. R., Hansbro, P. M., Chellappan, D. K., & Singh, S. K. (2022). Biomedical applications of metallic nanoparticles in cancer: Current status and future perspectives. *Biomedicine and Pharmacotherapy*, 150, 112951. DOI: 10.1016/j.biopha.2022.112951 PMID: 35447546

Kibria, M. G., Masuk, N. I., Safayet, R., Nguyen, H. Q., & Mourshed, M. (2023). Plastic waste: Challenges and opportunities to mitigate pollution and effective management. *International Journal of Environmental Research*, 17(1), 20. DOI: 10.1007/s41742-023-00507-z PMID: 36711426

Kiefer, J., Grabow, J., Kurland, H. D., & Müller, F. A. (2015). Characterization of nanoparticles by solvent infrared spectroscopy. *Analytical Chemistry*, 87(24), 12313–12317. DOI: 10.1021/acs.analchem.5b03625 PMID: 26593634

Kim, S. H., Lee, C. M., & Kafle, K. (2013). Characterization of crystalline cellulose in biomass: Basic principles, applications, and limitations of XRD, NMR, IR, Raman, and SFG. *Korean Journal of Chemical Engineering*, 30(12), 2127–2141. DOI: 10.1007/s11814-013-0162-0

Kim, Y., Kim, W., & Park, J. W. (2016). Principles and applications of force spectroscopy using atomic force microscopy. *Bulletin of the Korean Chemical Society*, 37(12), 1895–1907. DOI: 10.1002/bkcs.11022

Kiran, M. S., & Betageri, V. S., CR, R. K., Vinay, S. P., & Latha, M. S. (2020). Synthesis of gold nanoparticles using novel *Eucalyptus tereticornis* and their in-vitro antibacterial, antioxidant and anticancer studies. *Advances in Natural Sciences: Nanoscience and Nanotechnology*, 11(4), 045012.

Kiranmayee, M., Rajesh, N., Vidya Vani, M., Khadri, H., Mohammed, A., Chinni, S. V., Ramachawolran, G., Riazunnisa, K., & Moussa, A. Y. (2023). Green synthesis of Piper nigrum copper-based nanoparticles: In silico study and ADMET analysis to assess their antioxidant, antibacterial, and cytotoxic effects. *Frontiers in Chemistry*, 11, 1–22. DOI: 10.3389/fchem.2023.1218588 PMID: 37736256

Kirdar, S. S. (2015). Current and future applications of nanotechnology in the food industry. In Conference paper, ISITES2015, Valencia-Spain.

Kishanji, M., Mamatha, G., Obi Reddy, K., Varada Rajulu, A., & Madhukar, K. (2017). In situ generation of silver nanoparticles in cellulose matrix using Azadirachta indica leaf extract as a reducing agent. *IJPAC. International Journal of Polymer Analysis and Characterization*, 22(8), 734–740. DOI: 10.1080/1023666X.2017.1369612

Kisielowski, C., Liliental-Weber, Z., & Weber, E. R. (1996). Quantitative HRTEM: A novel approach towards application oriented basic research. *Brazilian Journal of Physics*, 26(1), 83–94.

Klaholz, B. P. (2019). Deriving and refining atomic models in crystallography and cryo-EM: The latest Phenix tools to facilitate structure analysis. *Acta Crystallographica. Section D, Structural Biology*, 75(10), 878–881. DOI: 10.1107/S2059798319013391 PMID: 31588919

Klinavičius, T., Khinevich, N., Tamulevičienė, A., Vidal, L., Tamulevičius, S., & Tamulevičius, T. (2024). Deep learning methods for colloidal silver nanoparticle concentration and size distribution determination from UV-Vis extinction spectra. *The Journal of Physical Chemistry. C, Nanomaterials and Interfaces*, 128(23), 9662–9675. DOI: 10.1021/acs.jpcc.4c02459

Kocazorbaz, E. (2021). Green Synthesis, Optimization, and Characterization of Silver Nanoparticles from Euphorbia rigida Leaf Extract and Investigation of Their Antimicrobal Activities. *Bilecik Şeyh Edebali Üniversitesi Fen Bilimleri Dergisi*, 8(2), 512–522. DOI: 10.35193/bseufbd.843005

Kokilavani, S., Janani, B., Balasurya, S., & Khan, S. S. (2023). Accumulation of Engineered Nanomaterials in Soil, Water, and Air. In Fernandez-Luqueno, F., & Patra, J. K. (Eds.), *Agricultural and Environmental Nanotechnology: Novel Technologies and their Ecological Impact* (pp. 551–582). Springer Nature., DOI: 10.1007/978-981-19-5454-2_21

Kokina, I., Plaksenkova, I., Jankovskis, L., Jermaļonoka, M., & Galek, R. (2024). New Insights on Biosynthesis of Nanoparticles Using Plants Emphasizing the Use of Alfalfa (Medicago sativa L.). *Journal of Nanotechnology*, 9721166, 14. DOI: 10.1155/2024/9721166

Kolesnikov, S. I., Timoshenko, A. N., Kazeev, K. S., Akimenko, Y. V., & Myasnikova, M. A. (2019). Ecotoxicity of copper, nickel, and zinc nanoparticles assessment on the basis of biological indicators of chernozems. *Eurasian Soil Science*, 52(8), 982–987. DOI: 10.1134/S106422931908009X

Kora, A. J., & Rastogi, L. (2018). Green synthesis of palladium nanoparticles using gum ghatti (Anogeissus latifolia) and its application as an antioxidant and catalyst. *Arabian Journal of Chemistry*, 11(7), 1097–1106. DOI: 10.1016/j.arabjc.2015.06.024

Kordy, M. G. M., Abdel-Gabbar, M., Soliman, H. A., Aljohani, G., BinSabt, M., Ahmed, I. A., & Shaban, M. (2022). Phyto-Capped Ag Nanoparticles: Green Synthesis, Characterization, and Catalytic and Antioxidant Activities. *Nanomaterials (Basel, Switzerland)*, 12(3), 373. DOI: 10.3390/nano12030373 PMID: 35159718

Kreuter, J. (1995). Nanoparticulate systems in drug delivery and targeting. *Journal of Drug Targeting*, 3(3), 171–173. DOI: 10.3109/10611869509015940 PMID: 8705247

Kreuter, J. (2007). Nanoparticles—A historical perspective. *International Journal of Pharmaceutics*, 331(1), 1–10. DOI: 10.1016/j.ijpharm.2006.10.021 PMID: 17110063

Krishna, D. N. G., & Philip, J. (2022). Review on surface-characterization applications of X-ray photoelectron spectroscopy (XPS): Recent developments and challenges. *Applied Surface Science Advances*, 12, 100332. DOI: 10.1016/j.apsadv.2022.100332

Krishna, G., Kumar, S. S., Pranitha, V., Alha, M., & Charaya, S. (2015). Biogenic synthesis of silver nanoparticles and their synergistic effect with antibiotics: A study against Salmonella sp. *International Journal of Pharmacy and Pharmaceutical Sciences*, 7(11), 84–88.

Krivorotova, T., Cirkovas, A., Maciulyte, S., Staneviciene, R., Budriene, S., Serviene, E., & Sereikaite, J. (2016). Nisin-loaded pectin nanoparticles for food preservation. *Food Hydrocolloids*, 54, 49–56. DOI: 10.1016/j.foodhyd.2015.09.015

Kuang, X., Wang, Z., Luo, Z., He, Z., Liang, L., Gao, Q. H., Li, Y., Xia, K., Xie, Z., Chang, R., Wang, Y., Liu, Y., Zhao, S., Su, J., Wang, Y., Situ, W., Chen, M., Zhao, Y., Chen, X., & Liu, H. (2022). Ag nanoparticles enhance immune checkpoint blockade efficacy by promoting of immune surveillance in melanoma. *Journal of Colloid and Interface Science*, 616, 189–200. DOI: 10.1016/j.jcis.2022.02.050 PMID: 35203032

Kučuk, N., Primožič, M., Knez, Ž., & Leitgeb, M. (2022). Sustainable Biodegradable Biopolymer- Based Nanoparticles for Healthcare Applications. *International Journal of Molecular Sciences*, 24(4), 3188. DOI: 10.3390/ijms24043188 PMID: 36834596

Kulkarni, D.-S. (2024). Combinations of Biotechnology and Nanotechnology in Industrial Wastewater Treatment. In Mishra, B. K. (Ed.), *Sustainable Science and Intelligent Technologies for Societal Development* (pp. 96–106). DOI: 10.4018/979-8-3693-1186-8.ch006

Kulkarni, D., Sherkar, R., Shirsathe, C., Sonwane, R., Varpe, N., Shelke, S., More, M. P., Pardeshi, S. R., Dhaneshwar, G., Junnuthula, V., & Dyawanapelly, S. (2023). Biofabrication of nanoparticles: Sources, synthesis, and biomedical applications. *Frontiers in Bioengineering and Biotechnology*, 11, 1159193. DOI: 10.3389/fbioe.2023.1159193 PMID: 37200842

Kumar Sur, U., Ankamwar, B., Karmakar, S., Halder, A., & Das, P. (2018). Green synthesis of Silver nanoparticles using the plant extract of Shikakai and Reetha. *Materials Today: Proceedings*, 5(1), 2321–2329. DOI: 10.1016/j.matpr.2017.09.236

Kumar, A., Gupta, V., Singh, P. P., Kujur, A., & Prakash, B. (2020). Fabrication of volatile compounds loaded-chitosan biopolymer nanoparticles: Optimization, characterization and assessment against Aspergillus flavus and aflatoxin B1 contamination. *International Journal of Biological Macromolecules*, 165, 1507–1518. DOI: 10.1016/j.ijbiomac.2020.09.257 PMID: 33038402

Kumar, B., Smita, K., Seqqat, R., Benalcazar, K., Grijalva, M., & Cumbal, L. (2016). In vitro evaluation of silver nanoparticles cytotoxicity on Hepatic cancer (Hep-G2) cell line and their antioxidant activity: Green approach for fabrication and application. *Journal of Photochemistry and Photobiology. B, Biology*, 159, 8–13. DOI: 10.1016/j.jphotobiol.2016.03.011 PMID: 27010841

Kumar, C. G., & Poornachandra, Y. (2015). Biodirected synthesis of Miconazole-conjugated bacterial silver nanoparticles and their application as antifungal agents and drug delivery vehicles. *Colloids and Surfaces. B, Biointerfaces*, 125, 110–119. DOI: 10.1016/j.colsurfb.2014.11.025 PMID: 25460601

Kumar, H., Bhardwaj, K., Kuča, K., Kalia, A., Nepovimova, E., Verma, R., & Kumar, D. (2020). Flower-based green synthesis of metallic nanoparticles: Applications beyond fragrance. *Nanomaterials (Basel, Switzerland)*, 10(4), 766. DOI: 10.3390/nano10040766 PMID: 32316212

Kumari, M., Pandey, S., Mishra, S. K., Giri, V. P., Agarwal, L., Dwivedi, S., & Mishra, A. (2020). Omics-based mechanistic insight into the role of bioengineered nanoparticles for biotic stress amelioration by modulating plant metabolic pathways. *Frontiers in Bioengineering and Biotechnology*, 8, 242. DOI: 10.3389/fbioe.2020.00242 PMID: 32363178

Kumari, S., & Rath, P. K. (2014). Extraction and characterization of chitin and chitosan from (Labeo rohit) fish scales. *Procedia Materials Science*, 6, 482–489. DOI: 10.1016/j.mspro.2014.07.062

Kumar, J. A., Krithiga, T., Manigandan, S., Sathish, S., Renita, A. A., Prakash, P., Prasad, B. S. N., Kumar, T. R. P., Rajasimman, M., Hosseini-Bandegharaei, A., Prabu, D., & Crispin, S. (2021). A focus to green synthesis of metal/metal based oxide nanoparticles: Various mechanisms and applications towards ecological approach. *Journal of Cleaner Production*, 324, 129–198. DOI: 10.1016/j.jclepro.2021.129198

Kumar, K. S., & Jebakumar, K. J.. (2021). Eco-friendly synthesis of zinc oxide nanoparticles using milk and their photocatalytic activity. *Journal of Nanoparticle Research*, 23(12), 312.

Kumar, N., Salar, R., Kumar, R., Prasad, M., Brar, B., & Nain, V. (2017). Green Synthesis of Silver Nanoparticles and its Applications—. *RE:view*.

Kumar, V., Dasgupta, N., & Ranjan, S. (2018). *Nanotoxicology: Toxicity Evaluation, Risk Assessment and Management*. CRC Press. DOI: 10.1201/b21545

Kumar, V., Guleria, P., & Kumar, V. (2020). Green synthesis of nanoparticles: A review. *Green Chemistry Letters and Reviews*, 13(1), 13–29.

Kumar, V., Yadav, S. C., & Yadav, S. K. (2010). Syzygium cumini leaf and seed extract mediated biosynthesis of silver nanoparticles and their characterization. *Journal of Chemical Technology and Biotechnology*, 85(10), 1301–1309. DOI: 10.1002/jctb.2427

Kuppusamy, P., Yusoff, M. M., Maniam, G. P., & Govindan, N. (2016). Biosynthesis of metallic nanoparticles using plant derivatives and their new avenues in pharmacological applications – An updated report. *Saudi Pharmaceutical Journal*, 24(4), 473–484. DOI: 10.1016/j.jsps.2014.11.013 PMID: 27330378

Kurhade, P., Kodape, S., & Choudhury, R. (2021). Overview on green synthesis of metallic nanoparticles. *Chemicke Zvesti*, 75(10), 5187–5222. DOI: 10.1007/s11696-021-01693-w

Kuriakose, S., & Philip, D. (2014). Green synthesis of ZnO nanoparticles using rubber leaf extract and its applications. *Materials Letters*, 89, 233–235.

Kus-Liśkiewicz, M., Fickers, P., & Ben Tahar, I. (2021). Biocompatibility and Cytotoxicity of Gold Nanoparticles: Recent Advances in Methodologies and Regulations. *International Journal of Molecular Sciences*, 22(20), 20. Advance online publication. DOI: 10.3390/ijms222010952 PMID: 34681612

Kwak, S. Y., Lew, T. T. S., Sweeney, C. J., Koman, V. B., Wong, M. H., Bohmert-Tatarev, K., & Strano, M. S. (2019). Chloroplast-selective gene delivery and expression in planta using chitosan-complexed single-walled carbon nanotube carriers. *Nature Nanotechnology*, 14(5), 447–455. DOI: 10.1038/s41565-019-0375-4 PMID: 30804482

Kwak, S. Y., Wong, M. H., Lew, T. T. S., Bisker, G., Lee, M. A., Kaplan, A., & Strano, M. S. (2017). Nanosensor technology applied to living plant systems. *Annual Review of Analytical Chemistry (Palo Alto, Calif.)*, 10(1), 113–140. DOI: 10.1146/annurev-anchem-061516-045310 PMID: 28605605

Kyobe, J. W., Khan, M. D., Kinunda, G., Mubofu, E. B., & Revaprasadu, N. (2020). Synthesis of CdTe quantum dots capped with castor oil using a hot injection solution method. *Materials Science in Semiconductor Processing*, 106, 104780. DOI: 10.1016/j.mssp.2019.104780

Labruère, R., Sona, A., & Turos, E. (2019). Anti–methicillin-resistant Staphylococcus aureus nanoantibiotics. *Frontiers in Pharmacology*, 10, 1121. DOI: 10.3389/fphar.2019.01121 PMID: 31636560

Lahiani, M. H., Dervishi, E., Chen, J., Nima, Z., Gaume, A., Biris, A. S., & Khodakovskaya, M. V. (2013). Impact of carbon nanotube exposure to seeds of valuable crops. *ACS Applied Materials & Interfaces*, 5(16), 7965–7973. DOI: 10.1021/am402052x PMID: 23834323

Lakkaboyana, S. K., Khantong, S., Asmel, N. K., Obaidullah, S., Kumar, V., Kannan, K., Venkateswarlu, K., Yuzir, A., & Wan Yaacob, W. Z. (2021). Indonesian Kaolin supported nZVI (IK-nZVI) used for the an efficient removal of Pb(II) from aqueous solutions: Kinetics, thermodynamics and mechanism. In Journal of Environmental Chemical Engineering (Vol. 9, Issue 6, p. 106483). Elsevier BV. DOI: 10.1016/j.jece.2021.106483

Lakshmanan, G., Sathiyaseelan, A., Kalaichelvan, P. T., & Murugesan, K. (2018). Plant-mediated synthesis of silver nanoparticles using fruit extract of Cleome viscosa L.: Assessment of their antibacterial and anticancer activity. In Karbala International Journal of Modern Science (Vol. 4, Issue 1, pp. 61–68). University of Kerbala. DOI: 10.1016/j.kijoms.2017.10.007

Lakshmipathy, R., Sarada, N. C., Chidambaram, K., & Pasha, S. K. (2015). One-step, low-temperature fabrication of CdS quantum dots by watermelon rind: a green approach. *International Journal of Nanomedicine, 10*(sup2), 183-188.

Lakshmipriya, T., & Gopinath, S. C. B. (2021). 1 - Introduction to nanoparticles and analytical devices. In S. C. B. Gopinath & F. B. T.-N. in A. and M. D. Gang (Eds.), *Nanoparticles in Analytical and Medical Devices* (pp. 1–29). Elsevier.

Lal, S., & Nayak, P. L. (2012). Green synthesis of gold nanoparticles using various extract of plants and spices. *International Journal of Science Innovations and Discoveries*, 2, 325.

Landriscina, A., Rosen, J., & Friedman, A. J. (2015). Biodegradable chitosan nanoparticles in drug delivery for infectious disease. *Nanomedicine (London)*, 10(10), 1609–1619. DOI: 10.2217/nnm.15.7 PMID: 26008195

Langer, K., Balthasar, S., Vogel, V., Dinauer, N., Von Briesen, H., & Schubert, D. (2003). Optimization of the preparation process for human serum albumin (HSA) nanoparticles. *International Journal of Pharmaceutics*, 257(1-2), 169–180. DOI: 10.1016/S0378-5173(03)00134-0 PMID: 12711172

Lanje, A. S., Sharma, S. J., Pode, R. B., & Ningthoujam, R. S. (2010). Synthesis and optical characterization of copper oxide nanoparticles *Advances in Applied Science Research 36-401*(2), 36–40.

Larkin, P. (2017). *Infrared and Raman spectroscopy: principles and spectral interpretation*. Elsevier.

Larsson, S., Jansson, M., & Boholm, Å. (2019). Expert stakeholders' perception of nanotechnology: Risk, benefit, knowledge, and regulation. *Journal of Nanoparticle Research*, 21(3), 1–17. DOI: 10.1007/s11051-019-4498-1

Lateef, A., Darwesh, O. M., & Matter, I. A. (2021). Microbial Nanobiotechnology: The Melting Pot of Microbiology, Microbial Technology and Nanotechnology. In Lateef, A., Gueguim-Kana, E. B., Dasgupta, N., & Ranjan, S. (Eds.), *Microbial Nanobiotechnology: Principles and Applications* (pp. 1–19). Springer., DOI: 10.1007/978-981-33-4777-9_1

Laurent, S., Bridot, J.-L., Vander Elst, L., & Muller, R. (2010). Magnetic Iron Oxide Nanoparticles for Biomedical Applications. *Future Medicinal Chemistry*, 2(3), 427–449. DOI: 10.4155/fmc.09.164 PMID: 21426176

Le Wee, J., Law, M. C., Chan, Y. S., Choy, S. Y., & Tiong, A. N. T. (2022). The potential of Fe-based magnetic nanomaterials for the agriculture sector. *ChemistrySelect*, 7(17), e202104603. DOI: 10.1002/slct.202104603

Lee, B., Park, J., Ryu, M., Kim, S., Joo, M., Yeom, J.-H., Kim, S., Park, Y., Lee, K., & Bae, J. (2017). Antimicrobial peptide-loaded gold nanoparticle-DNA aptamer conjugates as highly effective antibacterial therapeutics against Vibrio vulnificus. *Scientific Reports*, 7(1), 13572. DOI: 10.1038/s41598-017-14127-z PMID: 29051620

Lee, B., Yoon, S., Lee, J. W., Kim, Y., Chang, J., Yun, J., Ro, J. C., Lee, J.-S., & Lee, J. H. (2020). Statistical Characterization of the Morphologies of Nanoparticles through Machine Learning Based Electron Microscopy Image Analysis. *ACS Nano*, 14(12), 17125–17133. DOI: 10.1021/acsnano.0c06809 PMID: 33231065

Lee, K. X., Shameli, K., Yew, Y. P., Teow, S.-Y., Jahangirian, H., Rafiee-Moghaddam, R., & Webster, T. J. (2020). Recent Developments in the Facile Bio-Synthesis of Gold Nanoparticles (AuNPs) and Their Biomedical Applications. *International Journal of Nanomedicine*, 15, 275–300. DOI: 10.2147/IJN.S233789 PMID: 32021180

Leite, F. L. (2007). The Atomic Force spectroscopy as a tool to investigate surface forces: Basic principles and applications. A. Vilas M (Ed.), *Modern Research and Educational Topics in Microscopy*,(pp. 747-755). Formatex.

León, A., Reuquen, P., Garín, C., Segura, R., Vargas, P., Zapata, P., & Orihuela, P. A. (2017). FTIR and raman characterization of TiO_2 nanoparticles coated with polyethylene glycol as carrier for 2-methoxyestradiol. *Applied Sciences (Basel, Switzerland)*, 7(1), 1–9. DOI: 10.3390/app7010049

Letchumanan, D., Sok, S. P., Ibrahim, S., Nagoor, N. H., & Arshad, N. M. (2021). Plant-based biosynthesis of copper/copper oxide nanoparticles: An update on their applications in biomedicine, mechanisms, and toxicity. *Biomolecules*, 11(4), 564. DOI: 10.3390/biom11040564 PMID: 33921379

Lewczuk, B., & Szyryńska, N. (2021). Field-emission scanning electron microscope as a tool for large-area and large-volume ultrastructural studies. *Animals (Basel)*, 11(12), 1–21. DOI: 10.3390/ani11123390 PMID: 34944167

Liao, W., Liao, C., Lin, Y.-C., & Lien, C.-Y. (2024). Characterization of Gold Nanoparticles Synthesized with Zingiber zerumbet Extracts. *Journal of Chemistry*, 2024, 1–8. DOI: 10.1155/2024/1358495

Li, L. L., Ji, J., Fei, R., Wang, C. Z., Lu, Q., Zhang, J. R., Jiang, L.-P., & Zhu, J. J. (2012a). A facile microwave avenue to electrochemiluminescent two-color graphene quantum dots. *Advanced Functional Materials*, 22(14), 2971–2979. DOI: 10.1002/adfm.201200166

Lim, J., Petaroudi, M., Pearce, G., Yasakçi, V., & Unak, P. (2021). Effects of Fluorodeoxyglucose Conjugated and Unconjugated Iron Oxide Magnetic Nanoparticles on Macrophages: A Pilot Study. *Brazilian Archives of Biology and Technology*, 64, e21200736. DOI: 10.1590/1678-4324-2021200736

Lin, L., Xu, Y., Zhang, S., Ross, I. M., Ong, A. C., & Allwood, D. A. (2013). Fabrication and luminescence of monolayered boron nitride quantum dots. *Small (Weinheim an der Bergstrasse, Germany), 10*(1), 60-65.

Lin, H., Wang, C., Wu, J., Xu, Z., Huang, Y., & Zhang, C. (2015). Colloidal synthesis of MoS 2 quantum dots: Size-dependent tunable photoluminescence and bioimaging. *New Journal of Chemistry*, 39(11), 8492–8497. DOI: 10.1039/C5NJ01698C

Lin, Y.-S., Lin, Y.-F., Nain, A., Huang, Y.-F., & Chang, H.-T. (2021). A critical review of copper nanoclusters for monitoring of water quality. In *Sensors and Actuators Reports* (Vol. 3, p. 100026). Elsevier BV., DOI: 10.1016/j.snr.2021.100026

Li, P., & Hur, J. (2017). Utilization of UV-Vis spectroscopy and related data analyses for dissolved organic matter (DOM) studies: A review. *Critical Reviews in Environmental Science and Technology*, 47(3), 131–154. DOI: 10.1080/10643389.2017.1309186

Li, P., Xia, Y., Song, K., & Liu, D. (2024). The Impact of Nanomaterials on Photosynthesis and Antioxidant Mechanisms in Gramineae Plants: Research Progress and Future Prospects. *Plants*, 13(7), 984. DOI: 10.3390/plants13070984 PMID: 38611512

Li, R., Liu, Y., Li, Z., Shen, J., Yang, Y., Cui, X., & Yang, G. (2016). Bottom-up fabrication of single-layered nitrogen-doped graphene quantum dots through intermolecular carbonization arrayed in a 2D plane. *Chemistry (Weinheim an der Bergstrasse, Germany)*, 22(1), 272–278. DOI: 10.1002/chem.201503191 PMID: 26593633

Liu, F., Cheng, X., Xiao, L., Wang, Q., Yan, K., Su, Z., Wang, Y., Ma, C., & Wang, Y. (2021). Inside-outside Ag nanoparticles-loaded polylactic acid electrospun fiber for long-term antibacterial and bone regeneration. *International Journal of Biological Macromolecules*, 167, 1338–1348. DOI: 10.1016/j.ijbiomac.2020.11.088 PMID: 33232699

Liu, L., & Corma, A. (2018). Metal Catalysts for Heterogeneous Catalysis: From Single Atoms to Nanoclusters and Nanoparticles. *Chemical Reviews*, 118(10), 4981–5079. DOI: 10.1021/acs.chemrev.7b00776 PMID: 29658707

Liu, R., & Lal, R. (2014). Synthetic apatite nanoparticles as a phosphorus fertilizer for soybean (Glycine max). *Scientific Reports*, 4(1), 5686. DOI: 10.1038/srep05686 PMID: 25023201

Liu, S., Yu, J.-M., Gan, Y.-C., Qiu, X.-Z., Gao, Z.-C., Wang, H., Chen, S.-X., Xiong, Y., Liu, G.-H., Lin, S.-E., McCarthy, A., John, J. V., Wei, D.-X., & Hou, H.-H. (2023). Biomimetic natural biomaterials for tissue engineering and regenerative medicine: New biosynthesis methods, recent advances, and emerging applications. *Military Medical Research*, 10(1), 16. DOI: 10.1186/s40779-023-00448-w PMID: 36978167

Liu, W., Feng, Y., Yu, S., Fan, Z., Li, X., Li, J., & Yin, H. (2021). The Flavonoid Biosynthesis Network in Plants. *International Journal of Molecular Sciences*, 22(23), 12824. DOI: 10.3390/ijms222312824 PMID: 34884627

Liu, Y., Liang, Q., Liu, X., Raza, H., Ma, H., & Ren, X. (2022). Treatment with ultrasound improves the encapsulation efficiency of resveratrol in zein-gum Arabic complex coacervates. *Lebensmittel-Wissenschaft + Technologie*, 153, 112331. DOI: 10.1016/j.lwt.2021.112331

Li, Y., Zhao, Y., Cheng, H., Hu, Y., Shi, G., Dai, L., & Qu, L. (2012b). Nitrogen-doped graphene quantum dots with oxygen-rich functional groups. *Journal of the American Chemical Society*, 134(1), 15–18. DOI: 10.1021/ja206030c PMID: 22136359

Li, Z., Fu, Y., McClements, D. J., & Li, T. (2022). Impact of alginate block type on the structure and physicochemical properties of curcumin-loaded complex biopolymer nanoparticles. *Lebensmittel-Wissenschaft + Technologie*, 162, 113435. DOI: 10.1016/j.lwt.2022.113435

Loganathan, S., Valapa, R. B., Mishra, R. K., Pugazhenthi, G., & Thomas, S. (2017). Thermogravimetric analysis for characterization of nanomaterials. S. Thomas (Ed.), *Thermal and Rheological Measurement Techniques for Nanomaterials Characterization* (pp. 67-108). Elsevier. DOI: 10.1016/B978-0-323-46139-9.00004-9

Lomelí-Rosales, D. A., Zamudio-Ojeda, A., Reyes-Maldonado, O. K., López-Reyes, M. E., Basulto-Padilla, G. C., Lopez-Naranjo, E. J., Zuñiga-Mayo, V. M., & Velázquez-Juárez, G. (2022). Green Synthesis of Gold and Silver Nanoparticles Using Leaf Extract of Capsicum chinense Plant. *Molecules (Basel, Switzerland)*, 27(5), 1692. DOI: 10.3390/molecules27051692 PMID: 35268794

Lopes, I. S., Soares, J. K. C., de Medeiros, L. S., & Courrol, L. C. (2024). Evaluation of ALA-capped silver, copper, and silver–copper nanoparticles for controlling fungal plant pathogens. *Microbial Pathogenesis*, 191, 106672. DOI: 10.1016/j.micpath.2024.106672 PMID: 38705219

López-Lorente, Á. I., Cárdenas, S., & González-Sánchez, Z. I. (2019). Effect of synthesis, purification and growth determination methods on the antibacterial and antifungal activity of gold nanoparticles. *Materials Science and Engineering C*, 103, 109805. DOI: 10.1016/j.msec.2019.109805 PMID: 31349488

Lowry, G. V., Hill, R. J., Harper, S., Rawle, A. F., Hendren, C. O., Klaessig, F., Nobbmann, U., Sayre, P., & Rumble, J. (2016). Guidance to improve the scientific value of zeta-potential measurements in nanoEHS. *Environmental Science. Nano*, 3(5), 953–965. DOI: 10.1039/C6EN00136J

Lucas, W. J., & Lee, J. Y. (2004). Plasmodesmata as a supracellular control network in plants. *Nature Reviews. Molecular Cell Biology*, 5(9), 712–726. DOI: 10.1038/nrm1470 PMID: 15340379

Lu, J., Yan, M., Ge, L., Ge, S., Wang, S., Yan, J., & Yu, J. (2013). Electrochemiluminescence of blue-luminescent graphene quantum dots and its application in ultrasensitive aptasensor for adenosine triphosphate detection. *Biosensors & Bioelectronics*, 47, 271–277. DOI: 10.1016/j.bios.2013.03.039 PMID: 23587790

Lu, J., Yeo, P. S. E., Gan, C. K., Wu, P., & Loh, K. P. (2011). Transforming C60 molecules into graphene quantum dots. *Nature Nanotechnology*, 6(4), 247–252. DOI: 10.1038/nnano.2011.30 PMID: 21423185

Luo, Z., Qi, G., Chen, K., Zou, M., Yuwen, L., Zhang, X., Huang, W., & Wang, L. (2016). Microwave-Assisted Preparation of White Fluorescent Graphene Quantum Dots as a Novel Phosphor for Enhanced White-Light-Emitting Diodes. *Advanced Functional Materials*, 26(16), 2739–2744. DOI: 10.1002/adfm.201505044

Lu, X. Y., Wu, D. C., Li, Z. J., & Chen, G. Q. (2011). Polymer nanoparticles. *Progress in Molecular Biology and Translational Science*, 104, 299–323. DOI: 10.1016/B978-0-12-416020-0.00007-3 PMID: 22093222

Lv, J., Christie, P., & Zhang, S. (2019). Uptake, translocation, and transformation of metal-based nanoparticles in plants: Recent advances and methodological challenges. *Environmental Science. Nano*, 6(1), 41–59. DOI: 10.1039/C8EN00645H

Lv, Y.-C., Esmaeili Shahri, E., Mahmoudi, A., Keifi Naughabi, R., Abbaspour, S., & Tayebee, R. (2023). Bioinspired nickel oxide nanoparticle as an efficient nanocarrier in the delivery of doxorubicin as an anti-bladder cancer drug. *Inorganic Chemistry Communications*, 152, 110650. DOI: 10.1016/j.inoche.2023.110650

Madan, P. L. (1978). Microencapsulation I. Phase separation or coacervation. *Drug Development and Industrial Pharmacy*, 4(1), 95–116. DOI: 10.3109/03639047809055641

Madhumitha, G., & Elango, G. (2018). Green synthesis of ZnO nanoparticles using Tridax procumbens leaf extract and its applications. *Journal of Photochemistry and Photobiology. B, Biology*, 185, 150–158.

Madkour, L. H. (2018). *Biogenic– Biosynthesis Metallic Nanoparticles (MNPs) for Pharmacological, Biomedical and Environmental Nanobiotechnological Applications.*

Maensiri, S., Laokul, P., Klinkaewnarong, J., Phokha, S., Promarak, V., & Seraphin, S. (2008). Indium oxide (In_2O_3) nanoparticles using Aloe vera plant extract: Synthesis and optical properties. *Journal of Optoelectronics and Advanced Materials*, 10(3), 161–165.

Ma, H., Shieh, K. J., & Qiao, T. X. (2006). Study of transmission electron microscopy (TEM) and scanning electron microscopy (SEM). *Nature and Science*, 4(3), 14–22.

Mahalakhsmi, A., & Baskar, G. (2019). Green synthesis and characterization of cadmium-tellurium quantum dots using pomelo peel aqueous extract. *Journal of Electronic Materials*, 48(9), 5975–5979. DOI: 10.1007/s11664-019-07352-x

Maity, D., Gupta, U., & Saha, S. (2022). Biosynthesized metal oxide nanoparticles for sustainable agriculture: Next-generation nanotechnology for crop production, protection and management. *Nanoscale*, 14(38), 13950–13989. DOI: 10.1039/D2NR03944C PMID: 36124943

Majeed, H., Iftikhar, T., Nadeem, M. A., & Nazir, M. A. (2024). Green synthesis of *Eucalyptus globulus* zinc nanoparticles and its use in antimicrobial insect repellent paint formulation in bulk industrial production. *Heliyon*, 10(2), e24467. DOI: 10.1016/j.heliyon.2024.e24467 PMID: 38288019

Makarov, V. V., Love, A. J., Sinitsyna, O. V., Makarova, S. S., Yaminsky, I. V., Taliansky, M. E., & Kalinina, N. O. (2014). "Green" nanotechnologies: Synthesis of metal nanoparticles using plants. *Acta Naturae (English Ed.)*, 6(1), 35–44. DOI: 10.32607/20758251-2014-6-1-35-44 PMID: 24772325

Ma, L., Kohli, M., & Smith, A. (2013). Nanoparticles for Combination Drug Therapy. *ACS Nano*, 7(11), 9518–9525. DOI: 10.1021/nn405674m PMID: 24274814

Malik, S., Muhammad, K., & Waheed, Y. (2023). Nanotechnology: A Revolution in Modern Industry. *Molecules (Basel, Switzerland)*, 28(2), 661. DOI: 10.3390/molecules28020661 PMID: 36677717

Mali, S. C., Dhaka, A., Githala, C. K., & Trivedi, R. (2020). Green synthesis of copper nanoparticles using *Celastrus paniculatus* Willd. leaf extract and their photocatalytic and antifungal properties. *Biotechnology Reports (Amsterdam, Netherlands)*, 27, e00518. DOI: 10.1016/j.btre.2020.e00518 PMID: 32923378

Mallakpour, S., & Hussain, C. M. (2021). Characterization of nanoparticles by FTIR and FTIR-microscopy handbook of consumer nanoproducts. S. Mallakpour (Ed.), *Handbook of Consumer Nanoproducts* (pp. 978-981). DOI: 10.1007/978-981-15-6453-6

Mallin, M. A., & Cahoon, L. B. (2003). Industrialized animal production—A major source of nutrient and microbial pollution to aquatic ecosystems. *Population and Environment*, 24(5), 369–385. DOI: 10.1023/A:1023690824045

Manandhar, S., Sankhe, R., Harish Kumar, B., & Pai, K. S. R. (2021). Clinical protocols for nanopharmacokinetics assessment. In Thorat, N. D., & Kumar, N. (Eds.), *Nano-Pharmacokinetics and Theranostics* (pp. 193–220). Academic Press., DOI: 10.1016/B978-0-323-85050-6.00001-3

Manikandan, V., Velmurugan, P., Park, J.-H., Lovanh, N., Seo, S.-K., Jayanthi, P., Park, Y.-J., Cho, M., & Oh, B.-T. (2016). Synthesis and antimicrobial activity of palladium nanoparticles from Prunus × yedoensis leaf extract. *Materials Letters*, 185, 335–338. DOI: 10.1016/j.matlet.2016.08.120

Manimegalai, P., Selvam, K., Loganathan, S., Kirubakaran, D., Shivakumar, M. S., Govindasamy, M., Rajaji, U., & Bahajjaj, A. A. A. (2024). Green synthesis of zinc oxide (ZnO) nanoparticles using aqueous leaf extract of Hardwickia binata: Their characterizations and biological applications. *Biomass Conversion and Biorefinery*, 14(11), 12559–12574. DOI: 10.1007/s13399-023-04279-6

Manna, I., & Bandyopadhyay, M. (2019). A review on the biotechnological aspects of utilizing engineered nanoparticles as delivery systems in plants. *Plant Gene*, 17, 100167. DOI: 10.1016/j.plgene.2018.100167

Mansfield, E., Tyner, K. M., Poling, C. M., & Blacklock, J. L. (2014). Determination of nanoparticle surface coatings and nanoparticle purity using microscale thermogravimetric analysis. *Analytical Chemistry*, 86(3), 1478–1484. DOI: 10.1021/ac402888v PMID: 24400715

Maraveas, C., Bayer, I. S., & Bartzanas, T. (2021). Recent Advances in Antioxidant Polymers: From Sustainable and Natural Monomers to Synthesis and Applications. *Polymers*, 13(15), 2465. Advance online publication. DOI: 10.3390/polym13152465 PMID: 34372069

Mardina, V., Fadlly, T. A., Harmawan, T., Sufriadi, E., Iqramullah, M., Umar, H., & Ilyas, S. (2024). Green synthesis of gold nanoparticles from the aqueous extracts of Sphagneticola trilobata (L.) J.F Pruski as anti-breast cancer agents. *Journal of Advanced Pharmaceutical Technology & Research*, 15(2), 75–80. DOI: 10.4103/japtr.japtr_410_23 PMID: 38903547

Marín-Caba, L., Chariou, P. L., Pesquera, C., Correa-Duarte, M. A., & Steinmetz, N. F. (2019). Tobacco Mosaic Virus-Functionalized Mesoporous Silica Nanoparticles, a Wool-Ball-like Nanostructure for Drug Delivery. *Langmuir*, 35(1), 203–211. DOI: 10.1021/acs.langmuir.8b03337 PMID: 30576145

Marques, A. C., Rojas-Hernandez, R. E., & Almeida, R. M. (2021). Optical spectroscopy methods for the characterization of sol–gel materials. *Journal of Sol-Gel Science and Technology*, 100(1), 1–43. DOI: 10.1007/s10971-021-05592-0

Martin, C., Low, W., Gupta, A., Amin, M., Radecka, I., Britland, S., Raj, P., & Kenward, K. (2015). Strategies for antimicrobial drug delivery to biofilm. *Current Pharmaceutical Design*, 21(1), 43–66. DOI: 10.2174/1381612820666140905123529 PMID: 25189862

Martínez, C., J., A. Chequer, N., L. González, J., & Cordova, T. (. (2013). Alternative metodology for gold nanoparticles diameter characterization using PCA technique and UV-VIS spectrophotometry. *Nanoscience and Nanotechnology*, 2(6), 184–189. DOI: 10.5923/j.nn.20120206.06

Martínez, G., Merinero, M., Pérez-Aranda, M., Pérez-Soriano, E. M., Ortiz, T., Villamor, E., Begines, B., & Alcudia, A. (2020). Environmental impact of nanoparticles' application as an emerging technology: A review. *Materials (Basel)*, 14(1), 166. DOI: 10.3390/ma14010166 PMID: 33396469

Matouk, Z., Islam, M., Gutiérrez, M., Pireaux, J.-J., & Achour, A. (2022). X-ray Photoelectron Spectroscopy (XPS) Analysis of Ultrafine Au Nanoparticles Supported over Reactively Sputtered TiO2 Films. *Nanomaterials (Basel, Switzerland)*, 12(20), 3692. DOI: 10.3390/nano12203692 PMID: 36296882

Ma, X., Geiser-Lee, J., Deng, Y., & Kolmakov, A. (2010). Interactions between engineered nanoparticles (ENPs) and plants: Phytotoxicity, uptake and accumulation. *The Science of the Total Environment*, 408(16), 3053–3061. DOI: 10.1016/j.scitotenv.2010.03.031 PMID: 20435342

Mayeen, A., Shaji, L. K., Nair, A. K., & Kalarikkal, N. (2018). Morphological characterization of nanomaterials. S. Bhagyaraj (Ed.), *Characterization of Nanomaterials: Advances and key technologies* (pp. 335-364). Elsevier. DOI: 10.1016/B978-0-08-101973-3.00012-2

Mba, I. E., & Nweze, E. I. (2021). Nanoparticles as therapeutic options for treating multidrug-resistant bacteria: Research progress, challenges, and prospects. *World Journal of Microbiology & Biotechnology*, 37(6), 1–30. DOI: 10.1007/s11274-021-03070-x PMID: 34046779

Medici, S., Peana, M., Pelucelli, A., & Zoroddu, M. A. (2021). An updated overview on metal nanoparticles toxicity. *Seminars in Cancer Biology*, 76, 17–26. DOI: 10.1016/j.semcancer.2021.06.020 PMID: 34182143

Medina-Pérez, G., Fernández-Luqueño, F., Vazquez-Nuñez, E., López-Valdez, F., Prieto-Mendez, J., Madariaga-Navarrete, A., & Miranda-Arámbula, M. (2019). Remediating polluted soils using nanotechnologies: Environmental benefits and risks. *Polish Journal of Environmental Studies*, 28(3), 1013–1030. DOI: 10.15244/pjoes/87099

Mehmood, A., Murtaza, G., Bhatti, T., Kausar, R., & Ahmed, M. (2016). Biosynthesis, characterization and antimicrobial action of silver nanoparticles from root bark extract of Berberis lycium Royle. *Pakistan Journal of Pharmaceutical Sciences*, 29, 131–136. PMID: 26826826

Mehta, N., & Pai, R. (2022). Amalgamation of Nanoparticles within Drug Carriers: A Synergistic Approach or a Futile Attempt? *Pharmaceutical Nanotechnology*, 10(5), 354–367. DOI: 10.2174/2211738510666220902150449 PMID: 36056844

Meier, S., Moore, F., Morales, A., González, M. E., Seguel, A., Meriño-Gergichevich, C., & Mejías, J. (2020). Synthesis of calcium borate nanoparticles and its use as a potential foliar fertilizer in lettuce (Lactuca sativa) and zucchini (Cucurbita pepo). *Plant Physiology and Biochemistry*, 151, 673–680. DOI: 10.1016/j.plaphy.2020.04.025 PMID: 32353673

Mekuye, B., & Abera, B. (2023). Nanomaterials: An overview of synthesis, classification, characterization, and applications. *Nano Select*, 4(8), 486–501. DOI: 10.1002/nano.202300038

Menon, S., Agarwal, H., Kumar, S., & Kumar, V. (2017). Green synthesis of silver nanoparticles using medicinal plant Acalypha indica leaf extracts and its application as an antioxidant and antimicrobial agent against foodborne pathogens. *International Journal of Applied Pharmaceutics*, 9(5), 42. DOI: 10.22159/ijap.2017v9i5.19464

Menon, S., Rajeshkumar, S., & Venkat Kumar, S. (2017). A review on biogenic synthesis of gold nanoparticles, characterization, and its applications. *Resource-Efficient Technologies*, 3(4), 516–527. DOI: 10.1016/j.reffit.2017.08.002

Meshkatalsadat, M. H., Momeni, A., & Abdollahzadeh, M. R. (2023). Biosynthesis of Zinc Oxide Nanoparticles Using Punica granatum L. Waste Peel Extract, and Assessment of Antioxidant and Catalytic Activity. *Nano Biomedicine and Engineering*, 15(4), 378–388. DOI: 10.26599/NBE.2023.9290030

Mewada, A., Pandey, S., Shinde, S., Mishra, N., Oza, G., Thakur, M., Sharon, M., & Sharon, M. (2013). Green synthesis of biocompatible carbon dots using aqueous extract of Trapa bispinosa peel. *Materials Science and Engineering C*, 33(5), 2914–2917. DOI: 10.1016/j.msec.2013.03.018 PMID: 23623114

Meyer, R. C., Winter, A. R., & Weiser, H. H. (1959). Edible protective coatings for extending the shelf life of poultry. *Food Technology*, 13, 146–148.

Meyns, M., Dietz, F., Weinhold, C.-S., Züge, H., Finckh, S., & Gerdts, G. (2022). Multi-feature round silicon membrane filters enable fractionation and analysis of micro- and nanoplastics with Raman spectroscopy and nano-FTIR. *Analytical Methods*, 15(5), 606–617. DOI: 10.1039/D2AY01036D PMID: 36644945

Miao, T., Wang, J., Zeng, Y., Liu, G., & Chen, X. (2018). Polysaccharide-based controlled release systems for therapeutics delivery and tissue engineering: From bench to bedside. *Advancement of Science*, 5(4), 1700513. DOI: 10.1002/advs.201700513 PMID: 29721408

Mikhailova, E. O. (2020). Silver Nanoparticles: Mechanism of Action and Probable Bio-Application. *Journal of Functional Biomaterials*, 11(4), 84. DOI: 10.3390/jfb11040084 PMID: 33255874

Mikhailova, E. O. (2021). Gold Nanoparticles: Biosynthesis and Potential of Biomedical Application. *Journal of Functional Biomaterials*, 12(4), 70. DOI: 10.3390/jfb12040070 PMID: 34940549

Milan, J., Niemczyk, K., & Kus-Liśkiewicz, M. (2022). Treasure on the Earth—Gold Nanoparticles and Their Biomedical Applications. *Materials (Basel)*, 15(9), 3355. DOI: 10.3390/ma15093355 PMID: 35591689

Ming, X., Weber, G. H., Ayres, J. W., & Sandine, W. E. (1997). Bacteriocins applied to food packaging materials to inhibit Listeria monocytogenes on meats. *Journal of Food Science*, 62(2), 413–415. DOI: 10.1111/j.1365-2621.1997.tb04015.x

Minnelli, C., Moretti, P., Laudadio, E., Gerelli, Y., Pigozzo, A., Armeni, T., Galeazzi, R., Mariani, P., & Mobbili, G. (2022). Tuning curvature and phase behavior of monoolein bilayers by epigallocatechin-3-gallate: Structural insight and cytotoxicity. *Colloids and Surfaces. B, Biointerfaces*, 209, 112171. DOI: 10.1016/j.colsurfb.2021.112171 PMID: 34736221

Mino, L., Borfecchia, E., Segura-Ruiz, J., Giannini, C., Martinez-Criado, G., & Lamberti, C. (2018). Materials characterization by synchrotron x-ray microprobes and nanoprobes. *Reviews of Modern Physics*, 90(2), 025007. DOI: 10.1103/RevModPhys.90.025007

Mirabello, G., Lenders, J. J. M., & Sommerdijk, N. A. J. M. (2016). Bioinspired synthesis of magnetite nanoparticles. *Chemical Society Reviews*, 45(18), 5085–5106. DOI: 10.1039/C6CS00432F PMID: 27385627

Miri, A., Mahdinejad, N., Ebrahimy, O., Khatami, M., & Sarani, M. (2019). Zinc oxide nanoparticles: Biosynthesis, characterization, antifungal and cytotoxic activity. *Materials Science and Engineering C*, 104, 109981. DOI: 10.1016/j.msec.2019.109981 PMID: 31500056

Mirzaeifard, Z., & Shariatinia, Z. (2024). Economical, one-pot, and green synthesis of plant-based carbon quantum dots for efficient visible-light photocatalytic dye degradation and water purification. *Journal of the Taiwan Institute of Chemical Engineers*, 163, 105655. DOI: 10.1016/j.jtice.2024.105655

Mishra, R. K., Thomas, S., & Zachariah, A. K. (2017). Energy-dispersive X-ray spectroscopy techniques for nanomaterial. S. Thomas (Ed.), *Microscopy methods in nanomaterials characterization* (pp. 383-405). Elsevier. DOI: 10.1016/B978-0-323-46141-2.00012-2

Mishra, A., Goswami, R., & Bhatt, N. (2021). *Polymeric nanomaterials as broad-spectrum antimicrobial compounds Advances in Polymeric Nanomaterials for Biomedical Applications*. Elsevier.

Mishra, D., Rajurkar, S., Mishra, N., Jadhav, N., & Ballurkar, B. (2017). Green Synthesis of Gold Nanoparticles by Azadirachta indica Leaf Extract and Coating with Morinda citrifolia Fruit Extract: Their Characterization. *International Journal of Livestock Research*, 7, 205–219. DOI: 10.5455/ijlr.20170306091736

Mittal, A. K., Chisti, Y., & Banerjee, U. C. (2013). Synthesis of metallic nanoparticles using plant extracts. *Biotechnology Advances*, 31(2), 346–356. DOI: 10.1016/j.biotechadv.2013.01.003 PMID: 23318667

Mittal, D., Kaur, G., Singh, P., Yadav, K., & Ali, S. A. (2020). Nanoparticle-based sustainable agriculture and food science: Recent advances and future outlook. *Frontiers in Nanotechnology*, 2, 579954. DOI: 10.3389/fnano.2020.579954

Modena, M. M., Rühle, B., Burg, T. P., & Wuttke, S. (2019). Nanoparticle characterization: What to Measure? *Advanced Materials*, 31(32), 1–26. DOI: 10.1002/adma.201901556 PMID: 31148285

Mohamed, M. A., Jaafar, J., Ismail, A. F., Othman, M. H. D., & Rahman, M. A. (2017). Fourier transform infrared (FTIR) spectroscopy. N. Hilal, & A. F. Ismail (Eds.), *Membrane characterization* (pp. 3-29). Elsevier. DOI: 10.1016/B978-0-444-63776-5.00001-2

Mohamed, F. A. N., & Laraba-Djebari, F. (2016). Development and characterization of a new carrier for vaccine delivery based on calcium-alginate nanoparticles: Safe immunoprotective approach against scorpion envenoming. *Vaccine*, 34(24), 2692–2699. DOI: 10.1016/j.vaccine.2016.04.035 PMID: 27109567

Mohammadi Dehcheshmeh, I., Poursattar Marjani, A., Sadegh, F., & Ebrahim Soltani, M. (2024). Copper Application and Copper Nanoparticles in Chemistry. In Various Uses of Copper Material [Working Title]. IntechOpen. DOI: 10.5772/intechopen.1004068

Mohammadi, M., Shahisaraee, S. A., Tavajjohi, A., Pournoori, N., Muhammadnejad, S., Mohammadi, S. R., Poursalehi, R., & Delavari H, H. (2019). Green synthesis of silver nanoparticles using Zingiber officinale and Thymus vulgaris extracts: Characterisation, cell cytotoxicity, and its antifungal activity against Candida albicans in comparison to fluconazole. *IET Nanobiotechnology / IET*, 13(2), 114–119. DOI: 10.1049/iet-nbt.2018.5146 PMID: 31051440

Mohammed, A. E. (2015). Green synthesis, antimicrobial and cytotoxic effects of silver nanoparticles mediated by *Eucalyptus camaldulensis* leaf extract. *Asian Pacific Journal of Tropical Biomedicine*, 5(5), 382–386. DOI: 10.1016/S2221-1691(15)30373-7

Mohammed, A. E., Aldahasi, R. M., Rahman, I., Shami, A., Alotaibi, M., BinShabaib, M. S., ALHarthi, S. S., & Aabed, K. (2024). The antimicrobial activity of tea tree oil *(Melaleuca alternifolia)* and its metal nanoparticles in oral bacteria. *PeerJ*, 12, e17241. DOI: 10.7717/peerj.17241 PMID: 38854801

Mohan, M., Sankar, K., & Ghosh, A. (2020). Effect of zinc oxide nanoparticles on zinc uptake and growth of wheat (Triticum aestivum). *Journal of Plant Nutrition and Soil Science*, 183(3), 332–341. DOI: 10.1002/jpln.201900473

Mohanty, B., Aswal, V. K., Kohlbrecher, J., & Bohidar, H. B. (2005). Synthesis of gelatin nanoparticles via simple coacervation. *Journal of Surface Science and Technology*, 21(3/4), 149.

Mojaveryazdi, F. S., Muhamad, I. I., Rezania, S., & Benham, H. (2014). Importance of glucose and Pseudomonas in producing degradable plastics. *Jurnal Teknologi*, 69(5). Advance online publication. DOI: 10.11113/jt.v69.3194

Moloney, M. P., Govan, J., Loudon, A., Mukhina, M., & Gun'ko, Y. K. (2015). Preparation of chiral quantum dots. *Nature Protocols*, 10(4), 558–573. DOI: 10.1038/nprot.2015.028 PMID: 25741991

Montanha, G. S., Rodrigues, E. S., Marques, J. P. R., De Almeida, E., Dos Reis, A. R., & Pereira de Carvalho, H. W. (2020). X-ray fluorescence spectroscopy (XRF) applied to plant science: Challenges towards in vivo analysis of plants. *Metallomics*, 12(2), 183–192. DOI: 10.1039/c9mt00237e PMID: 31793600

Montes, A., Gordillo, M. D., Pereyra, C., & De La Ossa, E. M. (2011). Co-precipitation of amoxicillin and ethyl cellulose microparticles by supercritical antisolvent process. *The Journal of Supercritical Fluids*, 60, 75–80. DOI: 10.1016/j.supflu.2011.05.002

Moond, M., Singh, S., Sangwan, S., Devi, R., & Beniwal, R. (2022). Green synthesis and applications of silver nanoparticles: A systematic review. *AATCC Journal of Research*, 9(6), 272–285. DOI: 10.1177/24723444221119847

Moradi Dehaghi, S., Rahmanifar, B., Moradi, A. M., & Azar, P. A. (2014). Removal of Permethrin Pesticide from Water by Chitosan-Zinc Oxide Nanoparticles Composite as an Adsorbent. *Journal of Saudi Chemical Society*, 18(4), 348–355. DOI: 10.1016/j.jscs.2014.01.004

More, S., Bampidis, V., Franz, R., Bragard, C., Halldorsson, T., Hernández-Jerez, A., Hougaard Bennekou, S., Koutsoumanis, K., Lambré, C., Machera, K., Naegeli, H., Nielsen, S., Schlatter, J., & Schrenk, D., Silano (deceased), V., Turck, D., Younes, M., Castenmiller, J., Chaudhry, Q., … Schoonjans, R. (. (2021). Guidance on risk assessment of nanomaterials to be applied in the food and feed chain: Human and animal health. *EFSA Journal*, 19(8). Advance online publication. DOI: 10.2903/j.efsa.2021.6768 PMID: 34377190

Morrow, K. J.Jr, Bawa, R., & Wei, C. (2007). Recent advances in basic and clinical nanomedicine. *The Medical Clinics of North America*, 91(5), 805–843. DOI: 10.1016/j.mcna.2007.05.009 PMID: 17826104

Mouhamad, R. S., Al Khafaji, K. A., Al-Dharob, M. H., & Al-Abodi, E. E. (2022). Antifungal, antibacterial and anti-yeast activities evaluation of oxides of silver, zinc and titanium nanoparticles. *Chemistry International*, 8(4), 159–166. DOI: 10.5281/zenodo.7317423

Mourdikoudis, S., Pallares, R. M., & Thanh, N. T. (2018). Characterization techniques for nanoparticles: Comparison and complementarity upon studying nanoparticle properties. *Nanoscale*, 10(27), 12871–12934. DOI: 10.1039/C8NR02278J PMID: 29926865

Moustafa, H., Darwish, N. A., & Youssef, A. M. (2022). Rational formulations of sustainable polyurethane/chitin/rosin composites reinforced with ZnO-doped-SiO2 nanoparticles for green packaging applications. *Food Chemistry*, 371, 131193. DOI: 10.1016/j.foodchem.2021.131193 PMID: 34649200

Mo, Y., Tang, Y., Wang, S., Lin, J., Zhang, H., & Luo, D. (2015). Green synthesis of silver nanoparticles using *Eucalyptus* leaf extract. *Materials Letters*, 144, 165–167. DOI: 10.1016/j.matlet.2015.01.004

Muhamad, I. I., Sabbagh, F., & Abdul Karim, N. (2017). Plant Secondary Metabolites. In *Polyhydroxyalkanoates: A Valuable Secondary Metabolite Produced in Microorganisms and Plants* (1st ed., Vol. 3, pp. 205–234). Apple Academic Press.

Muhammed, A., Asere, T. G., & Diriba, T. F. (2024). Photocatalytic and antimicrobial properties of ZnO and Mg-Doped ZnO nanoparticles synthesized using *lupinus albus* leaf extract. *ACS Omega*, 9(2), 2480–2490. DOI: 10.1021/acsomega.3c07093 PMID: 38250416

Muir, V. G., & Burdick, J. A. (2020). Chemically modified biopolymers for the formation of biomedical hydrogels. *Chemical Reviews*, 121(18), 10908–10949. DOI: 10.1021/acs.chemrev.0c00923 PMID: 33356174

Mulvaney, S. P., & Keating, C. D. (2000). Raman spectroscopy. *Analytical Chemistry*, 72(12), 145–158. DOI: 10.1021/a10000155 PMID: 10882205

Munir, A., Salah, M. A., Ali, M., Ali, B., Saleem, M. H., Samarasinghe, K. G. B. A., De Silva, S. I. S., Ercisli, S., Iqbal, N., & Anas, M. (2024). Advancing Agriculture: Harnessing Smart Nanoparticles for Precision Fertilization. *BioNanoScience*. Advance online publication. DOI: 10.1007/s12668-024-01597-5

Murali, N., Rainu, S. K., Sharma, A., Siddhanta, S., Singh, N., & Betal, S. (2024). Remotely Controlled Surface Charge Modulation of Magnetoelectric Nanogenerators for Swift and Efficient Drug Delivery. *ACS Omega*, 9(26), 28937–28950. DOI: 10.1021/acsomega.4c03825 PMID: 38973906

Murdock, R. C., Braydich-stolle, L., Schrand, A. M., Schlager, J. J., Hussain, S. M., & Al, M. E. T. (2008). Characterization of nanomaterial dispersion in solution prior to in vitro exposure using dynamic light scattering technique. *Toxicological Sciences*, 101(2), 239–253. DOI: 10.1093/toxsci/kfm240 PMID: 17872897

Murei, A., Pillay, K., & Samie, A. (2021). Syntheses, characterization, and antibacterial evaluation of P. grandiflora extracts conjugated with gold nanoparticles. *Journal of Nanotechnology*, 2021(1), 8687627. DOI: 10.1155/2021/8687627

Muthiah, B., Muthukrishnan, L., Anita Lett, J., Sagadevan, S., Kesavan, S., Vennila, S., & Ahmad, N. (2020). *Eucalyptus concoction* mediated synthesis of gold nanoparticles and its bioactive role explored via antimicrobial and cytotoxic studies. *Journal of Nanoscience and Nanotechnology*, 20(10), 6326–6333. DOI: 10.1166/jnn.2020.17897 PMID: 32384982

Muthulakshmi, L., Rajini, N., Nellaiah, H., Kathiresan, T., Jawaid, M., & Rajulu, A. V. (2017). Preparation and properties of cellulose nanocomposite films with in situ generated copper nanoparticles using Terminalia catappa leaf extract. *International Journal of Biological Macromolecules*, 95, 1064–1071. DOI: 10.1016/j.ijbiomac.2016.09.114 PMID: 27984140

Naganthran, A., Verasoundarapandian, G., Khalid, F. E., Masarudin, M. J., Zulkharnain, A., Nawawi, N. M., Karim, M., Che Abdullah, C. A., & Ahmad, S. A. (2022). Synthesis, characterization and biomedical application of silver nanoparticles. *Materials (Basel)*, 15(2), 427. DOI: 10.3390/ma15020427 PMID: 35057145

Nagar, N., & Devra, V. (2018). Green synthesis and characterization of copper nanoparticles using Azadirachta indica leaves. *Materials Chemistry and Physics*, 213, 44–51. DOI: 10.1016/j.matchemphys.2018.04.007

Naiel, B., Fawzy, M., Halmy, M. W. A., & Mahmoud, A. E. D. (2022). Green synthesis of zinc oxide nanoparticles using Sea Lavender (Limonium pruinosum L. Chaz.) extract: Characterization, evaluation of anti-skin cancer, antimicrobial and antioxidant potentials. *Scientific Reports*, 12(1), 20370. DOI: 10.1038/s41598-022-24805-2 PMID: 36437355

Naimi-Shamel, N., Pourali, P., & Dolatabadi, S. (2019). Green synthesis of gold nanoparticles using Fusarium oxysporum and antibacterial activity of its tetracycline conjugant. *Journal de Mycologie Médicale*, 29(1), 7–13. DOI: 10.1016/j.mycmed.2019.01.005 PMID: 30709721

Nairm, J. G. (1995). 3 Coacervation-phase separation technology. []. Academic Press.]. *Advances in Pharmaceutical Sciences*, 7, 93–219. DOI: 10.1016/S0065-3136(06)80005-1

Nakum, J., & Bhattacharya, D. (2022). Various Green Nanomaterials Used for Wastewater and Soil Treatment: A Mini-Review. In Frontiers in Environmental Science (Vol. 9). Frontiers Media SA. DOI: 10.3389/fenvs.2021.724814

Nallusamy, S., & Manoj Babu, A. (2016). X-Ray differaction and fesem analysis for mixture of hybrid nano particles in heat transfer applications. *Journal of Nano Research*, 37, 58–67. . DOI: 10.4028/www .scientific.net/JNanoR.37.58

Nam, N. H., Luong, N. H. (2019). Nanoparticles: synthesis and applications. Materials for Biomedical Engineering. 211–40.

Nami, S., Aghebati-Maleki, A., & Aghebati-Maleki, L. (2021). Current applications and prospects of nanoparticles for antifungal drug delivery. *EXCLI Journal*, 20, 562–584. DOI: 10.17179/excli2020-3068 PMID: 33883983

Nan, C., Yue, W., Tao, L., & Yang, X. (2021). Fourier transform infrared nano-spectroscopy: Mechanism and applications. *Applied Spectroscopy Reviews*, 56(7), 531–552. DOI: 10.1080/05704928.2020.1830789

Nancy, B. A., & Elumalai, K. (2019). Synthesis of silver nanoparticles using Pelargonium graveolens essential Oil and anti-fungal activity. *International Journal of Pharmacy and Biological Sciences*, 9, 176–185.

Narasaiah, P., Kumar Mandal, B., & Sarada, N. C. (2017). Green synthesis of Pd NPs from Pimpinella tirupatiensis plant extract and their application in photocatalytic activity dye degradation. *IOP Conference Series. Materials Science and Engineering*, 263, 22013. DOI: 10.1088/1757-899X/263/2/022013

Narendhran, S., & Sivaraj, R. (2016). Biogenic ZnO nanoparticles synthesized using L. aculeata leaf extract and their antifungal activity against plant fungal pathogens. *Bulletin of Materials Science*, 39(1), 1–5. DOI: 10.1007/s12034-015-1136-0

Narware, J., Singh, S. P., Manzar, N., & Kashyap, A. S. (2023). Biogenic synthesis, characterization, and evaluation of synthesized nanoparticles against the pathogenic fungus Alternaria solani. *Frontiers in Microbiology*, 14, 1159251. DOI: 10.3389/fmicb.2023.1159251 PMID: 37138620

Naseer, M., Aslam, U., Khalid, B., & Chen, B. (2020). Green route to synthesize Zinc Oxide Nanoparticles using leaf extracts of Cassia fistula and Melia azadarach and their antibacterial potential. *Scientific Reports*, 10(1), 9055. DOI: 10.1038/s41598-020-65949-3 PMID: 32493935

Nasseri, M. A., Keshtkar, H., Kazemnejadi, M., & Allahresani, A. (2020). Phytochemical properties and antioxidant activity of Echinops persicus plant extract: Green synthesis of carbon quantum dots from the plant extract. *SN Applied Sciences*, 2(4), 1–12. DOI: 10.1007/s42452-020-2466-0

Nathanael, K., Cheng, S., Kovalchuk, N. M., Arcucci, R., & Simmons, M. J. H. (2023). Optimization of microfluidic synthesis of silver nanoparticles: A generic approach using machine learning. *Chemical Engineering Research & Design*, 193, 65–74. DOI: 10.1016/j.cherd.2023.03.007

Nath, D., & Banerjee, P. (2013). Green nanotechnology–a new hope for medical biology. *Environmental Toxicology and Pharmacology*, 36(3), 997–1014. DOI: 10.1016/j.etap.2013.09.002 PMID: 24095717

Navarro, E., Piccapietra, F., Wagner, B., Marconi, F., Kaegi, R., Odzak, N., & Behra, R. (2008). Toxicity of silver nanoparticles to Chlamydomonas reinhardtii. *Environmental Science & Technology*, 42(23), 8959–8964. DOI: 10.1021/es801785m PMID: 19192825

Naveenkumar, S., Kamaraj, C., Ragavendran, C., Vaithiyalingam, M., Sugumar, V., & Marimuthu, K. (2023). Gracilaria corticata red seaweed mediate biosynthesis of silver nanoparticles: Larvicidal, neurotoxicity, molecular docking analysis, and ecofriendly approach. *Biomass Conversion and Biorefinery*. Advance online publication. DOI: 10.1007/s13399-023-04026-x

Nawaz, S., & Bano, A. (2020). Effects of PGPR (*Pseudomonas* sp.) and Ag-nanoparticles on enzymatic activity and physiology of cucumber. *Recent Patents on Food, Nutrition & Agriculture*, 11(2), 124–136. DOI: 10.2174/2212798410666190716162340 PMID: 31322080

Nayab, D. E., & Akhtar, S. (2023). Green synthesized silver nanoparticles from *Eucalyptus* leaves can enhance shelf life of banana without penetrating in pulp. *PLoS One*, 18(3), e0281675. DOI: 10.1371/journal.pone.0281675 PMID: 36888584

Nazari, M., Ebrahimzadeh, F., Falaki, M., Hajimomeni, Y., Varma, R. S., & Fallahnia, N. (2022). Nanoparticles in the prevention, diagnosis, and treatment of the influenza virus. *Nanomedicine Research Journal*, 7(1), 19–36. DOI: 10.22034/nmrj.2022.01.002

Nazli, A., He, D. L., Liao, D., Khan, M. Z. I., Huang, C., & He, Y. (2022). Strategies and progresses for enhancing targeted antibiotic delivery. *Advanced Drug Delivery Reviews*, 189, 114502. DOI: 10.1016/j.addr.2022.114502 PMID: 35998828

Ndikau, M., Noah, N. M., Andala, D. M., & Masika, E. (2017). Green synthesis and characterization of silver nanoparticles using Citrullus lanatus fruit rind extract. *International Journal of Analytical Chemistry*, 2017(1), 8108504. DOI: 10.1155/2017/8108504 PMID: 28316627

Nejo, A., Adetona, A., & Lawal, A. (2024). Green Synthesis of Nickel Oxide Nanoparticles and its Application in the Degradation of Methyl Red. *Lafia Journal of Scientific and Industrial Research*, 2, 54–60. DOI: 10.62050/ljsir2024.v2n2.328

Newton, D. W. (1991). Coacervation: Principles and applications. In *Polymers for controlled drug delivery* (pp. 67-81). CRC Press.

Nguyen, M. D., Tran, H.-V., Xu, S., & Lee, T. R. (2021). Fe(3)O(4) Nanoparticles: Structures, Synthesis, Magnetic Properties, Surface Functionalization, and Emerging Applications. *Applied Sciences (Basel, Switzerland)*, 11(23), 11301. DOI: 10.3390/app112311301 PMID: 35844268

Niazi, S., & Lokesh, S. (2022). *Biopharmaceutical Manufacturing, Volume 2: Unit processes*. IOP Publishing. https://iopscience.iop.org/book/mono/978-0-7503-3179-1

Nikam, A., Pagar, T., Ghotekar, S., Pagar, K., & Pansambal, S. (2019). A review on plant extract mediated green synthesis of zirconia nanoparticles and their miscellaneous applications. *Journal of Chemical Reviews*, 1(3), 154–163. DOI: 10.33945/SAMI/JCR.2019.3.1

Ningthoujam, R., Sahoo, B., Ghosh, P., Shivani, A., Ganguli, P., & Chaudhuri, S. (2023). Green production of zero-valent iron nanoparticles using pomegranate peel extracts and its use in lindane degradation. In Nanotechnology for Environmental Engineering (Vol. 8, Issue 2, pp. 581–589). Springer Science and Business Media LLC. DOI: 10.1007/s41204-023-00313-0

Niraimathi, K. L., Sudha, V., Lavanya, R., & Brindha, P. (2013). Biosynthesis of silver nanoparticles using Alternanthera sessilis (Linn.) extract and their antimicrobial, antioxidant activities. *Colloids and Surfaces. B, Biointerfaces*, 102, 288–291. DOI: 10.1016/j.colsurfb.2012.08.041 PMID: 23006568

Nisar, P., Ali, N., Rahman, L., Ali, M., & Shinwari, Z. K. (2019). Antimicrobial activities of biologically synthesized metal nanoparticles: An insight into the mechanism of action. *Journal of Biological Inorganic Chemistry*, 24(7), 929–941. DOI: 10.1007/s00775-019-01717-7 PMID: 31515623

Nishimoto-Sauceda, D., Romero-Robles, L. E., & Antunes-Ricardo, M. (2022). Biopolymer nanoparticles: A strategy to enhance stability, bioavailability, and biological effects of phenolic compounds as functional ingredients. *Journal of the Science of Food and Agriculture*, 102(1), 41–52. DOI: 10.1002/jsfa.11512 PMID: 34460939

Niu, G., Zhao, L., Wang, Y., & Jiang, Y. (2023). PDA/gold nanorod-based nanoparticles for synergistic genetic and photothermal combination therapy for cancer treatment. *ChemPhysMater*, 2(1), 83–89. DOI: 10.1016/j.chphma.2022.07.001

Noma, S. A. A., Ulu, A., Acet, Ö., Sanz, R., Sanz-Pérez, E. S., Odabaşı, M., & Ateş, B. (2020). Comparative study of ASNase immobilization on tannic acid-modified magnetic Fe 3 O 4/SBA-15 nanoparticles to enhance stability and reusability. *New Journal of Chemistry*, 44(11), 4440–4451. DOI: 10.1039/D0NJ00127A

Oghenerume, P., Eduok, S., Ita, B., John, O., & Basssy, I. (2020). Impact of zinc oxide nanoparticles amended organic manure on *Arachis hypogaea* growth response and rhizosphere bacterial community. *International Journal of Plant and Soil Science*, 32(5), 24–35. DOI: 10.9734/ijpss/2020/v32i530279

Oh, S.-H., Jeong, S., Kim, I. S., Shon, H. K., & Jang, A. (2019). Removal behaviors and fouling mechanisms of charged antibiotics and nanoparticles on forward osmosis membrane. *Journal of Environmental Management*, 247, 385–393. DOI: 10.1016/j.jenvman.2019.06.070 PMID: 31254754

Olajire, A. A., & Mohammed, A. A. (2020). Green synthesis of nickel oxide nanoparticles and studies of their photocatalytic activity in degradation of polyethylene films. *Advanced Powder Technology*, 31(1), 211–218. DOI: 10.1016/j.apt.2019.10.012

Olawale, F., Ariatti, M., & Singh, M. (2021). Biogenic Synthesis of Silver-Core Selenium-Shell Nanoparticles Using Ocimum tenuiflorum L.: Response Surface Methodology-Based Optimization and Biological Activity. *Nanomaterials (Basel, Switzerland)*, 11(10), 10. Advance online publication. DOI: 10.3390/nano11102516 PMID: 34684956

Olivera, S., Muralidhara, H. B., Venkatesh, K., Guna, V. K., Gopalakrishna, K., & Kumar, K. Y. (2016). Potential Applications of Cellulose and Chitosan Nanoparticles/composites in Wastewater Treatment: A Review. *Carbohydrate Polymers*, 153, 600–618. DOI: 10.1016/j.carbpol.2016.08.017 PMID: 27561533

Ong, Y. T., Ahmad, A. L., Zein, S. H. S., & Tan, S. H. (2010). A review on carbon nanotubes in an environmental protection and green engineering perspective. *Brazilian Journal of Chemical Engineering*, 27(2), 227–242. DOI: 10.1590/S0104-66322010000200002

Ontiveros-Robles, J. A., Villanueva-Flores, F., Juarez-Moreno, K., Simakov, A., & Vazquez-Duhalt, R. (2023). Antibody-Functionalized Copper Oxide Nanoparticles with Targeted Antibacterial Activity. *ChemistryOpen*, 12(5), e202200241. DOI: 10.1002/open.202200241 PMID: 37226371

Onyancha, R. B., Aigbe, U. O., Ukhurebor, K. E., & Muchiri, P. W. (2021). Facile synthesis and applications of carbon nanotubes in heavy-metal remediation and biomedical fields: A comprehensive review. *Journal of Molecular Structure*, 1238, 130462. DOI: 10.1016/j.molstruc.2021.130462

Oppenheim, R. C. (1981). Solid colloidal drug delivery systems: Nanoparticles. *International Journal of Pharmaceutics*, 8(3), 217–234. DOI: 10.1016/0378-5173(81)90100-9

Oriola, A. O., Onwudiwe, D. C., & Oyedeji, A. O. (2021). Study on the synthesis and applications of metal nanoparticles in agriculture.

Oselusi, S. O., Sibuyi, N. R., Meyer, M., & Madiehe, A. M. (2024). Phytonanotherapeutic Applications of Plant Extract-Synthesized Silver Nanoparticles in Wound Healing—A Prospective Overview. *BioNanoScience*, 14(3), 1–21. DOI: 10.1007/s12668-024-01535-5

Osman, A. I., Zhang, Y., Farghali, M., Rashwan, A. K., Eltaweil, A. S., Abd El-Monaem, E. M., Mohamed, I. M. A., Badr, M. M., Ihara, I., Rooney, D. W., & Yap, P.-S. (2024). Synthesis of green nanoparticles for energy, biomedical, environmental, agricultural, and food applications: A review. *Environmental Chemistry Letters*, 22(2), 841–887. DOI: 10.1007/s10311-023-01682-3

Ouattara, B., Simard, R. E., Piette, G., Bégin, A., & Holley, R. A. (2000). Inhibition of surface spoilage bacteria in processed meats by application of antimicrobial films prepared with chitosan. *International Journal of Food Microbiology*, 62(1-2), 139–148. DOI: 10.1016/S0168-1605(00)00407-4 PMID: 11139014

Özkan Dağlioğlu, Y. (2021). Green Synthesis of Silver Nanoparticles Using Mandragora autumnalis; Its Characterization, Antioxidant and Antimicrobial Activities. *Erzincan Üniversitesi Fen Bilimleri Enstitüsü Dergisi*, 14(3), 1039–1054. DOI: 10.18185/erzifbed.915107

Padalia, H., Moteriya, P., & Chanda, S. (2015). Green synthesis of silver nanoparticles from marigold flower and its synergistic antimicrobial potential. *Arabian Journal of Chemistry*, 8(5), 732–741. DOI: 10.1016/j.arabjc.2014.11.015

Padhiary, M. (2024a). Membrane Technologies for Treating Wastewater in the Food Processing Industry: Practices and Challenges. In *Research Trends in Food Technology and Nutrition* (Vol. 27, pp. 37–62). AkiNik Publications. DOI: 10.22271/ed.book.2817

Padhiary, M. (2024b). Status of Farm Automation, Advances, Trends, and Scope in India. [IJSR]. *International Journal of Science and Research (Raipur, India)*, 13(7), 737–745. DOI: 10.21275/SR24713184513

Padhiary, M. (2024c). The Convergence of Deep Learning, IoT, Sensors, and Farm Machinery in Agriculture. In *Designing Sustainable Internet of Things Solutions for Smart Industries* (pp. 109–142). IGI Global., DOI: 10.4018/979-8-3693-5498-8.ch005

Padhiary, M., Barbhuiya, J. A., Roy, D., & Roy, P. (2024). 3D Printing Applications in Smart Farming and Food Processing. *Smart Agricultural Technology*, 9, 100553. DOI: 10.1016/j.atech.2024.100553

Padhiary, M., & Kumar, R. (2024). Enhancing Agriculture Through AI Vision and Machine Learning: The Evolution of Smart Farming. In Thangam, D. (Ed.), (pp. 295–324). Advances in Computational Intelligence and Robotics. IGI Global., DOI: 10.4018/979-8-3693-5380-6.ch012

Padhiary, M., & Roy, P. (2024). Advancements in Precision Agriculture: Exploring the Role of 3D Printing in Designing All-Terrain Vehicles for Farming Applications. *International Journal of Science and Research (Raipur, India)*, 13(5), 861–868. DOI: 10.21275/SR24511105508

Padhiary, M., Saha, D., Kumar, R., Sethi, L. N., & Kumar, A. (2024). Enhancing Precision Agriculture: A Comprehensive Review of Machine Learning and AI Vision Applications in All-Terrain Vehicle for Farm Automation. *Smart Agricultural Technology*, 8, 100483. DOI: 10.1016/j.atech.2024.100483

Padmavathi, S., & Bhadramaraju, R. (2024). 'A Review on Nanotechnology And Its Applications In The Field Of Pharmacy. *Educational Administration: Theory and Practice*, 30(1), 964–973. DOI: 10.53555/kuey.v30i1.5856

Pandurangan, A., & Jayaraman, R.. (2020). Green synthesis of zinc oxide nanoparticles using eggshells and their antibacterial and catalytic properties. *Journal of Environmental Chemical Engineering*, 8(2), 103797.

Panigrahi, S., Kundu, S., Ghosh, S., Nath, S., & Pal, T. (2004). General method of synthesis for metal nanoparticles. *Journal of Nanoparticle Research*, 6(4), 411–414. DOI: 10.1007/s11051-004-6575-2

Pan, L., Zou, J. J., Wang, S., Huang, Z. F., Yu, A., Wang, L., & Zhang, X. (2013). Quantum dot self-decorated TiO 2 nanosheets. *Chemical Communications*, 49(59), 6593–6595. DOI: 10.1039/c3cc42152j PMID: 23660603

Panwar, C., Phillips, A., Sangal, A., & Gupta, D. (2024). Synthesis of Gold Nanoparticles and Their Applications in Cancer Therapy. *Biointerface Research in Applied Chemistry*, 14, 12.

Pan, Z., Lin, Y., Sarkar, B., Owens, G., & Chen, Z. (2020). Green synthesis of iron nanoparticles using red peanut skin extract: Synthesis mechanism, characterization and effect of conditions on chromium removal. In *Journal of Colloid and Interface Science* (Vol. 558, pp. 106–114). Elsevier BV., DOI: 10.1016/j.jcis.2019.09.106

Pappas, P. G., Lionakis, M. S., Arendrup, M. C., Ostrosky-Zeichner, L., & Kullberg, B. J. (2018). Invasive candidiasis. *Nature Reviews. Disease Primers*, 4(1), 18026. DOI: 10.1038/nrdp.2018.26 PMID: 29749387

Paralikar, P., & Rai, M. (2018). Bio-inspired synthesis of sulphur nanoparticles using leaf extract of four medicinal plants with special reference to their antibacterial activity. *IET Nanobiotechnology / IET*, 12(1), 25–31. DOI: 10.1049/iet-nbt.2017.0079

Park, S. C., Kim, Y. M., Lee, J. K., Kim, N. H., Kim, E. J., Heo, H., Lee, M. Y., Lee, J. R., & Jang, M. K. (2017). Targeting and synergistic action of an antifungal peptide in an antibiotic drug-delivery system. *Journal of Controlled Release*, 256, 46–55. DOI: 10.1016/j.jconrel.2017.04.023 PMID: 28428067

Parsa, N. (2012). Environmental factors inducing human cancers. *Iranian Journal of Public Health*, 41(11), 1. PMID: 23304670

Pathak, L., Kanwal, A., & Agrawal, Y. (2015). Curcumin loaded self assembled lipid-biopolymer nanoparticles for functional food applications. *Journal of Food Science and Technology*, 52(10), 6143–6156. DOI: 10.1007/s13197-015-1742-2 PMID: 26396362

Patil, S. S., Shedbalkar, U. U., Truskewycz, A., Chopade, B. A., & Ball, A. S. (2016a). Nanoparticles for environmental clean-up: A review of potential risks and emerging solutions. *Environmental Technology & Innovation*, 5, 10–21. DOI: 10.1016/j.eti.2015.11.001

Patil, S., & Chandrasekaran, R. (2020). Biogenic nanoparticles: A comprehensive perspective in synthesis, characterization, application and its challenges. *Journal of Genetic Engineering and Biotechnology*, 18(1), 67. DOI: 10.1186/s43141-020-00081-3 PMID: 33104931

Patil, S., Singh, N., & Jagdale, P. (2018). Nanotechnology for occupational health and safety: Advances and Challenges. *Advances in Nanotechnology and the Environment*, 12, 57–78.

Patiño-Ruiz, D. A., Meramo-Hurtado, S. I., González-Delgado, Á. D., & Herrera, A. (2021). Environmental Sustainability Evaluation of Iron Oxide Nanoparticles Synthesized via Green Synthesis and the Coprecipitation Method: A Comparative Life Cycle Assessment Study. *ACS Omega*, 6(19), 12410–12423. DOI: 10.1021/acsomega.0c05246 PMID: 34056392

Patois, E., Capelle, M. A. H., Palais, C., Gurny, R., & Arvinte, T. (2012). Evaluation of nanoparticle tracking analysis (NTA) in the characterization of therapeutic antibodies and seasonal influenza vaccines: Pros and cons. *Journal of Drug Delivery Science and Technology*, 22(5), 427–433. DOI: 10.1016/S1773-2247(12)50069-9

Patra, J. K., & Baek, K.-H. (2015). Novel green synthesis of gold nanoparticles using Citrullus lanatus rind and investigation of proteasome inhibitory activity, antibacterial, and antioxidant potential. *International Journal of Nanomedicine*, 10, 7253–7264. PMID: 26664116

Patra, J. K., Das, G., & Baek, K.-H. (2016). Phyto-mediated biosynthesis of silver nanoparticles using the rind extract of watermelon (Citrullus lanatus) under photo-catalyzed condition and investigation of its antibacterial, anticandidal and antioxidant efficacy. *Journal of Photochemistry and Photobiology. B, Biology*, 161, 200–210. DOI: 10.1016/j.jphotobiol.2016.05.021 PMID: 27261701

Peng, H., Liu, P., Lin, D., Deng, Y., Lei, Y., Chen, W., Chen, Y., Lin, X., Xia, X., & Liu, A. (2016). Fabrication and multifunctional properties of ultrasmall water-soluble tungsten oxide quantum dots. *Chemical Communications*, 52(61), 9534–9537. DOI: 10.1039/C6CC03245A PMID: 27381501

Pennycook, S. J., Lupini, A. R., Varela, M., Borisevich, A., Peng, Y., Oxley, M. P., Benthem, K. V., Chisholm, M. F. (2006). Scanning Transmission Electron Microscopy for Nanostructure Characterization. *Scanning Microscopy for Nanotechnology*, 152–191. DOI: 10.1007/978-0-387-39620-0_6

Peralta-Videa, J. R., Huang, Y., Parsons, J. G., Zhao, L., Lopez-Moreno, L., Hernandez-Viezcas, J. A., & Gardea-Torresdey, J. L. (2016). Plant-based green synthesis of metallic nanoparticles: Scientific curiosity or a realistic alternative to chemical synthesis? *Nanotechnology for Environmental Engineering*, 1(1), 4. DOI: 10.1007/s41204-016-0004-5

Pérez-de-Luque, A. (2017). Interaction of nanomaterials with plants: What do we need for real applications in agriculture? *Frontiers in Environmental Science*, 5, 12. DOI: 10.3389/fenvs.2017.00012

Periakaruppan, R., Chen, X., Thangaraj, K., Jeyaraj, A., Nguyen, H. H., Yu, Y., Hu, S., Lu, L., & Li, X. (2021). Utilization of tea resources with the production of superparamagnetic biogenic iron oxide nanoparticles and an assessment of their antioxidant activities. *Journal of Cleaner Production*, 278, 123962. DOI: 10.1016/j.jclepro.2020.123962

Perotti, G. F., Tronto, J., Bizeto, M. A., Izumi, C., Temperini, M. L., Lugão, A. B., Parra, D. F., & Constantino, V. R. (2014). Biopolymer-clay nanocomposites: Cassava starch and synthetic clay cast films. *Journal of the Brazilian Chemical Society*, 25, 320–330. DOI: 10.5935/0103-5053.20130300

Petla, R. K., Vivekanandhan, S., Misra, M., Mohanty, A., & Satyanarayana, N. (2012). Soybean (Glycine Max) Leaf Extract Based Green Synthesis of Palladium Nanoparticles. *Journal of Biomaterials and Nanobiotechnology*, 03(1), 14–19. DOI: 10.4236/jbnb.2012.31003

Phaniendra, A., Jestadi, D. B., & Periyasamy, L. (2015). Free radicals: Properties, sources, targets, and their implication in various diseases. *Indian Journal of Clinical Biochemistry*, 30(1), 11–26. DOI: 10.1007/s12291-014-0446-0 PMID: 25646037

Philip, D. (2008). Synthesis and spectroscopic characterization of gold nanoparticles. *Spectrochimica Acta. Part A: Molecular and Biomolecular Spectroscopy*, 71(1), 80–85. DOI: 10.1016/j.saa.2007.11.012 PMID: 18155956

Philip, D. (2010). Green synthesis of gold and silver nanoparticles using Hibiscus rosa sinensis. *Physica E, Low-Dimensional Systems and Nanostructures*, 42(5), 1417–1424. DOI: 10.1016/j.physe.2009.11.081

Phung, X., Groza, J., Stach, E. A., Williams, L. N., & Ritchey, S. B. (2003). Surface characterization of metal nanoparticles. *Materials Science and Engineering A*, 359(1–2), 261–268. DOI: 10.1016/S0921-5093(03)00348-4

Pich, U., Fritsch, R., & Schubert, I. (1996). Closely related Allium species (Alliaceae) share a very similar satellite sequence. *Plant Systematics and Evolution*, 202(3-4), 255–264. DOI: 10.1007/BF00983386

Pinto, R. J., Lucas, J. M., Morais, M. P., Santos, S. A., Silvestre, A. J., Marques, P. A., & Freire, C. S. (2017). Demystifying the morphology and size control on the biosynthesis of gold nanoparticles using *Eucalyptus globulus* bark extract. *Industrial Crops and Products*, 105, 83–92. DOI: 10.1016/j.indcrop.2017.05.003

Pitek, A. S., Hu, H., Shukla, S., & Steinmetz, N. F. (2018). Cancer Theranostic Applications of Albumin-Coated Tobacco Mosaic Virus Nanoparticles. *ACS Applied Materials & Interfaces*, 10(46), 39468–39477. DOI: 10.1021/acsami.8b12499 PMID: 30403330

Poinern, G. E. J., Chapman, P., Shah, M., & Fawcett, D. (2013). Green biosynthesis of silver nanocubes using the leaf extracts from *Eucalyptus macrocarpa*. Nano Bulletin, 2(1).

Poloni, L. N., Zhong, X., Ward, M. D., & Mandal, T. (2016). Best Practices for Real-Time in Situ Atomic Force and Chemical Force Microscopy of Crystals. *Chemistry of Materials*, 29(1), 331–345. DOI: 10.1021/acs.chemmater.6b03082

Porcaro, F., Carlini, L., Ugolini, A., Visaggio, D., Visca, P., Fratoddi, I., Venditti, I., Meneghini, C., Simonelli, L., Marini, C., Olszewski, W., Ramanan, N., Luisetto, I., & Battocchio, C. (2016). Synthesis and Structural Characterization of Silver Nanoparticles Stabilized with 3-Mercapto-1-Propansulfonate and 1-Thioglucose Mixed Thiols for Antibacterial Applications. *Materials (Basel)*, 9(12), 1028. DOI: 10.3390/ma9121028 PMID: 28774148

Pourmortazavi, S. M., Taghdiri, M., Makari, V., & Rahimi-Nasrabadi, M. (2015). Procedure optimization for green synthesis of silver nanoparticles by aqueous extract of *Eucalyptus oleosa.Spectrochimica Acta. Part A: Molecular and Biomolecular Spectroscopy*, 136, 1249–1254. DOI: 10.1016/j.saa.2014.10.010 PMID: 25456666

Pourmortazavi, S. M., Taghdiri, M., Makari, V., Rahimi-Nasrabadi, M., & Batooli, H. (2017). Reducing power of *Eucalyptus oleosa* leaf extracts and green synthesis of gold nanoparticles using the extract. *International Journal of Food Properties*, 20(5), 1097–1103. DOI: 10.1080/10942912.2016.1203334

Prabha, A. S., Dorothy, R., Jancirani, S., Rajendran, S., Singh, G., & Kumaran, S. S. (2020). Recent advances in the study of toxicity of polymer-based nanomaterials. *Nanotoxicity*, 143-165. https://doi.org/ DOI: 10.1016/B978-0-12-819943-5.00007-5

Prada, Y., Guzmán, F., Ortíz, C., Cabanzo, R., Torres, R., & Mejía-Ospino, E. (2019). New synthetic peptides conjugated to gold nanoclusters: Antibiotic activity against Escherichia coli O157: H7 and methicillin-resistant Staphylococcus aureus (MRSA). *The Protein Journal*, 38(5), 506–514. DOI: 10.1007/s10930-019-09840-9 PMID: 31119600

Pradheesh, G., Suresh, S., Suresh, J., & Alexramani, V. (2020). Antimicrobial and Anticancer Activity Studies on Green Synthesized Silver Oxide Nanoparticles from the Medicinal Plant *Cyathea nilgiriensisHolttum.International Journal of Pharmaceutical Investigation*, 10(2), 146–150. DOI: 10.5530/ijpi.2020.2.27

Pramanik, S., & Chakraborty, P. (2014). Soil productivity analysis under different land uses of Sundarban delta. *Journal of Soil and Water Conservation*, •••, 13.

Prasad, P. R., Kanchi, S., & Naidoo, E. B. (2016). In-vitro evaluation of copper nanoparticles cytotoxicity on prostate cancer cell lines and their antioxidant, sensing and catalytic activity: One-pot green approach. *Journal of Photochemistry and Photobiology. B, Biology*, 161, 375–382. DOI: 10.1016/j.jphotobiol.2016.06.008 PMID: 27318296

Preetha, D., Arun, R., Kumari, P., & Aarti, C. (2013). Synthesis and characterization of silver nanoparticles using cannonball leaves and their cytotoxic activity against MCF-7 cell line. *Journal of Nanotechnology*, 2013(1), 1–5.

Preetha, P. S., & Balakrishnan, N. (2017). A review of nano fertilizers and their use and functions in soil. *International Journal of Current Microbiology and Applied Sciences*, 6(12), 3117–3133. DOI: 10.20546/ijcmas.2017.612.364

Priester, J. H., Ge, Y., Mielke, R. E., Horst, A. M., Moritz, S. C., Espinosa, K., & Holden, P. A. (2012). Soybean susceptibility to manufactured nanomaterials with evidence for food quality and soil fertility interruption. *Proceedings of the National Academy of Sciences of the United States of America*, 109(37), E2451–E2456. DOI: 10.1073/pnas.1205431109 PMID: 22908279

Prime, R. B., Bair, H. E., & Vyazovkin, S. Riga Alan, & Gallagher, P. K. (2009). Thermogravimetric analysis (TGA). J. D. Menczel & R. B. Prime (Eds.), *Thermal analysis of polymers:Fundamentals and applications* (pp. 241-317). Wily.

Priya, M., Venkatesan, R., Deepa, S., Sana, S. S., Arumugam, S., Karami, A. M., Vetcher, A. A., & Kim, S.-C. (2023). Green synthesis, characterization, antibacterial, and antifungal activity of copper oxide nanoparticles derived from Morinda citrifolia leaf extract. *Scientific Reports*, 13(1), 18838. DOI: 10.1038/s41598-023-46002-5 PMID: 37914791

Pryshchepa, O., Pomastowski, P., & Buszewski, B. (2020). Silver nanoparticles: Synthesis, investigation techniques, and properties. In *Advances in Colloid and Interface Science* (Vol. 284, p. 102246). Elsevier BV., DOI: 10.1016/j.cis.2020.102246

Puja, A. M., Rupa, E. J., Kim, Y. J., & Yang, D. C. (2023). Medicinal plant enriched metal nanoparticles and nanoemulsion for inflammation treatment: A narrative review on current status and future perspective. *Immuno*, 3(2), 182–194. DOI: 10.3390/immuno3020012

Punitha, U., & Mary Saral, A. (2024). Nickel oxide nanoparticles from Sargassumwightii: Synthesis,Characterization, and biomedical applications. *Results in Chemistry*, 7, 101289. DOI: 10.1016/j.rechem.2023.101289

Purushothaman, B., & Song, J. M. (2021). Ag 2 S quantum dot theragnostics. *Biomaterials Science*, 9(1), 51–69. DOI: 10.1039/D0BM01576H PMID: 33185212

Putro, J. N., Santoso, S. P., Ismadji, S., & Ju, Y. H. (2017). Investigation of heavy metal adsorption in binary system by nanocrystalline cellulose–bentonite nanocomposite: Improvement on extended Langmuir isotherm model. *Microporous and Mesoporous Materials*, 246, 166–177. DOI: 10.1016/j.micromeso.2017.03.032

Pyrzynska, K., & Sentkowska, A. (2021). Biosynthesis of selenium nanoparticles using plant extracts. In Journal of Nanostructure in Chemistry (Vol. 12, Issue 4, pp. 467–480). Springer Science and Business Media LLC. DOI: 10.1007/s40097-021-00435-4

Qasem, N. A. A., Mohammed, R. H., & Lawal, D. U. (2021). Removal of heavy metal ions from wastewater: a comprehensive and critical review. In npj Clean Water (Vol. 4, Issue 1). Springer Science and Business Media LLC. DOI: 10.1038/s41545-021-00127-0

Qiu, J., Li, D., Mou, X., Li, J., Guo, W., Wang, S., Yu, X., Ma, B., Zhang, S., Tang, W., Sang, Y., Gil, P. R., & Liu, H. (2016). Graphene Quantum Dots: Effects of Graphene Quantum Dots on the Self-Renewal and Differentiation of Mesenchymal Stem Cells (Adv. Healthcare Mater. 6/2016). *Advanced Healthcare Materials*, 5(6), 623–623. DOI: 10.1002/adhm.201670026 PMID: 26833812

Qu, D., Sun, Z., Zheng, M., Li, J., Zhang, Y., Zhang, G., & Xie, Z. (2015). Three colors emission from S, N Co-doped graphene quantum dots for visible light H2 production and bioimaging. *Advanced Optical Materials*, 3(3), 360–367.

Quirós, J., Gonzalo, S., Jalvo, B., Boltes, K., Perdigón-Melón, J. A., & Rosal, R. (2016). Electrospun cellulose acetate composites containing supported metal nanoparticles for antifungal membranes. *The Science of the Total Environment*, 563, 912–920. DOI: 10.1016/j.scitotenv.2015.10.072 PMID: 26524992

Raafat, D., Von Bargen, K., Haas, A., & Sahl, H. G. (2008). Insights into the mode of action of chitosan as an antibacterial compound. *Applied and Environmental Microbiology*, 74(12), 3764–3773. DOI: 10.1128/AEM.00453-08 PMID: 18456858

Rabeea, M. A., Owaid, M. N., & Muslim, R. F. (2021). Synthesis and characterization of silver nanoparticles by natural organic compounds extracted from *Eucalyptus* leaves and their role in the catalytic degradation of methylene blue dye. *Songklanakarin Journal of Science and Technology*, 43(1).

Rafiee, B., Ghani, S., Sadeghi, D., & Ahsani, M. (2018). Green synthesis of zinc oxide nanoparticles using *eucalyptus mellidora* leaf extract and evaluation of its antimicrobial effects. *Majallah-i Danishgah-i Ulum-i Pizishki-i Babul*, 20(10), 28–35.

Rahaiee, S., Shojaosadati, S. A., & Hashemi, M. (2023). An efficient ionic gelation based nano- delivery system to improve the stability and controlled release of saffron extracts. *Biocatalysis and Agricultural Biotechnology*, 52, 102831. DOI: 10.1016/j.bcab.2023.102831

Rahimi, H. R., & Doostmohammadi, M. (2019). Nanoparticle synthesis, applications, and toxicity. Applications of Nanobiotechnology, 10.

Rahman, M., Alam, K., Hafeez, A., Ilyas, R., & Beg, S. (2021). Metallic nanoparticles in drug delivery and cancer treatment. In *Nanoformulation strategies for cancer treatment* (pp. 107–119). Elsevier. DOI: 10.1016/B978-0-12-821095-6.00008-2

Rahmatyian, M., & Dadmehr, M. (2020). Biological synthesis of silver nanoparticle based on *Eucalyptus obliqua* and *Mentha spicata* plant extracts and evaluation of their physical properties. *Crop Biotechnology*, 10(31), 61–70.

Rai, M., Deshmukh, S., Ingle, A., & Gade, A. (2019). Silver nanoparticles: The powerful nanoweapon against multidrug-resistant bacteria. *Journal of Applied Microbiology*, 127(3), 701–712. PMID: 31216601

Rajakumar, G., Gomathi, T., Abdul Rahuman, A., Thiruvengadam, M., Mydhili, G., Kim, S.-H., Lee, T.-J., & Chung, I.-M. (2016). Biosynthesis and Biomedical Applications of Gold Nanoparticles Using Eclipta prostrata Leaf Extract. *Applied Sciences (Basel, Switzerland)*, 6(8), 222. DOI: 10.3390/app6080222

Rajchakit, U., & Sarojini, V. (2017). Recent developments in antimicrobial-peptide-conjugated gold nanoparticles. *Bioconjugate Chemistry*, 28(11), 2673–2686. DOI: 10.1021/acs.bioconjchem.7b00368 PMID: 28892365

Rajendran, V., & Aravindhan, R. S.. (2022). Sustainable synthesis of zinc oxide nanoparticles from human urine: Characterization and application in antimicrobial and photocatalytic activities. *Journal of Cleaner Production*, 340, 130784.

Rajeshkumar, S., Veena, P., & Santhiyaa, R. V. (2018). Synthesis and Characterization of Selenium Nanoparticles Using Natural Resources and Its Applications. In Nanotechnology in the Life Sciences (pp. 63–79). Springer International Publishing. DOI: 10.1007/978-3-319-99570-0_4

Rajeshkumar, S., & Bharath, L. V. (2017). Mechanism of plant-mediated synthesis of silver nanoparticles–a review on biomolecules involved, characterization and antibacterial activity. *Chemico-Biological Interactions*, 273, 219–227. DOI: 10.1016/j.cbi.2017.06.019 PMID: 28647323

Ramesh, A. V., Devi, D., Battu, G., & Basavaiah, K. (2018). A Facile plant mediated synthesis of silver nanoparticles using an aqueous leaf extract of Ficus hispida Linn. f. for catalytic, antioxidant and antibacterial applications. *South African Journal of Chemical Engineering*, 26, 25–34. DOI: 10.1016/j.sajce.2018.07.001

Ramesh, M., Anbuvannan, M., & Viruthagiri, G. (2015). Green synthesis of ZnO nanoparticles using 'Solanum nigrum' leaf extract and their antibacterial activity. *Spectrochimica Acta. Part A: Molecular and Biomolecular Spectroscopy*, 136, 864–870. DOI: 10.1016/j.saa.2014.09.105 PMID: 25459609

Ramezani, N., Ehsanfar, Z., Shamsa, F., Amin, G., Shahverdi, H. R., Esfahani, H. R. M., Shamsaie, A., Bazaz, R. D., & Shahverdi, A. R. (2008). Screening of medicinal plant methanol extracts for the synthesis of gold nanoparticles by their reducing potential. *Zeitschrift für Naturforschung. Section B. A Journal of Chemical Sciences*, 63(7), 903–908. DOI: 10.1515/znb-2008-0715

Rami, J. M., Patel, C. D., Patel, C. M., & Patel, M. V. (2020). Thermogravimetric analysis (TGA) of some synthesized metal oxide nanoparticles. *Materials Today: Proceedings*, 43, 655–659. DOI: 10.1016/j.matpr.2020.12.554

Ramrakhiani, L., & Ghosh, S. (2018). Metallic nanoparticle synthesised by biological route: Safer candidate for diverse applications. *IET Nanobiotechnology / IET*, 12(4), 392–404. DOI: 10.1049/iet-nbt.2017.0076 PMID: 29768220

Ramyadevi, J., Jeyasubramanian, K., Marikani, A., Rajakumar, G., & Rahuman, A. A. (2012). Synthesis and antimicrobial activity of copper nanoparticles. *Materials Letters*, 71, 114–116. DOI: 10.1016/j.matlet.2011.12.055

Ran, F., Li, C., Hao, Z., Zhang, X., Dai, L., Si, C., Shen, Z., Qiu, Z., & Wang, J. (2022). Combined bactericidal process of lignin and silver in a hybrid nanoparticle on E. coli. *Advanced Composites and Hybrid Materials*, 5(3), 1841–1851. DOI: 10.1007/s42114-022-00460-z PMID: 35434523

Rangaraj, S., Gopalu, K., Rathinam, Y., Periasamy, P., Venkatachalam, R., & Narayanasamy, K. (2014). Effect of silica nanoparticles on microbial biomass and silica availability in maize rhizosphere. *Biotechnology and Applied Biochemistry*, 61(6), 668–675. DOI: 10.1002/bab.1191 PMID: 24329970

Rani, M. Ankit, Yadav, J., & Shanker, U. (2022). Sustainable Green Doped Nanomaterials for Emerging Contaminants Removal. In Handbook of Green and Sustainable Nanotechnology: Fundamentals, Developments and Applications (pp. 1-30). Cham: Springer International Publishing.

Rani, H., Singh, S. P., Yadav, T. P., Khan, M. S., Ansari, M. I., & Singh, A. K. (2020). In-vitro catalytic, antimicrobial and antioxidant activities of bioengineered copper quantum dots using Mangifera indica (L.) leaf extract. *Materials Chemistry and Physics*, 239, 122052. DOI: 10.1016/j.matchemphys.2019.122052

Rani, N., Singh, P., Kumar, S., Kumar, P., Bhankar, V., & Kumar, K. (2023). Plant-mediated synthesis of nanoparticles and their applications: A review. *Materials Research Bulletin*, 163, 112233. DOI: 10.1016/j.materresbull.2023.112233

Rao, A., Schoenenberger, M., Gnecco, E., Glatzel, T., Meyer, E., Brändlin, D., & Scandella, L. (2007). Characterization of nanoparticles using atomic force microscopy. *Journal of Physics: Conference Series*, 61(1), 971–976. DOI: 10.1088/1742-6596/61/1/192

Rashid, M. I., Shahzad, T., Shahid, M., Ismail, I. M., Shah, G. M., & Almeelbi, T. (2017). Zinc oxide nanoparticles affect carbon and nitrogen mineralization of Phoenix dactylifera leaf litter in a sandy soil. *Journal of Hazardous Materials*, 324, 298–305. DOI: 10.1016/j.jhazmat.2016.10.063 PMID: 27810328

Rathod, S., Preetam, S., Pandey, C., & Bera, S. P. (2024). Exploring synthesis and applications of green nanoparticles and the role of nanotechnology in wastewater treatment. In *Biotechnology Reports* (Vol. 41, p. e00830). Elsevier BV., DOI: 10.1016/j.btre.2024.e00830

Rautela, A., & Rani, J., & Debnath (Das), M. (. (2019). Green synthesis of silver nanoparticles from Tectona grandis seeds extract: Characterization and mechanism of antimicrobial action on different microorganisms. *Journal of Analytical Science and Technology*, 10, 5. DOI: 10.1186/s40543-018-0163-z

Raval, N., Maheshwari, R., Kalyane, D., Youngren-Ortiz, S. R., Chougule, M. B., & Tekade, R. K. (2019). Importance of physicochemical characterization of nanoparticles in pharmaceutical product development. In *Basic fundamentals of drug delivery* (pp. 369–400). Academic Press. DOI: 10.1016/B978-0-12-817909-3.00010-8

Raveendran, P., Fu, J., & Wallen, S. L. (2003). Completely "Green" Synthesis and Stabilization of Metal Nanoparticles. *Journal of the American Chemical Society*, 125(46), 13940–13941. DOI: 10.1021/ja029267j PMID: 14611213

Read, D. S., Matzke, M., Gweon, H. S., Newbold, L. K., Heggelund, L., Ortiz, M. D., & Svendsen, C. (2016). Soil pH effects on the interactions between dissolved zinc, non-nano-and nano-ZnO with soil bacterial communities. *Environmental Science and Pollution Research International*, 23(5), 4120–4128. DOI: 10.1007/s11356-015-4538-z PMID: 25903189

Reddy, N. R., Sathe, S. K., & Salunkhe, D. K. (1982). Phytates in legumes and cereals. *Advances in Food Research*, 28, 1–92. DOI: 10.1016/S0065-2628(08)60110-X PMID: 6299067

Redondo-Gómez, C., Rodríguez Quesada, M., Vallejo Astúa, S., Murillo Zamora, J. P., Lopretti, M., & Vega-Baudrit, J. R. (2020). Biorefinery of biomass of agro-industrial banana waste to obtain high-value biopolymers. *Molecules (Basel, Switzerland)*, 25(17), 3829. DOI: 10.3390/molecules25173829 PMID: 32842473

Rehman, A., Naeem, A., Ahmad, I., Fozia, F., Almutairi, M. H., Aslam, M., Israr, M., Almutairi, B. O., & Ullah, Z. (2024). Synthesis of Plant-Mediated Iron Oxide Nanoparticles and Optimization of Chemically Modified Activated Carbon Adsorbents for Removal of As, Pb, and Cd Ions from Wastewater. *ACS Omega*, 9(1), 317–329. DOI: 10.1021/acsomega.3c05299 PMID: 38222602

Reimer, L. (2013). Transmission electron microscopy: physics of image formation and microanalysis. *Springer,* 36

Revathy, R., Sajini, T., Augustine, C., & Joseph, N. (2023). Iron-based magnetic nanomaterials: Sustainable approaches of synthesis and applications. *Results in Engineering*, 18, 101114. DOI: 10.1016/j.rineng.2023.101114

Rhyner, M. N., Smith, A. M., Gao, X., Mao, H., Yang, L., & Nie, S. (2006). Quantum dots and multifunctional nanoparticles: new contrast agents for tumor imaging.

Ribeiro, A. I., Dias, A. M., & Zille, A. (2022). Synergistic Effects Between Metal Nanoparticles and Commercial Antimicrobial Agents: A Review. *ACS Applied Nano Materials*, 5(3), 3030–3064. DOI: 10.1021/acsanm.1c03891 PMID: 36568315

Rizwan, M., & Ali, S., ur Rehman, M. Z., Malik, S., Adrees, M., Qayyum, M. F. & Ahmad, P. (2019). Effect of foliar applications of silicon and titanium dioxide nanoparticles on growth, oxidative stress, and cadmium accumulation by rice (Oryza sativa). *Acta Physiologiae Plantarum*, 41, 1–12.

Rocha, V., Ferreira-Santos, P., Aguiar, C., Neves, I. C., & Tavares, T. (2024). Valorization of plant by-products in the biosynthesis of silver nanoparticles with antimicrobial and catalytic properties. *Environmental Science and Pollution Research International*, 31(9), 14191–14207. DOI: 10.1007/s11356-024-32180-w PMID: 38278998

Rodrigues, M. L., de Oliveira, H. C., & Albuquerque, P. C. (2022). Antifungal development and the urgency of minimizing the impact of fungal diseases on public health. *ACS Bio & Med Chem Au*, 2(6), 506–511. DOI: 10.1021/acsbiomedchemau.2c00055

Rodrigues, M. L., & Nosanchuk, J. D. (2020). Fungal diseases as neglected pathogens: A wake-up call to public health officials. *PLoS Neglected Tropical Diseases*, 14(2), e0007964. DOI: 10.1371/journal.pntd.0007964 PMID: 32078635

Rodríguez-Gómez, F. D., Penon, O., Monferrer, D., & Rivera-Gil, P. (2023). Classification system for nanotechnology-enabled health products with both scientific and regulatory application. *Frontiers in Medicine*, 10, 1212949. Advance online publication. DOI: 10.3389/fmed.2023.1212949 PMID: 37601794

Rodríguez-León, E., Rodríguez-Vázquez, B. E., Martínez-Higuera, A., Rodríguez-Beas, C., Larios-Rodríguez, E., Navarro, R. E., López-Esparza, R., & Iñiguez-Palomares, R. A. (2019). Synthesis of Gold Nanoparticles Using Mimosa tenuiflora Extract, Assessments of Cytotoxicity, Cellular Uptake, and Catalysis. *Nanoscale Research Letters*, 14(1), 334. DOI: 10.1186/s11671-019-3158-9 PMID: 31654146

Rosales, T. K. O., & Fabi, J. P. (2023). Pectin-based nanoencapsulation strategy to improve the bioavailability of bioactive compounds. *International Journal of Biological Macromolecules*, 229, 11–21. DOI: 10.1016/j.ijbiomac.2022.12.292 PMID: 36586647

Roshmi, T., Soumya, K., Jyothis, M., & Radhakrishnan, E. (2015). Effect of biofabricated gold nanoparticle-based antibiotic conjugates on minimum inhibitory concentration of bacterial isolates of clinical origin. *Gold Bulletin*, 48(1-2), 63–71. DOI: 10.1007/s13404-015-0162-4

Roy, A., Pandit, C., Gacem, A., Alqahtani, M. S., Bilal, M., Islam, S., Hossain, M. J., & Jameel, M. (2022). Biologically Derived Gold Nanoparticles and Their Applications. *Bioinorganic Chemistry and Applications*, 2022(1), 8184217. DOI: 10.1155/2022/8184217 PMID: 35959230

Roy, S., & Das, T. (2015). Plant Mediated Green Synthesis of Silver Nanoparticles-A Review. *International Journal of Plant Biology & Research*, 3, 1044.

Rubina, M. S., Pigaleva, M. A., Naumkin, A. V., & Gromovykh, T. I. (2020). Bacterial Cellulose Film Produced by Gluconacetobacter hansenii as a Source Material for Oxidized Nanofibrillated Cellulose. *Doklady Physical Chemistry*, 493(2), 127–131. DOI: 10.1134/S0012501620080023

Rui, M., Ma, C., Hao, Y., Guo, J., Rui, Y., Tang, X., & Zhu, S. (2016). Iron oxide nanoparticles as a potential iron fertilizer for peanut (Arachis hypogaea). *Frontiers in Plant Science*, 7, 815. DOI: 10.3389/fpls.2016.00815 PMID: 27375665

Saadatkhah, N., Carillo Garcia, A., Ackermann, S., Leclerc, P., Latifi, M., Samih, S., Patience, G. S., & Chaouki, J. (2020). Experimental methods in chemical engineering: Thermogravimetric analysis—TGA. *Canadian Journal of Chemical Engineering*, 98(1), 34–43. DOI: 10.1002/cjce.23673

Saadh, M. J. (2022). Silver nanoparticle-conjugated antibiotics inhibit in vitro growth of Brucella melitensis. *Veterinary World*, 15(7), 1749–1752. DOI: 10.14202/vetworld.2022.1749-1752 PMID: 36185523

Saad, R., Nsour, M., Khader, Y., & Gunaid, M. (2021). Public Health Surveillance Systems in the Eastern Mediterranean Region: Bibliometric Analysis of Scientific Literature (Preprint). *JMIR Public Health and Surveillance*, 7(11), e32639. Advance online publication. DOI: 10.2196/32639 PMID: 34723831

Sadeghi, R., Mehryar, L., Karimi, M., & Kokini, J. (2016). Nanocapsule formation by individual biopolymer nanoparticles. In: Nanoencapsulation Technologies for the Food and Nutraceutical Industries, 404-446. https://doi.org/DOI: 10.1016/B978-0-12-809436-5.00011-2

Sadeghi, S., Madadlou, A., & Yarmand, M. (2014). Microemulsification–cold gelation of whey proteins for nanoencapsulation of date palm pit extract. *Food Hydrocolloids*, 35, 590–596. DOI: 10.1016/j.foodhyd.2013.07.021

Sadiq, M. U., Shah, A., Haleem, A., Shah, S. M., & Shah, I. (2023). *Eucalyptus globulus* mediated green synthesis of environmentally benign metal based nanostructures: A Review. *Nanomaterials (Basel, Switzerland)*, 13(13), 2019. DOI: 10.3390/nano13132019 PMID: 37446535

Saeed, M., Haq, A. U., Muneer, M., Usman, M., Naqvi, S. A. R., Adeel, M., & Nisar, A. (2021). assisted green synthesis of CoO and Ag-CoO and evaluation of their catalytic activities toward photodegradation of crystal violet dye. *Environmental Progress & Sustainable Energy*, 40, e13591. DOI: 10.1002/ep.13591

Sagar, P. V., Ramadevi, D., Basavaiah, K., & Botsa, S. M. (2024). Green synthesis of silver nanoparticles using aqueous leaf extract of *Saussureaobvallata* for efficient catalytic reduction of nitrophenol, antioxidant, and antibacterial activity. *Water Science and Engineering*, 17(3), 274–282. DOI: 10.1016/j.wse.2023.09.004

Şahin, S., Ünlü, C., & Trabzon, L. (2021). Affinity biosensors developed with quantum dots in microfluidic systems. *Emergent Materials*, 4(1), 187–209. DOI: 10.1007/s42247-021-00195-5 PMID: 33718778

Sahoo, M., Vishwakarma, S., Panigrahi, C., & Kumar, J. (2021). Nanotechnology: Current applications and future scope in food. *Food Frontiers*, 2(1), 3–22. DOI: 10.1002/fft2.58

Sahu, S. C. (2023). Impact of Nanotechnology in the Modern World. *Annals of Clinical Case Studies*, 5, 43–44.

Saidu, F. K., Mathew, A., Parveen, A., Valiyathra, V., & Thomas, G. V. (2019). Novel green synthesis of silver nanoparticles using clammy cherry (Cordia obliqua Willd) fruit extract and investigation on its catalytic and antimicrobial properties. *SN Applied Sciences*, 1(11), 1368. DOI: 10.1007/s42452-019-1302-x

Saif, S., Tahir, A., & Chen, Y. (2016). Green synthesis of iron nanoparticles and their environmental applications and implications. *Nanomaterials (Basel, Switzerland)*, 6(11), 209. DOI: 10.3390/nano6110209 PMID: 28335338

Sajid, M., & Płotka-Wasylka, J. (2020). Nanoparticles: Synthesis, characteristics, and applications in analytical and other sciences. *Microchemical Journal*, 154, 104623. DOI: 10.1016/j.microc.2020.104623

Saliem, A., Ibrahim, O., & Sarwa, I. (2018). Biosynthesis of Silver Nanoparticles using Cinnamon zeylanicum Plants Bark Extract . *Kufa Journal For Veterinary Medical Sciences*, 7(1), 51–63. DOI: 10.36326/kjvs/2016/v7i14294

Samadi, N., Hosseini, S., Fazeli, A., & Fazeli, M. (2010). Synthesis and antimicrobial effects of silver nanoparticles produced by chemical reduction method. *Daru : Journal of Faculty of Pharmacy, Tehran University of Medical Sciences*, 18(3), 168. PMID: 22615613

Sampatkumar, H. G., Antony, A. M., Trivedi, M., Sharma, M., Ghate, M., Baidya, M., Dateer, R. B., & Patil, S. A. (2024). In situ biosynthesis of palladium nanoparticles on banana leaves extract-coated graphitic carbon nitride: An efficient and reusable heterogeneous catalyst for organic transformations and antimicrobial agent. *Biomass Conversion and Biorefinery*, 14(9), 10045–10066. DOI: 10.1007/s13399-022-03222-5

Sánchez-Hernández, E., González-García, V., Correa-Guimarães, A., Casanova-Gascón, J., Martín-Gil, J., & Martín-Ramos, P. (2023). Phytochemical Profile and Activity against Fusarium Species of Tamarix gallica Bark Aqueous Ammonia Extract. In Agronomy (Vol. 13, Issue 2, p. 496). MDPI AG. DOI: 10.3390/agronomy13020496

Sandhu, R., Singh, N., Dhankhar, J., Kama, G., & Sharma, R. (2018). Dynamic light scattering (DLS) technique, principle, theoretical considerations and applications. *Nanotechnological and Biochemical Techniques for Assessing the Quality and Safety of Milk and Milk Products*, 135-137.

Sandhu, Z. A., Raza, M. A., Alqurashi, A., Sajid, S., Ashraf, S., Imtiaz, K., Aman, F., Alessa, A. H., Shamsi, M. B., & Latif, M. (2024). Advances in the Optimization of Fe Nanoparticles: Unlocking Antifungal Properties for Biomedical Applications. *Pharmaceutics*, 16(5), 645. DOI: 10.3390/pharmaceutics16050645 PMID: 38794307

Sangeetha, G., Rajeshwari, S., & Venckatesh, R. (2011). Green synthesis of zinc oxide nanoparticles by aloe barbadensis miller leaf extract: Structure and optical properties. *Materials Research Bulletin*, 46(12), 2560–2566. DOI: 10.1016/j.materresbull.2011.07.046

Sanità, G., Carrese, B., & Lamberti, A. (2020). Nanoparticle Surface Functionalization: How to Improve Biocompatibility and Cellular Internalization. *Frontiers in Molecular Biosciences*, 7, 587012. Advance online publication. DOI: 10.3389/fmolb.2020.587012 PMID: 33324678

Sankpal, P. S., Bandal, O. B., Bailgekar, A. V., Bilur, P., & Chavanpatil, S. (2020). Screening of in-vitro anthelmintic activity of silver nanoparticles of *Eucalyptus globulus* against *Pheretima posthuma. Global Journal for Research Analysis*, 9(2), 53–55.

Sanna, V., & Sechi, M. (2020). Therapeutic Potential of Targeted Nanoparticles and Perspective on Nanotherapies. *ACS Medicinal Chemistry Letters*, 11(6), 1069–1073. DOI: 10.1021/acsmedchemlett.0c00075 PMID: 32550978

Santos, N. C., & Domingues, M. M. (2018). Application of light scattering techniques to nanoparticle characterization and development. *Frontiers in Chemistry*, 6, 1–12. DOI: 10.3389/fchem.2018.00237 PMID: 29988578

Saraswat, P., Singh, S., Prasad, M., Misra, R., Rajput, V. D., & Ranjan, R. (2023). Applications of bio-based nanomaterials in environment and agriculture: A review on recent progresses. *Hybrid Advances*, 4, 100097. DOI: 10.1016/j.hybadv.2023.100097

Sarkar, B., Mahanty, A., Gupta, S. K., Choudhury, A. R., Daware, A., & Bhattacharjee, S. (2022). Nanotechnology: A next-generation tool for sustainable aquaculture. *Aquaculture (Amsterdam, Netherlands)*, 546, 737330. DOI: 10.1016/j.aquaculture.2021.737330

Saryanti, D., Nugraheni, D., & Astuti, N. S. (2020). Preparation and characterization of Betel Leaves (Piper betle Linn) extract nanoparticle with ionic gelation method. *Journal of Tropical Pharmacy and Chemistry*, 5(1), 15–20. DOI: 10.25026/jtpc.v5i1.224

Sasidharan, S., Chen, Y., Saravanan, D., Sundram, K. M., & Yoga Latha, L. (2010). Extraction, Isolation and Characterization of Bioactive Compounds from Plants' Extracts. *African Journal of Traditional, Complementary, and Alternative Medicines*, 8(1), 1–10. DOI: 10.4314/ajtcam.v8i1.60483 PMID: 22238476

Sathyanarayana, D. N. (2015). *Vibrational spectroscopy: theory and applications*. New Age International.

Satti, S. H., Raja, N. I., Javed, B., Akram, A., Mashwani, Z. U. R., Ahmad, M. S., & Ikram, M. (2021). Titanium dioxide nanoparticles elicited agro-morphological and physicochemical modifications in wheat plants to control Bipolaris sorokiniana. *PLoS One*, 16(2), 0246880. DOI: 10.1371/journal.pone.0246880 PMID: 33571310

Satyavani, K., Gurudeeban, S., Ramanathan, T., & Balasubramanian, T. (2011). Biomedical potential of silver nanoparticles synthesized from calli cells of *Citrullus colocynthis* (L.) Schrad. *Journal of Nanobiotechnology*, 9(1), 43. DOI: 10.1186/1477-3155-9-43 PMID: 21943321

Schaumlöffel, D. (2012). Nickel species: Analysis and toxic effects. In Journal of Trace Elements in Medicine and Biology (Vol. 26, Issue 1, pp. 1–6). Elsevier BV. DOI: 10.1016/j.jtemb.2012.01.002

Schmitt, C., & Turgeon, S. L. (2011). Protein/polysaccharide complexes and coacervates in food systems. *Advances in Colloid and Interface Science*, 167(1-2), 63–70. DOI: 10.1016/j.cis.2010.10.001 PMID: 21056401

Schneider, R. (2011). Electron Energy-Loss Spectroscopy (EELS) and Energy-Filtering Transmission Electron Microscopy (EFTEM). *Surface and Thin Film Analysis: A Compendium of Principles, Instrumentation, and Applications*, 67-91.

Schwab, F., Zhai, G., Kern, M., Turner, A., Schnoor, J. L., & Wiesner, M. R. (2016). Barriers, pathways and processes for uptake, translocation and accumulation of nanomaterials in plants–Critical review. *Nanotoxicology*, 10(3), 257–278. DOI: 10.3109/17435390.2015.1048326 PMID: 26067571

Scimeca, M., Bischetti, S., Lamsira, H. K., Bonfiglio, R., & Bonanno, E. (2018). Energy dispersive X-ray (EDX) microanalysis: A powerful tool in biomedical research and diagnosis. *European Journal of Histochemistry*, 62(1), 89–99. DOI: 10.4081/ejh.2018.2841 PMID: 29569878

Seleiman, M. F., Almutairi, K. F., Alotaibi, M., Shami, A., Alhammad, B. A., & Battaglia, M. L. (2020). Nano-fertilization as an emerging fertilization technique: Why can modern agriculture benefit from its use? *Plants*, 10(1), 2. DOI: 10.3390/plants10010002 PMID: 33375026

Selvakannan, P. R., Mandal, S., Phadtare, S., Gole, A., Pasricha, R., Adyanthaya, S. D., & Sastry, M. (2004). Water-dispersible tryptophan-protected gold nanoparticles prepared by the spontaneous reduction of aqueous chloroaurate ions by the amino acid. *Journal of Colloid and Interface Science*, 269(1), 97–102. DOI: 10.1016/S0021-9797(03)00616-7 PMID: 14651900

Senevirathna, S., Pitawala, J., Asela, U., & Attanayake, A. (2022). Biopolymers: Structure, properties, extraction methods and applications. *Sri Lankan Journal of Applied Sciences*, 1(01), 18–30.

Sengan, M., Subramaniyan, S. B., Prakash, S. A., Kamlekar, R., & Veerappan, A. (2019). Effective elimination of biofilm formed with waterborne pathogens using copper nanoparticles. *Microbial Pathogenesis*, 127, 341–346. DOI: 10.1016/j.micpath.2018.12.025 PMID: 30557585

Sen, I. K., Mandal, A. K., Chakraborti, S., Dey, B., Chakraborty, R., & Islam, S. S. (2013). Green synthesis of silver nanoparticles using glucan from mushroom and study of antibacterial activity. *International Journal of Biological Macromolecules*, 62, 439–449. DOI: 10.1016/j.ijbiomac.2013.09.019 PMID: 24076192

Seo, S., Lee, J. E., Lee, K., & Kim, H. N. (2022). Effects of microenvironmental factors on assessing nanoparticle toxicity. *Environmental Science. Nano*, 9(2), 454–476. DOI: 10.1039/D1EN00925G

Seralathan, J., Stevenson, P., Subramaniam, S., Raghavan, R., Pemaiah, B., Sivasubramanian, A., & Veerappan, A. (2014). Spectroscopy investigation on chemo-catalytic, free radical scavenging and bactericidal properties of biogenic silver nanoparticles synthesized using Salicornia brachiata aqueous extract. *Spectrochimica Acta. Part A: Molecular and Biomolecular Spectroscopy*, 118, 349–355. DOI: 10.1016/j.saa.2013.08.114 PMID: 24056313

Serpen, A., Capuano, E., Fogliano, V., & Gökmen, V. (2007). A new procedure to measure the antioxidant activity of insoluble food components. *Journal of Agricultural and Food Chemistry*, 55(19), 7676–7681. DOI: 10.1021/jf071291z PMID: 17708649

Serry, F. M., & Schmit, J. (2019). Characterization and Measurement of Microcomponents with the Atomic Force Microscope (AFM). In *Optical Inspection of Microsystems, Second Edition*, CRC Press, 155-176.

Seyedsharifi, R., & Khorramdel, S. (2015). Effects of nano-zinc oxide and seed inoculation by plant growth promoting rhizobacteria (PGPR) on yield, yield components and grain filling period of soybean (*Glycine max* l.). *Iranian Journal of Field Crops Research*, 13(4), 738–753.

Shafey, A. M. E. (2020). Green synthesis of metal and metal oxide nanoparticles from plant leaf extracts and their applications: A review. *Green Processing and Synthesis*, 9(1), 304–339. DOI: 10.1515/gps-2020-0031

Shahbaz, M., Akram, A., Raja, N. I., Mukhtar, T., Mashwani, Z. U. R., Mehak, A., Fatima, N., Sarwar, S., Haq, E. U., & Yousaf, T. (2022). Green synthesis and characterization of selenium nanoparticles and its application in plant disease management: A review. In Pakistan Journal of Phytopathology (Vol. 34, Issue 1, pp. 189–102). Pakistan Phytopathological Society. DOI: 10.33866/phytopathol.034.01.0739

Shahidi, F., Arachchi, J. K. V., & Jeon, Y. J. (1999). Food applications of chitin and chitosans. *Trends in Food Science & Technology*, 10(2), 37–51. DOI: 10.1016/S0924-2244(99)00017-5

Shahid, M., Ijaz, N., Shahid, B., Tufail, T., Ain, H. B. U., Hussain, M., Basharat, S., Ikram, A., & Al Jbawi, E. (2023). *Eucalyptus globulus* Labill. Mediated synthesis of ZnO nanoparticles, their Optimization and characterization. *Cogent Food & Agriculture*, 10(1), 2293332. DOI: 10.1080/23311932.2023.2293332

Shah, M., Fawcett, D., Sharma, S., Tripathy, S. K., & Poinern, G. E. J. (2015). Green synthesis of metallic nanoparticles via biological entities. *Materials (Basel)*, 8(11), 7278–7308. DOI: 10.3390/ma8115377 PMID: 28793638

Shah, M., Poinern, G. E. J., Sharma, S. B., & Fawcett, D. (2015). Biogenic synthesis of gold and silver nanoparticles using the leaf extract from *Eucalyptus macrocarpa.International Journal of Sciences*, 4(6), 27–33.

Shah, V., Collins, D., Walker, V. K., & Shah, S. (2014). The impact of engineered cobalt, iron, nickel and silver nanoparticles on soil bacterial diversity under field conditions. *Environmental Research Letters*, 9(2), 024001. DOI: 10.1088/1748-9326/9/2/024001

Shaik, M., Khan, M., Kuniyil, M., Al-Warthan, A., Alkhathlan, H., Siddiqui, M., Shaik, J., Ahamed, A., Mahmood, A., Khan, M., & Adil, S. (2018). Plant-Extract-Assisted Green Synthesis of Silver Nanoparticles Using Origanum vulgare L. Extract and Their Microbicidal Activities. In Sustainability (Vol. 10, Issue 4, p. 913). MDPI AG. DOI: 10.3390/su10040913

Shaikh, S., Rizvi, S. M. D., Shakil, S., Hussain, T., Alshammari, T. M., Ahmad, W., Tabrez, S., Al-Qahtani, M. H., & Abuzenadah, A. M. (2017). Synthesis and Characterization of Cefotaxime Conjugated Gold Nanoparticles and Their Use to Target Drug-Resistant CTX-M-Producing Bacterial Pathogens. *Journal of Cellular Biochemistry*, 118(9), 2802–2808. DOI: 10.1002/jcb.25929 PMID: 28181300

Shala, A. Y., & Gururani, M. A. (2021). Phytochemical properties and diverse beneficial roles of *Eucalyptus globulus* Labill.: A Review. *Horticulturae*, 7(11), 450. DOI: 10.3390/horticulturae7110450

Shamloo, H. B., Jafarzadeh, S., Jouyban, A., & Soleymani, J. (2022). Synthesis and application of concanavalin A-conjugated green luminescent gold nanoparticle/fluorescent polydopamine nanoparticles for specific differentiation of cancer cells from normal cells using glycan bioreceptors. *Materials Chemistry and Physics*, 288, 126344. DOI: 10.1016/j.matchemphys.2022.126344

Shammout, M., & Awwad, A. (2021). A novel route for the synthesis of copper oxide nanoparticles using Bougainvillea plant flowers extract and antifungal activity evaluation. *Chemistry International*, 7(1), 71–78. DOI: 10.5281/zenodo.4042902

Shang, Y., Hasan, M. K., Ahammed, G. J., Li, M., Yin, H., & Zhou, J. (2019). Applications of nanotechnology in plant growth and crop protection: A review. *Molecules (Basel, Switzerland)*, 24(14), 2558. DOI: 10.3390/molecules24142558 PMID: 31337070

Shankar, S. S., Rai, A., Ahmad, A., & Sastry, M. (2004). Rapid synthesis of Au, Ag, and bimetallic Au core–Ag shell nanoparticles using Neem (Azadirachta indica) leaf broth. *Journal of Colloid and Interface Science*, 275(2), 496–502. DOI: 10.1016/j.jcis.2004.03.003 PMID: 15178278

Shankar, S., Pangeni, R., Park, J. W., & Rhim, J. W. (2018). Preparation of sulfur nanoparticles and their antibacterial activity and cytotoxic effect. *Materials Science and Engineering C*, 92, 508–517. DOI: 10.1016/j.msec.2018.07.015 PMID: 30184776

Shanmugam, K., Subha, V., & Thulasimuthu, E. (2023). Environmental Aspects of Copper Nanoparticles Against Green Algae Water and Bactericidal Action From Ethanolic Root Extract of Asparagus Racemosus. MDPI AG. DOI: 10.20944/preprints202306.1501.v1

Shanmugam, P. (2022). Green route synthesis of alpinia calcarata functionalized gold nanoparticles for nonlinear optical applications. *Heliyon*, 8(8), e10409. DOI: 10.1016/j.heliyon.2022.e10409 PMID: 36082338

Shanwaz, M. M., & Shyam, P. (2022). Synthesis of Silver Nanoparticles from Vitex negundo Plant by Green Method and their Bactericidal Effects. In Letters in Applied NanoBioScience (Vol. 12, Issue 2, p. 59). AMG Transcend Association. DOI: 10.33263/LIANBS122.059

Shariatinia, Z. (2020). Biopolymeric Nanocomposites in Drug Delivery. In Nayak, A. K., & Hasnain, M. S. (Eds.), *Advanced Biopolymeric Systems for Drug Delivery* (pp. 233–290). Springer International Publishing., DOI: 10.1007/978-3-030-46923-8_10

Sharma, A., Goyal, A. K., & Rath, G. (2018). Recent advances in metal nanoparticles in cancer therapy. *Journal of Drug Targeting*, 26(8), 617–632. DOI: 10.1080/1061186X.2017.1400553 PMID: 29095640

Sharma, B., Tiwari, S., Kumawat, K. C., & Cardinale, M. (2023). Nano-biofertilizers as bio-emerging strategies for sustainable agriculture development: Potentiality and their limitations. *The Science of the Total Environment*, 860, 160476. DOI: 10.1016/j.scitotenv.2022.160476 PMID: 36436627

Sharma, D., Kanchi, S., & Bisetty, K. (2019). Biogenic synthesis of nanoparticles: A review. *Arabian Journal of Chemistry*, 12(8), 3576–3600. DOI: 10.1016/j.arabjc.2015.11.002

Sharma, D., Sabela, M. I., Kanchi, S., Bisetty, K., Skelton, A. A., & Honarparvar, B. (2018). Green synthesis, characterization and electrochemical sensing of silymarin by ZnO nanoparticles: Experimental and DFT studies. *Journal of Electroanalytical Chemistry (Lausanne, Switzerland)*, 808, 160–172. DOI: 10.1016/j.jelechem.2017.11.039

Sharma, G., Das, S., Kumar, P., & Stangar, U. L. (2024). Parametric effect of continuous synthesis of metal nanoparticles within microchannels and application: A review study. *Chemical Engineering Research & Design*, 209, 142–156. DOI: 10.1016/j.cherd.2024.07.058

Sharma, R. K., Gulati, S., & Mehta, S. K. (2016). Green synthesis of nanoparticles using plant extracts: A review. *Environmental Chemistry Letters*, 14(2), 223–238. DOI: 10.1007/s10311-015-0531-3

Sharma, R., Bisen, D. P., Shukla, U., & Sharma, B. G. (2012). X-ray diffraction: A powerful method of characterizing nanomaterials. *Recent Research in Science and Technology*, 4(8), 77–79.

Sharma, S., Parveen, R., & Chatterji, B. P. (2021). Toxicology of Nanoparticles in Drug Delivery. *Current Pathobiology Reports*, 9(4), 133–144. DOI: 10.1007/s40139-021-00227-z PMID: 34840918

Sharma, V. K., Yngard, R. A., & Lin, Y. (2015). Silver nanoparticles: Green synthesis and their antimicrobial activities. *Advances in Colloid and Interface Science*, 145(1-2), 83–96. DOI: 10.1016/j.cis.2008.09.002 PMID: 18945421

Sharma, V., & Kundu, P. P. (2006). Addition polymers from natural oils—A review. *Progress in Polymer Science*, 31(11), 983–1008. DOI: 10.1016/j.progpolymsci.2006.09.003

Sharma, V., Sharma, J., Kansay, V., Dutta, A., Raj, M., Singh, M., & Kumar, S. (2024). Green synthesis, characterization and drug-loaded iron oxide nanoparticles derived from Nerium oleander flower extract as a nanocarrier for in vitro antibacterial efficacy. *Nano Express*, 5(1), 015014. DOI: 10.1088/2632-959X/ad2997

Shcherbakova, E. N., Shcherbakov, A. V., Andronov, E. E., Gonchar, L. N., Kalenskaya, S. M., & Chebotar, V. K. (2017). Combined pre-seed treatment with microbial inoculants and Mo nanoparticles changes composition of root exudates and rhizosphere microbiome structure of chickpea (Cicer arietinum L.) plants. *Symbiosis*, 73(1), 57–69. DOI: 10.1007/s13199-016-0472-1

Shenashen, M., Derbalah, A., Hamza, A., Mohamed, A., & El Safty, S. (2017). Antifungal activity of fabricated mesoporous alumina nanoparticles against root rot disease of tomato caused by Fusarium oxysporium. *Pest Management Science*, 73(6), 1121–1126. DOI: 10.1002/ps.4420 PMID: 27558672

Shende, S., Ingle, A. P., Gade, A., & Rai, M. (2015). Green synthesis of copper nanoparticles by Citrus medica Linn. (Idilimbu) juice and its antimicrobial activity. *World Journal of Microbiology & Biotechnology*, 31(6), 865–873. DOI: 10.1007/s11274-015-1840-3 PMID: 25761857

Shin, D. O., Jeong, J. R., Han, T. H., Koo, C. M., Park, H. J., Lim, Y. T., & Kim, S. O. (2010). A plasmonic biosensor array by block copolymer lithography. *Journal of Materials Chemistry*, 20(34), 7241–7247. DOI: 10.1039/c0jm01319f

Shinde, U. A., & Nagarsenker, M. S. (2009). Characterization of gelatin-sodium alginate complex coacervation system. *Indian Journal of Pharmaceutical Sciences*, 71(3), 313. DOI: 10.4103/0250-474X.56033 PMID: 20490302

Shivaji, K., Mani, S., Ponmurugan, P., De Castro, C. S., Lloyd Davies, M., Balasubramanian, M. G., & Pitchaimuthu, S. (2018). Green-synthesis-derived CdS quantum dots using tea leaf extract: Antimicrobial, bioimaging, and therapeutic applications in lung cancer cells. *ACS Applied Nano Materials*, 1(4), 1683–1693. DOI: 10.1021/acsanm.8b00147

Shivani Rasool, H. I., Palanisamy, V., Mathisen, C., Schmidt, Ḱ. M., Wong, D. T., & Gimzewski, J. K. (2010). Structural-mechanical characterization and force spectroscopy saliva, using correlative AFM, FESEM, of nanoparticle exosomes in human. *American Chemical Society, 4*(4), 1921–1926.

Shnoudeh, A. J., Hamad, I., Abdo, R. W., Qadumii, L., Jaber, A. Y., Surchi, H. S., & Alkelany, S. Z. (2019). Synthesis, characterization, and applications of metal nanoparticles. In *Biomaterials and bionanotechnology* (pp. 527–612). Elsevier. DOI: 10.1016/B978-0-12-814427-5.00015-9

Shukla, A. K., Alam, J., Mishra, U., & Alhoshan, M. (2023). Effect of Cu-doped GO nanoparticles on polyphenylsulfone nanocomposite membrane surface and its application for the removal of organic pollutants and antibacterial analysis. In *Materials Today Communications* (Vol. 35, p. 106326). Elsevier BV., DOI: 10.1016/j.mtcomm.2023.106326

Siddiqi, K. S., Husen, A., & Rao, R. A. (2018). A review on biosynthesis of silver nanoparticles and their biocidal properties. *Journal of Nanobiotechnology*, 16(1), 1–28. DOI: 10.1186/s12951-018-0334-5 PMID: 29452593

Siddique, S., & Chow, J. C. (2020). Application of nanomaterials in biomedical imaging and cancer therapy. *Nanomaterials (Basel, Switzerland)*, 10(9), 1700. DOI: 10.3390/nano10091700 PMID: 32872399

Siegel, F. R., & Siegel, F. R. (2015). Population assessments: 2013–2050–2100: Growth, stability, contraction. Countering 21st Century Social-Environmental Threats to Growing Global Populations, 1-8.

Sila, M. J., Nyambura, M. I., Abong'o, D. A., Mwaura, F. B., & Iwuoha, E. (2019). Biosynthesis of silver nanoparticles from *Eucalyptus corymbia* leaf extract at optimized conditions. *Nano Hybrids and Composites*, 25, 32–45. DOI: 10.4028/www.scientific.net/NHC.25.32

Sillen, W. M., Thijs, S., Abbamondi, G. R., Janssen, J., Weyens, N., White, J. C., & Vangronsveld, J. (2015). Effects of silver nanoparticles on soil microorganisms and maize biomass are linked in the rhizosphere. *Soil Biology & Biochemistry*, 91, 14–22. DOI: 10.1016/j.soilbio.2015.08.019

Silva, A. A., Sousa, A. M. F., Furtado, C. R. G., & Carvalho, N. M. F. (2022). *Green magnesium oxide prepared by plant extracts: synthesis, properties and applications*. Materials Today Sustainability., DOI: 10.1016/j.mtsust.2022.100203

Singh, A. K., Pal, P., Gupta, V., Yadav, T. P., Gupta, V., & Singh, S. P. (2018). Green synthesis, characterization and antimicrobial activity of zinc oxide quantum dots using Eclipta alba. *Materials Chemistry and Physics*, 203, 40–48. DOI: 10.1016/j.matchemphys.2017.09.049

Singh, A., Goyal, V., Singh, J., Kaur, H., Kumar, S., Batoo, K. M., Gaur, J., Pal, M., Rawat, M., & Hussain, S. (2022). Structurally and morphologically engineered single-pot biogenic synthesis of NiO nanoparticles with enhanced photocatalytic and antimicrobial activities. *Journal of Cleaner Production*, 343, 131026. DOI: 10.1016/j.jclepro.2022.131026

Singh, H., Desimone, M. F., Pandya, S., Jasani, S., George, N., Adnan, M., Aldarhami, A., Bazaid, A. S., & Alderhami, S. A. (2023). Revisiting the Green Synthesis of Nanoparticles: Uncovering Influences of Plant Extracts as Reducing Agents for Enhanced Synthesis Efficiency and Its Biomedical Applications. *International Journal of Nanomedicine*, 18, 4727–4750. DOI: 10.2147/IJN.S419369 PMID: 37621852

Singh, H., Sharma, A., Bhardwaj, S. K., Arya, S. K., Bhardwaj, N., & Khatri, M. (2021). Recent advances in the applications of nano-agrochemicals for sustainable agricultural development. *Environmental Science. Processes & Impacts*, 23(2), 213–239. DOI: 10.1039/D0EM00404A PMID: 33447834

Singh, J., Dutta, T., Kim, K. H., Rawat, M., Samddar, P., & Kumar, P. (2018). 'Green' synthesis of metals and their oxide nanoparticles: Applications for environmental remediation. *Journal of Nanobiotechnology*, 16(1), 1–24. DOI: 10.1186/s12951-018-0408-4 PMID: 30373622

Singh, K. R., Natarajan, A., & Pandey, S. S. (2023). Bioinspired Multifunctional Silver Nanoparticles for Optical Sensing Applications: A Sustainable Approach. *ACS Applied Bio Materials*, 6(11), 4549–4571. DOI: 10.1021/acsabm.3c00669 PMID: 37852204

Singh, M., Thakur, V., Kumar, V., Raj, M., Gupta, S., Devi, N., Upadhyay, S. K., Macho, M., Banerjee, A., Ewe, D., & Saurav, K. (2022). Silver Nanoparticles and Its Mechanistic Insight for Chronic Wound Healing: Review on Recent Progress. *Molecules (Basel, Switzerland)*, 27(17), 5587. DOI: 10.3390/molecules27175587 PMID: 36080353

Singh, P., Kim, Y. J., Zhang, D., & Yang, D. C. (2016). Biological Synthesis of Nanoparticles from Plants and Microorganisms. *Trends in Biotechnology*, 34(7), 588–599. DOI: 10.1016/j.tibtech.2016.02.006 PMID: 26944794

Singh, P., Pandit, S., Mokkapati, V., Garg, A., Ravikumar, V., & Mijakovic, I. (2018). Gold nanoparticles in diagnostics and therapeutics for human cancer. *International Journal of Molecular Sciences*, 19(7), 1979. DOI: 10.3390/ijms19071979 PMID: 29986450

Singh, S. P., Subhadarshini, S., Dilnashin, H., Keswani, C., Sansinenea, E., & Shende, S. S. (2021). Biological synthesis of nanofertilizer and their effects on crop health. In *Microbial synthesis of nanomaterials* (pp. 201–216). Nova Science publishers.

Singh, S., Saikia, J. P., & Buragohain, A. K. (2013). A novel 'green'synthesis of colloidal silver nanoparticles (SNP) using *Dillenia indica* fruit extract. *Colloids and Surfaces. B, Biointerfaces*, 102, 83–85. DOI: 10.1016/j.colsurfb.2012.08.012 PMID: 23010109

Singh, T. A., Sharma, A., Tejwan, N., Ghosh, N., Das, J., & Sil, P. C. (2021). A state of the art review on the synthesis, antibacterial, antioxidant, antidiabetic and tissue regeneration activities of zinc oxide nanoparticles. *Advances in Colloid and Interface Science*, 295, 102495. DOI: 10.1016/j.cis.2021.102495 PMID: 34375877

Sinha, T., Adhikari, P. P., & Bhandari, V. M. (2022). Sustainable Fabrication of Copper Nanoparticles: A Potent and Affordable Candidate for Water Treatment, Water Disinfection, Antioxidant Activity and Theranostic Agent. In ChemistrySelect (Vol. 7, Issue 15). Wiley. DOI: 10.1002/slct.202103552

Siripireddy, B., & Mandal, B. K. (2017). Facile green synthesis of zinc oxide nanoparticles by *Eucalyptus globulus* and their photocatalytic and antioxidant activity. *Advanced Powder Technology*, 28(3), 785–797. DOI: 10.1016/j.apt.2016.11.026

Skłodowski, K., Chmielewska-Deptuła, S. J., Piktel, E., Wolak, P., Wollny, T., & Bucki, R. (2023). Metallic nanosystems in the development of antimicrobial strategies with high antimicrobial activity and high biocompatibility. *International Journal of Molecular Sciences*, 24(3), 2104. DOI: 10.3390/ijms24032104 PMID: 36768426

Slomberg, D. L., & Schoenfisch, M. H. (2012). Silica nanoparticle phytotoxicity to Arabidopsis thaliana. *Environmental Science & Technology*, 46(18), 10247–10254. DOI: 10.1021/es300949f PMID: 22889047

Smith, J. R., Breakspear, S., & Campbell, S. A. (2003). AFM in Surface Finishing: Part I - An Introduction. *Finishing*, 27(8), 35–40. DOI: 10.1080/00202967.2003.11871484

Sofi, W., Gowri, M., Shruthilaya, M., Rayala, S., & Venkatraman, G. (2012). Silver nanoparticles as an antibacterial agent for endodontic infections. *BMC Infectious Diseases*, 12(S1), 60. DOI: 10.1186/1471-2334-12-S1-P60

Soliman, A. S., El-Feky, S. A., & Darwish, E. (2015). Alleviation of salt stress on Moringa peregrina using foliar application of nanofertilizers. *Journal of Horticulture and Forestry*, 7(2), 36–47. DOI: 10.5897/JHF2014.0379

Soliman, M. K. Y., Salem, S. S., Abu-Elghait, M., & Azab, M. S. (2023). Biosynthesis of Silver and Gold Nanoparticles and Their Efficacy Towards Antibacterial, Antibiofilm, Cytotoxicity, and Antioxidant Activities. *Applied Biochemistry and Biotechnology*, 195(2), 1158–1183. DOI: 10.1007/s12010-022-04199-7 PMID: 36342621

Somya, A. (2020). Hybrid Ion Exchangers. In Hybrid Nanomaterials - Flexible Electronics Materials. IntechOpen. DOI: 10.5772/intechopen.92116

Somya, A. (2023). Metal Phosphates: Their Role as Ion Exchangers in Water Purification. In Metal Phosphates and Phosphonates (pp. 341–356). Springer International Publishing. DOI: 10.1007/978-3-031-27062-8_19

Somya, A., & Maheri, A. K. (2020). Studies on polyoxyethylene octyl phenyl ether supported thorium(IV) phosphates: a new cation exchange material. In Bulletin of Materials Science (Vol. 43, Issue 1). Springer Science and Business Media LLC. DOI: 10.1007/s12034-020-02153-z

Somya, A., & Singh, M. (2021). Studies on Sodium Lauryl Sulphate-supported Thorium (IV) Phosphate: A New Surfactant-supported Cation Exchange Resin, Useful in Water Purification. In Current Analytical Chemistry (Vol. 18, Issue 3, pp. 383–390). Bentham Science Publishers Ltd. DOI: 10.2174/15734110 17666210804102748

Somya, A., Rafiquee, M. Z. A., Varshney, K. G., & Al-Lohedan, H. A. (2014). Synthesis, characterization and analytical applications of N-dodecyl pyridinium chloride-cerium (IV) phosphate fibrous ion exchanger: selective for Hg(II) and its binary separations. In Desalination and Water Treatment (Vol. 56, Issue 3, pp. 638–645). Elsevier BV. DOI: 10.1080/19443994.2014.940388

Somya, A. (2019). Surfactant based hybrid ion exchangers. *Research Journal of Chemistry and Environment*, 23(3), 96.

Somya, A. (2024). An investigation study on water quality of few lakes of Bengaluru. In *Industry 4.0 with Modern Technology* (pp. 200–204). CRC Press., DOI: 10.1201/9781003450924-37

Somya, A., Khan, A., Hussein, M. A., Al-Romaizan, A. N., Malash, A. M., Topare, N. S., & Asiri, A. M. (2023). Biological and environmental studies on pectin-zirconium (IV) phosphate, a biopolymer doped hybrid cation exchanger. In *International Journal of Biological Macromolecules* (Vol. 250, p. 125803). Elsevier BV., DOI: 10.1016/j.ijbiomac.2023.125803

Somya, A., Rafiquee, M. Z. A., & Upadhyaya, V. (2018). Role of surfactants in the adsorption studies of n-butyl acetate-cerium (IV) phosphate for some alkaline earths and heavy metal ions. *Research Journal of Chemistry and Environment*, 22(6), 45.

Somya, A., & Upadhayaya, V. (2020). Synthesis and ion exchange characterization of n-butyl acetate based tin (IV) phosphate: A new intercalated hybrid ion exchange material. *Research Journal of Chemistry and Environment*, 24(7), 107.

Somya, A., & Upadhyaya, V. (2017). Synthesis, Characterization and Adsorption behaviour of N-dodecyl pyridinium chloride based Tin (IV) phosphate: A new surfactant based hybrid ion exchanger. *Research Journal of Chemistry and Environment*, 21(1), 18.

Song, J. Y., Kwon, E. Y., & Kim, B. S. (2010). Biological synthesis of platinum nanoparticles using Diopyros kaki leaf extract. *Bioprocess and Biosystems Engineering*, 33(1), 159–164. DOI: 10.1007/s00449-009-0373-2 PMID: 19701776

Sousa, M. G. C., Rezende, T. M. B., & Franco, O. L. (2021). Nanofibers as drug-delivery systems for antimicrobial peptides. *Drug Discovery Today*, 26(8), 2064–2074. DOI: 10.1016/j.drudis.2021.03.008 PMID: 33741497

Sparks, R. P., Jenkins, J. L., & Fratti, R. (2019). Use of surface plasmon resonance (SPR) to determine binding affinities and kinetic parameters between components important in fusion machinery. *SNAREs. Methods and Protocols*, •••, 199–210.

Spotti, M. L., Acosta, C. A., Perez, A. A., Spotti, M. J., Quintero-Cerón, J. P., Carrara, C. R., & Fioramonti, S. A. (2024). Whey protein nanoparticles obtained by water in oil emulsification followed by heating: Influence of surfactant type on the stability, rheology and interfacial properties. *Food Hydrocolloids*, 153, 109912. DOI: 10.1016/j.foodhyd.2024.109912

Spruit, S. L., van de Poel, I., & Doorn, N. (2016). Informed Consent in Asymmetrical Relationships: An Investigation into Relational Factors that Influence Room for Reflection. *NanoEthics*, 10(2), 123–138. DOI: 10.1007/s11569-016-0262-5 PMID: 27478516

Sre, P. R., Reka, M., Poovazhagi, R., Kumar, M. A., & Murugesan, K. (2015). Antibacterial and cytotoxic effect of biologically synthesized silver nanoparticles using aqueous root extract of *Erythrina indica lam.Spectrochimica Acta. Part A: Molecular and Biomolecular Spectroscopy*, 135, 1137–1144. DOI: 10.1016/j.saa.2014.08.019 PMID: 25189525

Srihasam, S., Thyagarajan, K., Korivi, M., Lebaka, V. R., & Mallem, S. P. (2020). Phytogenic Generation of NiO Nanoparticles Using Stevia Leaf Extract and Evaluation of Their In-Vitro Antioxidant and Antimicrobial Properties. *Biomolecules*, 10(1), 89. DOI: 10.3390/biom10010089 PMID: 31935798

Srikar, S. K., Giri, D. D., Pal, D. B., Mishra, P. K., & Upadhyay, S. N. (2016). Green synthesis of silver nanoparticles: A review. *Green and Sustainable Chemistry*, 6(01), 34–56. DOI: 10.4236/gsc.2016.61004

Srivastav, A., Ganjewala, D., Singhal, R. K., Rajput, V. D., Minkina, T., Voloshina, M., Srivastava, S., & Shrivastava, M. (2021). Effect of ZnO nanoparticles on growth and biochemical responses of wheat and maize. *Plants*, 10(12), 2556. DOI: 10.3390/plants10122556 PMID: 34961025

Srivastava, S. (2023). Ethical issues regarding the use of nanobiotechnology-based products. In Pratap Singh, R., Adetunji, C. O., Singh, R. L., Singh, J., Solanki, P. R., & Singh, K. R. B. (Eds.), *Nanobiotechnology for the Livestock Industry* (pp. 435–473). Elsevier., DOI: 10.1016/B978-0-323-98387-7.00005-7

Stetefeld, J., Mckenna, S. A., & Patel, T. R. (2016). Dynamic light scattering : A practical guide and applications in biomedical sciences. *Biophysical Reviews*, 8(4), 409–427. DOI: 10.1007/s12551-016-0218-6 PMID: 28510011

Stevie, F. A., & Donley, C. L. (2020). Introduction to X-ray photoelectron spectroscopy. *Journal of Vacuum Science & Technology. A, Vacuum, Surfaces, and Films*, 38(6), 063204. DOI: 10.1116/6.0000412

Stewart, P. L. (2017). Cryo-electron microscopy and cryo-electron tomography of nanoparticles. *Wiley Interdisciplinary Reviews. Nanomedicine and Nanobiotechnology*, 9(2), 1–16. DOI: 10.1002/wnan.1417 PMID: 27339510

Stryer, L., Berg, J. M., & Tymoczko, J. L. (2002). *Biochemistry and Study Guide* (5th ed.). W.H. Freeman.

Suazo-Hernández, J., Arancibia-Miranda, N., Mlih, R., Cáceres-Jensen, L., Bolan, N., & Mora, M. D. L. L. (2023). Impact on some soil physical and chemical properties caused by metal and metallic oxide engineered nanoparticles: A review. *Nanomaterials (Basel, Switzerland)*, 13(3), 572. DOI: 10.3390/nano13030572 PMID: 36770533

Sudarshan, N. R., Hoover, D. G., & Knorr, D. (1992). Antibacterial action of chitosan. *Food Biotechnology*, 6(3), 257–272. DOI: 10.1080/08905439209549838

Sultan, M., Hafez, O. M., Saleh, M. A., & Youssef, A. M. (2021). Smart edible coating films based on chitosan and beeswax–pollen grains for the postharvest preservation of Le Conte pear. *RSC Advances*, 11(16), 9572–9585. DOI: 10.1039/D0RA10671B PMID: 35423446

Sun, W. (2018). Principles of atomic force microscopy. *Atomic Force Microscopy in Molecular and Cell Biology*, 1-28.

Sun, C., Lux, S., Müller, E., Meffert, M., & Gerthsen, D. (2020). Versatile application of a modern scanning electron microscope for materials characterization. *Journal of Materials Science*, 55(28), 13824–13835. DOI: 10.1007/s10853-020-04970-3

Sundar, S., Kundu, J., & Kundu, S. C. (2010). Biopolymeric nanoparticles. *Science and Technology of Advanced Materials*, 11(1), 014104. Advance online publication. DOI: 10.1088/1468-6996/11/1/014104 PMID: 27877319

Sunder, S. S. (1993, October). The ecological, economic and social effects of *Eucalyptus*. In Proceedings of the regional expert consultation on eucalyptus (Vol. 1). FAO Rome Italy.

Sun, H., Peng, Q., Guo, J., Zhang, H., Bai, J., & Mao, H. (2022). Effects of short-term soil exposure of different doses of ZnO nanoparticles on the soil environment and the growth and nitrogen fixation of alfalfa. *Environmental Pollution*, 309, 119817. DOI: 10.1016/j.envpol.2022.119817 PMID: 35872284

Sun, Q., Cai, X., Li, J., Zheng, M., Chen, Z., & Yu, C. P. (2014). Green synthesis of silver nanoparticles using tea leaf extract and evaluation of their stability and antibacterial activity. *Colloids and Surfaces. A, Physicochemical and Engineering Aspects*, 444, 226–231. DOI: 10.1016/j.colsurfa.2013.12.065

Sun, Z., Chen, Y., Ke, Q., Yang, Y., & Yuan, J. (2002). Photocatalytic degradation of a cationic azo dye by TiO2/bentonite nanocomposite. *Journal of Photochemistry and Photobiology A Chemistry*, 149(1-3), 169–174. DOI: 10.1016/S1010-6030(01)00649-9

Suriyakalaa, U., Antony, J. J., Suganya, S., Siva, D., Sukirtha, R., Kamalakkannan, S., & Achiraman, S. (2013). Hepatocurative activity of biosynthesized silver nanoparticles fabricated using Andrographis paniculata. *Colloids and Surfaces. B, Biointerfaces*, 102, 189–194. DOI: 10.1016/j.colsurfb.2012.06.039 PMID: 23018020

Susan Johnson, M., Nair, R. S., & Nambiar, A. (2023). Electrical conductivity of green synthesized copper nanoparticles by P.granatum. *Materials Today: Proceedings*, •••, 1–5. DOI: 10.1016/j.matpr.2023.02.175

Susanti, D., Haris, M. S., Taher, M., & Khotib, J. (2022). Natural Products-Based Metallic Nanoparticles as Antimicrobial Agents. *Frontiers in Pharmacology*, 13, 895616. DOI: 10.3389/fphar.2022.895616 PMID: 35721199

Su, Y., Ashworth, V., Kim, C., Adeleye, A. S., Rolshausen, P., Roper, C., White, J., & Jassby, D. (2019). Delivery, uptake, fate, and transport of engineered nanoparticles in plants: A critical review and data analysis. *Environmental Science. Nano*, 6(8), 2311–2331. DOI: 10.1039/C9EN00461K

Suzuki, N., Wang, Y., Elvati, P., Qu, Z. B., Kim, K., Jiang, S., Baumeister, E., Lee, J., Yeom, B., Bahng, J. H., Lee, J., Violi, A., & Kotov, N. A. (2016). Chiral graphene quantum dots. *ACS Nano*, 10(2), 1744–1755. DOI: 10.1021/acsnano.5b06369 PMID: 26743467

Swain, S., Barik, S. K., Behera, T., Nayak, S. K., Sahoo, S. K., Mishra, S. S., & Swain, P. (2016). Green Synthesis of Gold Nanoparticles Using Root and Leaf Extracts of Vetiveria zizanioides and Cannabis sativa and its Antifungal Activities. *BioNanoScience*, 6(3), 205–213. DOI: 10.1007/s12668-016-0208-y

Sykes, E., Dai, Q., Sarsons, C., Chen, J., Rocheleau, J., Hwang, D., Zheng, G., Cramb, D., Rinker, K., & Chan, W. (2016). Tailoring nanoparticle designs to target cancer based on tumor pathophysiology. *Proceedings of the National Academy of Sciences of the United States of America,* 113. DOI: 10.1073/pnas.1521265113

Szczyglewska, P., Feliczak-Guzik, A., & Nowak, I. (2023). Nanotechnology–General Aspects: A Chemical Reduction Approach to the Synthesis of Nanoparticles. *Molecules (Basel, Switzerland)*, 28(13), 4932. DOI: 10.3390/molecules28134932 PMID: 37446593

Taha, M., Hassan, M., Essa, S., & Tartor, Y. (2013). Use of Fourier transform infrared spectroscopy (FTIR) spectroscopy for rapid and accurate identification of yeasts isolated from human and animals. *International Journal of Veterinary Science and Medicine*, 1(1), 15–20. DOI: 10.1016/j.ijvsm.2013.03.001

Tahvildari, K., & Mojahedi, H. (2016). Trends. *Life Sciences*, 5, 2319.

Takma, D. K., Bozkurt, S., Koç, M., Korel, F., & Nadeem, H. Ş. (2023). Characterization and encapsulation efficiency of zein nanoparticles loaded with chestnut fruit shell, cedar and sweetgum bark extracts. *Food Hydrocolloids for Health*, 4, 100151. DOI: 10.1016/j.fhfh.2023.100151

Tamil Elakkiya, V., Meenakshi, R. V., Senthil Kumar, P., Karthik, V., Ravi Shankar, K., Sureshkumar, P., & Hanan, A. (2022). Green synthesis of copper nanoparticles using Sesbania aculeata to enhance the plant growth and antimicrobial activities. *International Journal of Environmental Science and Technology*, 19(3), 1313–1322. DOI: 10.1007/s13762-021-03182-9

Tamilselvan, S., Ashokkumar, T., & Govindaraju, K. (2017). Microscopy based studies on the interaction of bio-based silver nanoparticles with Bombyx mori Nuclear Polyhedrosis virus. *Journal of Virological Methods*, 242, 58–66. DOI: 10.1016/j.jviromet.2017.01.001 PMID: 28065747

Tanaka, N., & Saitoh, K. (2015). Basics of STEM. *In Scanning Transmission Electron Microscopy of Nanomaterials: Basics of Imaging And Analysis* (pp. 41-108).

Tang, C. Y., & Yang, Z. (2017). Transmission electron microscopy (TEM). N. Hilal, A. F. Ismail (Eds.), *Membrane Characterization* (pp. 145-159). Elsevier. DOI: 10.1016/B978-0-444-63776-5.00008-5

Tan, G., Tevlek, A., & Aydin, H. M. (2023). Comparison of garlic and onion extract-derived gold nanoparticles: Characterization and anticancer activity. *Journal of Drug Delivery Science and Technology*, 84, 104542. DOI: 10.1016/j.jddst.2023.104542

Tang, D., Liu, J., Yan, X., & Kang, L. (2016). Graphene oxide derived graphene quantum dots with different photoluminescence properties and peroxidase-like catalytic activity. *RSC Advances*, 6(56), 50609–50617. DOI: 10.1039/C5RA26279H

Tang, L., Ji, R., Cao, X., Lin, J., Jiang, H., Li, X., Teng, K. S., Luk, C. M., Zeng, S., Hao, J., & Lau, S. P. (2012). Deep ultraviolet photoluminescence of water-soluble self-passivated graphene quantum dots. *ACS Nano*, 6(6), 5102–5110. DOI: 10.1021/nn300760g PMID: 22559247

Tang, Y., Su, Y., Yang, N., Zhang, L., & Lv, Y. (2014). Carbon nitride quantum dots: A novel chemiluminescence system for selective detection of free chlorine in water. *Analytical Chemistry*, 86(9), 4528–4535. DOI: 10.1021/ac5005162 PMID: 24655009

Tanhaei, B., Ayati, A., Lahtinen, M., & Sillanpää, M. (2015). Preparation and Characterization of a Novel chitosan/Al2O3/magnetite Nanoparticles Composite Adsorbent for Kinetic, Thermodynamic and Isotherm Studies of Methyl Orange Adsorption. *Chemical Engineering Journal*, 259, 1–10. DOI: 10.1016/j.cej.2014.07.109

Taniguchi, N. (1983). Current status in, and future trends of, ultraprecision machining and ultrafine materials processing. *CIRP Annals*, 32(2), 573–582. DOI: 10.1016/S0007-8506(07)60185-1

Tan, N. P. B., & Lee, C. H. (2017). Environment-friendly approach in the synthesis of metal/polymeric nanocomposite particles and their catalytic activities on the reduction of p-nitrophenol to p-aminophenol. In *Green chemical processing and synthesis*. IntechOpen.

Tao, C. (2018). Antimicrobial activity and toxicity of gold nanoparticles: Research progress, challenges and prospects. *Letters in Applied Microbiology*, 67(6), 537–543. DOI: 10.1111/lam.13082 PMID: 30269338

Tao, R., Kang, K., Li, X., Li, R., Huang, R., Jin, Y., Qiu, L., & Zhang, W. (2021). Controlled Synthesis of Palladium Nanoparticles with Size-Dependent Catalytic Activities Enabled by Organic Molecular Cages. *Inorganic Chemistry*, 60(16), 12517–12525. DOI: 10.1021/acs.inorgchem.1c01723 PMID: 34320317

Tarannum, N., Divya, D., & Gautam, Y. K. (2019). Facile green synthesis and applications of silver nanoparticles: a state-of-the-art review. In RSC Advances (Vol. 9, Issue 60, pp. 34926–34948). Royal Society of Chemistry (RSC). https://doi.org/DOI: 10.1039/C9RA04164H

Taş, R., Köroğlu, E., & Celebioglu, H. U.Extract and Antibacterial Activity. (2021). Green Synthesis of Nickel Nanoparticles Using Peumus Boldus Koch. Extract and Antibacterial Activity. *International Journal of Innovative Engineering Applications*, 5(2), 152–155. DOI: 10.46460/ijiea.929625

Tavares, I. S., Caroni, A. L. P. F., Neto, A. D., Pereira, M. R., & Fonseca, J. L. C. (2012). Surface charging and dimensions of chitosan coacervated nanoparticles. *Colloids and Surfaces. B, Biointerfaces*, 90, 254–258. DOI: 10.1016/j.colsurfb.2011.10.025 PMID: 22078924

Tem, S. D. (2008). Characterization and simulation technologies of nanomaterial. M. Nayfeh (Ed.), *Fundamentals and Applications of Nano Silicon in Plasmonics and Fullerines* (pp. 153–167). Elsevier. DOI: 10.1016/B978-0-323-48057-4.00007-4

Thakkar, K. N., Mhatre, S. S., & Parikh, R. Y. (2010). Biological synthesis of metallic nanoparticles. *Nanomedicine; Nanotechnology, Biology, and Medicine*, 6(2), 257–262. DOI: 10.1016/j.nano.2009.07.002 PMID: 19616126

Thakur, K., Bala, I., Rajeshwer, D. M., & Bhatt, A. K. (2017). Evaluation of effectiveness of biologically synthesized silver nanoparticles of *Eucalyptus globules* leaf extract against pathogenic and acne-inducing bacteria. *Journal of Nanomedicine & Nanotechnology*, 8(3), 443. DOI: 10.4172/2157-7439.1000443

Thema, F. T., Manikandan, E., Gurib-Fakim, A., & Maaza, M. (2016). Single phase Bunsenite NiO nanoparticles green synthesis by Agathosma betulina natural extract. *Journal of Alloys and Compounds*, 657, 655–661. DOI: 10.1016/j.jallcom.2015.09.227

Thesing, A., Nascimento, J. E. R. D., Jacob, R. G., & Ferreira, J. (2018, February). *Eucalyptus* oil-mediated synthesis of gold nanoparticles. *Journal of Chemistry and Chemical Engineering. New York.*, 12(2), 52–59.

Thilakan, D., Patankar, J., Khadtare, S., Wagh, N. S., Lakkakula, J., El-Hady, K. M., Islam, S., Islam, M. R., Khan, M. S., Alafaleq, N. O., & Tarique, M. (2022). Plant-Derived Iron Nanoparticles for Removal of Heavy Metals. *International Journal of Chemical Engineering*, 2022, 1517849. DOI: 10.1155/2022/1517849

Thirumoorthy, G., Balasubramanian, B., George, J. A., Nizam, A., Nagella, P., Srinatha, N., Pappuswamy, M., Alanazi, A. M., Meyyazhagan, A., Rengasamy, K. R. R., & Veerappa Lakshmaiah, V. (2024). Phytofabricated bimetallic synthesis of silver-copper nanoparticles using *Aerva lanata* extract to evaluate their potential cytotoxic and antimicrobial activities. *Scientific Reports*, 14(1), 1270. DOI: 10.1038/s41598-024-51647-x PMID: 38218918

Thomas, R., Jishma, P., Snigdha, S., Soumya, K., Mathew, J., & Radhakrishnan, E. (2020). Enhanced antimicrobial efficacy of biosynthesized silver nanoparticle based antibiotic conjugates. *Inorganic Chemistry Communications*, 117, 107978. DOI: 10.1016/j.inoche.2020.107978

Tian, X., Nejadnik, M. R., Baunsgaard, D., Henriksen, A., Rischel, C., & Jiskoot, W. (2016). A comprehensive evaluation of nanoparticle tracking analysis (NanoSight) for characterization of proteinaceous submicron particles. *Journal of Pharmaceutical Sciences*, 105(11), 3366–3375. DOI: 10.1016/j.xphs.2016.08.009 PMID: 27663383

Timilsena, Y. P., Akanbi, T. O., Khalid, N., Adhikari, B., & Barrow, C. J. (2019). Complex coacervation: Principles, mechanisms and applications in microencapsulation. *International Journal of Biological Macromolecules*, 121, 1276–1286. DOI: 10.1016/j.ijbiomac.2018.10.144 PMID: 30352231

Titus, D., James Jebaseelan Samuel, E., & Roopan, S. M. (2018). Nanoparticle characterization techniques. A. K. Shukla and S. Iravani (Eds.), *Green Synthesis, Characterization and Applications of Nanoparticles* (pp. 303-319). Elsevier. DOI: 10.1016/B978-0-08-102579-6.00012-5

Titus, D., James Jebaseelan Samuel, E., & Roopan, S. M. (2019). Chapter 12 - Nanoparticle characterization techniques. In A. K. Shukla & S. B. T.-G. S. Iravani Characterization and Applications of Nanoparticles (Eds.), *Micro and Nano Technologies* (pp. 303–319). Elsevier.

Toksha, B., Sonawale, V. A. M., Vanarase, A., Bornare, D., Tonde, S., Hazra, C., & Chatterjee, A. (2021). Nanofertilizers: A review on synthesis and impact of their use on crop yield and environment. *Environmental Technology & Innovation*, 24, 101986. DOI: 10.1016/j.eti.2021.101986

Tokudome, Y., Morimoto, T., Tarutani, N., Vaz, P. D., Nunes, C. D., Prevot, V., Stenning, G. B. G., & Takahashi, M. (2016). Layered double hydroxide nanoclusters: Aqueous, concentrated, stable, and catalytically active colloids toward green chemistry. *ACS Nano*, 10(5), 5550–5559. DOI: 10.1021/acsnano.6b02110 PMID: 27124717

Torney, F., Trewyn, B. G., Lin, V. S. Y., & Wang, K. (2007). Mesoporous silica nanoparticles deliver DNA and chemicals into plants. *Nature Nanotechnology*, 2(5), 295–300. DOI: 10.1038/nnano.2007.108 PMID: 18654287

Tran, T.-K., Nguyen, M.-K., Lin, C., Hoang, T.-D., Nguyen, T.-C., Lone, A. M., Khedulkar, A. P., Gaballah, M. S., Singh, J., Chung, W. J., & Nguyen, D. D. (2024). Review on fate, transport, toxicity and health risk of nanoparticles in natural ecosystems: Emerging challenges in the modern age and solutions toward a sustainable environment. *The Science of the Total Environment*, 912, 169331. DOI: 10.1016/j.scitotenv.2023.169331 PMID: 38103619

Tsai, G. J., & Su, W. H. (1999). Antibacterial activity of shrimp chitosan against Escherichia coli. *Journal of Food Protection*, 62(3), 239–243. DOI: 10.4315/0362-028X-62.3.239 PMID: 10090242

Tsilo, P. H., Basson, A. K., Ntombela, Z. G., Dlamini, N. G., & Pullabhotla, R. V. S. R. (2024). Green synthesis and characterisation of silver nanoparticles utilising a bioflocculant obtained from *Pichia kudriavzevii* isolated from kombucha tea SCOBY. Advances in Materials and Processing Technologies, 1-17.

Tudi, M., Daniel Ruan, H., Wang, L., Lyu, J., Sadler, R., Connell, D., Chu, C., & Phung, D. T. (2021). Agriculture Development, Pesticide Application and Its Impact on the Environment. In International Journal of Environmental Research and Public Health (Vol. 18, Issue 3, p. 1112). MDPI AG. DOI: 10.3390/ijerph18031112

Tugarova, A. V., Mamchenkova, P. V., Dyatlova, Y. A., & Kamnev, A. A. (2018). FTIR and Raman spectroscopic studies of selenium nanoparticles synthesised by the bacterium Azospirillum thiophilum. *Spectrochimica Acta. Part A: Molecular and Biomolecular Spectroscopy*, 192, 458–463. DOI: 10.1016/j.saa.2017.11.050 PMID: 29220816

Turgeon, S. L., Schmitt, C., & Sanchez, C. (2007). Protein–polysaccharide complexes and coacervates. *Current Opinion in Colloid & Interface Science*, 12(4-5), 166–178. DOI: 10.1016/j.cocis.2007.07.007

Tyagi, P. K., Gola, D., Tyagi, S., Mishra, A. K., Kumar, A., Chauhan, N., Ahuja, A., & Sirohi, S. (2020). Synthesis of zinc oxide nanoparticles and its conjugation with antibiotic: Antibacterial and morphological characterization. *Environmental Nanotechnology, Monitoring & Management*, 14, 100391. DOI: 10.1016/j.enmm.2020.100391

Utsunomiya, S., & Ewing, R. C. (2003). field scanning transmission electron microscopy, scanning transmission electron microscopy-energy dispersive X-ray spectrometry, and energy-filtered transmission. *Environmental Science & Technology*, 37(4), 786–791. https://pubs.acs.org/doi/abs/10.1021/es026053t. DOI: 10.1021/es026053t PMID: 12636280

Vahabi, S., Nazemi Salman, B., & Javanmard, A. (2013). Atomic force microscopy application in biological research: A review. *Iranian Journal of Medical Sciences*, 38(2), 76–83. PMID: 23825885

Van Loo, R. (2019). Regulatory Monitors: Policing Firms in the Compliance Era. *Columbia Law Review*, 119(2), 369–444.

Vanlalveni, C., Lallianrawna, S., Biswas, A., Selvaraj, M., Changmai, B., & Rokhum, S. L. (2021). Green synthesis of silver nanoparticles using plant extracts and their antimicrobial activities: A review of recent literature. *RSC Advances*, 11(5), 2804–2837. DOI: 10.1039/D0RA09941D PMID: 35424248

Vanti, G. L., Nargund, V. B., N, B. K., Vanarchi, R., Kurjogi, M., Mulla, S. I., Tubaki, S., & Patil, R. R. (2019). Synthesis of Gossypium hirsutum-derived silver nanoparticles and their antibacterial efficacy against plant pathogens. *Applied Organometallic Chemistry*, 33(1), 4630. DOI: 10.1002/aoc.4630

Varshney, K. G., Rafiquee, M. Z. A., & Somya, A. (2007). Effect of surfactants on the adsorption behaviour of cerium (IV) phosphate, cation exchanger for alkaline earths and heavy metal ions. In Colloids and Surfaces A: Physicochemical and Engineering Aspects (Vol. 301, Issues 1–3, pp. 69–72). Elsevier BV. DOI: 10.1016/j.colsurfa.2006.12.025

Varshney, K. G., Rafiquee, M. Z. A., & Somya, A. (2007). Effect of surfactants on the adsorption behaviour of tin(IV) phosphate, cation exchanger for alkaline earths and heavy metal ions. In Colloids and Surfaces A: Physicochemical and Engineering Aspects (Vol. 301, Issues 1–3, pp. 224–228). Elsevier BV. DOI: 10.1016/j.colsurfa.2006.12.061

Varshney, K. G., Rafiquee, M. Z. A., & Somya, A. (2008). Triton X-100 based cerium(IV) phosphate as a new Hg(II) selective, surfactant based fibrous ion exchanger: Synthesis, characterization and adsorption behaviour. In Colloids and Surfaces A: Physicochemical and Engineering Aspects (Vol. 317, Issues 1–3, pp. 400–405). Elsevier BV. DOI: 10.1016/j.colsurfa.2007.11.012

Vaseghi, Z., Nematollahzadeh, A., & Tavakoli, O. (2018). Green methods for the synthesis of metal nanoparticles using biogenic reducing agents: A review. *Reviews in Chemical Engineering*, 34(4), 529–559. DOI: 10.1515/revce-2017-0005

Vasireddy, R., Paul, R., & Mitra, A. K. (2012). Green synthesis of silver nanoparticles and the study of optical properties. *Nanomaterials and Nanotechnology*, 2, 8. DOI: 10.5772/52329

Vasoya, N. H. (2023). Revolutionizing Nano Materials Processing through IoT-AI Integration: Opportunities and Challenges. *Journal of Materials Science Research and Reviews*, 6(3), 3.

Vassallo, A., Silletti, M. F., Faraone, I., & Milella, L. (2020). Nanoparticulate antibiotic systems as antibacterial agents and antibiotic delivery platforms to fight infections. *Journal of Nanomaterials*, 2020(1), 6905631. DOI: 10.1155/2020/6905631

Vasudev, S., Saxena, J., Srivastava, B., & Reddy, P. (2023). *Advancements in Immunology, Nanodrugs and Public health.*

Vecchio, M. G., Loganes, C., & Minto, C. (2016). Beneficial and healthy properties of *Eucalyptus* plants: A great potential use. *The Open Agriculture Journal*, 10(1), 52–57. DOI: 10.2174/1874331501610010052

Verma, A., & Mehata, M. S. (2016). Controllable synthesis of silver nanoparticles using Neem leaves and their antimicrobial activity. *Journal of radiation Research and applied sciences, 9*(1), 109-115.

Verma, A. K., & Kumar, P. (2022). On recent developments in biosynthesis and application of Au and Ag nanoparticles from biological systems. *Journal of Nanotechnology*, 2022(1), 5560244. DOI: 10.1155/2022/5560244

Verma, G., & Mishra, M. (2018). Development and optimization of UV-Vis spectroscopy - A review. *World Journal of Pharmaceutical Research*, 7(11), 1170–1180. DOI: 10.20959/wjpr201811-12333

Verma, M. L., Dhanya, B. S., Thakur, M., Jeslin, J., & Jana, A. K. (2021). Chapter Eight - Plant derived nanoparticles and their biotechnological applications. In Verma, S. K., & Das, A. K. B. T.-C. A. C. (Eds.), *Biosynthesized Nanomaterials* (Vol. 94, pp. 331–362). Elsevier. DOI: 10.1016/bs.coac.2021.01.011

Vida-simiti, I., Cluj-napoca, U. T., Chicinas, I., Cluj-napoca, U. T., Batin, G., & Cluj-napoca, U. T. (2004). Applications of Scanning Electron Microscopy (Sem). *Romanian Journal of Physics*, 49, 955–965.

Vijayakumar, S., Vaseeharan, B., Malaikozhundan, B., Gopi, N., Ekambaram, P., Pachaiappan, R., Velusamy, P., Murugan, K., Benelli, G., Suresh Kumar, R., & Suriyanarayanamoorthy, M. (2017). Therapeutic effects of gold nanoparticles synthesized using *Musa paradisiaca* peel extract against multiple antibiotic resistant *Enterococcus faecalis* biofilms and human lung cancer cells (A549). *Microbial Pathogenesis*, 102, 173–183. DOI: 10.1016/j.micpath.2016.11.029 PMID: 27916691

Vijayaraghavan, K., & Ashokkumar, T. (2017). Plant-mediated biosynthesis of metallic nanoparticles: A review of literature, factors affecting synthesis, characterization techniques and applications. *Journal of Environmental Chemical Engineering*, 5(5), 4866–4883. DOI: 10.1016/j.jece.2017.09.026

Vijayaram, S., Razafindralambo, H., Sun, Y.-Z., Vasantharaj, S., Ghafarifarsani, H., Hoseinifar, S. H., & Raeeszadeh, M. (2023). Applications of Green Synthesized Metal Nanoparticles — a Review. In Biological Trace Element Research (Vol. 202, Issue 1, pp. 360–386). Springer Science and Business Media LLC. DOI: 10.1007/s12011-023-03645-9

Vijilvani, C., Bindhu, M. R., Frincy, F. C., AlSalhi, M. S., Sabitha, S., Saravanakumar, K., Devanesan, S., Umadevi, M., Aljaafreh, M. J., & Atif, M. (2020). Antimicrobial and catalytic activities of biosynthesized gold, silver and palladium nanoparticles from Solanum nigurum leaves. *Journal of Photochemistry and Photobiology. B, Biology*, 202, 111713. DOI: 10.1016/j.jphotobiol.2019.111713 PMID: 31760373

Vinodhini, S., Vithiya, B. S. M., & Prasad, T. A. A. (2022). Green synthesis of palladium nanoparticles using aqueous plant extracts and its biomedical applications. *Journal of King Saud University. Science*, 34(4), 102017.

Viswanathan, V. K., Manoharan, S. R. R., Subramanian, S., & Moon, A. (2019). Nanotechnology in spine surgery: A current update and critical review of the literature. *World Neurosurgery*, 123, 142–155. DOI: 10.1016/j.wneu.2018.11.035 PMID: 30447449

Vitta, Y., Figueroa, M., Calderon, M., & Ciangherotti, C. (2020). Synthesis of iron nanoparticles from aqueous extract of *Eucalyptus robusta* Sm and evaluation of antioxidant and antimicrobial activity. *Materials Science for Energy Technologies*, 3, 97–103. DOI: 10.1016/j.mset.2019.10.014

Vodyashkin, A. A., Kezimana, P., Vetcher, A. A., & Stanishevskiy, Y. M. (2021). Biopolymeric Nanoparticles–Multifunctional Materials of the Future. *Polymers*, 14(11), 2287. DOI: 10.3390/polym14112287 PMID: 35683959

Waghmode, S. R., Dudhane, A. A., & Mhaindarkar, V. P. (2021). *Syzygium cumini* mediated green synthesis of silver nanoparticles for reduction of 4-nitrophenol and assessment of its antibacterial activity. In *Nanocomposite Materials for Biomedical and Energy Storage Applications*. IntechOpen.

Waheed, I., Ali, A., Tabassum, H., Khatoon, N., Lai, W. F., & Zhou, X. (2024). Lipid-based nanoparticles as drug delivery carriers for cancer therapy. *Frontiers in Oncology*, 14(04), 1–17. DOI: 10.3389/fonc.2024.1296091 PMID: 38660132

Walemba, E. M. (2020). Synthesis and Characterization of Hemocompatible Antibiotic Nanoparticles.

Wang, C., Zhang, Z., Wang, J., Wang, Q., & Shang, L. (2022). Biohybrid materials: Structure design and biomedical applications. *Materials Today. Bio*, 16, 100352. DOI: 10.1016/j.mtbio.2022.100352 PMID: 35856044

Wang, H., Han, S., Sun, J., Fan, T., Tian, C., & Wu, Y. (2012). Amphiphilic dextran derivatives nanoparticles for the delivery of mitoxantrone. *Journal of Applied Polymer Science*, 126(S1), E35–E43. DOI: 10.1002/app.36534

Wang, H., Wang, Z., Yue, R., Gao, F., Ren, R., Wei, J., Wang, X., & Kong, Z. (2020). Functional group-rich hyperbranched magnetic material for simultaneous efficient removal of heavy metal ions from aqueous solution. In *Journal of Hazardous Materials* (Vol. 384, p. 121288). Elsevier BV., DOI: 10.1016/j.jhazmat.2019.121288

Wang, H., Yuan, X., Wang, H., Chen, X., Wu, Z., Jiang, L., Xiong, W., & Zeng, G. (2016). Facile synthesis of Sb2S3/ultrathin g-C3N4 sheets heterostructures embedded with g-C3N4 quantum dots with enhanced NIR-light photocatalytic performance. *Applied Catalysis B: Environmental*, 193, 36–46. DOI: 10.1016/j.apcatb.2016.03.075

Wang, P., Lombi, E., Zhao, F. J., & Kopittke, P. M. (2016). Nanotechnology: A new opportunity in plant sciences. *Trends in Plant Science*, 21(8), 699–712. DOI: 10.1016/j.tplants.2016.04.005 PMID: 27130471

Wang, S., Gao, Y., Jin, Q., & Ji, J. (2020). Emerging antibacterial nanomedicine for enhanced antibiotic therapy. *Biomaterials Science*, 8(24), 6825–6839. DOI: 10.1039/D0BM00974A PMID: 32996490

Wang, S., Wang, F., Gao, S., & Wang, X. (2016). Heavy metal accumulation in different rice cultivars as influenced by foliar application of nano-silicon. *Water, Air, and Soil Pollution*, 227(7), 1–13. DOI: 10.1007/s11270-016-2928-6

Wang, T., Jin, X., Chen, Z., Megharaj, M., & Naidu, R. (2014). Green synthesis of Fe nanoparticles using *Eucalyptus* leaf extracts for treatment of eutrophic wastewater. *The Science of the Total Environment*, 466, 210–213. DOI: 10.1016/j.scitotenv.2013.07.022 PMID: 23895784

Wang, T., & Liu, W. (2022). Emerging investigator series: metal nanoparticles in freshwater: transformation, bioavailability and effects on invertebrates. *Environmental Science. Nano*, 9(7), 2237–2263. DOI: 10.1039/D2EN00052K PMID: 35923327

Wang, W., Jimmy, C. Y., Shen, Z., Chan, D. K., & Gu, T. (2014). gC 3 N 4 quantum dots: Direct synthesis, upconversion properties and photocatalytic application. *Chemical Communications*, 50(70), 10148–10150. DOI: 10.1039/C4CC02543A PMID: 25052166

Wang, W., Yang, Z., Zhang, A., & Yang, S. (2021). Water retention and fertilizer slow release integrated superabsorbent synthesized from millet straw and applied in agriculture. *Industrial Crops and Products*, 160, 113126. DOI: 10.1016/j.indcrop.2020.113126

Wang, X., Wang, Y., He, H., Chen, X., Sun, X., Sun, Y., Zhou, G., Xu, H., & Huang, F. (2016). Steering graphene quantum dots in living cells: Lighting up the nucleolus. *Journal of Materials Chemistry. B*, 4(4), 779–784. DOI: 10.1039/C5TB02474A PMID: 32262959

Wang, Y., & Nowack, B. (2018). Environmental risk assessment of engineered nano-SiO2, nano iron oxides, nano-CeO2, nano-Al2O3, and quantum dots. *Environmental Toxicology and Chemistry*, 37(5), 1387–1395. DOI: 10.1002/etc.4080 PMID: 29315795

Wang, Z. L. (1999). Transmission Electron Microscopy and Spectroscopy of Nanoparticles. *Characterization of Nanophase Materials*, 18, 37–80. DOI: 10.1002/3527600094.ch3

Wang, Z., Liu, X., Duan, Y., & Huang, Y. (2022). Infection microenvironment-related antibacterial nanotherapeutic strategies. *Biomaterials*, 280, 121249. DOI: 10.1016/j.biomaterials.2021.121249 PMID: 34801252

Watt, J., Huber, D. L., & Stewart, P. L. (2019). Soft matter and nanomaterials characterization by cryogenic transmission electron microscopy. *MRS Bulletin*, 44(12), 942–948. DOI: 10.1557/mrs.2019.285

Weijie, M., Chongnv, W., Xuming, P., Weixin, J., Yuhang, W., & Benhui, S. (2020). TiO2 nanoparticles and multi-walled carbon nanotubes monitoring and bioremediation potential using ciliates *Pseudocohnilembus persalinus.Ecotoxicology and Environmental Safety*, 187, 109825. DOI: 10.1016/j.ecoenv.2019.109825 PMID: 31677570

Wei, X., Cao, P., Wang, G., Liu, Y., Song, J., & Han, J. (2021). CuO, ZnO, and γ-Fe2O3 nanoparticles modified the underground biomass and rhizosphere microbial community of *Salvia miltiorrhiza* (Bge.) after 165-day exposure. *Ecotoxicology and Environmental Safety*, 217, 112232. DOI: 10.1016/j.ecoenv.2021.112232 PMID: 33864980

Welch, E. C., Powell, J. M., Clevinger, T. B., Fairman, A. E., & Shukla, A. (2021). Advances in biosensors and diagnostic technologies using nanostructures and nanomaterials. *Advanced Functional Materials*, 31(44), 2104126. DOI: 10.1002/adfm.202104126

Wen, J., Fu, W., Ding, S., Zhang, Y., & Wang, W. (2022). Pyrogallic acid modified nanoscale zerovalent iron efficiently removed Cr(VI) by improving adsorption and electron selectivity. In Chemical Engineering Journal (Vol. 443, p. 136510). Elsevier BV. DOI: 10.1016/j.cej.2022.136510

Weng, X., Guo, M., Luo, F., & Chen, Z. (2017). One-step green synthesis of bimetallic Fe/Ni nanoparticles by *Eucalyptus* leaf extract: Biomolecules identification, characterization and catalytic activity. *Chemical Engineering Journal*, 308, 904–911. DOI: 10.1016/j.cej.2016.09.134

Wesołowska, M., Rymarczyk, J., Góra, R., Baranowski, P., Sławiński, C., Klimczyk, M., & Schimmelpfennig, L. (2021). New slow-release fertilizers-economic, legal and practical aspects: A Review. *International Agrophysics*, 35(1), 11–24. DOI: 10.31545/intagr/131184

Woodley, C. M., Amado, P. S., Cristiano, M. L., & O'Neill, P. M. (2021). Artemisinin inspired synthetic endoperoxide drug candidates: Design, synthesis, and mechanism of action studies. *Medicinal Research Reviews*, 41(6), 3062–3095. DOI: 10.1002/med.21849 PMID: 34355414

World Health Organization. (2022). *WHO fungal priority pathogens list to guide research, development and public health action*. World Health Organization. https://www.who.int/publications/i/item/9789240060241

Woźniak-Budych, M. J., Przysiecka, Ł., Langer, K., Peplińska, B., Jarek, M., Wiesner, M., Nowaczyk, G., & Jurga, S. (2017). Green synthesis of rifampicin-loaded copper nanoparticles with enhanced antimicrobial activity. *Journal of Materials Science. Materials in Medicine*, 28(3), 1–16. DOI: 10.1007/s10856-017-5857-z PMID: 28150115

Wu, H., Tito, N., & Giraldo, J. P. (2017). Anionic cerium oxide nanoparticles protect plant photosynthesis from abiotic stress by scavenging reactive oxygen species. *ACS Nano*, 11(11), 11283–11297. DOI: 10.1021/acsnano.7b05723 PMID: 29099581

Wu, J., Zheng, Y., Song, W., Luan, J., Wen, X., Wu, Z., Chen, X., Wang, Q., & Guo, S. (2014). In situ synthesis of silver-nanoparticles/bacterial cellulose composites for slow-released antimicrobial wound dressing. *Carbohydrate Polymers*, 102, 762–771. DOI: 10.1016/j.carbpol.2013.10.093 PMID: 24507345

Wu, S., Shanmugam, R., Madasamy, M., & Vanaja, M. (2020). Green synthesis of copper nanoparticles using Cissus vitiginea and its antioxidant and antibacterial activity against urinary tract infection pathogens. *Artificial Cells, Nanomedicine, and Biotechnology*, 48(1), 1153–1158. DOI: 10.1080/21691401.2020.1817053 PMID: 32924614

Xiao, Z., Liu, W., Zhu, G., Zhou, R., & Niu, Y. (2014). A review of the preparation and application of flavour and essential oils microcapsules based on complex coacervation technology. *Journal of the Science of Food and Agriculture*, 94(8), 1482–1494. DOI: 10.1002/jsfa.6491 PMID: 24282124

Xu, J., Luo, X., Wang, Y., & Feng, Y. (2018). Evaluation of zinc oxide nanoparticles on lettuce (Lactuca sativa L.) growth and soil bacterial community. *Environmental Science and Pollution Research International*, 25(6), 6026–6035. DOI: 10.1007/s11356-017-0953-7 PMID: 29238929

Xulu, J. H., Ndongwe, T., Ezealisiji, K. M., Tembu, V. J., Mncwangi, N. P., Witika, B. A., & Siwe-Noundou, X. (2022). The Use of Medicinal Plant-Derived Metallic Nanoparticles in Theranostics. *Pharmaceutics*, 14(11), 2437. DOI: 10.3390/pharmaceutics14112437 PMID: 36365255

Xu, Q., Li, W., Ding, L., Yang, W., Xiao, H., & Ong, W. J. (2019). Function-driven engineering of 1D carbon nanotubes and 0D carbon dots: Mechanism, properties and applications. *Nanoscale*, 11(4), 1475–1504. DOI: 10.1039/C8NR08738E PMID: 30620019

Xu, S., Li, D., & Wu, P. (2015). One-pot, facile, and versatile synthesis of monolayer MoS2/WS2 quantum dots as bioimaging probes and efficient electrocatalysts for hydrogen evolution reaction. *Advanced Functional Materials*, 25(7), 1127–1136. DOI: 10.1002/adfm.201403863

Xu, Y., Liu, J., Gao, C., & Wang, E. (2014). Applications of carbon quantum dots in electrochemiluminescence: A mini review. *Electrochemistry Communications*, 48, 151–154. DOI: 10.1016/j.elecom.2014.08.032

Xu, Y., Wang, X., Zhang, W. L., Lv, F., & Guo, S. (2018). Recent progress in two-dimensional inorganic quantum dots. *Chemical Society Reviews*, 47(2), 586–625. DOI: 10.1039/C7CS00500H PMID: 29115355

Yaashikaa, P. R., Senthil Kumar, P., & Varjani, S. (2022). Valorization of agro-industrial wastes for biorefinery process and circular bioeconomy: A critical review. *Bioresource Technology*, 343, 126126. DOI: 10.1016/j.biortech.2021.126126 PMID: 34673193

Yadav, S., Lal, S., Yadav, S., Laxman, J., Verma, B., Sushma, M., Choudhary, R., Singh, P., Singh, S., Sharma, V., & Singh, B. (2019). *Use of Nanotechnology in Agri-food Sectors and Apprehensions: An Overview.*

Yadi, M., Mostafavi, E., Saleh, B., Davaran, S., Aliyeva, I., Khalilov, R., . . . Panahi, Y. (2018). Current developments in green synthesis of metallic nanoparticles using plant extracts: a review. *Artificial cells, nanomedicine, and biotechnology, 46*(sup3), 336-343.

Yakura, K., & Tanifuji, S. (1983). Molecular cloning and restriction analysis of Eco RI-fragments of Vicia faba rDNA. *Plant & Cell Physiology*, 24(7), 1327–1330.

Yang, J. S., Martinez, D. A., & Chiang, W. H. (2017). Synthesis, characterization and applications of graphene quantum dots. *Recent Trends in Nanomaterials: Synthesis and Properties*, 65-120.

Yang, J., Song, Z., Ma, J., & Han, H. (2020). Toxicity of molybdenum-based nanomaterials on the soybean–rhizobia symbiotic system: Implications for nutrition. *ACS Applied Nano Materials*, 3(6), 5773–5782. DOI: 10.1021/acsanm.0c00943

Yang, Z., & Zhu, M. (2023). Case Studies: Ultraviolet-Visible (UV-Vis) Spectroscopy. In *Springer Handbook of Advanced Catalyst Characterization* (pp. 265–283). Springer International Publishing. DOI: 10.1007/978-3-031-07125-6_12

Yan, H., Li, P., Wen, F., Xu, Q., Guo, Q., & Su, W. (2023). Green synthesis of carbon quantum dots from plant turmeric holds promise as novel photosensitizer for in vitro photodynamic antimicrobial activity. *Journal of Materials Research and Technology*, 22, 17–34. DOI: 10.1016/j.jmrt.2022.11.090

Yanik, S., Emre, D., Alp, M., Algi, F., Yilmaz, S., Bilici, A., & Ozkan-Ariksoysal, D. (2023). A novel electrochemical biosensor based on palladium nanoparticles decorated on reduced graphene oxide-polyaminophenol matrix for the detection and discrimination of mitomycin C-DNA and acyclovir-DNA interaction. *Journal of Pharmaceutical and Biomedical Analysis*, 234, 115524. DOI: 10.1016/j.jpba.2023.115524 PMID: 37320972

Yano, J., & Yachandra, V. K. (2009). X-ray absorption spectroscopy. *Photosynthesis Research*, 102(2-3), 241–254. DOI: 10.1007/s11120-009-9473-8 PMID: 19653117

Yao, F., Zhu, P., Chen, J., Li, S., Sun, B., Li, Y., Zou, M., Qi, X., Liang, P., & Chen, Q. (2023). Synthesis of nanoparticles via microfluidic devices and integrated applications. *Mikrochimica Acta*, 190(7), 256. DOI: 10.1007/s00604-023-05838-4 PMID: 37301779

Yasmin, H., Giwa, S. O., Noor, S., & Aybar, H. Ş. (2023). Influence of Preparation Characteristics on Stability, Properties, and Performance of Mono- and Hybrid Nanofluids: Current and Future Perspective. *Machines (Basel)*, 11(1), 112. DOI: 10.3390/machines11010112

Yassin, M. T., Mostafa, A. A.-F., Al-Askar, A. A., & Al-Otibi, F. O. (2022). Synergistic antifungal efficiency of biogenic silver nanoparticles with itraconazole against multidrug-resistant candidal strains. *Crystals*, 12(6), 816. DOI: 10.3390/cryst12060816

Ye, D., & Peramo, A. (2014). Implementing tissue engineering and regenerative medicine solutions in medical implants. *British Medical Bulletin*, 109(1), 3–18. DOI: 10.1093/bmb/ldt036 PMID: 24357734

Yeh, T. F., Teng, C. Y., Chen, S. J., & Teng, H. (2014). Nitrogen-doped graphene oxide quantum dots as photocatalysts for overall water-splitting under visible light Illumination. *Advanced Materials*, 26(20), 3297–3303. DOI: 10.1002/adma.201305299 PMID: 24677453

Yetisgin, A. A., Cetinel, S., Zuvin, M., Kosar, A., & Kutlu, O. (2020). Therapeutic Nanoparticles and Their Targeted Delivery Applications. *Molecules (Basel, Switzerland)*, 25(9), 2193. DOI: 10.3390/molecules25092193 PMID: 32397080

Yip, L. X., Wang, J., Xue, Y., Xing, K., Sevencan, C., Ariga, K., & Leong, D. T. (2024). Cell-derived nanomaterials for biomedical applications. *Science and Technology of Advanced Materials*, 25(1), 2315013. DOI: 10.1080/14686996.2024.2315013 PMID: 38476511

Youssef, A. M., El-Sayed, H. S., El-Sayed, S. M., Fouly, M., & El-Aziz, M. A. (2023). Novel bionanocomposites based on cinnamon nanoemulsion and TiO2-NPs for preserving fresh chicken breast fillets. *Food and Bioprocess Technology*, 16(2), 356–367. DOI: 10.1007/s11947-022-02934-w

Youssef, A. M., El-Sayed, H. S., Islam, E. N., & El-Sayed, S. M. (2021). Preparation and characterization of novel bionanocomposites based on garlic extract for preserving fresh Nile tilapia fish fillets. *RSC Advances*, 11(37), 22571–22584. DOI: 10.1039/D1RA03819B PMID: 35480459

You, T., Liu, D., Chen, J., Yang, Z., Dou, R., Gao, X., & Wang, L. (2018). Effects of metal oxide nanoparticles on soil enzyme activities and bacterial communities in two different soil types. *Journal of Soils and Sediments*, 18(1), 211–221. DOI: 10.1007/s11368-017-1716-2

Yuan, H., Guo, H., Luan, X., He, M., Li, F., Burnett, J., Truchan, N., & Sun, D. (2020). Albumin nanoparticle of paclitaxel (Abraxane) decreases while taxol increases breast cancer stem cells in treatment of triple negative breast cancer. *Molecular Pharmaceutics*, 17(7), 2275–2286. DOI: 10.1021/acs.molpharmaceut.9b01221 PMID: 32485107

Yusefi-Tanha, E., Fallah, S., Rostamnejadi, A., & Pokhrel, L. R. (2020). Zinc oxide nanoparticles (ZnONPs) as a novel nanofertilizer: Influence on seed yield and antioxidant defense system in soil grown soybean (Glycine max cv. Kowsar). *The Science of the Total Environment*, 738, 140240. DOI: 10.1016/j.scitotenv.2020.140240 PMID: 32570083

Zafar, A., Alruwaili, N. K., Imam, S. S., Alsaidan, O. A., Ahmed, M. M., Yasir, M., Warsi, M. H., Alquraini, A., Ghoneim, M. M., & Alshehri, S. (2021). Development and Optimization of Hybrid Polymeric Nanoparticles of Apigenin: Physicochemical Characterization, Antioxidant Activity and Cytotoxicity Evaluation. *Sensors (Basel)*, 22(4), 1364. DOI: 10.3390/s22041364 PMID: 35214260

Zainab, S., Hamid, S., Sahar, S., & Ali, N. (2022). Fluconazole and biogenic silver nanoparticles-based nano-fungicidal system for highly efficient elimination of multi-drug resistant Candida biofilms. *Materials Chemistry and Physics*, 276, 125451. DOI: 10.1016/j.matchemphys.2021.125451

Zaini, M. S., Ying Chyi Liew, J., Alang Ahmad, S. A., Mohmad, A. R., & Kamarudin, M. A. (2020). Quantum confinement effect and photoenhancement of photoluminescence of PbS and PbS/MnS quantum dots. *Applied Sciences (Basel, Switzerland)*, 10(18), 6282. DOI: 10.3390/app10186282

Zareshahrabadi, Z., Karami, F., Taghizadeh, S., Iraji, A., Amani, A. M., Motamedi, M., & Zomorodian, K. (2021). Green synthesis of silver nanoparticles using aqueous extract of Lamium album and their antifungal properties. *Journal of Nano Research*, 67, 55–67. . DOI: 10.4028/www.scientific.net/JNanoR.67.55

Zarghani, M., Nemati-Anaraki, L., Sedghi, S., Chakoli, A. N., & Rowhani-Farid, A. (2023). Iranian researchers' perspective about concept and effect of open science on research publication. *BMC Health Services Research*, 23(1), 437. DOI: 10.1186/s12913-023-09420-9 PMID: 37143102

Zayed, M. F., Mahfoze, R. A., El-Kousy, S. M., & Al-Ashkar, E. A. (2020). *In-vitro* antioxidant and antimicrobial activities of metal nanoparticles biosynthesized using optimized *Pimpinella anisum* extract. *Colloids and Surfaces. A, Physicochemical and Engineering Aspects*, 585, 124–167. DOI: 10.1016/j.colsurfa.2019.124167

Zazo, H., Colino, C. I., & Lanao, J. M. (2016). Current applications of nanoparticles in infectious diseases. *Journal of Controlled Release*, 224, 86–102. DOI: 10.1016/j.jconrel.2016.01.008 PMID: 26772877

Zhang, J., Niu, Y., Zhou, Y., Ju, S., & Gu, Y. (2022). Green preparation of nano-zero-valent iron-copper bimetals for nitrate removal: Characterization, reduction reaction pathway, and mechanisms. In Advanced Powder Technology (Vol. 33, Issue 11, p. 103807). Elsevier BV. DOI: 10.1016/j.apt.2022.103807

Zhang, C., Cui, Y., Song, L., Liu, X., & Hu, Z. (2016). Microwave assisted one-pot synthesis of graphene quantum dots as highly sensitive fluorescent probes for detection of iron ions and pH value. *Talanta*, 150, 54–60. DOI: 10.1016/j.talanta.2015.12.015 PMID: 26838381

Zhang, F., Liu, F., Wang, C., Xin, X., Liu, J., Guo, S., & Zhang, J. (2016). Effect of lateral size of graphene quantum dots on their properties and application. *ACS Applied Materials & Interfaces*, 8(3), 2104–2110. DOI: 10.1021/acsami.5b10602 PMID: 26725374

Zhang, H., Huang, M., Zhang, W., Gardea-Torresdey, J. L., White, J. C., Ji, R., & Zhao, L. (2020). Silver nanoparticles alter soil microbial community compositions and metabolite profiles in unplanted and cucumber-planted soils. *Environmental Science & Technology*, 54(6), 3334–3342. DOI: 10.1021/acs.est.9b07562 PMID: 32088952

Zhang, L., Peng, D., Liang, R. P., & Qiu, J. D. (2015). Nitrogen-doped graphene quantum dots as a new catalyst accelerating the coordination reaction between cadmium (II) and 5, 10, 15, 20-tetrakis (1-methyl-4-pyridinio) porphyrin for cadmium (II) sensing. *Analytical Chemistry*, 87(21), 10894–10901. DOI: 10.1021/acs.analchem.5b02450 PMID: 26427520

Zhang, M., Li, M., Zhang, D., Yu, Y., Zhu, K., Zang, X., & Liu, D. (2024). Preparation and Investigation of Sustained-Release Nanocapsules Containing Cumin Essential Oil for Their Bacteriostatic Properties. *Foods*, 13(6), 947. DOI: 10.3390/foods13060947 PMID: 38540937

Zhang, S., Zhang, Z., & Vardhanabhuti, B. (2014). Effect of charge density of polysaccharides on self-assembled intragastric gelation of whey protein/polysaccharide under simulated gastric conditions. *Food & Function*, 5(8), 1829–1838. DOI: 10.1039/C4FO00019F PMID: 24920131

Zhang, X. F., Liu, Z. G., Shen, W., & Gurunathan, S. (2016). Silver nanoparticles: Synthesis, characterization, properties, applications, and therapeutic approaches. *International Journal of Molecular Sciences*, 17(9), 1534. DOI: 10.3390/ijms17091534 PMID: 27649147

Zhang, X., Xie, H., Liu, Z., Tan, C., Luo, Z., Li, H., Lin, J., Sun, L., Chen, W., Xu, Z., Xie, L., Huang, W., & Zhang, H. (2015). Black phosphorus quantum dots. *Angewandte Chemie International Edition*, 54(12), 3653–3657. DOI: 10.1002/anie.201409400 PMID: 25649505

Zhao, J., Zhao, L., Lan, C., & Zhao, S. (2016). Graphene quantum dots as effective probes for label-free fluorescence detection of dopamine. *Sensors and Actuators. B, Chemical*, 223, 246–251. DOI: 10.1016/j.snb.2015.09.105

Zhao, R., Xiang, J., Wang, B., Chen, L., & Tan, S. (2022). Recent advances in the development of noble metal NPs for cancer therapy. *Bioinorganic Chemistry and Applications*, 2022(1), 2444516. DOI: 10.1155/2022/2444516 PMID: 35126483

Zhao, X., Ma, X., Sun, J., Li, D., & Yang, X. (2016). Enhanced catalytic activities of surfactant-assisted exfoliated WS2 nanodots for hydrogen evolution. *ACS Nano*, 10(2), 2159–2166. DOI: 10.1021/acsnano.5b06653 PMID: 26785941

Zhou, W., Apkarian, R., Wang, Z. L., & Joy, D. (2007). Fundamentals of scanning electron microscopy (SEM). W. Zhou, Z. L. Wang (Eds.), *Scanning Microscopy for Nanotechnology: Techniques and applications* (pp. 1–40). DOI: 10.1007/978-0-387-39620-0_1

Zhu, S., Yu, T., Xu, T., Chen, H., Dustdar, S., Gigan, S., Gunduz, D., Hossain, E., Jin, Y., Lin, F., Liu, B., Wan, Z., Zhang, J., Zhao, Z., Zhu, W., Chen, Z., Durrani, T. S., Wang, H., Wu, J., … Pan, Y. (2023). Intelligent Computing: The Latest Advances, Challenges, and Future. *Intelligent Computing, 2*, 0006. DOI: 10.34133/icomputing.0006

Zhuang, Z., Li, Z., Gong, G., Li, Q., Zhang, Y., Huang, C., & Guo, Z. (2023). Two-dimensional superlattice films of gold nanoparticle-polystyrene composites: A bioactive platform for bone homeostasis maintenance. *Advanced Composites and Hybrid Materials*, 6(5), 166. DOI: 10.1007/s42114-023-00738-w

Zhu, H., Liu, A., Shan, F., Yang, W., Zhang, W., Li, D., & Liu, J. (2016). One-step synthesis of graphene quantum dots from defective CVD graphene and their application in IGZO UV thin film phototransistor. *Carbon*, 100, 201–207. DOI: 10.1016/j.carbon.2016.01.016

Zhu, W., Hu, C., Ren, Y., Lu, Y., Song, Y., Ji, Y., Han, C., & He, J. (2021). Green synthesis of zinc oxide nanoparticles using Cinnamomum camphora (L.) Presl leaf extracts and its antifungal activity. *Journal of Environmental Chemical Engineering*, 9(6), 106659. DOI: 10.1016/j.jece.2021.106659

Zhu, X., & Loh, X. J. (2015). Layer-by-layer assemblies for antibacterial applications. *Biomaterials Science*, 3(12), 1505–1518. DOI: 10.1039/C5BM00307E PMID: 26415703

Zhu, Y., & Durr, H. (2015). The future of electron microscopy. *Physics Today*, 68(4), 32–38. DOI: 10.1063/PT.3.2747

Zielińska, A., Skwarek, E., Zaleska, A., Gazda, M., & Hupka, J. (2009). Preparation of silver nanoparticles with controlled particle size. *Procedia Chemistry*, 1(2), 1560–1566. DOI: 10.1016/j.proche.2009.11.004

Zohra, E., Ikram, M., Omar, A. A., Hussain, M., Satti, S. H., Raja, N. I., Mashwani, Z.-U.-R., & Ehsan, M. (2021). Potential applications of biogenic selenium nanoparticles in alleviating biotic and abiotic stresses in plants: A comprehensive insight on the mechanistic approach and future perspectives. In Green Processing and Synthesis (Vol. 10, Issue 1, pp. 456–475). Walter de Gruyter GmbH. DOI: 10.1515/gps-2021-0047

Zohri, M., Alavidjeh, M. S., Haririan, I., Ardestani, M. S., Ebrahimi, S. E. S., Sani, H. T., & Sadjadi, S. K. (2010). A comparative study between the antibacterial effect of nisin and nisin-loaded chitosan/alginate nanoparticles on the growth of Staphylococcus aureus in raw and pasteurized milk samples. *Probiotics and Antimicrobial Proteins*, 2(4), 258–266. DOI: 10.1007/s12602-010-9047-2 PMID: 26781321

Zolnik, B. S., González-Fernández, Á., Sadrieh, N., & Dobrovolskaia, M. A. (2010). Nanoparticles and the Immune System. *Endocrinology*, 151(2), 458–465. DOI: 10.1210/en.2009-1082 PMID: 20016026

Zou, L. K., Wang, H. N., Pan, X., Tian, G. B., Xie, Z. W., Wu, Q., Chen, H., Xie, T., & Yang, Z. R. (2008). Expression, purification and characterization of a phyA m-phyCs fusion phytase. *Journal of Zhejiang University. Science. B.*, 9(7), 536–545. DOI: 10.1631/jzus.B0720006 PMID: 18600783

Zuhrotun, A., Oktaviani, D. J., & Hasanah, A. N. (2023). Biosynthesis of Gold and Silver Nanoparticles Using Phytochemical Compounds. *Molecules (Basel, Switzerland)*, 28(7), 3240. DOI: 10.3390/molecules28073240 PMID: 37050004

Zulfiqar, F., Navarro, M., Ashraf, M., Akram, N. A., & Munné-Bosch, S. (2019). Nanofertilizer use for sustainable agriculture: Advantages and limitations. *Plant Science*, 289, 110270. DOI: 10.1016/j.plantsci.2019.110270 PMID: 31623775

About the Contributors

Susanta Das is a seasoned educator with a proven record of teaching and mentoring diverse students, working at multiple institutions, and providing service to the various initiatives of organizations. He received his Ph.D. and M.A. degrees from Western Michigan University (WMU), USA, and M.Sc. degree from Banaras Hindu University (BHU), India all in Physics. He continued his post-doctoral research as a Marie-Curie post-doctoral fellow at Stockholm University Stockholm, Sweden on beam diagnostics for the DESIREE (Double ElectroStatic Ion Ring ExpEriment) in the project DITANET (Diagnostic Techniques for particle Accelerators – a Marie-Curie initial training NETwork), at the Indian Institute of Science Education and Research-Kolkata, India on high-pressure physics as a Project Scientist-B, and at the University of Electro-Communications, Tokyo, Japan on ion-surface interactions as a Post-doctoral fellow. Throughout his career, Dr. Das worked and collaborated with many researchers, Ph.D., master, and visiting students from different countries. He further visited many countries to discuss research and an international conference participant (UK, Italy, Germany, Greece, Belgium, Bulgaria, Romania, Brazil etc.). He co-authored several research articles in WoS/Scopus indexed international journals, conference proceedings, and scholarly book chapters published by renowned international publishing houses. Beyond his research endeavors, Dr. Das has a storied history of service to academic and administrative committees, exhibiting his commitment to the growth and development of educational institutions. His experience includes tenure at Central University South Bihar, Sri Sri University, and P.K. University before assuming his current role at Ajeenkya DY Patil University. He received the Marie-Curie post-doctoral fellowship in Sweden, Science Academies' Summer Research Fellowship in India, Gwen Frostic Doctoral Fellowship, Department Graduate Research and Creative Scholar Award by WMU, and Leo R. Parpart Doctoral Fellowship by Dept. of Physics, WMU, among many others, throughout his academic journey. At present, Dr. Das continues to delve into cutting-edge fields, with a keen interest in nanotechnology, quantum computing, and data science. His multifaceted contributions, spanning teaching, research, and administrative leadership, showcase a dedicated professional who is instrumental in advancing the vision and mission of the institutions he serves.

Shankar Khade completed M.Sc. in Biotechnology from Swami Ramanand Teerth Marathwada University, Nanded, Maharashtra, and received his M. Tech. in Biotechnology from National Institute of Technology (NIT), Rourkela. He earned his Ph.D. in Biochemical Engineering from IIT, Varanasi. Along with being with association with ADYPU, he is also an Editorial Board Member of the journal "American Journal of Chemical and Biochemical Engineering" and Reviewer of "Preparative Biochemistry and Biotechnology" and "Signa Vitae" Journals. He has presented his Ph.D. research work at an international conference organised by NTU and AIChE in Singapore and many more National conferences. He has also published 12 peer-reviewed research articles in reputed journals and an edited book in IGI Global, USA. He has a core interest in the field of Microbial Biochemicals of therapeutic importance, Bioenergy (2G and

3G ethanol), and bioremediation. His doctoral studies were focused on the "production and Purification of uricase enzyme from microbial sources and its application in the treatment of gout disease". During M. Tech (2011 – 13) and Ph.D. (2013 – 18), he received an MHRD fellowship, Govt. of India. Later, he joined as a Project Associate at R&D, Hindustan Petroleum Corporation Limited (HPCL), Bangalore, and worked in the Bioenergy area.

Debanjali Barman Roy has completed her Ph.D. Degree in Physics from ICFAI University, Dehradun. She received Kalpana Chawla Academic excellence award and Kalpana Chawla scholarship in the year 2013 from Late Mr. Chawla (Father of Late Kalpana Chawla). She has published numerous research articles in reputed international journals, books, and conference proceedings. Furthermore, she guided several undergraduate and post-graduate dissertations on nanoparticles synthesis and characterization. In addition, she has a teaching experience of more than 8 years as an Assistant Professor of Physics in different educational organization in India. She is also involved in mentoring the students and mentored more than 500 students approximately so far. Apart from teaching, she was the In-Charge of Cultural club in ICFAI University, Dehradun. Currently she is working as an Assistant Professor of Physics in Ajeenkya DY Patil University, Pune and also an active member of Student Activity committee, School of Engineering.

KhushbuTrivedi is an Assistant Professor in Computer Science & Engineering at Ajeenkya D.Y. Patil University, Pune. She is Pursuing her Ph.D. from College of Engineering Pune (COEP). She completed her B.Tech from Rajasthan Technical University and M.Tech from State Government College, Rajasthan. She qualified national level UGC-NET examination with more than 99 percentile. She got "Best Paper Award" for her research work in "Mass Estimation and Density Estimation" in IEEE conference. She had submitted an idea in "Smart India Hackathon 2017" under Department of Rural Development Ministry. She has attended National Institute of Technical Teachers and Training and Research (NITTTR), Chandigarh workshops on various technologies. She has many papers published in international journals. She has more than 6 years of teaching experience in various institutes and universities in Pune and Rajasthan. She also served the role of Teaching assistant in Govt. College Engineering, Ajmer (Rajasthan). She Published Various research Papers and book chapters in Scopus and IEEE journals and conferences. Her core interest areas include Artificial intelligence, Data Mining, Machine Learning, Algorithms, Python and Blockchain.

***

Nehir Arik received a B.Sc. in Bioengineering from Ege University in 2015. She did her M.Sc. in Department of Biocomposite Engineering at the Izmir Katip Celebi University in 2017 and Ph.D. studies in Department of Molecular Biology and Genetics at the Middle East Technical University in 2022. She completed her Erasmus internship at National Hellenic Research Foundation (NHRF, Athens-Greece) in 2014. She worked at Akdeniz University and Middle East Technical University as a research assistant. She works as an industrial property assistant expert at Turkish Patent and Trademark Office since 2021. She is currently conducting her postdoc studies in Department of Materials at the Manchester University. Her current research interests are using of biomaterials as biosensors in wastewater systems, as well as their use in wound healing and skin cancer treatment.

Alka Bali is Professor of Pharmaceutical Chemistry at University Institute of Pharmaceutical Sciences (UIPS), Panjab University, Chandigarh. She is the Coordinator of the M. Pharm. (Pharmaceutical Analysis) section and also the Placement In-charge of the Institute. She did her B. Pharm., M. Pharm. (Pharmaceutical Chemistry) and Ph. D. from the same institute. She has an academic and research experience spanning 29 years of undergraduate and postgraduate teaching. She has expertise in the fields of Pharmaceutical analysis and Medicinal chemistry. Her research areas of interest include Pharmaceutical analysis with focus on method development and validation, spectroscopic characterization of impurities in drugs and drug products, synthetic medicinal chemistry encompassing computer-aided drug designing (CADD), QSAR, synthesis and pharmacological evaluation of new chemical entities in the area of antipsychotics, anti-inflammatory agents and angiotensin II antagonists. She has guided 50 M. Pharm students and currently, four M. Pharm and two PhD students are pursuing their research degrees under her guidance. She has authored three books, 14 book chapters, 47 international and 5 national publications in reputed journals and she has 70 conference papers to her credit.

Madhura Bodkhe is a final year passout student (2024) from chemical engineering department of UICT, Jalgaon, Maharashtra, India

Tejas Chalke is a final year passout student (2024) from chemical engineering department of UICT, Jalgaon, Maharashtra, India.

Joydeb Dey have pursued his PhD from the Department of Life Science & Bioinformatics, Assam University, Silchar, under the supervision of Dr. Himanshu Kishore Prasad, Associate Professor, Assam University. His thesis is entitled 'Arginine catabolism and its role in pathogenesis of C. albicans'. His work was focused on functional characterization and purification of three potential Ureohydrolase superfamily members from C. albicans. He is interested in delving into the study of medicinal plant-derived metallic nanoparticles imparting antifungal actions, to enlighten the scope of this field for combating the anti-fungal resistance issues against current pharmaco-drugs available.

Poulami Dey is an M. Tech student specializing in Food Process Engineering at the Indian Institute of Technology Kharagpur. With a strong background in Agricultural Engineering, she focuses on developing innovative solutions to improve food quality and sustainability. Driven by a passion for enhancing food safety and efficiency, she actively participates in academic projects and workshops, contributing to advancements in the field. Her commitment to excellence reflects her aspiration to make a meaningful impact on engineering and technology.

Rimpi Foujdar is currently working in the capacity of Assistant Professor in the Department of Food Technology, Haldia Institute of Technology.

Ajaygiri Goswami is currently director of UICT Jalgaon. He has 25 years of experience in teaching and research.

Samrin Gul is an Assistant professor, College of Agriculture, University of Sargodha, Punjab, Pakistan.

Ayesha Hameed is a dedicated researcher and educator with an MPhil in Biotechnology from the University of Gujrat, Pakistan. With a strong focus on agricultural biotechnology, she has a significant understanding of resistance in wheat varieties of Pakistan towards cereal aphids. Ayesha is experienced in proteomic analysis, gene expression studies, and stress physiology in crops. With two years of teaching experience, she has instructed undergraduate students in the Department of Biochemistry and Biotechnology at both the University of Gujrat and Saint Mary's College Gujranwala, Pakistan. Her contributions to the field include a chapter on Plant-Mediated Synthesis of Biocompatible Metal Nanoparticles for Sustainable Agriculture, which showcases her commitment to advancing sustainable agricultural practices.

Hammad Ali Khan, an esteemed researcher, earned his BS in Chemistry, specializing in analytical chemistry, from Government College University Faisalabad, followed by an M.Phil. in Analytical Chemistry from the same institution. His research spans nanocatalysis for enhancing chemical transformations, innovative analytical methods for precise characterization, and the synthesis of heterocyclic compounds with applications in medicine and pharmaceuticals. Khan's investigation into the optoelectronic properties of these compounds aims to drive advancements in high-tech applications. With a substantial portfolio of published research papers, Khan currently serves as a dedicated researcher at the Department of Chemistry, Government College University Faisalabad, where he continues to advance the frontiers of analytical chemistry and inspire future scientific endeavors.

Sunil J. Kulkarni has done his PhD in Chemical Engineering with focus on adsorption for wastewater treatment. He has done his Masters in chemical engineering from Tatyasaheb Kore Institute of Engineering and Technology, Warananagar. He has worked as an Assistant Professor in Chemical Engineering Department of Datta Meghe College of Engineering, Airoli, Navi Mumbai, India for 13 years. He has total 25 years of teaching and research experience. Dr. Sunil is involved with different academic activities across the universities in Maharashtra. His area of investigation in doctoral studies is phenol and heavy metal removal from waste water. He has also worked on kinetics and isotherms of various sorption operations. His area of research includes adsorption, clean technology and environmental biotechnology. He is on the reviewer board of many international journals and reviewed many international papers. He is also governing body member, promotional editor and executive member of many reputed international journals. Currently he is working as Associate Professor in chemical engineering and and Dean Research and Development at GIT, Lavel, India.

Nawneet Kumar Kurrey is an accomplished cell biologist, possesses diverse expertise in molecular biology, cancer biology, genomics, drug discovery, and nutraceuticals. He is a valuable team player with excellent communication skills and is proficient in assay development and troubleshooting complex systems. With over sixteen years of experience in research, teaching, and training, He has made significant contributions to academic and government biotechnology R&D institutions, showcasing exceptional team management and collaboration skills.

Abdul Malik did his M.Sc. (Hons) Agriculture - Plant Breeding and Genetics. He is a Researcher at University of Agriculture, Faisalabad.

Mohammed Shabab holds a Ph.D. from a joint program at the National Chemical Laboratory, Pune, India, and the Max Planck Institute for Plant Breeding Research, Cologne, Germany. With fifteen years of combined teaching and research experience as a staff scientist and faculty member at various institutes, he is deeply engrossed in exploring the biotic interactions of living systems. Following his postdoctoral tenure, he joined the Massachusetts Institute of Technology (MIT), USA, as a staff scientist, delving into signaling and bactericidal cysteine-rich peptides. He continues his research on signaling in S.meliloti-Medicago Interactions and Antimicrobial Peptides. Beyond research, he passionately teaches undergraduate courses, including Analytical Techniques, Foundation in Biology and Biotechnology, and Genetic Engineering.

Nutan Napte was born in 1997 in Pune. Maharashtra, and had her secondary and higher secondary education from Vidyadham Prashala Shikrapur. She completed her graduation in Agriculture from the College of Agriculture, Dhule and her M.Sc. (Agriculture) in Plant Pathology with distinction from the College of Agriculture, Pune, Mahatma Phule Krishi Vidyapeeth Rahuri, Ahmednagar, Maharashtra, India. She is currently pursuing her Ph. D from Post Graduate Institute, Mahatma Phule Krishi Vidyapeeth Rahuri, Ahmednagar, Maharashtra, India. She has qualified ASRB- NET (Plant Pathology) in 2023. She has contributed to publications including 5 research articles, 1 review paper, 2 popular articles and 3 book chapters. She also accomplished one international and several national training programs.

Mrutyunjay Padhiary is a distinguished Assistant Professor in the Department of Agricultural Engineering at Assam University, Silchar, India, with a deep academic background. He possesses 18 years of committed academic experience. His previous positions included significant teaching and research duties at the Faculty of Technology, UBKV in West Bengal, and Tribhuvan University in Nepal, as well as research experience at IIT Kharagpur. He has been awarded multiple awards, including an International NUS Research Fellowship and an excellent score on the 2001 GATE Exam, demonstrating his proficiency in precision agriculture. He has written numerous articles in esteemed journals and is a lifetime member of the Indian Society of Agricultural Engineers. He has made substantial contributions to biofuel production, artificial intelligence vision, aquaculture, precision agriculture, agricultural automation, and artificial intelligence. Additionally, he has engaged in academic travel abroad, where he conducted extensive biofuel research at the National University of Singapore. He is currently working in several research projects, which range from autonomous surface vessels to Precision Agriculture All-Terrain Vehicles.

Himanshu Kishore Prasad is a faculty member in the Department of Life Science and Bioinformatics at Assam University, Silchar, Assam. His research interests encompass antifungal mechanisms and microbial biotechnology.

Swati Ray is currently working in the capacity of Assistant Professor Grade III in the Department of Food Technology, Haldia Institute of Technology for the past 15 years.

Anjali Roy is currently a second year Ph.D research scholar at the Department of Life Science & Bioinformatics, Assam University, Silchar. Her field of research is associated with microbial biotechnology, dealing with yeasts. She received a bachelor's degree with honors in botany from

Gurucharan University, Silchar and a master's degree in botany as well from the Department of Life Science and Bioinformatics, Assam University.

Dipak Roy is a B.Tech student in Electronics and Communication Engineering at Tezpur University and an active NCC cadet. Currently working on an Image Processing project, he has developed strong skills in programming and embedded systems. With a keen interest in research and technology, he is dedicated to exploring innovative solutions and contributing to advancements in the field. His academic pursuits, combined with practical experience, reflect a strong commitment to excellence and a passion for engineering and technological development.

Sujoy Saha graduated from Bidhan Chandra Krishi Viswavidyalaya and after completion of Ph. D. in Plant Pathology from the ICAR-Indian Agricultural Research Institute, New Delhi served as an Assistant Plant Pathologist with the Department of Agriculture, Govt. of West Bengal, the main area of research being disease management of cereal crops, potatoes and vegetables from 2000-2008. Served as Senior Scientist, Plant Pathology at the ICAR-Indian Institute of Vegetable Research, Varanasi where the primary area of research was new generation fungicides, PGPRs and disease dynamics about climate change from 2008-2015. Identified by APEDA to develop package of practices for exportable vegetables. Presently serving as a Principal Scientist at the ICAR-National Research Centre for Grapes, Pune where he is entrusted to look into key matters of export, food safety and biosecurity. Coordinates several research projects with academic institutions, and government bodies as well as with the pesticides industry. Published around 140 research papers in journals of national and international repute, one compendium, 23 book chapters, 15 manuals, 92 research abstracts, 20 leaflets/Folders and more than 100 popular articles. Being life members of APIV, IMS, INS, ISVS and ISCA has delivered several keynote and lead lectures in national and international conferences and functions as a reviewer of several journals of national and International repute. Promulgated three technologies duly recognized by ICAR and was nominated as an expert by ICAR for TBT and SPS issues to assist Ministry of Commerce, GOI regarding pesticide residues and CODEX. He has guided 5 PhD students and 19 MSc dissertations.

Santosh Kumar Sethi is working as Head, School of Biotechnology, Gangadhar Meher University, Sambalpur, Odisha. He has received his Doctoral degree in Biotechnology especially in the field of Microbial genetics under stress and their perspective. He has awarded with the National Post Doctoral Fellowship from UGC, CSIR and ICAR, Govt. of India. His research areas focused on Biocontrol, Biofertilizer and mitigation potential of GHGs from rice ecosystem through potential microbes. He has well versed with the process and protocols for working with people in rural areas for transfer of technology to beneficiaries. Dr. Sethi is an active life member of prestigious organizations of India like AMI (Association of Microbiologists of India), Society of Biotechnology and Bioinformatics (SBB)

Sharmistha Sharma is currently a first year Ph.D research scholar at the Department of Life Science and Bioinformatics, Assam University, Silchar. Her interest lies in the field of research associated with microbial biotechnology, that deals primarily with fungi and bacteria. She received a bachelor's degree with honors in botany from Gurucharan University, Silchar and a master's degree in botany as well from the Department of Life Science and Bioinformatics, Assam University.

Khushbu Siddiqui is working as Project Assistant in the School of engineering at Ajeenkya D Y Patil University. She has completed her master's in biotechnology from Veer Bahadur Singh Purvanchal university and has qualified GATE in Lifesciences with all India rank of 331.

Jyoti Srivastava is pursuing PhD on Discovery of antimicrobial compounds for combating infectious diseases at Ajeenkya D Y Patil University, Pune. She holds a M.Sc. degree in Biotechnology from Utkal University, Bhubaneswar, Odisha. She has completed Master's thesis from NIT Rourkela. She holds 12+ years of experience in in various reputed organizations. She worked as Research Scientist at IIT (BHU), Corporate experience in Pharma company and in academic. She has published 6 research articles and book chapters in reputed publishers.

Qasim Waheed studied from Kohat University of Science and Technology, KPK Pakistan. Qasim Waheed Born on January 9, 2000, Qasim Waheed is a young and ambitious individual pursuing a degree in Biotechnology and Genetic Engineering. With a strong passion for science and innovation, Qasim is driven to make a positive impact in the field. Outside of academics, Qasim is an avid sports enthusiast with a love for cricket and football. He enjoys staying active, competing with friends, and following professional sports. As Qasim continues his educational journey, he is excited to explore the possibilities that biotechnology and genetic engineering have to offer. With a strong foundation in science and a keen interest in sports, Qasim is poised to achieve great things in his future endeavors.

Saurabh Yadav is a faculty in the Department of Biotechnology in Hemwati Nandan Bahuguna Garhwal University, Srinagar, India. His research areas are Plant Biotechnology,, Plant-Pathogen interactions and Intellectual Property Rights, Legal Biotechnology.

Index

E

F

G

H

I

M

N

P

S

T

V

W

X

Z

www.ingramcontent.com/pod-product-compliance
Lightning Source LLC
LaVergne TN
LVHW062106300926
849161LV00010B/2381